for *Physical Geography: Science and Systems of the Human Environment*, 4th Edition, Canadian Version

Check with your instructor to find out if you have access to *WileyPLUS!*

Study More Effectively with a Multimedia Text

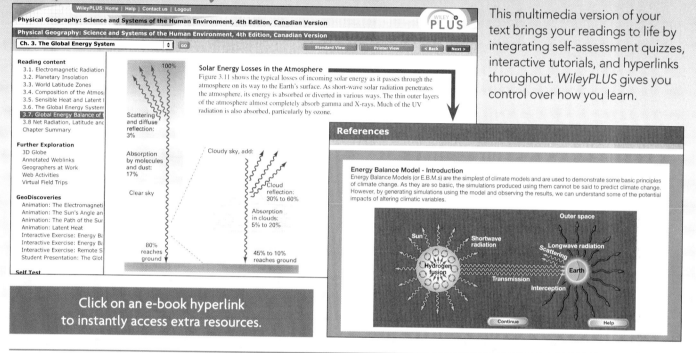

This multimedia version of your text brings your readings to life by integrating self-assessment quizzes, interactive tutorials, and hyperlinks throughout. *WileyPLUS* gives you control over how you learn.

Click on an e-book hyperlink to instantly access extra resources.

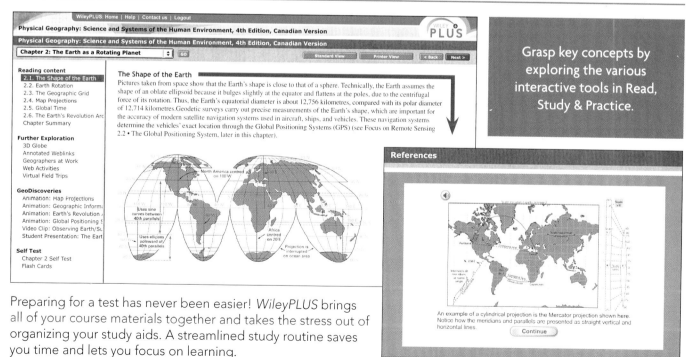

Grasp key concepts by exploring the various interactive tools in Read, Study & Practice.

Preparing for a test has never been easier! *WileyPLUS* brings all of your course materials together and takes the stress out of organizing your study aids. A streamlined study routine saves you time and lets you focus on learning.

John Wiley & Sons, Inc.

for *Physical Geography: Science and Systems of the Human Environment*, 4th Edition, Canadian Version

Complete and Submit Assignments On-line Efficiently

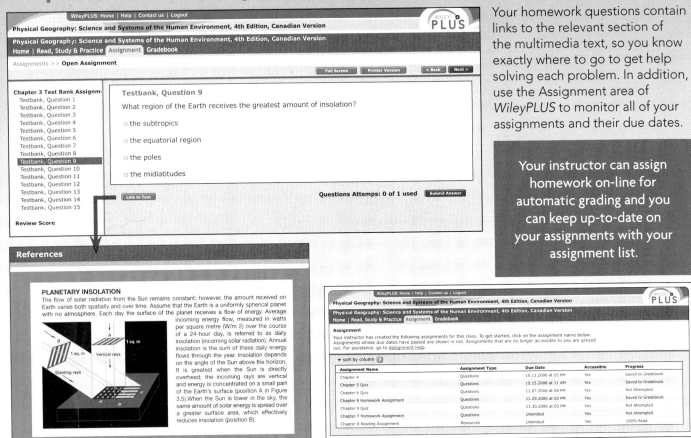

Your homework questions contain links to the relevant section of the multimedia text, so you know exactly where to go to get help solving each problem. In addition, use the Assignment area of *WileyPLUS* to monitor all of your assignments and their due dates.

Your instructor can assign homework on-line for automatic grading and you can keep up-to-date on your assignments with your assignment list.

Keep Track of Your Progress

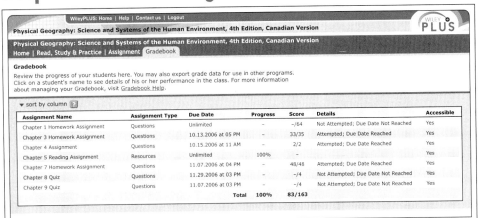

Your personal Gradebook lets you review your answers and results from past assignments as well as any feedback your instructor may have for you.

Keep track of your progress and review your completed questions at any time.

Technical Support: http://higheredwiley.custhelp.com
Student Resource Centre: http://www.wileyplus.com

For further information regarding *WileyPLUS* and other Wiley products, please visit www.wiley.ca.

PHYSICAL

Canadian Version

Fourth Edition

GEOGRAPHY

Science and Systems of the Human Environment

WILEY

John Wiley & Sons, Inc.

PHYSICAL GEOGRAPHY

Canadian Version
Fourth Edition

Science and Systems of the Human Environment

ALAN STRAHLER
Boston University

O.W. ARCHIBOLD
University of Saskatchewan

In collaboration with
Christopher R. Burn
Carleton University

Robert Gilbert
Queen's University

Acquisitions Editor	Michael Valerio
Editorial Manager	Karen Staudinger
Director of Publishing Services	Karen Bryan
Development Editors	Leanne Rancourt, Sarah Wolfman-Robichaud
Media Editor	Elsa Passera-Berardi
Editorial Assistant	Sheri Coombs
Photo Researchers	Francine Geraci, Christina Beamish, Darren Yearsley
Marketing Manager	Anne-Marie Seymour
Designer	Adrian So
Cover Image	GeoStock/Getty Images
Printing & Binding	Quebecor World Inc.

This book was typset in 10/12 Times New Roman by Prepare, Inc. and printed and bound by Quebecor. The cover was printed by Quebecor.

The paper was manufactured by a mill whose forest management programs include sustained yield harvesting of its timberlands. Sustained yield harvesting principles ensure that the number of trees cut each year does not exceed the amount of new growth.

This book is printed on acid-free paper.

Library of Congress Cataloguing in Publication Data:
Strahler, Alan, H.
Archibold, O.W.
Physical Geography: Science and Systems of the Human Environment, Fourth Edition Canadian Version/ Alan Strahler, O.W. Archibold

ISBN-13: 978-0-470-12555-7
ISBN-10: 0-470-12555-1

Printed in the United States of America.

1 2 3 4 5 QW 11 10 09 08 07

PREFACE

Welcome to the fully revised Canadian version of *Physical Geography: Science and Systems of the Human Environment* Fourth Edition. Although physical geography is a truly global science, Canada presents an environment with many unique geographic characteristics. In the third edition we introduced, for the first time, a greater focus on Canadian content within the text. Our goal was to provide a balance between local content without sacrificing the treatment of global processes. This edition makes extensive use of illustrations and examples from across Canada to provide a greater focus on local content that will be of particular relevance to both students and physical geography instructors in Canadian colleges and universities. Using reviewer feedback as our guide, we have further accentuated the coverage of Canadian geographical features to provide a strong pedagogical framework to stimulate interest in the global processes studied in physical geography.

HALLMARK FEATURES OF THIS TEXT

Systems-Based Approach

In keeping with the instructional goals of previous editions of *Physical Geography*, the primary focus of this new edition is to provide a view of physical geography as a key discipline in understanding the Earth's diverse environments and how they are modified by global change. This has been done by taking a systems-based approach, particularly flow systems of energy and matter as a unifying theme. In this way we provide a paradigm for understanding the underlying scientific principles and common elements of the many processes that constitute the science of physical geography.

Quantitative Features

As the text develops chapter by chapter, quantitative material is integrated through the use of *Working It Out* boxes. This material has been separated in order to allow the instructor flexibility over when to use it. The problems have been selected to help show how scientific methodology is employed and how it helps geographers better understand the environments they study, rather than as a test of arithmetic skills. In this way, we hope to open the door for students, in a population that is traditionally less oriented toward mathematics, to realize how quantitative methods can contribute to the understanding of scientific principles.

End of Chapter Materials

Each chapter concludes with a number of aids to facilitate student learning. The *Chapter Summary* is worded like a scientific abstract, succinctly covering all the major concepts of the chapter. The list of *Key Terms* indicates the most important concepts to study to understand the chapter material. The *Review Questions* are designed as oral or written exercises that require description or explanation of important ideas and concepts. *Visualizing Exercises* utilize sketching or graphing as a way of motivating students to visualize key concepts. *Essay Questions* and *Problems* require more synthesis or the reorganization of knowledge in a new context. Last of all, *A Closer Look* boxes focus on the development of key geographical skills in an applied and student-friendly manner.

Pedagogical Effectiveness

Today, science must be taught in an open, accessible manner to reach students, and written material needs to communicate directly and simply. Moreover, today's students are accustomed to visual learning and to an interactive methodology. The text and illustrations of *Physical Geography* have been devised to be accessible and inviting and to provide clear descriptions of scientific principles. Every map and line drawing is carefully designed and styled to make sure that its message is clear, obvious, and direct. The photos have been selected not only to provide a fine illustration of the relevant science concept, but also to demonstrate it effectively and strikingly. In summary, we continue to strive to produce an illustrated text of which we are very proud.

NEW TO THIS EDITION

Canadian Content

Using reviewer feedback as our guide, we have emphasized the use of Canadian geographical features as bridges to a better comprehension of the global processes that we study in physical geography. Below is a list of some of the major chapter revisions that incorporate Canadian content.

Chapter 10: **Midlatitude and High-Latitude Climates** now has a greater focus on the nature and impact of different climate types across Canada. Examples include recent ice storms in Quebec, drought in the Prairies, and flood and slope instability hazards in British Columbia. The role of fire is discussed in the context of boreal climates, and the impact of global warming in the Arctic is considered.

Chapter 13: **Volcanic and Tectonic Landforms** contains new information on volcanic activity in Canada including a more complete treatment of hot springs and earthquake activity.

Chapter 14: **Weathering and Mass Wasting** provides a revised section on *Processes and Landforms of Permafrost Regions* and a comprehensive introduction to the unique features of Canada's periglacial environment.

Chapter 15: **The Cycling of Water on the Continents** continues to provide information on *Lakes* that is particularly applicable to Canada, a country where lakes are a conspicuous feature of the landscape.

Chapter 16: **Fluvial Processes and Landforms** is a fully revised chapter that uses Canadian examples to describe current ideas in channel form and processes with more emphasis on the physics of hydraulic processes.

Chapter 18: **The Work of Wind and Waves** has been restructured and commences with a description of the physical natures of wind movement and particle entrainment at the land surface and the terrestrial features this produces. Waves and coastal landforms are then discussed with examples.

Chapter 20: **Soil Systems** incorporates a fully revised account of the Canadian system of soil classification and descriptions of characteristic soils in Canada.

Simplified Presentation

The other principle aim of the new edition is to make the text easier to use for one semester courses. Increasingly in Canada introductory physical geography courses are taught in one semester. To better reflect this, the content has been rewritten to make the subject matter more manageable in a shorter teaching timeframe. Great care, however, was taken to uphold the reputation of previous editions that *Physical Geography* helps the instructor teach the rigor of the discipline in a pedagogically stimulating environment without oversimplification. In addition, we have revised many of the previous edition's features in order to provide a more user friendly and systematic treatment. Among these are:

- *Eye on Global Change* features examine the human impact on predicted global climate change.
- *Focus on Remote Sensing* features showcase how satellite technology is helping geographers learn more about the Earth.

- *A Closer Look* features provide enhanced coverage of emerging and important topics that impact physical geography, including natural hazards and environmental impacts.
- Each chapter contains revised *Eye on the Landscape* features that show dynamic landscapes, and help the reader to see them from a geographer's perspective.
- *Geographers at Work* features provide an interesting look at some of the noteworthy Canadian geographers, what they practise in the field, and its relevancy to the changing human and environmental needs.

FlyBy Coordinates

Since the last edition, there have been many new developments in computer software applications of interest to students and instructors of geography. Open source software programs such as Google Earth and World Wind by NASA provide the user access to virtual globes. To visit selected locations mentioned in the text, the map coordinates provided serve as a tool to navigate these virtual globes.

ADDITIONAL RESOURCES

The accompanying website (www.wiley.com/canada/strahler) provides access to a full range of resources for the instructor and student, including a fully revised Image Bank for instructors, and student information on how to access the accompanying WileyPLUS version of this text.

WileyPLUS

Also available is WileyPLUS, an online suite of resources that includes a multimedia version of the text that will help students visualize key geographical concepts, be better prepared for lectures, and allow instructors to track student progress throughout the course more easily. Students can take advantage of tools such as self-assessment quizzes and animated tutorials to help them study more effectively. WileyPLUS is designed to provide instant feedback for students as they practise on their own. Instructors in turn, can create assignments and automate the assigning and grading of homework or quizzes using the gradebook feature. They can also create class presentations using PowerPoint slides and interactive simulations, or they can upload their own material to create a customized presentation.

ACKNOWLEDGEMENTS

Preparation of the Canadian version of this text has been greatly aided by numerous Canadian colleagues and reviewers. Firstly, we would especially like to thank Professor Christopher R. Burn of Carleton University for contributing to the revision of the coverage of periglacial phenomena, and Professor Robert Gilbert of Queen's University for his revision on lakes. Both Professors Burn and Gilbert willingly lent their expertise, and their insight has greatly helped in creating a better text.

We would also like to thank professors from the previous edition who helped lay the foundation of this text. Their insight helped establish a greater demand for Canadian content in a physical geography text, and as a result set the wheels in motion for this edition. They are as follows: Professors Daryl Dagesse, Brock University; Catherine Moore, Concordia University; and Lori Daniels, University of British Columbia.

Reviewer feedback played a key role in the revision strategy of the new edition of *Physical Geography*. Instructors from across the country were asked to provide detailed feedback on all chapters. This call was answered by many, and the final result is the text you have here. We therefore take this opportunity to thank the following instructors who willingly gave their time to help improve this text:

Chris Ayles, Camosun College
Doug Banting, Ryerson University
Jim Bowers, Langara College
Darryl Dagesse, Brock University
Dirk DeBoer, University of Saskatchewan
Brett Eaton, University of British Columbia
Sarah Finkelstein, University of Toronto
David Gregory, Athabasca University
Xulin Guo, University of Saskatchewan
John Iacozza, University of Manitoba
David Kemp, Lakehead University
Norma Kerby, Northwest Community College
Wim Kok, Northern Lights College
Colin Laroque, Mount Allison University
Merrin Macrae, University of Waterloo
Steven Marsh, University College of Fraser Valley
Shawn Marshall, University of Calgary
William Osei, Algoma University College
David Rowbotham, Nipissing University
Anne Marie Ryan, Dalhousie University
Jan Thompson, Kwantlen University College
Hannah Wilson, Malaspina University College
Kathy Young, York University

Special thanks also go to the contributors to our *Geographers at Work* series for the time and effort they spent in introducing their work to beginning students. They are:

Konrad Gajewski, University of Ottawa
Xulin Guo, University of Saskatoon
Marilyne Jollineau, Brock University
Colin P. Laroque, Mount Allison University
Shawn Marshall, University of Calgary
Ian McKendry, University of British Columbia
Joel Mortyn, University of British Columbia
Trisalyn A. Nelson, University of Victoria
Tarmo Remmel, York University
Paul Treitz, Queen's University
James A. Voogt, University of Western Ontario
Will Wilson, Lakehead University

It is with particular pleasure that we thank the staff at Wiley Canada for their careful work and encouragement in the preparation and production of the fourth edition of *Physical Geography*: they include Acquisitions Editor Michael Valerio, Developmental Editors Leanne Rancourt and Sarah Wolfman-Robichaud, Photo Researchers Francine Geraci, Christina Beamish, and Darren Yearsley, Copyeditor Alison Arnot, Proofreader Laurel Hyatt, and Indexer Edwin Durbin. Design and production thanks go to Karen Bryan, Adrian So, and Prepare Inc.

Alan Strahler   O.W. Archibold
Boston, Massachusetts   *Saskatoon, Saskatchewan*
May 2007   May 2007

ABOUT THE AUTHORS

Alan Strahler received his B.A. degree in 1964 and his Ph.D. degree in 1969 from The Johns Hopkins University, Department of Geography and Environmental Engineering. He has held academic positions at the University of Virginia, the University of California at Santa Barbara, and Hunter College of the City University of New York, and has been Professor of Geography at Boston University since 1988. With the late Arthur Strahler, he is a co-author of eight textbook titles with eleven revised editions on physical geography and environmental science. He has published over 250 articles in the refereed scientific literature, largely on the theory of remote sensing of vegetation, and has also contributed to the fields of plant geography, forest ecology, and quantitative methods. His research has been supported by over $6 million in grant and contract funds, primarily from NASA. In 1993, he was awarded the Association of American Geographers/Remote Sensing Specialty Group Medal for Outstanding Contributions to Remote Sensing. In 2000, he received the honorary degree Doctorem Scientiarum Honoris Causa (D.S.H.C) from the Université Catholique de Louvain, Belgium, for his academic accomplishments in teaching and research. In 2004, he was honoured by election to the rank of Fellow in the American Association for the Advancement of Science.

Bill Archibold received his B.A. degree in Geography and Biology from the University of Keele, in 1968. He completed an M.Sc. in Natural Resource Survey at the University of Sussex in 1969, and received a Ph.D. degree from Simon Fraser University in 1975. He has held a teaching position in the Department of Geography at the University of Saskatchewan since 1972, and is currently the head of department. His main teaching areas have been in physical geography including biogeography, meteorology and field techniques. His research has focused on a variety of ecological and environmental topics, such as the effects of forest fire and logging in the boreal forest, the effects of air pollution on plant communities, and various studies in microclimate. In addition, he has carried out extensive studies on the ecology of wild rice in Saskatchewan, and is currently investigating the feasibility of green roof technology in a prairie climate. He has published numerous articles in scientific journals and is the author of *Ecology of World Vegetation* a comprehensive textbook that describes Earth's major terrestrial, aquatic and marine biomes.

BRIEF CONTENTS

Part 1 Introduction 2

Chapter 1 Introducing Physical Geography 4

Part 2 Weather and Climate Systems 18

Chapter 2 The Earth as a Rotating Planet 20

Chapter 3 The Global Energy System 21

Chapter 4 Air Temperature and Air Temperature Cycles 68

Chapter 5 Winds and the Global Circulation System 98

Chapter 6 Atmospheric Moisture and Precipitation 128

Chapter 7 Weather Systems 152

Chapter 8 The Global Scope of Climate 176

Chapter 9 Low-Latitude Climates 200

Chapter 10 Mid-Latitude and High-Latitude Climates 224

Part 3 Systems and Cycles of the Solid Earth 264

Chapter 11 Earth Materials and the Cycle of Rock Transformation 266

Chapter 12 The Lithosphere and the Tectonic System 286

Chapter 13 Volcanic and Tectonic Landforms 310

Part 4 Systems of Landform Evolution 334

Chapter 14 Weathering and Mass Wasting 336

Chapter 15 The Cycling of Water on the Continents 364

Chapter 16 Fluvial Processes and Landforms 394

Chapter 17 Landforms and Rock Structure 418

Chapter 18 The Work of Wind and Waves 436

Chapter 19 Glacier Systems and the Late-Cenozoic Ice Age 462

Part 5 Systems and Cycles of Soils and the Biosphere 490

Chapter 20 Soil Systems 492

Chapter 21 Systems and Cycles of the Biosphere 518

Chapter 22 Biogeographic Processes 542

Chapter 23 The Earth's Terrestrial Biomes 568

CONTENTS

Part 1 | Introduction   2

1 | Introducing Physical Geography   4

The Discipline of Geography   5
Physical Geography   5
Tools in Geography   7
Physical Geography, Environment, and Global Change   7
Global Climate Change   9
The Carbon Cycle   9
Biodiversity   9
Pollution   9
Extreme Events   10
Organizing Information in Physical Geography   10
Earth Realms   10
Scales in Physical Geography   10
Systems in Physical Geography   11
Open and Closed Flow Systems   12
Feedback and Equilibrium in Flow Systems   13
Time Cycles   14

Part 2 | Weather and Climate Systems   18

2 | The Earth as a Rotating Planet   20

The Shape of the Earth   21
Earth Rotation   21
Environmental Effects of Earth Rotation   21
The Geographic Grid   22
Parallels and Meridians   22
Latitude and Longitude   22
Working It Out 2.1 • Distances from Latitude and Longitude   23
Map Projections   24
Focus on Remote Sensing 2.2 • The Global Positioning Systems   24

Polar Projection   25
Mercator Projection   25
Goode Projection   26
Global Time   26
Standard Time   27
*Geographer's Tools 2.3 • Geographic Information
   Systems   28*
World Time Zones   29
Daylight Saving Time   29
Geographers at Work • Tarmo Remmel, Ph. D   30
International Date Line   30
The Earth's Revolution around the Sun   31
Tilt of the Earth's Axis   31
Solstice and Equinox   31
Equinox Conditions   32
Solstice Conditions   33
*A Closer Look: Geographer's Tools 2.4 • Focus
   on Maps   34*

3 | The Global Energy System   40

Electromagnetic Radiation   41
Radiation and Temperature   42
Focus on Systems 3.1 • Forms of Energy   43
Solar Radiation   44
Working It Out 3.2 • Radiation Laws   44
Characteristics of Solar Energy   45
Long-Wave Radiation from the Earth   45
Planetary Insolation   46
Daily Insolation through the Year   46
World Latitude Zones   47
Composition of the Atmosphere   48
Earth's Primordial Atmosphere   48
The Present Atmosphere   48
Ozone in the Upper Atmosphere   49
*Eye on Global Change 3.3 • The Ozone Layer – Shield
   to Life   50*
Air Pollutants   50

Sensible Heat and Latent Heat Transfer   52
The Global Energy System   53
Solar Energy Losses in the Atmosphere   53
Albedo   54
Counter-Radiation and the Greenhouse Effect   54
**Global Energy Balance of the Earth-Atmosphere
   System   55**
Incoming Short-Wave Radiation   55
*Focus on Remote Sensing 3.4 • CERES – Clouds and
   the Earth's Radiant Energy System   56*
Surface Energy Flows   56
Energy Flows to and from the Atmosphere   57
Climate and Environmental Change   57
**Net Radiation, Latitude, and the
   Energy Balance   58**
Solar Energy   59
*A Closer Look: Geographer's Tools 3.5 • Remote
   Sensing for Physical Geography   60*

**4 | Air Temperature and Air Temperature
   Cycles   68**

Surface Temperature   70
Air Temperature   70
Measurement of Air Temperature   70
*Focus on Systems 4.1 • The Surface Energy Balance
   Equation   72*
The Daily Cycle of Air Temperature   72
Daily Insolation and Net Radiation   72
Daily Temperature   73
Temperatures Close to the Ground   74
Environmental Contrasts: Rural and Urban
   Temperatures   74
Geographers at Work • Dr. James A. Voogt   76
Temperature Structure of the Atmosphere   76
Troposphere   76
Stratosphere and Upper Layers   77
High Mountain Environments   78
Temperature Inversion   79
The Annual Cycle of Air Temperature   79
Net Radiation and Temperature   80
Land and Water Contrasts   80
World Patterns of Air Temperature   82
Factors Controlling Air Temperature Patterns   83
World Air Temperature Patterns for January
   and July   83
The Annual Range of Air Temperatures   85
**Global Warming and the Greenhouse
   Effect   86**
Factors Influencing Climatic Warming
   and Cooling   86
Working It Out 4.2 • Exponential Growth   87
*Eye on Global Change 4.3 • Carbon Dioxide – On the
   Increase   88*
The Temperature Record   **90**
Future Scenarios   91
*A Closer Look: Eye on Global Change 4.4 • The IPCC
   Report of 2001   92*

5 | Winds and the Global Circulation System 98

Atmospheric Pressure 99
Measuring Atmospheric Pressure 99
How Air Pressure Changes with Altitude 100
Wind 100
Measurement of Wind 100
Winds and Pressure Gradients 100
The Coriolis Effect and Winds 101
Working it Out 5.1 • Pressure and Density in the Oceans and Atmosphere 102
Surface Winds on an Idealized Earth 102
Focus on Systems 5.2 • A Simple Convective Wind System 104
Global Wind and Pressure Patterns 104
Subtropical High-Pressure Belts 104
The ITCZ and the Monsoon Circulation 105
Wind and Pressure Features of Higher Latitudes 107
Cyclones and Anticyclones 108
Local Winds 109
Winds in the Upper Atmosphere 111
Eye on the Environment 5.3 • Wind Power, Wave Power, and Current 112
The Geostrophic Wind 112
Global Circulation at Upper Levels 114
Rossby Waves, Jet Streams, and the Polar Front 115
Temperature Layers of the Ocean 118
Ocean Currents 118
Surface Currents 118
Deep Currents and Thermohaline Circulation 120
Eye on Global Change 5.4 • El Niño 122

6 | Atmospheric Moisture and Precipitation 128

Three States of Water 129
The Hydrosphere and the Hydroligic Cycle 130
Working it Out 6.1 • Energy and Latent Heat 130
Focus on Systems 6.2 • The Global Water Balance as a Matter Flow System 132
Humidity 132
Relative Humidity 133
The Adiabatic Process 134
Dry Adiabatic Rate 135
Wet (Saturated) Adiabatic Rate 135
Working it Out 6.3 • The Lifting Condensation Level 136
Clouds 136
Cloud Forms 137
Fog 138
Focus on Remote Sensing 6.4 • Observing Clouds from GOES 139
Precipitation 140
Precipitation Processes 142
Orographic Precipitation 143
Convergent Precipitation 144
Convectional Precipitation 144

Atmospheric Stability 145
Thunderstorms 146
Focus on Systems 6.5 • The Thunderstorm as a Flow System 148
Microbursts 148

7 | Weather Systems 152

Air Masses 153
North American Air Masses 155
Travelling Cyclones and Anticyclones 156
Wave Cyclones 157
Weather Changes within a Wave Cyclone 159
Cyclone Tracks and Cyclone Families 161
The Tornado 163
Tropical and Equatorial Weather Systems 164
Easterly Waves and Weak Equatorial Lows 164
Polar Outbreaks 164
Tropical Cyclones 164
Impacts of Tropical Cyclones 167
Poleward Transport of Heat and Moisture 168
Atmosphere Heat and Moisture Transport 168
Oceanic Heat Transport 169
Cloud Cover, Precipitation, and Global Warming 169
Eye on the Environment 7.1 • Hurricane Katrina 170
Geographers at Work • Dr. Ian McKendry 172

8 | The Global Scope of Climate 176

Keys to Climate 177
Temperature Regimes 178
Focus on Systems 8.1 • Time Cycles of Climate 179
Working it Out 8.2 • Averaging in Time Cycles 180
Global Precipitation 181
Seasonality of Precipitation 185
Geographers at Work • Dr. Colin P. Laroque 187

Climate Classification 187
Overview of the Climates 189
Dry and Moist Climates 191
A Closer Look: Special Supplement 8.3 • The Köppen Climate System 192

9 | Low-Latitude Climates 200
Overview of the Low-Latitude Climates 201
The Wet Equatorial Climate (Köppen *AF*) 202
The Monsoon and Trade-Wind Coastal Climate (Köppen *AF, AM*) 204
The Low-Latitude Rainforest Environment 207
The Wet-Dry Tropical Climate (Köppen *AW, CWA*) 207
The Savanna Environment 208
Working it Out 9.1 • Cycles of Rainfall in the Low Latitudes 210
Eye on Global Change 9.2 • Drought and Land Degradation in the African Sahel 212
The Dry Tropical Climate (Köppen *BWH, BSH*) 215
The Tropical Desert Environment 217
Highland Climates of Low Latitudes 219

10 | Midlatitude and High-Latitude Climates 224
Overview of the Midlatitude Climates 225
Working it Out 10.1 • Standard Deviation and Coefficient of Variation 226
The Dry Subtropical Climate 226
The Subtropical Desert Environment 227
The Moist Subtropical Climate 228
The Moist Subtropical Environment 230
The Mediterranean Climate 232
The Mediterranean Climate Environment 233
The Marine West-Coast Climate 234
The Marine West-Coast Environment 235
Geographers at Work • Dr. Trisalyn A. Nelson 236

Focus on Systems 10.2 • California Rainfall Cycles and El Niño 238
The Dry Midlatitude Climate 240
The Dry Midlatitude Environment 241
The Moist Continental Climate 242
The Moist Continental Forest and Prairie Environment 242
Eye on the Environment 10.3 • Drought on the Prairies 244
Overview of the High-Latitude Climates 247
The Boreal Forest Climate 247
The Boreal Forest Environment 247
The Tundra Climate 248
The Arctic Tundra Environment 249
Arctic Permafrost 250
The Ice Sheet Climate 251
The Ice Sheet Environment 251
Our Changing Climate 251
A Closer Look: Eye on Global Change 10.4 • Regional Impacts of Climate Change on North America 252

Part 3 | Systems and Cycles of the Solid Earth 264

11 | Earth Materials and the Cycle of Rock Transformation 266
The Crust and its Composition 267
Rocks and Minerals 268
Igneous Rocks 270
Common Igneous Rocks 271
Intrusive and Extrusive Igneous Rocks 272
Chemical Alteration of Igneous Rocks 274
Sediments and Sedimentary Rocks 274
Clastic Sedimentary Rocks 275
Chemically Precipitated Sedimentary Rocks 276
Hydrocarbon Compounds in Sedimentary Rocks 279
Metamorphic Rocks 279
The Cycle of Rock Transformation 281
Focus on Systems 11.1 • Powering the Cycle of Rock Transformation 282

12 | The Lithosphere and the Tectonic System 286
The Earth's Structure 287
The Earth's Interior 288
The Lithosphere and Asthenosphere 290
The Geologic Time Scale 290
Major Relief Features of the Earth's Surface 291
Relief Features of the Continents 291
Working It Out 12.1 • Radioactive Decay and Radiometric Dating 292
Relief Features of the Ocean Basins 295

Midoceanic Ridge and Ocean Basic Features 295
The Global System of Lithospheric Plates 297
Plate Tectonics 297
Plate Motions and Interactions 297
Tectonic Processes 300
Subduction Tectonics 301
Orogens and Collisions 302
Continental Rupture and New Ocean Basins 303
Continents of the Past 304

13 | Volcanic and Tectonic Landforms 310

Landforms 311
Volcanic Activity 312
Stratovolcanoes 312
Shield Volcanoes 314
Volcanic Activity across the Globe 316
Volcanic Eruptions as Environmental Hazards 316
Landforms of Tectonic Activity 317
Fold Belts 317
Focus on Systems 13.1 • The Life Cycle of a Hotspot Volcano 318
Eye on the Environment 13.2 • Geothermal Energy Sources 320
Faults and Fault Landforms 321
Focus on Remote Sensing 13.3 • Remote Sensing of Volcanoes 322
Earthquakes 325
Working it Out 13.4 • Earthquake Magnitude 326
Earthquakes and Plate Tectonics 327
Seismic Sea Waves 330

Part 4 | Systems of Landform Evolution 334

14 | Weathering and Mass Wasting 336
Physical Weathering 338
Frost Action 338
Salt-Crystal Growth 339
Unloading 340
Other Physical Weathering Processes 340
Chemical Weathering and its Landforms 340
Hydrolysis and Oxidation 340
Acid Action 340
Mass Wasting 341
Slopes 342
Soil Creep 342
Earth Flow 343
Environmental Impact of Earth Flows 343
Mudflow and Debris Flood 344
Landslide 344

Induced Mass Wasting 345
Working it Out 14.1 • The Power of Gravity 346
Induced Earth Flows 346
Scarification of the Land 346
Processes and Landforms of the Permafrost Regions 348
Permafrost 348
Permafrost Temperature 350
The Active Layer 351
Forms of Ground Ice 351
Focus on Systems 14.2 • Permafrost as an Energy Flow System 352
Thermokarst Lakes 355
Geographers at Work • Konrad Gajewski, Ph.D. 356
Forest Fires and Permafrost 356
Retrogressive Thaw Slumps 357
Patterned Ground and Solifluction 358
Alpine Tundra 359
Environmental Problems of Permafrost 359
Climate Change in the Arctic 360

15 | The Cycling of Water on the Continents 364

Groundwater 366
The Water Table Surface 366
Aquifers 367
Limestone Solution by Groundwater 368
Limestone Caverns 368
Karst Landscapes 368
Problems of Groundwater Management 369
Water Table Depletion 370
Contamination of Groundwater 370
Surface Water 371
Overland Flow and Stream Flow 371
Stream Discharge 372
Drainage Systems 373
Stream Flow 373
Focus on Systems 15.1 • Stream Flow and its Measurement 374
How Urbanization Affects Stream Flow 376
The Annual Flow Cycle of a Large River 376
River Floods 376
Flood Prediction 377
Working it Out 15.2 • Magnitude and Frequency of Flooding 378
The Red River Flood of 1997 380
Lakes 380
Origins of Lakes 382
Water Balance of a Lake 382
Eye on the Environment 15.3 • The Great Lakes 384
Saline Lakes and Salt Flats 386

Lake Temperature and Circulation 386
Seasonal Circulation in Lakes 387
Sediment in Lakes 388
Surface Water as a Natural Resource 389

16 | Fluvial Processes and Landforms 394

Fluvial Processes and Landforms 395
Erosional and Depositional Landforms 395
Slope Erosion 396
Accelerated Erosion 397
Rilling and Gullying 398
The Work of Streams 399
Stream Erosion 399
Stream Transportation 400
Capacity of a Stream to a Transport Load 402
Stream Deposition 402
Aggradation and Flood Plain Development 402
Working it Out 16.1 • River Discharge and Suspended Sediment 403
Stream Gradation 406
Landscape Evolution of a Graded Stream 408
Waterfalls 409
Focus on Systems 16.2 • Stream Networks as Trees 410
Dams 411
Entrenched Meanders 412
Fluvial Processes in an Arid Climate 413
Alluvial Fans 414

17 | Landforms and Rock Structure **418**

Rock Structure as a Landform Control 419
Strike and Dip 420
*Focus on Remote Sensing 17.1 • Remote Sensing of
Rock Structures 421*
**Landforms of Horizontal Strata and Coastal
Plains 422**
Arid Regions **422**
Drainage Patterns on Horizontal Strata **423**
Coastal Plains **423**
Landforms of Folded Rock Layers 424
Sedimentary Domes 424
Fold Belts 425
*Working it Out 17.2 • Properties of Stream
Networks 426*
**Landforms Developed on Other Land-Mass
Types 428**
Erosion Forms on Fault Structures 428
Exposed Batholiths 428
Deeply Eroded Volcanoes 429
Impact Structures 430
**The Geographic Cycle of Land-Mass
Denudation 431**

18 | The Work of Wind and Waves **436**

**The Work of Wind on Terrestrial
Environments 436**
Transport by Wind 438
Dust Storms 439
Erosion by Wind 440
**Depositional Landforms Associated
with Wind 441**
Sand Dunes 442
Types of Sand Dunes 442
Coastal Foredunes 444
Loess 444
The Work of Wind on Coastal Environments 445

Wind and Waves 445
The Work of Waves 446
Marine Erosion 446
Marine Transport and Deposition 448
Littoral Drift and Shore Protection 449
The Tides 449
The Physics of Tides 449
Tide Cycles and Amplitudes 451
Tidal Currents 452
Types of Coastlines 452
Coastlines of Submergence 452
Emergent Coastlines 453
Neutral Coastlines 453
*A Closer Look: Eye on Global Change 18.1 • Global
Change and Coastal Environments 454*
Geographers at Work • Dr. Marilyne Jollineau 456
Organic Coastlines 457
Fault Shorelines and Marine Terraces 458
Rising Sea Level 458

19 | Glacier Systems and the Late-
Cenozoic Ice Age **462**

Glaciers 463
Alpine Glaciers 465
*Focus on Systems 19.1 • A Glacier as a Flow System of
Matter and Energy 466*
*Focus on Remote Sensing 19.2 • Remote Sensing
of Glaciers 468*
Landforms Made by Alpine Glaciers 468
Glacial Troughs and Fjords 469
*Eye on Global Change 19.3 • Glacial Retreat – A Global
Overview 472*
Ice Sheets of the Present 472
Sea Ice and Icebergs 473
The Late-Cenozoic Ice Age 475
Glaciation during the Ice Age 476
Landforms Made by Ice Sheets 477

Erosion by Ice Sheets   477
Development of the Great Lakes   478
Deposits Left by Ice Sheets   479
Moraines   480
Outwash and Eskers   482
Drumlins and Till Plains   482
Marginal Lakes and Their Deposits   482
Geographers at Work • Dr. Shawn Marshall   483
*Eye on Global Change 19.4 • Ice Sheets and Global
   Warming   484*
Investigating the Late-Cenozoic Ice Age   484
Causes of the Late-Cenozoic Ice Age   485
Causes of Glaciation Cycles   486
Holocene Environments   486

Part 5 | Systems and Cycles of Soils and the Biosphere   490

20 | Soil Systems   492

The Nature of the Soil   494
Soil Colour and Texture   494
Soil Structure   495
Soil Minerals   496
Soil Colloids   496
Soil Acidity and Alkalinity   497
Soil Moisture   498
The Soil Water Balance   498
A Simple Soil Water Budget   499
Soil Development   499
Soil Horizons   000
*Working it Out 20.1 • Calculating A Simple Soil Water
   Budget   500*
Soil-Forming Processes   501
Soil Temperature   502
Surface Configuration   503
Biological Processes   504
The Global Scope of Soils   504
Soils of Canada   504
Soils of the World   510

21 | Systems and Cycles of the Biosphere   518

Energy Flow in Ecosystems   519
The Food Web   519
Photosynthesis and Respiration   520
Net Photosynthesis   521
Net Primary Production   522
Net Production and Climate   523
Biomass as an Energy Source   524
Biogeochemical Cycles in the Biosphere   524
Geographers at Work • Joel Mortyn   525
Nutrient Elements in the Biosphere   525
The Carbon Cycle   526

The Oxygen Cycle   526
The Nitrogen Cycle   527
*Eye on Global Change • Human Impact on the
   Carbon Cycles   528*
The Sulphur Cycle   530
Sedimentary Cycles   532
*A Closer Look: Focus on Remote Sensing 21.2 •
   Monitoring Global Productivity from Space   534*

22 | Biogeographic Processes   542

Ecological Biogeography   543
Water Need   544
Temperature   546
Other Climatic Factors   546
Bioclimatic Frontiers   547
Geomorphic and Edaphic (Soil) Factors   547
Disturbance   548
Interactions among Species   550
Ecological Succession   551
Geographers at Work • Dr. Xulin Guo   552
Succession, Change, and Equilibrium   552
Historical Biogeography   554
Evolution   554
Speciation   555
Extinction   556
Dispersal   556
Distribution Patterns   557
*Working it Out 22.1 • Island Biogeography: The
   Species-Area Curve   558*
Biogeographic Realms   560
Biodiversity   562

23 | The Earth's Terrestrial Biomes 568

Natural Vegetation 569
Structure and Life Form of Plants 570
Terrestrial Ecosystems – The Biomes 570
Geographers at Work • Dr. Paul Treitz 571
*Focus on Remote Sensing 23.1 • Mapping Global
   Land Cover by Satellite 572*
Forest Biome 575
*Eye on the Environment 23.2 • Exploitation of the
   Low-Latitude Rain Forest Ecosystem 578*
Savanna Biome 587
Geographers at Work • Dr. Will Wilson 589
Grassland Biome 589
Desert Biome 591
Tundra Biome 592
Climatic Gradients and Vegetation Types 594
Altitude Zones of Vegetation 596

Appendix 1 | Climate Definitions and
   Boundaries 600

Appendix 2 | Conversion Factors 602

**Answers to *Working it Out*
Problems 603**

Glossary 607

Photo Credits 633

Index 636

Part 1

CHAPTER IN PART 1

1 Introducing Physical Geography

INTRODUCTION

Thin layers of fog are dispersing in the morning sun to reveal the glaciated peaks of the Canadian Rockies and the conifer forests that cover the lower slopes.

Geography is concerned with the places of the Earth, as well as the physical and human processes that differentiate those places and make them unique. Geography captures the spatial connections among human activities as they occur on the Earth's physical landscape. ■ The ability to model and predict human and natural spatial phenomena makes geography a vital discipline in today's world. Human impact on the environment increases inexorably as the world's population grows and makes increasing demands on the Earth's natural resources. In order to live in harmony with the environment, the population will have to make difficult choices in the future, and understanding geography can help with those decisions. ■ Part 1, our introduction to geography, provides an orientation to the discipline and explains the role physical geography plays in understanding global change. We also present some important ideas about systems of energy and matter flow in physical geography, which we will use in many later chapters.

Chapter 1

EYE ON THE LANDSCAPE

Vancouver, British Columbia enjoys a spectacular setting on the Strait of Georgia, flanked by the Pacific and Vancouver Island Ranges.
What else would the geographer see? . . . Answers at the end of the chapter.

INTRODUCING PHYSICAL GEOGRAPHY

The Discipline of Geography
Physical Geography
Tools in Geography
Physical Geography, Environment, and Global Change
Global Climate Change
The Carbon Cycle
Biodiversity
Pollution
Extreme Events
Organizing Information in Physical Geography
Earth Realms
Scales in Physical Geography
Systems in Physical Geography
Open and Closed Flow Systems
Feedback and Equilibrium in Flow Systems
Time Cycles

THE DISCIPLINE OF GEOGRAPHY

Geography is the study of the evolving character and organization of the Earth's surface. It investigates how, why, and where human and natural activities occur and how these activities are interconnected. Differentiating the Earth's surface into unique **places** is covered under the general theme of **regional geography**. For example, what makes Vancouver, British Columbia unique? Is it the city's spectacular setting where the Coast Mountains meet the Pacific Ocean? The marine west-coast climate that provides its mild and rainy winters and blue summer skies? Its position as a seaport gateway to Asia? In fact, these are only some of the attributes that contribute to making Vancouver the unique place that it is.

Although each place is unique, the physical, economic, and social processes that operate there may be quite similar to those found in other regions. Thus, geographers are concerned with discovering, understanding, and modelling the processes that differentiate places on the Earth's surface. Why are pineapples cheap in Hawaii and expensive in Toronto? Lobster cheap in Halifax but expensive in Calgary. These are examples of a simple principle of **economic geography**—that prices include transportation costs and when goods travel a longer distance, they are usually more expensive. Discovering such principles and extending them to model and predict spatial phenomena is the domain of **systematic geography**. Thus, geographers study both the characteristics that define a place and the connections between places.

PHYSICAL GEOGRAPHY

As a field of study, systematic geography is often divided into two broad areas—**human geography**, which deals with social, economic, and behavioural processes that differentiate places, and **physical geography**, which covers the atmospheric, terrestrial, and marine environments on local, regional, and global scales. It focuses primarily on factors and processes that are part of the human environment, and provides an explanation for many common natural phenomena. The main fields of physical and human geography are shown in Figure 1.1. Aspects of physical geography, listed on the upper left side of the diagram, form the basis of this book.

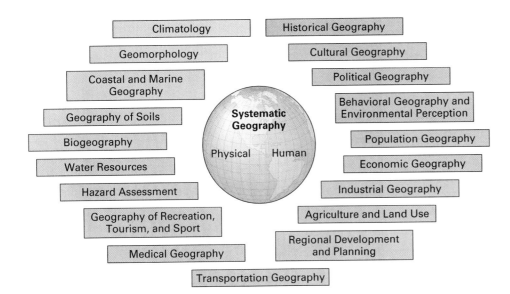

1.1 Fields of systematic geography

Meteorology and **climatology** are both concerned with the atmosphere. Meteorology deals primarily with the processes that cause short-term fluctuations in the atmosphere's measurable properties, which form the basis of daily weather reports. Climatology describes the results of these processes in terms of their variability in space and time. In general, climate is the average variability in weather at different places around the world. Chapters 2–10 discuss the processes that control weather and climate. Climatology is also concerned with climate change, both past and future. This aspect of climatology relies heavily on computer-based global climate models (GCMs) to predict how human activities, such as releasing carbon dioxide (CO_2) from fossil fuel burning, might affect global climates.

Geomorphology is the science of Earth surface processes and landforms (Figure 1.2). The combined influence of human and natural factors is constantly altering the Earth's surface. The work of gravity in the collapse and movement of Earth materials, as well as flowing water, blowing wind, breaking waves, and moving ice, acts to weaken, transport, and deposit rock and sediment. This sculpts details onto a surface that is constantly being renewed though volcanic activity and crustal movements. Chapters 11–19 describe these geomorphic processes and the basic geology of the rock materials involved. Modern geomorphology also focuses on predictive landform models. For example, in **oceanography** and **marine geography**, computer models might be generated to predict the rate and impact of coastal erosion in relation to sea level changes that are likely to occur as a result of global warming.

Geography of soils (Chapter 20) includes the study of the distribution of soil types and properties and the processes of soil formation. It is related to both geomorphic processes of rock breakup and weathering, and biological processes associated with plants and organisms living in the soil. Since both geomorphic and biologic processes are influenced by the surface temperature and availability of moisture, broad-scale soil patterns are often related to climate.

Biogeography is the study of the distributions of organisms and the processes that produce these patterns. Local distributions of plants and animals typically depend on the suitability of the habitat that supports them (Chapter 21-22). In this application, biogeography is closely aligned with *ecology*. Over broader scales and time periods, the migration, evolution,

1.2 McFarlane Falls, Saskatchewan Water is an inexorable geomorphic agent, and over time the waterfall and bedrock remnants will be gradually eroded and the rapids will disappear.

1.3 Deciduous forest, Ontario. The diversity of the forest becomes more apparent as the trees take on their autumn hues in response to shortening days and cooler temperatures caused by the revolution of Earth around the Sun.

and extinction of plants and animals are key processes that determine their spatial distribution patterns (Chapter 22). Thus, biogeographers often try to reconstruct past patterns of plant and animal communities from various kinds of fossil evidence. In light of the increasing human impact on the environment, *biodiversity*—the assessment of the richness of life and life forms on Earth—is a biogeographic topic of considerable importance (Figure 1.3). The present global-scale distribution of life forms in the Earth's major biomes provides the context for biodiversity (Chapter 23).

In addition to these fields of physical geography, **water resources** and **hazards assessment**, are important subfields in applied physical geography. Water resources couples the basic study of the location, distribution, and movement of water, for example in river systems, with the study of water quality and human use. This field involves many aspects of human geography, including regional development and planning, political geography, and agriculture and land use. Chapters 15 and 16 review water resources in a discussion of wells, dams, and water quality.

Hazards assessment also blends physical and human geography. What are the risks of living next to a river and how do inhabitants perceive those risks? What is the government's role in protecting citizens from floods or assisting them in recovery from flood damages? Answering questions like these requires knowledge of not only how physical systems work, but also how humans interact with their environment, both as individuals and as societies. An understanding of physical processes, such as floods, earthquakes, and landslides, provides the background for assessing the impact of natural hazards.

Most of the remaining fields of human geography shown in Figure 1.1 have links to physical geography. For example, climatic and biogeographic factors may determine the spread of disease-carrying mosquitoes (medical geography). Mountain barriers may isolate populations and increase the cost of transporting goods from one place to another (cultural geography, transportation geography, and economic geography). Unique landforms and landscapes may draw tourism (geography of recreation, tourism, and sport).

TOOLS IN GEOGRAPHY

Because geographers deal with spatial phenomena, they commonly use maps to represent spatial information. A **map** shows point, line, or area data—that is, locations, connections, and regions (Figure 1.4a). The map's scale links the true distance between places with the distance on the map. The art and science of map-making is called **cartography**. Chapter 2 provides more information on this important subject.

Maps are very useful for storing information, but they have limitations. In the past two decades, advances in data collection, storage, analysis, and display have led to the development of **geographic information systems (GIS)**. These spatial databases rely on computers for analysis, manipulation, and display of spatial data (Figure 1.4b). Chapter 2 provides more information on GIS.

Another important technique for acquiring spatial information is **remote sensing**, in which aircraft or spacecraft provide images of the Earth's surface (Figure 1.4c). Depending on the scale of the remotely sensed image, the information obtained can range from fine local detail—such as the arrangement of trees in a woodland—to a large area picture—for example, the "greenness" of vegetation for an entire continent. Chapter 3 discusses remote sensing further.

Tools in geography also include **mathematical modelling** and **statistics**. Using mathematics and computers to model geographic processes is a powerful approach to understanding both natural and human phenomena. Statistics provides methods to analyze geographic data to assess differences, trends, and patterns.

PHYSICAL GEOGRAPHY, ENVIRONMENT, AND GLOBAL CHANGE

Physical geography is concerned with the natural world around us. Because natural processes are constantly active, the Earth's environments are always changing. Sometimes the changes are slow and subtle, as when rivers slowly cut into their bedrock channels.

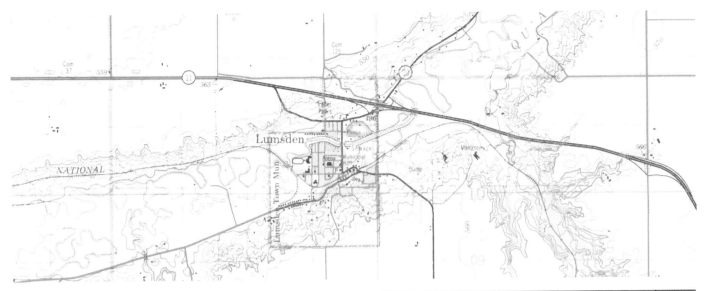

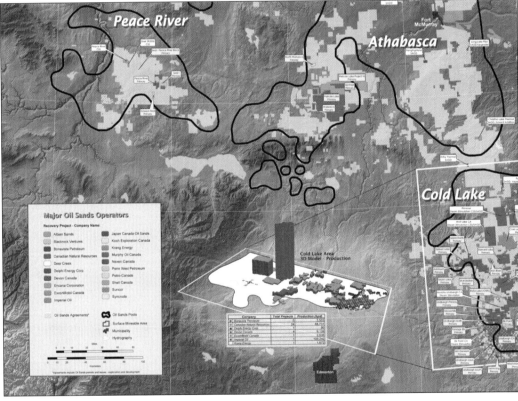

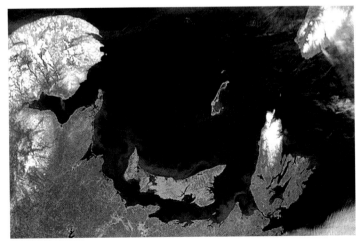

1.4 Tools of Physical Geography a. **Cartography** Maps display geographic information in printed form. b. **Geographic information systems (GIS)** Computer programs that store and manipulate geographic data are essential to modern applications of geography. c. **Remote sensing** Remote sensing includes observing the Earth from the perspective of an aircraft or spacecraft. This scene shows algal blooms in the sea around Prince Edward Island (SeaWiFS).

1.5 Frank slide The town of Frank lies at the foot of Turtle Mountain in southern Alberta. Turtle Mountain is comprised of limestone thrust over weaker sediments including coal. In the early morning hours of April 29, 1903, more than 80 million tons (30 million cubic metres) of rock collapsed from the north face and buried part of the town.

1.6 Athabasca tar sands The Athabasca oil sands in northern Alberta consist of oil-rich bitumen mixed with sand and clays. The near-surface deposits, which are about 50 metres thick, occur at a depth of about 75 metres of overburden in an area of about 3400 square kilometres. *Fly By: 57° 01' N; 111° 38' W*

At other times, the changes are rapid and dramatic, for example, during a landslide (Figure 1.5).

Environmental change is now produced not only by the natural processes that have acted on the planet for millions of years, but also by human activity. The human race has populated the planet so completely that few places remain free of some form of human impact (Figure 1.6). Physical geography is the key to understanding the effect this has on the environment. Special sections throughout this book discuss this aspect of physical geography.

GLOBAL CLIMATE CHANGE

Over the past decade, many scientists have come to the opinion that human activity has begun to change the Earth's climate. How has this happened? The answer lies in the greenhouse effect, a natural phenomenon that is necessary for life on Earth. Gases such as carbon dioxide, methane, and nitrous oxide, which are released by human activities, block heat radiation from leaving the Earth. This induced intensification of the greenhouse effect is widely implicated in global warming. This book discusses many of the potential impacts of **climate change**, such as increased frequency of extreme weather events and the expected rise in sea level.

THE CARBON CYCLE

One way to reduce the human impact on the greenhouse effect is to slow the release of CO_2 from fossil fuel burning. Because modern civilization depends on the energy of fossil fuels to carry out almost every task, reducing fossil fuel consumption to stabilize CO_2 concentration in the atmosphere is not easy. However, some natural processes reduce atmospheric CO_2. Plants withdraw CO_2 from the atmosphere by absorbing it during photosynthesis. In addition, CO_2 is soluble in sea water. These two important pathways, by which carbon flows from the atmosphere to lands and oceans, are part of the **carbon cycle**.

BIODIVERSITY

There is a growing awareness that the diversity of Earth's plants and animals is an immensely valuable resource. One important reason for preserving this **biodiversity** is that, over time, species have evolved natural biochemical defences against diseases and predators. Some of these have proven useful in agriculture and medicine, where they have been used to develop such things as natural pesticides and cancer drugs.

Another important reason for maintaining biodiversity is that complex ecosystems with many species tend to be more stable and respond better to environmental change. If human activities inadvertently reduce biodiversity significantly, there is more risk of unexpected and irreversible degradation of natural environments. Chapters 22 and 23 present the processes that create and maintain biodiversity in the Earth's many natural habitats.

POLLUTION

Unchecked human activity can degrade environmental quality. In addition to releasing CO_2, fuel consumption can release gases that are hazardous to health, especially when they react with each other to form toxic compounds, such as ozone and nitric acid, in photochemical smog. Water pollution from fertilizer runoff, toxic industrial production waste, and acid mine drainage can severely degrade water quality. This pollution affects not only river ecosystems, but also the human populations that depend on them as sources of water and food. Chapter 15 discuss the causes and effects of **environmental pollution** in the context of air and water.

EXTREME EVENTS

Catastrophic events, such as floods, fires, hurricanes, and earthquakes, have great and long-lasting impacts on human and natural systems. As Earth responds to current changes in climate, models predict that weather extremes will become more severe and frequent. Droughts, and consequent wildfires and crop failures, will occur more often, as will extreme rains and floods. In the last decade there have been many examples of extreme weather events, such as Hurricane Katrina in 2005. Although the human impact on natural systems is implicated in these **extreme events**, a lot of questions remain unanswered.

Some extreme events, such as earthquakes and volcanic eruptions, are produced by forces deep within the Earth that are not affected by human activity. But as the human population continues to expand and rely increasingly on complex technological infrastructures, the more damaging and disruptive the impact of such events will be. The chapters that follow discuss many types of extreme events, including tornadoes, hurricanes, and tsunami.

ORGANIZING INFORMATION IN PHYSICAL GEOGRAPHY

In all sciences, recurring principles and ideas organize the wealth of knowledge that has accumulated. In physical geography, one approach is to focus on the Earth's **realms**—the major components of the planet, each with its own unique properties. Also important is the **scale** on which processes operate. A more integrative idea is that of **systems**, which involves the flow of energy and matter. Finally, **cycles** describe the regular changes in a system's energy and matter flows that recur with predictable regularity.

1.7 The mountainous scenery along the Icefields Parkway in Jasper National Park, Alberta provides fine examples of the union of Earth's four great realms.

EARTH REALMS

The atmosphere, lithosphere, hydrosphere, and biosphere are the four great realms that form the basis of physical geography (Figure 1.7).

The **atmosphere** is a gaseous layer that surrounds the Earth. It receives heat and moisture from the surface, redistributes them globally, and returns some heat and all the moisture to the surface. The atmosphere also supplies vital elements—carbon, hydrogen, and oxygen—that are needed to sustain life forms.

The outermost solid layer of the Earth is the **lithosphere**, the surface of which is sculpted into landforms, such as mountains, hills, and plains. Over most of the continents, the outermost layer of the lithosphere is a shallow layer of soil in which nutrients are available to organisms.

The water in the world's oceans is the largest component of the **hydrosphere**, which also includes lakes, rivers, and glacier ice, as well as the water present in the atmosphere, as gaseous vapour, liquid droplets, and solid ice crystals. In the lithosphere, water is found in the uppermost layers in soils and in ground water reservoirs.

The **biosphere** encompasses all of the Earth's living organisms. Life forms on Earth use the atmosphere's gases, the hydrosphere's water, and the lithosphere's nutrients, and so the biosphere depends on all three of the other great realms. Most of the biosphere is contained in the **life layer**, which includes the surface of the continents and the upper 100 metres or so of the ocean.

SCALES IN PHYSICAL GEOGRAPHY

Natural processes act over a wide range of scales in time and space. It may take millions of years for Earth forces to produce a massive chain of mountains like the Rockies; the fury of a hurricane may last for only a few days; while an earthquake can cause widespread damage in a few minutes.

On a **global scale**, Earth is a nearly spherical planet (it is a bit flattened at the poles and bulges at the equator). Its shape affects how its land and water surfaces absorb the sun's energy. Unequal solar heating produces currents of air and water that are incorporated into the global atmospheric and oceanic circulation system. The resulting weather and climate patterns can be readily distinguished on a **continental scale**.

Observations on a **regional scale** might be discerning vegetation patterns, while more detailed assessments of species distributions could be made on a **local scale**, where differences between, say, a steep slope and an adjacent valley floor become apparent. At the finest level are **individual-scale** landscape features, such as a grassy sand dune on a beach (Figure 1.8).

1.8 Different scales of time and space contribute to this landscape in Death Valley, California.

energy will be in motion. For example, a system that describes the flow of solar energy from the sun to the Earth and its atmosphere is termed an *energy flow system*.

Flow systems have a **structure** of interconnected **pathways**. For example, the reflection of solar radiation from the top of a cloud is a pathway in which a flow of solar energy is turned back toward space. The term *components* refers to a system's parts, such as the pathways, their connections, and the types of matter and/or energy that flow within the system.

A flow system needs some sort of **power source**. Sources of power in *natural flow systems* include energy flowing from the sun to the Earth, and the outward flow of heat from the Earth's interior. Flow systems will also have **inputs** and **outputs**.

A river drainage basin is an example of a flow system in physical geography. It consists of water in a set of connected stream channels (Figure 1.9). In this system, the stream channels—pathways—are connected in a structure—the channel network—that organizes the flow of water and sediment from high lands to the ocean. The input is in the form of rain or snow. The energy source comes from gravity, and the output might be in the form of a delta where the sediment is deposited.

SYSTEMS IN PHYSICAL GEOGRAPHY

The processes that interact within the four realms to shape the life layer and differentiate global environments are varied and complex. Many of these processes are interconnected and are more readily understood by adopting a **systems approach**, which emphasizes how and where matter and energy flow in natural systems. A system is a set or collection of things that are somehow related or organized. An example is the solar system—a collection of planets that revolve around the sun.

Physical geography uses specific types of systems. For example, **flow systems** describe how matter and energy move through time from one location to another. Understanding flow systems in physical geography is important because it helps to explain how things are connected. In some physical geography systems,

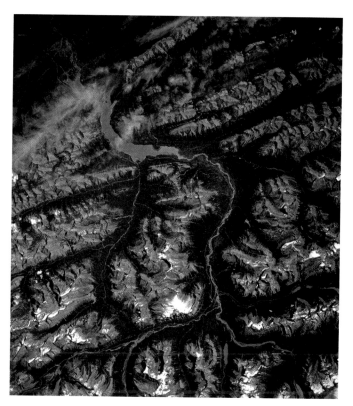

1.9 River channels as flow system pathways The North Saskatchewan River provides an example of a natural flow system. Several tributary streams flow into the North Saskatchewan River as it flows northeast through the Rocky Mountains west of Red Deer, Alberta.

OPEN AND CLOSED FLOW SYSTEMS

A river system is an **open flow system** because there are inputs and outputs of energy and matter. Another type of matter flow system has no inputs or outputs. Instead, the system's flowing materials move endlessly in a series of interconnected paths or loops in what is termed a **closed flow system**. This type of system is also known as a **material cycle**.

Most natural closed flow systems have complicated structures consisting of many interconnected looping pathways. They also require a flow of energy to sustain themselves. Although a material cycle may be closed, energy flow systems that feed them will always be open.

The **hydrologic cycle**, in which water circulates between the biosphere, atmosphere, lithosphere, and hydrosphere, is an example of a closed flow system in physical geography (Figure 1.10). It is maintained by condensation, precipitation, infiltration, runoff, and evapotranspiration. These processes occur simultaneously and continuously on a global scale. The loops in the hydrologic cycle are flow paths of water in gaseous, liquid, and solid forms—liquid form as water flowing in rivers, gaseous form as moist air currents in the atmosphere, and solid form as slowly advancing glaciers. The loops are interconnected and ultimately maintained by the sun's energy flow system.

Whether a matter flow system is open or closed depends partly on its boundary. In Figure 1.11a the system boundary has been drawn around a single river network. Water enters the network when it runs off the land into the river system. Water exits from the network at a point along its course. Defined in this way, the river system is an open flow system in which water enters and leaves.

When the system boundary is redrawn to include the entire Earth and its atmosphere (Figure 1.11b), a new pathway must be added—the return flow of water from the oceans to the atmosphere by evaporation. There is no input or output because water does not leave the Earth's atmosphere or enter it from space. Thus, the system, which now represents the hydrologic cycle, is closed. Indeed, on a global scale, any matter flow system is closed, since with few exceptions (for example, meteorites), nothing enters or leaves the planet and its atmosphere.

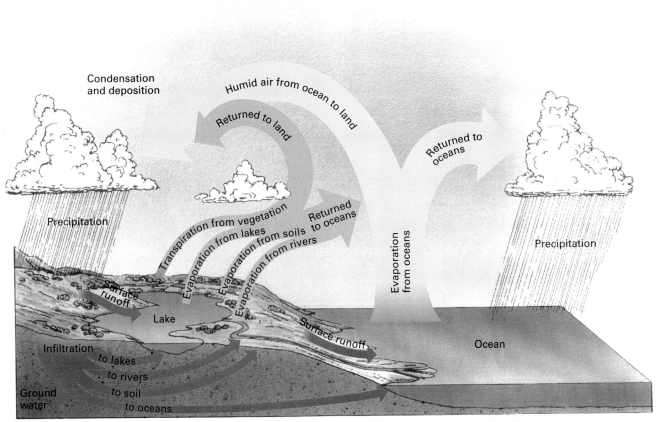

1.10 The hydrologic cycle The hydrologic cycle traces the various paths of water from oceans, through the atmosphere, to land, and its return to oceans.

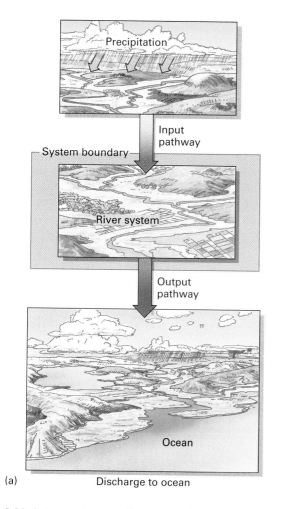

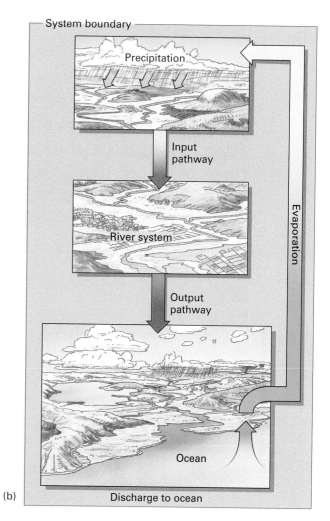

1.11 A river system as a flow system The river is an open flow system in (a) because water enters and leaves at specific points in the network. When the boundary is moved to enclose the entire Earth and its atmosphere (b), the system becomes a closed, global flow system—the hydrologic cycle.

However, energy flow systems are always open. Earth absorbs some portion of the radiant energy it receives from the sun, so there is always an energy input. However, the Earth is warmer than the depths of space and thus emits radiant energy. A portion of that energy ultimately leaves the Earth and its atmosphere, so there is always an output energy flow, even when the system boundary encloses the Earth and its atmosphere.

FEEDBACK AND EQUILIBRIUM IN FLOW SYSTEMS

Two other important concepts associated with flow systems are **feedback** and **equilibrium**. Feedback occurs when the flow in one pathway acts either to reduce or increase the flow in another pathway. Negative feedback helps to dampen external changes imposed on the system. For example, as water evaporates from a soil, the soil particles hold the remaining water more tightly, and so the rate of evaporation slows. Positive feedback occurs when some force or factor creates a situation that increases the effect. For example, wa-

ter falling on a slope may start to erode small channels or rills. With each rainstorm the rills capture more water and so become enlarged, directing more and more water into the channel, leading to the creation of large gullies (Figure 1.12).

Equilibrium refers to a steady condition in which the flow rates in a system's various pathways do not change significantly. A lake within a closed basin is an example of an equilibrium system. This type of lake has no stream outlets and would dry up completely if not fed by nearby streams. Water enters the lake from rivers and streams that feed it; it leaves the lake by evaporation.

If the climate becomes wetter and the input from rivers that feed the lake increases, then the water level rises and the area of the lake expands. Because of the greater area, evaporation is greater. Eventually, the level rises to the point where the increased evaporation rate equals the increased inflow rate. That is, the level reaches an equilibrium. The combination of input, surface area, and evaporation is a negative feedback. Systems that come to an equilibrium are normally stabilized by negative feedback loops.

The effect of clouds on the global climate system illustrates the role of negative and positive feedback on an equilibrium system. Low clouds reflect sunlight back to space much more efficiently than dark land or ocean surfaces. They provide an energy flow pathway in which a portion of the solar energy flow is turned backward and redirected toward space. This pathway tends to cool the surface, and so it acts as a negative feedback. High clouds are different, however. They tend to absorb the outgoing flow of heat from the Earth to space and redirect it earthward. Thus, high clouds provide a positive feedback that warms the surface (Figure 1.13).

A small increase in the Earth's surface temperature will evaporate more water from the oceans and moist land surfaces, and more clouds will form. Global climate models predict that this will increase high cloud cover. Thus, the effect will be a positive feedback that will make the surface even warmer.

TIME CYCLES

Any system, whether open or closed, can undergo a change in the flow rates of energy or matter within its pathways; flow rates may accelerate or slow down. These changes in activity can also be reversed. Consequently, a rate can alternately speed up and slow down at various intervals in what is called a **time cycle**.

1.13 Clouds provide both positive and negative feedback controls on global temperature.

Many natural systems have a rhythm of increasing and decreasing flow. The annual revolution of the Earth around the sun generates a time cycle of energy flow in many natural systems. This produces the rhythm of the seasons. The rotation of the Earth on its axis produces the night-and-day cycle of darkness and light. The moon, in its monthly orbit around the Earth, generates the cycle of tides (Figure 1.14).

1.12 Rilling and gullying, as seen here at Drumheller, Alberta, is a positive feedback process that accelerates as more water is captured by the developing channel system.

1.14 The undercut form of this marine stack at Hopewell Rocks, New Brunswick, illustrates the range of the tidal cycle on the Bay of Fundy.

Other time cycles with durations of tens to hundreds of thousands of years describe the alternate growth and shrinkage of the great ice sheets. Still others, with durations of millions of years, describe cycles in which super continents form, break apart, and reform anew.

A Look Ahead

This chapter has presented an introduction to physical geography and to some of the tools and approaches that geographers use in studying the landscape. It has also noted some key environmental and global change topics. In addition, it has presented some broad-scale concepts that cut across all of physical geography. These include the great Earth *realms* that physical geographers study; the *scales*, from local to global, that characterize Earth processes; the flow *systems* that power the natural processes of physical geography; and the *cycles* of processes that repeat in time and space. The chapters that follow explore these concepts further.

CHAPTER SUMMARY

- Geography is the study of the evolving character and organization of the Earth's surface. Regional geography is concerned with how the Earth's surface is differentiated into unique places, while systematic geography is concerned with the processes that differentiate places in time and space.
- Within systematic geography, human geography deals with social, economic, and behavioural processes that differentiate places, and physical geography examines the natural processes occurring at the Earth's surface. Climatology and meteorology are concerned with the variability in space and time of the heat and moisture states of the Earth's surface. Geomorphology focuses on Earth surface processes and landforms. Oceanography and marine geography combine the study of geomorphic processes that shape shores and coastlines with their application to coastal development and marine resource use. Geography of soils includes the study of the distribution of soil types, their properties, and the processes that formed them. Biogeography is the study of the distributions of organisms on varying spatial and temporal scales, as well as the processes that produce these distribution patterns. Water resources and hazards assessment are applied fields that blend both physical and human geography.
- Important tools for studying the fields of physical geography include maps, geographic information systems, remote sensing, mathematical modelling, and statistics.
- Physical geography is concerned with the natural world, which is constantly changing. The effect of human activities on natural environments is also of interest especially in the way it is implicated in global climate change. Maintaining global biodiversity is important both for ecosystem stability and for preserving a potential resource of bioactive compounds for human benefit. Unchecked human activity can degrade environmental quality and create environmental pollution. Extreme events—storms and droughts, for example—may be more frequent with global warming caused by human activity.
- Realms, scales, systems, and cycles are four overarching themes that appear in physical geography. The four great Earth realms are the atmosphere, hydrosphere, lithosphere, and biosphere. The life layer is the shallow surface layer where lands and oceans meet the atmosphere and where most forms of life are found. The systems of interaction between the realms can be examined on different scales—global, continental, regional, local, and individual.
- A systems approach to physical geography helps in understanding interconnections between natural processes by considering them as flow systems. A distinction can be made between matter flow systems and energy flow systems. Flow systems are composed of pathways of energy and/or matter that are interconnected in a structure. All flow systems have a power source.
- Open flow systems have inputs and outputs, while closed flow systems do not. Closed matter flow systems are also called material cycles in which materials move in an endless series of interconnected pathways. The hydrologic cycle and the carbon cycle are examples of cycles in physical geography. Although *matter flow systems* may be open or closed, *energy flow systems* are always open.
- Feedback in a flow system occurs when the flow in one pathway affects the flow in another pathway. *Positive feedback* enhances or increases the flow within a pathway, while negative feedback reduces it. A system with *negative feedback* loops or pathways tends to be self-stabilizing and moves toward an equilibrium—a steady state in which flow rates in system pathways remain about the same.
- Systems may undergo periodic, repeating changes in flow rates that constitute time cycles. Important time cycles in physical geography range in length from hours to millions of years.

KEY TERMS

atmosphere
biodiversity
biogeography
biosphere
carbon cycle
cartography
climate change
climatology
closed flow system
continental scale
cycles
economic geography
environmental pollution
equilibrium

extreme events
feedback
flow systems
geographic information
    system (GIS)
geography
geography of soils
geomorphology
global scale
hazards assessment
human geography
hydrologic cycle
hydrosphere
individual scale

inputs
life layer
lithosphere
local scale
map
marine geography
material cycle
mathematical modelling
meteorology
oceanography
open flow system
outputs
pathways
physical geography

place
power source
realms
regional geography
regional scale
remote sensing
scale
statistics
structure
systematic geography
systems
systems approach
time cycle
water resources

REVIEW QUESTIONS

1. What is geography? Regional geography? Systematic geography?
2. Identify and define three important fields of science within physical geography.
3. Identify and describe three tools that geographers use to acquire and display spatial data.
4. Why is the loss of biodiversity a concern of biogeographers and ecologists?
5. Give examples of extreme events and discuss how they might be influenced by human activity.
6. Name and describe each of the four great physical realms of Earth. What is the life layer?
7. Provide two examples of processes or systems that operate on each of the following scales: global, continental, regional, local, and individual.

8. What is a flow system? Provide an example. Distinguish two types of flow systems.
9. What distinguishes an open flow system from a closed flow system?
10. What is a cycle (material cycle)? Why are current research efforts focused on the carbon cycle?
11. Describe the concepts of feedback and equilibrium as applied to systems. Provide an example drawn from a natural system.
12. What is a time cycle as applied to a system? Give an example of a time cycle evident in natural systems.

EYE ON THE LANDSCAPE

Chapter Opener *Vancouver, British Columbia* With its deep embayments and steep topography, this is a coastline of submergence (**A**). Much of the terrain was carved into steep slopes and wide valleys by glacial ice descending from nearby peaks during the late-Cenozoic Ice Age, when the sea level was as much as 100 metres lower than it is today. With the melting of ice sheets, the sea level rose, drowning the landscape. Also note the snowcapped peaks in the far distance (**B**). Rugged landscapes like this often indicate recent plate tectonic activity in which immense crustal plates collide to push up mountain chains like B.C.'s Coast Mountains. Snow persists only on the higher peaks of the Coast Mountains (**C**) in this mild, moist marine west coast climate, but high annual precipitation favours the growth of dense conifer forests, remnants of which are preserved within the city (**D**).

River meandors. The Vermillion River near Sudbury, Ontario meanders through stands of spruce and aspen (**A**) wich are characteristic of the mixed wood section of the southern boreal forest

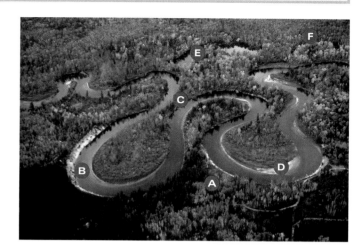

(see Chapter 23). The outer nank of a meander (**B**) is comparately steep where it collapses as it is undercut by the river (see Chapter 16). A narrow neck of land may be left between adjacent meanders (**C**). The inner bank of a meander slopes more gently to the river where sediments are deposited in the slower moving water. Abandoned channes may remain as oxbow lakes (**E**) or becoming progressively infilled with sediment and peat (**F**).

Part 2

CHAPTERS IN PART 2

2   The Earth as a Rotating Planet
3   The Global Energy System
4   Air Temperature and Air Temperature Cycles
5   Winds and the Global Circulation System
6   Atmospheric Moisture and Precipitation
7   Weather Systems
8   The Global Scope of Climate
9   Low-Latitude Climates
10 Mid-Latitude and High-Latitude Climates

WEATHER AND CLIMATE SYSTEMS

Late-winter sunsent over the Churchill River, northern Saskatechewan.

The flow of energy from the Sun to the Earth powers a vast and complex system of energy and matter flows within the atmosphere and oceans, and at the land surface. Part 2 explores how these flows are linked to weather and climate. ■ The six chapters (2-7) look at the forces and factors that influence the measurable properties of the atmosphere—those that are associated with general weather phenomena. They start with a discussion of the Earth's rotation on its axis and its revolution around the Sun and how this induces the daily and seasonal rhythms in energy. They discuss the link between energy flows and global temperature, then review global pressure and circulation patterns. The atmospheric components of the global water cycle are discussed with particular emphasis on evaporation, precipitation, and storm systems. ■ The last three chapters in Part 2 are devoted to climate and describe the general weather patterns that different regions of the world experience during a typical year.

Chapter 2

EYE ON THE LANDSCAPE

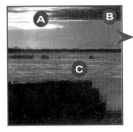

January in the Prairies *A CN freight train heading west near Saskatoon, Saskatchewan.* **What else would the geographer see? . . . Answers at the end of the chapter.**

THE EARTH AS A ROTATING PLANET

The Shape of the Earth
Earth Rotation
Environmental Effects of Earth Rotation
The Geographic Grid
Parallels and Meridians
Latitude and Longitude
Map Projections
*Working It Out 2.1 • Distances from Latitude
    and Longitude*
*Focus on Remote Sensing 2.2 • The Global
    Positioning System*
Polar Projection
Mercator Projection
Goode Projection
*Geographer's Tools 2.3 • Geographic
    Information Systems*
Global Time
Standard Time
World Time Zones
Daylight Saving Time
Geographers at Work • Tarmo Remmel, Ph. D.
International Date Line
The Earth's Revolution around the Sun
Tilt of the Earth's Axis
Solstice and Equinox
Equinox Conditions
Solstice Conditions
A Closer Look
Geographer's Tools 2.4 • Focus on Maps

The systematic study of physical geography starts with a discussion of the motions of the Earth and Sun, and their implications for global location, world time zones, and the changing seasons.

THE SHAPE OF THE EARTH

Pictures taken from space show that the Earth's shape is close to that of a sphere. Technically, the Earth assumes the shape of an *oblate ellipsoid* because it bulges slightly at the equator and flattens at the poles, due to the centrifugal force of its rotation. Thus, the Earth's equatorial diameter is about 12,756 kilometres, compared with its polar diameter of 12,714 kilometres. *Geodetic surveys* carry out precise measurements of the Earth's shape, which are important for the accuracy of modern satellite navigation systems used in aircraft, ships, and vehicles. These navigation systems determine the vehicles' exact location through the Global Positioning Systems (GPS) (see *Focus on Remote Sensing 2.2 • The Global Positioning System*, later in this chapter).

EARTH ROTATION

Earth rotation refers to the counter-clockwise turning of the Earth on its axis, an imaginary straight line passing through the centre of the planet and joining the North and South poles. The Earth completes one rotation with respect to the Sun every day (Figure 2.1). Each rotation defines the solar day and is the basis of conventional time systems. The axis of rotation also serves as a reference in setting up the system of latitude and longitude used for global positioning.

ENVIRONMENTAL EFFECTS OF EARTH ROTATION

One effect of the Earth's rotation is that it imposes a daily, or diurnal, rhythm on how it receives solar energy. This in turn influences air temperature, air humidity, and air motion. A second effect of the Earth's rotation is that it causes the winds and ocean currents to turn consistently in a sideways direction. In the northern hemisphere, flows are deflected to the right, and in the southern hemisphere flows are deflected to the left. This important phenomenon is termed the *Coriolis effect* (see Chapter 5). Additionally, the Earth's rotation is linked to the diurnal rise and fall of the ocean tides (see Chapter 18).

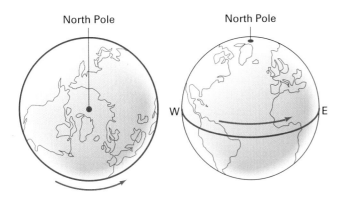

2.1 Direction of Earth rotation The direction of rotation of the Earth can be thought of as (a) counter-clockwise at the North Pole, or (b) eastward (from left to right) at the equator.

THE GEOGRAPHIC GRID
PARALLELS AND MERIDIANS

The **geographic grid** is a spherical coordinate system used to determine the locations of features on the Earth's surface. It is constructed from two sets of intersecting circles (Figure 2.2). One set of circles, termed **parallels**, is arranged perpendicular to the axis of rotation. These intersect the second set of circles, known as **meridians**, at right angles.

The **equator** lies midway between the North Pole and South Pole, and is the Earth's longest parallel. The other parallels run parallel to the equator and their positions refer to the angular distance from the Earth's centre. The 49th parallel, which forms much of the border between Canada and the United States, is a well-known example. Meridians run from the North Pole to the South Pole and represent half of an imaginary circle drawn around the surface of the Earth.

Meridians and parallels define geographic directions. Meridians run in a true north-south direction; parallels are east-west lines. Every point on the globe is associated with a unique combination of one parallel and one meridian; their intersection defines the position of the point.

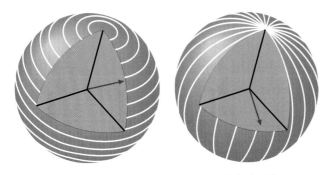

2.2 Parallels and meridians (a) Parallels of latitude divide the globe crosswise into rings. (b) Meridians of longitude divide the globe from pole to pole.

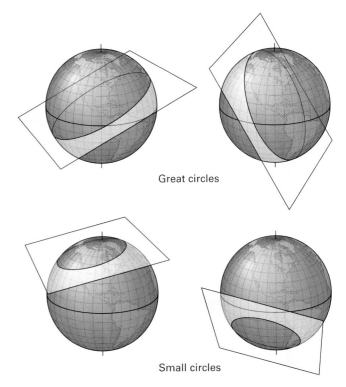

2.3 Great and small circles A great circle is created when a plane passes through the Earth, intersecting the Earth's centre. Small circles are created when a plane passes through the Earth but does not intersect the centre point.

Two types of circles—great circles and small circles—can be drawn on the surface of a globe. With a **great circle** the plane of intersection passes through the centre of the globe and so divides it into two equal halves. With a **small circle** the plane of intersection passes through the surface of the globe, but not its centre (Figure 2.3). Thus, meridians are actually halves of great circles, while all parallels, except the equator, are small circles. Great circles can be aligned in any direction on the globe and can be drawn to pass through any two points on its surface. Note that the arc of the great circle between the two points represents the shortest distance between them.

LATITUDE AND LONGITUDE

Locations on the Earth's surface are represented by latitude and longitude. Parallels of **latitude** measure the angular distance north or south of the equator, ranging from lat. 0° at the equator to 90° N at the North Pole and 90° S at the South Pole. Meridians of **longitude** measure angular distance east or west of the prime meridian, which runs through the Royal Observatory in Greenwich, England (Figure 2.4), and is commonly referred to as the Greenwich meridian. It has the value long. 0°. The longitude of a meridian is measured eastward or westward from the prime meridian and ranges from 0° to 180° E or 180° W.

For greater precision, latitude and longitude angles can be subdivided into *minutes* and *seconds*. A minute is 1/60 of a

Table 2.1 Distance in kilometres of 1 degree of longitude at various latitudes

Latitude	Length of 1 Degree of Longitude
0	111.32 km
15	107.55 km
30	96.45 km
45	78.85 km
60	55.80 km
75	28.90 km
90	0 km

2.4 The prime meridian This photograph, taken at dusk at the old Royal Observatory at Greenwich, England, shows the prime meridian, which has been marked as a stripe on the pavement in front. *Fly By: 51° 28' 37" N; 0° 00' 02" W*

degree, and a second is 1/60 of a minute, or 1/3,600 of a degree. Degrees of latitude and longitude can also measure distance. A degree of latitude, which measures distance in a north-south direction, is equal to about 111 kilometres. However, the distance associated with a degree of longitude varies, because meridians converge toward the poles. Thus, the length of one degree of longitude varies from approximately 111 kilometres at the equator to 0 kilometres at the poles (see Table 2.1). *Working It Out 2.1 • Distances from Latitude and Longitude* shows how to convert positional information from latitude and longitude coordinates into distance.

Latitude and longitude can now be determined quickly and accurately with the help of the **Global Positioning System (GPS)**. This satellite-based system constantly sends radio signals to Earth

WORKING IT OUT | 2.1 Distances from Latitude and Longitude

Statements of latitude and longitude do not describe distances in kilometres or miles directly. However, for latitude, the conversions from degrees into kilometres can be estimated quite easily. One degree of latitude is approximately equivalent to 111 kilometres of surface distance in the north-south direction. For example, Saskatoon at lat. 52° N is about 52 × 111 = 5,772 kilometres north of the equator.

East-west distances cannot be converted so easily from degrees of longitude into kilometres or miles because the meridians converge toward the poles. Only at the equator is a degree of longitude equivalent to 111 kilometres (see Table 2.1). At lat. 60° N or S, the distance between meridians is about half

that at the equator, or about 56 kilometres. At the poles, the distance is zero.

The formula to determine the length of a degree of longitude is

$$L_{LONG} = \cos (lat) \times L_{LAT}$$

where L_{LONG} is the length of a degree of longitude; cos is the trigonometric cosine function; *lat* is the latitude, in degrees, of the location at which the length is to be calculated; and L_{LAT} is the length of a degree of latitude, that is, 111 kilometres.

For example, a degree of longitude at 20° lat. has a length of

$$L_{LONG} = \cos (lat) \times L_{LAT}$$
$$= \cos (20°) \times 111 \text{ km}$$
$$= 0.866 \times 111 \text{ km}$$
$$= 96.1 \text{ km}$$

This principle can be used to find the distance between two cities located at the same latitude. For example, Vancouver and Winnipeg, at 49°N, are separated by 26° of longitude.

First, determine the length of a degree of longitude at 49° lat. That is,

$$L_{LONG} = \cos (lat) \times L_{LAT}$$
$$= \cos (49°) \times 111 \text{ km}$$
$$= 0.656 \times 111 \text{ km}$$
$$= 72.8 \text{ km}$$

Then,

$$26° \text{ long} \times 72.8 \text{ km} = 1,892.8 \text{ km}$$

2.5 GPS receiver Latitude and longitude can be rapidly and accurately determined using a hand-held GPS receiver.

with information that allows a GPS receiver (Figure 2.5) to calculate its position on the Earth's surface. *Focus on Remote Sensing 2.2 • The Global Positioning System* provides more information on the GPS and how it works.

MAP PROJECTIONS

The geographic grid is the basis of *cartography*, or map making, which provides techniques for accurately displaying the locations of continents, rivers, cities, islands, and other geographic features on maps and in atlases. Because the Earth's shape is nearly spherical it is impossible to represent it on a flat sheet of paper without cutting, stretching, or otherwise distorting the curved surface in some way. The main properties affected are direction, distance, shape, and area. A map preserves direction when the bearing between two points is correct anywhere on the map. Similarly, a map preserves shape when shape and area on the map are uniformly proportional to the real world locations they represent.

There are various methods to change the actual geographic grid of curved parallels and meridians into a flat coordinate system. This transformation is achieved mathematically using various **map projections**. There are many different map projections, but no single projection can preserve, simulataneously, all of the properties that are affected. The purpose of the map, therefore, determines which projection to use. Three common ones are the polar projection, Mercator projection, and Goode projection. The interchapter feature at the end of the chapter, *A Closer Look: Geographer's Tools 2.4 • Focus on Maps*, provides more information about how maps are made and how they display information.

FOCUS ON REMOTE SENSING | 2.2   The Global Positioning System

The latitude and longitude coordinates of a point on the Earth's surface describe its position exactly. But how are these coordinates determined? For the past few hundred years, latitude was calculated from the position of the stars, and longitude was determined with the aid of an accurate clock set to Greenwich Mean Time.

The Global Positioning System (GPS), which can provide location information to an accuracy of about 10 metres horizontally and 15 metres vertically, has now superseded this. The system uses 24 satellites that orbit the Earth every 12 hours, continuously broadcasting their position and a precise time signal.

To determine location, a receiver listens simultaneously to signals from four or more satellites. The receiver compares the time readings transmitted by each satellite with the receiver's own clock to determine how long it took for each signal to reach the receiver. Since the radio signal travels at a known rate, the receiver can convert the signal travel time into the distance between the receiver and the satellite. By combining the distance to each satellite with the position of the satellite in its orbit at the time of the transmission, the receiver calculates its position on the ground.

Several types of errors can affect the accuracy of the location. First, unpredictable events such as solar particle showers cause small perturbations in the orbits of the satellites. Another source of error is small variations in the atomic clock that each satellite carries. A larger source of error, however, is the effect of the atmosphere on the satellite signal. The layer of charged particles at the outer edge of the atmosphere (ionosphere) and water vapour slow the radio waves. Since these conditions change continuously, the speed of the radio waves varies in an unpredictable way. Another problem is that the radio waves may bounce off local obstructions or be blocked in mountainous terrain.

Two GPS units can determine a more accurate location, to within about one metre horizontally and two metres vertically. One unit acts as a stationary base station, and the other is moved to the desired locations. The base station unit is placed at a position that is known with high accuracy. By comparing its exact position with that calculated from the satellite signals, it corrects any errors and then broadcasts that information to the GPS field unit at the desired location. Because this method compares two sets of signals, it is known as *differential GPS*. Differential GPS is now widely used for coastal navigation and in aircraft landing systems.

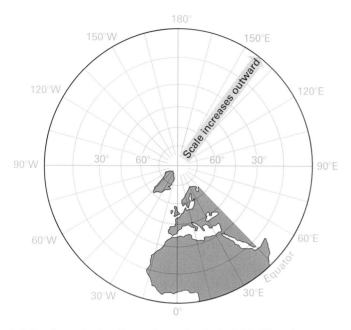

2.6 A polar projection The map is centred on the North Pole. All meridians are straight lines radiating from the centre point, and all parallels are concentric circles. Scale increases toward the periphery.

POLAR PROJECTION

The **polar projection** (Figure 2.6) can be centred on either the North or South Pole. Meridians are straight lines radiating outward from the pole, and parallels are nested circles centred on the pole. The space between the parallels increases outward from the centre. The equator usually forms the boundary of the map. Because the parallels with the meridians intersect at right angles, this projection shows the true shapes of small areas, such as islands. However, because scale increases away from the centre, shapes look disproportionately larger toward the edge of the map.

MERCATOR PROJECTION

The **Mercator projection** (Figure 2.7) is a rectangular grid with meridians shown as straight vertical lines, and parallels as straight horizontal lines. Meridians are evenly spaced, but the spacing between parallels increases with latitude and at 60° is double that at the equator. Closer to the poles, the spacing increases even more, and the map must be cut off at some arbitrary parallel; typically about 80° N and 65° S are used. This change of scale enlarges features near the poles. For example, Greenland appears to be nearly the size of Africa. In fact, the surface area of Greenland is about

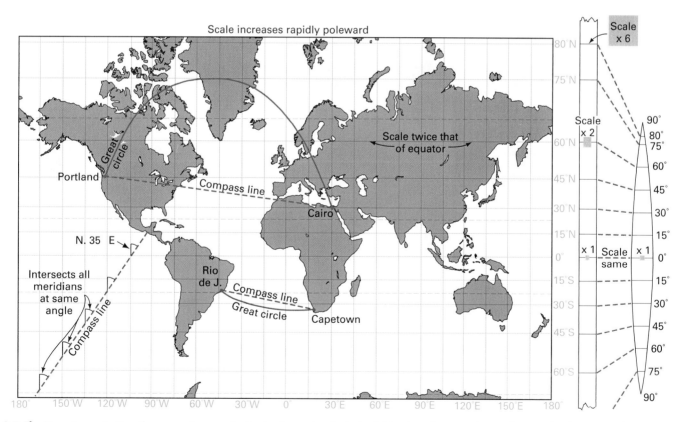

2.7 The Mercator projection The compass line connecting two locations, such as Portland and Cairo, shows the compass bearing of a course directly connecting them. However, the shortest distance between them lies on a great circle, which is a longer, curving line on this map projection. The diagram at the right shows how rapidly the map scale increases at higher latitudes. At lat. 60°, the scale is double the equatorial scale. At lat. 80°, the scale is six times greater than at the equator. For this reason, the map is typically truncated at about 80° N and 65° S.

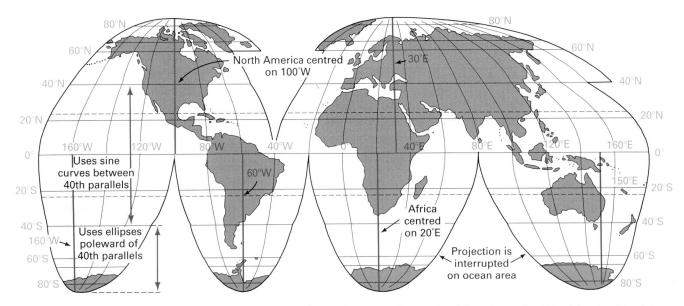

2.8 The Goode projection The meridians in this projection follow sine curves between lat. 40° N and lat. 40° S, and ellipses between lat. 40° and the poles. Although the shapes of continents are distorted, their relative surface areas are preserved. (Copyright © by the University of Chicago. Used by permission of the Committee on Geographical Studies, University of Chicago.)

2,175,000 square kilometres, compared with about 30,000,000 square kilometres for Africa.

The Mercator projection has several special properties that make it indispensable for navigation. One is that a straight line drawn anywhere on the map is a line of constant compass direction. A navigator can therefore simply draw a line between any two points on the map and measure the *bearing,* or direction of the line, with respect to a nearby meridian on the map. Since the meridian is a true north-south line, the angle will give the compass bearing to follow. Once heading in that compass direction, a ship or airplane would eventually reach its destination.

However, this route will not necessarily follow the shortest actual distance between two points. As noted earlier, the shortest path between two points on the globe is always a portion of a great circle. On a Mercator projection, a great circle line curves (except on the equator) and can falsely seem to represent a much longer distance than a compass line.

Because the Mercator projection shows the true compass direction of any straight line on the map, it is used to show many types of straight-line features, such as the flow lines of winds and ocean currents, and global temperature and pressure patterns.

GOODE PROJECTION

The **Goode projection** uses two sets of mathematical curves to form its meridians. Between the 40th parallels, it uses sine curves, and beyond the 40th parallel, toward the poles, it uses ellipses. Since the ellipses converge to meet at the pole, the entire globe can be shown. The straight, horizontal parallels make it easy to scan across the map at any given level.

The Goode projection represents areas in correct proportion to the Earth's surface. Because of this, the Goode map is used to show regional distributions of geographical features, such as soils and vegetation that occur in various parts of the world. However, it distorts the shapes of areas, particularly in high latitudes. This effect is minimized by separating the map into sectors, each centred on a different vertical meridian (Figure 2.8). This type of split map is called an *interrupted projection.* Although the interrupted projection greatly reduces shape distortion, it does have the drawback of separating parts of the Earth's surface that are actually close together.

Maps are in wide use today for many applications as a simple and efficient way to compile and store spatial information— information associated with a specific location or area of the Earth's surface. However, in the past two decades, maps have been supplemented by more powerful computer-based methods for acquiring, storing, processing, analyzing, and outputting spatial data. *Geographer's Tools 2.3 • Geographic Information Systems* describes the basic concepts of these geographic information systems (GISs) and how they work.

GLOBAL TIME

Maps and map projections are a practical application of the Earth's geographic grid. Global time systems are also derived from the geographic grid, in conjunction with the Earth's rotation.

The global time system is based on the Sun and its apparent passage across the sky. In the morning, the Sun is low on the eastern horizon, and as the day progresses, it rises higher, until **solar noon,** when it reaches its highest point in the sky or *zenith.* After

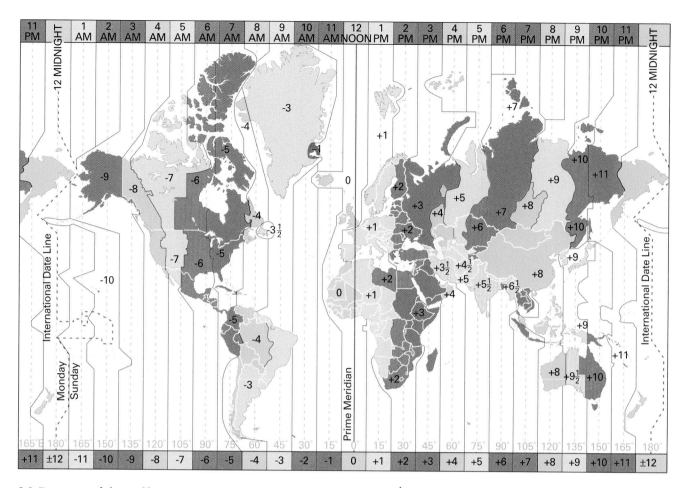

2.9 Time zones of the world Dashed lines represent 15° meridians, and bold lines represent 7½° meridians. Alternate zones appear in colour. (After U.S. Navy Oceanographic Office)

solar noon, the Sun's elevation in the sky decreases. By late afternoon, the Sun appears low in the sky, and at sunset, it rests on the western horizon. In most locations, the transition from daylight to darkness occurs with each daily rotation of the Earth. The daily cycle is not synchronous—it may be a sunny morning in Vancouver but night has come to Delhi.

The starting point for global time systems is the noon meridian—the north-south meridian along which at any given moment the sun is at its zenith. The noon meridian sweeps around the Earth with every 24-hour rotation. Each 15° of longitude therefore equates to one hour of time, and because of the direction of rotation, places to the east are always ahead.

STANDARD TIME

The world adopted standard time zones in 1884, largely as a result of the efforts of Canadian railway engineer, Sir Sandford Fleming. In the **standard time system**, the Earth is divided into 24 **time zones**. All inhabitants within a zone keep time according to a *standard meridian* that passes through their zone. Since the standard meridians are usually 15 degrees apart, the difference in time between adjacent zones is normally one hour.

Theoretically, the standard time system should consist of belts exactly 15° wide, extending to meridians 7½° east and west of each standard meridian. However, this is inconvenient if the boundary meridians divide a state, county, or city into two different time zones. As a result, time zone boundaries are often routed to follow agreed-upon natural or political boundaries. Figure 2.9 shows the 24 principal standard time zones of the world. It also shows the time of day in each zone when it is noon at the Greenwich meridian. The country spanning the greatest number of time zones from east to west is Russia, with eleven zones, but these are grouped into eight standard time zones. China spans five time zones but runs on a single national time using the standard meridian of Beijing.

Figure 2.10 shows the time zones for Canada in summer and winter. Some of the time zone boundaries are conveniently located along provincial boundaries, but there are several exceptions. For example, the Kootenay region of British Columbia sets its clocks to Mountain Standard Time in keeping with the neighbouring province of Alberta. Ontario, Quebec, and Labrador each have two time zones. Most of Nunavut uses the Eastern Time Zone; however, Kangiqliniq (Ranklin Inlet) uses Central Time

GEOGRAPHER'S TOOLS | 2.3   Geographic Information Systems

What Is a GIS?

A **geographic information system** contains software to acquire, process, store, query, create, analyze, and display spatial data—pieces of information that are in some way associated with a specific location or area of the Earth's surface. A simple example of a GIS is a map overlay system. Imagine that a planner is deciding how to divide a tract of land into building lots. Appropriate inputs might include a topographic map showing land elevations, a vegetation map showing the type of existing plant cover, a map of existing roads and trails, a map of streams and watercourses, a map of wetland areas, a map of power transmission lines and gas pipelines crossing the area, and so forth. From these, the planner could identify steep slopes, conservation areas, and other attributes, before laying out roads, designing bridges, and planning for other essential amenities.

To do this efficiently, the planner needs to be able to overlay the various maps. That's where computer-based geographic information systems come in; they allow the planner to manipulate spatial data for a variety of applications. GIS technology can alleviate problems that might arise, for example, when the input maps are drawn at different scales or when map data need further analysis, such as to identify areas of steep slopes.

Spatial Objects in Geographic Information Systems

Geographic information systems are designed to manipulate spatial objects. A *spatial object* is a geographic area to which some information is attached. This information may be as sim-

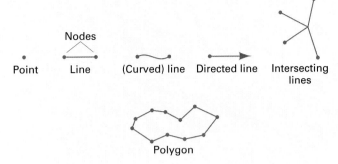

Spatial objects Spatial objects in a GIS can include points, lines of various types, intersecting lines, and polygons.

ple as a place name or as complicated as a large data table. A spatial object will normally have a boundary, described as a *polygon*, that outlines the object.

Spatial objects include points and lines. A *point* can be considered a special type of spatial object with no area. A *line* is also a spatial object with no area, but it has two points associated with it, one at each end of the line. These special points are referred to as *nodes*. Lines can be straight or curved. If the two nodes marking the ends of the line are differentiated as the start and end, then the line has a direction and its two sides can be distinguished—for example, land on one side and water on the other. Lines connect to other lines when they share a common node. A series of connected lines that form a closed chain is a polygon.

By defining spatial objects in this way, computer-based geographic information systems can perform many different

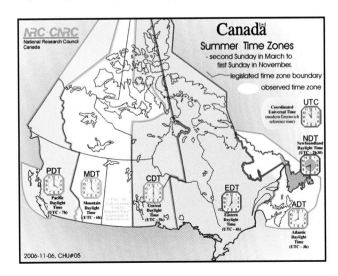

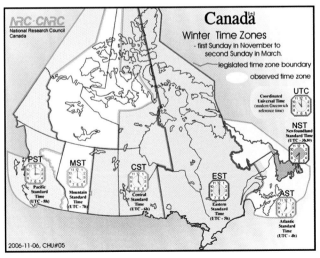

2.10 Time zones of Canada This figure shows the time zones for Canada in winter and summer. Several time zones do not follow provincial boundaries, and not all places observe Daylight Saving Time, hence the boundaries vary between summer and winter.

types of operations to compare objects and generate new objects. For example, suppose a GIS data layer composed of conservation land in a region is represented as polygons, and another layer containing the location of pre-existing water wells is represented as points within the region. The GIS can identify the wells located on conservation land and produce a new data layer showing this. The GIS could also compare the conservation layer to a layer of polygons showing vegetation type, and it could tabulate the amount of conservation land in forest, grassland, brush, and so on. Similarly, it could calculate distance zones around a spatial object in order to create a map of buffer zones located within, say, 100 metres of conservation land.

Key Elements of a GIS

A geographic information system consists of five elements: data acquisition, preprocessing, data management, data manipulation and analysis, and product generation. Each is a component or process needed to ensure the functioning of the system as a whole. In the *data acquisition* process, data are gathered together for the particular application. These may include maps, air photos, or information tables. In *preprocessing*, the GIS converts the assembled spatial data into forms that are compatible with the system to produce data layers of spatial objects and their associated information. The *data management* component creates, stores, retrieves, and modifies data layers and spatial objects, which is essential to the proper functioning of all parts of the GIS. The key feature of the GIS is the *manipulation and analysis* component, with which the user asks and answers questions about spatial data

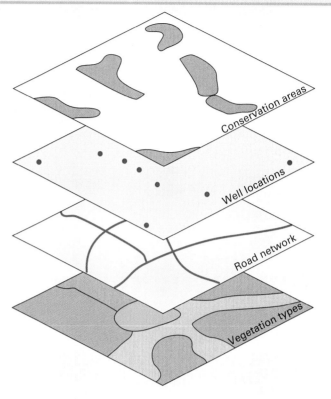

Data layers in a GIS A GIS allows easy overlay of spatial data layers for such queries as "Identify all wells on conservation land."

and creates new data layers of derived information. The last component of the GIS, *product generation*, produces output in the form of maps, graphics, tabulations, and statistical reports, which are the end products the users desire.

and Southampton Island only Standard Time. Newfoundland Standard Time is adjusted by 30 minutes to coincide with the half-hour meridian at the provincial capital St. John's. Here, time on the clocks is closer to local solar time. Most of Saskatchewan is in the Central Time Zone, except the city of Lloydminster, which observes Mountain Standard Time.

WORLD TIME ZONES

World time zones are numbered to indicate the number of hours' difference between time in a zone and time in Greenwich. A number of −7, for example, indicates that local time is seven hours behind Greenwich Time, while a +2 indicates that local time is two hours ahead of Greenwich Time. Because of the historical importance of the Greenwich Observatory, world time was

traditionally referenced to Greenwich Mean Time (GMT). Recently **Coordinated Universal Time (UTC)** has superceded GMT as the global reference. UTC is the legal time standard recognized by all nations and administered by the Bureau International de l'Heure, located near Paris. UTC is a high-precision atomic time standard that is periodically adjusted by *leap seconds* to compensate for discrepancies in the Earth's rotation. It is precisely related to the Earth's angular rotation, rather than to the passage of a 24-hour period.

DAYLIGHT SAVING TIME

Especially in urban areas, many human activities begin well after sunrise and continue long after sunset. Therefore, in many parts of the world clocks are adjusted during the part of the

GEOGRAPHERS AT WORK

The Use of Geographic Information Systems in the Boreal Forests

by Tarmo Remmel, Ph.D., York University

The predominantly coniferous boreal forest wraps across the northern reaches of Canada, Europe, and Eurasia forming a nearly continuous forest growing on nutrient-poor soils and which has adapted to a relatively dry and cold climate. This zone provides natural habitat to numerous

Tarmo Remmel conducting a tree survey in a southern Ontario woodlot for a comparison study.

fauna and oxygen-producing flora, while functioning as a filter for the air that we breathe. However, aside from being a seemingly untouched wilderness, the boreal forest is also a vast reserve of natural resources that are periodically disturbed by large natural wildfires that represent a regular ecological mechanism driving regeneration, nutrient cycling, habitat change, diversification, and carbon release to the atmosphere.

Since forestry-related harvesting operations are also prevalent in the boreal forest, the compounded disturbance effect poses several management complications that yield complex questions. Should natural wildfires be suppressed to protect economically viable forest resources? If wildfires are a natural process, should forest harvesting operations emulate the patterns of natural disturbance? Although some jurisdictions have policies regarding the emulation of natural disturbances by anthropogenic

processes, these systems are not clearly understood; their impact on the landscape and ultimately the climate are unknown.

Measuring and quantifying spatial patterns as recorded during field campaigns, from satellite image interpretation, and GIS analyses are Tarmo Remmel's specialty. He is currently working to quantify the spatial behaviour of these large boreal wildfires, such that probabilistic models can be built to predict the spatial occurrence probabilities of vegetation residuals and to determine differences in forest recovery following natural versus anthropogenic disturbances. By better understanding of spatial forest processes involving disturbance and recovery, forest management practices might be improved to ensure the sustainability of boreal forest resources and to minimize greenhouse gas releases to the atmosphere that could be influencing the global climate.

year with a longer daylight period to correspond more closely with the pace of society. This adjusted time system, called **Daylight Saving Time**, is obtained by setting all clocks ahead by one hour. The effect of the time change is to transfer the early morning daylight period, theoretically wasted while schools, offices, and factories are closed, to the early evening, when most people are awake and busy. Daylight Saving Time also yields a considerable savings in power used for electric lights. In Canada, Daylight Saving Time comes into effect on the second Sunday in March and is discontinued on the first Sunday of November, although it is not observed in some areas. This is the case throughout most of Saskatchewan, with the exception of Lloydminster in the west of the province and Denare Beach and Creighton in the east. Other places that do not alter their clocks include Fort St. John, Dawson Creek in British Columbia, Creston in the East Kootenays of B.C., and Southampton Island in Nunavut.

INTERNATIONAL DATE LINE

Counting eastward from the Greenwich meridian, places on the 180th meridian are 12 hours ahead. Counting westward from the Greenwich meridian, places on the 180th meridian are 12 hours behind. Since this is the same meridian, an adjustment must be made to resolve this apparent paradox.

Imagine that it is exactly midnight on the 180th meridian on, say, June 26. At that instant, the same 24-hour calendar day covers the entire globe. To the east it will be the early morning of June 26, while to the west it is late in the evening of June 26. It is the same calendar day on both sides of the meridian but 24 hours apart in time.

Repeating the calculations an hour later, at 1 a.m., to the east it is the early morning of June 26, but to the west, midnight of June 26 has passed, and it is now the early morning of June 27. So on the west side of the 180th meridian, it is also 1 a.m., but one day later than on the east side. For this reason, the 180th meridian serves as the **International Date Line**. This means that travelling westward across the date line, calendars must advance by one day. When travelling eastward, calendars turn back by a day. For example, flying westward from Vancouver to Tokyo, you may depart on a Tuesday evening and arrive on a Thursday morning after a flight that lasted only 14 hours. On an eastward flight, you may actually arrive the day before you took off.

The International Date Line does not follow the 180th meridian exactly. Like many time zone boundaries, it deviates from the meridian for practical reasons. As shown in Figure 2.9, it has a zigzag offset between Asia and North America, as well as an eastward offset in the South Pacific to keep clear of New Zealand and several island groups.

THE EARTH'S REVOLUTION AROUND THE SUN

So far, we have discussed the importance of the Earth's rotation on its axis. Another important motion of the Earth is its **revolution**, its movement in orbit around the Sun.

The Earth completes a revolution around the Sun in 365.242 days—about one quarter day more than the calendar year of 365 days. Every four years, the extra quarter days add up to about one whole day. The 29th day in February in leap years corrects the calendar, although further minor corrections are necessary to perfect the system.

The Earth traces a slightly elliptical orbit around the Sun. Thus, the distance between the Earth and the Sun varies by about three percent during each revolution. The Earth is nearest to the Sun at *perihelion*, which occurs on or about January 3, and farthest from the Sun at *aphelion*, on or about July 4.

The direction of the Earth's revolution around the Sun is counter-clockwise, the same direction as the Earth's rotation on its axis (Figure 2.11).

2.11 Revolution of the Earth and Moon Viewed from a point over the Earth's North Pole, the Earth both rotates and revolves in a counter-clockwise direction. From this viewpoint, the Moon also rotates counter-clockwise.

2.12 The phases of the moon The Moon orbits Earth every 28 days in a counter-clockwise direction. The new moon occurs when the Moon is between the Sun and Earth. It waxes through first quarter to full moon when it is on the side away from the Sun, then wanes through last quarter to new moon again, and then the cycle repeats.

The Moon also rotates on its axis and revolves around the Sun in a counter-clockwise direction. However, the Moon's rate of rotation is such that one side of the Moon is always directed toward the Earth. Thus, astronomers had never seen the far side of the Moon until a Soviet spacecraft passing the Moon transmitted photos back to Earth in 1959. The phases of the Moon are determined by its position in orbit around the Earth during the month (Figure 2.12).

TILT OF THE EARTH'S AXIS

The seasons are related to the orientation of the Earth's axis of rotation as it revolves around the Sun. The Earth's axis is tilted with respect to the **plane of the ecliptic**—the plane circumscribed by the Earth's orbit around the Sun. Figure 2.13 shows the plane of the ecliptic as it intersects the Earth. The axis of the Earth's rotation is tilted at an angle of $66\frac{1}{2}°$ to the plane of the ecliptic. Note that the direction toward which the axis points is fixed in space—it aims toward Polaris, the north star. Neither the angle nor the direction of the axis change as the Earth revolves around the Sun.

SOLSTICE AND EQUINOX

Because the direction of the Earth's axis of rotation is fixed, the North and South Poles are alternately tilted toward and away from the Sun at different points on the orbit. The Earth is always divided into two hemispheres with respect to the Sun's rays. One hemisphere is illuminated by the Sun, and the other lies in darkness (Figure 2.14). The *circle of illumination*, which separates the illuminated hemisphere from the non-illuminated hemisphere, essentially represents the twilight period at dawn and dusk.

On or about December 22, the Earth is positioned so that the North Pole is inclined at the maximum angle of $23\frac{1}{2}°$ away from the Sun, and the South Pole is inclined at the same angle toward the Sun. In Canada, this event is usually called the **winter solstice**. But while it is winter in the northern hemisphere, it is summer in the southern hemisphere, so the preferred term is *December solstice* to avoid any confusion. At this time of the year, the southern hemisphere is tilted toward the Sun and receives strong solar heating.

Six months later, on or about June 21, the Earth is at the opposite point on its orbit. At this date, known as the **summer solstice** (or *June solstice*), the North Pole is tilted at $23\frac{1}{2}°$ toward the Sun and the South Pole is tilted away from the Sun. Theoretically, this is the time of maximum energy gain for the northern hemisphere, and minimum energy gain for the southern hemisphere.

The equinoxes occur midway between the solstice dates. At an **equinox**, the Earth's axial tilt is neither toward nor away from the Sun. The *vernal (spring) equinox* occurs on or about March 21, and the *autumnal (fall) equinox* occurs on or about September 22.

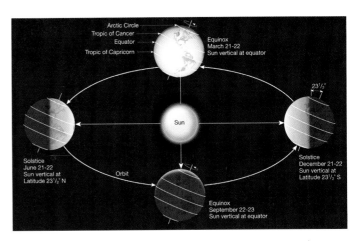

2.14 The seasons The spring, summer, autumn, and winter seasons occur because the Earth's tilted axis keeps a constant orientation in space as the Earth revolves around the Sun. In summer the northern hemisphere is inclined toward the Sun, and in winter away from the Sun. Both hemispheres are illuminated equally at the vernal (spring) and autumnal equinoxes.

The orientation of the Earth with the Sun is identical on the two equinoxes. Note that the date of the solstices and equinoxes in a particular year may vary by a day or so, since the revolution period is not exactly 365 days.

EQUINOX CONDITIONS

At the equinoxes, the circle of illumination passes through the North and South poles (Figure 2.15), and the Sun's rays just touch the Earth's surface at either pole. On these dates, the *subsolar point*, the point on the Earth's surface where the Sun at noon is directly overhead, falls on the equator. Here, the angle between the Sun's rays and the Earth's surface is 90°, and solar radiation theoretically is at its most intense. At an intermediate latitude, such as 40° N, the rays of the Sun at noon strike the surface at a lesser angle. This *zenith angle*—the elevation of the Sun above the horizon at noon—is equal to 90° minus the latitude, in this example 90° − 40° = 50°. A sextant, a common navigation instrument used by sailors, uses this principle by taking a site on the Sun when it is at its zenith to fix the ship's latitude. At the equinoxes, the latitude is equal to 90° minus the zenith angle of Sun. At any other time of the year, a correction has to be applied to compensate for the Sun's movement north and south of the equator with the changing seasons.

In the 24-hour period at equinox, during which the Earth completes one rotation, every point on the surface receives 12 hours of darkness and 12 hours of daylight. This is because the circle of illumination passes through the poles, dividing every parallel exactly in half. The word "equinox" is derived from the Latin *aequus* (equal) and *nox* (night). Note that we have carefully

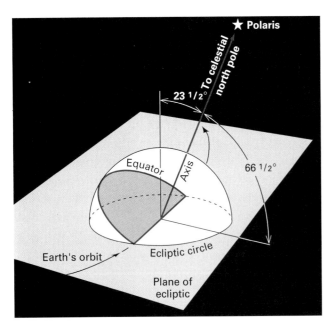

2.13 The tilt of the Earth's axis of rotation with respect to its orbital plane is $66\frac{1}{2}°$ As the Earth moves in its orbit on the plane of the ecliptic around the Sun, its rotational axis remains pointed toward Polaris, the north star. Note the angle between the equator and the ecliptic plane is $23\frac{1}{2}°$.

used the term daylight to describe the period of the day during which the sun is above the horizon. When the sun is not too far below the horizon, scattering of solar rays by atmospheric particles still lights the sky and we observe twilight. At high latitudes during the polar night, a twilight period of some hours length provides enough illumination for many human activities.

SOLSTICE CONDITIONS

In both the summer and winter solstices the circle of illumination passes from the Arctic Circle to the Antarctic Circle (Figure 2.16). Note that during the summer solstice the Earth's orientation puts the entire region north of the **Arctic Circle** in 24 hours of daylight. The situation is reversed in the southern hemisphere, where regions beyond the **Antarctic Circle** are in continuous darkness. During the summer solstice the subsolar point is at lat. $23\frac{1}{2}°$ N, the parallel known as the **Tropic of Cancer**. Because the Sun is directly over the Tropic of Cancer at this time, solar energy is most intense there.

At the winter solstice, conditions are exactly reversed from those of the summer solstice. Everywhere south of lat. $66\frac{1}{2}°$ S lies under the Sun's rays for 24 hours, and the subsolar point is now positioned over the **Tropic of Capricorn** at lat. $23\frac{1}{2}°$ S.

The solstices and equinoxes represent conditions at only four times of the year. Between these times, the subsolar point travels northward and southward in an annual cycle between the Tropics of Cancer and Capricorn. The latitude of the subsolar point is referred to as the Sun's **declination**. Nautical almanacs contain tables listing these values for every day of the year. Sailors use these tables, which show what is referred to as the *Sun's ephemeris*, to establish latitudes using a sextant.

As the seasonal cycle progresses in polar regions, areas of 24-hour daylight or 24-hour night shrink as the Sun begins to move

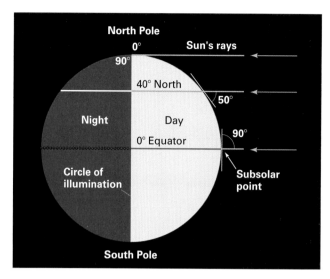

2.15 Equinox conditions At this time, the Earth's axis of rotation is exactly at right angles to the direction of solar illumination. The subsolar point lies on the equator. The Sun touches the horizon at both poles.

out of their respective hemispheres and grow as it returns. At other latitudes, the length of daylight changes progressively from one day to the next. The difference in duration of daylight and darkness increases with latitude. Only the equator experiences approximately 12 hours of daylight and 12 hours of darkness each day of the year.

A Look Ahead

The flow of solar energy to the Earth powers most of the natural processes that occur each day, from changes in the weather to the work of streams in carving the landscape. The next chapter examines in detail solar energy and its interaction with the Earth's atmosphere and surface.

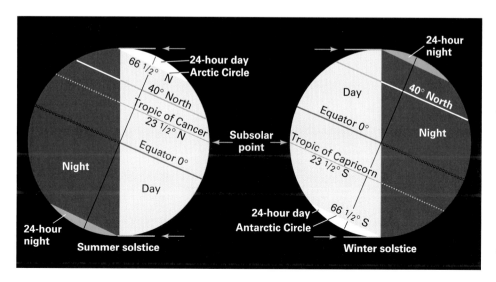

2.16 Solstice conditions Because of the tilt of the Earth's axis, polar regions experience either 24-hour daylight or 24-hour darkness. The subsolar point lies on the Tropic of Cancer (lat. $23\frac{1}{2}°$ N) in June and on the Tropic of Capricorn (lat. $23\frac{1}{2}°$ S) in December.

A CLOSER LOOK   Geographer's Tools 2.4   Focus on Maps

The purpose of this special supplement is to provide additional information on the subject of cartography, the science of maps and their construction.

More about Map Projections

A *map projection* is an orderly system of parallels and meridians that is used as a base for transferring the features of the spherical Earth to a flat surface. All map projections distort the Earth's shape in some way as a result of cutting, stretching, or other manipulation of the true spatial properties.

A simple map projection might consist of a grid of squares or rectangles with horizontal lines representing parallels and vertical lines representing the meridians. Modern computer-generated world maps often use this type of grid to display data that consist of a single number for each square (see Figure 1). A grid of this kind can represent the approximate spacing of the parallels, but it fails to show how the meridians converge toward the poles.

Early attempts to find satisfactory map projections used a simple concept. Imagine the continents and a grid of meridians and parallels drawn on a transparent sphere. A light source placed at the centre of the sphere will cast an image of the continents and grid on a surface outside the sphere. Three basic surfaces can be used: flat (or planar), conical, and cylindrical (Figure 2).

First is a flat paper disk balanced on the North Pole. The shadow of the grid on this plane surface appears as a combination of concentric circles (parallels) and radial straight lines (meridians). This produces a polar (or zenithal) projection. Second is a cone resting point-up on the sphere. The cone can be unrolled to produce a map that is part of a full circle. This is called a conic projection. Parallels

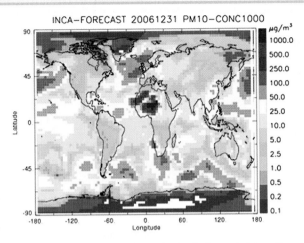

Figure 1 A computer-generated map showing the fine particulate matter concentration in the atmosphere on November 28, 2006.

are arcs of circles, and meridians are radiating straight lines. Third, a cylinder touching the sphere at the equator when unrolled produces a cylindrical projection, with a rectangular grid. The Mercator projection uses this principle.

Note that none of these three projection methods can show the entire Earth grid, no matter how large a sheet of paper is used to draw the image. To show the entire Earth grid, a quite different system must be devised using various mathematical techniques. The *Goode projection* is an example.

Scales of Globes and Maps

All globes and maps depict the Earth's features at a much smaller size than they really are. Globes are basically scale models of the Earth with the *scale* of a globe rep-

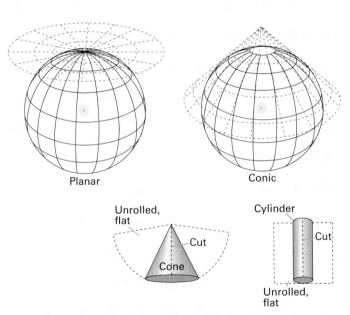

Figure 2 Rays from a central light source would cast shadows of a transparent, spherical Earth on different surfaces. The conical and cylindrical surfaces can be unrolled to make flat maps. (A. N. Strahler)

resenting the ratio between the size of the globe and the size of the Earth. Size in this case refers to some measure of length or distance. For example, a globe 20 centimetres in diameter would have a scale of 20 centimetres to 13,000 kilometres (the approximate diameter of the Earth), which reduces to 1 cm = 650 km. This relationship holds true for distances between any points on the globe.

Scale is more usefully stated as a simple fraction, termed the *representative fraction*, obtained by reducing both the Earth and globe distances to the same unit of measure. The advantage of the representative fraction is that it is entirely free of any specified units of measure, such as metre, kilometre, or mile. Thus, in the previous example, the size of the Earth is converted to centimetres, and the representative fraction is obtained as 20/1,300,000,000, which gives 1/65,000,000 or 1:65,000,000. The fractional notation in which a colon separates the numerator from the denominator is the most common way to express this ratio.

Being a true-scale model of the Earth, a globe has a constant scale everywhere on its surface, but this is not true of a map projection drawn on a flat surface. All map projections stretch the Earth's surface in a non-uniform manner, so that the map scale changes from place to place. It is, however, possible to select a meridian or parallel—the equator, for example—for which a scale fraction can be given.

Small-Scale and Large-Scale Maps

When geographers refer to small-scale and large-scale maps, they refer to the value of the representative fraction. For example, a global map at a scale of 1:65,000,000 has a scale value of 0.00000001524, which is obtained by dividing 1 by 65,000,000. A hiker's topographic map at a scale of 1:25,000 has a scale value of 0.00004. Since the global scale value is smaller, it is a small-scale map, while the hiker's map is a large-scale map, even though it represents only a small area (Figure 3). Most large-scale maps carry a *graphic scale*, which is a line marked off into units representing kilometres (Figure 4). Graphic scales make it easy to measure ground distances by marking the positions of two places on the edge of a piece of paper, laying the paper on the graphic scale, and reading the distance directly from the scale.

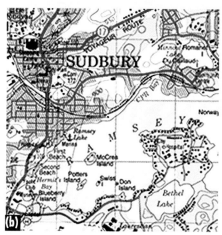

Figure 3 In the current National Topographic Series, the Canadian government produces maps at two scales: (a) 1:50,000 and (b) 1:250,000.

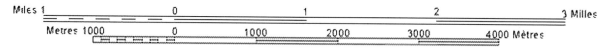

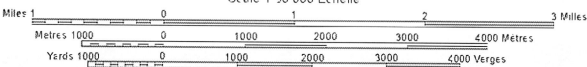

Figure 4 Graphic scales used on Canadian topographic maps.

Informational Content of Maps

Multipurpose, large-scale maps "zoom in" so they are capable of portraying a considerable amount of geographic information (See Figure 3a). They must show this information in a convenient and effective manner using a variety of symbols.

Map symbols associate information with points, lines, and areas. Point symbols consist of various types of dots, such as a closed circle, open circle, letter, numeral, or a little picture of the object it represents. Linear features like roads are represented by lines of varying widths and styles. Specific areas can be represented by patches, shown simply by a line marking their edge or a distinctive pattern or solid colour. Patterns are highly varied and might consist of tiny dots or parallel, intersecting, and wavy lines. The inside back cover of the book shows the symbols used on Canadian National Topographic Series (NTS) maps.

The size and shape of map symbols in relation to map scale is of prime importance in cartography. Large-scale maps can show objects in their true outline form. As map scale decreases, representation becomes more and more generalized. In physical geography, an excellent example is the depiction of a river, such as the South Saskatchewan. Figure 5 shows the river at three scales, starting with a detailed plan, progressing to a double-line symbol that generalizes the channel form, and ending with a single-line symbol. As map symbols become more generalized, the details of the river banks and channel bends simplify as well. The level of depiction of fine details is termed *resolution*. Large-scale maps have much higher resolution than small-scale maps.

Presenting Numerical Data on Thematic Maps

In physical geography, much of the information collected about particular areas is in numerical form, such as air temperatures and amounts of rainfall. Another information category consists merely of the presence or absence of an attribute, such as a particular type of soil. In such cases, a dot can mean "present," so that an area of scattered dots develops as the locations are entered (Figure 6).

Some measurements are taken at predetermined locations, such as weather stations, and although the numbers and locations may be accurate, it may be difficult to see the spatial pattern present in the data. For this reason, cartographers often simplify arrays of point values into *isopleth* maps, which use lines of equal values. In drawing an isopleth, the line is routed

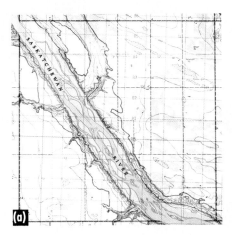 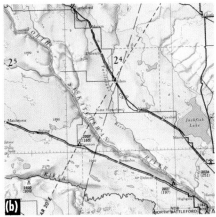 

Figure 5 Maps of the Saskatchewan River on three scales: (a) 1:20,000 (channel contours give depth below mean water level); (b) 1:250,000 (only waterline shown to depict channel); (c) 1:2,000,000 (channel shown as a solid line symbol).

CHAPTER SUMMARY

- The rotation of the Earth on its axis and the revolution of the Earth in its orbit around the Sun are fundamental topics in physical geography. The Earth is nearly spherical. It rotates on its *axis* once in 24 hours. The axis of rotation marks the *North* and *South poles*. The direction of rotation is counter-clockwise.
- The Earth's rotation provides the first great rhythm of our planet—the daily alternation of sunlight and darkness.

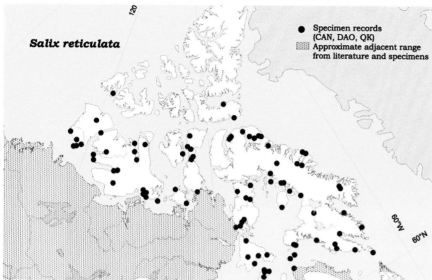

among the points in a way that best indicates the available data. Colour often enhances the patterns, as shown in the annual precipitation map of Alberta in Figure 7. In contrast to the isopleth map is the *choropleth* map, which identifies information in categories. The map of soils of Alberta (Figure 8) is an example of a thematic choropleth map.

Figure 6 A dot map showing the distribution of soils of the Alfisols order in the United States. (From P. Gersmehl, Annals of the Assoc. of Amer. Geographers, vol. 67. Copyright © Association of American Geographers. Used with permission.)

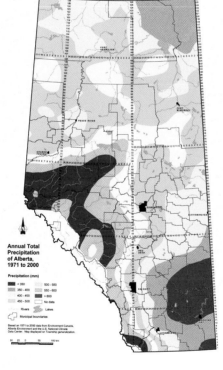

Figure 7 The boundaries between adjacent colours on this precipitation map of Alberta represent isohyets; they are based on data collected from weather stations throughout the province. Colours enhance the precipitation pattern, with blue signifying wetter areas and red the drier areas.

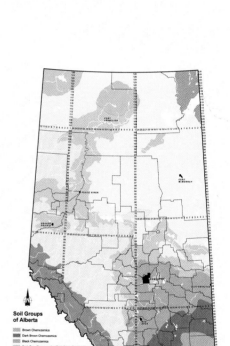

Figure 8 The coloured areas on this map represent the different soils found in Alberta. The boundaries have no numerical values, but simply denote where one soil replaces another.

- The Earth's axis of rotation provides a reference for the system of location on the Earth's surface—the geographic grid, which consists of meridians and parallels. This system is indexed by the latitude and longitude system.
- We require a map projection to display the Earth's curved surface on a flat map. The polar projection is centred on either pole. The Mercator projection converts the curved geographic grid into a flat, rectangular grid and best displays directional features. The Goode projection distorts the shapes of continents and coastlines but preserves the areas of land masses in their correct proportion.
- The Earth's rotation is the basis of time. The Earth rotates by 15° each hour. In the standard time system, time is defined by standard meridians that are normally 15° apart. Daylight Saving Time is commonly used to advance the clock by one hour. At

the International Date Line, the calendar day changes, advancing forward one day for westward travel, and dropping back a day for eastward travel.

- The seasons, the second great rhythm of the Earth, arise from the revolution of the Earth in its orbit around the Sun combined with the tilt of the Earth's axis with respect to its orbital plane. The solstices and equinoxes mark the cycle of this revolution. At the summer (June) solstice, the northern hemisphere is tilted toward the Sun. At the winter (December) solstice, the southern hemisphere is tilted toward the Sun. At the equinoxes, day and night are of equal length.

KEY TERMS

Antarctic Circle
Arctic Circle
Coordinated Universal Time (UTC)
Daylight Saving Time
declination
Earth rotation
equator
equinox

geographic grid
geographic information system (GIS)
Global Positioning System (GPS)
Goode projection
great circle
International Date Line
latitude

longitude
map projection
Mercator projection
meridian
parallel
plane of the ecliptic
polar projection
revolution
small circle

solar noon
standard time system
summer solstice
time zone
Tropic of Cancer
Tropic of Capricorn
winter solstice

REVIEW QUESTIONS

1. What is the approximate shape of the Earth? How is this known? What is the Earth's true shape?
2. What is meant by Earth rotation? Describe three environmental effects of the Earth's rotation.
3. Describe the geographic grid, including parallels and meridians.
4. How do latitude and longitude determine position on the globe? In what units are they measured?

5. Name three types of map projections and describe each briefly. Give reasons why a different map projection might be chosen to display different types of geographical information.
6. Explain the global time-keeping system. Define and use the terms *standard time*, *standard meridian*, and *time zone* in your answer.
7. What is meant by the "tilt of the Earth's axis"? How is the tilt responsible for the seasons?

GEOGRAPHER'S TOOLS 2.3   Geographic Information Systems

1. What is a geographic information system?
2. Identify and describe three types of spatial objects.

3. What are the key elements of a GIS?

A CLOSER LOOK   Geographer's Tools 2.4   Focus on Maps

1. Can the scale of a flat map be uniform everywhere on the map? Do large-scale maps show large areas or small areas?
2. What types of symbols are found on maps, and what types of information do they carry?

3. How are numerical data represented on maps? Identify three types of isopleths. What is a choropleth map?

VISUALIZATION EXERCISES

1. Sketch a diagram of the Earth at an equinox. Show the North and South Poles, the equator, and the circle of illumination. Indicate the direction of the Sun's incoming rays and shade the night portion of the globe.

2. Sketch a diagram of the Earth at the summer (June) solstice, showing the same features. Also include the Tropics of Cancer and Capricorn, and the Arctic and Antarctic Circles.

ESSAY QUESTION

1. Suppose that the Earth's axis were tilted at 40° instead of $22\frac{1}{2}°$. What would be the global effects of this change? How would the seasons change at your location?

WORKING IT OUT 2.1 | Distances from Latitude and Longitude

1. Toronto and Quito, Ecuador, are located on about the same meridian, but they are about 44° lat. apart. What is the approximate distance between these cities?

2. Saskatoon and Oxford, England, are both located at approximately 52° N, but their longitudes are 106° W and 1° W, respectively. What is the approximate distance between the two cities?

3. A map of a region close to the equator shows an area of 1° of latitude by 1° of longitude. About how many square kilometres does the map cover? How does this compare with a 1° by 1° area at lat. 60° N near Churchill, Manitoba?

Find answers at the back of the book

EYE ON THE LANDSCAPE

Chapter Opener The sun is low in the sky at this time of year (**A**). Incoming solar energy is further reduced by a layer of altostratus cloud (**B**), and by the albedo of the snow-covered landscape (**C**) that has been molded and packed by the wind.

Shadows in a winter woodland (right). The low sun angle during winter at higher latitudes is evident from shadows cast on the snow. The camera points towards north and given the orientation of the shadows the photograph was taken before non while the sun was in the southeast (**A**). The openness of the forest is suggested from the variations in light intensity on the snow (**B**), and the absence of tree crown shadows suggests that

the trees are deciduous. Slight variations in snow accumulation around some of the tree trunks (**C**) suggests that prevailing winds are from the west-northwest.

Chapter 3

EYE ON THE LANDSCAPE

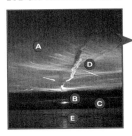

Sun rise. The high cirrus clouds are illuminated by the rising sun, but Earth's surfaced remains in partial shadow. Because of the low sun angle, the dispersing contrail left by a passing aircraft is casting a shadow on the clouds above.

What else would the geographer see? . . . Answers at the end of the chapter.

THE GLOBAL ENERGY SYSTEM

Electromagnetic Radiation
Focus on Systems 3.1 • *Forms of Energy*
Radiation and Temperature
Working It Out 3.2 • *Radiation Laws*
Solar Radiation
Characteristics of Solar Energy
Long-Wave Radiation from the Earth
Planetary Insolation
Daily Insolation through the Year
World Latitude Zones
Composition of the Atmosphere
Earth's Primordial Atmosphere
The Present Atmosphere
Ozone in the Upper Atmosphere
Air Pollutants
Eye on Global Change 3.3 • *The Ozone
   Layer—Shield to Life*
Sensible Heat and Latent Heat Transfer
The Global Energy System
Solar Energy Losses in the Atmosphere
Albedo
Counter-Radiation and the Greenhouse Effect
Focus on Remote Sensing 3.4 • *CERES—
   Clouds and the Earth's Radiant Energy System*
**Global Energy Balance of the
Earth–Atmosphere System**
Incoming Short-Wave Radiation
Surface Energy Flows
Energy Flows to and from the Atmosphere
Climate and Environmental Change
**Net Radiation, Latitude, and the Energy
Balance**
Solar Energy
A Closer Look:
Geographer's Tools 3.5 • *Remote Sensing for
   Physical Geography*

The constant flow of energy from the Sun drives nearly all of the natural processes that shape the Earth. It is the source of energy for wind, waves, rivers, and ocean currents and is reflected in the rich diversity of climates.

This chapter explains how solar radiation is intercepted by the Earth, flows through the atmosphere, and interacts with the land surfaces and the oceans. Solar energy is not received uniformly over the surface because of the Earth's spherical shape, the tilt of the polar axis, and its constant rotation and revolution. Regional differences in energy flow rates affect temperature and pressure, which drive the global wind system. Ocean currents are also linked to the redistribution of energy. Although more energy is received in low-latitudes than at the poles, the redistribution of energy in the ocean and atmosphere circulation results in a global energy balance in which the flows, or fluxes, reaching the Earth are the same as those that are returned to space.

ELECTROMAGNETIC RADIATION

The study of the global energy system begins with **electromagnetic radiation**—the principal form in which energy is transported from the Sun to the Earth. Electromagnetic radiation consists of electric and magnetic waves that oscillate together, at right angles to each other. They travel in a direction that is perpendicular to the direction of their oscillation (Figure 3.1).

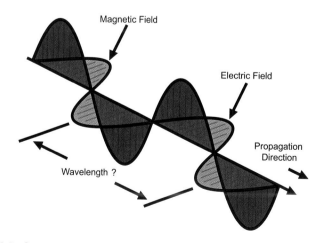

3.1 Electromagnetic radiation is a collection of energy waves with different wavelengths. Wavelength λ is the crest-to-crest distance between successive wave crests.

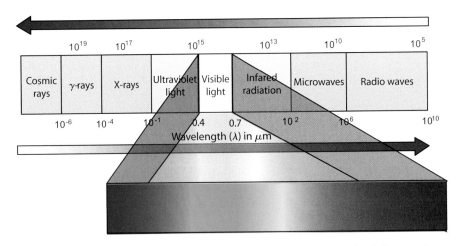

3.2 The electromagnetic spectrum Electromagnetic radiation can exist at any wavelength. By convention, names are assigned to a specific wavelength region.

Light and heat are two familiar forms of electromagnetic radiation. Light is radiation that is visible to our eyes. Heat radiation, though not visible, is easily felt when you hold your hand near a warm object, such as an oven. *Wavelength* λ describes the distance separating adjacent wave crests; the unit of measurement is the *micrometre* (μm), which is 10^{-6} metres.

Electromagnetic waves of different wavelengths form the **electromagnetic spectrum** (Figure 3.2). At the short wavelength end of the spectrum are *gamma rays* and *X-rays*. Their wavelengths are normally expressed in *nanometres* (nm = 10^{-9} m). Gamma and X-rays have high energies and can be hazardous to health. Gamma and X-ray radiation grade into **ultraviolet (UV) radiation**, which begins at about 10 nanometres and extends to 400 nanometres (or 0.4 μm). Like gamma and X-rays, ultraviolet radiation can be damaging to living tissues.

The **visible light** portion of the spectrum starts at about 0.4 micrometres, depicted here in violet. Colours then grade through blue, green, yellow, orange, and red, reaching the end of the visible spectrum at about 0.7 micrometres. Next is *near-infrared* radiation, with wavelengths from 0.7 to 1.2 micrometres. This radiation is similar to visible light in that most of it comes from the Sun. However, human eyes are not sensitive to radiation beyond about 0.7 micrometres, so it is not visible. Healthy plant leaves and green tissues reflect near-infrared light strongly, a fact that is used in remote sensing. (See the feature, *Geographer's Tools 3.5 • Remote Sensing for Physical Geography*, at the end of the chapter).

Short-wave infrared radiation lies in the range of 1.2 to 3 micrometres. Like near-infrared radiation, it comes mostly from the Sun. In remote sensing, this radiation can distinguish clouds from snow. It is also useful for differentiating rock types. From 3 micrometres to about 6 micrometres is *middle-infrared* radiation. It comes from the Sun as well, but is also emitted by fires burning at the Earth's surface, such as forest fires or gas well flares.

Thermal infrared radiation is between 6 and about 300 micrometres and includes the heat radiated from objects at temperatures normally found at Earth's surface. Most heat radiation is in the range of 8 to 12 micrometres. Thermal infrared radiation is no different in principle from visible light radiation. Although photographic film is not sensitive to thermal radiation, it is possible to acquire a thermal infrared image using special sensors (see Figure 3.3). Beyond the infrared wavelengths are microwaves, radar, radio waves, and television transmissions.

Electromagnetic radiation is one of several fundamental forms of energy. *Focus on Systems 3.1 • Forms of Energy* discusses other forms that are important in physical geography.

RADIATION AND TEMPERATURE

Two important physical principles concern the emission of electromagnetic radiation. First, there is an inverse relationship between the range of wavelengths that an object emits and the surface temperature of the object. For example, the Sun, a very hot object, emits radiation predominantly in shorter wavelengths. In contrast, the Earth, a much cooler object, emits radiation at longer wavelengths. Second, hot objects radiate more energy than cooler objects. The flow of radiant energy from the

FOCUS ON SYSTEMS | 3.1 Forms of Energy

The term *energy* was first used in 1807, but the idea is much older, going back to Sir Isaac Newton's laws of motion, published in 1687. In 1695 Gottfried Leibnitz proposed the existence of a "vital force." Today this is termed **kinetic energy**—the energy of mass (or "matter") in motion, and is expressed mathematically as

$$1/2 \ mv^2$$

where m = mass, and v = velocity

Energy is defined as the ability to do work, which in this sense is equal to the change in kinetic energy in a body. Energy "moves" or "travels" with (or within) forms of matter. For example, the gasoline engine in a car transfers the energy of fuel combustion into forward motion. The entire moving object—the car—then acquires kinetic energy. Kinetic energy is one of the forms of energy that comes under the general heading of **mechanical energy**. Should the car crash head-on into a massive power pole, it will demonstrate its ability to do work upon its own body, upon its passengers, and upon the pole. The energy released in the collision will increase in direct proportion to the mass of the car, and it will also increase with the square of the car's speed, since kinetic energy is equal to $1/2 \ mv^2$.

Now that the car is motionless it no longer has kinetic energy. Energy cannot be simply destroyed and disappear, but it can be transformed. At the point of impact, the car body would have become heated for an instant, as kinetic energy of motion was turned into **heat energy**.

In this form of energy, the atoms and molecules within a solid (or liquid, or gas) are in rapid motion. The hotter the substance, the faster the molecules vibrate. This form of kinetic energy is termed *sensible heat* because it can be sensed by touch, as well as be measured by a thermometer. Sensible heat is extremely important in the study of temperature changes in air, water, and soil.

Radiant energy is the form of energy that is emitted from the Sun and is the Earth's principal form of energy. In this form, energy travels through space and easily penetrates transparent substances, such as air, water, and glass.

The force of gravity is a form of energy that arises when one object is attracted to another. Imagine rolling a heavy object, such as a boulder, to the top of a cliff, then nudging it over the edge and watching it drop to the ground below. Moving the boulder to the top of the hill requires the expenditure of energy to overcome the ever-present downward pull of gravity. Thus the boulder is endowed with a certain amount of **potential energy**, also referred to as **stored energy**. With each metre of vertical distance the boulder falls, it converts a quantity of that potential energy into kinetic energy. (We can disregard the small energy loss due to friction with the surrounding air.) Upon impact, the kinetic energy is converted to heat energy, which quickly dissipates.

Potential energy, present because of gravity, must always be evaluated in terms of a given reference level, or *base*

level. The standard base level is the average level of the ocean, called "mean sea level." In nature, any kind of matter, on the land or in the atmosphere, holds potential energy with respect to sea level. Examples are raindrops and hailstones formed in a storm cloud or the mineral particles of ash and cinder spewed from an erupting volcano. As these substances fall toward the Earth, their potential energy is converted to kinetic energy. Potential energy can also be converted to kinetic energy as substances descend along sloping and winding pathways. The flow of water in a river or ice in a glacier are two examples.

Chemical energy, which is absorbed or released by matter when chemical reactions take place, is also important. Chemical reactions involve the coming together of atoms to form simple molecules, the recombining of simple molecules into new and more complex compounds, and the reverse changes back into simpler forms. Most of the time, chemical energy is stored energy held within the molecule. An important example is photosynthesis, in which green plants absorb radiant energy from the Sun to produce complex organic carbohydrates. Chemical energy is also stored for long periods of geologic time in the hydrocarbon compounds that are used as fuels, such as coal, oil, and natural gas. As these fuels are burned, their molecules combine with atmospheric oxygen, releasing heat energy as well as many simpler molecules, such as carbon dioxide and water.

surface of an object is directly related to its absolute temperature raised to the fourth power. Absolute temperature is temperature measured on the Kelvin (K) scale, with absolute zero (0 K, which is equivalent to -273 °C) representing the absence of all heat.

Because of this relationship, a small increase in temperature results in a large increase in the rate at which radiation is emitted. *Working It Out 3.2 • Radiation Laws* describes these two principles in more detail.

3.3 A thermal infrared image This suburban scene was imaged at night in the thermal infrared spectral region. Red tones indicate the highest temperatures and black tones the coldest. Windows appear red because they are warm and radiate more intensely. House walls are intermediate in temperature and appear blue. The purple tones of the roads, driveways, and trees show they are relatively cool. The black ground and sky are coldest.

SOLAR RADIATION

The Sun is a ball of constantly churning gases that are heated by continuous nuclear reactions. It is of average size compared with other stars, and has a surface temperature of about 6,000°C. Like all objects, it emits energy in the form of electromagnetic radiation. The energy travels away from the Sun at a speed of about 300,000 kilometres per second. At that rate, it takes the energy about 8.3 minutes to travel the 150-million-kilometre distance to the Earth.

None of the solar radiation is lost as it travels through space. However, the rays spread apart as they move away from the Sun so the more distant planets receive less radiation per square metre than those nearer the Sun. Earth intercepts less than one-billionth of the Sun's total energy output.

WORKING IT OUT | 3.2   **Radiation Laws**

The relationship between the flow of energy from an object and that object's surface temperature is described by the **Stefan-Boltzmann Law** as

$$M = \sigma T^4$$

where M is the energy flow from the surface, in watts per square metre (W/m^{-2}); σ is the Stefan-Boltzmann constant (5.67 × 10^{-8} W/m^{-2} K^{-4}); and T is the temperature of the surface, in kelvins. The watt (W) is a measure of energy flow. The symbol K denotes absolute temperature in kelvins.

Wien's Law states that the wavelength at which radiation emitted by a surface is greatest is related to temperature as

$$\lambda_{MAX} = b/T$$

where λ_{MAX} is the wavelength at which radiation is at a maximum, in micrometres; b is a constant equal to 2,898 micrometre; and T is the temperature of the surface in kelvins.

Both of these laws apply only to a perfectly radiating surface known as a **blackbody**; that is, one that emits energy at the same rate as it absorbs it. Most natural surfaces are good radiators and do not differ too much from a blackbody.

The following calculation uses these formulae to determine the flow of energy emitted by a surface and then to determine the wavelength at which radiation from the surface is greatest. Assume the temperature of the surface is 450 K (177°C, a common oven temperature for baking). Using the Stefan-Boltzmann equation, substituting 450 K for T gives:

$$M = \sigma T^4 = (5.67 \times 10^{-8} \text{ W/m}^{-2} \text{ K}^{-4}) \times (450 \text{ K})^4$$
$$= (5.67 \times 10^{-8} \text{ W/m}^{-2} \text{ K}^{-4}) \times (4.10 \times 1010 \text{ K}^4)$$
$$= 2.325 \times 10^3 \text{ W/m}^{-2}$$
$$= 2,325 \text{ W/m}^{-2}$$

Applying Wien's Law, then

$$\lambda_{MAX} = b/T = 2,898 \; \mu m \text{ K}/450 \text{ K} = 6.44 \; \mu m$$

In comparison, a surface at 293 K (20°C, approximately room temperature) gives an emission rate of 418 watts per square metre and a wavelength of maximum emission of 9.89 micrometres. Thus, the oven surface emits more than five times more energy than the room temperature surface, and the wavelength of maximum emission for the oven is shorter.

The Sun's interior is the source that generates solar energy. Here, hydrogen is converted to helium at extremely high temperatures and pressures. This process of nuclear fusion produces a vast quantity of energy, which finds its way to the Sun's surface. Because the energy production rate is nearly constant, the output of solar radiation varies only slightly. The term **solar constant** describes the more or less uniform rate at which the Earth intercepts this energy. The solar constant is measured beyond the outer limits of Earth's atmosphere, and has a value of about 1,376 watts per square metre (W/m^{-2}).

The **watt** (W) describes a rate of energy flow. When used to measure the intensity of a flow of radiant energy, a unit surface area of one square metre (1 m^2) is normally specified. Thus, the measure of intensity of received (or emitted) radiation is given as watts per square metre (W/m^2).

CHARACTERISTICS OF SOLAR ENERGY

Figure 3.4 shows details of the electromagnetic radiation that reaches Earth from the Sun. The wavelengths of solar energy received at the surface range from about 0.3 to 3 micrometres, and

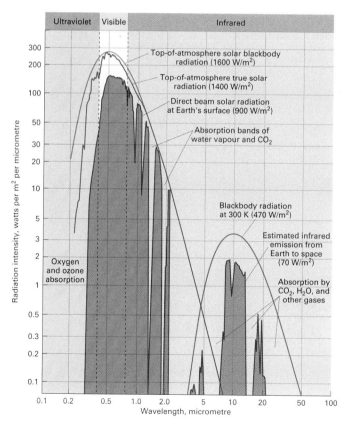

3.4 Spectra of solar and Earth radiation This figure plots both short-wave radiation, which comes from the Sun (left side), and long-wave radiation, which is emitted by the Earth's surface and the atmosphere (right side). Note that both scales are logarithmic. (After W. D. Sellers, *Physical Climatology*, University of Chicago Press. Used with permission.)

are termed **short-wave radiation**. The uppermost curve for solar radiation shows how the Sun would supply solar energy at the top of the atmosphere if it were radiating as a *blackbody*. The actual output of the Sun, as measured at the top of the atmosphere, is quite close to the blackbody condition, except for UV wavelengths, where the Sun emits less energy than is theoretically possible. Note that the Sun's output peaks in the visible part of the spectrum. Thus, human vision is adjusted to the wavelengths where solar light energy is most abundant.

The line showing solar radiation reaching the Earth's surface is quite different. As solar radiation passes through the atmosphere, it is affected by absorption and scattering. **Absorption** occurs when molecules and particles in the atmosphere intercept radiation at particular wavelengths. Absorption is shown by the dips in the graph—for example, at about 1.3 and 1.9 micrometres. At these wavelengths, molecules of water vapour and carbon dioxide strongly absorb solar radiation and prevent nearly all of it from reaching the Earth's surface. Note also that oxygen and ozone absorb almost all of the UV radiation at wavelengths shorter than about 0.3 micrometres. Atmospheric absorption is one of the flows of energy in the global energy balance and is important because it warms the atmosphere directly.

Solar radiation passing through the atmosphere can also be scattered. **Scattering** refers to the change in direction of radiation caused by a molecule or particle. Scattered rays may go back toward space or continue toward the surface. The line on Figure 3.4 is for direct beam energy only, and does not include any scattered radiation that makes its way to the surface.

LONG-WAVE RADIATION FROM THE EARTH

Since the Earth's surface and atmosphere are much colder than the sun's surface, our planet radiates less energy than the sun, and the energy emitted has longer wavelengths. The right side of Figure 3.4 shows the Earth's emission spectrum. The upper line is the radiation of a blackbody at a temperature of about 300 K (23°C), which is a good approximation for Earth as a whole. At this temperature, radiation ranges from about 3 to 30 micrometres and peaks at about 10 micrometres in the thermal infrared region; these emissions are considered **long-wave radiation**.

Beneath the Earth's theoretical blackbody curve is an irregular line that shows upwelling energy emitted by the Earth and its atmosphere, as measured at the top of the atmosphere. Compared with the blackbody curve, much of the Earth's surface radiation is absorbed by the atmosphere, especially at about 6–8 micrometres, 14–17 micrometres, and above 21 micrometres. Water and carbon dioxide are the primary absorbers of this terrestrial long-wave energy. Such absorption is an important part of the *greenhouse effect*, which helps to warm the Earth's surface. These absorption regions are flanked by three regions in which outgoing energy flow is significant—4–6 micrometres, 8–14 micrometres, and 17–21 micrometres. These are *windows* through which long-wave radiation leaves the Earth and flows to space.

Remote sensing instruments can detect the radiant energy that passes upward through the atmosphere and use it to compose images of the Earth's surface. *A Closer Look: Geographer's Tools 3.5 • Remote Sensing for Physical Geography* at the end of the chapter provides more information on these techniques.

PLANETARY INSOLATION

The flow of solar radiation from the Sun remains constant; however, the amount received on Earth varies both spatially and over time. Assume that the Earth is a uniformly spherical planet with no atmosphere. Each day the surface of the planet receives a flow of energy. Average incoming energy flow, measured in watts per square metre (W/m^{-2}) over the course of a 24-hour day, is referred to as daily **insolation** (**in**coming **sol**ar radi**ation**). Annual insolation is the sum of these daily energy flows through the year.

Insolation depends on the angle of the Sun above the horizon. It is greatest when the Sun is directly overhead; the incoming rays are vertical and energy is concentrated on a small part of the Earth's surface (position A in Figure 3.5). When the Sun is lower in the sky, the same amount of solar energy is spread over a greater surface area, which effectively reduces insolation (position B).

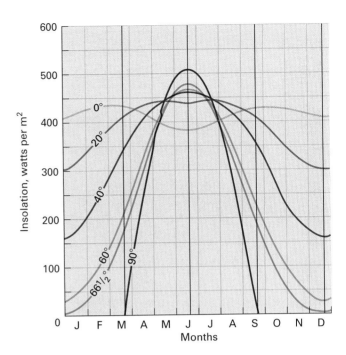

3.6 Daily insolation at the top of the atmosphere through the year at various latitudes (northern hemisphere) Black lines mark the equinoxes and solstices. Latitudes between the Equator (0°) and Tropic of Cancer (23 ½° N) show two maxima; elsewhere only one maximum occurs. Between the Arctic Circle (66 ½° N) and the North Pole, insolation is zero for at least some of the year. (Copyright © A. N. Strahler. Used with permission.)

In the absence of an atmosphere, daily insolation at a location depends on two factors: (1) the angle at which the Sun's rays strike the Earth, and (2) the length of time the location is directly exposed to the rays. Both of these factors are controlled by the latitude of the location and the time of year. For example, in midlatitude locations in summer, daylight may persist for 16 hours or more, the midday Sun rises high in the sky, and surface heating is quite intense, thus insolation is greater. During the equinoxes the subsolar point lies on the Equator, and at this time of the year daylight everywhere is limited to 12 hours, so insolation is less at these midlatitude locations.

DAILY INSOLATION THROUGH THE YEAR

The angle of the solar rays varies with the Sun's position in the sky. Insolation at a location increases as the Sun gets higher above the horizon, until it is at its maximum at the summer solstice in each hemisphere. For the northern hemisphere this would be in June, and for the southern hemisphere in December. Figure 3.6 shows the variation in daily insolation at selected latitudes in the northern hemisphere over the course of the year.

Following the curve for 40° N (about 2° south of Windsor, Ontario), the daily average insolation ranges from about 160 watts per square metre at the December solstice to about 460

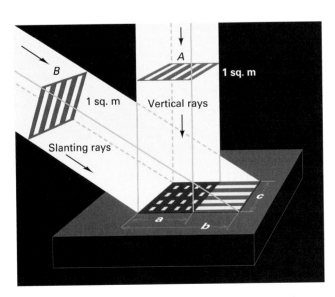

3.5 Insolation and Sun angle The angle of the Sun's rays determines the intensity of insolation on the ground. The energy contained in one square metre of solar rays perpendicular to the Earth's surface (for example, at the Tropic of Cancer during the summer solstice, as illustrated by A) is concentrated in one square metre on the ground (a × c). However, the same energy in the slanting rays in B is spread over a larger surface area (b × c).

watts per square metre at the June solstice. Equinox values are around 350 watts per square metre. Inuvik, NWT, is situated just north of the Arctic Circle, and conditions here are approximately the same as those shown by the $66\frac{1}{2}°$ N line. At this latitude direct insolation is 0 watts per square metre during the December solstice, rising to about 175 watts per square metre by the March equinox, and reaching 475 watts per square metre at the time of the June solstice. Note that direct insolation during June is potentially higher at Inuvik than Windsor, and both locations would receive more insolation than the Equator, which at this time of the year would expect about 380 watts per square metre. These values confirm that insolation, in the absence of any effect from atmosphere, depends on how high the Sun rises during the day, and how long it stays above the horizon. Indeed, at the June solstice, insolation at the North Pole exceeds 500 watts per square metre and is higher than at any other location.

Note that for the Equator, there are two insolation maxima—one at each equinox—and two minima—one at each solstice. In fact, two daily insolation maxima occur at all latitudes between the Tropic of Cancer ($23\frac{1}{2}°$ N) and the Tropic of Capricorn ($23\frac{1}{2}°$ S). However, as either tropic is approached, the two maxima get closer and closer in time, and then merge into a single maximum. The line representing latitude 20° N (approximate latitude of Mexico City) shows two maxima about two months apart. Note that the two insolation peaks that occur at the Equator are not quite equal and are not exactly at the equinoxes. These small variations arise because the Earth's orbit is not exactly circular.

Based on the information in Figure 3.6, the following conclusions can be drawn regarding daily insolation:

- Latitudes between the Tropic of Cancer and Arctic Circle, and between the Tropic of Capricorn and the Antarctic Circle, show a wavelike pattern of greater daily insolation at the summer solstice and lower daily insolation at the winter solstice.
- Between the Arctic and Antarctic circles and the North and South poles respectively, the Sun is below the horizon for at least part of the year, during which daily insolation drops to zero.
- Daily insolation is greatest at either the North or South Pole during the respective summer solstice in each hemisphere.
- There are two maxima and two minima in daily insolation at the Equator, occurring at the equinoxes and solstices, respectively.
- Moving away from the Equator toward the two tropics, there are also two maxima and minima daily insolation values; however, as the tropic is approached, the two maxima get closer in time, merging into a single maximum at the tropic.

These differences in daily insolation are important because they measure how much solar energy is available to heat the Earth's surface. Thus, they are the most important factor in determining air temperature. The seasonal change in daily insolation at a location is, therefore, a major determinant of climate.

Annual insolation varies smoothly from the Equator to the poles and is greater at lower latitudes (Figure 3.7). High latitudes still receive a considerable flow of solar energy, with annual insolation at the poles about 40 percent of that at the Equator.

WORLD LATITUDE ZONES

The seasonal pattern of daily insolation can be used as a basis for dividing the globe into broad latitude zones (Figure 3.8). The limits shown in the figure and specified below are generalized and provide a convenient nomenclature to identify general geographic zones.

The *Equatorial zone* encompasses the Equator and covers the latitude belt roughly 10° N to 10° S. Here the Sun provides intense insolation throughout most of the year, and daylight and nighttime are of roughly equal length. Spanning the Tropics of Cancer and Capricorn are the *tropical zones*, ranging from latitudes 10° to 25° N and S. A marked seasonal cycle exists in these zones, combined with high annual insolation.

Moving beyond the tropical zones are transitional regions called the *subtropical zones*. These may be conveniently delineated by the latitude belts 25° to 35° N and S. These zones have a strong seasonal cycle combined with annual insolation similar to that of the tropical zones.

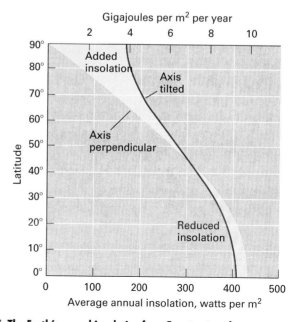

3.7 The Earth's annual insolation from Equator to poles

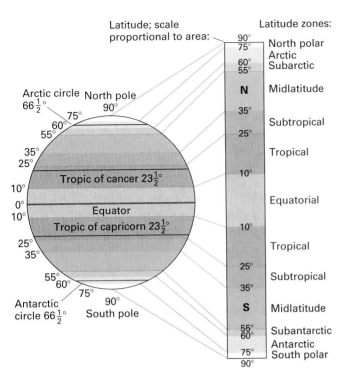

3.8 World latitude zones A system of latitude zones, based on the seasonal patterns of daily insolation observed over the globe.

The *midlatitude zones* lie between 35° and 55° N and S latitude. Here the Sun's zenith angle changes markedly during the year. There are pronounced differences in day length from winter to summer, and seasonal contrasts in insolation are quite strong. Consequently, these regions can experience a large annual range in temperature.

Bordering the midlatitude zones on the poleward side are the *subarctic* and *subantarctic zones* at 55° to 60° N and S latitudes. Astride the Arctic and Antarctic Circles from latitudes 60° to 75° N and S lie the *Arctic* and *Antarctic zones*. All of these zones have an extremely large yearly variation in day length, resulting in enormous contrasts in insolation from solstice to solstice.

The northern and southern *polar zones* occupy the areas between about 75° latitude and the poles. Here the polar regime that approaches six months of daylight followed by six months of darkness is predominant. These zones experience the greatest seasonal contrasts of insolation.

COMPOSITION OF THE ATMOSPHERE

The flows of energy within the atmosphere and between the atmosphere and the Earth's surface have a profound influence on the energy budget. The effects of these energy flows are dependent on the nature and properties of the gases and other atmospheric constituents that envelope the Earth.

EARTH'S PRIMORDIAL ATMOSPHERE

It is believed that the Earth was formed by an aggregation of gas and dust that was part of a **nebula**. The gravitational pull of the growing planet was probably too small to retain any gaseous envelope; therefore, the thinking is that the Earth's atmosphere evolved from self-contained secondary sources. Initially, the atmosphere probably consisted of helium and hydrogen, with traces of ammonia and methane. A hydrogen and helium atmosphere is similar to those found on the outer planets and the Sun. With time, carbon dioxide (CO_2) began to accumulate, with lesser amounts of nitrogen, sulphur dioxide, and water vapour. These gases would have been emitted from ancient volcanoes. This CO_2-dominated atmosphere is similar to present-day conditions on Mars and Venus.

No oxygen is, or presumably ever has been, released from volcanoes. However, volcanoes do expel water vapour, and in ancient times this would have resulted in the formation of clouds and rain of high acidity, which caused rapid weathering of rocks. Geologic evidence suggests that such conditions had developed by about 3.8 billion years ago. With time, oxygen became more prevalent; evidence of an oxidizing atmosphere is seen in the formation of ancient limestone and other rocks.

The change from an oxygen-poor to an oxygen-rich atmosphere resulted from the photo-dissociation of water molecules by UV radiation. UV radiation could penetrate to the Earth's surface because there was no ozone, which is itself a derivative of oxygen. Atmospheric oxygen levels produced in this way would have been about 0.001 percent of present concentrations. The subsequent rise in oxygen is attributed to photosynthesis.

For primitive life forms to develop, cells had to be protected from the lethal UV radiation that penetrated the primitive atmosphere. This protection was possible a few metres below the ocean surface. Thus, primitive algae could survive in a relatively thin ocean layer, perhaps to a depth of 30 metres, at which point nearly all visible light is absorbed. Once oxygen concentrations reached about 10 percent of current atmospheric levels, it is postulated that sufficient ozone was present to shield land surfaces from harmful UV radiation. This likely occurred some 400 million years ago, and allowed organisms to rapidly spread on land.

Terrestrial photosynthesis led to a rapid increase in oxygen. At the same time atmospheric CO_2 levels declined, as the gas was incorporated into organic materials and deposited in the ancient coal seams and carbonate rocks. The Earth's atmosphere has remained essentially unchanged since this time, although concerns are now being raised over the impact pollution is having.

THE PRESENT ATMOSPHERE

The Earth's atmosphere is a mixture of gases, not a single chemical compound. It extends to a height of approximately 10,000 kilometres and is held to the Earth by gravity. Because air is highly compressible, the lower layers of atmosphere are much denser than those above. Consequently, about 97 percent of the mass of the atmosphere lies within 30 kilometres of the Earth's surface. From the surface to an altitude of about 80 kilometres, the chemical composition of air is uniform in terms of the proportions of constituent gases. Pure, dry air consists largely of nitrogen, about 78 percent by volume, and oxygen, about 21 percent (see Table 3.1). Other gases, such as argon and carbon dioxide, as well as water vapour and various pollutants, account for the remaining one percent.

Nitrogen (N_2) is an inert gas that does not readily combine with other substances. It is removed from the atmosphere in precipitation during lightning storms and by specialized nitrogen fixing bacteria in the soil for use by plants. Unlike nitrogen, *oxygen* (O_2) is chemically active and combines readily with other elements in the oxidation process. Combustion of fuels represents a rapid form of oxidation, while in certain forms of rock weathering oxidation is very slow. Living tissues require oxygen to convert foodstuffs into energy.

The remaining one percent of dry air is mostly argon (Ar), an inactive gas of little importance in natural processes. In addition, there is a small amount of *carbon dioxide* (CO_2), which is particularly efficient at absorbing long-wave energy radiated from the Earth's surface. This warms the lower atmosphere, which then reradiates some heat back to the surface through a process called the *greenhouse effect*. Green plants also convert carbon dioxide into carbohydrates in photosynthesis.

Another important component of the atmosphere is *water vapour* (H_2O), the gaseous form of water. Individual water vapour molecules mix freely throughout the atmosphere, just like the other gases. Usually, water vapour makes up less than one percent of the atmosphere, but its concentration varies and can exceed four percent under some conditions. Since water vapour, like carbon dioxide, is a good absorber of long-wave radiation, it also plays a major role in warming the lower atmosphere.

The atmosphere also contains many other gases, most of which are pollutants, such as sulphur dioxide and halogens. Dusts and fine particles that originate from volcanic eruptions or other natural processes absorb and scatter radiation. Many of these particulates also play an important role in the formation of precipitation by acting as **condensation nuclei** around which water droplets and ice crystals can form.

OZONE IN THE UPPER ATMOSPHERE

Another small, but important, constituent of the atmosphere is **ozone** (O_3), a form of oxygen in which three oxygen atoms are bonded together. Ozone absorbs UV radiation, thus shielding plants and animals from its harmful effects. The concentration of ozone in the atmosphere is measured in **Dobson units** (DU). If the ozone in a column of air stretching from the Earth's surface to the top of the atmosphere were concentrated in a single layer at standard temperature and pressure (i.e., at 0°C and 101.3 kilopascals), the thickness of that layer, measured in hundredths of a millimetre, would be the concentration of ozone expressed in Dobson units (Figure 3.9). Ozone is found mostly in the stratosphere, a layer of the atmosphere that lies about 14 to 50 kilometres above the surface (see Chapter 4).

Gaseous chemical reactions in the stratosphere produce ozone. Oxygen molecules (O_2) absorb UV energy and split into two oxygen atoms (O + O). A free oxygen atom (O) then combines with an O_2 molecule to form ozone (O_3). Once formed, ozone can also be destroyed by UV radiation, thus splitting to form O_2 + O. The net effect is that ozone (O_3), molecular oxygen (O_2), and atomic oxygen (O) are constantly formed, destroyed, and reformed in the ozone layer, absorbing UV radiation with each transformation.

Table 3.1 Average composition of the atmosphere up to an altitude of 25 kilometres

Gas	Chemical Formula	Percent Volume
Nitrogen	N_2	78.08%
Oxygen	O_2	20.95%
*Water	H_2O	0 to 4%
Argon	Ar	0.93%
*Carbon dioxide	CO_2	0.0360%
Neon	Ne	0.0018%
Helium	He	0.0005%
*Methane	CH_4	0.00017%
Hydrogen	H_2	0.00005%
*Nitrous oxide	N_2O	0.00003%
*Ozone	O_3	0.000004%

* Denotes gases that vary in concentration both spatially and seasonally.

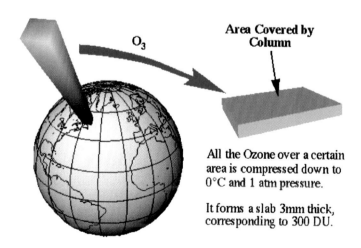

3.9 Ozone in the atmosphere is measured in Dobson units (DU) If the ozone in an air column, compressed to standard temperature and pressure (0°C at 101.3 kPa), formed a slab three millimetres thick, it would be equivalent to 300 DU.

EYE ON GLOBAL CHANGE | 3.3   The Ozone Layer—Shield to Life

The release of **chlorofluorocarbons (CFCs)** into the atmosphere poses a serious threat to the ozone layer. CFCs are synthetic industrial chemical compounds containing chlorine, fluorine, and carbon atoms. Although household uses of CFCs in aerosol sprays were banned in most parts of the world in the 1970s, CFCs are still in wide use as cooling fluids in existing refrigeration systems. When these appliances are disposed of, CFCs can be released into the air.

Molecules of CFCs in the atmosphere are very stable. They move upward by diffusion without chemical change until they eventually reach the ozone layer. There, they absorb UV radiation and decompose into chlorine oxide (CIO) molecules. The chlorine oxide molecules in turn break down molecules of ozone, converting them into oxygen molecules. This reduces the concentration of ozone in the ozone layer, which adversely affects its ability to absorb UV radiation. Other gaseous molecules, including nitrogen oxides, bromine oxides, and hydrogen oxides can also reduce the concentration of ozone in the stratosphere.

However, not all threats to the ozone layer can be attributed to human activity. Aerosols inserted into the stratosphere by volcanic activity can also reduce ozone concentrations. The June 1991 eruption of Mount Pinatubo in the Philippines reduced the average concentration of global ozone in the stratosphere by four percent the following year, with reductions over midlatitudes of up to nine percent.

Studies of the ozone layer based on satellite data during the 1980s showed a substantial decline in total global ozone. By late in the decade, scientists had agreed that ozone reduction was occurring much faster than had been predicted. Compounding the problem of global ozone decrease was the discovery in the mid-1980s of a "hole" in the ozone layer over Antarctica. The ozone hole is defined as the area over Antarctica in which the total ozone concentration is less than 220 Dobson units. The ozone hole has steadily increased in area

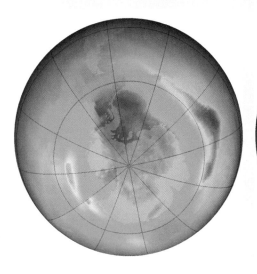 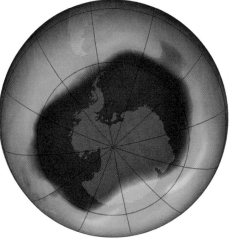

Maps of ozone concentration for the South Pole region on September 24, 1979 and 2006 Prior to 1979, values were in the range of 250 to 300 DU, but plunged to as little as 85 DU in some parts of Antarctica in 2006. Data were acquired by the Total Ozone Mapping Spectrometer (TOMS) on NASA's Earth Probe TOMS-EP Satellite. (Courtesy of NASA)

If the concentration of ozone is reduced, fewer transformations occur, reducing the amount of UV absorption. If UV radiation were to reach the Earth's surface at full intensity, exposed bacteria on the Earth's surface would be destroyed, and unprotected plant and animal tissues would be severely damaged. Of particular concern is the potential increase in skin cancer. Other possible effects include a reduction in crop yields and damage to some forms of aquatic life.

The presence of the ozone layer is thus essential to maintaining a viable environment for life on this planet. Some forms of air pollution can reduce ozone concentrations substantially in some regions of the atmosphere at certain times of the year. *Eye on Global Change 3.3 • The Ozone Layer—Shield to Life* describes this threat to the ozone layer.

AIR POLLUTANTS

An **air pollutant** is an unwanted substance injected into the atmosphere from the Earth's surface by either natural or human activities. Air pollutants include gases, aerosols, and particulates. **Aerosols** are extremely small particles that float freely with normal

and intensity over the past two decades. Seasonal thinning of the Antarctic ozone layer occurs during the southern hemisphere spring, with ozone concentration typically reaching a minimum in late September and October of each year.

This thinning occurs after the formation of a polar vortex in the stratosphere in the winter. During the long polar night, this vast whirlpool of trapped air is kept out of the Sun, preventing ozone formation. The air in the vortex is very cold, and clouds of crystalline water (ice) or other water-containing compounds form within it. The crystals are important because they provide a surface on which chemical reactions can take place. One of these reactions converts chlorine from a stable form to chlorine oxide (ClO), which can rapidly destroy ozone in the presence of sunlight. As the southern hemisphere spring approaches, the Sun slowly illuminates the polar vortex, and the ClO reacts with ozone, reducing ozone concentrations and forming the ozone hole.

On September 24, 2006, the area of the Antarctic ozone hole reached 29.5 million square kilometres (see Figure to the left), which equaled the previous record single-day value of September 9, 2000. In addition, the concentration of ozone was only 85 DU in some parts of eastern Antarctica, and was practically absent in the ozone layer between 13–21 kilometres above the surface. In this critical layer, a record low of only 1.2 DU was recorded, compared with an average reading of 125 DU earlier in the year.

In the northern hemisphere, conditions for the formation of an ozone hole are not as favourable. An Arctic polar vortex forms, but it is much weaker than the Antarctic vortex and is less stable. As a result, an early-spring ozone hole is not usually observed in the Arctic. However, in 1993, 1996, 1997, and 1999, notable Arctic ozone holes occurred, and atmospheric computer models have predicted more of these events in the period 2010–2019.

Although the most dramatic reductions in stratospheric ozone levels have been associated with Arctic and Antarctic ozone holes, reductions have been noted elsewhere as well—including the northern midlatitudes. A number of studies during the 1990s showed reductions of six to eight percent in stratospheric ozone concentrations occurring over North America.

As the global ozone layer thins, the amount of incoming UV solar radiation reaching the Earth's surface should increase. Recent studies of satellite data have documented this effect. Since 1978, the average annual exposure to UV radiation increased by 6.8 percent per decade along the 55° N parallel. At 55° S, the increase was 9.8 percent per decade.

Responding to the global threat of ozone depletion and its anticipated impact on the biosphere, 20 nations signed the Vienna Convention for the Protection of the Ozone Layer in 1985. In the Montreal Protocol of 1987, they agreed to specific reductions in CFCs, with developed nations phasing out CFC production by 2000. Halting the use of CFCs in developing nations, scheduled for 2010, is the next step. In 1999, developed nations pledged $1 billion to help developing nations switch to safer alternatives. As a result of the Montreal Protocol, the concentrations of ozone-depleting substances in the stratosphere are decreasing from a peak reached in 2001 by about 0.1 to 0.2 percent annually. However, many of these substances stay in the atmosphere for more than 40 years. The present estimate is that the ozone hole will not fully recover until about 2065.

---

air movements. **Particulates** are larger, heavier particles that eventually fall back to Earth. Most pollutants generated by human activity arise in two ways. The first is through day-to-day activities, such as driving automobiles. The second is through industrial activities, such as fossil fuel combustion or smelting of mineral ores.

One group of pollutants is known collectively as greenhouse gases. Although relatively scarce in the atmosphere, they are of special importance because of their ability to trap long-wave radiation. (Chapter 4 discusses the link between greenhouse gases and global warming.) The most significant greenhouse gas is CO_2, which although naturally present in the atmosphere, is released in great quantities through combustion of fossil fuels. Methane (CH_4) is produced by anaerobic processes in bogs and swamps, such as those found throughout the Canadian boreal forest. However, methane is also produced in rice paddies, by livestock, and by biomass burning done when clearing land for agriculture. Nitrous oxide (N_2O) is released in automobile exhaust and from chemical fertilizers. Chlorofluorocarbons (CFCs), which include aerosol propellants and refrigerants, are also important greenhouse gases.

In addition to N₂O, nitric oxide (NO), nitrogen dioxide (NO₂), and nitrates (−NO₃) are also present in the atmosphere; this mixture is usually referred to as NOₓ. NOₓ reacts in the presence of sunlight with volatile organic compounds, such as those released from petroleum refineries. These **photochemical reactions** can cause brown hazes and also produce ozone. Unlike stratospheric ozone, the ozone found in the lowest layers of the atmosphere is an air pollutant and is damaging to lungs and other living tissues.

Similarly, sulphur oxides (SO₂ and SO₃) are referred to as SOₓ. These gases, together with NO₂, readily combine with oxygen and water in the presence of sunlight to form sulphuric and nitric acid aerosols. These aerosols serve as condensation nuclei and acidify the tiny water droplets that coalesce to form acid precipitation. Sulphur gases are largely produced in smelters and electricity generating stations that burn coal or lower-grade fuel oils. These types of facilities also supply most of the unwanted particulate matter in the atmosphere. Some of it is coarse soot particles—fly ash—that settle quite close to the source as **fallout**. Particles too small to settle are later swept down to Earth by precipitation in a process called **washout**.

Acidic particles that are removed from the atmosphere through **dry deposition** form thin dust layers on plants and soils, which can have severe environmental consequences (Figure 3.10).

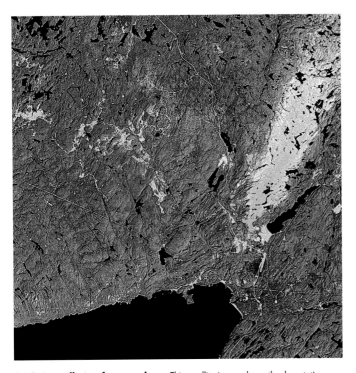

3.10 Air pollution from smelters This satellite image shows the devastating effects of pollutants emitted from smelting operations near Wawa, Ontario. Healthy vegetation is shown in red to pink tones. The blue streak on the right-hand side of the image is barren of vegetation. The smelting facilities are located at the narrowest point of the blue plume. *Fly By: 48° 03′ N; 84° 45′ W*

This dust acidifies any precipitation that subsequently falls. In winter, acid particles can mix with snow as it accumulates, and when the snow melts, a surge of acid water is released into soils and streams.

An important factor in the level of impact of acid deposition on the environment is the ability of the soil and surface water to absorb and neutralize acid. In dry climates, such as the Canadian Prairies, acidic deposition can be readily neutralized because surface waters are normally somewhat alkaline. Areas where soil water is naturally acidic, such as the forest regions of Canada, are most sensitive to acid deposition. Many lakes in eastern Canada have experienced increases in fish mortality, attributed to acidification.

SENSIBLE HEAT AND LATENT HEAT TRANSFER

The transfer of short-wave energy from the Sun across the vacuum of space is accomplished primarily by radiation, and occurs without the involvement of a physical substance. In the same way, long-wave energy from the Earth and its atmosphere can escape into space. In addition to radiation, energy transfer in the form of sensible and latent heat is also important in maintaining the global energy balance.

Sensible heat is heat energy held by an object or substance that can be measured by a thermometer. The store of sensible heat increases as the object's temperature rises. When two objects of unlike temperature contact each other, heat energy moves by **conduction** from the warmer to the cooler. This type of heat flow is referred to as **sensible heat transfer**. Sensible heat transfer can also occur by **convection**. In this process, a fluid is heated by a warm surface, expands, and rises, creating an upward flow. This flow moves heat away from the surface. Air acts as a fluid in much the same way as water, and convection is an important factor underlying winds and other dynamic processes in the atmosphere (see Chapter 5).

Unlike sensible heat, a thermometer cannot measure **latent heat**. Rather it is heat that is taken up and stored in the form of molecular motion when a substance changes state from a solid to a liquid, from a liquid to a gas, or from a solid directly to a gas. For example, energy is required to convert liquid water to water vapour. As water evaporates, it absorbs energy from the environment, which accelerates the random motion of the water molecules. The water absorbs the energy in the form of latent heat and uses it to overcome the molecular forces of attraction in the liquid. Because the energy is latent heat, the change in state from liquid to vapour occurs without a rise in temperature. In this way, energy is transferred to the atmosphere, carried in the form of randomly moving water vapour molecules.

Although a thermometer cannot measure the latent heat of water vapour in air, energy is stored there just the same. When water vapour reverts to a liquid or solid, the latent heat is released, making the surroundings warmer. In the Earth's atmosphere, **latent heat transfer** occurs when water evaporates from a moist land surface or open water surface. This process transfers heat from the surface to the atmosphere. On a global scale, latent heat transfer by movement of moist air provides an important mechanism for transporting large amounts of energy from one region of the Earth to another. *Working It Out 5.1 • Energy and Latent Heat* in Chapter 5 provides more information on the amount of energy acquired or released in these processes.

THE GLOBAL ENERGY SYSTEM
SOLAR ENERGY LOSSES IN THE ATMOSPHERE

Figure 3.11 shows the typical losses of incoming solar energy as it passes through the atmosphere on its way to the Earth's surface. As short-wave solar radiation penetrates the atmosphere, its energy is absorbed or diverted in various ways. The thin outer layers of the atmosphere almost completely absorb gamma and X-rays. Much of the UV radiation is also absorbed, particularly by ozone.

As radiation moves through deeper and denser atmospheric layers, gas molecules cause some radiation to be scattered. Radiation that is scattered sideways in all directions is termed **diffuse radiation**. It is unchanged except for the direction it travels. Dust and other particles in the air can also cause scattering. Some energy returns to space as a result of scattering, and some eventually flows down to the surface. Radiation that is turned back to space is termed **diffuse reflection**, which, under clear sky conditions, amounts to about three percent of the original incoming solar energy.

Gas molecules and particles of dust and other solids can absorb radiation as it passes through the atmosphere. Carbon dioxide and water are the two primary absorbers, but because the air's water vapour content can vary greatly, absorption also varies from one location to another. Absorption results in a rise in air temperature. In this way, incoming solar radiation causes some direct heating of the lower atmosphere. Absorption accounts for about 17 percent of the incoming solar radiation, while scattering and diffuse reflection account for about 3 percent. Thus, under clear skies, diffuse reflection and absorption total about 20 percent, leaving as much as 80 percent of the solar radiation to reach the ground.

The presence of clouds can greatly increase the amount of incoming solar radiation reflected back to space. The bright, white surfaces of thick, low clouds are extremely good reflectors of short-wave radiation. Cloud reflection can account for the direct return to space of 30 to 60 percent of incoming radiation (Figure 3.12). Clouds also absorb as much as 5 to 20 percent of the radiation, depending on cloud type and thickness. Under heavy cloud cover, as little as 10 percent of incoming solar radiation may actually reach the ground. The amount and distribution of global cloud cover are important factors affecting the global energy budget.

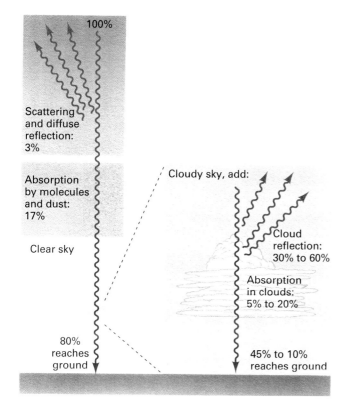

3.11 Fate of incoming solar radiation Losses of incoming solar energy are much lower with clear skies (left) than with cloud cover (right). (Copyright © A. N. Strahler. Used with permission.)

3.12 Effect of cloud cover on the amount of energy received at the surface The marked colour contrast between the upper and lower parts of clouds illustrates how much energy is reflected and absorbed.

ALBEDO

The proportion of short-wave radiant energy scattered upward by a surface is termed **albedo**. For example, a surface that reflects 40 percent of the short-wave radiation it receives has an albedo of 0.40. Albedo is an important property of a surface because it measures how much incident solar energy it will absorb (see Table 3.2). A surface with a high albedo, such as snow or ice (0.50 to 0.85), reflects much or most of the solar radiation, thereby considerably reducing the amount of energy it absorbs. A surface with a low albedo, such as black pavement (0.03), absorbs nearly all the incoming solar energy. Since the energy absorbed by a surface can warm the air immediately above it by conduction and convection, surface temperatures will be warmer over a low albedo surface (pavement) than over a high albedo surface (snow or ice).

The albedo of a water surface varies with the angle of incoming radiation. It is very low (0.03) for nearly vertical rays on calm water. However, when the Sun shines on a water surface at a low angle, much of the radiation is directly reflected. For fields, forests, and bare ground, albedos are of intermediate value, ranging from as low as 0.05 to as high as 0.30.

Certain orbiting satellites are equipped with instruments to measure the energy levels of short-wave and long-wave radiation at the top of the atmosphere. Data from these satellites have been used to estimate Earth's average albedo. This albedo includes reflection by the Earth's atmosphere as well as all its surfaces, so it includes diffuse reflection by clouds, dust, and atmospheric molecules. The albedo values obtained in this way vary between 0.29 and 0.34. This means that the Earth–atmosphere system directly returns to space about one third of the solar radiation it receives. *Focus on Remote Sensing 3.4 • CERES—Clouds and the Earth's Radiant Energy System* describes how satellite instruments map and monitor the reflection of solar short-wave radiation and the emission of long-wave radiation by the Earth–atmosphere system.

Table 3.2 Albedo (reflectivity) of various surfaces

Surface	Percent
Fresh snow	80–85
Old snow	50–60
Sand	20–30
Grass	20–25
Dry earth	15–25
Wet earth	10
Forest	5–10
Water (sun near horizon)	50–80
Water (sun near zenith)	3–5
Thick cloud	70–80
Thin cloud	25–50
Earth and atmosphere	30

COUNTER-RADIATION AND THE GREENHOUSE EFFECT

The amount of short-wave energy absorbed by a surface is an important determinant of its temperature. However, a surface is also warmed significantly by long-wave radiation emitted by the atmosphere and absorbed by the ground. Figure 3.13 shows energy flows between the surface, atmosphere, and space. On the left is the flow of short-wave radiation from the Sun to the surface. Some of this radiation is reflected back to space (as measured by the surface's albedo), but much of it is absorbed, warming the surface.

The surface emits long-wave radiation upward as flows, labelled A and B. Some of this radiation escapes directly to space (A); the atmosphere absorbs the remainder (B). Although the atmosphere is colder than the surface, it too emits long-wave radiation, as shown by flows C and D. It emits this radiation in all directions, some upward to space (C), and some toward the Earth's surface (D). Since this downward flow is the opposite direction to long-wave radiation leaving the surface, it is termed **counter-radiation**. It replaces some of the heat emitted by the surface.

The amount of counter-radiation depends mainly on the presence of carbon dioxide and water vapour in the atmosphere. Much of the long-wave radiation emitted upward from the Earth's surface is absorbed by these two atmospheric constituents. This absorbed energy raises the temperature of the atmosphere, thereby causing it to emit more counter-radiation. Thus, the lower atmosphere, with its long-wave-absorbing gases, acts like a

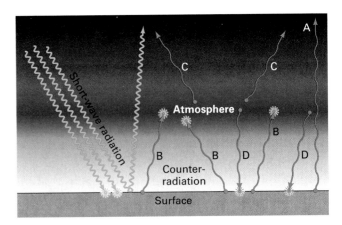

3.13 Counter-radiation and the greenhouse effect Short-wave radiation (left) passes through the atmosphere and is absorbed at the surface, warming the surface. The surface emits long-wave radiation. Some of this flow passes directly to space (A), but most is absorbed by the atmosphere (B). In turn, the atmosphere radiates long-wave energy back to the surface as counter-radiation (D) and also to space (C). The return of outbound long-wave radiation by counter-radiation constitutes the greenhouse effect.

blanket that traps heat. Because liquid water is also a strong absorber of long-wave radiation, cloud layers, which are composed of tiny water droplets, are even more effective than carbon dioxide and water vapour in producing an insulating effect.

This mechanism, in which the atmosphere traps long-wave radiation and returns it to the surface, is termed the **greenhouse effect**. The heating of air in a car parked in the Sun with the windows closed illustrates the same phenomenon. The window glass permits entry of short-wave energy but absorbs and blocks the exit of long-wave energy.

GLOBAL ENERGY BALANCE OF THE EARTH–ATMOSPHERE SYSTEM

Earth constantly absorbs short-wave solar radiation and emits long-wave radiation, and these flows of energy from the Sun to the Earth and then back out into space form a complex system involving not only radiant energy, but also energy storage and energy transport in various forms. The pathways of energy flow between the Sun, the Earth's surface, and the atmosphere integrate all the dynamic processes that shape the planet. Ultimately, the amount of energy received by the Earth–atmosphere system equals the amount of energy returned to space. Therefore, on a global scale, there is a radiation balance.

The global radiation balance is an example of an energy flow system, and can be calculated from the incoming short-wave and outgoing long-wave radiation fluxes. The input flow of solar radiation to Earth is determined from the flow rate and the area that re-

ceives the flow. The rate is described by the solar constant, which has a value of 1,376 watts per square metre (1.367 kilowatts per square metre). The Earth's radius is about 6,400 kilometres. Thus, Earth presents a 6,400-kilometre-wide disk to the Sun's rays, with a surface area of 1.29×10^{14} square metres. Therefore, the energy flow from the Sun to the Earth is

$$\begin{array}{c} \text{Flow per} \\ \text{Unit Area} \end{array} \times \begin{array}{c} \text{Area of Earth} \\ \text{Presented to Sun} \end{array} = \begin{array}{c} \text{Sun–Earth} \\ \text{Energy Flow} \end{array}$$

$$1.367 \text{ kW/m}^2 \times 1.29 \times 10^{14} \text{ m}^2 = 1.77 \times 10^{14} \text{ kW}$$

Energy outflows take the form of reflected short-wave radiation and emitted long-wave radiation. Earth reflects about one third of the solar radiation that it receives, or about $1.77 \times 10^{14} \text{ kW} \div 3 = 0.59 \times 10^{14} \text{ kW}$. The remaining two thirds ($1.77 - 0.59 = 1.18 \times 10^{14} \text{ kW}$) is absorbed by the Earth and its atmosphere and is ultimately emitted as long-wave radiation.

An important principle of physics is that energy is neither created nor destroyed, but only transformed. It is therefore possible to follow the initial stream of solar energy and account for its diversion into different system pathways and its conversion into different energy forms. A full accounting of all the energy flows among the Sun, the atmosphere, the Earth's surface, and space forms the **global energy balance**.

INCOMING SHORT-WAVE RADIATION

Figure 3.14 presents the global energy balance for the Earth–atmosphere system. The fate of the downward flow of solar radiation is shown on the left. Reflection by gas molecules and

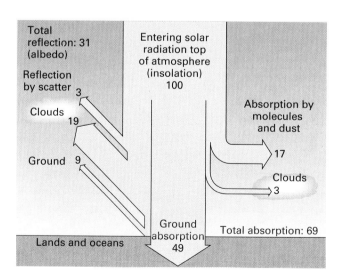
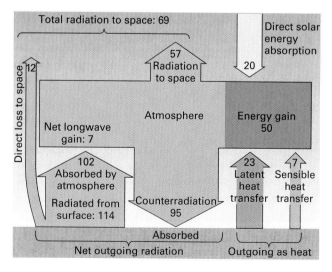

3.14 The global energy balance The left diagram shows the fate of incoming solar radiation. The right diagram shows long-wave energy flows occurring between the surface and atmosphere and space. Also shown are the transfers of latent heat, sensible heat, and direct solar absorption, which balance the budget for the Earth–atmosphere system. Values are given as percentages of a total insolation of 100. (Data from Kiehl and Trenberth, 1997)

The Earth's global **radiation balance** is the primary determinant of long-term surface temperature. However, this balance is increasingly sensitive to human activities, such as the conversion of forests to pasture or the release of greenhouse gases into the atmosphere. Thus, it is important to monitor the Earth's radiation budget over time as accurately as possible.

For nearly 20 years, the Earth's global radiation balance has been the subject of NASA missions, which have launched radiometers—radiation-measuring devices—into orbit around the Earth. These devices scan Earth and measure the amount of short-wave and long-wave radiation at the top of the atmosphere. An ongoing NASA experiment entitled CERES—Clouds and the Earth's Radiant Energy System—is placing a new generation of these instruments in space to monitor the global radiation balance.

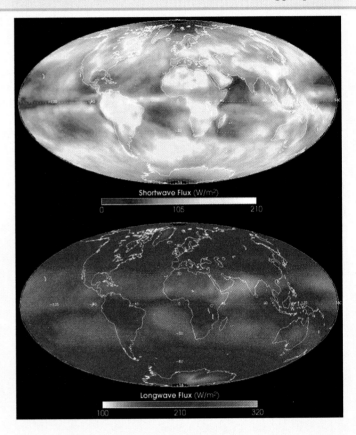

Global short-wave and long-wave energy fluxes from CERES These images show average short-wave and long-wave energy flows from Earth for March 2000, as measured by the CERES instrument on NASA's Terra satellite platform. (Courtesy of NASA)

dust, clouds, and the surface (including the oceans) totals 31 percent (0.31) and represents the combined albedo of the Earth and the atmosphere. The right side of the diagram shows values for short-wave energy absorption in the atmosphere. The combined losses through absorption by molecules, dust, and clouds average 20 percent. With 31 percent of the incoming solar energy flow reflected and 20 percent absorbed in the atmosphere, the remaining 49 percent is absorbed by the Earth's land and water surfaces.

SURFACE ENERGY FLOWS

The right part of Figure 3.14 shows the components of outgoing long-wave radiation for the Earth's surface and the atmosphere. The large arrow on the left shows that total long-wave radiation leaving the Earth's land and ocean surface is 114 percent of the total incoming solar radiation. This value is higher than 100 percent because the surface also receives a flow of 95 units of long-wave counter-radiation from the atmosphere. This is in addition to the

49 units of short-wave radiation that are absorbed by the surface (see left diagram). So the surface actually receives 95 (long-wave) plus 49 (short-wave) to equal 144 units.

The two smaller arrows on the far right of the diagram show the flow of energy away from the surface as latent heat (23 units) and sensible heat (7 units). Evaporation of water from moist soil or oceans transfers latent heat from the surface to the atmosphere. Sensible heat transfer occurs when heat is directly conducted from the surface to the adjacent air layer. When air is warmed by direct contact with a surface and rises, it transports its heat with it by convection. In this process, heat flows from the surface to the atmosphere. This leaves 144 − 23 (latent heat) − 7 (sensible heat) = 114 units that must be accounted for at the surface.

The long, thin arrow on the extreme left indicates that 12 units are lost to space, leaving 102 units absorbed by the atmosphere. (These flows are equivalent to flows A and B in Figure 3.13). Total energy loss through radiation from the surface is therefore equivalent to 12 (direct loss to space) + 102 (absorbed by atmosphere) = 114 percent.

The CERES instruments scan Earth from horizon to horizon, measuring outgoing energy flows of reflected solar radiation (0.3 to 5.0 micrometres), out-going long-wave radiation (8 to 12 micrometres), and outgoing total radiation (0.3 to 100 micrometres). Daily global maps of the Earth's upwelling radiation fields are prepared from these observations. The figure to the left shows global reflected solar energy and emitted long-wave energy averaged over the month of March 2000, as obtained from the CERES instrument aboard NASA's Terra satellite platform.

The top image shows average short-wave flux ranging from 0 to 210 watts per square metre. The largest flows occur over regions of thick cloud cover near the Equator, where much of the solar radiation is reflected back to space. In the midlatitudes, persistent cloudiness during this month also shows up as light tones, as do bright tropical deserts, such as the Sahara. Snow and ice surfaces in polar regions are quite reflective, but in March, at about the time of the equinox, sun angle is still quite low. Therefore, the amount of radiation the polar regions receive is much less than at the equator or midlatitudes. Consequently, the polar regions don't appear bright in this image. Oceans, especially where skies are clear, absorb solar radiation and thus show low short-wave fluxes.

The bottom image shows long-wave flux on a scale from 100 to 320 watts per square metre. Cloudy equatorial regions have low values, showing the blanketing effect of thick clouds that trap long-wave radiation beneath them. Warm tropical oceans in regions of clear sky emit the most long-wave energy. Toward the poles, surface and atmospheric temperatures drop, so long-wave energy emission also drops significantly.

Clouds are important determiners of the global radiation balance. A primary goal of the CERES experiment is to learn more about the Earth's cloud cover, which changes from hour to hour. This knowledge can improve global climate models that predict the impact of human and natural change on the Earth's climates. The most important contribution of CERES, however, is continuous monitoring of the Earth's radiant energy flows. In this way, small, long-term changes induced by human or natural processes can be detected, despite the large spatial and temporal variations in energy flows associated with cloud cover.

The latent heat flow (23 units) and sensible heat flow (7 units) are not part of the radiation balance, since they are not in the form of radiation. However, they are an important part of the total energy budget of the surface, which includes all forms of energy. Taken together, these two flows account for 30 units leaving the surface. With their contribution, the surface energy balance is complete. Total gains are 95 (long-wave) + 49 (short-wave) = 144. Total losses are 114 (long-wave) + 23 (latent heat) + 7 (sensible heat) = 144.

ENERGY FLOWS TO AND FROM THE ATMOSPHERE

Like the Earth's surface, the atmosphere gains energy by absorption of short-wave radiation, amounting to 20 units (17 units by gas molecules and dust and 3 units by clouds). The atmosphere also gains energy from latent heat transfer (23 units) and sensible heat transfer (7 units). In long-wave radiation, the atmosphere absorbs 102 units emitted by the surface. The sum of these energy units is 152.

A loss of 57 units of long-wave energy is radiated from the atmosphere to space (flows labelled C in Figure 3.13) and a loss of 95 units occurs as counter-radiation to the surface (flows labelled D in Figure 3.13). Together these total 152 units, which balance the atmosphere's energy budget.

This analysis illustrates the atmosphere's vital role in trapping heat through the greenhouse effect. Without the 95 units of counter-radiation from the atmosphere, the surface would have only 49 units of absorbed short-wave radiation to emit. The temperature needed to radiate so few units is well below freezing. So without the greenhouse effect, Earth would be much less hospitable.

CLIMATE AND ENVIRONMENTAL CHANGE

The preceding analysis helps to illustrate how global change might affect the Earth's climate. Suppose that clearing forests for agriculture and turning agricultural lands into urban and suburban areas decreases surface albedo. The ground would absorb more energy, raising its temperature. That, in turn, would increase the flow of surface long-wave radiation to the atmosphere. This radiation would be absorbed and then boost counter-radiation, which would probably amplify the warming through the greenhouse effect.

But what if air pollution causes more, low, thick clouds to form? Since low clouds increase short-wave reflection back to space, the effect will be to cool the surface and atmosphere. Conversely, an increase in high altitude clouds would result in greater absorption of long-wave energy compared with the amount of short-wave energy they reflect. This should make the atmosphere warmer and boost counter-radiation, increasing the greenhouse effect. The energy flow linkages between the Sun, surface, atmosphere, and space are critical components of the climate system, and understanding them helps to better appreciate the complexities of global climate change.

NET RADIATION, LATITUDE, AND THE ENERGY BALANCE

The Earth intercepts solar energy, and when it absorbs this energy, the planet's temperature tends to rise. This is offset by the reduction in temperature that occurs when energy is radiated to space.

Over time, these incoming and outgoing radiation flows must balance for the Earth as a whole. However, incoming and outgoing flows do not have to balance instantaneously at a specific location. At night, for example, there is no incoming radiation, yet the darkened Earth's surface and the atmosphere still emit outgoing radiation. Furthermore, in some locations, more radiant energy can be lost than gained; for example, in the polar regions. This is offset by tropical regions, which gain more radiant energy than they lose.

Net radiation is the difference between all incoming radiation and all outgoing radiation. In places where radiant energy flows in faster than it flows out, net radiation is positive, providing an energy surplus. In other places, where radiant energy flows out faster than it flows in, net radiation is negative, yielding an energy deficit. On an annual basis the net radiation balance is zero for the entire Earth–atmosphere system.

Solar energy input varies strongly with latitude. Between about 40° N lat. and 40° S lat. there is a net radiant energy gain, or energy surplus. Here incoming solar radiation exceeds outgoing long-wave radiation throughout the year (Figure 3.15). From 40° N and 40° S toward the poles, net radiation is negative, providing an energy deficit. In these regions outgoing long-wave radiation exceeds incoming short-wave radiation. The area on the graph labelled "surplus" is equal in size to the

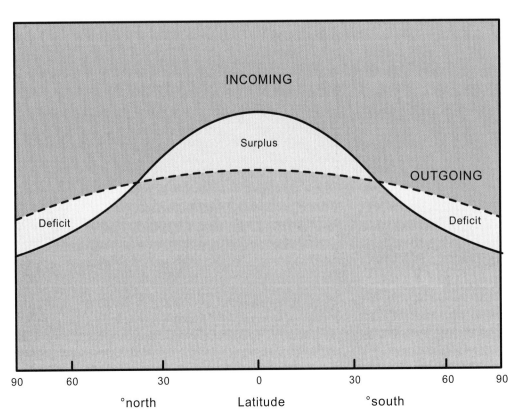

3.15 Annual surface net radiation from pole to pole Where net radiation is positive, incoming solar radiation exceeds outgoing long-wave radiation, there is an energy surplus, and energy moves toward the poles as latent and sensible heat. Where net radiation is negative, there is an energy deficit, and latent and sensible heat are lost in the form of outgoing long-wave radiation.

combined areas labelled "deficit." This confirms that net radiation for the Earth as a whole is zero and that global incoming short-wave radiation exactly balances global outgoing long-wave radiation.

The pattern of energy surplus at low latitudes and energy deficit at high latitudes creates a flow of energy from low latitudes to high. This energy flow, in the form of sensible and latent heat, occurs through the poleward movement of warm water and warm, moist air in global ocean and atmospheric circulation. At the same time cooler water and cooler, drier air moves toward the Equator. Without this broad-scale circulation, low latitudes would heat up and high latitudes would cool down until a radiative balance is achieved, resulting in much more extreme temperature contrasts.

SOLAR ENERGY

The Earth intercepts solar energy at the rate of 1.77×10^{14} kilowatts per year. This quantity of energy is about 28,000 times as much as all the energy currently being consumed each year by human society. Thus, an enormous energy source is near at hand waiting to be used. In addition to its abundance, solar energy, unlike fossil fuels, does not release carbon dioxide or other pollutants.

Solar energy can be converted directly into electricity using photovoltaic cells. Its use has been limited because of the relatively high cost of manufacturing these solar cells, but the technology is now widely used for roadside emergency telephones and similar applications. Using current technology, arrays of solar cells would need to occupy large areas to produce enough energy outputs for general use. Electricity storage is another problem, since the solar cells will not generate power at night.

A second common application of solar energy is heating buildings. Most systems use solar collectors placed on roofs. A typical collector consists of a network of aluminum or copper tubes carrying circulating water. The tubes, painted black, absorb solar energy and use it to heat the water to a temperature of about 65°C. Several advanced installations have been constructed, which use arrays of computer-controlled mirrors to focus the Sun's rays onto boilers to produce steam for use in power generators.

Commercial-scale projects have generally employed arrays of flat, moveable mirrors called heliostats to focus the Sun's energy onto a central **solar power tower**, such as Solar 1, which was constructed in 1981 near Barstow, California. It provided steam for five years, delivering 10 megawatts of electric power. In 1994–95, the plant was modified by adding more heliostats and converting it to use molten nitrate salt, which is heated to 565°C (Figure 3.16). This stored heat was used during the night to produce steam to generate a continuous supply of electricity. The design, with modifications, was used for Solar Tres in Spain, which has 2,493 heliostats, with a total reflective area of 240,000 square metres.

A Look Ahead

Earth's energy balance is sensitive to a number of factors that determine how energy is transmitted and absorbed. The net effect is reflected in global temperature. Chapter 4 presents a discussion of present temperature patterns and how and why they vary from region to region, as well as the causes of seasonal and daily patterns. Chapter 5 covers the link between the energy balance and global circulation.

3.16 Solar II This solar power plant was built as a demonstration project in 1994–95. The heliostats were moved by computer-controlled motors to focus the Sun's energy on a reservoir of molten nitrate salt located atop the central tower. Heat from the salt was exchanged to create steam that was used to generate power.

A CLOSER LOOK   GEOGRAPHER'S TOOLS 3.5   Remote Sensing for Physical Geography

Remote sensing refers to gathering information from great distances and over broad areas, usually through instruments mounted on aircraft or orbiting space vehicles. These instruments measure electromagnetic radiation coming from the Earth's surface and atmosphere. The data acquired by remote sensors are typically displayed as images, but are often processed further to provide other types of outputs, such as maps of albedo, vegetation condition or extent, or land-cover class. Solar radiation reaching the Earth's surface is concentrated in wavelengths from about 0.3 to 2.1 micrometres in visible, near-infrared, and short-wave infrared wave bands, and remote sensors are commonly constructed to measure radiation in all or part of this range. For remote sensing of emitted energy, the object or substance itself is the source of the radiation, which typically is related to its temperature.

Colours and Spectral Signatures

Most objects or substances at the Earth's surface appear coloured to the human eye. The various colours mean that they reflect different wavelengths of light in the visible spectrum. Figure 1 shows typical reflectance spectra for water, vegetation, and soil in the solar wavelength range as they might be viewed through the atmosphere. Note that water vapour in the atmosphere strongly absorbs radiation at wavelengths from about 1.2 to 1.4 micrometres and 1.75 to 1.9 micrometres, so it is not possible for a remote sensor to detect surface features at those wavelengths.

Water surfaces are always dark, but are slightly more reflective in the blue and green regions of the visible spectrum. Thus, water appears blue or blue-green to our eyes. Beyond the visible region, water absorbs nearly all short-wave radiation it receives and so looks black in images acquired in the near-infrared and short-wave infrared regions.

Vegetation appears dark green to the human eye, which means that it reflects more energy in the green portion of the visible spectrum while reflecting somewhat less in the blue and red portions. But vegetation also reflects strongly in near-infrared wavelengths, which the human eye cannot see. Because of this property, vegetation will be bright in near-infrared images. This distinctive behaviour of vegetation—appearing dark in visible bands and bright in the near-infrared—is the basis for most vegetation remote sensing.

The soil spectrum shows a steady increase of reflectance across the visible and near-infrared spectral regions. Soil is brighter overall than vegetation and is somewhat more reflective in the orange and red portions. Thus, it often appears brown. (Note that this is just a "typical" spectrum—soil colour can actually range from black to bright yellow or red.)

The pattern of relative brightness within spectral bands is referred to as the *spectral signature* of an object or substance. Spectral signatures can be used to recognize objects or surfaces in remotely sensed images in much the same way that we recognize objects by their colours. In computer

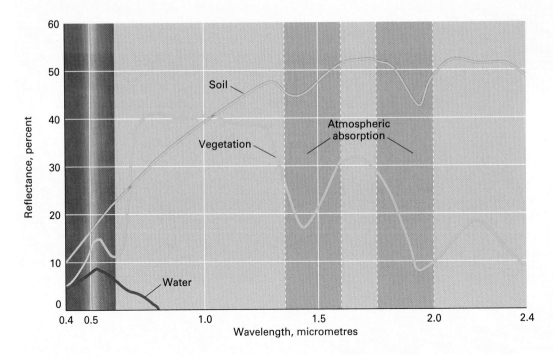

Figure 1 Reflectance spectra of vegetation, soil, and water

processing of remotely sensed images, spectral signatures can be used to make classification maps showing, for example, water, vegetation, and soil.

Aerial Photography

Aerial photography is the oldest form of remote sensing. Air photos have been in wide use since before World War II. Common practice is to have the field of one photograph overlap the next along the plane's flight path so that the photographs can be viewed stereoscopically for a three-dimensional effect. Because of its high resolution (degree of sharpness) and low cost, aerial photography is commonly used in remote sensing.

Aerial photography often uses colour infrared film. This special film is sensitive to near-infrared wavelengths in addition to visible wavelengths. Red colour in the film is produced as a response to near-infrared light, green colour is produced by red light, and blue colour by green light. Because healthy, growing vegetation reflects much more strongly in the near-infrared than in the red or green regions of the spectrum, vegetation has a characteristic red appearance (Figure 2).

Thermal Infrared Sensing

Radiation leaving the Earth's surface is concentrated in the thermal infrared spectral region, ranging from about 8 to 12 micrometres. Besides absolute temperature, the intensity of infrared emission depends on the *emissivity* of an object or substance. Emissivity is the ratio of the actual energy emitted to that of a blackbody at the same temperature. For most natural Earth surfaces, emissivity is comparatively high—between 0.85 and 0.99. Differences in emissivity affect thermal images. For example, two different surfaces might be at the same temperature, but the one with the higher emissivity will look brighter because it emits more energy. Rock types can be distinguished and mapped using thermal infrared images from several wavelengths (Figure 3).

Radar

There are two classes of remote sensor systems: passive and active. *Passive systems* acquire images without providing a source of wave energy, in much the same way as a camera acquires an image. *Active systems*, such as radar, use a beam of energy as a source, sending the beam toward an object. Part of the energy is reflected back to the source, where it is recorded by a detector.

Radar systems in remote sensing use the *microwave* portion of the electromagnetic spectrum. An advantage of radar systems is that they use wavelengths that are not significantly absorbed by liquid water. This means that radar systems

Figure 2 This colour infrared aerial photograph shows part of Bow River Irrigation District near Vauxhall, Alberta. The various red tones represent different types of crops, several of which are irrigated with centre-pivot systems that give the distinctive round patterns.

Figure 3 Images of the Saline Valley area of California, acquired by the Advanced Spaceborne Thermal Emission and Reflection Radiometer (ASTER). Image (a) displays visible and near infrared bands. Vegetation appears red, snow and dry salt lakes are white, and exposed rocks are brown, grey, yellow, and blue. Rock colors mainly reflect the presence of iron minerals, and variations in albedo. Image (b) displays short wavelength infrared bands. In this wavelength region, clay, carbonate, and sulphate minerals have unique absorption features, resulting in distinct colours in the image. For example, limestones are yellow-green, and purple areas are kaolinite-rich clay mineral. Image (c) displays thermal infrared bands. In these wavelengths, variations in quartz content are more or less red; carbonate rocks are green, and volcanic rocks with high proportions of magnesium and iron are purple.

can penetrate clouds to provide images of the Earth's surface in any weather. At some wavelengths, however, microwaves are scattered by water droplets and can produce a return signal sensed by the radar apparatus. This effect is the basis for weather radars, which can detect rain and hail and are used in local weather forecasting.

Figure 4 shows a radar image of folded strata in the Appalachian Mountains in south-central Pennsylvania. It is produced by an airborne radar instrument that sends pulses of radio waves at an angle to the Earth's surface. Surfaces that are more perpendicular to the radar beam will

return the strongest echo and therefore appear lightest in tone. In contrast, those surfaces facing away from the beam will appear darkest. This produces an image resembling a three-dimensional model of the landscape illuminated by a light.

Digital Imaging

Modern remote sensing relies heavily on computer processing to extract and enhance information from remotely sensed data. This requires the data to be in the form of a *digital image*. In this format, the picture consists of a large

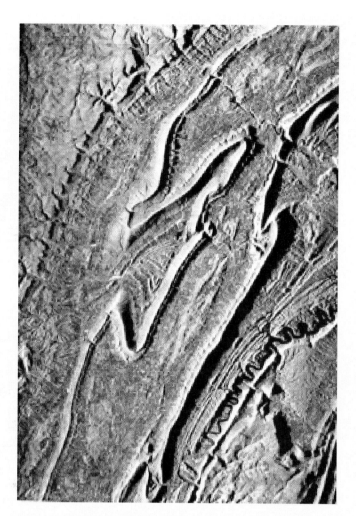

Figure 4 Side-looking radar image from south-central Pennsylvania. The image shows a portion of folded strata in the Appalachians with zigzag ridges and intervening valleys. The area shown is about 40 kilometres wide. (SAR image courtesy of Intera Technologies Corporation, Calgary, Alberta)

number of grid cells arranged in rows and columns. Each grid cell records a brightness value and is referred to as a *pixel*. Normally, low pixel values represent dark (low reflectance), and high pixel values represent light (high reflectance). The brightness values are normally viewed as a false colour image that is hard to distinguish from a colour photograph.

The great advantage of digital images over images on photographic film is that a computer can process them, for example, to increase contrast, sharpen edges, or assign distinctive colour classes. *Image processing* refers to the manipulation of digital images to extract, enhance, and display the information they contain. In remote sensing, image processing is a broad field that includes many methods and techniques for processing remotely sensed data.

Orbiting Earth Satellites

With the development of orbiting Earth satellites, remote sensing has expanded into a major branch of geographic research. Because orbiting satellites can image and monitor large geographic areas or even the entire Earth, global and regional studies have become possible. The two general types of Earth-observing satellites follow either sun-synchronous or geostationary orbits.

Sun-synchronous polar orbiting environmental satellites have an orbit that passes close to the North and South Poles (Figure 5). Because the Earth is not a perfect sphere, the force of gravity is slightly greater at the Equator. When the orbit crosses the Equator, the difference in gravity pushes the satellite slightly eastward. This keeps the satellite in time with the Sun, so that its overpasses occur at the same time of day. Typical Sun-synchronous satellites take 90 to 100 minutes to circle the Earth and are located at heights of about 700 to 800 kilometres above the Earth. As the Earth rotates to the east at a rate of 15° longitude per hour, the satellite's orbit moves to the west on each suc-

Figure 5 Earth track of a Sun-synchronous satellite With the Earth track inclined at about 80° to the Equator, the orbit slowly swings eastward at about 30° longitude per month, maintaining its relative position with respect to the Sun. (The height of the orbit above the Earth's surface is greatly exaggerated.)

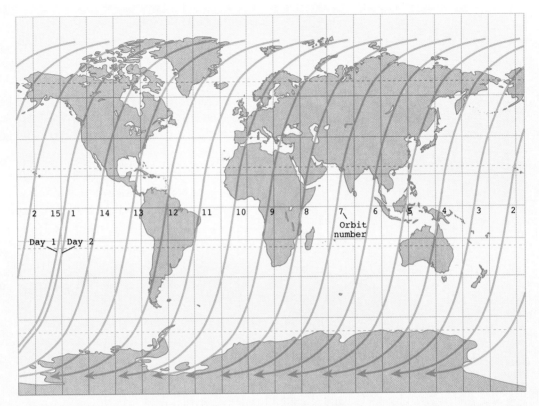

Figure 6 Ground paths of successive orbits during one day for a near-polar orbiting satellite The atmosphere normally contains fewer clouds early in the morning, so Earth-observing satellites, such as Landsat and SPOT, are often set to pass over at about 9:30–10:30 a.m. local time.

cessive pass (Figure 6). In this way, a polar-orbiting satellite completes 14 to 16 orbits per day, and covers the entire Earth's surface every 16 days.

Satellites in *geostationary orbit*, such as the GOES weather satellites, constantly revolve above the Equator (Figure 7). The orbit height, about 35,800 kilometres, is set so that the satellite makes one revolution in exactly 24 hours in the same direction that the Earth turns. Thus, the satellite always remains above the same point on the Equator. From its high vantage point, the geostationary orbiter provides a view of nearly half of the Earth and its atmosphere.

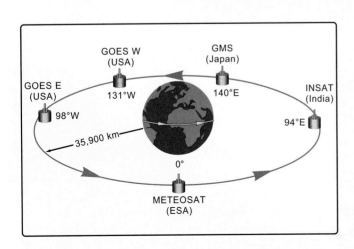

Figure 7 Geostationary satellites remain in the same position above the Earth The GOES weather satellites located above the Equator and the Atlantic and Pacific oceans provide continuous images of weather conditions for eastern and western North America.

CHAPTER SUMMARY

- Electromagnetic radiation is a form of energy emitted by all objects. The wavelength of the radiation determines its characteristics. The hotter an object, the shorter the wavelengths of the radiation and the greater the amount of radiation it emits.
- Radiation emitted by the Sun includes *ultraviolet, visible, near-infrared, and short-wave infrared radiation. Thermal infrared radiation*, which is emitted by Earth surfaces, is familiar as heat. The atmosphere absorbs and scatters radiation in certain wavelength regions. Radiation flows are measured in watts per square metre.
- The Earth continuously absorbs solar short-wave radiation and emits long-wave radiation. In the long run, the gain and loss of radiant energy remain in balance, and the Earth's average temperature remains constant.
- Insolation, the rate of solar radiation flow available at a location at a given time, is greater when the Sun is higher in the sky. *Daily insolation* is also greater when the period of daylight is longer. Between the tropics and poles, the Sun rises higher in the sky and stays longer in the sky at the summer solstice than at the equinox; daylight hours are shortest during the winter solstice. Near the Equator, daily insolation is greater at the equinoxes than at the solstices. Annual insolation is greatest at the Equator and least at the poles. However, the poles still receive 40 percent of the radiation the Equator receives.
- The pattern of annual insolation with latitude leads to a natural naming convention for latitude zones: *Equatorial, Tropical, Subtropical, Midlatitude, Subarctic (Subantarctic), Arctic (Antarctic)*, and *Polar*.
- The Earth's atmosphere is dominated by *nitrogen* and *oxygen gases. Carbon dioxide* and *water vapour*, though lesser constituents by volume, are important because they absorb long-wave radiation and enhance the greenhouse effect.
- Ozone (O_3) is a small but important constituent of the atmosphere and is concentrated in a layer of the upper atmosphere. It is formed from oxygen (O_2) by chemical reactions that absorb ultraviolet radiation, thus sheltering the organisms of the Earth's surface from its damaging effects.
- Latent heat and sensible heat are additional forms of energy. Latent heat is taken up or released when a change of state occurs. Sensible heat is contained within a substance. It can be transferred to another substance by *conduction* or *convection*.
- The global energy system includes a number of important pathways by which insolation is transferred and transformed. Part of the solar radiation passing through the atmosphere is absorbed or scattered by molecules, dust, and larger particles. Some of the scattered radiation returns to space as *diffuse reflection*. Land surfaces, ocean surfaces, and clouds also reflect some solar radiation back to space.
- The percentage of radiation that a surface absorbs is termed its albedo. The albedo of the Earth and atmosphere together is about 30 percent.
- The atmosphere absorbs long-wave energy emitted by the Earth's surface, causing the atmosphere to counter-radiate some of that long-wave radiation back to Earth, thereby creating the greenhouse effect. Because of this heat trapping, the Earth's surface temperature is considerably warmer than it would be without an atmosphere.
- Net radiation describes the balance between incoming and outgoing radiation. At latitudes lower than 40 degrees, annual net radiation is positive. It is negative at higher latitudes. This imbalance creates poleward heat transport of latent and sensible heat in the movement of warm water and warm, moist air, and thus provides the power that drives ocean currents and broad-scale atmospheric circulation patterns.

KEY TERMS

absorption	conduction	electromagnetic spectrum	latent heat
aerosol	convection	fallout	latent heat transfer
air pollutant	counter-radiation	global energy balance	long-wave radiation
albedo	diffuse radiation	greenhouse effect	mechanical energy
blackbody	diffuse reflection	heat energy	net radiation
chemical energy	Dobson unit	infrared radiation	ozone
chlorofluorocarbons (CFCs)	dry deposition	insolation	particulates
condensation nuclei	electromagnetic radiation	kinetic energy	photochemical reaction

potential or stored energy   sensible heat   solar power tower   washout

radiant energy   sensible heat transfer   Stefan-Boltzmann's Law   watt

radiation balance   short-wave radiation   ultraviolet (UV) radiation   Wien's Law

scattering   solar constant   visible light

REVIEW QUESTIONS

1. What is electromagnetic radiation? How is it characterized? Identify the major regions of the electromagnetic spectrum.
2. How does an object's temperature influence the nature and amount of electromagnetic radiation it emits?
3. What is the solar constant? What is its value? What units are used to measure it?
4. How does solar radiation received at the top of the atmosphere differ from solar radiation received at the Earth's surface? What are the roles of absorption and scattering?
5. Compare the terms *short-wave radiation* and *long-wave radiation*. What are their sources?
6. How does the atmosphere affect the flow of long-wave energy from the Earth's surface to space?
7. What is the Earth's global energy balance, and how are short-wave and long-wave radiation involved?
8. How does the sun's path in the sky influence daily insolation at a location?
9. What influence does latitude have on the annual cycle of daily insolation? On annual insolation?
10. Identify the three largest components of dry air. Why are carbon dioxide and water vapour important atmospheric constituents?
11. How does the ozone layer protect the life layer?
12. What is energy? What distinguishes the following forms of energy: kinetic, potential, mechanical, radiant, and chemical?
13. Describe latent heat transfer and sensible heat transfer.
14. Define albedo and give two examples.
15. Describe the counter-radiation process and how it relates to the greenhouse effect.
16. What important physical principle permits the construction of budgets of energy and matter flow?
17. Discuss the energy balance of the Earth's surface. Identify the types and sources of energy flows that the surface receives. Do the same for energy flows that it loses.
18. Discuss the energy balance of the atmosphere. Identify the types and sources of energy flows that the atmosphere receives. Do the same for energy flows that it loses.
19. What is net radiation? How does it vary with latitude?
20. What is the role of poleward heat transport in balancing the net radiation budget by latitude?
21. Why is solar power in principle more desirable than power produced by burning fossil fuels?

A CLOSER LOOK   EYE ON GLOBAL CHANGE 3.3   The Ozone Layer—Shield to Life

1. What are CFCs, and how do they affect the ozone layer?
2. When and where have ozone reductions been reported? Have corresponding reductions in ultraviolet radiation been noted?
3. What is the outlook for the future health of the ozone layer?

A CLOSER LOOK   GEOGRAPHER'S TOOLS 3.5   Remote Sensing for Physical Geography

1. What is remote sensing? What is a remote sensor?
2. Compare the reflectance spectra of water, vegetation, and a typical soil. How do they differ in visible wavelengths? In infrared wavelengths?
3. What colour is vegetation on colour infrared film? Why?
4. What is emissivity, and how does it affect the amount of energy radiated by an object?
5. Is radar an example of an active or passive remote sensing system? Why?
6. What is a digital image? What advantage does a digital image have over a photographic image?
7. How does a Sun-synchronous orbit differ from a geostationary orbit? What are the advantages of each type?

VISUALIZATION EXERCISES

1. Sketch the world latitude zones on a circle representing the globe and give their approximate latitude ranges.

2. Sketch a simple diagram of the Sun above a layer of atmosphere and the Earth's surface. Draw arrows to indicate flows of energy among Sun, atmosphere, and surface. Label each arrow according to the process it represents.

ESSAY QUESTION

1. Discuss how the atmosphere will influence a beam of either (a) short-wave solar radiation entering the Earth's atmosphere heading toward the surface, or (b) a beam of long-wave radiation emitted from the surface heading toward space.

PROBLEMS   WORKING IT OUT 3.2   Radiation Laws

1. Assuming that the Sun behaves as a blackbody, what is the flow rate of energy leaving its surface if the surface is at a temperature of 5,950 K (5,677°C)? At what wavelength will the emitted radiation be greatest?

2. The average surface temperature of the Earth is approximately 15.4°C. Assuming that the surface radiates perfectly, what is the flow rate of energy emitted by the surface? At what wavelength will the radiation be greatest?

3. Using the answers to problems 1 and 2, determine the ratio of the flow rate of energy emitted by the Sun's surface to the flow rate of energy emitted by the Earth's surface.

4. Venus has a radius of about 6,050 kilometres. Because it is closer to the Sun than the Earth, the solar radiation flow it receives is 1.92 times stronger than that received by the Earth. The atmosphere of Venus consists of dense clouds of carbon dioxide vapour, which reflect about 65 percent of the incoming solar radiation. Draw a diagram of the energy flow system for Venus and calculate the rates of inflow and outflow of short-wave and long-wave radiation.

Find answers in the back of the book

EYE ON THE LANDSCAPE

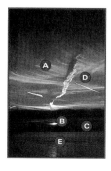

Chapter Opener   Sun rise. The high cirrus clouds (**A**) are illuminated by the rising sun (**B**), but the bank of low stratus cloud (**C**) and Earth's surface remains in partial shadow. Because of the low sun angle, the dispersing contrail (**D**) left by a passing aircraft is casting a shadow on the clouds above. Sunlight is reflected in the smooth ocean surface, but the effect is diminished where the waves are bigger (**E**).

Inco refinery, Sudbury Ontario Constructed in 1972, the 380 m superstack, at the Inco refinery in Sudbury, Ontario remains the second tallest in the world. The chimney disperses gases, including sulphur dioxide into faster moving air currents aloft (**A**) al-

lowing them to be dispersed many kilometers from the source. Scattering of incoming radiation increases at low sun angles when the solar rays travel through a greater thickness of atmosphere. The orange glow (**B**) at sunset can be enhanced by impurities in the air.

Chapter 4

EYE ON THE LANDSCAPE

Winter in Jasper National Park, Alberta.
What else would the geographer see? . . . Answers at the end of the chapter.

AIR TEMPERATURE AND AIR TEMPERATURE CYCLES

Surface Temperature

Focus on Systems 4.1 • *The Surface Energy Balance Equation*

Air Temperature

Measurement of Air Temperature

The Daily Cycle of Air Temperature

Daily Insolation and Net Radiation

Daily Temperature

Temperatures Close to the Ground

Environmental Contrasts: Rural and Urban Temperatures

Geographers at Work • *Dr. James A. Voogt*

Temperature Structure of the Atmosphere

Troposphere

Stratosphere and Upper Layers

High Mountain Environments

Temperature Inversion

The Annual Cycle of Air Temperature

Net Radiation and Temperature

Land and Water Contrasts

World Patterns of Air Temperature

Factors Controlling Air Temperature Patterns

World Air Temperature Patterns for January and July

The Annual Range of Air Temperatures

Global Warming and the Greenhouse Effect

Factors Influencing Climatic Warming and Cooling

Working It Out 4.2 • *Exponential Growth*

The Temperature Record

Eye on Global Change 4.3 • *Carbon Dioxide— On the Increase*

Future Scenarios

A Closer Look:

Eye on Global Change 4.4 • *The IPCC Report of 2001*

Air temperature affects many aspects of human life, from the clothing we wear to the fuel we pay for. Air temperature, along with precipitation, is a key determinant of climate. Thus, it is an important environmental factor that affects many landscape processes including rock weathering and erosion, soil formation, and the type of plant and animal life that is present in a region.

Is air temperature changing? It is commonly assumed that the Earth is becoming increasingly warmer, and that as a result, sea level is rising, climate boundaries are shifting, and severe weather is becoming more frequent. Global warming is a controversial issue, but it is generally agreed that human activities, such as forest clearance and air pollution, have had an impact on global temperatures. This is discussed later in the chapter, in the context of the Earth's natural long-term cycles.

Air temperature, as presented in weather reports, is measured at a height of 1.25 metres above the ground surface. This fundamental property of the atmosphere is influenced by seven important factors:

1. **Insolation.** The two motions of the Earth—its daily rotation on its axis and its annual revolution in its solar orbit—create the daily and annual cycles of insolation. In turn, these produce the cycles of air temperature that distinguish day and night, winter and summer.

2. **Latitude.** Annual insolation decreases toward the poles, so less energy is available to heat the air, and temperatures generally fall. Similarly, seasonal differences in insolation are more pronounced at higher latitudes, resulting in a corresponding increase in the annual temperature range.

3. **Cloud cover.** Cloud cover can influence air temperature by altering the amount of energy received and retained at the Earth's surface.

4. **Surface type.** Drier surfaces tend to heat more rapidly than moister surfaces because less energy is dissipated through evaporation. Surface albedo will also affect how much energy is absorbed or reflected.

5. **Coastal versus interior location.** Water heats and cools more slowly than land, so air temperatures over water bodies tend to be less extreme than over land surfaces.

6. **Elevation.** At high elevations, the atmosphere is less dense and more long-wave energy can escape to space. Lower air density also permits more solar energy to reach the surface during the day. Consequently, the daily temperature range in mountainous areas is generally quite pronounced.

7. **Slope and Aspect.** Radiation intensity and surface warming increase with slope angle to the point at which incoming solar rays are perpendicular to the surface. The slope angle that corresponds to maximum insolation will vary seasonally with the Sun's apparent movement north and south of the equator. There is also a diurnal component due to the Sun's passage across the sky each day. Slope aspect—the direction the slope faces—also influences the amount of radiation received; south-facing slopes are normally warmer than north-facing slopes.

SURFACE TEMPERATURE

Temperature represents the amount of thermal energy available in a system, and is conveniently defined as a measure of the average kinetic energy associated with the motion of atoms and molecules in a gaseous, liquid, or solid substance. Heat, or loss thereof, is energy that is transferred from one object to another because of their temperature difference. When a substance receives a flow of energy, its temperature rises. Conversely, if a substance loses energy, its temperature falls.

Net radiation is the balance between incoming short-wave radiation and outgoing long-wave radiation. It produces a radiant energy flow that can heat or cool a substance. During the day, incoming solar radiation normally exceeds outgoing long-wave radiation, so the net radiation balance is positive and the Earth's surface warms. At night, net radiation is negative, and the surface temperature falls as long-wave energy is radiated to space.

A second form of energy transfer is **conduction**. Conduction is the principal method of heat transfer in solids where atoms are in constant contact. For example, heat transfer in soils is by conduction and is always directed from a higher to a lower temperature. The rate at which heat flows within solids, such as soil, is its *thermal conductivity*. It varies according to the temperature gradient and the *specific heat* of the substance. Specific heat is the amount of heat per unit mass required to raise the temperature of a substance by one degree Celsius. Dry sand, for example, has a specific heat of $0.8 \text{ J g}^{-1}\text{K}^{-1}$ (joules per gram per Kelvin). Thermal conductivity in soils is determined primarily by how porous and moist the soil is, and its organic matter content.

Latent heat transfer is also important. When water evaporates at a surface, the change of state from liquid to vapour removes stored heat, thus cooling the surface. This is the latent heat of vaporization and amounts to $2.501 \times 10^6 \text{ J kg}^{-1}$. When water condenses at a surface, an equivalent amount of energy is released, warming the surface. This is the latent heat of condensation.

Another form of energy transfer is **convection**, in which heat is distributed in a fluid by mixing. For example, if air in contact with a heated soil surface is warmed, the gas molecules spread out, causing it to become less dense than the surrounding, unheated air. The warmed air will subsequently rise and transfer heat by convection. Upward and downward air currents can warm or cool the surface and generate turbulence in the atmosphere. Convection is an important energy transfer process in both the atmosphere and the oceans.

Since energy is neither created nor destroyed—just transformed—the energy flows occurring at a surface must be in balance, what is termed the **energy balance**. *The surface energy balance equation* describes how net radiation, latent heat, and sensible heat flows are balanced through conduction and convection. *Focus on Systems 4.1 • The Surface Energy Balance Equation* discusses this equation more fully. As well as explaining the temperature of the surface, the latent heat transfer term in the equation also describes how water evaporates from the soil.

AIR TEMPERATURE

Air temperature is probably the most commonly reported piece of weather information. Reported air temperatures closely follow changes and trends experienced at the ground surface. Because this surface effect varies with height, recordings are made at a standard level of 1.25 metres above the ground (at least until the winter snows come), preferably above short grass.

MEASUREMENT OF AIR TEMPERATURE

The basic instrument for recording temperature is the *thermometer*. The traditional design uses a capillary tube partially filled with mercury. The column of mercury expands and contracts as the temperature changes, and the top of the column corresponds to a calibrated scale. Because of concern for human safety and the environment since mercury is toxic, most thermometers today use an alcohol-based liquid; this has an advantage since it does not freeze at low temperatures, such as those encountered in the Canadian Arctic.

For standard measurements thermometers are housed in *instrument shelters* (also called Stevenson screens). These are louvred boxes that hold thermometers and other weather instruments at a standardized height, while shading them from the direct rays of the Sun (Figure 4.1). Air circulates freely through the louvres, ensuring that temperatures inside the shelter are the same as the outside air.

4.1 Weather recording instruments An instrument shelter housing a pair of thermometers. The shelter is constructed with louvred sides for ventilation and painted white to reflect solar radiation.

Liquid-filled thermometers are now being replaced by electronic, thermistor-based instruments for the routine measurement of temperature. A thermistor is a temperature sensitive resistor constructed of semiconductor material, which exhibits a large change in resistance proportional to a small change in temperature. There are several advantages to this technology. For example, information can be transmitted automatically from remote locations without the need for human observation, and the data can be input directly into computer programs for processing and use in weather forecasting. Many weather stations are now equipped with this type of automatic temperature measurement system.

Although some weather stations report temperatures hourly, most stations report only the highest and lowest temperatures recorded during a 24-hour period. Temperature measurements collected by national agencies, such as the Meteorological Service of Canada, are used to generate weather reports and forecasts (Figure 4.2). This agency has daily, monthly, and annual temperature statistics for each reporting station in Canada. For many stations, hourly temperature readings are archived, as well as summary data, such as daily maximum, minimum, and mean temperature—the average of the maximum and minimum daily values. Monthly summaries of maximum, minimum, and mean temperatures are also available. With sufficient record length (usually at least 30 years), these statistics, along with others such as daily precipitation, are used to describe the climate at the station and its surrounding area.

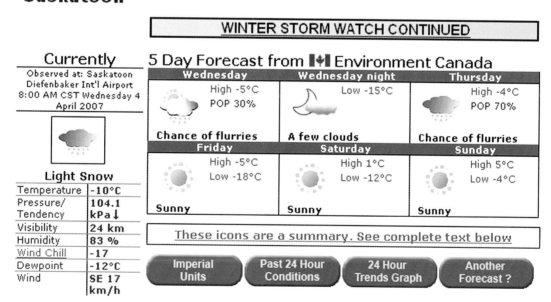

4.2 The Environment Canada Meteorological Service of Canada website (http://www.weatheroffice.ec.gc) provides current forecasts as well as archived data for all weather stations in Canada, including historical data from stations that are no longer active.

FOCUS ON SYSTEMS | 4.1   The Surface Energy Balance Equation

The surface energy balance equation considers the surface to be a thin boundary layer positioned between the atmosphere and the ground. For daytime conditions it is written as

$$Q^* \downarrow = Q_G \downarrow + Q_H \uparrow + Q_E \uparrow$$

where

Q^* is net wave radiation;

Q_G is soil heat flux;

Q_H is sensible heat flux; and

Q_E is latent heat flux.

Arrows indicate the direction of the positive fluxes and would be reversed at night.

The surface is also considered to be an energy transfer layer that changes one type of energy flow to another, but is too thin to hold any heat itself. Thus, the surface energy balance equation can be rewritten as

$$R_{LONG} - R_{SHORT} + H_{LATENT} + H_{SENSIBLE} + H_{SOIL} = 0$$

where

R_{LONG} is the net flow of long-wave radiation between the surface and the atmosphere;

R_{SHORT} is the flow of short-wave radiation absorbed by the surface (note that outgoing flows are considered positive and incoming flows are negative);

H_{LATENT} is the latent heat flow between the surface and the atmosphere;

$H_{SENSIBLE}$ is the sensible heat flow between the surface and the atmosphere; and

H_{SOIL} is the sensible heat flow between the surface and the soil.

From the principle of energy conservation, incoming energy flow must balance outgoing energy flow, so the terms of the equation must add up to zero.

The equation includes two radiation terms. R_{SHORT} is the flow of short-wave radiation to the surface from the Sun. This flow includes only radiation that is absorbed by the surface and does not include short-wave radiation that is reflected directly back toward the atmosphere. Because the normal convention is to consider energy flows to the surface as negative, R_{SHORT} is negative during the day. At night, this flow will drop to zero. R_{LONG} is the net long-wave radiation. Because it is a net term, it represents the balance between outgoing long-wave radiation emitted by the surface and incoming long-wave radiation from the atmosphere above. Because the surface is usually warmer than the atmosphere, even at night, the flow will normally be away from the surface, which is taken as positive. The sum of R_{SHORT} and R_{LONG} is the net radiation flow ($Q^* \downarrow$).

H_{LATENT} is the latent heat flow and represents the energy associated with the evaporation of water or ice sublimation. Latent heat flow ($Q_E \uparrow$) occurs when water vapour diffuses into the atmosphere following a phase change. During the day, latent heat flow will be positive as soil water evaporates. The latent heat flux is regulated by the moisture gradient away from the surface. Without vertical motion, the air immediately above the surface would quickly become saturated and energy transfer would cease. Transfer of latent heat increases during turbulent, windy conditions, when moist air near the surface is continually replaced by drier air. At night, condensation or deposition may occur, yielding dew or frost. If so, latent heat will be released at the surface, providing a heat flow to the surface (negative flow).

$H_{SENSIBLE}$ is the sensible heat flux ($Q_H \uparrow$) that arises when the surface

THE DAILY CYCLE OF AIR TEMPERATURE

Because the Earth rotates on its axis, incoming solar energy at a location can vary widely throughout the 24-hour period, while outgoing long-wave energy remains more constant. During the day, net radiation is positive, and the surface gains heat. At night, net radiation is negative, and the surface loses heat. This is reflected in the daily cycle of rising and falling air temperatures.

DAILY INSOLATION AND NET RADIATION

Figure 4.3 shows idealized curves of daily insolation, net radiation, and air temperature for a typical observation station in southern Ontario at lat. 45° N. The time scale is set so that noon occurs when the Sun is at its highest elevation in the sky.

Graph (*a*) shows daily insolation. At the equinox, insolation begins at about sunrise (6 a.m.), rises to a peak value at noon, and declines to zero at sunset (6 p.m.). At the June solstice, insolation begins about two hours earlier (4 a.m.) and ends about two hours later (8 p.m.). The June peak is much greater than at the equinox, and the total insolation for the day is also much greater. At the December solstice, insolation begins about two hours later than at the equinox (8 a.m.) and ends about two hours earlier (4 p.m.). Both the peak intensity and daily total insolation are greatly reduced in the winter.

Graph (*b*) shows net radiation for the surface. When net radiation is positive, the surface gains heat, and when net radiation

conducts heat to the atmosphere in the boundary layer. Sensible heat is then carried upward by convection. It will be positive when the air is warmed by the surface, which is the normal condition during the day. At night, the

surface can become colder than the air in contact with it, so the heat flow may be negative, from the air to the surface.

H_{SOIL} is the flow of heat by conduction from the surface to the soil below

$(Q_G\downarrow)$. This flow will normally be away from the surface (positive) during the day, as heat is conducted from the warm surface into the soil. At night, however, the surface will normally cool sufficiently for heat to be conducted upward (negative).

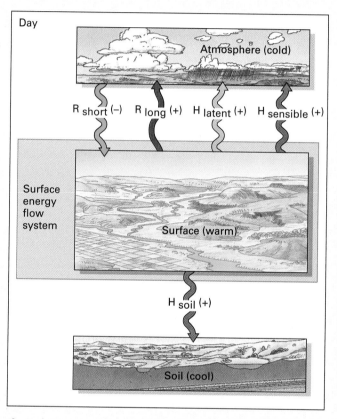
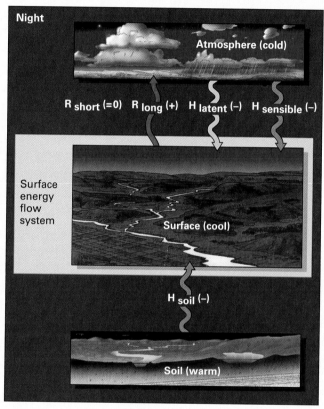

The surface energy balance equation for typical day and night conditions

is negative, it loses heat. The net radiation curves generally resemble those for insolation. At midnight, net radiation is negative. This deficit continues into the early morning hours. Net radiation becomes positive shortly after sunrise and rises sharply to a peak at noon. In the afternoon, net radiation decreases and reaches zero shortly before sunset. With no incoming insolation, net radiation then becomes negative and a deficit persists until morning.

All net radiation curves show the same general daily pattern, but they differ in magnitude. For the June solstice, the positive values are quite large, and the surplus period is much longer than the deficit period. Thus, for the day as a whole, net radiation is positive. At the December solstice, the surplus period is short and

the surplus is small. The total deficit covers nearly 18 hours, and net radiation for that day is negative. This pattern of positive daily net radiation in summer and negative daily net radiation in winter is reflected in the annual temperature.

DAILY TEMPERATURE

Graph (c) shows the typical, or average, air temperature cycle for a 24-hour day. The minimum daily temperature usually occurs about half an hour after sunrise. Since net radiation has been negative during the night, heat has flowed from the ground surface, and the ground has cooled the surface air layer to its lowest temperature. As net radiation becomes positive, the surface warms quickly and transfers heat to the air above. Air temperature rises

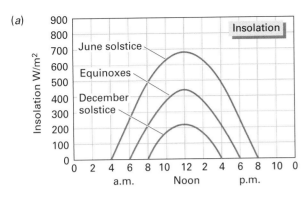

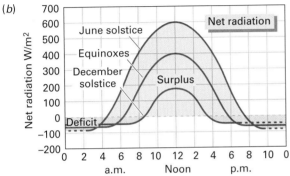

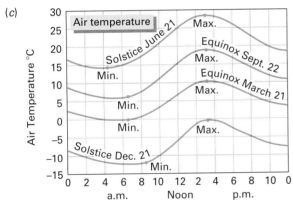

4.3 Daily cycles of insolation, net radiation, and air temperature These three graphs show idealized daily cycles for a midlatitude station in southern Ontario. Insolation (a) is a strong determinant of net radiation. (b) Air temperatures (c) respond by generally increasing when net radiation is positive and decreasing when it is negative.

sharply in the morning hours and continues to rise long after the noon peak of net radiation.

Although air temperature might be expected to rise as long as net radiation is positive, another process begins in the early afternoon on a sunny day. Convection currents develop and mix with the air several hundred metres above the surface, carrying heated air aloft and bringing cooler air down toward the surface. Therefore, the temperature peak usually occurs in mid-afternoon. The figure shows the peak at about 3 p.m., but it can typically occur between 2 and 4 p.m., depending on local conditions. Air temperature tends to fall quite rapidly until sunset, then at a slower rate through the night.

As expected, daily temperatures in summer are higher than in winter. However, the September equinox temperatures are higher than those in March. Even though net radiation is the same for the two equinoxes, the temperature curves differ because each reflects conditions in the preceding season. The summer energy surplus persists into the autumn, whereas warming in the spring may be delayed by the reflective properties of the remaining snow cover and the thermal inertia of the frozen soils.

TEMPERATURES CLOSE TO THE GROUND

Although air and ground surface temperatures generally show the same trends, the surface temperatures are likely to be more extreme. Figure 4.4 presents a series of generalized temperature profiles from about 30 centimetres below the surface to a height of 1.5 metres at five times during an autumn day. At 8 a.m. air and soil temperatures are uniform, resulting in a vertical line on the graph. By noon the surface is considerably warmer than the air temperature at 1.25 metres, and the soil below the surface has also warmed. By 3 p.m., the soil surface is much warmer than the air. The soil cools rapidly during the evening, and by 5 a.m., it is colder than the air.

Thus, daily temperature variation is greatest at the surface, while air temperature at the standard recording height is less variable. Note also that within the soil, the daily cycle weakens with depth to a point where the temperature does not change during the 24-hour cycle.

ENVIRONMENTAL CONTRASTS: RURAL AND URBAN TEMPERATURES

In actively growing vegetation, water is taken up by plant roots and moved to the leaves, where it is lost to the atmosphere through **transpiration**. This cools the leaf surfaces, which in turn tends to lower the temperature of the air. The cooling effect of vegetation is even greater in a forest. Not only are transpiring leaves abundant, but solar radiation is intercepted by the tall leafy canopy and the understory plants. Thus, solar radiation is not concentrated intensely at the ground surface, but is used to warm the whole forest layer. Water also evaporates directly from the soil, again cooling the surface and moderating air temperatures. Transpiration and evaporation are collectively termed **evapotranspiration**.

In contrast, water drains from the impervious roofs, sidewalks, and streets of a city. It is channelled into storm sewer systems, where it flows directly into rivers, lakes, or oceans, rather than soaking into the ground beneath the city. Little energy is dissipated in evaporation, so these impervious surfaces can warm quickly on sunny days. The moderating effect of evapotranspiration is limited to ornamental trees, lawns, parks, and bare patches of soil.

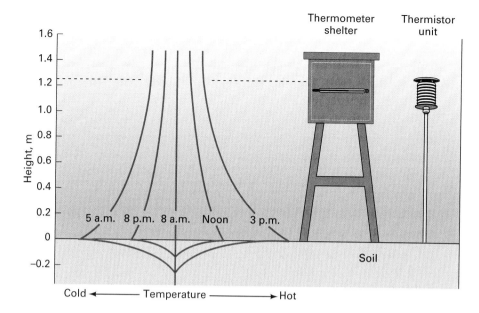

4.4 Daily temperature profiles close to the ground This simplified diagram shows the temperature profile close to the ground surface at five times of day. The soil surface becomes hot by mid-afternoon, but cools greatly at night.

City surfaces are also darker and more absorbent than rural surfaces. Asphalt paving and roofing absorb about twice as much solar energy as vegetation. Heat absorption is also enhanced by the many vertical surfaces in cities, which reflect radiation from one to another. Since some radiation is absorbed with each reflection, the network of vertical surfaces tends to trap solar energy and radiant heat more effectively than a simple horizontal surface. Concrete, asphalt, and other materials conduct and store heat better than soil, even when the soil is dry. At night, this heat is conducted back to the surface, keeping nighttime temperatures warmer.

Another important factor in warming the city is fuel consumption and waste heat. In summer, city temperatures increase through the use of air conditioning. This equipment pumps heat out of buildings, releasing it to the air. The power used to run the air conditioning systems also generates heat. In winter heat is directly lost from poorly insulated buildings. Vehicles contribute heat as well as greenhouse gases, which can also increase heat gain.

As a result of these effects, air temperatures in the central region of a city are typically several degrees warmer than those of the surrounding suburbs and countryside. The elevated temperatures represent an urban **heat island** (Figure 4.5). In large cities the heat island persists through the night because of the slow release of heat stored during the daytime hours.

The urban heat island effect has important environmental consequences. Higher temperatures demand more air conditioning and electric power in the summer. The fossil fuel burned to generate this power contributes carbon dioxide and pollutants to the air. The increased temperatures are also more conducive to smog formation, which is unhealthy and damaging to materials. Cities are now taking steps to reduce the heat island

effect by planting more vegetation and encouraging the use of more reflective surfaces. In desert climates, heat islands may not develop because evapotranspiration from irrigated vegetation can lower city temperatures compared with the surrounding area.

4.5 The urban heat island of Brisbane, Australia. Temperatures are highest in the downtown area; parks and rural areas are coolest. The Brisbane River also stands out as a cool line in the image.

GEOGRAPHERS AT WORK

How Cities Alter Weather and Climate
by Dr. James A. Voogt, University of Western Ontario

Are you an urban dweller? If so, you probably know that the climate of your city is different from the climate of the suburbs and rural areas nearby. For one thing, it's hotter. And air quality is generally lower. But did you know that city air is also typically drier?

Urban climatology is a new and expanding field of geography that focuses on how human activities affect urban climate. Urban climatologists are concerned with a range of problems, including atmospheric processes in an urban environment; urban climate and health; improvement of urban climate by adding parks and green spaces; urban climate and urban planning; and use of remote sensing to observe urban areas.

Urban climatology is the specialty of James Voogt, shown at work in the photo at left. He focuses on the key issue of measuring urban temperatures accurately using infrared radiometers—devices that remotely measure surface temperature by detecting the longwave radiation coming from surfaces, such as streets, plazas, and lawns, that is a function of their surface temperature. Through the day and night, different types of surfaces will have different temperatures, and a radiometer looking down on a city from atop a tower will see a mixture of different surfaces and temperatures that varies with the look angle. By sorting out these effects, James can study how urban surface temperatures relate to urban air temperature measured by thermometers; how radiometers on aircraft and orbiting satellites sense the urban heat island; and how to plan urban areas to consider the impact of surface temperatures on the local climates that affect residents. Read more about James's work on our web site.

Dr. Voogt setting up a thermal scanner to monitor the surface temperatures in an urban street canyon. The meteorological tower used to measure the heat fluxes, turbulence and many other atmospheric characteristics from within the street canyon to well above roof level can be seen on the left.

TEMPERATURE STRUCTURE OF THE ATMOSPHERE

The atmosphere is mainly warmed from the surface below, so temperatures can be expected to decrease with altitude. This general condition is termed a **lapse rate**. The **normal temperature lapse rate** measures the drop in temperature in stationary air, averaged for the entire Earth over a long period. On average, temperature drops with altitude at a rate of 6.4°C per 1,000 metres (Figure 4.6). For example, when the air temperature near the surface is 20°C, the air at an altitude of 8 kilometres should be about −30°C. Keep in mind that the normal temperature lapse rate is an average value, and that on any given day the observed lapse rate might be quite different. The actual change in temperature with altitude at a specific time and specific location is termed the **environmental temperature lapse rate**.

Figure 4.6 shows another important feature of the atmosphere. For the first 12 kilometres or so, temperature falls with increasing altitude. However, at about 14 kilometres (depending on latitude) the temperature stops decreasing, the trend reverses and the temperature slowly increases with altitude. This property distinguishes two different layers in the lower atmosphere—the troposphere and the stratosphere.

TROPOSPHERE

The **troposphere** is the lowest layer of the atmosphere and contains about 80 percent of its total mass. Most weather phenomena are restricted to this layer. Temperature decreases with increasing altitude in the troposphere. The troposphere is thickest in the equatorial and tropical regions, where it extends to about 16 kilometres; at the poles it thins to about 6 kilometres.

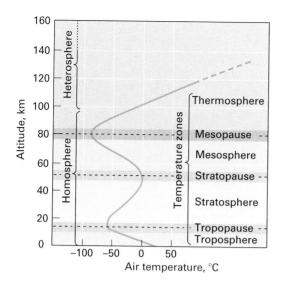

4.6 Temperature structure of the atmosphere Temperature decreases at a mean rate of 6.4°C/1,000 metres in the troposphere. This value is termed the normal temperature lapse rate. Temperature increases with altitude in the stratosphere. Above the stratosphere are the mesosphere and thermosphere. The homosphere, in which air's chemical components are well mixed, ranges from the surface to nearly 100 kilometres.

This is partly due to the rotation of the Earth on its axis, but is also linked to thermal expansion in the warmer zones. One important feature of the troposphere is that it contains practically all of the water vapour present in the atmosphere. However, because warm air holds more water vapour than cold air, it is mainly concentrated at lower altitudes. (Figure 4.7).

The troposphere gives way to the stratosphere at the *tropopause*. Traditionally the tropopause is defined as the lowest level where the lapse rate falls below 2°C per kilometre and remains below for up to two kilometres above. Above the tropopause, the temperature starts to increase with altitude. The thickness of the troposphere decreases during the winter season due to thermal contraction of the air column (Figure 4.8).

STRATOSPHERE AND UPPER LAYERS

Above the tropopause lies the **stratosphere**, in which the air becomes slightly warmer as altitude increases. The stratosphere extends to an altitude of roughly 50 kilometres above the Earth's surface. The strong, persistent winds—the jet streams—that blow from west to east in the lower stratosphere are linked to the formation and movement of pressure systems in the troposphere. However, there is little mixing of air between the troposphere and stratosphere, so water vapour and dust are practically absent from the stratosphere.

One important feature of the stratosphere is the ozone layer, which shields organisms by absorbing the intense, ultraviolet radiation emitted by the Sun. The warming of the stratosphere with altitude is caused largely by the absorption of solar energy by ozone molecules.

Above the stratosphere are two other layers—the mesosphere and thermosphere. In the *mesosphere*, temperature falls with elevation. This layer begins at the *stratopause*, and ends at the *mesopause*, where it gives way to the *thermosphere*, a layer of

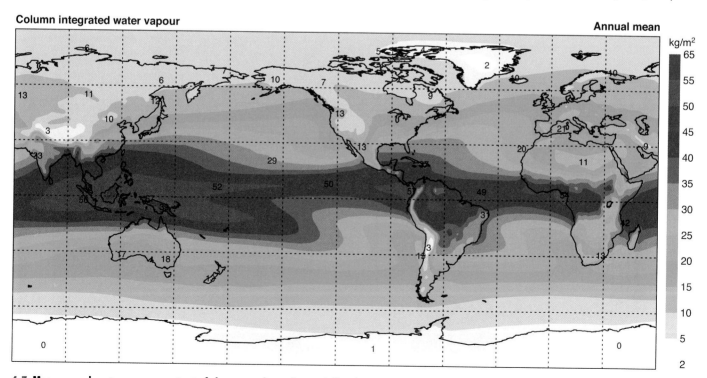

Column integrated water vapour

Annual mean

4.7 Mean annual water vapour content of the atmosphere. Moisture held in the atmosphere is equivalent to about 10 days supply of precipitation globally. Average water vapour content is highest in lower latitudes where air temperatures are warm throughout the year

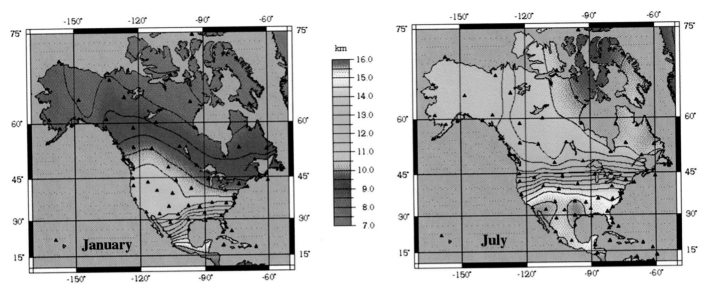

4.8 Mean height of the tropopause over North America in January and July. Summer warming and expansion of the atmosphere increases the altitude of the tropopause over Canada by about 2000 metres.

increasing temperature. However, at the altitude of the thermosphere, the air is so thin that it holds very little heat.

Apart from important differences in water vapour and ozone, the composition of the atmosphere is uniform for about the first 100 kilometres of altitude, which includes the troposphere, stratosphere, mesosphere, and lower portion of the thermosphere. This region is referred to as the *homosphere*. Above 100 kilometres, gas molecules tend to become increasingly sorted into layers by molecular weight and electric charge. This region is referred to as the *heterosphere*.

HIGH MOUNTAIN ENVIRONMENTS

Various environmental changes occur with increasing altitude because the air is less dense. With fewer air molecules and aerosol particles to scatter and absorb the Sun's light, incoming radiation is more intense. Also, there is less carbon dioxide and water vapour, reducing the greenhouse effect. This generally results in cooler temperatures, especially at night when heat loss is rapid. Air pressure is also lower because there are fewer gas molecules per unit volume of air.

Elevation has a pronounced effect on the daily temperature cycle (Figure 4.9). Mean temperatures clearly decrease with ele-

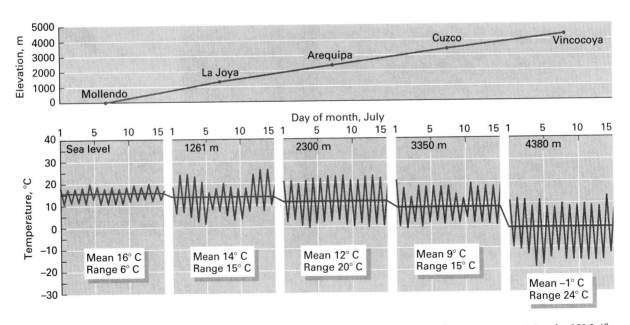

4.9 The effect of elevation on air temperature cycles Daily maximum and minimum air temperatures for mountain stations in Peru, lat. 15° S. All data cover the same 15-day observation period in July. As elevation increases, the mean daily temperature decreases and the temperature range increases.

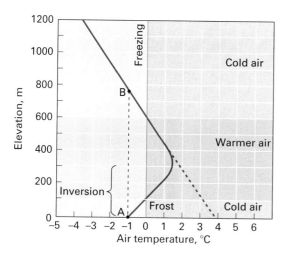

4.10 A ground inversion with frost While air temperature normally decreases with altitude (dashed line), in a ground inversion, temperature increases for some distance above the ground.

vation, from 16°C at sea level to −1°C at 4,380 metres. The daily range also increases with elevation, except for in Cuzco, Peru. This large city does not experience nighttime temperatures as low as expected because of its urban heat island.

TEMPERATURE INVERSION

During the night, the ground surface radiates long-wave energy to the atmosphere, net radiation becomes negative, and the surface cools. The air near the surface will also be cooled. Under clear, calm conditions, a cool, dense air will accumulate at the ground surface to produce a **temperature inversion**. Figure 4.10 shows an example of a **ground inversion** where air temperature at the surface (point A) has dropped to −1°C. Temperature becomes warmer above the surface, up to about 300 metres. Here the lapse rate reverses and the normal decrease in temperature with altitude resumes. As air cools and becomes denser, it can flow gently downward into low lying depressions. Areas that are susceptible to this type of imperceptible cold air drainage are termed *frost hollows*.

Low-level temperature inversions often occur over snow-covered surfaces in winter. Inversions of this type are intense and can extend thousands of metres into the air. They build up over many long nights in arctic and polar regions, where the solar heat of the short winter day cannot completely compensate for nighttime cooling. Inversions can also result when a warm air layer overlies a colder one. A sure sign of a temperature inversion is smoke that spreads across the sky as a thin layer (Figure 4.11). Temperature inversions suppress vertical mixing of the air and can seal air pollution within a thin surface layer.

THE ANNUAL CYCLE OF AIR TEMPERATURE

Earth's revolution around the Sun, in combination with the tilt of the Earth's axis, is the most important determinant of the annual cycle of net radiation. This, in turn, causes an annual cycle of

4.11 Low-level temperature inversion. Smoke from a power station in Saskatoon, Saskatchewan spreads out at top of the inversion later on a cold day in February.

mean monthly air temperatures, although other factors, such as maritime or continental location and cloud cover, also have an important influence.

NET RADIATION AND TEMPERATURE

Figure 4.12 (*a*) shows the yearly cycle of net radiation for four stations spanning a latitudinal range from near the equator to almost the Arctic Circle. Graph (*b*) shows mean monthly air temperatures for these same stations.

At Manaus, located nearly on the equator, the average net radiation rate is strongly positive in every month. However, there are two minor peaks. These coincide approximately with the equinoxes, when the Sun is nearly directly overhead. The temperature graph for Manaus shows uniform air temperatures, averaging about 27°C for the year. The annual temperature range—the difference between the highest and lowest mean monthly temperatures—is only 1.7°C. On the basis of temperature, this climate has no seasons.

Aswan, Egypt, is a desert location on the Nile River at lat. 24° N. The net radiation rate curve shows a large surplus in every month, with values ranging from about 35 watts per square metre in December to about 125 watts per square metre in June. The temperature graph shows a corresponding cycle, reaching 33°C during June, July, and August, and dropping to 16°C in December and January. The annual temperature range is 17°C.

Hamburg, Germany, is located at lat. 54° N. Here the net radiation rate cycle is strongly developed. The rate is positive from February to October, but a deficit occurs from November to January. The temperature cycle reflects the general reduction in total insolation at this latitude. Summer months reach a maximum of just over 16°C, while in winter temperatures drop to about freezing (0°C). The annual temperature range is about 17°C, the same as at Aswan.

At Yakutsk, Russia, lat. 62° N, the net radiation rate is negative during the long, dark winters and the radiation deficit lasts about six months. Mean temperatures in winter are between −35 and −45°C. In mid-summer, when daylight lasts almost 24 hours each day, the net radiation rate rises to a strong peak and exceeds those of the other three stations. Air temperature rises quickly after the winter solstice and reaches 17°C by June. Because of Yakutsk's high latitude and continental interior location, its annual temperature range is considerable—more than 60°C.

LAND AND WATER CONTRASTS

The surface layer of any extensive, deep body of water heats and cools more slowly than the surface layer of a large body of land when both are subjected to the same intensity of insolation. Consequently, daily and annual air temperature cycles are different between coastal and interior locations.

(*a*) **Net radiation**

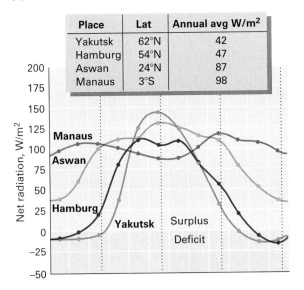

Place	Lat	Annual avg W/m²
Yakutsk	62°N	42
Hamburg	54°N	47
Aswan	24°N	87
Manaus	3°S	98

(*b*) **Monthly mean air temperature**

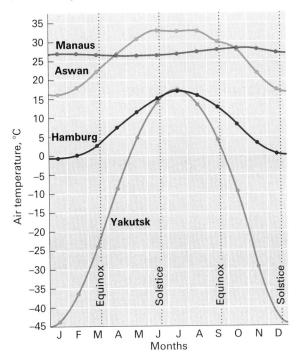

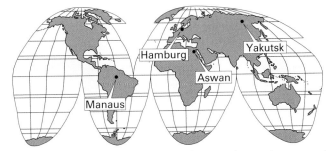

4.12 Net radiation and temperature cycles Net radiation and temperature for Aswan (Egypt), Hamburg (Germany), and Yakutsk (Russia) all show strong annual cycles with summer maxima and winter minima. In contrast, Manaus (Brazil), near the equator, shows two net radiation peaks and nearly uniform temperatures throughout the year. (Data courtesy of David H. Miller)

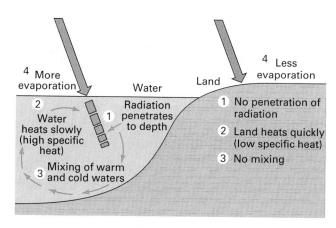

4.13 Land–water contrasts These four differences illustrate why a land surface heats more rapidly and more intensely than the surface of a deep water body. As a result, locations near the ocean have more uniform air temperatures.

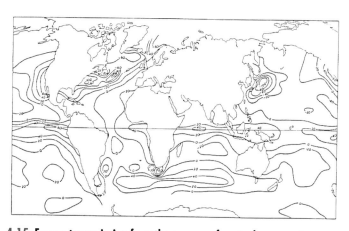

4.15 Energy transmission from the ocean surface to deeper water. Negative values indicate where energy is brought to the surface. (Note: units are in kilolangleys. To convert to W/m2, multiply by 1.328.)

Four important thermal differences between land and water surfaces account for the land–water contrast (Figure 4.13). First, much of the incoming solar radiation penetrates water, distributing heat throughout a substantial water layer. In contrast, solar radiation does not penetrate soil or rock, so its heating effect is concentrated at the surface. Consequently, the radiation will warm the water only slightly, while it warms a land surface more intensely.

A second factor is the different specific heats of water and land. Water has a comparatively high specific heat of 4.184 J $g^{-1}K^{-1}$, while a typical value for rock and soil is about 0.8 J $g^{-1}K^{-1}$. Thus, it takes about five times as much heat to raise the temperature of water by one degree as it does to raise the temperature of rock. The same will be true for cooling—after losing the same amount of heat, the temperature of water falls less than the temperature of rock.

A third difference is the amount of energy that is dissipated through evaporation of water. This is equivalent to 2,259 J $g^{-1}K^{-1}$. Water continually evaporates from the surface of the ocean and so reduces the amount of energy that is available for warming. Land surfaces can also be cooled by evaporation, but only if water is pres-

ent, as is the case for moist soils. When the surface dries, evaporation ceases, and the energy is then converted to heat. Figure 4.14 shows the global surface heat flux, representing the quantity of energy from oceans and continents used for evaporation.

Finally, a warm surface layer in the oceans can mix with cooler water below to produce a more uniform temperature throughout. For open water, the mixing is produced by wind-generated waves and density currents. Clearly, no such mixing occurs on land surfaces. In some localized areas, as much as 150 watts per square metre is transferred from the surface oceans in this way, but generally the process accounts for less than 50 watts per square metre. Elsewhere energy is brought in to the surface, notably in the North Pacific and North Atlantic Oceans (Figure 4.15).

Figure 4.16 shows the temperature record for Prince Rupert, on the west coast of British Columbia, for the period of July 16–22, 2006. Temperatures increase slowly over the week, with the daily maxima between 15°C and 22°C. The daily range is about 5–6°C. Temperatures are higher at Penticton, in the interior of the province, with daily maxima of 25–35°C. The daily range is about 15°C.

Figure 4.17 shows the annual temperature cycles for four Canadian locations. Tofino, located on Vancouver Island in B.C., is exposed to the onshore flow of air from the Pacific Ocean. In winter, mean temperatures fall to about 5°C, and in summer they rise to 15°C. The annual temperature range at Tofino is 10°C. In contrast, Winnipeg, Manitoba, experiences a continental climate with mean winter temperatures of −18°C and mean summer temperatures of 19°C. At Winnipeg, the annual temperature range is 37°C.

Summerside, P.E.I., is also a maritime location. However, temperatures are not as mild as at Tofino. The influence of the Atlantic Ocean at Summerside is reduced because airflow is principally from west to east across the continent. Mean winter temperatures are about −7°C, and in summer about 19°C. The annual temperature range is 26°C at Summerside, compared with 10°C at Tofino. Sault

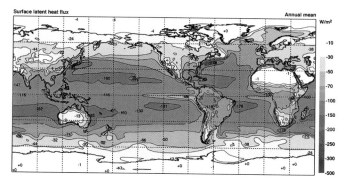

4.14 Surface latent heat flux Energy used in evaporation is greatest in the subtropical oceans and negligible in the polar regions and extreme deserts.

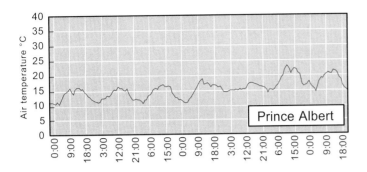

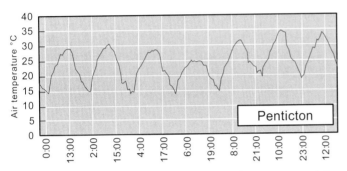

4.16 Maritime and continental temperatures Hourly temperature records for Prince Rupert and Penticton in British Columbia for the period of July 16–22, 2006. The daily temperature cycle at Prince Rupert on the Pacific Coast is less pronounced than at Penticton in the province's interior.

Ste. Marie, Ontario, is located at the east end of Lake Superior, with Lake Michigan and Lake Huron close by. The Great Lakes are sufficiently large to moderate the climate of the continental interior. With winter temperatures of −11°C and summer temperatures about 17°C, the annual range at Sault Ste. Marie is 28°C, compared

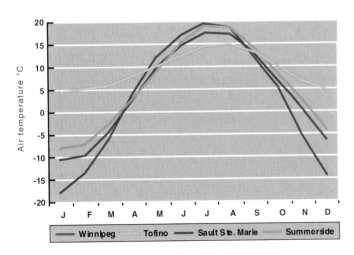

4.17 Mean monthly temperatures for Tofino, B.C., Winnipeg, Manitoba, Sault Ste. Marie, Ontario, and Summerside, P.E.I. The moderating of the nearby ocean is less pronounced at Summerside compared with Tofino. Conversely, the annual temperature range at Sault Ste. Marie is less than at Winnipeg, because of the influence of the Great Lakes.

with 37°C at Winnipeg. Another contrast between maritime and continental locations is the timing of maximum and minimum temperatures. Insolation reaches a maximum at summer solstice, but it is still strong for a long period afterward. This means that heat energy continues to flow into the ground for several weeks after the solstice. Therefore, the hottest month of the year for interior regions is generally July, the month following the solstice. Similarly, the coldest month of the year for large land areas is January, the month after the winter solstice. This is because the ground continues to lose heat even after insolation begins to increase. At coastal locations, maximum and minimum air temperatures are generally reached later than in the continental interiors, because water bodies heat and cool more slowly.

This effect is not always apparent because other factors, such as cloud cover, may affect temperature patterns. The total hours of bright sunshine is greatest at Winnipeg, with mean monthly totals ranging from 96 hours to 317 hours, totalling approximately 2,400 hours annually. The totals at Sault Ste. Marie and Summerside are similar, at 1,945 and 1,920, respectively. Tofino receives about 1,680 hours annually, ranging from 55 hours in December to 220 hours in July (Figure 4.18).

WORLD PATTERNS OF AIR TEMPERATURE

The distribution of air temperatures on a map is normally shown by **isotherms**—lines drawn to connect locations having the same temperature. In Figure 4.19 the recorded air temperatures have been placed at their proper locations. These may be single readings, such as a daily maxima or minima, or they may be averages of many years of records for a particular day or month of a year, depending on the

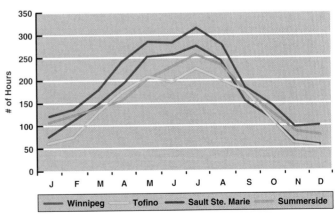

4.18 Total hours of bright sunshine at Tofino, B.C., Winnipeg, Manitoba, Sault Ste. Marie, Ontario, and Summerside, P.E.I. Less cloud cover in the drier climate of the Prairies accounts for the high number of sunshine hours at Winnipeg compared to other regions in Canada.

purpose of the map. The isotherms are constructed by drawing smooth lines through and between the points. Usually, isotherms represent 5- or 10-degree differences, but they can be drawn at any convenient temperature interval. Isothermal maps depict broad temperature patterns from which *temperature gradients*—directions along which temperature changes—can be discerned.

FACTORS CONTROLLING AIR TEMPERATURE PATTERNS

Three main factors—latitude, marine-continental location, and altitude—determine the patterns of isotherms on a global scale. As latitude increases, mean annual insolation decreases, and there is a corresponding decrease in temperatures. Latitude affects seasonal variation, while marine-continental influences also produce seasonal temperature contrasts. Marine locations, where prevailing winds come from the oceans, have more uniform temperatures—cooler in summer and warmer in winter. Continental stations show a much larger annual temperature range. Ocean currents also keep coastal waters, and consequently air temperatures, somewhat warmer or cooler than expected. Similarly, mountain ranges are cooler than surrounding regions because temperature decreases with elevation.

WORLD AIR TEMPERATURE PATTERNS FOR JANUARY AND JULY

Seven important points can be discerned from the January and July world temperature maps shown in Figure 4.20.

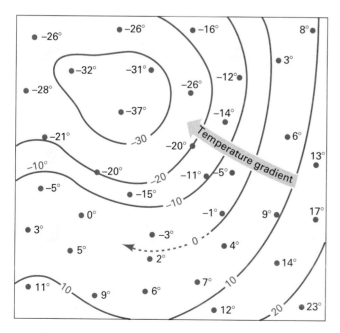

4.19 Isotherms Isotherms are used to plot temperature maps. Each line connects points having the same temperature. Where temperature changes along one direction, a temperature gradient exists.

1. **Temperatures decrease from the equator to the poles.** The latitudinal decrease in temperature corresponds closely with the decrease in mean annual insolation. The effect is most pronounced in the southern hemisphere where land areas are less extensive.

2. **Large land masses located in the subarctic and Arctic zones develop centres of extremely low temperatures in winter.** In January, low-temperature centres are strongly developed over North America and Siberia. The high albedo of the snow cover is an important factor keeping winter temperatures low in these regions. Greenland is also a region of low temperature. In addition to the permanent ice cover, the general height of Greenland's ice surface rises to more than 3,000 metres.

3. **Temperatures in equatorial regions change little from January to July.** The northern and southern positions of the 25°C isotherms are separated by a broad zone in which temperatures remain above 25°C in both January and July. Although latitudinal position of the two isotherms moves with the seasons, the equator, with a few minor exceptions, always falls between them. Relatively constant insolation in the region primarily causes this uniformity of equatorial temperatures.

4. **Isotherms make a large north-south shift from January to July over continents in the midlatitude and subarctic zones.** In the winter, isotherms generally move toward the equator, while in the summer, they arch toward the poles. The effect is especially pronounced in North America and Eurasia. For example, the 10°C isotherm borders the Gulf of Mexico in January, but in July it reaches to the Arctic in northwestern Canada. The isotherms over oceans shift much less. This difference is due to the contrast between oceanic and continental surface properties, which cause continents to heat and cool more rapidly than oceans.

5. **Highlands are always colder than surrounding lowlands. This effect is clearly illustrated by the Andes Mountains in South America, and in the highlands of Ethiopia.** Similarly, in North America, the −5°C and −10°C isotherms dip southwards around the Rocky Mountains in winter, indicating that the centre of the range is colder than the adjacent areas. In summer, the 20°C and 25°C isotherms show a similar pattern because temperature decreases with elevation.

6. **Areas of perpetual ice and snow are always intensely cold.** Greenland and Antarctica contain Earth's two great ice sheets. They are cold for two reasons. First, their surfaces are high, rising to more than 3,000 metres at their centres. Second, the snow surfaces have a high albedo. Since little solar energy is absorbed, the snow surface remains cold and chills the air above it. The Arctic Ocean, with its cover of floating ice, also maintains its cold temperatures throughout the year (Figure 4.21). However, the cold is much less intense in January in the Arctic than on the Greenland ice sheet, since ocean water underneath the ice acts as a heat reservoir and moderates the temperature of the ice above.

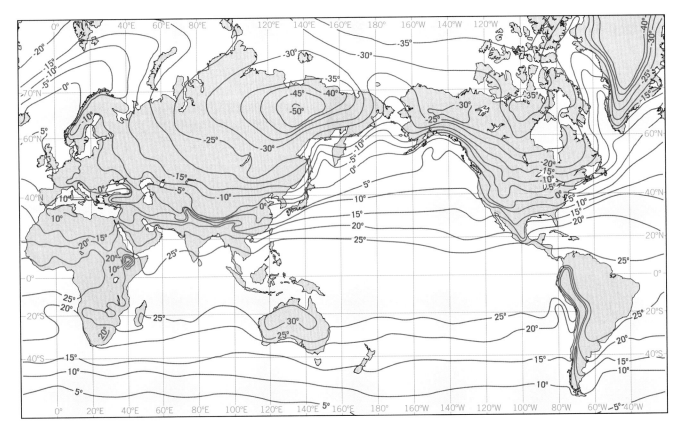

JANUARY

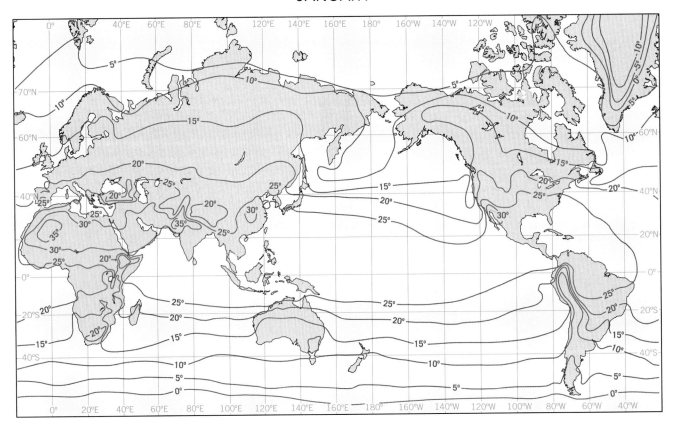

JULY

4.20 Mean monthly air temperatures (°C) for January and July. Temperatures remain relatively constant in tropical regions, but marked seasonal contrasts occur at higher latitudes.

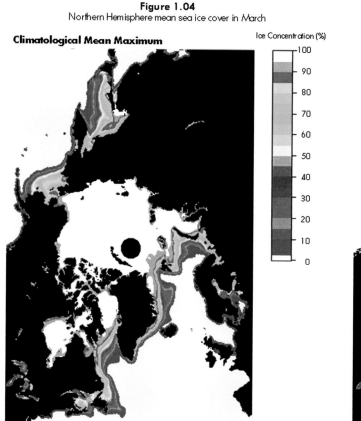

Figure 1.04
Northern Hemisphere mean sea ice cover in March

Climatological Mean Maximum

Ice Concentration (%)

Figure 1.03
Northern Hemisphere mean sea ice cover in September

Climatological Mean Minimum

Ice Concentration (%)

4.21 Arctic sea ice. The extent of sea ice cover in the Arctic changes with the seasons, reaching a maximum in March and shrinking to a minimum in September.

7. **Isothermal patterns associated with ocean currents.** The northward bend in the 20°C and 25°C isotherms in January off the west coast of South America is associated with the cold Peru Current. A similar effect is seen off the coast of southwest Africa where the Benguela Current brings cold water north from the seas off Antarctica. The 5°C January isotherm is positioned farther north over the Atlantic Ocean compared with the Pacific Ocean. This can be attributed to the warm waters carried by the North Atlantic Drift from the general vicinity of the Gulf of Mexico.

THE ANNUAL RANGE OF AIR TEMPERATURES

Figure 4.22 shows the annual range of air temperatures; the lines resembling isotherms show the difference between the January and July monthly mean temperatures. Five characteristic patterns emerge.

1. **The annual range increases with latitude, especially over northern hemisphere continents.** This trend is most clearly shown for North America and Asia. This is due to the more pronounced contrast between summer and winter insolation at higher latitudes.

2. **The greatest ranges occur in the subarctic and arctic zones of Asia and North America.** The map shows two very strong centres of large annual range — one in northeast Siberia and the other in northwest Canada–eastern Alaska. In these regions, summer insolation is nearly the same as at the equator, while winter insolation is very low.

3. **The annual range is moderately large on land areas near the Tropics of Cancer and Capricorn.** These are principally desert regions, for example the Sahara in North Africa, the Kalahari in southern Africa, and the Gibson Desert in Australia. Dry air and the absence of clouds and moisture allow these continental locations to cool strongly in winter and warm strongly in summer, even though insolation contrasts with the season are not as great as at higher latitudes.

4. **The annual range over oceans is less than that over land at the same latitude.** This can be seen by following a parallel of lat. 60° N, for example. Starting near the west coast of Europe, the range is about 10°C over the Atlantic but increases to 55°C in central Asia. In the Pacific, the range falls to 20°C off Alaska and increases to more than 40°C in North America. Again, these major differences are due to the contrast between land and water surfaces.

5. **The annual range is very small over oceans in the tropical zone.** The temperature range over tropical oceans is less than 3°C since there is little seasonal variation in insolation near the equator, and water heats and cools more slowly.

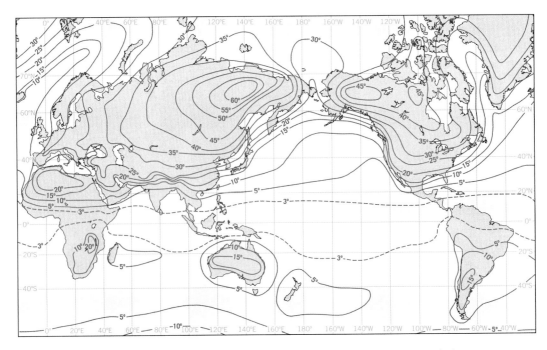

4.22 Annual range of air temperature in degrees Celsius Data show difference between January and July mean temperatures.

GLOBAL WARMING AND THE GREENHOUSE EFFECT

It is now nearly certain that global temperatures are rising as a result of human activities. Carbon dioxide (CO_2), which is produced by many human activities, is generally cited as a major cause of climate warming. This gas is released to the atmosphere in large quantities by fossil fuel burning. The increase in CO_2 with time follows an *exponential growth* curve, which *Working It Out 4.2 • Exponential Growth* describes more fully.

The buildup of CO_2 is a problem because it accentuates the greenhouse effect, which increases atmospheric absorption of outgoing long-wave radiation. A portion of the absorbed energy is subsequently counter-radiated back toward the Earth's surface. For this reason, the greenhouse effect is escalating and global temperatures are rising—or soon will rise—in response. (See *Eye on Global Change 4.3 • Carbon Dioxide—On the Increase.*)

Also of concern are other gases that are normally present in much smaller concentrations—methane (CH_4), chlorofluorocarbons (CFCs), tropospheric ozone (O_3), and nitrous oxide (N_2O). Taken together with CO_2, they are known as **greenhouse gases**. Though less abundant, these gases are better absorbers of long-wave radiation than CO_2, and so are more effective, molecule per molecule, at enhancing the greenhouse effect. This comparative efficiency is termed the **global warming potential** (GWP). The GWP of a gas is based on its heat-absorbing capacity, as well as the rate at which it is removed from the atmosphere over a given number of years. The GWP of CO_2 is 1. The time interval, which is normally 100 years, is an essential part of the rating scheme (Table 4.1.)

FACTORS INFLUENCING CLIMATIC WARMING AND COOLING

Figure 4.23 shows how a number of important factors have influenced global warming since about 1850. The five bars on the

Table 4.1 Global warming potential (GWP) of the main greenhouse gases

Greenhouse Gas	Atmospheric Lifetime (Years)	GWP	Pre-Industrial Concentration	Concentration in 1994
Carbon dioxide (CO_2)	Variable	1	278,000 ppbv	358,000 ppbv
Methane (CH_4)	12, 2 ± 3	21	700 ppbv	1,721 ppbv
Nitrous oxide (N_2O)	120	310	275 ppbv	311 ppbv
CFC-12 (CCl_2F_2)	102	6,200–7,100	0	0.503 ppbv
HCFC-22 ($CHClF_2$)	12, 1	1,300–1,400	0	0.105 ppbv
Perfluoromethane (CF_4)	50,000	6,500	0	0.070 ppbv
Sulphur hexa-fluoride (SF_6)	3,200	23,900	0	0.032 ppbv

| WORKING IT OUT | 4.2 | **Exponential Growth** |

Exponential growth occurs when something increases by a constant percentage during each growth period. The important thing about this type of growth is that it is compounded. For example, money in a bank account will grow according to the interest rate, such as 5 percent per year. A dollar will earn $.05 in the first year, but the interest on the second year applies to the $1.05 accumulated from the first year, not on the original $1 amount.

The figure on the right shows an example of exponential growth at two rates—3 percent and 6 percent per time period. The vertical axis shows the multiplier that applies to the original amount. For example, after 20 time periods, the multiplier for the 6 percent growth rate is slightly more than 3, meaning the original quantity will have more than tripled in that time. In contrast, the multiplier for the 3 percent growth rate is less than 2.

Here is an approximate formula for exponential growth:

$$M = e^{(R \times T)}$$

where M is the multiplier; R is the percentage rate, expressed in decimal points (i.e., $R = 0.04$ for 4 percent per year); and T is the number of time periods (i.e., years) that elapse.

The symbol e stands for the base of natural logarithms, which has a value of 2.718. To evaluate the expression, first find ($R \times T$), then raise e to that power.

For example, suppose the CO_2 concentration in the atmosphere is 360 parts per million (ppm) and is increasing at a rate of 4 percent per year. What concentration can be expected in 20 years if the present growth rate continues?

$$M = e^{(R \times T)} = e^{(0.04 \times 20)} = 2.718^{0.80} = 2.226$$

Thus, the concentration of 360 ppm will grow to about $360 \times 2.226 = 801$ ppm.

Sometimes it is convenient to think in terms of a doubling; that is, the time it will take for a quantity to double given that it is growing exponentially at a fixed percentage rate. The figure to the left shows graphically the doubling time for growth rates of 3 and 6 percent. It is the time associated with the multiplier 2; doubling times for 3 and 6 percent are 11.6 and 23.3, respectively. There is a handy rule to figure out the doubling time from the growth rate: divide 70 by the percentage growth rate, and the result will be the doubling time. So, a 4 percent annual growth rate would have a doubling time of 70/4 = 17.5 years.

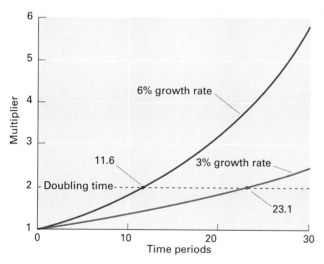

Exponential growth The multiplier for a quantity growing at a constant percentage rate increases with time. The higher the growth rate, the more quickly the multiplier increases.

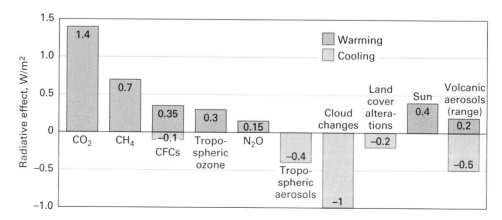

4.23 Factors affecting global warming and cooling Greenhouse gases act primarily to enhance global warming, while aerosols, cloud changes, and land-cover alterations caused by human activity act to retard global warming. Natural factors may be either positive or negative. (After Hansen et al., 2000, Proc. Nat. Acad. Sci., used with permission)

EYE ON GLOBAL CHANGE | 4.3   Carbon Dioxide—On the Increase

In the centuries before global industrialization, the level of CO_2 concentration in the atmosphere was slightly less than 300 parts per million (ppm) by volume. During the last hundred years or so, that amount has increased substantially due to fossil fuel burning. When fuels like coal, oil, or natural gas are burned, they yield water vapour and carbon dioxide. The release of water vapour is not a significant problem because a large amount is normally present in the global atmosphere. However, the natural amount of CO_2 is small. The present concentration of 360 ppm represents a 20 percent increase and is directly linked to the use of fossil fuels. Studies of atmospheric gases trapped in glacial ice indicate that this is the highest level attained in the last 420,000 years and nearly double the amount of CO_2 that was present during the most recent ice age.

The figure below shows how CO_2 has increased with time since 1860. Atmospheric concentration remained nearly stable until about 1940, but has been increasing at progressively faster rates over the past several decades. Even with concerted global action to reduce CO_2 emissions in the future, scientists estimate that levels will stabilize at a value not lower than about 550 ppm by the late twenty-first century.

Predicting the buildup of CO_2 is difficult because not all the gas emitted to the atmosphere remains there. Instead it circulates between the major Earth realms—the atmosphere,

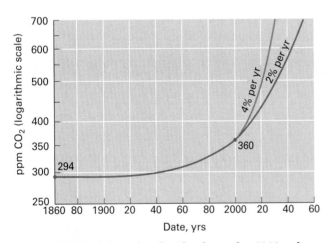

Increase in atmospheric carbon dioxide, observed to 2000 and predicted into the twenty-first century

biosphere, hydrosphere, and lithosphere—in a complex series of processes that combine to form the **carbon cycle** (see figure on opposite page). Green plants use CO_2 in photosynthesis, the process by which they use light energy to build their tissues. Photosynthesis removes about 121 gigatonnes (Gt) of CO_2 from the atmosphere annually. This represents about 20 percent of the total CO_2 stored in plant tissues. Some CO_2 is returned to the atmosphere through plant respiration, but the majority is taken out of circulation until the plants die and their remains decompose. In natural systems, decomposition and respiration each contribute about 60 Gt, which balances the annual uptake by photosynthesis.

Human activity can affect this balance by clearing and burning the vegetation cover to open up more land for development. This practice increases the amount of CO_2 in the atmosphere. However, when land is allowed to return to its natural state, CO_2 is removed from the air by the trees. At present, midlatitude forests in the northern hemisphere are growing more rapidly than they are being destroyed. However, this may be outweighed by the tropical deforestation that is taking place in South America, Africa, and Asia. (See *Eye on Global Change 23.1 • Exploitation of the Low-Latitude Rainforest Ecosystem*.)

The ability of plants to temporarily remove CO_2 from the atmosphere has led to the development of an international system of "carbon credits". The first such scheme, the European Union's Emissions Trading Scheme, started in January 2005. It was developed as a way to meet the 2012 CO_2 emission reduction obligations that were agreed to by member countries in the Kyoto Protocol. Carbon credits trading involves the sale of permits to emit carbon dioxide in exchange for land set aside for reforestation and afforestation. For example, a company in Europe could offset its CO_2 emissions by funding a reforestation project in Africa. Proponents of the scheme see it as a way to stabilize atmospheric CO_2 levels while concurrently reversing global deforestation and reducing environmental deterioration that accompanies the loss of forest cover. The decision to sign on to the Kyoto Protocol on April 29, 1998 committed Canada to a 240-megatonne reduction in annual

left show the principal greenhouse gases. Although carbon dioxide is the largest, the other four together contribute about the same amount of warming. Taken as a whole, the total enhanced energy flow to the surface produced by greenhouse gases is about three watts per square metre, which is about 1.25 percent of the solar energy absorbed by the Earth and atmosphere.

The decay of organic matter in wetlands naturally releases methane (CH_4). However, human activity generates about double that amount through rice cultivation, farm animal wastes, bacterial decay in sewage and landfills, fossil fuel extraction, transportation, and biomass burning. Chlorofluorocarbons (CFCs) have both a warming and cooling effect. The warming effect re-

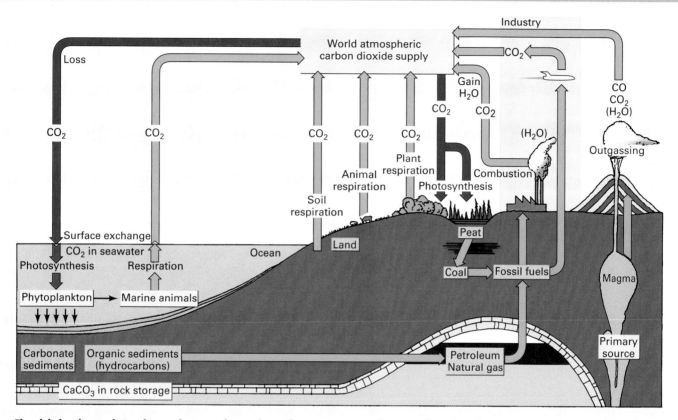

The global carbon cycle involves carbon transfers and transformations among the atmosphere, biosphere, hydrosphere, and lithosphere.

emissions of CO_2 and other greenhouse gases between 2008 and 2012. It is expected that Canada can meet this through a combination of improved industrial practices and carbon sequestration, particularly in the boreal forest.

Another part of the carbon cycle involves the oceans. The ocean's surface layer contains microscopic plant life (phytoplankton) that takes in about 92 gigatonnes of carbon annually. The CO_2 in the ocean water initially comes from the atmosphere and is mixed into the ocean by surface waves. When the plankton die, they sink to the ocean bottom. There they decompose and release CO_2, enriching the water near the ocean floor.

This CO_2 eventually returns to the surface through a system of global ocean currents in which CO_2-poor waters sink in the northernmost Atlantic and CO_2-rich waters rise to the surface in the northernmost Pacific. In fact, the ocean acts like a slow conveyor belt, moving CO_2 from the surface to ocean depths and releasing it again in a cycle lasting about 1,500 years (see Chapter 6). It is estimated that ocean surface waters absorb more CO_2 than they release. Therefore, CO_2 may be accumulating in ocean depths. However, current studies of global climate computer simulations indicate that the oceans may not be as effective in removing excess CO_2 as previously thought.

Although there is a great deal of uncertainty about the movements and buildup rate of excess CO_2 released to the atmosphere by fossil fuel burning, one thing is certain. Without conversion to solar, nuclear, and hydroelectric power, fuel consumption will continue to release carbon dioxide, and its effect on climate will continue to increase.

sults because these compounds are good absorbers of long-wave energy. The cooling effect occurs because CFCs destroy ozone in the stratosphere, and ozone contributes to warming. (See *Eye on Global Change 3.3 • The Ozone Layer—Shield to Life.*) Ozone warming also occurs in the troposphere, where O_3 is created by air pollution. Nitrous oxide (N_2O) is released by bacteria acting on nitrogen fertilizers in soils and runoff water. Motor vehicles also emit significant amounts of N_2O.

The next three factors—tropospheric aerosols, cloud changes, and land cover alterations—are primarily a result of human activity, and all tend to cool the Earth–atmosphere system. Tropospheric aerosols are produced mainly by fossil fuel burning and are considered a potent form of air pollution. They include sulphate particles, fine soot, and organic compounds. Aerosols act to scatter solar radiation back to space, thus reducing the flow of solar energy available to warm the surface. In addition, they enhance the formation of low, bright clouds that also reflect solar radiation back to space. These, and other changes in cloud cover caused directly or indirectly by human activity, lead to a significant cooling effect. Land-cover alteration, which has also induced cooling, includes conversion of forested lands to cropland and pastures, which are brighter and reflect more solar energy.

The last two factors are natural in origin. Solar output has increased slightly, causing warming. Volcanic aerosols, however, have at times caused warming and at other times cooling.

Taken together, the warming effect of the greenhouse gases has outweighed the cooling effect of other factors, and the result is a net warming effect of about 1.6 watts per square metre, which is about 0.23 percent of the total solar energy flow absorbed by the Earth and atmosphere. Has this enhanced energy flow caused global temperatures to rise? If not, will it warm the Earth in the near future? Before addressing these important questions, it is necessary to review the record of the Earth's surface temperature over the past few centuries.

THE TEMPERATURE RECORD

Figure 4.24 shows the Earth's mean annual surface temperature from 1866 to 2002, as obtained from surface air temperature measurements. The temperature is expressed as a difference from the average annual temperature for the period 1951–80. Two curves are shown: yearly data and data smoothed over a five-year

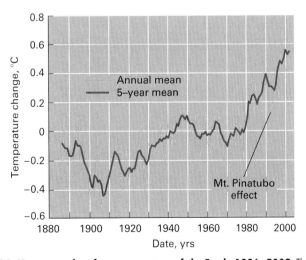

4.24 Mean annual surface temperature of the Earth, 1886–2002 The vertical scale shows departures in degrees from a zero line of reference representing the average for the years 1951–80. The yellow line shows the mean for each year. The red line shows a running five-year average. Note the effect of the Mount Pinatubo eruption in 1992–93. (James Hansen/NASA Goddard Institute for Space Studies)

4.25 Eruption of Mount Pinatubo, Philippines, April 1991 Volcanic eruptions like this can inject particles and gases into the stratosphere, influencing climate for several years afterwards. *Fly By: 15° 08′ N; 120° 21′ E*

period. Although temperature has increased, especially toward the end, there have been wide swings in the mean annual surface temperature.

Some of the variation is due to volcanic activity, which can propel particles and gases—especially sulphur dioxide (SO_2)—into the stratosphere, forming stratospheric aerosols. Strong winds spread the aerosols quickly throughout the entire layer. The aerosols have a cooling effect because they reflect incoming solar radiation. For example, the eruption of Mount Pinatubo in the Philippines in the spring of 1991 lofted 15 to 20 million tonnes of sulphuric acid aerosols into the stratosphere (Figure 4.25). The aerosol layer produced by the eruption contributed to the 2–3 percent reduction in solar radiation reaching the Earth's surface following the blast. In response, global temperatures fell about 0.3°C in 1992 and 1993.

Although direct air temperature measurement only began in the middle of the nineteenth century, the record can be extended further back by using **dendrochronology** (tree-ring analysis), a method based on the annual growth rings that develop in trees. In climates where the seasons are distinct, annual rings are usually clearly defined (Figure 4.26). If growing conditions are good, the annual ring is wide. If they are poor, the annual ring is narrow. For trees along the timberline in North America, the width is related to temperature—the trees grow better when temperatures are warmer. Since only one ring is usually formed each year, the date of each ring is easy to determine by counting backward from the present. Tree species with long life-spans can be used to develop temperature records that extend back many centuries. In the southwest United States, for example, living and dead bristlecone pines (*Pinus longaeva*) have been used to develop a chronology that extends back 8,500 years.

4.26 Tree rings in a cross-section taken from a 50-year-old Douglas fir (*Pseudotsuga menziesii*) tree. Careful measurement of the width of tree rings can be used to assess long term temperature patterns.

Figure 4.27 shows a reconstruction of northern hemisphere temperatures from 1700 using tree-ring analysis. Values are expressed as a difference from the mean temperature for 1950–65. From 1880 to present, the temperatures reconstructed from tree-ring analysis seem to fit the observed temperatures in Figure 4.24 quite well. Similar analysis of an earlier period shows another cycle of temperature increase and decline. The low point, around 1840, marks a historic cold event during which European alpine glaciers became more active and advanced. Other evidence indicates that the two cycles in Figure 4.24 are part of a natural global cycle of temperature warming and cooling lasting about 150 to 200 years. These cycles have occurred regularly over the last thousand years and may be linked to sun spot activity.

Again, a number of theories have considered the cause of these cycles, but no consensus has been reached. Still longer cycles of temperature change are evident in the fossil record, inducing the advance and retreat of continental and mountain glaciers during the late-Cenozoic Ice Age (see Chapter 20). Thus, the temperature record shows that the Earth's climate is naturally quite variable, responding to many different influences on many different time scales.

FUTURE SCENARIOS

The year 2005 was the warmest on record, with an average temperature of 14.7°C, which was 0.75°C above the global mean for 1950–80. It was also reported to be the warmest year of the past millennium, according to temperature reconstructions using tree rings and glacial ice cores. While a record year, 2005 was only slightly warmer than 1998, when worldwide temperatures were 0.71°C above the 1950–80 mean. The third highest year was 2002, followed in sequence by 2003, 2004, 1990, and 1995.

In 1995 the Intergovernmental Panel on Climate Change (IPCC), a United Nations-sponsored group of more than 2,000 scientists, concluded for the first time that human activity has caused climatic warming. This judgement was based largely on computer simulations of global climate that account for the release of CO_2 and SO_2 from fossil fuel burning since the turn of the century. The simulations agreed with the patterns of warming observed over that period, leading to the conclusion that, despite natural variability, human influence has been felt in the climate record of the twentieth century. Reports issued in 2001 and 2007 cited increased confidence in this conclusion. *A Closer Look: Eye on Global Change 4.4 • The IPCC Report of 2001* provides a summary of observed climate change and predictions of climate change for the remainder of the twenty-first century.

Using various projections of the continuing release of greenhouse gases coupled with computer climate models, IPCC scientists projected that global temperatures will warm between 1.4°C and 5.8°C by the year 2100. This warming trend will be accompanied by other environmental changes.

Of particular concern is a rise in sea level, as glaciers and sea ice melt in response to the warming. Current predictions call for a rise of between 9 and 88 centimetres in sea level by the year 2100. This would place as many as 92 million people at risk from annual flooding. Climate change could also promote the spread of insect-borne diseases, such as malaria. Furthermore, climate boundaries may shift, making some regions wetter, while others become drier. Thus, agricultural patterns could change, displacing large human populations as well as natural

4.27 Three centuries of northern hemisphere temperatures A reconstruction of the departures of northern hemisphere temperatures from the 1950–65 mean, based on analyses of tree rings sampled along the northern tree limit of North America. Original data are from 1700 to 1975; for completeness, data from Figure 4.24 are shown from 1975 to the present in a different colour. (Courtesy of Gordon C. Jacoby of the Tree-Ring Laboratory of the Lamont-Doherty Geological Observatory of Columbia University)

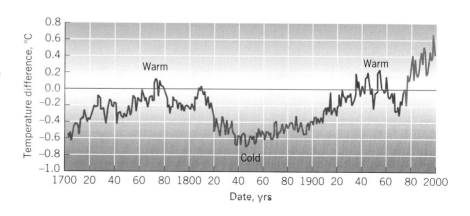

In 2001, the United Nations Intergovernmental Panel on Climate Change (IPCC), a body of scientists nominated by countries throughout the world, issued three major reports on human-induced changes in global climate. The reports concerned the scientific basis for anticipated climate change, the projected impacts of the change, and strategies for mitigating these impacts.

To estimate the degree of climate change in time throughout the world, IPCC used several complex global climate models—mathematical models that predict the state of the atmosphere and land and water surfaces at short time intervals over long periods. The models were driven by predicted releases of greenhouse gases under different scenarios of global economic growth and social evolution through the end of the twenty-first century. The two graphs accompanying this review show global temperature change and sea-level rise as modelled under these scenarios. Although there is variation between outcomes, it is clear that both global temperature and sea level will rise significantly by 2100 based on this analysis.

Here are some of the panel's more specific findings, taken from *Climate Change 2001: The Scientific Basis, A Report of the Working Group I of the Intergovernmental Panel on Climate Change*, Summary for Policymakers (Geneva, Switzerland: IPCC, 2001).

Recent Climate Change

- Global average surface temperature increased about 0.6°C during the twentieth century. The 1990s was the warmest decade and 1998 the warmest year since 1861. Nighttime daily minimum air temperatures rose about twice as fast as daytime maximum temperatures. Since 1940, the frequency of extreme low temperatures has decreased. The frequency of extreme high temperatures has increased a smaller amount.

- The lowest eight kilometres of the atmosphere has warmed at a rate of about 0.04°C per decade since at least 1979.

- Snow and ice cover has decreased by about 10 percent since the late 1960s. The duration of snow cover has reduced by about two weeks in the mid- and high latitudes of the northern hemisphere. The extent of sea ice in spring and summer in the northern hemisphere has decreased by 10 to 15 percent since the 1950s. Summer and fall sea ice is thinner.

- Global average sea level rose between 10 and 20 centimetres during the twentieth century. Global ocean heat content has increased since at least the late 1950s.

- Precipitation increased by 0.5 to 1 percent per decade in the twentieth century over the mid- and high latitudes of the northern hemisphere. Rainfall decreased in subtropical regions. There was an increase of 2 to 4 percent in the frequency of heavy precipitation events in the latter half of the twentieth century. The frequency and intensity of droughts in some parts of Africa and Asia have also increased in recent decades.

- Cloud cover increased by about 2 percent in the mid- and high latitudes of the northern hemisphere during the twentieth century.

- Since 1970, El Niño episodes have been more frequent, intense, and persistent, compared with those in the previous 100 years.

- Some important aspects of climate have not changed. Some parts of the southern hemisphere oceans and parts of Antarctica have not warmed in recent decades. No significant changes have occurred in Antarctic sea ice coverage since 1978. Tropical and extratropical cyclone frequencies have not changed.

- Concentrations of greenhouse gases in the atmosphere have increased as a result of human activities. CO_2 has increased by 31 percent since 1750. For the past two decades, CO_2 concentration has increased by about 1.5 parts per million (ppm) per year. Methane (CH_4) concentration has increased by 151 percent since 1750; nitrous oxide (N_2O) by about 17 percent; and ozone (O_3) by about 36 percent.

- There is newer and stronger evidence that human activities are responsible for most of the warming observed since 1950.

Climate Change in the Twenty-First Century

The IPCC reports also offer these predictions for the rest of the twenty-first century. Although they may not all come true, they are considered likely given the extent of present knowledge and current levels of confidence in global climate model predictions, as well as reasonable scenarios for economic growth, development, and emission control.

- Atmospheric composition will continue to change. CO_2 from fossil fuel burning will continue to increase. As this occurs, land and ocean will take up a decreasing fraction of CO_2 released, accentuating the increase. By 2100, CO_2 concentration will have increased by 90 to 250 percent above the 1750 value, depending on the scenario of economic growth, development, and CO_2 release.

- Other greenhouse gas concentrations are likely to change, increasing under most scenarios. Aerosols could contribute significantly to warming if their release is unabated.

- Global average temperature will rise between 1.4 and 5.8°C from 1990 to 2100. The projected rate of warming is much larger than that of the late twentieth century and is likely to be greater than any warming episode in the last 10,000 years.

- Land areas will warm more rapidly than the global average, especially northern latitudes in the winter.

- Precipitation will increase in northern mid- to high latitudes and in Antarctica in winter. Larger year-to-year variations in precipitation are likely. At low latitudes, there will be decreases in precipitation in some areas and increases in others.

- Over nearly all land areas, there will be higher maximum temperatures and more hot days; higher minimum temperatures and fewer cold days and frost days; reduced daily temperature range; more intense precipitation events; increased risk of summer drought; and an increase in peak wind and precipitation intensities of tropical cyclones.

- Snow and ice cover will decrease further, and ice caps and glaciers will continue their widespread retreat observed in the late twentieth century. The Antarctic Ice Sheet will gain mass from increased precipitation, while the Greenland Ice Sheet will thin.

- Sea level will rise from 9 to 88 centimetres depending on the scenario, due to thermal expansion and a gain in volume from melting ice.

- Human-generated climate change will persist for many centuries into the future. Even if concentrations of greenhouse gases are stabilized, oceans will continue to warm, sea level will continue to rise, and ice sheets and glaciers will continue to melt. Models suggest that a local warming of 3°C over the Greenland ice cap, if sustained over 10 centuries, would completely melt the ice cap, raising the sea level by about seven metres.

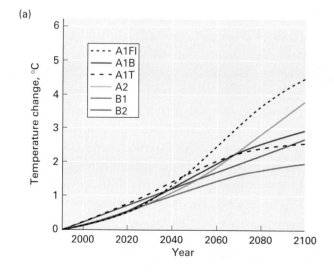

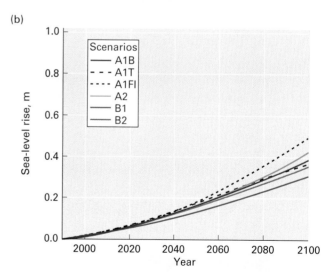

Increases in temperature and sea-level rise modelled by the IPCC These graphs show how temperature (a) and sea level (b) are predicted to rise between now and 2100, based on global climate models. Curves show means for several models under six emissions scenarios. The **A1** scenarios describe a world of rapid growth with economic and social convergence among regions, leading to a more uniform world. Scenario **A1F1** projects heavy reliance on fossil fuels; **A1T** projects a reliance on non-fossil energy sources; and **A1B** a balance across all sources. The **A2** scenario is a more heterogeneous world with less convergence and greater regional isolation. The **B** scenarios are similar to **A1** but move toward a service and information economy that emphasizes social and environmental sustainability. **B1** assumes more global convergence, while **B2** assumes more independence among regions. (From IPCC, *Climate Change Report 2001: Synthesis Report,* copyright IPCC 2001, used with permission of Cambridge University Press)

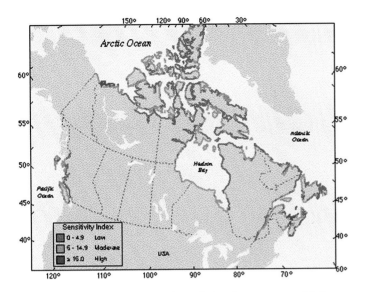

4.28 Generalized sensitivity analysis of Canadian coasts to global sea-level rise. The Maritime provinces are particularly susceptible to potential sea level rise.

ecosystems. In Canada, the Maritime provinces are considered to be one of the areas that are most sensitive to predicted sea level change (Figure 4.28).

A recently discovered effect of climatic warming is the enhancement of climate variability. Events such as very high 24-hour precipitation—extreme snowstorms, rainstorms, sleet, and ice storms, for example—appear to be occurring more frequently since 1980. More frequent and more intense spells of hot and cold weather may also be related to climatic warming.

The world has become acutely aware of the problem of the buildup in CO_2 and other greenhouse gases. Although efforts such as the Kyoto Protocol will slow the buildup of greenhouse gases, the ultimate solution will certainly involve greater reliance on solar, wind, and geothermal energy sources, which produce power without releasing CO_2. Energy conservation and the development of new methods of using fossil fuels that reduce CO_2 emissions also will be important.

A Look Ahead

In this chapter, air temperature and temperature cycles were described, along with the factors that influence them: insolation, latitude, cloud cover, surface type, continental-maritime location, elevation, and aspect. Air temperature is closely linked to atmospheric pressure and global wind patterns, as well as evaporation and condensation of moisture in the atmosphere. The following chapters discuss these important elements of weather and climate.

CHAPTER SUMMARY

- Air temperature—measured at 1.25 metres above the surface—is influenced by insolation, latitude, cloud cover, surface type, location, elevation, slope, and aspect. The energy balance of the ground surface is determined by net radiation, conduction to the soil, latent heat transfer, and convection to and from the atmosphere.

- Air temperature is measured using a *thermometer* or *thermistor*. Weather stations make daily minimum, maximum, and mean temperature measurements. Some measure temperature hourly.

- The two major cycles of air temperature—daily and annual—are controlled by the cycles of insolation produced by the rotation and revolution of the Earth. These cycles induce cycles of net radiation at the surface. When net radiation in the daily cycle is positive, air temperatures increase; when it is negative, air temperatures decrease. This principle operates through the seasonal cycle, which produces annual temperature variations.

- Temperatures of air and soil at or close to the ground surface are more variable than air temperature measured at the standard height of 1.25 metres.

- Surface characteristics also affect temperatures. Rural surfaces are generally moist and slow to heat and cool, while urban surfaces are dry and absorb and give up heat readily. This difference creates an urban heat island effect.

- Air temperature normally falls with altitude in the troposphere. The average value of decrease with altitude is the normal temperature lapse rate, 6.4°C per 1,000 metres. The temperature decrease measured on a specific day may differ from the average rate and is known as the environmental temperature lapse rate. At the *tropopause*, this decrease stops. In the overlying stratosphere, temperatures increase slightly with altitude.

- Air temperatures observed at mountain locations decrease with elevation, but day–night temperature differences increase with elevation.

- When air temperature increases with altitude, a temperature inversion is present. This can develop on clear nights when the surface loses long-wave radiation to space.

- Annual air temperature cycles are influenced by the annual pattern of net radiation, which depends largely on latitude and cloud cover.

- Maritime or continental location is another important factor. Ocean temperatures vary less than land temperatures because water heats more slowly, absorbs energy throughout a surface layer, and can mix and evaporate freely. Maritime locations that receive oceanic air therefore show smaller ranges of both daily and annual temperature.

- Global temperature patterns for January and July show the effects of latitude, maritime-continental location, and elevation. Equatorial temperatures vary little from season to season. Toward the poles, temperatures decrease with latitude, and continental surfaces in polar regions can become very cold. At higher elevations, temperatures are always colder.

- Isotherms over continents move over a wide latitudinal range with the seasons, while isotherms over oceans move through a much smaller range. The annual range in temperature increases with latitude and is greatest in northern hemisphere continental interiors.

- Earth's global temperature changes from year to year. Within the last few decades, global temperatures have been increasing. CO_2 released by fossil fuel burning is important in causing warming, but so are the other greenhouse gases, CH_4, CFCs, O_3, and N_2O. Aerosols scatter sunlight back to space and induce more low clouds, so they tend to lower global temperatures. Solar output and volcanic activity also influence global temperatures.

- Nearly all scientists agree that the human-induced buildup of greenhouse gases has begun to affect global climate. However, natural cycles, such as variations in the Sun's output, still provide strong influences. The continued release of large quantities of greenhouse gases at increasing rates is expected to result in a significant rise in global temperatures that will cause climate zones to shift and raise sea level.

KEY TERMS

air temperature	environmental temperature	ground inversion	normal temperature lapse rate
carbon cycle	lapse rate	heat island	stratosphere
conduction	evapotranspiration	isotherm	temperature inversion
convection	exponential growth	lapse rate	transpiration
dendrochronology	global warming potential	latent heat transfer	troposphere
energy balance	greenhouse gases	net radiation	

REVIEW QUESTIONS

1. Identify seven important factors in determining air temperature and air temperature cycles.
2. What factors influence the temperature of a surface?
3. How are mean daily air temperature and mean monthly air temperature determined?
4. How does the daily temperature cycle measured within a few centimetres of the surface differ from the air temperature cycle at the standard measurement height?
5. Compare the characteristics of urban and rural surfaces and describe how the differences affect urban and rural air temperatures. Include a discussion of the urban heat island.
6. What are the two lowest layers of the atmosphere and what is the zone that separates them? How are they distinguished? Name the two upper layers.
7. How and why are the temperature cycles of high mountain stations different from those of lower elevations?
8. Explain how latitude affects the annual cycle of air temperature through net radiation by comparing Manaus, Aswan, Hamburg, and Yakutsk

9. Why do large water bodies heat and cool more slowly than land masses? What effect does this have on daily and annual temperature cycles for coastal and interior locations?
10. What three factors are most important in explaining the world pattern of isotherms? Explain how and why each factor is important, and what effect it has.
11. Turn to the January and July world temperature maps shown in Figure 4.20. Make seven important observations about the patterns and explain why each occurs.
12. Turn to the world map of annual temperature range in Figure 4.22. What five important observations can you make about the annual temperature range patterns? Explain each.
13. Identify the important greenhouse gases and rank them in terms of their warming effect. What human-influenced factors act to cool global temperature? How?
14. Describe how global air temperatures have changed in the recent past. Identify some factors or processes that influence global air temperatures on this time scale.

FOCUS ON SYSTEMS 4.1 | The Surface Energy Balance Equation

1. Consider the surface energy balance during the day. What happens if the surface falls under the shadow of a cloud? How will this affect energy flows to and from the surface? What would you expect to happen to the temperature of the surface?

2. Suppose that on a hot sunny day, a surface layer of soil dries out so that water is no longer available for evaporation. How will this affect energy flows to and from the surface? What will happen to the surface temperature?

3. Air temperature is measured at a standard height (1.25 metres) above the surface. Would you expect this air temperature to be warmer or colder than the surface during the day? Why? What about at night? Why?

EYE ON GLOBAL CHANGE 4.3 | Carbon Dioxide—On the Increase

1. Why has the atmospheric concentration of CO_2 increased in recent years?

2. How does plant life affect the level of atmospheric CO_2?

3. What is the role of the oceans in influencing atmospheric levels of CO_2?

A CLOSER LOOK | Eye on Global Change 4.4 | The IPCC Report of 2001

1. What global climate changes have been noted for the last half of the twentieth century in global temperature, snow and ice cover, precipitation, and greenhouse gas concentrations?

2. How is climate predicted to change by the end of the twenty-first century with respect to temperature, precipitation, snow and ice cover, and sea level?

VISUALIZATION EXERCISES

1. Sketch graphs showing how insolation, net radiation, and temperature might vary during a 24-hour cycle at a mid-latitude station such as Toronto.

2. Sketch a graph of air temperature with height showing a ground inversion. Where and when is such an inversion likely to occur?

ESSAY QUESTIONS

1. Prince Rupert, on B.C.'s north Pacific coast, and Edmonton, Alberta, are at about the same latitude. Sketch the annual temperature cycle you would expect for each location. How do they differ and why? Select one season, summer or winter, and sketch a daily temperature cycle for each location. Again, describe how they differ and why.

2. Most scientists have concluded that human activities are acting to raise global temperatures. What human processes are involved? How do they relate to natural processes? Are global temperatures increasing now? What other effects could be influencing global temperatures? What are the consequences of global warming?

PROBLEMS | Working It Out 4.2 | Exponential Growth

1. If the present concentration of CO_2 is 360 ppm and it increases at a rate of 2 percent per year, what will the concentration be in 50 years? What will the concentration be if CO_2 is increasing at a rate of 3 percent?

2. The population of Singapore is about 2.8 million, and it is increasing at an annual rate of 1.3 percent. The Republic of

Congo has a population of about 2.4 million, increasing at an annual rate of 3 percent. What is the doubling time of each population? How large will the populations of these two countries be in 25 years if growth continues at the current rates?

Find answers in the back of the book

EYE ON THE LANDSCAPE

Chapter Opener Winter in Jasper National Park, Alberta. Moisture condenses in the cold air above the Athabasca River to form an ice fog (**A**) which interferes with the incoming solar beam to create an optical effect similar to a sundog (**B**). Albedo is high where ice covers the river (**C**) but low on the adjacent roadway (**D**). Differential absorption of solar energy in this way can affect the rate at which the debris laden winter snowpack melts (**E**)

Toronto city skyline from Lake Ontario. Radiation and temperature are greatly affected by surface properties. Urban structures vary considerably in terms of their thermal characteristics, but generally absorb shortwave energy readily. The canyon-like form of the modern city (**A**) also tends to trap heat. Vegetated surfaces (**B**) will help to moderate temperatures through higher albedo as well as greater use of energy for evaporation and transpiration. The intervening waters of Lake Ontario (**C**) will provide a thermostatic character to the local microclimate by providing a cooling effect in the summer and remaining warmer in winter until freeze-up.

Chapter 5

EYE ON THE LANDSCAPE

What else would the geographer see? . . . Answers at the end of the chapter.

WINDS AND THE GLOBAL CIRCULATION SYSTEM

Atmospheric Pressure
Measuring Atmospheric Pressure
How Air Pressure Changes with Altitude
Wind
Measurement of Wind
Winds and Pressure Gradients
The Coriolis Effect and Winds
Working It Out 5.1. • Pressure and Density in the Oceans and Atmosphere
Surface Winds on an Idealized Earth
Focus on Systems 5.2 • A Simple Convective Wind System
Global Wind and Pressure Patterns
Subtropical High-Pressure Belts
The ITCZ and the Monsoon Circulation
Wind and Pressure Features of Higher Latitudes
Cyclones and Anticyclones
Local Winds
Eye on the Environment 5.3 • Wind Power, Wave Power, and Current Power
Winds in the Upper Atmosphere
The Geostrophic Wind
Global Circulation at Upper Levels
Rossby Waves, Jet Streams, and the Polar Front
Temperature Layers of the Ocean
Ocean Currents
Surface Currents
Deep Currents and Thermohaline Circulation
Eye on Global Change 5.4 • El Niño

The air is always in motion. Why does the air move? What are the forces that cause winds to blow? Why do winds blow more often in some directions than others? What is the pattern of wind flow in the upper atmosphere, and how does it affect our weather? How does the global wind pattern produce ocean currents? These are some of the questions investigated in this chapter.

Winds are caused by unequal heating of the Earth's atmosphere, which results in differences in pressure and causes air at the surface to move toward the warmer location. For example, sea breezes bring cooler air onshore during the day. This effect also produces global wind motions, since solar heating of the atmosphere in equatorial and tropical regions is more intense than in mid- and higher latitudes. In response to this difference in heating, global-scale pressure gradients move vast bodies of warm air toward the poles, and cool air shifts toward the equator. The direction of these global wind motions is also influenced by the Earth's rotation.

Another feature of the global circulation system is ocean currents, which are largely driven by surface winds. Ocean currents, like global wind motions, act to move heat energy across parallels of latitude, and are also affected by the Earth's rotation.

ATMOSPHERIC PRESSURE

The Earth is surrounded by a deep layer of air that presses on the solid or liquid surface beneath it. The force per unit area exerted by the weight of a column of air above a particular location is referred to as pressure. Pressure occurs because air molecules have mass and are constantly being pulled toward the Earth by gravity. The standard unit of pressure is the *pascal* (Pa), but for meteorological purposes pressure is normally measured in **kilopascals** (kPa). At sea level, the average pressure of the atmosphere is 101.3 kilopascals (101,320 pascals). Note that 101.3 kilopascals is equivalent to 1,013 millibars (mb)—an older unit of pressure that is still commonly used.

MEASURING ATMOSPHERIC PRESSURE

Atmospheric pressure is measured by a **barometer**. The *mercury barometer* has become the standard instrument for weather measurements because of its high accuracy. It works on the principle that atmospheric pressure forces liquid mercury to rise in a tube from which air has been evacuated.

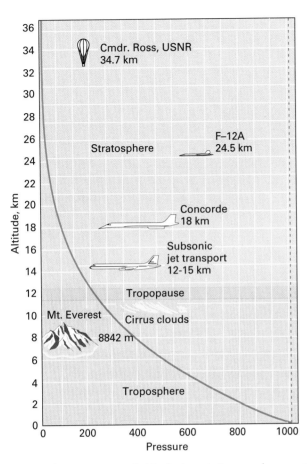

5.1 Atmospheric pressure and altitude Atmospheric pressure decreases with increasing altitude above the Earth's surface.

Another type of barometer is the *aneroid barometer*. It uses a thin-walled, sealed canister from which some air has been removed to establish a partial vacuum. The canister flexes as air pressure changes, and through a mechanical linkage a needle moves across the scale.

Atmospheric pressure at a location varies by only a small proportion from day to day. On a cold, clear winter day, the barometric pressure at sea level might be as high as 103 kilopascals, while in the centre of a storm system, pressure might drop to 98 kilopascals, a difference of about 5 percent.

HOW AIR PRESSURE CHANGES WITH ALTITUDE

Figure 5.1 shows how atmospheric pressure decreases with altitude. Moving upward from the Earth's surface, the decrease in pressure is initially quite rapid. At higher altitudes, the decrease is much slower. Aircraft cabins are pressurized for passenger comfort to about 80 kilopascals, which corresponds to an elevation of about 1,800 metres.

Working It Out 5.1 • Pressure and Density in the Oceans and Atmosphere explains the physical principles associated with change in atmospheric pressure with height above the surface, as well as the increase of water pressure in the ocean with depth.

WIND

Wind is air motion across the Earth's surface. Wind movement is predominantly horizontal. Air motion that is vertical is known by other terms, such as updrafts or downdrafts.

MEASUREMENT OF WIND

Air motion is characterized by a direction and velocity. Wind direction can be determined by a *wind vane*. The wind vane faces into the wind, and wind direction is always given as the direction from which the wind is coming. So, a west wind is one that comes from the west and moves to the east. Wind speed is measured by an *anemometer*. The most common type is the cup anemometer. The cups rotate with a speed proportional to that of the wind and are calibrated in metres per second. Modern wind vanes and anemometers are connected to data storage modules and their measurements are fed into computers for analysis.

WINDS AND PRESSURE GRADIENTS

Wind is caused by differences in atmospheric pressure from place to place. Air tends to move from high to low pressure (Figure 5.2) due to the simple physical principle that any fluid (such as air) subjected to gravity will move until the pressure is uniform. The lines on the map in Figure 5.2, called **isobars**, connect locations of equal pressure. At Edmonton, the pressure is high (H) and the barometer (adjusted to sea-level pressure) reads 102.8 kilopascals; at Saskatoon, the pressure is low (L), with a barometer reading of 99.7 kilopascals. The resulting **pressure gradient** produces the *pressure gradient force* that moves air from Edmonton toward Saskatoon. The greater the pressure difference between the two locations, the greater this force will be and the stronger the wind.

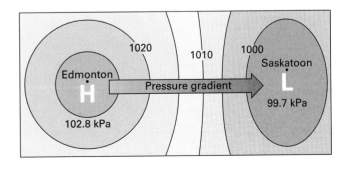

5.2 Isobars and a pressure gradient High pressure is centred at Edmonton and low pressure is centred at Saskatoon.

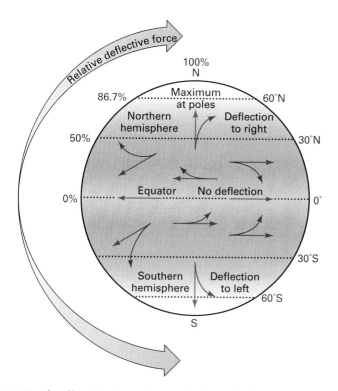

5.3 Coriolis effect direction and strength The Coriolis effect acts to deflect the paths of winds or ocean currents to the right in the northern hemisphere and to the left in the southern hemisphere, as viewed from the starting point. Blue arrows show the direction of initial motion, and red arrows show the direction of motion apparent to the observer on the Earth.

5.4 The Coriolis effect The path of a rocket launched from New York toward the North Pole will appear to be deflected to the right by the Earth's rotation. If launched from the North Pole toward New York, its path also will appear to be deflected to the right

THE CORIOLIS EFFECT AND WINDS

The pressure gradient force tends to move air from high pressure to low pressure; however, on a global scale, the direction of air motion is also affected by the Earth's rotation. This is termed the **Coriolis effect**, which causes air currents to follow gently curving paths as they blow across the Earth's surface. The air moves as though a force were pulling it sideways. The apparent deflection is to the right in the northern hemisphere and to the left in the southern hemisphere. Deflection is strongest near the poles and decreases to zero at the equator (Figure 5.3). The apparent force does not depend on direction of motion—deflection occurs whether the object is moving toward the north, south, east, or west. The Coriolis effect is a sideways-turning force that always acts at right angles to the direction of motion. The magnitude of the Coriolis effect increases with the speed of the motion of the air current. The Coriolis effect similarly affects ocean currents.

The Coriolis effect is a result of the Earth's rotation. Imagine that a rocket is launched from New York toward the North Pole, following the 74°W longitude meridian (Figure 5.4). Because of the Earth's rotation, a point on the Earth's surface at the latitude of New York (about 40°N) moves eastward at about 1,300 kilometres per hour. Although the rocket is aimed properly along the meridian, its motion will have an eastward component of 1,300 km/hr. Although it travels toward the pole along a straight path, it will appear to curve to the right, because the Earth rotates beneath it.

Air in motion near the surface is also subjected to a frictional force due to surface roughness. The frictional force is proportional to the wind speed and always acts opposite to the direction of motion. The combination of the pressure gradient force, the force of Coriolis effect, and the frictional force produces a direction of motion that is toward low pressure but at an angle to the pressure gradient (Figure 5.5).

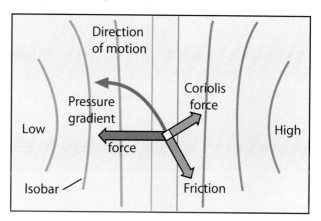

5.5 Balance of forces on a parcel of surface air in the Northern Hemisphere Although the pressure gradient will push a parcel of air toward low pressure, it will be deflected by the force of the Coriolis effect and slowed by friction with the surface. The direction of air motion will be the result of the three forces acting together.

WORKING IT OUT | 5.1   Pressure and Density in the Oceans and Atmosphere

Although it is sometimes useful to think of the Earth's surface as the bottom of a vast "ocean of air," the atmosphere is quite different from the oceans; the oceans are composed of a nearly incompressible liquid, whereas the atmosphere is composed of a mixture of readily compressible gases. Recall that both liquids and gases are classified as fluids.

The pressure at any level in a fluid is created by the mass of the fluid above that level. In the oceans, it's relatively simple to determine that weight. Pressing down on one square metre of ocean at a given depth (D) is the weight of $D \times 1$ m^3 of water. Each cubic metre of water weighs about 1,000 kilograms. So as a quick rule of thumb, the pressure at any depth in the ocean is

$$P_{OCEAN} = 1,000D = D \times 10^3 \text{ kg/m}^2$$

where P is the pressure in kilograms per square metre and D is the depth in metres.

For example, the pressure at 500 metres depth is about 500,000 kg/m^2, which equals 500×10^3 kg/m^2. At 5 kilometres, or 5,000 metres, it is about 5,000,000 kg/m^2, which equals $5,000 \times 10^3$ kg/m^2.

These values are based on a constant density of 1,000 kilograms per cubic metre. To be precise, there is a slight variation in the density of sea water with pressure, temperature, and salinity. That is, colder, saltier, and deeper water will be denser. At the surface, the density of ocean water is 1,028 kilograms per cubic metre, while at 5,000 metres depth, the density is 1,051 kilograms per cubic metre, an increase of about 2.2 percent. The left graph in the figure on the facing page, which plots pressure with depth in the ocean, uses a constant density value of 1,035 kilograms per cubic metre.

Note that at any depth in the ocean, there will be not only the weight of the water above, but also the weight of the atmosphere above the water. Sea-level pressure is about 100 kilopascals, which converts to 1,000 kilograms per square metre. Thus, the atmosphere adds the pressure of only about one more metre of depth.

Though small, the density differences in water with temperature and salinity can affect the pressure at a given depth enough to create pressure gradients and therefore induce movement of water from higher to lower pressure. An example is the convection loop circulation that occurs when water is heated from below.

In contrast to the oceans, the atmosphere is readily compressible. What is the weight of a cubic metre of air? There is no rule-of-thumb answer to that question because density varies with pressure. When a given volume of air is compressed, more molecules of the gases that compose air are present, and so the volume will have a greater mass and weigh more. At the Earth's surface, the density of a cubic metre of air is about 1.225 kilograms per cubic metre, while at 11 kilometres, the typical cruising altitude of an airplane, it is about 0.364 kilograms per cubic metre, or a little less than one-third of the surface value.

The difference is nearly all due to pressure, but temperature is also important. At a given pressure, warmer air will be less dense than colder air. Temperature differences create pressure gradients that induce air movements. The perpetual contrast between the warm atmosphere near the equator and cold atmosphere near the poles sustains the winds that constantly move and mix in the troposphere.

Atmospheric pressure is plotted against altitude in the graph on the right. Note that the graph curves smoothly, whereas the graph of pressure against ocean depth is straight. The atmospheric pressure graph curves because air is compressible and its density decreases with altitude.

SURFACE WINDS ON AN IDEALIZED EARTH

Consider wind and pressure patterns on a hypothetical Earth—one without a complicated pattern of land and water and no seasonal changes (Figure 5.6).

Recall that when air is heated, a convection loop can form. Since insolation is strongest when the Sun is directly overhead, the surface and atmosphere at the equator will be heated more strongly than other places on this hypothetical Earth. The result will be two convection loops, the **Hadley cells**, which form in the northern and southern hemispheres. In each Hadley cell, air rises over the equator and is drawn toward the poles by the pressure gradient. In addition, the air is turned by the force of the Corio-

lis effect and so heads westward as well as poleward. The uplifted air descends at about 30° latitude, completing a convection loop.

Since air rises at the equator, a zone of surface low pressure will result. This zone is known as the *equatorial trough*. The wind arrows show that air in both hemispheres moves toward the equatorial trough, where it converges and moves into the atmosphere as part of the Hadley cell circulation. This zone of convergence is termed the **intertropical convergence zone (ITCZ)**. Winds in the ITCZ are light and variable, because the air is generally rising.

Where the Hadley cell circulation descends, surface pressures are high. This produces **subtropical high-pressure belts**, centred at about 30° latitude in the northern and southern hemispheres.

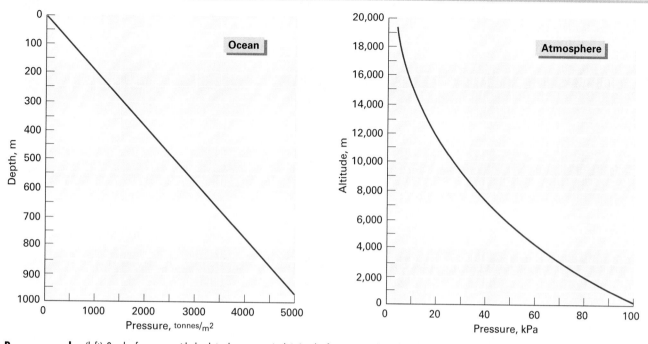

Pressure graphs (left) Graph of pressure with depth in the oceans. (right) Graph of pressure with altitude in the atmosphere.

Atmospheric pressure at any given altitude can be approximated by the following formula:

$$PZ = 101.3[1 - 0.0226Z]^{5.26}$$

where PZ is the pressure in kilopascals at height Z in kilometres. This formula takes into account the fact that both temperature and pressure decrease with altitude. Note that the value of 101.3 is the surface atmospheric pressure in kilopascals for standard conditions.

For example, let's find the pressure at 12 kilometres, which is approximately the altitude of a passenger jet on a transcontinental flight. Substituting for Z, we have

$$\begin{aligned}PZ &= 101.3 \times [1 - (0.0226 \times 12)]^{5.26}\\ &= 101.3 \times (1 - 0.271)^{5.26}\\ &= 101.3 \times (0.729)^{5.26}\\ &= 101.3 \times 0.189\\ &= 19.2 \text{ kPa}\end{aligned}$$

The subtropical high-pressure belts comprise a series of large and stable high-pressure centres.

Winds around the subtropical high-pressure centres spiral outward and move toward equatorial as well as midlatitudes. The winds moving toward the equator are the strong and dependable *trade winds*. The *northeast trades* occur in the northern hemisphere; the *southeast trades* south of the equator. Air spiralling outward from the subtropical high-pressure centres toward the poles produces southwesterly winds in the northern hemisphere and northwesterly winds in the southern hemisphere.

Between about 30° and 60° latitude, the pressure and wind pattern becomes more complex. A conflict zone, delineated by the **polar front**, occurs where colder polar air meets the warmer subtropical air. Masses of cool, dry air, termed *polar outbreaks*, move into this zone from the Arctic and Antarctic regions. Pressures and winds can be quite variable in the midlatitudes. On average, however, winds are more often from the west, so the region is said to have *prevailing westerlies*. Many of the weather systems that affect Canada and Western Europe originate along the polar front. The resulting low-pressure systems usually bring cloudy, unsettled weather and precipitation as they travel from west to east through these regions (see Chapter 7). At the poles, the air is intensely cold, which creates centres of high pressure. Winds spiralling outward around a polar anticyclone (a high-pressure

Pressure gradients develop because of unequal heating of the atmosphere. The simple convective wind system shown in the figure depicts a uniform atmosphere overlying the ground (part *a*). The lines mark isobaric surfaces. Imagine that the air is heated over *Y* (perhaps by sunlight) and cooled over *X* and *Z* (perhaps by radiation loss), as in part (*b*). As the air at *Y* expands in response to heating, it pushes the isobaric surfaces upward. At a constant height in the atmosphere, for example, along line *X'-Y'-Z'*, a pressure gradient is induced; at *Y'*, the pressure is between 98 and 97 kilopascals, while at *X'* and *Z'*, it is less than 97 kilopascals. Thus, there will be a pressure gradient from *Y'* to *X'* and from *Y'* to *Z'*.

The pressure gradient causes air to move away from *Y'* (part *c*). However, as soon as this air moves, it changes the surface pressure at *X*, *Y*, and *Z*. Since there is less air above *Y*, pressure at the ground falls. Meanwhile, there is more air above points *X* and *Z*, so pressure there rises. This creates a new pressure gradient at the surface that moves air from *X* and *Z* to *Y* (part *d*). The resulting circulation has produced two convection loops (shown by the arrows) in which air at the surface converges at the location of heating and moves away, or diverges, at higher altitude.

The principle that heating of the air produces a pressure gradient that causes air to move is important and is associated with the establishment of surface thermal high- and low-pressure zones. In general, low surface pressure is associated with warm air, and high surface pressure with cool air. On a global scale, the process is seen in the tendency for deserts to have low atmospheric pressure, while regions of persistent ice and snow have high pressure. However, it is important to note that not all surface pressure conditions develop this way.

High- and low-pressure systems can also develop as a result of convergence and divergence associated with general global circulation of the atmosphere.

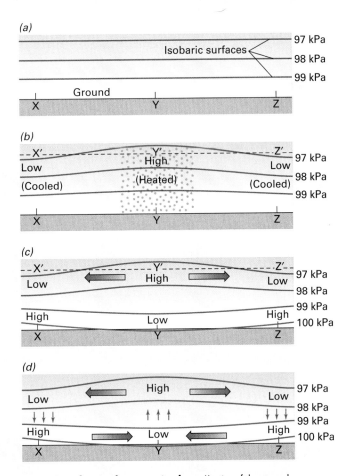

Formation of a simple convection loop. Heating of the atmosphere over point Y creates a pressure gradient that causes a convection loop.

centre) create surface winds generally from an easterly direction, known as the *polar easterlies*. In reality, this situation exists only in the south polar region. In the north polar region, winds tend to have an eastward component, but there is too much variation in direction to consider polar easterlies as the dominant winds there.

GLOBAL WIND AND PRESSURE PATTERNS

Figure 5.7 shows average wind and pressure conditions for January and July. Data are corrected for the elevation of the recording station, so that pressures are shown for sea level. Average barometric pressure is 101.3 kilopascals; values greater than this are "high" (red lines), while lower values are "low" (green lines).

SUBTROPICAL HIGH-PRESSURE BELTS

The most prominent features of the maps are the subtropical high-pressure belts, created by the Hadley cell circulation. The high-pressure belt in the southern hemisphere conforms well to the hypothetical pattern of Figure 5.6. It has three large high-pressure cells that persist year round over oceans. A fourth, weaker high-

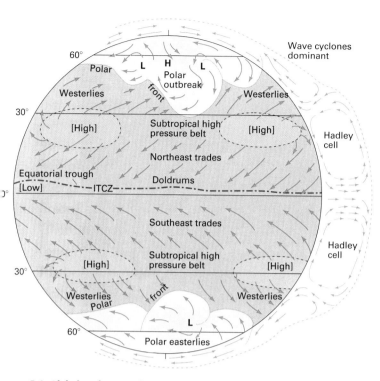

5.6 Global surface winds on an idealized Earth Global surface winds and pressures are shown on an Earth with no disruption from oceans and continents and no seasonal variation. The cross section at the right shows winds in the atmosphere.

pressure cell forms over Australia in July, as the continent cools during the southern hemisphere winter.

In the northern hemisphere, the subtropical high-pressure belt shows two large anticyclones (see *Cyclones and Anticyclones* later in this chapter) centred over oceans—the Hawaiian High in the Pacific and the Azores High in the Atlantic. From January to July, these intensify, move northward, and have a dominant influence on summer weather in North America.

On the east coast, warm, moist air from the Azores High is brought onshore from the Caribbean Sea, producing generally hot, humid weather for the central and eastern United States. On the west coast, dry subsiding air moves out of the continent, bringing fair weather and rainless conditions to places like California. In winter, the two semi-permanent anticyclones weaken and move to the south, allowing colder winds and air masses from the north and west to penetrate more deeply into the continent.

THE ITCZ AND THE MONSOON CIRCULATION

Recall from Chapter 3 that insolation is most intense when the Sun is directly overhead. Remember, too, that the latitude at which the Sun is directly overhead changes with the seasons, migrating between the tropics of Cancer and Capricorn. Since the Hadley cell circulation is driven by this heating, the general global pressure and wind belts also shift with the seasons.

In the western hemisphere, the ITCZ moves a few degrees north from January to July over the oceans (Figure 5.7). In South America, the ITCZ lies across the Amazon in January and swings northward by about 20° to the northern part of the continent. But in Africa and Asia, the shift is more pronounced. In January, the ITCZ runs south across eastern Africa and crosses the Indian Ocean to northern Australia at a latitude of about 15° S. In July, it swings north across Africa along the southern margin of the Sahara, and then passes farther north to lie along the south rim of the Himalayas, at a latitude of about 25° N. This represents a shift of about 40 degrees of latitude.

Why does such a large shift occur in Asia? In January, an intense high-pressure system, the Siberian High, forms in central Asia due to the extreme cold. In July, this high-pressure centre is replaced by a low centred over the deserts of the Middle East, which develops because of intense summer heating.

The movement of the ITCZ and the change in the pressure pattern with the seasons create a reversing wind pattern in Asia known as the **monsoon** (Figure 5.8). In the winter, there is a strong outflow of dry, continental air from the north across China, Southeast Asia, India, and the Middle East. During this *winter monsoon*, dry conditions prevail. In the summer, warm, humid air from the Indian Ocean and the southwestern Pacific moves northward and northwestward into Asia, passing over India, Indochina, and China. This airflow is known as the *summer monsoon* and is accompanied by heavy rainfall in southeastern Asia.

An additional element that accentuates the monsoonal airflows in Asia is the diversion of high level air streams north and south of the Himalayas. Also implicated is intense radiation heating and cooling of the Tibetan Plateau, which lies at a general elevation of 5,500–6,000 metres. In addition, the seasonal shift of the ITCZ between the southern to northern hemispheres causes a reversal in the Coriolis effect on the tropical winds. Deflection is reversed as the southeast trade winds cross from the southern hemisphere into the northern hemisphere. The resulting southwesterly winds from the Indian Ocean supply moisture for the wet monsoon season.

North America does not have the remarkable extremes of monsoon winds experienced in Asia. Even so, in summer, warm, moist air originating in the Gulf of Mexico tends to move northward across the central and eastern part of the United States into Canada. At times, moist air from the Gulf of California also invades the desert southwest, causing widespread scattered thundershowers and creating the "Arizona monsoon." In winter, the airflow pattern across North America changes, and dry, continental air moves out from the Arctic, bringing occasional frosts as far south as Florida.

The north–south movement of air across North America is accentuated by the continent's topography. The central lowlands

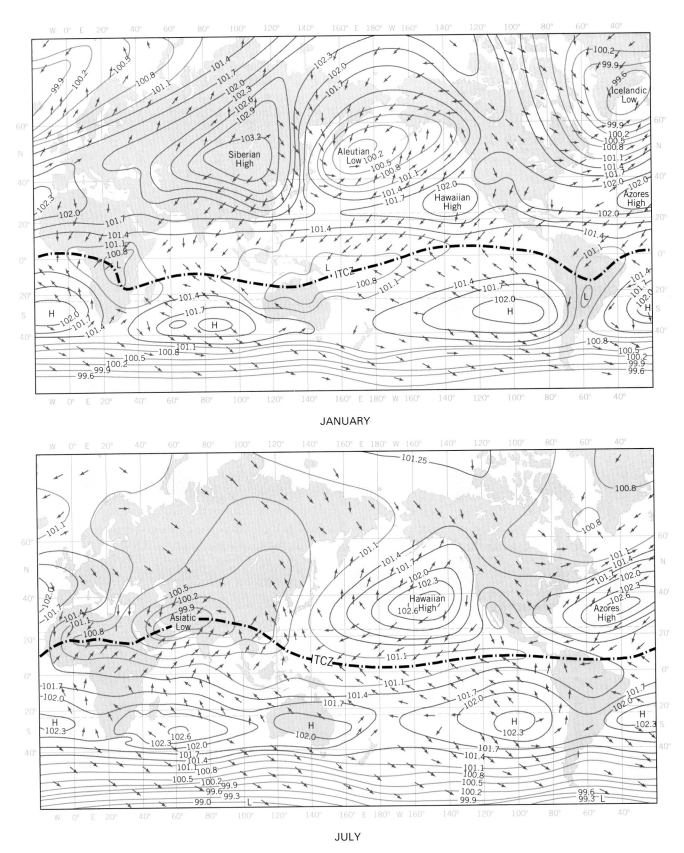

JANUARY

JULY

5.7 Atmospheric pressure maps The maps show mean monthly atmospheric pressure and prevailing surface winds for January and July. Pressure units are kilopascals reduced to sea level. (Data compiled by John E. Oliver)

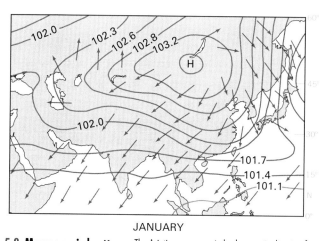

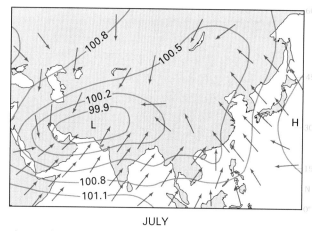

5.8 Monsoon wind patterns The Asiatic monsoon winds alternate in direction from January to July, responding to reversals of barometric pressure (in kilopascals) over the large continent.

present few barriers to meridional (north–south) air flow. Conversely, the high mountains in the west tend to block the air flow from the Pacific Ocean. Generally air from the Pacific is modified by passing over the mountains, where much of the moisture is lost. Similarly, the Appalachian Mountains limit airflow from the Atlantic Ocean. However, they do influence the passage of weather systems, deflecting them to the northeast through the Gulf of St. Lawrence (Figure 5.9). Some of the heavy rainstorms and snowfalls that affect eastern Canada can be attributed to this.

WIND AND PRESSURE FEATURES OF HIGHER LATITUDES

The northern and southern hemispheres are quite different in their geography (Figure 5.10). The northern hemisphere has two

large continental masses separated by oceans, with an ocean also at the pole. The southern hemisphere is essentially a large ocean with a cold, glacier-covered continent at the centre. These contrasting land–water patterns strongly influence the development of seasonal high- and low-pressure centres.

Recall from Chapter 4 that continents will be cold in winter and warm in summer, compared with oceans at the same latitude. Cold air is generally associated with surface high pressure and warm air with surface low pressure. Thus, continents typically show high pressure in winter and low pressure in summer.

This pattern is quite apparent in the northern hemisphere. In winter, the strong Siberian High is found in Asia, and the weaker Canadian High is found in North America. From these high-pressure centres, air spirals outward, bringing cold air to the south. Over the oceans, two large centres of low pressure are established—the Icelandic Low and the Aleutian Low. These low-pressure centres are regions where winter storm systems are spawned, the sizes and positions of which are quite variable.

In summer, the pattern reverses. The continents show generally low surface pressure, while high pressure builds over the

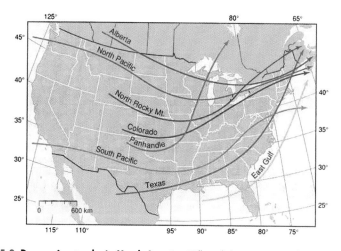

5.9 Depression tracks in North America Midlatitude low pressure weather systems move predominantly from west to east and are directed by upper atmosphere circulation patterns as well as surface topography. The are usually distinguished by region of origin and my receive names such as the Alberta clippers which in winter bring unusually cold weather and snow to eastern Canada and the United States.

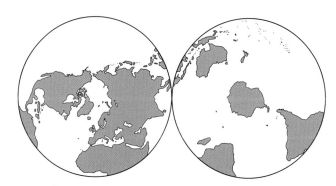

5.10 Land–ocean contrasts by hemisphere The northern and southern hemispheres are quite different in their distribution of lands and oceans.

oceans. The Asiatic Low is strongly developed. Winds spiralling inward form part of the monsoon cycle and bring warm, moist air from the Indian Ocean over India and Southeast Asia. A smaller low forms over the deserts of southwestern United States and northwestern Mexico. The two subtropical highs, the Hawaiian High and the Azores High, strengthen and dominate the Pacific and Atlantic oceans. Winds spiralling outward keep the west coasts of North America and Europe warm and dry, and the east coasts of North America and Asia warm and moist.

The higher latitudes of the southern hemisphere consist of a polar continent surrounded by a large ocean. Since Antarctica is covered by a glacial ice sheet and is cold at all times, the permanent South Polar High is centred there. Easterly winds spiral outward from the high-pressure centre. Surrounding the high is a band of deep low pressure with strong, inward-spiralling westerly winds that intensify toward the pole.

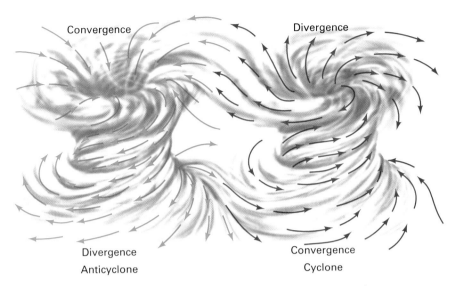

5.12 Dynamic convection loop A cyclone and anticyclone linked together in a convection loop.

CYCLONES AND ANTICYCLONES

Low- and high-pressure centres are common features of the daily weather maps in Canada. Unlike the semi-permanent highs and lows shown on the January and July world maps (Figure 5.7), the daily maps show the distribution pressure systems that normally move across a region and bring changing weather conditions. Low-pressure centres, known as **cyclones**,[1] are often associated with cloudy or rainy weather, and high-pressure centres, or **anticyclones**, usually bring fair weather.

These low- and high-pressure centres develop as air converges and diverges in response to regional pressure gradients (Figure 5.11). In a low-pressure system the pressure gradient is inward. The Coriolis effect and friction with the surface cause the surface air to move at an angle across the pressure gradient. This creates a convergent, inward spiralling motion. In the northern hemisphere, the cyclonic spiral is counter-clockwise because the Coriolis effect acts to the right. In the southern hemisphere, the cyclonic spiral is clockwise because the Coriolis effect acts to the left. For anticyclones, the pressure gradient is out from the centre; this creates a divergent, outward spiralling motion.

Cyclones and anticyclones are large features of the lower atmosphere—often 1,000 kilometres, or more, across. The vertical motion of air in adjacent low- and high-pressure centres is generally connected in a convection loop (Figure 5.12). In the anticyclone, air converges in the upper atmosphere and descends in a

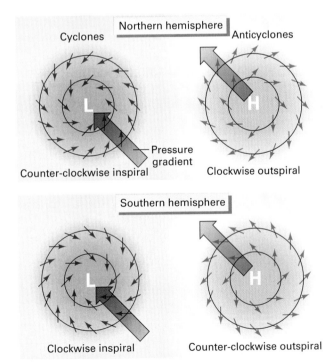

5.11 Air motion in cyclones and anticyclones Surface winds spiral inward toward the centre of a cyclone but outward and away from the centre of an anticyclone. Because the Coriolis effect deflects moving air to the right in the northern hemisphere and to the left in the southern hemisphere, the imposed curvature reverses from one hemisphere to the other.

[1] Note that cyclone is also the name given to intense low-pressure, tropical storms (hurricanes) in Australia, Japan, and the eastern Pacific. The term **depression** is often preferred when describing a low-pressure centre, particularly, those that bring cloudy conditions and precipitation in midlatitudes (see Chapter 7).

spiralling motion. Near the surface, the flow diverges, fanning out and moving away from the centre of high pressure. In the cyclone, air converges near the surface and spirals upward; in the atmosphere, the air diverges and spirals outward.

The upward and downward flows are linked as a portion of the moving air flows from the anticyclone to the cyclone near the surface, and another portion flows from the cyclone to the anticyclone in the atmosphere. The loop is not perfect because air escapes and enters from the sides of the flow.

LOCAL WINDS

Although global winds are important in determining the prevailing winds in a given area, local weather conditions may be influenced by winds that develop in response to small-scale or short-term pressure gradients. Local winds can be divided into two groups. The first is a wind that develops only in one area because a body of water or the local topography causes specific air movements. The second is a wind that an area's inhabitants consider to be distinctive despite the fact that it may have originated many hundreds of kilometres away.

Sea and *land breezes* are examples of how heating and cooling produce convection loops to create local winds. Recall from Chapter 4 that land surfaces heat and cool more rapidly than water, so a coastline is a location where temperature contrasts can easily develop. During the day, the air over the land is warmed by the surface, and a convection loop forms (Figure 5.13). The *sea breeze*, which usually begins to blow onshore on a summer's af-

ternoon, brings cool marine air toward land, while in the atmosphere the flow is toward the ocean. This cool air off the water effectively lowers temperatures. At night, radiation cooling over land creates a reversed convection loop, and a land breeze develops that blows out to sea. Air flow in land and sea breezes is generally about 5–10 metres per second. The effect is limited to the lowest 1,000 metres or so of the atmosphere and to a narrow zone, 15–20 kilometres wide, straddling the shoreline.

This effect is not restricted to ocean beaches, but can be detected around large lakes, such as the Great Lakes. The narrow strip of land that separates the largest of the Great Lakes—Superior, Michigan, and Huron—experiences another effect from the onshore air flow. In summer the warm air can pick up moisture from the lakes, and a little way inland where the ground is warmer, it rises in gentle convection currents. This can sometimes produce afternoon clouds and light rain (Figure 5.14).

Sea and *land* breezes are examples of *local wind* systems that affect relatively small areas and are brought about by localized differences in atmospheric heating and cooling. Mountain and valley winds, drainage winds, and chinooks, described below, are also usually classed as local winds because their effects are relatively confined.

Mountain and *valley winds* alternate in direction in a similar way as land and sea breezes. During the morning, as valley slopes begin to heat up, a convection loop forms across the valley (Figure 5.15).

The air rises up the valley sides, spreads across the valley, and sinks back to the surface near the valley's centre. The descending air is warmed by compression, which can help disperse mists and fogs that might have formed in the lowest terrain during the night (Figure 5.16).

Later in the morning as the valley floor gets warmer, a longitudinal flow component develops that moves air slowly along the valley toward its head. This is the valley wind. In the evening the higher slopes begin to cool by radiation, which, in turn, chills

5.13 Sea and land breezes The contrast between the land surface, which can heat and cool rapidly each day, and the ocean surface, which has a more uniform temperature, induces pressure gradients to create sea and land breezes.

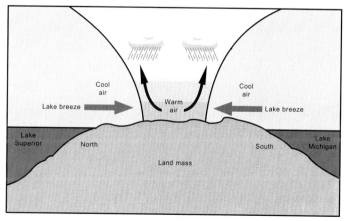

5.14 Shower development in Great Lakes region Convergence of moist air brought onshore by lake breezes rises, cools and condenses to form rainshowers further inland.

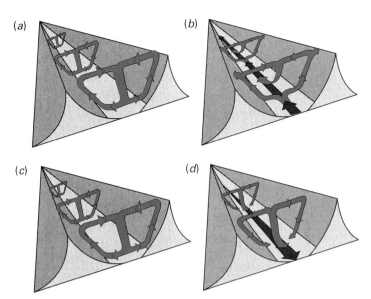

(a)

(b)

(c)

(d)

5.15 Mountain and valley winds During the day, air rises over the heated valley slopes (a) and will flow towards the head of the valley (b). At night as the air cools the circulation reverses (c) and a down-valley flow commences (d).

5.16 Valley mist Mists and fog which form as cold air settles in a valley floor begin to dissipate within a few hours of sunrise. Clearing usually commences in the centre of the valley where local convection currents descend and the air is warmed adiabatically.

the air. This reverses the convection loop and the cool dense air begins to move down into the valley. Air movement is relatively gentle, and small undulations, even bands of forest, can temporarily dam the flow. This may induce a pulsing flow to the mountain winds as they move down the slope. During the night the coldest air will settle at the lowest elevations, and by morning a temperature inversion often develops. The gentle air currents up and down slopes that are associated with mountain and valley winds help insects move into alpine meadows during the day, where they feed and pollinate the plants before returning to more hospitable sites at night.

Drainage, or *katabatic*, *winds* are similar to mountain winds, except that the air flows under the influence of gravity from higher to lower regions. They typically develop over high plateaus and snowfields; this accentuates cooling and creates reservoirs of cold, dense air. The cold air eventually spills over low divides or through passes, down into adjacent lowlands as a strong, cold, dry wind. Drainage winds occur in many of the world's mountainous regions and go by various local names. The *mistral* of the Rhône Valley in southern France is a well-known example. In extreme environments, such as on the ice sheets of Greenland and Antarctica, powerful drainage winds move across the ice surface and are funnelled through coastal valleys. Picking up loose snow, these winds produce blizzard-like conditions that last for days at a time.

A *Santa Ana* is a type of local wind that occurs when the outward flow of dry air from an anticyclone is combined with the local effects of mountainous terrain. The Santa Ana is a hot, dry easterly wind that sometimes blows from the interior desert region of southern California across coastal mountain ranges to reach the Pacific coast. This wind is funnelled through local

mountain gaps or canyon floors, where it gains great force, with speeds frequently exceeding 80 kilometres per hour. At times the Santa Ana wind carries large amounts of dust. Because this wind is dry, hot, and strong, it can easily fan wildfires in brush or forest out of control (Figure 5.17).

A *Bora* is a cold, dry wind that brings unseasonable conditions to the coastal regions of the Adriatic Sea. The outflow of

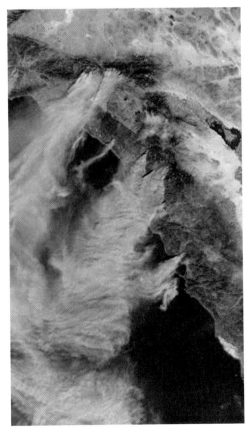

5.17 Santa Ana. This satellite image shows smoke from brush fires in being carried offshore by the Santa Ana winds that blow out from a high pressure centre over California.

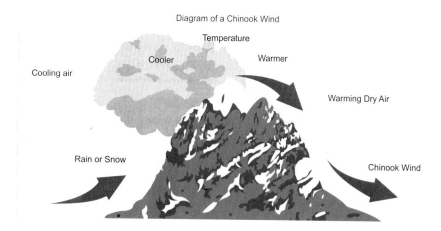

Diagram of a Chinook Wind

5.18 Chinook winds. Chinook winds can form when moist air is forced over mountain ranges. If the air loses moisture through condensation, it will warm at a faster rate as it descends on the lee side creating the warm, dry Chinook winds that commonly develop in winter in southern Alberta.

air is associated with a polar high-pressure area positioned over the snow-covered mountains of the interior plateau of Croatia. Thus the effect is most pronounced in winter. Speeds in excess of 200 kilometres per hour have been recorded during these events.

A *chinook* is a warm wind that results when air currents pass over a mountain range and descend on the sheltered side (Figure 5.18). The descending air is warmed and dried, which can lead to a dramatic rise in temperature. The effect is most pronounced in the winter and can lead to rapid sublimation of snow.

In Canada, chinooks occur most frequently in southern Alberta (Figure 5.19), but the effects are occasionally felt in Saskatchewan. One of the best examples of a chinook in Canada occurred in 1962 at Pincher Creek, Alberta, where temperatures rose by 33°C in one hour; in 1966, also in Pincher Creek, a 21°C rise in temperature in four minutes was recorded. Lethbridge, Alberta, has experienced chinook winds of hurricane force; exceptional gusts of 170 kilometres per hour were reported in 1962. In Europe, a similar wind experienced in the Alps is termed the *foehn*; and in Argentina it is known as the *zonda*.

Surface winds are generated when pressure gradient forces cause air to move. Energy of motion—kinetic energy—is stored in the air motion. The stored energy can be extracted by windmills or wind turbines to do other work, such as generate electric power. *Eye on the Environment 5.3 • Wind Power, Wave Power, and Current Power* describes this source of energy as well as those of waves and ocean currents.

WINDS IN THE UPPER ATMOSPHERE

How does air move at the higher levels of the troposphere? And how do pressure gradients arise at upper levels? Recall that pressure decreases less rapidly with height in warmer air than in

colder air (Figure 5.1). This principle is derived from fundamental physical laws describing the temperature, density, and pressure relationships within a thick layer of gas. Also recall that the Earth's insolation is greatest near the equator and least near the poles. Thus, there is a temperature gradient from the equator to the poles.

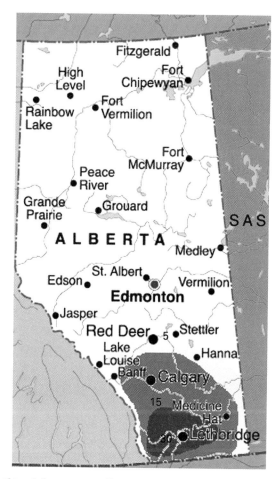

5.19 Chinook frequency in Alberta. Southern Alberta normally experiences about 30 chinooks each year. Few chinooks occur north of Red Deer and only rarely are they experienced as far east as Saskatchewan.

EYE ON THE ENVIRONMENT | 5.3   *Wind Power, Wave Power, and Current Power*

Wind power is an indirect form of solar energy that has been used for centuries. In the low countries of Europe (like the Netherlands), the windmill played a major role in pumping water from the polders as they were reclaimed from tidal land. In these low, flat areas, streams could not be adapted to waterpower, and the windmill was also used to grind grain.

The design of new forms of windmills has intrigued inventors for many decades. The total supply of wind energy is enormous. The World Meteorological Organization has estimated that the combined electrical-generating wind power of favourable sites throughout the world totals about 20 million megawatts.

Wind turbines, each with a generating capacity ranging from a few hundred kilowatts up to 4.2 megawatts for large-scale production facilities, have been assembled in groups at favourable locations to form "wind farms." Arranged in rows along ridge crests, the turbines intercept local winds of exceptional frequency and strength. Currently in Canada wind generators have the capacity to produce about 1,200 megawatts of electricity. One of the first installations was at Pincher Creek, Alberta. It began operating in 1993 with the construction of a 19-megawatt wind farm that captured the persistent wind flow descending from the Rocky Mountains. There are now more than 100 wind turbines in the area. Alberta, which generates about 350 megawatts of electricity, leads Canada in wind power capacity. Ontario generates about 300 megawatts, Quebec about 200 megawatts, and Saskatchewan about 175 megawatts. The largest wind farm in Canada is located near Swift Current, Saskatchewan, with a capacity of 149.4 megawatts. However, this will be exceeded by the Prince Project (190 megawatts) in Ontario when its second phase is completed.

The figure to the right shows the potential for wind energy generation in Canada. The majority of wind turbines currently in use are located at sites where the average wind speed is at least six metres per second (22 kilometres per hour). Wind energy has become competitive with conventional sources of energy, and government support for wind

Modern windmills at the Cap Chat, Quebec windfarm

Figure 5.20 shows a simple generalized cross section of the atmosphere from the pole on the left (*X*) to about 30° latitude on the right (*Y*). Because of temperature differences, the isobaric surfaces slope downward from the low latitudes to the pole. This creates a pressure gradient force. For example, at height H_1, the pressure at *X* is about 94 kilopascals, while at *Y* it is about 95.5 kilopascals, creating a pressure gradient of 1.5 kilopascals. However, the pressure difference becomes greater with elevation. At H_2, the polar pressure is about 82.5 kilopascals, while the low-latitude pressure is about 89 kilopascals. Here the pressure gradient is 6.5 kilopascals. At H_3, the gradient is about 10 kilopascals. These values are just examples, but they show that the pressure gradient force increases with altitude, with accompanying strong winds.

THE GEOSTROPHIC WIND

Given that a pressure gradient force exists pushing air toward the poles, how will this force produce wind and what will be the wind direction? Recall that any wind motion is subjected to the Coriolis effect, which will turn it to the right in the northern hemisphere and to the left in the southern hemisphere. Thus, air motion toward the poles will be to the east, creating west winds in both hemispheres.

power continues to increase. In Canada, wind energy is promoted through the Wind Power Production Incentive (WPPI), a program designed to reduce greenhouse gas emissions. Its goal is to increase wind power generation to 4,000 megawatts.

Wave energy is another indirect form of solar energy. Nearly all ocean waves are produced by wind blowing over the sea surface. Water moves in vertical orbits in waves, and energy is extracted from this motion with the use of a floating object tethered to the sea floor. As the floating mass rises and

falls, an attached mechanism drives a generator. Typically, these are pneumatic systems that use pressure changes caused by the rise and fall of the water.

Harnessing the vast power of ocean currents, such as the Gulf Stream or the Kuroshio (a similar current that flows along the east coast of Japan), is a prospect that has not gone unnoticed. One plan proposes tethering a large number of current-driven turbines to the ocean floor. Each turbine would operate below the ocean surface. They would be 170 metres in diameter and capable of generating 83 megawatts of power.

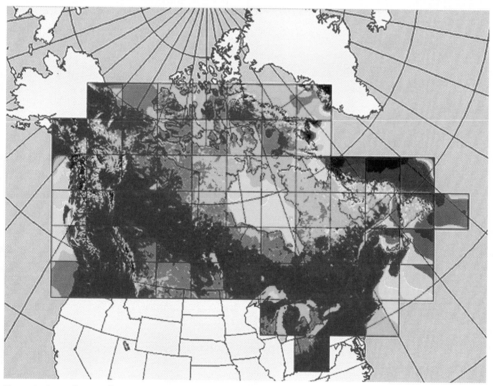

Numerical simulation of potential wind energy in Canada

Figure 5.21 (*a*) shows how the Coriolis effect acts on an air parcel at high altitudes. Compared with Figure 5.5, which shows the forces affecting air moving close to the surface, the force of friction is absent in the upper atmosphere because the Earth's surface has no effect on the air. Thus, only two forces act on the air parcel—the pressure gradient force and the Coriolis effect. Imagine a still parcel of air that begins to move toward the poles in response to the pressure gradient force (*b*). At first it travels in the direction of the pressure gradient, but as it accelerates, the Coriolis effect pulls it increasingly toward the right. As its velocity increases, the parcel turns increasingly to the right until the

Coriolis effect balances the gradient force. At that point, the sum of forces on the parcel is zero, so its speed and direction remain constant. The *term geostrophic* wind is used to identify this type of airflow. It occurs at upper levels in the atmosphere and is parallel to the isobars.

Figure 5.22 shows an upper-air map of North America in late June. The contours show the height of a 50-kilopascal pressure level, which dips down toward the Earth's surface where the upper-air pressure is lower. Similarly, the pressure level rises up when the upper-air pressure is higher. Thus low altitude indicates low pressure, and high altitude indicates high pressure.

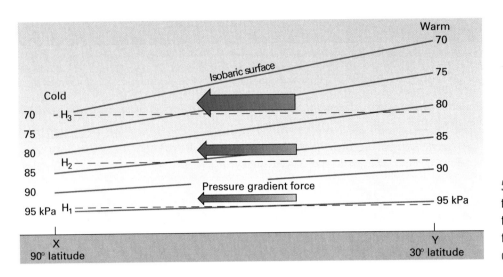

5.20 Upper-air pressure gradient Because the atmosphere is warmer near the equator than at the poles, a pressure gradient force acts to push air toward the poles. The gradient force increases with altitude.

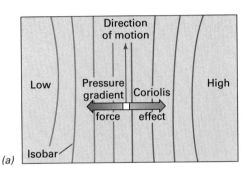

The map shows a large, upper-air low over Baffin Island. Low pressure in the upper atmosphere indicates cold air, as would be expected at this latitude. The upper-air low centred over the Great Lakes is a mass of cold, Arctic air that has moved southward to dominate the eastern part of the continent. A large, but weaker, upper-air high is centred over the southwestern desert. This will be a mass of warm air, heated by intense surface insolation on the desert below.

The winds generally follow the contours of the 50-kilopascal surface, showing geostrophic flow. Wind speeds are faster where the contours are closest and the pressure gradient force is strongest. The winds around the upper-air low tend to spiral inward and converge on the low before descending to the surface. Similarly, the winds around the upper-air high spiral outward and diverge away from the centre. This outflow is fed by the inward and upward spiral of winds from a surface low pressure system.

GLOBAL CIRCULATION AT UPPER LEVELS

The general pattern of airflows at higher levels in the troposphere comprised four major features—weak equatorial easterlies, tropical high-pressure belts, upper-air westerlies, and a polar low (Figure 5.23).

The pressure gradient force that generates westerly winds in the upper atmosphere results from the permanent temperature differences between the tropics and the poles. The *upper-air westerlies* blow in a complete circuit around the Earth, from about 25° lat. almost to the poles. At high latitudes the westerlies form a huge circumpolar spiral, circling a great polar low-pressure centre. Toward lower latitudes, atmospheric pressure rises steadily,

5.21 Geostrophic wind (a) At upper levels in the atmosphere, a parcel of air is subjected to a pressure gradient force and the Coriolis effect. (b) As a parcel of air moves in response to a pressure gradient, it is turned progressively sideways until the gradient force and Coriolis effect balance, producing the geostrophic wind.

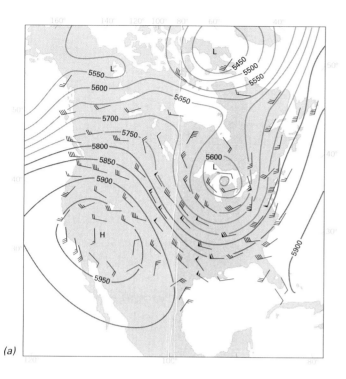

(a)

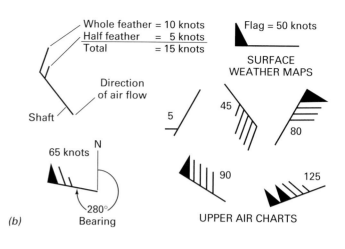

(b)

5.22 **Upper-air wind map** (a) An upper-air map for a late June day. Lines are height contours (metres) for the 50-kilopascal surface. (b) Explanation of wind arrows. 1 knot (nautical mile per hour) = 0.514 metres per second.

forming a *tropical high-pressure belt* at 15° to 20° N and S lat. This is the zone of high-altitude divergence associated with the rising limbs of the Hadley cells. Between the high-pressure ridges are zones of lower pressure in which the winds are light but generally easterly. These winds are called the *equatorial easterlies*.

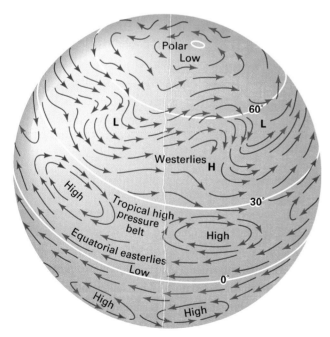

5.23 **Global upper-level winds** Global circulation in the upper troposphere at mid- and high latitudes is dominated by strong west winds. These westerly winds sweep to the north or south around centres of high and low pressure in the upper atmosphere. In the equatorial region, a weak easterly wind pattern prevails. (Copyright © A. N. Strahler)

ROSSBY WAVES, JET STREAMS, AND THE POLAR FRONT

The path of the upper-air westerlies periodically experiences undulations, called **Rossby waves** (Figure 5.24). The number of Rossby waves ranges from three to seven. They form in a contact zone between cold polar air and warm tropical air, called the *polar front*. Although the polar front is a common feature on surface weather maps (see Chapter 6), this boundary zone extends from the Earth's surface to the upper troposphere.

For several days or weeks, the flow may be fairly smooth. Then an undulation develops, and warm air pushes toward the poles while a tongue of cold air is brought to the south. Eventually, the tongue is pinched off, leaving a pool of cold air at a latitude far south of its point of origin. This cold pool may persist for some days or weeks. Because of its cold centre, it will show low pressure and convergence higher in the atmosphere. Air will descend in its core and diverge at the surface, creating a surface high pressure. Conversely, a high-pressure warm air pool will diverge in the upper atmosphere because it is fed by rising air that has converged in a low-pressure system at the surface.

Because the Rossby wave circulation brings warm air toward the poles and cold air toward the equator, it is a primary mechanism to transport heat toward the poles. It is also the reason weather in the midlatitudes is often so variable, as pools of warm, moist air and cold, dry air alternately invade midlatitude land masses. In addition, upper air flow around the Rossby waves helps to direct the movement of surface pressure systems. Of particular significance is the creation of an *omega block*, which can occur

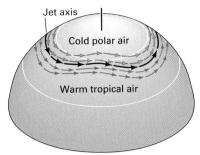

The jet stream begins to undulate.

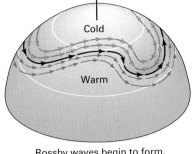

Rossby waves begin to form.

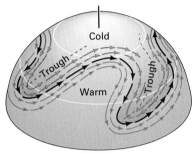

Waves are strongly developed. The cold air occupies troughs of low pressure.

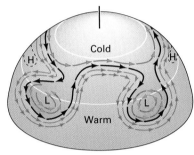

When the waves are pinched off, they form cyclones of cold air.

5.24 Rossby wave cycle Rossby waves form in the upper-air westerlies of the northern and southern hemisphere at the boundary between cold polar air and warm tropical air. This cycle shows the formation of large waves in the northern hemisphere that are pinched off, leaving pools of polar air at lower latitudes. (Copyright © A. N. Strahler)

when the Rossby waves develop pronounced lobes. As warm air pushes into higher latitudes, it creates a well-defined high-pressure region at the surface, which forces surface winds to blow around it. Persistent omega blocks, which can remain stationary for several weeks, may cause drought conditions in the Canadian Prairies if moist airflow from the Pacific is carried north of the agricultural regions (Figure 5.25).

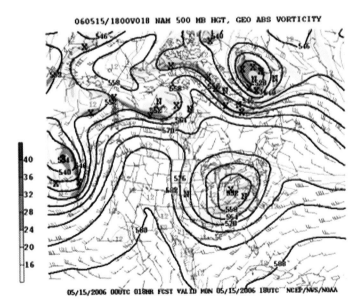

5.25 Omega block. Dry spells in the Prairies can sometimes occur when weather systems are deflected to higher latitudes around a sharply curved Rossby wave pattern such as the one developing over Idaho and Wyoming.

Air flow has a pronounced north–south—or *meridional*—component when the Rossby waves follow strongly meandering paths, such as in an omega block situation. However, at the beginning of each cycle, air flow is predominately from west to east—or *zonal*. Such variation in flow paths influences the time it takes for surface weather systems to pass through a region and will affect how quickly weather changes at midlatitudes.

Jet streams are important features of upper-air circulation. They are narrow zones at high altitude in which wind flow can reach great speeds. Jet streams occur at the approximate altitude of the tropopause (10–12 kilometres), where atmospheric pressure gradients are strong. This accounts for their development along the polar front. The polar jet streams essentially mark the boundary of the Rossby waves in the northern and southern hemisphere, so their position progressively changes over the course of a few weeks. Because upper air flow is closely linked to the movement of surface pressure systems, Canadian weather maps usually include the location of the polar front jet stream (Figure 5.26).

The meandering paths of the polar front jet stream typically lie between 35° and 65° latitude and their presence is often marked by a band of high cloud (Figure 5.27). The greatest wind speeds occur in the centre of the jet stream and range from 75 to 125 metres per second (about 270 to 450 kilometres per hour). Wind speeds are generally lower in summer, sometimes falling below 30 metres per second (100 kilometres per hour). Aircraft often use the polar jet stream to increase ground speed and reduce fuel consumption when flying eastward, say from Toronto to London, England. In the westward direction, flight paths are chosen to avoid the strong headwinds.

As well as the two westerly polar jet streams, there are two smaller *subtropical jet streams* closer to the equator (Figure 5.28). These occupy a position at the tropopause above the subtropical high-pressure cells at about 20° to 40° latitude in the northern and southern hemispheres. Here, westerly wind speeds reach maximum values of 100 to 110 metres per second (about 360 to 400 kilometres per hour). This jet stream is associated with the increase in velocity induced by conservation of angular momentum, which occurs as an air parcel moves toward the poles from the equator. The quantity of angular momentum of an air parcel rotating at the equator is based on the product of the turning rate of the Earth (approximately 1,700 kilometres per hour at the equator, dropping to 0 at the poles) and the distance between the parcel and the axis of rotation (the tropopause is about 6,400 kilometres away from the Earth's axis at the equator, compared with about 10 kilometres at the poles). As the parcel moves poleward, it gets closer to the Earth's axis of rotation, and by conservation of angular momentum, its turning rate must increase—in the same way that ice skaters increase their spinning speed by drawing their arms and legs inward. For air parcels, the effect is to increase their eastward velocity. The result is a flow of swiftly moving air that feeds into the descending currents over the subtropical highs.

5.27 Polar jet stream The course of the jet stream is indicated by the band of high cloud as it travels eastwards over Newfoundland where some winter snow cover and ice floes can be seen.

A third type of jet stream is found at even lower latitudes. Known as the *tropical easterly jet stream*, it runs from east to west—opposite in direction to that of the polar front and subtropical jet streams. The tropical easterly jet stream occurs only in the summer and is limited to the northern hemisphere over Southeast Asia, India, and Africa.

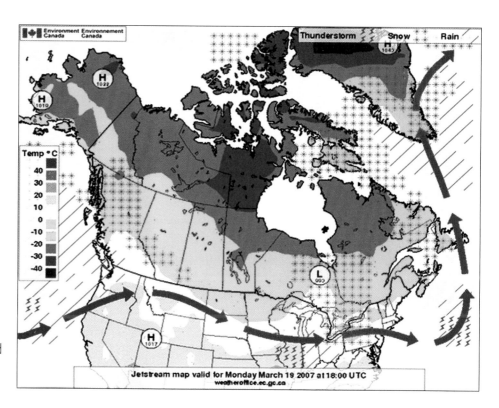

5.26 Environment Canada's "Weather at a Glance" map series. The position of the jet stream is marked by red arrows on the daily weather maps produced by Environment Canada. They can be accessed at http://weatheroffice.ec.gc.ca/jet_stream/index_e.html

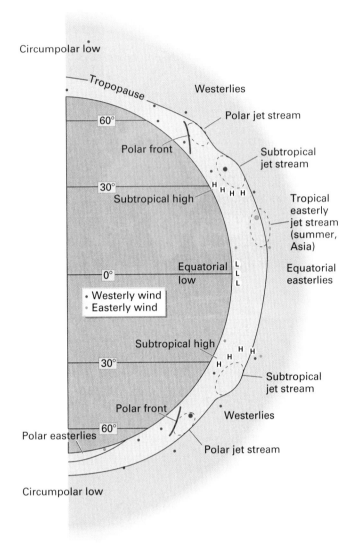

5.28 Upper-level circulation cross section A schematic diagram of wind directions and jet streams along an average meridian from pole to pole. The polar and subtropical jets are westerly in direction, in contrast to the tropical easterly jet that is confined to the northern hemisphere. (Copyright © A. N. Strahler)

TEMPERATURE LAYERS OF THE OCEAN

The ocean, like the atmosphere, has a layered structure. Ocean layers are recognized in terms of temperature, which generally are highest at the surface and decrease with depth. This trend is not surprising, since the heat sources that warm the ocean are solar insolation and heat supplied by the overlying atmosphere, both of which act at or near the surface.

Figure 5.29 shows the ocean's layered temperature structure. At low latitudes throughout the year and in midlatitudes in the summer, a warm surface layer develops. Here wave action mixes heated surface water with the water below it to form a warm layer

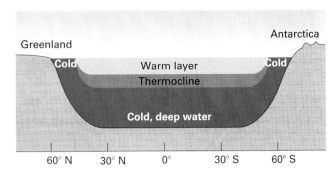

5.29 Ocean temperature structure A schematic north–south cross section of the world's oceans shows that the warm surface water layer disappears in Arctic and Antarctic latitudes, where very cold water lies at the surface. The thickness of the warm layer and the thermocline is greatly exaggerated.

that may be as thick as 500 metres with a temperature of 20° to 25° C in oceans near the equator. Below the warm layer, temperatures drop rapidly in a zone known as the *thermocline*. Very cold water extends from the thermocline to the deep ocean floor. Temperatures near the deep ocean floor range from 0° C to 5° C. In the Arctic and Antarctic regions, the warm layer and thermocline are absent.

OCEAN CURRENTS

Just as the atmosphere has a circulation pattern, so too do the oceans. An *ocean current* is any persistent, dominantly horizontal flow of ocean water, which can occur at the surface or at any depth. Surface currents are driven by prevailing winds. Deep currents are powered by changes in temperature and density occurring in surface waters, causing them to sink. Current systems act to exchange heat between low and high latitudes and are essential in sustaining the global energy balance.

SURFACE CURRENTS

The patterns of surface ocean currents are strongly related to prevailing surface winds (Figure 5.30). Energy is transferred from wind to water by the friction of the air blowing over the water surface. Because of the Coriolis effect, the actual direction of water drift is deflected about 45° from the direction of the driving wind. Ocean currents move warm waters toward the poles and cold waters toward the equator, and consequently, they are important regulators of air temperatures.

The circulation includes large circular movements, called **gyres**, which are centred approximately at latitudes 30°. These gyres track the movements of air around the subtropical high-pressure cells. An *equatorial current* with westward flow marks the belt of the trade winds. Although the trade winds blow to the southwest and northwest, at an angle across the parallels of latitude, the water movement follows the parallels. As these equato-

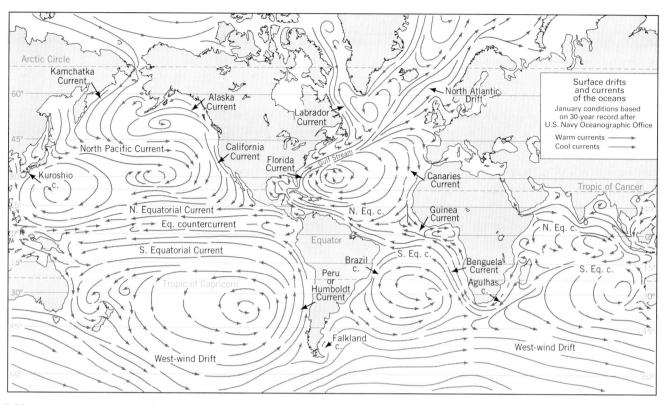

5.30 January ocean currents Surface drifts and currents of the oceans in January. (Based on data from U.S. Navy Oceanographic Office. Redrawn and revised by A. N. Strahler.)

rial currents approach land, they are turned poleward along the west sides of the oceans, forming warm currents paralleling the coast. Examples are the Kuroshio Current of Japan and the North Atlantic Drift of eastern North America. The North Atlantic Drift transports warm water from the Gulf of Mexico to Europe. The Russian port of Murmansk, on the Arctic Circle, remains ice-free year round because of this influx of warm water.

The equatorial currents are separated by an equatorial counter-current. A slow, eastward movement of water over the zone of the westerlies is named the *west-wind drift*. It covers a broad belt between lat. 35° and 45° in the northern hemisphere and between lat. 30° and 60° in the southern hemisphere. As the slow, eastward-moving waters of the *west-wind drift* approach the western sides of the continents, they are deflected toward the equator along the coast as cool currents.

These currents are often accompanied by *upwelling* along continental margins. In this process, colder water from greater depths rises to the surface. Examples of upwelling are the Peru (or Humboldt) Current, off the coast of Chile and Peru; the Benguela Current, off the coast of southern Africa; and the California Current. Upwelling is important for returning nutrients to the illuminated waters at the ocean surface, where phytoplankton can use them to sustain the marine food cycle. In addition, fogs brought

onshore from the Peru and Benguela currents provide much needed moisture to the deserts of Chile and Namibia.

The west-wind drifts also move water toward the poles to join Arctic and Antarctic circulations. Such is the case in the northeastern Atlantic Ocean, where the North Atlantic Drift spreads around the British Isles, into the North Sea, and along the Norwegian coast. In the northern hemisphere, where the polar sea is largely landlocked, cold currents flow toward the equator along the east sides of continents. For example, the Labrador Current flows south along the east coast of Canada. The mixing of cool air associated with the Labrador Current with air warmed by the North Atlantic Drift produces frequent fogs over the Grand Banks off Newfoundland (see Chapter 6). In addition, the Labrador Current carries icebergs from Greenland into eastern coastal waters until early summer (Figure 5.31).

Figure 5.32 shows a satellite image of ocean temperatures along the east coast of North America for a week in April. The Gulf Stream stands out as a tongue of red and yellow (warm) colour, moving along the coast and heading in a northeasterly direction. Cooler water from the Labrador Current, in green and blue tones, hugs the northern Atlantic coast. This current heads south and then turns to follow the Gulf Stream to the northeast. Instead of mixing, the two flows remain quite distinct. The

5.31 Icebergs off of Newfoundland. Icebergs in eastern Canadian waters mostly originate from the glaciers of West Greenland where more than 30,000 are produced each year. They usually are seen from March until July, but are most abundant in late April.

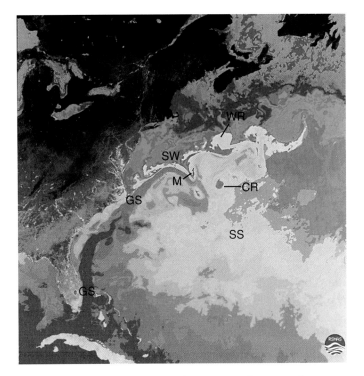

5.32 Sea-surface temperatures in the western Atlantic A satellite image showing sea-surface temperature for a week in April from data acquired by the NOAA-7 orbiting satellite. Cold water appears in green and blue tones; warm water in red and yellow tones. The image shows the Gulf Stream (GS) and its interactions with cold water of the Continental Slope (SW), brought down by the Labrador Current, and the warm water of the Sargasso Sea (SS). Other features include a meander (M), a warm-core ring (WR), and a cold-core ring (CR). The image was made from data collected by the NOAA-7 orbiting satellite.

boundary between them shows a wavelike flow, much like Rossby waves in the atmosphere. Warm and cold bodies of water are cut off to float freely, forming warm-core and cold-core rings.

One of the most dramatic phenomena associated with ocean surface currents is *El Niño*. During an El Niño event, Pacific surface currents shift into an unusual pattern. Pacific upwelling along the Peruvian coast ceases, trade winds weaken, and a weak equatorial eastward current develops. Global patterns of precipitation also change during El Niño events, bringing floods to some regions and droughts to others. In contrast to El Niño is *La Niña*, in which normal Peruvian coastal upwelling is enhanced, trade winds strengthen, and cool water is carried far westward in an equatorial plume.

Figure 5.33 shows two satellite images of sea-surface temperature observed during El Niño and La Niña years. During La Niña conditions (left), cold, upwelling water (dark green) is brought to the surface along the Peruvian coast. It is moved northward by the Peruvian Current, and then carried westward into the Pacific by the South Equatorial Current. During an El Niño year (right), the eastward motion of warm water acts as a barrier to the Peruvian Current. Some slight upwelling occurs (yellow), but the amount is greatly reduced compared with La Niña years. *Eye on Global Change 5.4 • El Niño* provides a fuller description of the El Niño–La Niña phenomenon.

DEEP CURRENTS AND THERMOHALINE CIRCULATION

Deep currents move water in a slow circuit across the floors of the world's oceans. They are generated when surface waters become denser and slowly sink downward. Coupled with these deep currents are broad and slow surface currents on which the more rapid surface currents described above are superimposed. This slow flow pattern, which links all of the world's oceans, is commonly referred to as the *great ocean conveyor* (Figure 5.34). It is a *thermohaline circulation* process that depends on the change in density of sea water caused by temperature and salinity differences in North Atlantic Ocean waters.

Warm Atlantic surface water slowly moves northward through the equatorial and tropical zones. As this surface layer is warmed, evaporation occurs and the layer becomes saltier, slightly increasing its density. As the water reaches higher latitudes, it loses heat to the atmosphere, and so it becomes colder and still denser. Eventually, along the northern boundary of the North Atlantic, the surface layer becomes dense enough to sink.

Carried along the bottom of the North and South Atlantic oceans, the cold, dense water eventually reaches the Southern Ocean. Here, upwelling and mixing occur, and the deep waters are brought to the surface of the Indian and southern Pacific oceans. A coupled circulation loop moves surface water from the Pacific

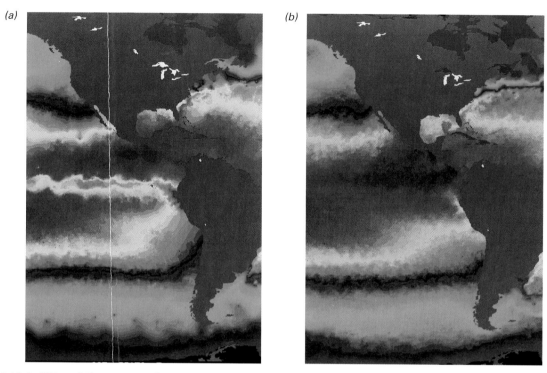

(a)   *(b)*

5.33 La Niña and El Niño sea-surface temperature These striking images shows sea-surface temperatures in the eastern tropical Pacific during (a) La Niña and (b) El Niño years as mesured by the NOAA-7 satellite. Green tones indicate cooler temperatures, while red tones indicate warmer temperatures, (Otis B. Brown/Univ. of Miami and Gene Carl Feldman, NASA/GSFC.)

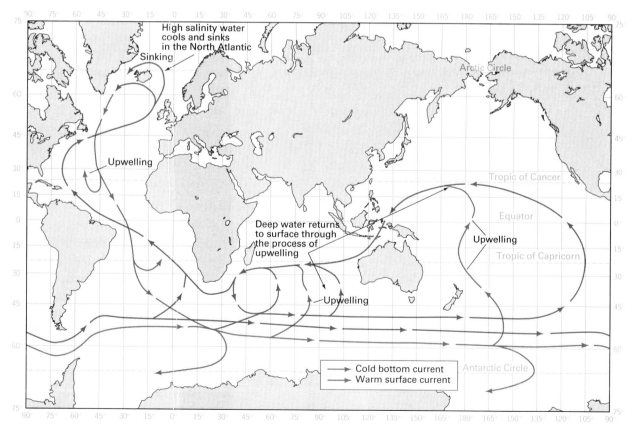

5.34 Ocean circulation Deep ocean currents, generated by the sinking of cold, salty water in the northern Atlantic, circulate sea water in slowly moving coupled loops involving the Atlantic, Pacific, Indian, and Southern oceans. (After A. J. Gordon, Nature 382:399–400, August 1996. Used with permission.)

EYE ON GLOBAL CHANGE | 5.4   El Niño

At intervals of about three to eight years, a remarkable disturbance of ocean and atmosphere occurs. It begins in the eastern Pacific Ocean and spreads its effects widely over the globe. This disturbance lasts more than a year, bringing droughts, heavy rainfalls, severe spells of heat and cold, or a high incidence of cyclonic storms to various parts of the Pacific and its eastern coasts. This phenomenon is called *El Niño*. The expression comes from Peruvian fishermen, who refer to the *Corriente del Niño*, or the "Current of the Christ Child," in describing an invasion of warm surface water that occurs once every few years around Christmas time, greatly depleting their catch of fish. El Niño occurs at irregular intervals and with varying degrees of intensity. Notable El Niño events occurred in 1891, 1925, 1940–41, 1965, 1972–73, 1982–83, 1989–90, 1991–92, 1994–95, 1997–98, and 2002–03.

Normally, the cool Peru (Humboldt) Current flows northward off the South American coast, then near the equator it turns westward across the Pacific as the South Equatorial Current (see Figure 5.30). The Peru Current is fed by upwelling of cold, deep water, bringing with it nutrients that serve as food for marine life. With the onset of El Niño, upwelling ceases, the cool water is replaced by warm, sterile water from the west, and the abundant marine life disappears.

In an El Niño year, a major change in barometric pressure occurs across the entire stretch of the equatorial zone as far west as southeastern Asia. Normally, high pressure prevails in the eastern Pacific, with low pressure over northern Australia, the East Indies, and New Guinea, where the largest and warmest body of ocean water can be found (see part a in figure to the right). Abundant rainfall occurs in this area during December, which is the high-sun season in the southern hemisphere.

During an El Niño event, the low-pressure system over the western Pacific is less intense and drought conditions replace the heavy rains. Air pressure drops in the equatorial zone of the mid-Pacific, and the high-pressure region in the eastern Pacific weakens. Heavy rainfall accompanies the strengthening of the equatorial trough in the central and eastern Pacific (part b). The shift in barometric pressure patterns is known as the *Southern Oscillation* (part c). The Southern Oscillation is calculated from the monthly differences in air pressure between Tahiti and Darwin, as follows:

$$SOI = 10 \frac{[\text{Pdiff} - \text{Pdiffav}]}{SD(\text{Pdiff})}$$

where

Pdiff = (average Tahiti MSLP for the month)-(average Darwin MSLP for the month)

Pdiffav = long-term average of Pdiff for the month in question

SD(Pdiff) = long-term standard deviation of Pdiff for the month in question

The multiplication by 10 is a convention. Using this convention, the Southern Oscillation ranges from about −35 to about +35, and the value can be quoted as a whole number. The Southern Oscillation is usually computed on a monthly basis.

Surface winds and currents also change with this change in pressure. During normal conditions, the strong prevailing trade winds blow westward, causing warm ocean water to move to the western Pacific and "pile up" near the western equatorial low. The characteristic airflow driven westward by the pressure gradient, known as the Walker circulation, results in the accumulation of warm ocean water in the western Pacific (see figure to the right). The ocean surface is some 60 centimetres higher in the western Pacific as a result of this motion. This westward motion causes the normal upwelling along the South American coast, as bottom water is carried up to replace the water dragged to the west.

During an El Niño event, the easterly trade winds die with the change in atmospheric pressure. A weak westerly wind flow sometimes occurs, completely reversing the normal wind direction. Without the pressure of the trade winds to hold them back, warm waters surge eastward. Sea-surface temperatures and sea level rise off the tropical western coasts of the Americas.

The major change in sea-surface temperatures that accompanies an El Niño can also shift weather patterns across large regions of the globe. Recurring, large-scale weather anomalies caused by changes in pressure and circulation patterns are termed teleconnections (see figures to the right). During El Niño events torrential rains bring relief to the arid coastal regions of South America, and drier conditions prevail in Australia and the East Indies. Above normal rainfall is also reported in east Africa. In North America, winter storms are generally more intense along the Pacific coast, but winters are milder and snowfalls lighter across the Prairies.

A somewhat rarer phenomenon also capable of altering global weather patterns is La Niña (the girl child), a condition roughly opposite to El Niño. During a La Niña period, sea-surface temperatures in the central and western Pacific Ocean fall to lower than average levels. This happens because the South Pacific subtropical high becomes strongly developed during the high-sun season. The result is abnormally strong southeast trade winds. The force of these winds drags a more-than-normal amount of warm surface water westward, which enhances upwelling along western continental coasts.

El Niño and La Niña show how dynamic the Earth really is. As grand-scale, global phenomena, El Niño and La Niña illustrate how the circulation patterns of the ocean and atmosphere are coupled by energy exchange and interact to provide teleconnections capable of producing extreme events affecting millions of people throughout the world.

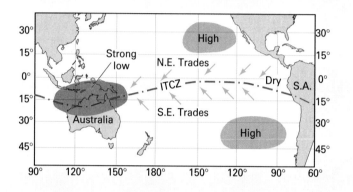

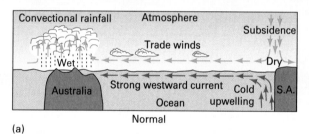

(a)

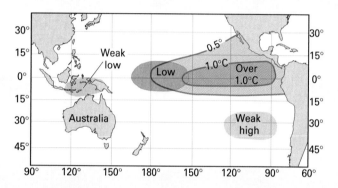

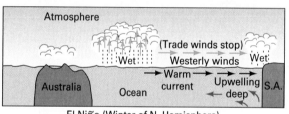

El Niño (Winter of N. Hemisphere)

(b)

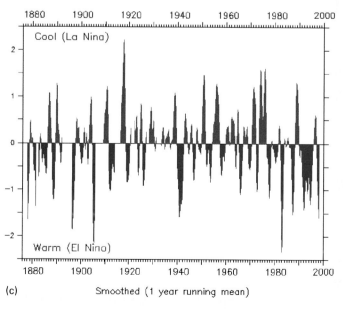

(c) Smoothed (1 year running mean)

El Niño maps The above maps show pressure conditions in the tropical Pacific and eastern Indian Ocean in November during normal and El Niño years. In a normal year (a), low pressure dominates in Malaysia and northern Australia. In an El Niño year (b), low pressure moves eastward to the central part of the western Pacific, and sea-surface temperatures become warmer in the eastern central Pacific. This shift in barometric pressure (c) is termed the Southern Oscillation. (Parts (a) and (b), copyright © A. N. Strahler)

(Continued)

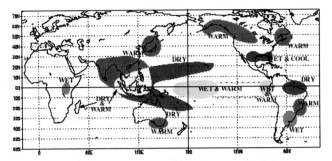

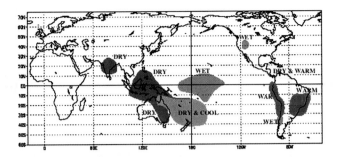

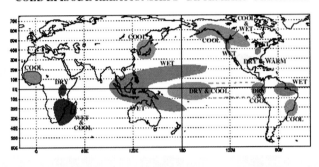

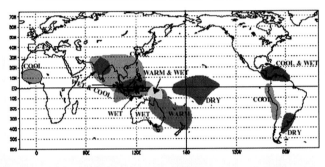

(a) El Niño conditions

(b) La Niña conditions

Weather anomalies associated with El Nino and La Nina events (teleconnections)

through the Indonesian seas, past Australia, and into the Indian Ocean. The flow continues around the southern tip of Africa, and enters the South Atlantic to complete the circuit.

Thermohaline circulation plays an important role in the carbon cycle by moving CO_2-rich surface waters into the ocean depths. Deep ocean circulation provides a system for storage and release of CO_2 in a cycle lasting about 1,500 years. (See Chapter 4, *Eye on Global Change 4.4 • Carbon Dioxide—On the Increase.*) This allows the ocean to moderate rapid changes in atmospheric CO_2 concentration, such as those produced by fossil fuel burning.

Some scientists have observed that inputs of fresh water into the North Atlantic could slow or stop thermohaline circulation. The fresh water would decrease the density of the ocean water, preventing it from sinking. Without sinking, the circulation would stop. In turn, this would interrupt a major flow pathway for the transfer of heat from equatorial regions to the northern midlatitudes. Inputs of fresh water from the sudden drainage of large meltwater lakes at the end of the Cenozoic Ice Age (see Chapter 19) may explain the periodic cycles of warm and cold temperatures noted over the past 12,000 years.

A Look Ahead

The energy flows associated with atmospheric pressure conditions and winds are closely linked to global precipitation patterns. Zones of convergence create rising air currents that characteristically result in clouds and precipitation. In areas of divergence, the skies are normally clear and dry weather prevails. Chapter 6 discusses the connection between atmospheric circulation and precipitation processes. Surface pressure systems are steered by the global wind systems, and further influenced by the movement of air upward in the atmosphere. Chapter 7 discusses how this affects the development of weather systems and storms.

CHAPTER SUMMARY

- The term *atmospheric pressure* describes the weight of air pressing on a unit of surface area. Atmospheric pressure is measured by a barometer. Atmospheric pressure decreases rapidly as altitude increases.
- Wind occurs when air moves with respect to the Earth's surface. Air motion is produced by pressure gradients that are formed when air in one location is heated to a temperature that is warmer than another. Heating creates high pressure in the upper atmosphere, which moves high-level air away from the area of heating. This motion induces low pressure at the surface, pulling surface air toward the area of heating, and forming a convection loop.
- The Earth's rotation strongly influences atmospheric circulation through the Coriolis effect. The force of the *Coriolis effect* deflects wind motion.
- Because the equatorial and tropical regions are heated more intensely than the higher latitudes, two convection loops develop—the Hadley cells. These loops drive the *northeast* and *southeast trade winds*, the convergence and lifting of air at the intertropical convergence zone (ITCZ), and the sinking and divergence of air in the subtropical high-pressure belts.
- The most persistent features of the global pattern of atmospheric pressure are the subtropical high-pressure belts, which are generated by the Hadley cell circulation. They intensify during their high-sun season.
- The monsoon circulation of Asia responds to a reversal in atmospheric pressure over the continent with the seasons. A *winter monsoon* flow of cool, dry air from the northeast alternates with a *summer monsoon* flow of warm, moist air from the southwest.
- In the midlatitudes and toward the poles, westerly winds prevail. In winter, continents develop high pressure, and intense oceanic low-pressure centres are found off the Aleutian Islands and near Iceland in the northern hemisphere. In the summer, the continents develop low pressure as oceanic subtropical high-pressure cells intensify and move toward the poles.
- Cyclones are centres of low pressure in which surface airflow spirals inwards. They represent regions of *convergence*. Anticyclones are centres of high pressure; they are regions of divergence where surface airflow spirals outwards.
- *Sea* and *land breezes* are examples of *convection loops* formed from unequal heating and cooling of the land surface compared with a nearby water surface.
- *Local winds* are generated by local pressure gradients. The sea and land breezes, as well as *mountain* and *valley winds*, are examples of local winds caused by local surface heating. Other local winds include *drainage winds*, *Santa Ana*, and *chinook winds*.
- Winds higher in the atmosphere are dominated by a global pressure gradient force between the tropics and pole in each hemisphere that is generated by the hemispheric temperature gradient from warm to cold. Coupled with the Coriolis effect, the gradient generates strong westerly *geostrophic winds* in the upper atmosphere. In the equatorial region, weak easterlies dominate the upper-level wind pattern.
- Rossby waves develop in the *upper-air westerlies*, bringing cold, polar air toward the equator and warmer air toward the poles. The *polar-front* and *subtropical* jet streams are concentrated westerly wind streams with high wind speeds. The *tropical easterly jet stream* is weaker and limited to Southeast Asia, India, and Africa.
- Oceans show a warm surface layer, a *thermocline*, and a deep cold layer. Near the poles, the warm layer and thermocline are absent.
- *Ocean* surface *currents* are dominated by huge gyres driven by the global surface wind pattern. *Equatorial currents* move warm water westward and then poleward along the east coasts of continents. Return flows bring cold water toward the equator along the west coasts of continents.
- *El Niño* events occur when an unusual flow of warm water in the equatorial Pacific moves eastward to the coasts of Central and South America, suppressing the normal northward flow of the Peru Current. This greatly reduces upwelling along the Peruvian coast. El Niño events normally occur on a three- to eight-year cycle and affect global patterns of precipitation in many regions.
- Slow, deep ocean currents are driven by the sinking of cold, salty water in the northern Atlantic. This *thermohaline circulation* pattern involves nearly all the Earth's ocean basins, and also acts to moderate the buildup of atmospheric CO_2 by moving CO_2-rich surface waters to ocean depths.

KEY TERMS

anticyclone	gyre	jet stream	Rossby waves
barometer	Hadley cell	kilopascal	subtropical high-pressure belts
Coriolis effect	intertropical convergence	monsoon	wind
cyclone	zone (ITCZ)	polar front	
depression	isobar	pressure gradient	

REVIEW QUESTIONS

1. Explain atmospheric pressure. Why does it occur? How is atmospheric pressure measured and in what units? What is the normal value of atmospheric pressure at sea level? How does atmospheric pressure change with altitude?

2. Describe a simple convective wind system, explaining how air motion arises from a pressure gradient force induced by heating.

3. Describe land and sea breezes. How do they illustrate the concepts of pressure gradient and convection loop?

4. What is the Coriolis effect, and why is it important? What produces it? How does it influence the motion of wind and ocean currents in the northern hemisphere? In the southern hemisphere?

5. Define cyclone and anticyclone. How does air move within each? What is the direction of circulation of each in the northern and southern hemispheres? What type of weather is associated with each and why?

6. What is the Asian monsoon? Describe the features of this circulation in summer and winter. How is the ITCZ involved? How is the monsoon circulation related to the high- and low-pressure centres that develop seasonally in Asia?

7. Compare the winter and summer patterns of high and low pressure that develop in the northern hemisphere with those that develop in the southern hemisphere.

8. What are drainage winds? What local names are applied to them?

9. How does global scale heating of the atmosphere create a pressure gradient force that increases with altitude?

10. What is the geostrophic wind, and what is its direction with respect to the pressure gradient force?

11. Describe the basic pattern of global atmospheric circulation at upper levels.

12. What are Rossby waves? Why are they important?

13. Identify five jet streams. Where do they occur? In which direction do they flow?

14. What is the general pattern of ocean surface current circulation? How is it related to global wind patterns?

15. How does thermohaline circulation induce deep ocean currents?

EYE ON THE ENVIRONMENT 5.3   Wind Power, Wave Power, and Current Power

1. Discuss wind power as a source of energy. Include indirect use in tapping waves and ocean currents for energy.

EYE ON GLOBAL CHANGE 5.4   El Niño

1. Compare the normal pattern of wind, pressure, and ocean currents in the equatorial Pacific with the pattern during an El Niño event.

2. What are some of the weather changes reported for El Niño events?

3. What is La Niña, and how does it compare with the normal pattern?

VISUALIZATION EXERCISES

1. Sketch an idealized Earth (without seasons or ocean–continent features) and its global wind system. Label the following on your sketch: equatorial trough, Hadley cell, ITCZ, northeast trades, polar easterlies, polar front, polar outbreak, southeast trades, subtropical high-pressure belts, and westerlies.

2. Draw four spiral patterns showing outward and inward flow in clockwise and counter-clockwise directions. Label each as appropriate to cyclonic or anticyclonic circulation in the northern or southern hemisphere.

ESSAY QUESTION

1. An airline pilot is planning a non-stop flight from Vancouver to Sydney, Australia. What general wind conditions can the pilot expect to find in the upper atmosphere? What jet streams will be encountered? Will they slow or speed the aircraft on its way?

2. You are planning to take a round-the-world cruise, leaving Montreal in October. Your vessel's route will take you through the Mediterranean Sea to Cairo, Egypt, in early December. Then you will pass through the Suez Canal and Red Sea to the Indian Ocean, calling at Mumbai, India, in January. From Mumbai, you will sail to Djakarta, Indonesia, and then go directly to Perth, Australia, arriving in March. Rounding the southern coast of Australia, your next port of call is Auckland, New Zealand, which you will reach in April. From Auckland, you head directly to Vancouver, your final destination, arriving in June. Describe the general wind and weather conditions you will experience on each leg of your journey.

PROBLEMS WORKING IT OUT 5.1 Pressure and Density in the Oceans and Atmosphere

1. A diving pool is five metres deep. What is the pressure on a person swimming at the bottom of the pool, including atmospheric pressure? What fraction of the pressure is due to the water, and what fraction is due to the atmosphere? Suppose the swimmer is now a deep-sea diver at a depth of 100 metres. What are the fractions in this case?

2. The Yellowhead Pass in the Canadian Rockies is at an elevation of 1,110 metres compared with the summit of nearby Mt. Robson at 3,954 metres. Assuming a sea level pressure of 101.4 kilopascals, what would you expect a barometer to read at these two locations? What percentage of sea-level pressure is each value?

3. In the course of the passage of a hurricane, a weather observer notes a change in barometric pressure from 101.6 kilopascals to 97.9 kilopascals, a difference of 3.7 kilopascals. Using the formula for atmospheric pressure with altitude, at what altitude would you expect to find a pressure of 97.9 kilopascals under normal conditions?

Find answers at the back of the book

EYE ON THE LANDSCAPE |

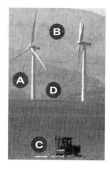

Chapter Opener Dunes and drifting Wind farm at Pincher Creek, Alberta. The wind turbines (**A**) are mounted several meters above the ground to reduce the effect of friction on air movement. The estimated average wind speed is about 9 m/s. The westerly airflow is funneled through mountainous terrain (**B**). Winter temperatures are cold but are moderated by frequent Chinooks created by the adjacent mountains. The region also lies in a rainshdow (see Chapter 6), but receives sufficient moisture for arable crops and forage (**C**), Forests of lodgepole pine and other conifers clothe the lower mountain slopes (**D**) (see Chapter 23).

Dunes and drifting snow. Unconsolidated materials, such as sand, can be readily moved by the wind to form surface features such as ripples (**A**) Small mounds (**B**) will accumulate where wind velocity is reduced in the lee of obstructions such as plants. The position of these mounds suggests the prevailing wind is from the

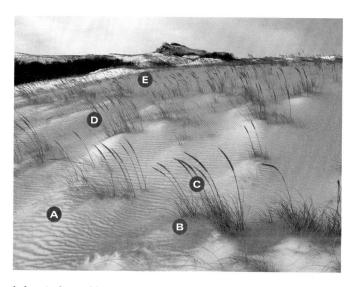

left as indicated by the curvature of the grass stems (**C**). The light covering of snow accentuates these patterns and processes with small drifts in the lee of the grass (**D**) and deeper accumulations where airflow is interrupted by shrubs (**E**).

Chapter 6

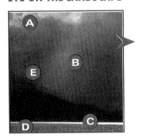

EYE ON THE LANDSCAPE

Rain over Johnson Strait, British Columbia.
What else would the geographer see? . . . Answers at the end of the chapter.

ATMOSPHERIC MOISTURE AND PRECIPITATION

Three States of Water
Working It Out 6.1 • *Energy and Latent Heat*
The Hydrosphere and the Hydrologic Cycle
Focus on Systems 6.2 • *The Global Water Balance as a Matter Flow System*
Humidity
Relative Humidity
The Adiabatic Process
Dry Adiabatic Rate
Wet (Saturated) Adiabatic Rate
Working It Out 6.3 • *The Lifting Condensation Level*
Clouds
Cloud Forms
Fog
Focus on Remote Sensing 6.4 • *Observing Clouds from GOES*
Precipitation
Precipitation Processes
Orographic Precipitation
Convergent Precipitation
Convectional Precipitation
Atmospheric Stability
Thunderstorms
Microbursts
Focus on Systems 6.5 • *The Thunderstorm as a Flow System*

This chapter focuses on water in the atmosphere, in its vapour, liquid, and solid forms. Most water enters the atmosphere by evaporation from the oceans. When water evaporates it takes up latent heat, which is later released during condensation. Thus, water plays a key role in the Earth's energy flows. Water returns to the Earth's surface in liquid or solid forms of precipitation. When it falls over land, precipitation provides the water or ice that carves the distinctive landforms that shape the continents.

Whenever air moves upward in the atmosphere, the decrease in pressure with altitude causes the air to expand and cool. Air rises as a result of four major situations:

- when winds move air over a mountain barrier;
- when unequal heating at the ground surface creates a bubble of air that is warmer and less dense than the air that surrounds it;
- when convergence takes place, which is especially important along the Intertropical Convergence Zone (ITCZ), and leads to heavy rainfall in the tropics; and
- when warm air rises over a mass of cooler, denser air in areas of low pressure.

This chapter focuses on the first three causes of air rising, or uplift. The fourth is associated with the distinctive weather systems that develop along the polar front, and is discussed in Chapter 7.

THREE STATES OF WATER

Water can exist in three states—solid (ice), liquid (water), and gas (water vapour). A change of state from solid to liquid, liquid to gas, or solid to gas requires the input of heat energy (Figure 6.1). This energy, termed *latent heat*, is drawn in from the surroundings and stored within the water molecules. The reverse changes–from liquid to solid, gas to liquid, or gas to solid–requires the release of latent heat.

Each type of transition is known by a specific name. Melting, freezing, evaporation, and condensation are all familiar terms. **Sublimation** is the direct transition from solid to vapour. A common example would be the way old ice cubes appear to shrink when stored for a long time because of the constant circulation of cold, dry air in the freezer. The Meteorological Service of Canada calls the reverse process, when water vapour crystallizes as ice, sublimation, as well; this often occurs on a winter's night and

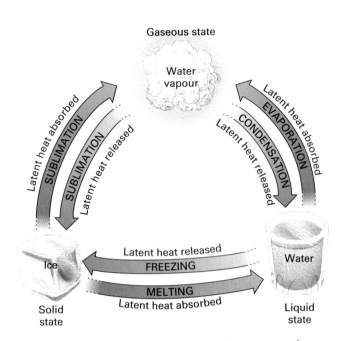

6.1 Three states of water Arrows show the ways that any one state of water can change into either of the other two states. Heat energy is absorbed or released, depending on the direction of change.

causes frost to form on car windshields. *Working It Out 6.1 • Energy and Latent Heat* presents the amount of latent heat taken up and released with changes of state of water.

THE HYDROSPHERE AND THE HYDROLOGIC CYCLE

The **hydrosphere** includes water on the Earth in all its forms. About 97.2 percent of the hydrosphere consists of ocean water. The remaining 2.8 percent is fresh water. The largest reservoir of fresh water is the ice stored in the world's ice sheets and mountain glaciers. This water accounts for 2.15 percent of total global water.

Fresh liquid water is found both on top of and beneath the Earth's land surfaces. *Subsurface water* occupies openings in soil and rock. Most of it is held in deep storage as *groundwater*, at a level where plant roots cannot access it. Groundwater makes up 0.63 percent of the hydrosphere, leaving 0.02 percent of the remaining fresh water.

This small remaining portion of the Earth's water is distributed in lakes, rivers, the soil, and the atmosphere. Although small in proportion to the total amount of water in the hydrosphere, it is important because it includes the water available for plants, animals, and human use. *Soil water*, which is found in the soil within reach of plant roots, comprises 0.005 percent of the global total. *Surface water* is found in streams, lakes, marshes, and swamps. Although most of this surface water is in lakes, about 50 percent would be technically classed as saline. An extremely small proportion—about 0.0001 percent—is found in the streams and rivers that flow toward the sea or inland lakes.

WORKING IT OUT | 6.1   Energy and Latent Heat

In the metric system, energy is measured in joules. The *joule* (J) is defined as one newton-metre, which is the energy expended by a force of one newton acting through a distance of one metre. A newton is the force produced by an acceleration of one metre per second squared (m/s²) applied to a mass of one kilogram. The joule thus has units of force multiplied by distance, which is a measure of energy. The joule is related to the watt (W), in that one watt is defined as the flow of one joule of energy per second. From this definition, the energy consumed by a 100-watt light bulb in one second is 100 joules.

The amount of latent heat taken up or released by water when a change of state occurs depends on the type of change and the temperature of the water. For melting and freezing, the energy

amounts to about 335 kilojoules per kilogram (kJ/kg) when water changes from a liquid at 0°C to a solid at the same temperature. For evaporation and condensation, the energy required or released is about 2,260 kilojoules per kilogram, at 100°C. Energy is also required to heat the water from 0°C to 100°C—4.19 kilojoules per kilogram for each Celsius degree, or 419 kilojoules per kilogram for 100°. These values depend partly on atmospheric pressure and apply to sea-level conditions.

To determine the amount of energy required for sublimation from solid to gas, add the amount of heat required first for melting, then for warming to the boiling point, and finally, for evaporating. That is, 335 + 419 + 2,260 = 3,014 kilojoules per kilogram. This will also be the

amount of energy released when the reverse sublimation process (deposition) occurs, depositing frost.

Evaporation can, of course, occur at temperatures well below 100°C. For example, the energy required to evaporate a kilogram of liquid water at 25°C will be the energy required to bring the water to the evaporation point added to the latent heat required to convert it to vapour. For the first quantity, each degree below 100°C requires 4.19 kilojoules per kilogram. Thus, 100 − 25 = 75°C, hence 4.19 × 75 = 314 kilojoules per kilogram are needed. To this add the 2,260 kilojoules per kilogram necessary to bring about a change of state from liquid to vapour. Therefore, the total energy required to evaporate a kilogram of liquid water at 25°C is 314 + 2,260 = 2,574 kilojoules per kilogram.

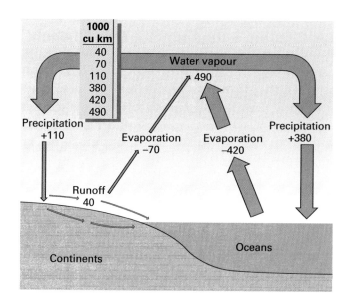

6.2 The global water balance The figures above give average annual water flows in and out of land areas and oceans. Values are in thousands of cubic kilometres. Global precipitation equals global evaporation. (Based on data of John. R. Mather)

The quantity of water held as vapour and cloud water droplets in the atmosphere is about 0.001 percent of the hydrosphere. This is approximately equivalent to the amount of precipitation that normally falls on Earth over 10 days. Though small, this reservoir of water is of enormous importance. It provides the supply of precipitation that replenishes all freshwater stocks on land. In addition, the transport of water vapour from warm tropical oceans to cooler regions provides a global flow of energy, in the form of latent heat, from low to high latitudes.

The movements of water among the great global reservoirs constitute the **hydrologic cycle**. This cycle is summarized in Figure 6.2 where the global water balance is discussed more fully. In the hydrologic cycle, water moves from land and ocean to the atmosphere as water vapour and returns as precipitation. Because precipitation over land exceeds evaporation, water also runs off the land to the oceans.

The cycle begins with **evaporation**. In this process water from ocean or land surfaces changes state from liquid to vapour and enters the atmosphere. Total evaporation is about six times greater over oceans than land. This is mainly due to the immense area of the oceans, which cover more than 70 percent of Earth's surface. In addition, land surfaces are not always wet enough to yield much evaporated water. The highest evaporation rates occur over tropical oceans. Evaporation is also high from wet tropical landmasses, such as the Amazon and Congo basins, where it is supplemented by transpiration from the forest cover. Evaporation is low over tropical deserts, where the rate is similar to that in cold polar regions.

Once in the atmosphere, water vapour can condense into droplets, or directly form ice crystals through sublimation. The water eventually falls to Earth as **precipitation** in the form of rain, snow, or hail. Total precipitation over the oceans is nearly four times greater than precipitation over land. Precipitation is generally heaviest along the ITCZ, where the moist trade winds converge to form the rising limb of the Hadley cells (Figure 6.3). Precipitation is much lower in the subtropics, where air descends in the zone of semi-permanent high pressure. In the mid-latitudes, precipitation is mainly associated with centres of low pressure that

6.3 Satellite image taken from the Global Precipitation Climatology Project (GPCP) Satellite estimates of rain are obtained from infrared radiance from geostationary and polar orbiting satellites and from passive microwave sensors on polar orbiting satellites. Image provided by Dr. Robert Adler, Goddard Space Flight Center, NASA (Adler et al., 2001).

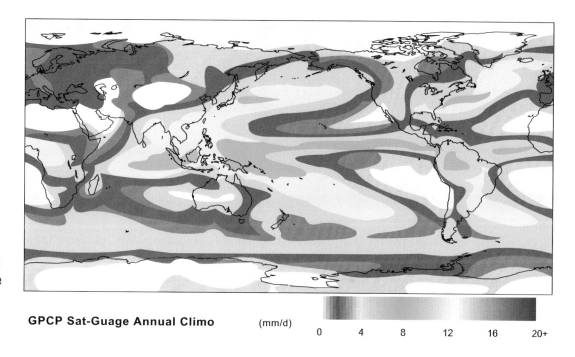

GPCP Sat-Guage Annual Climo   (mm/d)   0   4   8   12   16   20+

FOCUS ON SYSTEMS | 6.2   The Global Water Balance as a Matter Flow System

The Earth and its atmosphere contain a finite amount of water. Consequently, the exchange of water among the atmosphere, oceans, and continents must be in balance on the global scale. In this analysis, assume that the volume of ocean waters and the overall volume of fresh water in surface and subsurface water remain constant from year to year. This is probably quite reasonable, unless climate changes rapidly.

Water moves through the hydrologic cycle primarily as liquid and solid water in precipitation and as water vapour in evaporation. On the continents, water is present in the form of liquid water, and in solid form as snow and ice. The ocean surface is mainly liquid, but the Arctic Ocean and some parts of the ocean near Antarctica have a cover of sea ice that may be capped with snow.

Flows of water link the atmosphere and the two surface components. For budgeting purposes, the convention is to describe flows to the surface (land or oceans) as positive and flows leaving the surface as negative.

Since the cycle is in balance, flows into and out of the atmosphere add to zero. (See Figure 6.2 for flow values.) That is,

$$P_{(LAND+LAKE)} + E_{(LAND+LAKE)} + P_{OCEAN} + E_{OCEAN} = 0$$
$$(+110,000) + (-70,000) + (+380,000) + (-420,000) = 0$$

Here, $P_{(LAND+LAKE)}$ and P_{OCEAN} are precipitation flows from the atmosphere to the land and oceans, and $E_{(LAND+LAKE)}$ and E_{OCEAN} are evaporation flows from the land and oceans to the atmosphere. The term *precipitation* includes flows of water as rain and snow, as well as the direct sublimation of water vapour as frost. Similarly, evaporation includes sublimation of snow and ice to water vapour.

Although the global balance for the atmosphere is zero, there is a positive balance for land (+44,800), indicating more precipitation than evaporation, and a negative balance for oceans (−44,800), indicating more evaporation than precipitation. Thus, each year land and lakes gain 44,800 cubic kilometres of water, while oceans lose 44,800 cubic kilometres of water. Without a link between land and the oceans, the oceans would eventually empty and all the water would accumulate over land. This doesn't happen because a flow pathway connects land and oceans—runoff of water from rivers, glaciers, and ice sheets into the ocean. For the oceans to stay at the same level, and for the same amount of water to be stored on the land year after year, the flow in runoff must equal 44,800 cubic kilometres.

All matter flow systems need a power source. For the water flow system the power source is the Sun. Solar energy evaporates water and moves it into the atmosphere, where later it falls to the Earth as precipitation. Gravity plays a further role in moving streams of water and ice off the lands and into the oceans as runoff.

develop along the polar front, although convection is also important. Cold polar air holds little moisture, and precipitation is generally low in the high latitudes. The effects of mountain barriers, which may increase or decrease precipitation, are superimposed on these general regional patterns. Chapter 8 discusses global precipitation in more detail.

Precipitation that falls directly into the ocean can return immediately to the atmosphere through evaporation. However, on land surfaces, precipitation can follow three pathways. First, it can evaporate and return to the atmosphere as water vapour. Second, it can sink into the soil and then into the surface rock layers below, where it may be temporarily stored as groundwater reserves. Third, precipitation can flow from the land, concentrating in streams and rivers that carry it to a lake or the ocean. This flow of water is known as surface *runoff*. Chapter 16 discusses the terrestrial components of the hydrologic cycle.

Focus on Systems 6.2 • The Global Water Balance as a Matter Flow System describes how the evaporation, precipitation, and runoff pathways in the hydrologic cycle are interconnected in a closed matter flow system.

HUMIDITY

The amount of water vapour present in the air varies widely from place to place and over time. It is negligible in the cold, dry air of Arctic regions in winter, but can be as much as 4 or 5 percent of a given volume of air in the warm wet regions near the Equator. The amount of water vapour present in the air is commonly reported as **relative humidity**, but this is just one of many terms used.

In meteorology, the term **vapour pressure** refers to the contribution that water vapour makes to the pressure exerted by the atmosphere. The quantity of water vapour present in the atmosphere depends on air temperature, and increases as temperature increases. **Saturation vapour pressure (SVP)** is the maximum amount of water vapour that can be present at a

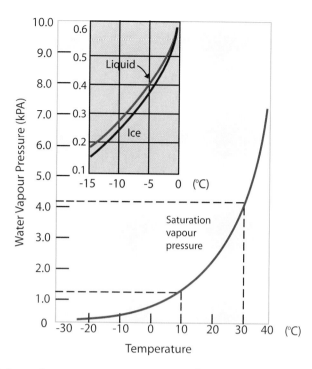

6.4 Saturation vapour pressure increases with temperature. At temperatures below freezing, the SVP is greater over super-cooled water than over ice, because more energy is needed to break the molecular bonds in an ice crystal.

6.5 Frost point Hoarfrost on trees in the Yukon forms when the temperature of the dew is below freezing.

RELATIVE HUMIDITY

Relative humidity compares the amount of water vapour present with the maximum amount that the air can hold at that temperature. It is expressed as a percentage. For example, if the air holds half the moisture possible for a given temperature, then relative humidity is 50 percent. When the humidity is 100 percent, the air holds the maximum amount possible; it is saturated, and will be at the dew-point temperature.

A change in relative humidity of the atmosphere can happen in one of two ways. The first is through direct gain or loss of water vapour. For example, if an exposed water surface or wet soil is present, additional water vapour can enter the air, raising the humidity. This process is slow because the water vapour must diffuse upward from the surface into the air above.

The second way is by lowering the temperature of the air. Even though no water vapour is added, relative humidity rises because the air's capacity to hold water vapour is lowered. The existing amount of water vapour then represents a higher percentage of the total capacity. For this reason relative humidity is generally highest at night, when air temperatures are lower, as shown for Kelowna, B.C. in Figure 6.6.

It is useful to look at the relationship between the various parameters that are associated with humidity. At 2 p.m. on October 2, air temperature at Kelowna was 16.3°C, relative humidity 35 percent, and the dew-point temperature 0.6°C. By 7 a.m. the following morning, these values had changed to 3.9°C, 93 percent, and 2.7°C, respectively. A graph, such as Figure 6.5, or a table can be used to determine the SVP at any given air temperature. Thus, at 3.9°C, the SVP is 8.1 grams per kilogram; and at 16.3°C, it is 18.3 grams per kilogram. SVPs for the recorded dew-point temperatures of 0.6°C and 2.7°C are 6.4 grams per kilogram and 7.5 grams per kilogram, respectively. Relative humidity can be calculated as

given temperature (Figure 6.4). For example, the SVP of air is 1.23 kilopascals at 10°C and rises to 4.25 kilopascals at 30°C. Note that a water molecule, comprising two hydrogen atoms and one oxygen atom, has a molecular mass of 18.02. Dry air, composed mainly of nitrogen and oxygen, has an equivalent molecular mass of 28.57. Thus, moist air actually exerts less pressure than dry air.

The actual mass of water vapour held by a parcel of air is its **specific humidity**. This is the mass of water vapour that is available for precipitation. Specific humidity is defined as the mass of water vapour contained in a given mass of air; it is expressed as grams of water vapour per kilogram of air (g/kg). *Saturation specific humidity* is the maximum mass of water that can be present at a given temperature. For example, saturation specific humidity at 10°C is 7 grams of water vapour per kilogram of air and rises to 26 grams per kilogram at 30°C.

Another way to describe the water vapour content of air is by its **dew-point temperature**, which is the temperature at which the water vapour in air condenses into water. If air is slowly chilled, it will eventually reach saturation. At this temperature, the air holds the maximum amount of water vapour possible. If cooling continues, condensation will begin and dew will form. The temperature at which saturation occurs is therefore known as the dew-point temperature. If the dew-point temperature is below freezing, then it is referred to as the **frost point**, and hoarfrost forms (Figure 6.5)

Kelowna October 1-3, 2006

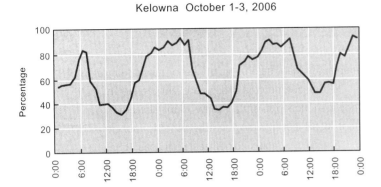

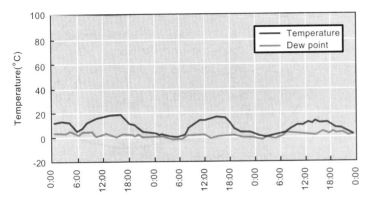

6.6 Relative humidity, air temperature, and dew-point temperature for Kelowna, B.C., for October 1–3, 2006 Relative humidity changes with temperature because the capacity of warm air to hold water vapour is greater than that of cold air, even though the amount of water vapour in the air remains more or less constant.

$$\text{Relative Humidity} = \frac{\text{SVP of dew-point temperature}}{\text{SVP of air temperature}} \times 100$$

Thus relative humidity at 2 p.m. on October 2

$$= \frac{6.4}{18.3} \times 100 = 35\%$$

and at 7 a.m. on October 3 $= \frac{7.5}{8.1} \times 100 = 93\%$

There was little change in the amount of water vapour present in the air during this three-day period, as indicated by the dew-point temperature. However, the capacity of the air to hold water vapour increased during the warmer daytime hours, causing relative humidity to fall. With lower air temperature at night, relative humidity increased, and the cycle repeated itself.

Humidity can be measured in various ways. The traditional method uses two matched thermometers mounted together side by side in an instrument called a wet-and-dry-bulb thermometer. The wet-bulb thermometer is covered with a cotton sleeve that is kept moist by distilled water that is drawn up the cotton wick from a reservoir. The dry-bulb thermometer is a standard thermometer. Evaporation cools the wet bulb, so as long as the air is not saturated, the drier the air, the greater the cooling. If the air is saturated, then evaporation cannot occur. Relative humidity is read from a sliding scale set to the wet-bulb and dry-bulb temperatures.

The *sling psychrometer* is a variant of the wet-and-dry-bulb thermometer. It is whirled in the open air and then the temperatures on the two thermometers are read. A *hair hygrograph* was the standard instrument for recording humidity over a period of time. The sensor in this instrument is a bundle of human hair. It works on the principle that hair expands by as much as 3 percent as air changes from dry to saturated.

Electronic instruments that read relative humidity directly are now used extensively. The most common are *capacitive sensors*, which consist of a thin layer of water absorbent polymer or inorganic material that is coated onto a conductive base. The material absorbs water vapour in an amount that depends on the relative humidity. The water vapour affects the ability of the metal film to hold an electric charge. An electronic circuit senses this ability and converts it to a direct reading of relative humidity. The advantage of these electronic sensors is that they provide automatic weather stations with data that computers can easily process.

THE ADIABATIC PROCESS

If there is sufficient water vapour present in a mass of air, and it can be cooled to the dew point, some of the water vapour will begin to form water droplets or ice crystals. For example, radiative heat loss at night cools the ground surface and the air in contact with it. If the air's temperature falls below the dew point, condensation or sublimation will give rise to dew or frost. In this same way, small patches of mist often develop over lakes and ponds (Figure 6.7). However, this process cannot produce the copious condensation necessary for widespread precipitation. Precipitation is formed only when a substantial mass of air experiences a steady drop in temperature below the dew point. This requires a parcel of air to be lifted to a higher level in the atmosphere.

6.7 Temperatures below dew point Mists over lakes and rivers often form through radiative cooling during the night.

DRY ADIABATIC RATE

An important principle of physics is that when a gas is allowed to expand, its temperature drops. Conversely, when a gas is compressed, its temperature rises. Physicists use the term **adiabatic process** to refer to a heating or cooling process that occurs solely as a result of pressure change. Because atmospheric pressure decreases with altitude, the pressure on a parcel of air that is rising also decreases. Consequently, the air parcel will expand. The molecules of air must do work as they expand. This will affect the air parcel's temperature, which is a measure of the kinetic energy of its molecules. During expansion, the total amount of energy in the parcel remains the same—none is added or lost. The air parcel can either use this energy to do the work of expansion or to maintain its temperature, but it can't use it for both. Thus, when the parcel expands, its temperature drops. Conversely, when a parcel of air sinks, higher pressure closer to the Earth's surface compresses and warms it.

If the parcel of air is unsaturated, then its temperature changes when displaced vertically at the **dry adiabatic lapse rate**. If the air parcel rises through the atmosphere it will *cool* at a rate of about 10°C/1,000 metres. If it sinks, it will *warm* at the same rate.

The dry adiabatic rate applies to a mass of air that is moving up or down through the atmosphere. It is always constant and is determined by physical laws. It is quite different from the normal and environmental temperature lapse rates discussed in Chapter 4. These lapse rates refer to the change in temperature of air with altitude in the absence of any vertical motion. The environmental temperature lapse rate will vary from time to time and from place to place, depending on the state of the atmosphere, with the average condition given by the normal temperature lapse rate.

WET (SATURATED) ADIABATIC RATE

Assume that an air parcel starts at a temperature of 20°C and is forced to rise upward through the atmosphere (Figure 6.8). As it rises, its temperature drops at the dry adiabatic rate, 10°C/1,000 metres, and when it reaches 1,000 metres, its temperature has fallen to 10°C. At this altitude, the air has cooled to its dew-point temperature, and condensation starts to occur. The altitude at which the air reaches this condition is called the **lifting condensation level**.

The dew-point temperature also changes slightly with altitude at a rate known as the *dew-point lapse rate*, which is 1.8°C/1,000 metres. Given that the initial dew-point temperature of the air mass was 11.8°C, at 1,000 metres the dew-point temperature will be 10°C. That is also the temperature of the parcel described above, so condensation will begin at that level. The lifting condensation level is thus determined by the initial temperature of the air and its dew point. *Working It Out 6.3 • The Lifting Condensation Level* shows how to calculate the lifting condensation level, given these two values.

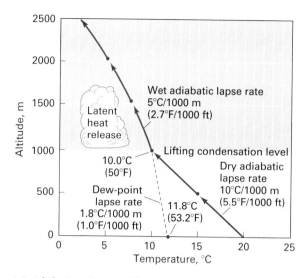

6.8 Adiabatic cooling An adiabatic decrease of temperature in a rising parcel of air leads to condensation of water vapour and the formation of clouds. (Copyright © A. N. Strahler)

If cooling continues, water droplets will form, producing a cloud. If the parcel of saturated air continues to rise, however, a new principle comes into effect—latent heat release. When condensation occurs, latent heat is released at the rate of 2,260 kilojoules per kilogram, which warms the rising air. Thus, adiabatic cooling occurs at the same time the air is heated through the release of latent heat by condensation.

Because more energy is released through adiabatic cooling than is gained by condensation, air will continue to cool as it rises. However, the release of latent heat slows the cooling down. This cooling rate for saturated air is called the **wet (saturated) adiabatic lapse rate** and ranges between 4 and 9°C per 1,000 metres. Unlike the dry adiabatic lapse rate, which remains constant, the wet adiabatic lapse rate is variable because it depends on the temperature and pressure of the air and its moisture content. For most situations, however, a value of 5.5°C/1,000 metres can be used. Because specific humidity increases with temperature, warm saturated air produces more liquid water than cold saturated air, and so releases more latent heat. Consequently, the wet adiabatic lapse rate is much less than the dry adiabatic lapse rate if rising air is warm, but is similar if rising air is cold (Table 6.1). Note, in

Table 6.1 Wet adiabatic lapse rates for different air temperatures at selected altitudes

Pressure (kPa)	Temperature (°C)				
	40	20	0	−20	−40
100 (100m)	9.5	8.6	6.4	4.3	3.0
80 (1,950m)	9.4	8.3	6.0	3.9	2.8
60 (4,200m)	9.3	7.9	5.4	3.5	2.6
40 (7,200m)	9.1	7.3	4.6	3.0	2.4
20 (12,000m)	8.6	6.0	3.4	2.5	2.0

WORKING IT OUT | 6.3   The Lifting Condensation Level

When a parcel of air moves upward, it cools at the dry adiabatic rate of 10°C/1,000 metres. This cooling occurs because the parcel is subject to lower atmospheric pressure as it rises, and under the adiabatic principle, it expands and cools. The change in pressure also affects the dew-point temperature, which falls at a rate of 1.8°C/1,000 metres. When the temperature of the cooling air parcel reaches the dew-point temperature, condensation will begin to occur. The elevation at which condensation begins is referred to as the *lifting condensation level*.

Suppose an air parcel is 20°C and is raised 500 metres at the dry adiabatic lapse rate. Its temperature will drop by 500/1,000 × 10 = 5°C, resulting in a temperature of 20 − 5 = 15°C. In equation form, this is

$$T = T_0 - (H \times R_{DRY})$$

where T is the temperature of the parcel; T_0 is the starting temperature at a height of 0 metres; H is the height in metres; and R_{DRY} is the dry adiabatic rate, 10°C/1,000 metres.

Assume that the dew point of that same parcel of air is 11°C. At 500 metres, the dew-point temperature will fall by 500/1000 × 1.8 = 0.9°C, and the resulting dew-point temperature will be 11 − 0.9 = 10.1°C. Similarly, we can write

$$T_D = T_{DEW} - (H \times R_{DEW})$$

where T_D is the dew-point temperature of the parcel at height H; T_{DEW} is the starting dew-point temperature; and R_{DEW} is the dew-point lapse rate, 1.8°C/1,000 metres.

Condensation will occur when the parcel's temperature reaches the dew point; that is, at the height at which $T = T_D$. This level can be found from

$$T_0 - (H \times R_{DRY}) = T_{DEW} - (H \times R_{DEW})$$

After some algebraic rearrangement the formula becomes

$$H = \frac{T_0 - T_{DEW}}{R_{DRY} - R_{DEW}}$$

By substituting the values for the dry adiabatic rate and dew-point lapse rate, the formula becomes

$$H = 1,000 \times (T_0 - T_{DEW})/8.2$$

This gives the lifting condensation level directly in metres. The lifting condensation level for the air parcel described above is

$$H = 1,000 \times (20 - 11)/8.2 = 1,000 \times 9/8.2 = 1,098 \text{ metres}$$

To find the temperature of the air parcel at the lifting condensation level, the equation becomes

$$T = T_0 - H \times R_{DRY} = 20 - 1,098 \times 10/1,000 = 9.0°C$$

Figure 6.9, the wet adiabatic rate is shown as a slightly curving line to indicate that its value increases with altitude.

CLOUDS

Views of the Earth from space show that clouds cover about half of the Earth at any given time. Low clouds reflect solar energy, thus cooling the Earth-atmosphere system, while high clouds absorb outgoing long-wave radiation, causing the Earth-atmosphere system to warm.

A **cloud** is made up of water droplets or ice particles suspended in air. These particles range in diameter from 20 to 50 micrometres (about 0.002–0.005 millimetres). Each cloud particle forms around a tiny centre of solid matter, called a *condensation nucleus*. Such nuclei have diameters in the range 0.1 to 1 micrometre.

An important source of condensation nuclei is the surface of the sea. When winds create waves, droplets of spray from the crests of the waves are carried upward in turbulent air. Evaporation of sea water droplets leaves a tiny residue of crystalline salt suspended in the air. This aerosol is hygroscopic—it strongly attracts water molecules. Another source of nuclei is the heavy load of dust carried by polluted air over cities, which can aid condensation and the formation of clouds and fog. But even clean air contains enough condensation nuclei for the formation of clouds.

The condensation of water vapour on condensation nuclei initiate cloud droplets, and their growth continues as long as water vapour condenses onto their surfaces. Droplet growth depends on the SVP (e_s) of the droplet and its surrounding environment. If the SVP in the environment is greater than the SVP in the droplet, the droplet grows due to condensation. If the SVP in the environment is less than the SVP in the droplet, the droplet shrinks through evaporation. After their initial growth, droplets have different SVPs based on their *size* and *purity*. These two droplet properties affect subsequent growth through the **curvature effect** and the **solute effect**.

The curvature effect is related to the size of the droplet and arises because the surface of a small sphere (or droplet) has greater curvature than the surface of a large sphere. The SVP in-

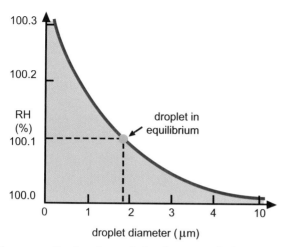

6.9 The curvature effect limits the growth of small, pure water droplets to conditions of supersaturation.

creases as the droplet size becomes smaller, and for small droplets to grow the air must be supersaturated (Figure 6.9). Consequently, it is easier for water to evaporate from, and harder for water vapour to condense to, smaller droplets than larger ones. Because smaller droplets lose water at a faster rate than larger droplets, a higher degree of suspersaturation is required for them to form and grow.

If a cloud is composed of pure water droplets of varying sizes, the smaller ones might evaporate. However, the solute effect compensates for most of the curvature effect, and allows droplets to develop when air is below saturation. The solute effect occurs because cloud droplets form around condensation nuclei, such as salt crystals, and are rarely composed of pure water. Equilibrium

SVP for solutions is lower than for pure water. Hence, condensation begins on these hygroscopic particles before air is saturated. For small droplets, the SVP is much less than that for pure water. For example, sodium chloride can initiate condensation at relative humidities as low as 80 percent, which can produce haze. Because growth is dependent on droplet chemistry, those with a higher salt concentration will grow faster. Thus, a range of droplet sizes develops within a cloud, which is important for the development of precipitation.

Typically, water at the Earth's surface turns to ice when the temperature falls below 0°C. However, when water is dispersed as tiny droplets in clouds, it can remain in the liquid, *super-cooled* state at temperatures far below freezing, although freezing will occur if ice nuclei are present in the atmosphere. Fine particles of kaolinite clay are a common form of ice nuclei; these become active at −10°C. Hence clouds may consist mostly of water droplets at temperatures down to about −10°C; however, below that, a mix of water droplets and ice crystals occurs. The coldest clouds, with temperatures below −40°C, occur at altitudes of 6 to 12 kilometres and are formed entirely of ice particles.

CLOUD FORMS

Clouds come in many shapes and sizes, from the small, white puffy clouds often seen in summer to the dark layers that produce heavy rain. Meteorologists classify clouds into four families, arranged by height—the high, middle, and low clouds, and clouds with strong vertical development (Figures 6.10). The high, middle, and low clouds are further classified on the basis of their general shape or form.

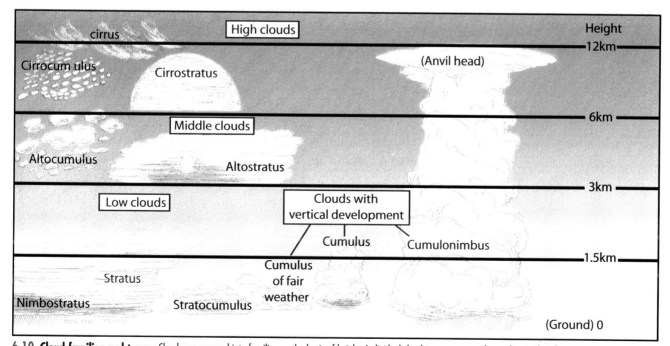

6.10 Cloud families and types Clouds are grouped into families on the basis of height. Individual cloud types are named according to their form.

6.11 Cloud photos (a) Rolls of altocumulus clouds, also known as a "mackerel sky." (b) High cirrus clouds lie above puffy cumulus clouds in this scene from Holland. (c) Fibrous cirrus clouds are drawn out in streaks by high-level winds to form "mare's tails." (d) Upper layer: lenticular clouds, a type of altostratus formed when moist air is uplifted as it crosses a range of hills. Lower layer: cumulus clouds, also triggered by the uplift.

Two major classes are recognized on the basis of form—stratiform and cumuliform. *Stratiform* clouds are extensive, layered clouds that generally cover large areas (Figure 6.11). A common type is low-altitude *stratus*, which forms when air layers are forced to rise gradually. This can happen when one air layer overrides another. As the overriding layer rises and cools, condensation occurs and a blanket-like cloud cover forms. If the overriding layer is quite moist and rising continues, dense, thick stratiform clouds result that can produce prolonged rain or snow.

Cumuliform clouds are globular masses associated with rising parcels of air. The air rises because it is warmer than the surrounding air. As it is buoyed upward, it is cooled adiabatically. Once it reaches condensation level, it begins to form a cloud. The most common cloud of this type is the *cumulus* cloud. Sometimes the upward movement yields dense, tall clouds that produce thunderstorms. This form of cloud is the *cumulonimbus*.

Individual cloud forms provide visible clues to air motions that are otherwise not apparent. On a continental or global scale, cloud patterns also serve to mark air motions and reveal major features of air circulation. One of the most important tools for observing cloud patterns remotely is the GOES series of geosta-tionary satellites. *Focus on Remote Sensing 6.4 • Observing Clouds from GOES* shows some examples of GOES images.

FOG

Fog is simply a cloud layer at or very close to the surface. Fog is a major environmental hazard, especially for transportation on highways and airports. For centuries, fog at sea has been a navigational hazard increasing the danger of ship collisions and groundings. In addition, polluted fog—**smog**, like that in London, England in the early part of the twentieth century—can cause lethal respiratory problems.

One type of fog, known as *radiation fog*, is formed at night when the temperature of the air layer at ground level falls below the dew point. This kind of fog is associated with low-level temperature inversions and typically forms upwards from the ground surface as the night progresses. *Valley fog* is similar to radiation fog in that cold air drains downward and collects in low-lying areas, chilling the overlying moist air. *Precipitation fog* forms when rain falls into cold air and evaporates; it often occurs along warm fronts. *Steam fog*, a common feature of Arctic coastlines, forms when water evaporates from a warm body of water and condenses in the cold air above.

FOCUS ON REMOTE SENSING | 6.4   **Observing Clouds from GOES**

Some of the most familiar images of Earth from space are those acquired by the Geostationary Operational Environmental Satellite (GOES) system. Images from the GOES series of satellites and its predecessors have been in constant use by meteorologists and weather forecasters since 1974. The primary mission of the GOES series is to view cloud patterns and track weather systems by providing frequent images of the Earth from a consistent viewpoint in space. This capability assists with forecasting storms and severe weather.

A key feature of the GOES series is its geostationary orbit that keeps the satellites above the same point on the equator. From this vantage point they can acquire a constant stream of images of the Earth below. A geostationary satellite is ideal for viewing clouds and tracking cloud patterns.

Altogether, 13 GOES satellites have been placed in orbit since 1975. The latest, GOES-13, was launched on May 24, 2006. GOES-9 and GOES-10 are still in limited use, and GOES-11 and GOES-12 operate as GOES-West and GOES-East, respectively, providing images from two

points on the equator, at longitudes 75° W and 135° W. From 135° W, the GOES-West satellite can track Pacific storm systems as they approach North America and move across western Canada and the United States. From 75° W, the GOES-East satellite observes weather systems in the eastern part of the continent. GOES-East also observes the tropical Atlantic, identifying tropical storms and hurricanes as they form and move toward the Caribbean Sea and along the east coast.

The present generation of GOES platforms carries two primary instruments—the GOES Imager and the GOES Sounder. The Imager acquires data in five spectral bands, ranging from the visible red to the thermal infrared. It acquires images of clouds, water vapour, surface temperature, winds, albedo, and fires and smoke. The Sounder uses 19 spectral channels to observe atmospheric profiles of temperature, moisture, ozone, and cloud height and cover.

The image below was acquired from the GOES-12 on September 22, 2006. In this near-noon image, the entire side of the Earth nearest to the

satellite is illuminated. The brown and green tones in the image are derived from visible band data. Vegetated areas appear green, while semi-arid and desert landscapes appear yellow-brown. The dark blue primary colour is derived from a thermal infrared band, but is scaled inversely so that cold areas are light and warm areas are dark. Since clouds are bright white in visible wavelengths and are colder than surface features, they appear white. The oceans appear blue because they are dark in the visible bands but are still somewhat warm.

For weather forecasting, an important tool of the latest generation of GOES imagers is the water vapour image. The image at the right shows a global water vapour image also for November 27, 2006. It is constructed from GOES-East and -West images, as well as those of two other geostationary satellites (MeteoSat at 0° long. and GMS at 140° W). The brightest areas show regions of active precipitation—note the numerous convective storms in the equatorial zone. Dark areas show low water vapour content.

Earth from GOES-12 The GOES-12 geostationary satellite acquired this image on September 22, 2006.

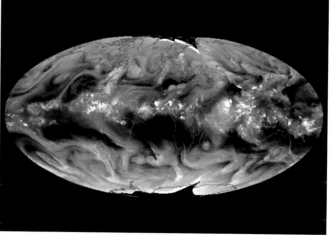

Water vapour composite image acquired November 27, 2006 Areas of high atmospheric water vapour content are bright in this image, prepared by merging data from four geostationary satellite instruments.

6.12 Sea fog A layer of sea fog along the coast of California

Another type of fog, *advection fog*, results when a warm, moist air layer moves over a cold surface. As the warm air layer loses heat to the surface, its temperature drops below the dew point, and condensation sets in. Advection fog commonly occurs over oceans where warm and cold currents flow side by side. When warm, moist air above the warm current moves over the cold current, condensation occurs. The fogs of the Grand Banks off Newfoundland are formed in this way because the cold Labrador Current comes in contact with the warmer waters of the Gulf Stream. Advection fog can also develop on land when a cold air mass mixes with warmer, moister air. A similar process gives rise to the *sea fog* that forms along the west coasts of South America and Africa. They are caused when warm, moist tropical air is cooled by the Peruvian and Benguela currents that bring cold water from the southern oceans around Antarctica. Similar fogs develop off the coast of California where the air is chilled by the cold California Current (Figure 6.12).

PRECIPITATION

The radial growth of the droplets in clouds slows down as they get bigger and competition for available moisture increases. The droplets' surface area gets progressively larger so more water must condense. The need for more water also slows the condensation rate, which is limited by the speed at which latent heat can be released. Once droplets grow large enough they begin to fall through the cloud. As they fall, they can collide with other droplets and coalesce to form larger droplets. Larger droplets also drag smaller droplets down behind them. This is the **collision-wake capture** process, which arises because of the range of droplet sizes naturally present in clouds.

The rate at which the droplet descends depends on its size. For example, in still air a droplet with a diameter of 20 micrometres has a terminal velocity of 0.01 metres per second, compared with 6.5 metres per second for a droplet 4,000 micrometres in diameter. However, larger droplets can also break apart as they collide, rapidly increasing the number of precipitation-size droplets. The longer a droplet remains in the cloud, the greater the opportunity it has to interact with other droplets. Droplet size can therefore increase if strong vertical updrafts develop in the cloud.

In colder clouds, especially in mid- and high latitudes, much of the precipitation originates from ice crystals through the **Bergeron (Bergeron-Findeisen) process**. This process depends on the presence of super-cooled water droplets in the cloud. Small ice needles will form directly from super-cooled water droplets if the temperature in the cloud falls to about −35°C. Freezing occurs at much higher temperatures if freezing nuclei are present. Once ice crystals have started to form, super-cooled water droplets will automatically freeze to them. An ice crystal will also grow as water vapour is converted directly to ice by deposition. Ice crystals will aggregate into larger snowflakes as they fall through the cloud in a process similar to the collision-wake capture process. If temperatures in the lower atmosphere are near or below freezing, the snowflakes will remain in solid form; otherwise, they melt to produce rain or sleet. Precipitation can develop quickly through the Bergeron process; it is the dominant process in large cumulus and cumulonimbus clouds, which usually extend well above the freezing level, even in the tropics.

Cloud droplets grow by condensation and sublimation to diameters of 50 to 100 micrometres. Droplets formed by coalescence, either in liquid or solid forms, can become much larger. A droplet diameter of about 500 micrometres is characteristic of *drizzle*. For *rain* the average droplet size is about 1,000 to 2,000 micrometres, but droplets can reach a maximum diameter of about 7,000 micrometres (about 7 millimetres). If they grow bigger than this, they become unstable and disintegrate into smaller drops while falling. The largest droplets are usually associated with thundershower activity and the warm moist clouds in the equatorial and tropical zones.

Most precipitation begins as ice crystals and snowflakes, which melt as they fall, reaching the Earth's surface as rain. Conversely, raindrops that fall through a cold air layer can produce pellets or grains of ice, or a mixture of snow and rain commonly known as *sleet*. *Ice storms* occur when the ground is frozen and the temperature in the lowest air layer is also below freezing. Under these conditions, rain falling through the layer is chilled and freezes onto ground surfaces as a clear glaze. Ice storms cause great damage, especially to telephone and power lines and to tree limbs due to the weight of the ice. In addition, the slippery glaze makes roads and sidewalks extremely hazardous. In Canada, freezing rain tends to occur most frequently in southern Ontario, the Maritimes, and Newfoundland (Figure 6.13).

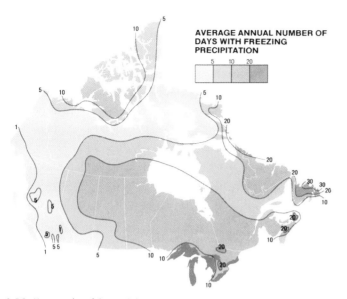

6.13 Mean number of days with freezing rain in Canada

6.15 Streaks of rain that evaporate before reaching the ground are termed virga.

Hail, another form of precipitation, consists of lumps of ice averaging 5–50 micrometres in diameter, although it can be much larger when associated with severe thunderstorms. The largest hailstone recorded in Canada fell in 1973 near Cedoux, Saskatchewan; it weighed 290 grams and was 114 millimetres in diameter. In 2003, a hailstone with a diameter of 178 millimetres was reported in Nebraska, U.S.A. Hailstones are formed by the accumulation of ice layers on ice pellets that are suspended in the strong updrafts of a thunderstorm. The maximum growth rate occurs at about –13°C; growth is uncommon below –30°C, as super-cooled water droplets are not abundant at that temperature. Thus, hail is most common in midlatitudes during early summer when surface temperatures are warm enough to promote thunderstorm activity, but the atmosphere is still relatively cool. Hail is less common in low latitudes, despite frequent thunderstorms, because the tropical atmosphere tends to be warmer. In Canada, hail occurs most frequently in Alberta, British Columbia, and Saskatchewan (Figure 6.14).

On some occasions, wisps or streaks of rain or snow fall from a cloud, but evaporate before reaching the ground. This phenomenon, called *virga*, is commonly observed during summer in the Canadian Prairies (Figure 6.15).

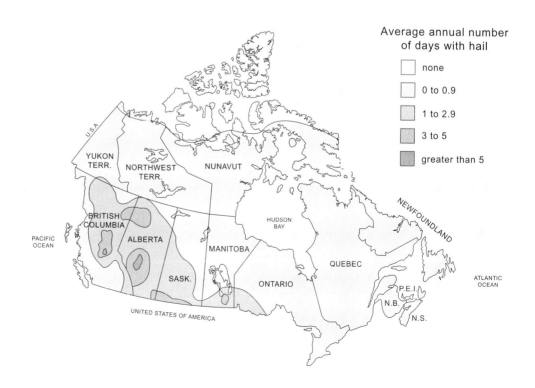

6.14 Mean number of days with hail in Canada

A *rain gauge* measures rainfall; the standard design consists of a narrow, calibrated cylinder with a funnel at the top. These are usually read once each day to provide the total precipitation amount for that day. An alternative is a tipping bucket rain gauge, which consists of two containers mounted on a pivot, beneath a funnel. When one container is filled, it tips and empties, and the second one begins to fill. Each container (or bucket) is calibrated to tip when it contains 0.1 millimetre of rain. With each tip, a signal is sent to a recording device and a record is compiled of the time and intensity of the rain.

Snowfall can be measured in a similar way by melting the snow as it falls into a heated rain gauge. This reduces snowfall to an equivalent rainfall, allowing rainfall and snowfall records to be combined in a single record of precipitation. Ordinarily, a 10-centimetre layer of snow is assumed to be equivalent to 10 millimetres of rain. Because snow can persist on the ground for much of the winter, the accumulated depth of snow is often reported for each month. Traditionally snow stakes—poles calibrated like rulers—have been used to read snow depths, but electronic instruments are now available. Snow depth measurements can be problematic; wind can blow the snow away in some areas, and elsewhere drifting can heap it up to unrepresentative depths. In addition, the snowpack ages over time through settling and melting or sublimation.

The snow season begins in August in Canada's Arctic islands, and by early December, most of the country is covered. In some areas, such as coastal British Columbia and southern Ontario, the snow cover may last for only a few weeks, but it will persist until June or July at high latitudes (Figure 6.16).

The Meteorological Service of Canada defines *blowing snow* as when horizontal visibility is restricted to 10 kilometres or less because the wind has raised snow particles to a height greater than two metres. This occurs at wind speeds of 35–39 kilometres per hour. Blowing snow occurs on about 20 days in the Prairies and other parts of southern Canada, but in northern communities 60–90 days of blowing snow are not uncommon (Figure 6.17). Wind speeds below 35 kilometres per hour are associated with *drifting snow*, which moves at a height of less than two metres. Blizzard conditions develop at wind speeds above 40 kilometres per hour, when visibility is reduced to less than 1 kilometre and temperatures are below –10°C.

PRECIPITATION PROCESSES

Air that is moving upward will be chilled by the adiabatic process to the saturation point and then to condensation. If the air is sufficiently moist, precipitation will occur. Air can be forced into the atmosphere in four ways:

- It can be forced to rise over a range of hills or mountains, leading to **orographic precipitation**.
- It is forced to rise when there is a net inflow due to horizontal **convergence**, as occurs with the northeast and southeast trade winds along the ITCZ.
- During **convection**, a parcel of air is heated by warm ground, so it becomes less dense than the air around it and rises.
- In midlatitudes, air masses of contrasting types meet along frontal zones where colder air tends to be forced under warmer air, forcing it to rise. This process leads to the formation of low pressure systems, and characteristically produces cloudy skies and precipitation. This type of precipitation is known as *cyclonic (or frontal) precipitation*, and is discussed further in Chapter 7.

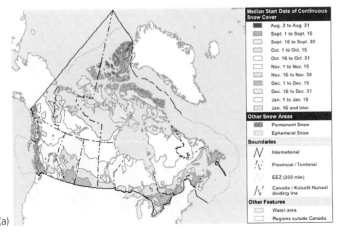

(a)

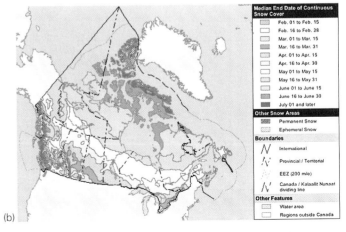

(b)

6.16 Median dates for the start (a) and end (b) of snow cover in Canada. The data are based on the first and last dates with 14 consecutive days of snow cover greater than two centimetres in depth.

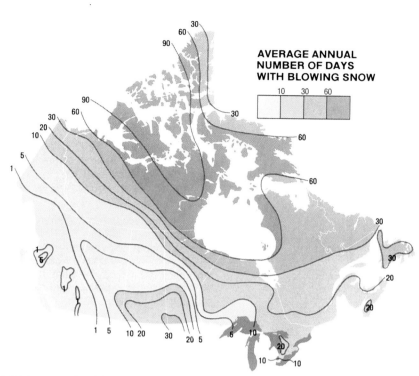

6.17 Average number of days with blowing snow in Canada

OROGRAPHIC PRECIPITATION

In orographic precipitation, winds move moist air up and over high terrain. In the example shown in Figure 6.18, moist air arrives at the coast (1) and rises on the windward side of the mountain range. As the air rises, it is cooled at the dry adiabatic rate until it reaches the lifting condensation level. Here condensation sets in, and clouds begin to form (2). Cooling now proceeds at the wet adiabatic rate. Eventually, precipitation begins and continues to fall as the air continues up the slope.

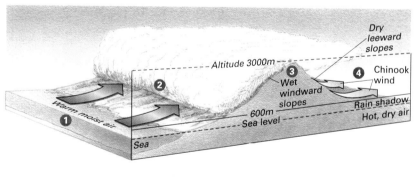

6.18 Orographic precipitation The forced ascent of a warm, moist oceanic air over a mountain barrier produces precipitation and a rain shadow effect. As the air moves up the mountain barrier, it loses moisture through precipitation. As it descends the far slope, it is warmed and may lead to chinook conditions (see chapter 5).

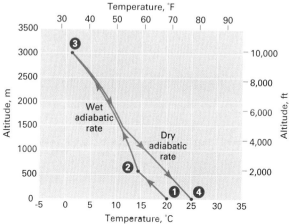

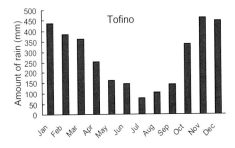

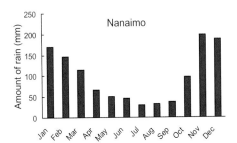

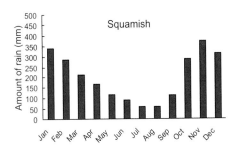

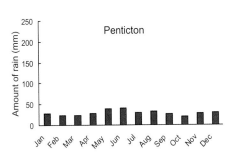

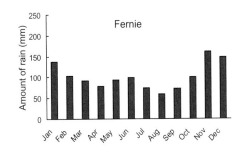

6.19 Orographic precipitation in British Columbia The effect of mountain ranges on precipitation is strong in British Columbia because of the prevailing flow of moist oceanic air from west (e.g., Vancouver) to east (e.g., Fernie). Centres of high precipitation coincide with the western slopes of mountain ranges. Rain shadows occur in the intervening valleys, such as the Okanagan.

After passing over the mountain summit, the air begins to descend down the leeward slopes of the range (3). As it descends, it is compressed and warmed, and cloud droplets and ice crystals evaporate or sublimate. Initially this takes up latent heat so the descending air warms at the wet adiabatic rate, but eventually the sky clears, and warming continues at the dry adiabatic rate. At the base of the mountain (4), the air is now warmer and drier, since much of its moisture was removed in precipitation. This effect creates a *rain shadow*, and under some conditions in winter, chinook winds can occur.

Precipitation patterns in British Columbia are clearly influenced by the orographic process (Figure 6.19). Annual precipitation totals almost 1,600 millimetres at Vancouver, where the onshore winds begin their ascent over the Coast Mountains. Penticton, in the Okanagan Valley, receives about 330 millimetres of precipitation annually. Precipitation rates rise again further inland at places like Fernie, as the air continues its ascent over the Rocky Mountains.

CONVERGENT PRECIPITATION

Convergence is the net addition of masses of air at a given point, and normally arises from opposing airflows. Vertical motions associated with convergence are typically rather weak and will usually produce cirrostratus and similar clouds with little vertical development. However, along the ITCZ, warm, moist tropical air contributes to the rising limbs of the Hadley cells. Once uplift is initiated, energy released through condensation will augment the process, and usually considerable cloud cover and rain develops. Hurricanes are also convergent systems that generate precipitation.

CONVECTIONAL PRECIPITATION

The convection process starts when a surface is heated unequally, resulting in localized pockets of warm air that rise because they are less dense than the surrounding, cooler air. As the air pocket rises, it is cooled adiabatically, but will continue to rise as long as

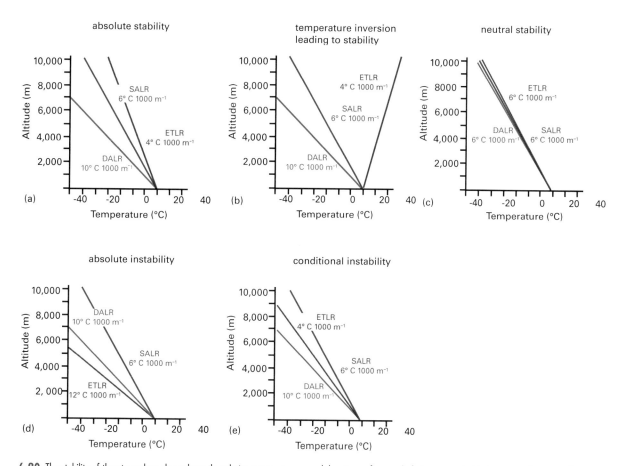

6.20 The stability of the atmosphere depends on the relative temperature — and density — of a parcel of air compared with the surrounding atmosphere.

it remains warmer than the air around it. When it reaches the dew point, condensation occurs, and the rising air becomes visible as a cumulus cloud. The flat base of cumulus clouds marks the lifting condensation level at which condensation begins. As the cloud grows it can shade the ground, which may reduce surface heating and limit further development. However, activity will begin again if the cloud encounters another warm area as it drifts downwind. In addition, if the air is moist, sufficient latent heat can be released to ensure that the uplift will continue.

ATMOSPHERIC STABILITY

A parcel of air that rises may settle back down to the same altitude that it started from, once the force lifting it has been removed. If this occurs, the parcel of air is said to be **stable**. It is denser than the surrounding air. Alternatively, the parcel of air may continue to rise, leading to a condition of instability. The parcel of air is unstable because it is less dense than the surrounding air. The stability of the atmosphere is determined by comparing the temperature of the air at various heights with the temperature of an air parcel as it rises through it. The environmental temperature lapse rate (ETLR) determines the temperature of the surrounding air, while the dry or saturated adiabatic lapse rates (DALR or SALR) determine the temperature of the rising parcel of air.

A condition of **absolute stability** develops when the temperature of the surrounding air is warmer than the air parcel (Figure 6.20a). If stable air is forced to rise, it tends to spread out horizontally, and if clouds form, they tend to be relatively thin layers, such as stratus or altostratus clouds. Inversion layers (Figure 6.20b) are absolutely stable because at any altitude the surrounding air is always warmer than air that is forced upward into the atmosphere. **Neutral stability** (Figure 6.20c) occurs when the ETLR equals the DALR or SALR; a parcel of air that is forced to rise will always be the same temperature as its surroundings. Common ways in which a stable atmosphere can develop include raising the temperature of the air in the upper atmosphere through warm advection or cooling the surface air by nighttime radiation or cold advection.

Absolute instability (Figure 6.20d) develops when the rising air remains warmer and less dense than the surrounding air. The air will continue to rise after the initial lifting force is removed. Absolute instability may occur in shallow surface air layers on hot sunny days. **Conditional instability** (Figure 6.20e) arises when the ETLR is between the DALR and SALR and is associated with moist air. If the rising air parcel remains unsaturated, it is stable unless it rises above the condensation level. At this altitude the air becomes saturated and unstable. Thunderstorms can be triggered when conditionally unstable air rises over mountains. Some general ways in which the atmosphere can become unstable include cooling of the air in the atmosphere by cold advection or radiation cooling, warming of the air at the surface due to daytime heating, or the influx of warm air at the surface.

The terms "stability" and "instability" apply to air that is displaced from its original altitude. Thus, air that is forced down through the atmosphere would be considered stable if later it rises and returns to its original altitude. Conversely, if the air continues to sink, it would be considered unstable.

Figure 6.21 provides an example of how the stability of an air parcel is determined from the ETLR and adiabatic lapse rates. The temperature of the surrounding air at ground level is 26°C and it has an ETLR of 12°C/1,000 metres. Energy conducted from the underlying warm ground heats a parcel of air by 1°C to 27°C, and the air begins to rise. At first, this air parcel cools at the DALR. At 500 metres, the parcel is 22°C, while the surrounding air is 20°C. Since it is still warmer than the surrounding air, it continues to rise. At 1,000 metres, it reaches the lifting condensation level. The temperature of the parcel is 17°C.

As the parcel rises above the condensation level, it cools at the SALR, which here is assumed to be 5°C/1,000 metres. Now the parcel cools more slowly as it rises. At 1,500 metres the parcel is 14.5°C, while the surrounding air is 8°C. Since the parcel is still warmer than the surrounding air, it continues to rise. Note that the difference in temperature between the rising parcel and the surrounding air actually increases with altitude. This means that the parcel will be buoyed upward ever more strongly, forcing even more condensation and precipitation.

The key to the convectional precipitation process is latent heat. When water vapour condenses into cloud droplets or forms ice crystals, it releases latent heat to the rising air parcel. By keeping the parcel warmer than the surrounding air, this latent heat fuels the convection process, driving the parcel ever higher. When the parcel reaches a high altitude, most of its water will have condensed. As adiabatic cooling continues, less latent heat will be released. As a result, the uplift will weaken and eventually cease, and the parcel will dissipate into the surrounding air.

THUNDERSTORMS

Thunderstorms are intense local storms associated with tall, dense cumulonimbus clouds in which there are very strong updrafts of air. They typically consist of several individual convection cells, each with a distinct life cycle that lasts about 30 minutes (Figure 6.22). Air rises within each cell as a succession of bubble-like air parcels, and intense adiabatic cooling produces precipitation. Precipitation can be in the form of water at lower levels, and solid forms at high levels where cloud temperatures are coldest. The uplift rate slows toward the top of the cloud, which in extreme cases may exceed 12 kilometres. A distinctive anvil shape develops where strong winds drag out the top of the thunderstorm cloud.

Ice particles falling from the top of the cloud act as nuclei for freezing and sublimation at lower levels. Ice crystals can be kept in the upper atmosphere by the strong updrafts, but eventually they fall and coalesce into large raindrops as they melt. The falling raindrops drag the air downward, which feeds a

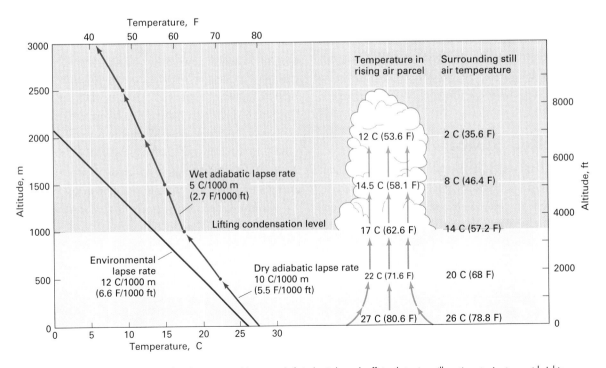

6.21 Convection in unstable air When the air is unstable, a parcel of air that is heated sufficiently to rise will continue to rise to great heights.

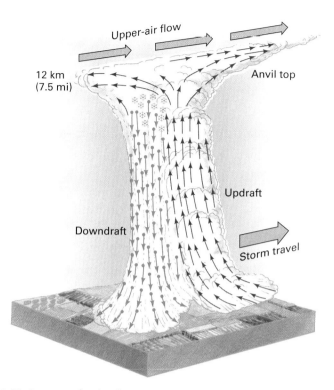

6.22 Anatomy of a thunderstorm A thunderstorm usually consists of several cells in which bubbles of moist, condensing air rise upward. Their upward movement creates a corresponding downdraft, expelling rain, hail, and cool air from the storm as it moves forward.

downdraft within the convection cell. As moisture in the cloud is used up, the supply of latent heat is diminished, updrafts become weaker, downdrafts begin to dominate, and the cell dissipates. However, on reaching the surface, the cold downdraft may force warm, moist surface air upward. This rising air then condenses and builds into a new cell. In this way a thunderstorm can continue for several hours. In some cases the downdrafts emerge from the cloud base with great force to form a gust front or plough wind.

Thunderstorms characteristically produce heavy, localized rainfall. In Canada, the highest rainfall intensity on record was reported at Buffalo Gap, Saskatchewan, in 1961, where 250 millimetres of rain fell in less than an hour. In addition to heavy rainfall, many thunderstorms produce hail. Hailstones have a minimum diameter of 5 millimetres. If they are smaller than that, they are defined as ice pellets. Hailstones are formed by the accumulation of clear and opaque ice layers on ice pellets suspended in the thunderstorm's strong updrafts. The onion-like layering depends on the size and abundance of super-cooled water droplets present in the cloud. Opaque ice forms when small, super-cooled liquid water droplets freeze rapidly on impact and trap air bubbles within the ice. When freezing is slower, air bubbles are able to escape and the ice layer is clear.

Economic losses from hailstorms can be considerable in both urban and rural areas. For example, the hailstorm that hit Calgary on September 7, 1991 caused more than $400 million in damage. Losses to crops are equally devastating. Hail insurance is a necessary expense for many Prairie farmers. On average, hail destroys roughly 3 percent of Prairie crops each year, mostly during a few severe storms. Damage swaths are typically 3–20 kilometres wide and 50–150 kilometres long. To reduce damage from hail, several cloud seeding experiments have been initiated. Typically, aircraft inject silver iodide crystals into the clouds in an attempt to produce smaller and softer hailstones that might melt as they fall to the ground. The success of these attempts at weather modification is still unclear.

Lightning is another phenomenon that results from convection cell activity (Figure 6.23). This occurs when updrafts and downdrafts cause positive and negative static charges to accumulate within different regions of the cloud. Lightning is an electric discharge passing between differently charged parts of the cloud mass or between the cloud and the ground. During a lightning discharge, a current of as much as 100,000 amperes may develop. This heats the air intensely, and makes it expand very rapidly, which generates sound waves that are heard as thunder. Most lightning discharges occur within the cloud, but a significant proportion strike land. Most do little or no damage, although electrical supplies are frequently interrupted. However, lightning is responsible for about 50 percent of the forest fires in Canada and economic losses are considerable.

A thunderstorm is a good example of a flow system of matter and energy. *Focus on Systems 6.5 • The Thunderstorm as a Flow System* examines the thunderstorm from the systems perspective.

6.23 Lightning strikes the CN Tower in Toronto.

FOCUS ON SYSTEMS | 6.5   The Thunderstorm as a Flow System

A thunderstorm is a dramatic and spectacular example of a system in which matter, in the form of air and water, moves from one location to another under the influence of a coupled flow of energy. The diagram of these flows shows (a) the structure of the storm, (b) how water flows and changes state within the storm, and (c) the flow of energy within the storm.

Water flows in a thunderstorm system (b) start as rising cells of warm, moist air. They serve to bring in water vapour at low altitude, which condenses to liquid and solid forms as it moves upward. At the top of the cloud, high altitude, through-flowing winds carry away some ice particles. The remaining ice crystals and water droplets descend in

another part of the cell and strike the ground as rain, hail, or even snow. Thus, water flows upward as vapour, then downward as precipitation.

As the convection cells within the storm cloud move moist air into the atmosphere, condensation converts latent heat to sensible heat, enhancing the convective uplift (part c). At the top of the storm, high-level winds carry a portion of the sensible heat forward and away from the storm. As solid and liquid water particles are carried *downward*, some evaporation and sublimation occur, which convert a portion of the sensible heat released back into latent heat. At the surface, the net effect is that a larger quantity of latent heat is lost than gained, resulting in generally cooler air.

In the atmosphere, the air remains warmer than before the storm, owing to the gain in sensible heat.

The power source for the storm is the latent heat in the water vapour that moves upward. By changing from the vapour to the liquid or solid state, the vapour releases the energy that powers the storm. Some of that energy is recovered when solid or liquid water is converted to vapour in the downdraft cell, but a larger portion is dissipated at higher altitudes. Thus, the storm moves heat from the surface to upper altitudes. The ultimate power source for the storm is the Sun, which heated the surface and brought about the evaporation that created the moist air.

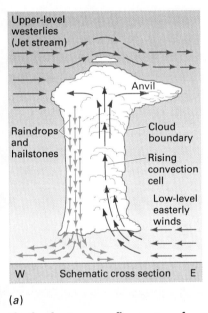

(a)

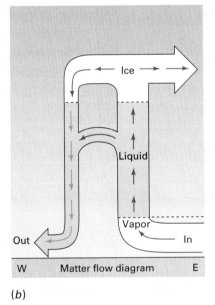

(b)

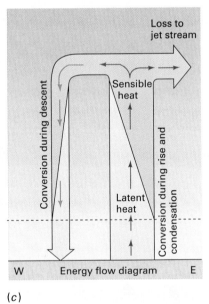

(c)

The thunderstorm as a flow system of matter and energy

MICROBURSTS

The downdraft that accompanies a thunderstorm can sometimes be very intense—so intense that it is capable of causing low-flying aircraft to crash. Such intense downdrafts are called **microbursts** (Figure 6.24). The downward-moving air flows outward in all directions and is often, but not always, accompanied by rain.

An aircraft flying through a microburst first encounters strong headwinds, which may cause a bumpy ride but do not interfere with the airplane's ability to fly. However, as it passes through the far side of the microburst, the aircraft encounters a strong tailwind. The lift of the aircraft's wings depends on the speed of the air flowing across them, and the tailwind greatly reduces this speed. If the tailwind is strong enough, the aircraft may be unable to maintain its altitude and could crash.

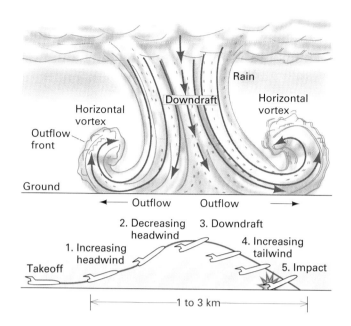

A Look Ahead

This chapter has shown that no matter what process causes precipitation, latent heat is always released. Because a significant amount of oceanic evaporation results in precipitation over land, there is a significant flow of latent heat from oceans to land. In addition, global air circulation patterns move this latent heat toward the poles. This creates the conditions necessary for another precipitation process—that associated with the low-pressure centres that move across the midlatitudes. The following chapter discusses these dynamic weather systems.

6.24 Anatomy of a microburst (above) Schematic cross section showing a downdraft reaching the ground and producing a horizontal outflow. (below) An aircraft taking off through a microburst suffers a loss of lift and could crash. (Adapted from diagrams by Research Applications Program, National Center for Atmospheric Research, Boulder, Colorado)

CHAPTER SUMMARY

- Precipitation is the fall of liquid or solid water from the atmosphere to the Earth's land or ocean surface. Evaporating, condensing, melting, freezing, and sublimation describe water's changes of state.
- Water moves freely between ocean, atmosphere, and land in the hydrologic cycle. The global water balance describes these flows. The fresh water in the atmosphere and on land in lakes, streams, rivers, and *groundwater* is only a small portion of the total water in the hydrosphere.
- Humidity describes the amount of water vapour present in air. The ability of air to hold water vapour depends on temperature. Warm air can hold much more water vapour than cold air.
- Vapour pressure refers to the contribution that water vapour makes to the pressure exerted by the atmosphere. Saturation vapour pressure (SVP) is the maximum amount of water vapour that can be present at a given temperature.
- Specific humidity measures the mass of water vapour in a mass of air, in grams of water vapour per kilogram of air. Relative humidity measures water vapour in the air as the percentage of the maximum amount of water vapour that can be held at the given air temperature. The dew-point temperature, at which condensation occurs, also measures the moisture content of air.
- The *adiabatic principle* states that when a gas is compressed, it warms, and when a gas expands, it cools. When an air parcel moves upward in the atmosphere, it encounters a lower pressure and so expands and cools.
- The dry adiabatic lapse rate describes the rate of cooling with altitude. If the air is cooled below the dew point, condensation or sublimation occurs and latent heat is released. This heat reduces the air parcel's cooling rate with altitude. When condensation or sublimation is occurring, the cooling rate is described as the wet (or saturated) adiabatic lapse rate.
- Clouds are composed of water droplets or ice crystals that form on *condensation nuclei*. Clouds typically occur in layers, as *stratiform* clouds, or in globular masses, as *cumuliform* clouds. *Fog* occurs when a cloud forms at ground level.
- Precipitation from clouds occurs as *rain, hail, snow,* and *sleet*. When *super-cooled* rain falls on a surface at temperatures below freezing, it produces an *ice storm*.
- There are four types of precipitation processes: orographic, convergent, convectional, and *cyclonic*. In orographic precipitation, air moves up and over a mountain barrier. As it moves up, it is cooled adiabatically and rain forms. As it descends the far side of the mountain, it is warmed, producing a *rain shadow* effect.
- Convergent precipitation occurs when moist air flows come together and then rise, as is the case along the ITCZ. Convergent precipitation is also associated with hurricanes.
- In convectional precipitation, unequal heating of the surface causes an air parcel to become warmer and less dense than the surrounding air. Because it is less dense, it rises. As it moves upward, it cools, and condensation with precipitation may occur.
- Stability and instability refer to the tendency for air, if displaced vertically, to return to its former altitude. This air is stable. Air that, when displaced, continues to move away from its original position is considered unstable. Comparing the air parcel's temperature with that of the surrounding air determines its stability.
- Under conditions of unstable air, thunderstorms can form, yielding hail and lightning. (Chapter 8 describes cyclonic precipitation.)

KEY TERMS

absolute instability
absolute stability
adiabatic process
Bergeron process
cloud
collision-wake capture
conditional instability
convectional precipitation
convergent precipitation

curvature effect
dew-point temperature
dry adiabatic lapse rate
evaporation
fog
frost point
hydrologic cycle
hydrosphere
ice nuclei

lifting condensation level
microburst
neutral stability
orographic precipitation
precipitation
relative humidity
saturation vapour pressure
   (SVP)
smog

solute effect
specific humidity
stable air
sublimation
thunderstorm
vapour pressure
wet (or saturated) adiabatic
   lapse rate

REVIEW QUESTIONS

1. Identify the three states of water and the terms used to describe possible changes of state.
2. What is the hydrosphere? Where is water found on the planet? In what amounts? How does water move in the hydrologic cycle?
3. Define saturation vapour pressure. How is the moisture content of air influenced by air temperature?
4. Define relative humidity. How is relative humidity measured? Sketch a graph showing relative humidity and temperature through a 24-hour cycle.
5. Use the terms *saturation*, *dew point*, and *condensation* to describe what happens when a parcel of moist air is chilled.
6. What is the adiabatic process? Why is it important?
7. Distinguish between dry and wet adiabatic lapse rates. In a parcel of air moving upward in the atmosphere, when do

they apply? Why is the wet adiabatic lapse rate less than the dry adiabatic rate? Why is the wet adiabatic rate variable?
8. How are clouds classified? Name four cloud families, two broad types of cloud forms, and three specific cloud types.
9. What is fog? Explain how radiation fog and advection fog form.
10. How is precipitation formed? Describe the process for warm and cool clouds.
11. Describe the orographic precipitation process. What is a rain shadow? Provide an example of the rain shadow effect.
12. What is unstable air? What are its characteristics?
13. Describe the convectional precipitation process. What is the energy source that powers this source of precipitation? Explain.

FOCUS ON SYSTEMS 6.2   The Global Water Balance as a Matter Flow System

1. Sketch a diagram showing five pathways of matter flow for water in the global water balance flow system. Indicate their magnitudes on the diagram.
2. Suppose that global climate warming increases evaporation over land and oceans. Provided that the cycle remains in balance

(that is, $P_{(LAND + LAKE)} + E_{(LAND + LAKE)} + P_{OCEAN} + E_{OCEAN} = 0$), what will be the effect on precipitation?
3. At present, oceans cover about 71 percent of the Earth's surface. Suppose this value were 50 percent. How will the global water balance be affected?

FOCUS ON SYSTEMS 6.5   The Thunderstorm as a Flow System

1. Compare and contrast the energy and matter flows in updraft cells and downdraft cells within a convective storm.
2. How would the convective precipitation process be affected by stable surrounding air?

VISUALIZATION EXERCISES

1. Draw a diagram showing the main features of the hydrologic cycle. Include water flows connecting land, ocean, and atmosphere. Label the flow paths.
2. Draw a graph of the temperature of an air parcel as it moves up and over a mountain barrier, producing precipitation.
3. Sketch the anatomy of a thunderstorm cell. Show rising bubbles of air, updraft, downdraft, precipitation, and other features.

ESSAY QUESTIONS

1. Water in the atmosphere is important for understanding weather and climate. Organize an essay or oral presentation on this topic, focusing on the following questions: What part of the global water supply is atmospheric? Why is it important? What is its global role? How does the capacity of air to hold water vapour vary? How is the moisture content of air measured? Clouds and fog visibly demonstrate the presence of atmospheric water. How do they form?

2. Compare and contrast orographic and convectional precipitation. Begin with a discussion of the adiabatic process and the generation of precipitation within clouds. Then compare the two processes, paying special attention to the conditions that create uplift. Can convectional precipitation occur in an orographic situation? Under what conditions?

PROBLEMS | WORKING IT OUT 6.1 Energy and Latent Heat

1. Figure 6.2 shows that about 490 cubic kilometres of water evaporate from land and water surfaces each year. If all this evaporation occurs from a water surface at 15°C, how much energy is required? (1 cubic metre of water has a mass of about 103 kilograms.)

PROBLEMS | WORKING IT OUT 6.3 The Lifting Condensation Level

1. What is the lifting condensation level for a parcel of air at an initial temperature of 25°C and a dew-point temperature of 18°? What is the temperature of the air parcel at that level?

2. Suppose that the same parcel of air is warmed to 30°C before it begins its ascent. What will be the lifting condensation level for the warmed parcel? What will be its temperature at that level?

Find answers at the back of the book

EYE ON THE LANDSCAPE |

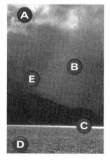

Chapter Opener Rain over Johnson Strait, B.C. The tops of the storm clouds (**A**) show white against the dark sky. A shaft of sunlight (**B**) illuminates a patch of water (**C**) and reflects from the small wavelets (**D**). Heavy rain and cloud obscure the mountains in the background (**E**).

Steam fog rises off of Hamilton Harbour, Ontario. Steam fog forms as water vapour from the warmer, unfrozen sections of Lake Ontario (**A**) condenses in the colder air above. The fog layer is more or less complete near the lake surface where the air is fully saturated (**B**), but begins to dissipate in the drier air above (**C**). The fog plumes rise almost vertically in the calm air near the lake surface, but are slowly drifting away in the faster moving air aloft (**D**).

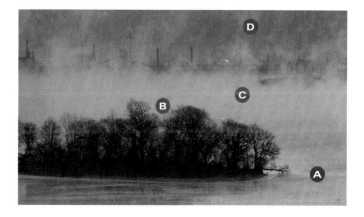

Chapter 7

EYE ON THE LANDSCAPE

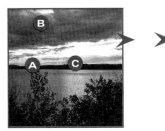

Stormy skies at sunset.
What else would the geographer see? . . . Answers at the end of the chapter.

WEATHER SYSTEMS

Air Masses
North American Air Masses
Travelling Cyclones and Anticyclones
Wave Cyclones
Weather Changes within a Wave Cyclone
Cyclone Tracks and Cyclone Families
The Tornado
Tropical and Equatorial Weather Systems
Easterly Waves and Weak Equatorial Lows
Polar Outbreaks
Tropical Cyclones
Impacts of Tropical Cyclones
Eye on the Environment 7.1 • *Hurricane Katrina*
Geographers at Work • *Dr. Ian McKendry*
Poleward Transport of Heat and Moisture
Atmospheric Heat and Moisture Transport
Oceanic Heat Transport
Cloud Cover, Precipitation, and Global Warming

The Earth's atmosphere is in constant motion, driven by pressure gradients and steered by the Coriolis effect. Air that has acquired characteristic temperature and humidity properties in one location is moved horizontally by the wind to new locations. Vertical motion in the atmosphere favours cloud formation and precipitation when air is lifted and cooled adiabatically; however, when air subsides, moisture returns to the vapour state through compression and adiabatic warming. In this way, the Earth's wind system influences the day-to-day weather.

Some patterns of wind circulation occur regularly, and give rise to recurring weather conditions. For example, travelling low-pressure centres often bring clouds and precipitation. Outbreaks of polar air typically give cold, clear weather in winter. These recurring circulation patterns and the conditions associated with them are called **weather systems**.

Weather systems range in size from a thousand kilometres or more, in the case of a large travelling anticyclone, down to one kilometre or so, in the case of a tornado. They may last for hours or weeks, depending on their size and strength. Some forms of weather systems—thunderstorms, for example, and less frequently, tornadoes and hurricanes—involve strong winds and heavy rainfall and can be destructive to life and property.

AIR MASSES

An **air mass** is a large body of air with fairly uniform temperature and moisture characteristics. It can be several thousand kilometres across and can extend upward to the top of the troposphere. A given air mass is characterized by a distinctive combination of surface temperature, environmental temperature lapse rate, and surface-specific humidity. Air masses acquire their characteristics in *source regions* where the air moves slowly or stagnates. For example, an air mass over a tropical ocean is characteristically warm and has a high water vapour content. Over a large tropical desert, slowly subsiding air forms a hot air mass with low humidity. Snow-covered land surfaces in the Arctic give rise to very cold air masses with low water vapour content.

Air masses move from one region to another under the influence of pressure gradients and upper-level wind patterns, and are sometimes pushed or blocked by high-level jet stream winds. When an air mass moves to a new area, its temperature and moisture properties will begin to change because they are influenced by the new surface environment.

Air masses are classified on the basis of the latitude where they originate and the nature of the underlying surface of their source regions. Latitude primarily determines surface temperature and the environmental temperature lapse rate of the air mass, while the nature of the underlying surface—continent or ocean— usually determines the moisture content. Five types of air masses are distinguished with respect to latitude. These are referred to as Arctic (A), Antarctic (AA), Polar (P), Tropical (T), and Equatorial (E). For the type of underlying surface, two subdivisions are used: maritime (m) and continental (c).

Air Mass	Symbol	Source Region
Arctic	A	Arctic Ocean and lands on its fringes
Antarctic	AA	Antarctica
Polar	P	Continents and oceans, lat. 50–60° N and S
Tropical	T	Continents and oceans, lat. 20–35° N and S
Equatorial	E	Oceans close to the equator
Maritime	m	Oceans
Continental	c	Continents

Six important types of air masses can be identified from these descriptive labels (Table 7.1). Table 7.1 lists the general locations of the source regions together with the typical values of surface temperature and specific humidity, although these can vary seasonally. Air mass temperature can range from – 46°C for continental arctic (cA) air masses, to 27°C for maritime equatorial air masses (mE). Specific humidity shows a high range—from 0.1 grams per kilogram for the cA air mass to as much as 19 grams per kilogram for the mE air mass. Thus, maritime equatorial air can hold about 200 times more moisture than continental arctic air.

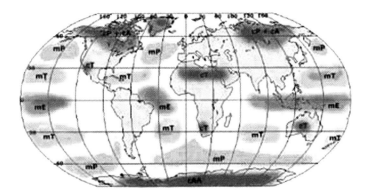

7.1 Global air masses and source regions Tropical (T) and equatorial (E) source regions provide warm or hot air masses, while polar (P), Arctic (A), and Antarctic (AA) source regions provide colder air masses of low specific humidity. Maritime (m) air masses originating over oceans are moister than those associated with continental (c) land surfaces.

Figure 7.1 shows the generalized global distribution of source regions of these air masses. Note that the polar air masses (mP, cP) originate in the subarctic latitude zone, not in the polar latitude zone.

The maritime tropical air mass (mT) and maritime equatorial air mass (mE) originate over warm oceans in the tropical and equatorial zones. They are quite similar in temperature and water vapour content. With their high specific humidity values, both are capable of heavy precipitation. The continental tropical air mass (cT) has its source region over subtropical deserts on the continents. Although this air mass will contain water vapour, it tends to be stable and has low relative humidity when heated strongly during the daytime.

The maritime polar air mass (mP) originates over midlatitude oceans. Since the quantity of water vapour it holds is not as large as maritime tropical air masses, an mP air mass typically yields less precipitation. Much of this precipitation is orographic, and occurs

Table 7.1 Properties of typical air masses

Air Mass	Symbol	Source Region	Properties	°C	Specific Humidity (g/kg)
Maritime equatorial	mE	Warm oceans in the equatorial zone	Warm, very moist	27°	19
Maritime tropical	mT	Warm oceans in the tropical zone	Warm, moist	24°	17
Continental tropical	cT	Subtropical deserts	Warm, dry	24°	11
Maritime polar	mP	Midlatitude oceans	Cool, moist (winter)	4°	4.4
Continental polar	cP	Northern continental interiors	Cold, dry (winter)	−11°	1.4
Continental arctic (and continental antarctic)	cA (cAA)	Regions near North and South, Poles	Very cold, very dry (winter)	−46°	0.1

over mountain ranges on the western coasts of continents. Where prevailing winds bring a steady supply of mP air to an area, the amount of precipitation can be substantial. For example, average annual precipitation at Prince Rupert, on the northwest coast of British Columbia, exceeds 3,000 millimetres. The continental polar air mass (cP) originates over North America and Eurasia in the subarctic zone. It has low specific humidity and is very cold in winter. Finally, the continental Arctic (and continental Antarctic) air mass type (cA, cAA), is extremely cold and holds almost no water vapour.

NORTH AMERICAN AIR MASSES

Figure 7.2 shows the source regions of air masses that have a strong influence on the weather in North America. Continental polar (cP) air masses originate over north-central Canada. These air masses form tongues of cold, dry air that periodically extend south and east from the source region to produce anticyclones accompanied by cool or cold temperatures and clear skies. Arctic air masses (cA) that develop in winter over the Arctic Ocean and its bordering lands are extremely cold and stable. When an Arctic air mass moves southward, it produces a severe cold wave that occasionally will penetrate as far south as the Gulf of Mexico.

Maritime polar air masses (mP) originate over the North Pacific and Bering Strait, in the region of the persistent Aleutian low-pressure centre. These air masses are characteristically cool and moist, with a tendency in winter to become unstable, giving

heavy precipitation over the coastal ranges in Alaska, British Columbia, and Washington. Other maritime polar air masses originate over the North Atlantic Ocean. They, too, are cool and moist, and are felt in Newfoundland and Labrador, the Maritimes, and New England, especially in winter.

Maritime tropical air masses (mT) from the Gulf of Mexico most commonly affect the central and eastern United States. As they move northward, they bring warm, moist, unstable air to the eastern part of the continent. Because of these air masses, weather conditions in the summer are often hot and uncomfortably humid; they also produce many thunderstorms. Maritime tropical air masses from the Atlantic Ocean also are brought onshore by air flowing out of the subtropical (Azores) high pressure cell; these air masses mostly affect the eastern seaboard of the United States. Many of the hurricanes that move into the continent originate in this source region.

In the Pacific Ocean, the source region for maritime tropical air masses lies at the eastern margins of the semi-permanent Hawaiian high pressure cell. Occasionally, in summer, this moist, unstable air mass penetrates the southwestern desert region, bringing severe thunderstorms to southern California and southern Arizona. In winter, a tongue of mT air frequently reaches the California coast, bringing heavy rainfall that is intensified when forced to rise over coastal mountain ranges.

A hot, dry continental tropical air mass (cT) originates over northern Mexico, western Texas, New Mexico, and Arizona during the summer. Although this air mass does not travel widely, it occasionally reaches into the Canadian Prairies and brings short spells of unusually hot, dry weather.

The properties of air masses are modified as they move away from their source regions. The degree of modification depends on their speed and also the general routes that they follow. For example, a continental polar air mass that originates over the Yukon may retain its general characteristics if it moves into the Prairies through northern Saskatchewan. However, if it moves eastwards and eventually passes over Hudson Bay into Ontario, its temperature and moisture properties could be greatly altered depending on the speed at which it moves and the season. For example, in late autumn a polar air mass may pass from a cold, snow-covered land surface to a warm water surface, and in turn become warmer and moister.

An air mass usually has a well-defined boundary between itself and a neighbouring air mass. These transitional zones are termed **fronts**, across which there are marked differences in atmospheric conditions. Fronts are three-dimensional in that they extend vertically into the upper troposphere as well as horizontally across a region. In North America, the boundary between polar and tropical air masses produces the well-defined *polar front*, which is located below the axis of the jet stream.

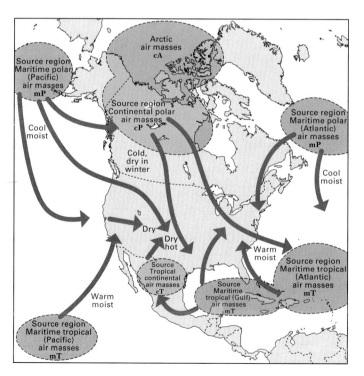

7.2 North American air mass source regions and trajectories *Air masses acquire temperature and moisture characteristics in their source regions, and then move across the continent. (Data from the U.S. Weather Service)*

TRAVELLING CYCLONES AND ANTICYCLONES

Air masses are set in motion by the pressure gradients caused by uneven heating or cooling of the atmosphere. They may be driven along by the general global wind system, or by the spiralling motion that develops due to pressure gradients associated with low pressure and high pressure centres. Air spirals inward and converges in a *cyclone* (low pressure), while air spirals outward and diverges in an *anticyclone* (high pressure). Most cyclones and anticyclones are large features that extend for hundreds of kilometres and move slowly across the Earth's surface. They are especially characteristic of the midlatitudes, where *travelling cyclones* and *travelling anticyclones* bring changing weather conditions as they pass across the region.

In a cyclone, convergence and upward motion cause air to rise and be cooled adiabatically. If the air is moist, condensation or sublimation can occur. This is **cyclonic precipitation**. Many cyclones are weak and pass overhead with little more than a period of cloud cover or light precipitation. However, when pressure gradients are steep and the air is laden with water vapour, strong winds and heavy rain or snow can accompany the cyclone. In this case, the disturbance is called a **cyclonic storm**.

Travelling cyclones fall into three types. First is the wave cyclone of the midlatitude, Arctic, and Antarctic zones. This type of cyclone ranges in intensity from a weak disturbance to a powerful storm. Second is the tropical cyclone associated with tropical and subtropical zones. This type ranges in intensity from a mild disturbance to the highly destructive hurricane or typhoon. A third type is the tornado, a small, intense cyclone of enormously powerful winds. The tornado is much smaller in size than other cyclones and is related to exceptionally strong convectional activity.

In an anticyclone, divergence and settling cause air to descend and be warmed adiabatically. Thus, condensation does not occur. Skies are clear, except for occasional puffy cumulus clouds that sometimes develop in a moist surface air layer due to weak local convection. Because of these characteristics, anticyclones are often termed *fair-weather systems*. Toward the centre of an anticyclone, the pressure gradient is weak, and winds are light and variable. Travelling anticyclones are found in the midlatitudes. They are typically associated with clear, dry air that moves eastward and toward the equator. Figure 7.3 is a geostationary satellite image of eastern North America. A large anticyclone is centred over the area, bringing fair weather and cloudless skies.

7.3 Anticyclonic conditions This geostationary satellite image of eastern North America shows a large anticyclone centred over the area, bringing fair weather and cloudless skies. The boundary between clear sky and clouds running across the Gulf of Mexico and the Florida peninsula delineates the leading edge of the cool, dry air mass. The cloud edge at the top of the photo marks the cold fronts of two air masses advancing eastward and southward.

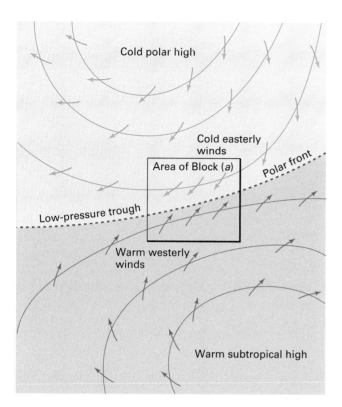

7.4 Conditions for the formation of a wave cyclone Two anticyclones, one with warm subtropical air and the other with cold polar air, are in contact at the polar front. The shaded area in block (a) of Figure 7.5 shows the early stage of development of a wave cyclone.

WAVE CYCLONES

In mid- and high latitudes, the dominant form of weather system is the **wave cyclone**, a large mass of inward-spiralling air that repeatedly forms, intensifies, and dissolves along the polar front. Figure 7.4 shows a situation favourable to the formation of a wave cyclone. Two large anticyclones are in contact on the polar front. The boundary constitutes a *low pressure trough* between the two high pressure cells. One contains a cold, dry polar air mass, and the other a warm, moist maritime air mass. Air flow converges from opposite directions toward the trough. Air flow from the subtropical high is from the southwest, and winds from the polar high are predominantly from the northeast. The converging air flows create an unstable situation leading to the development of a wave cyclone.

Figure 7.5 (a-d) shows the life cycle of a wave cyclone. In the early stage (Figure 7.5a), the polar-front region shows a wave beginning to form. Cold air is turned in a southerly direction and warm air in a northerly direction, producing a notch on the polar front in which the two air masses begin to advance on each other. As these frontal motions develop, warm air is lifted over the cold air, leading to cloud formation and precipitation.

7.5 Development of a wave cyclone In (a), a wave motion begins at a point along the polar front. The wave along the cold and warm fronts deepens and intensifies in (b). In (c), the cold front overtakes the warm front, producing an occluded front in the centre of the cyclone. Later, the polar front is re-established with a mass of warm air isolated in the atmosphere (d).

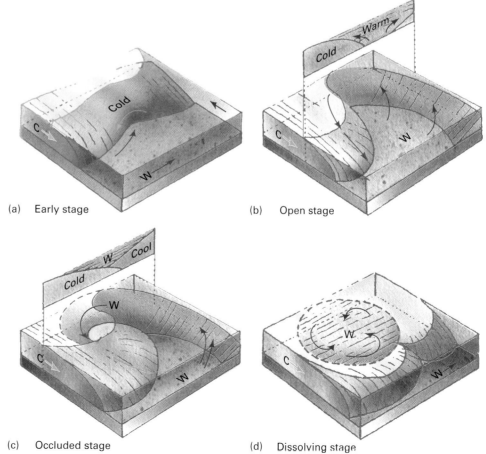

(a) Early stage

(b) Open stage

(c) Occluded stage

(d) Dissolving stage

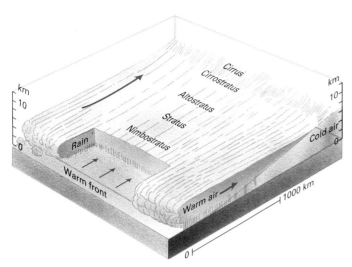

7.6 Warm front In a warm front, warm air advances toward cold air and rides up and over it. A notch of cloud is cut away to show where rain is falling from the dense stratus cloud layer. (Drawn by A.N. Strahler)

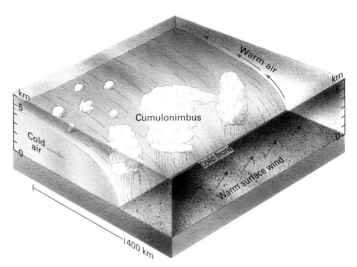

7.7 Cold front At a cold front, a cold air mass lifts a warm air mass into the atmosphere. The upward motion can set off a line of thunderstorms. Note that the frontal boundary is actually much less steep than is shown in this schematic drawing. (Drawn by A.N. Strahler)

In the *open stage* (Figure 7.5b), the wave disturbance along the polar front has deepened and intensified. Warm air actively moves northeastward along a warm front, and cold air pushes southward along a cold front.

Figure 7.6 shows a **warm front** in which warm air moves into a region of colder air. The cold air mass remains in contact with the ground because it is denser. Consequently, the moving mass of warm air is forced to rise over the cold air below. The gradient on a warm front is usually about 1:300, and results from the cold, denser air in the leading sector being feathered out by friction at the surface as it moves across the region. The rising motion is therefore quite gradual and a characteristic sequence of clouds tends to build up. When the warm air contains a lot of moisture, cloud formation commences, with high cirrus clouds becoming progressively thicker and lower until the sky is obscured with low stratus clouds. Precipitation often follows, which, although not intense, will often last for several hours or even days.

A **cold front** develops when a cold air mass invades a zone occupied by a warm air mass (Figure 7.7). Because the colder air mass is denser, it remains in contact with the ground. The gradient of a cold front is typically 1:50, so it is much steeper than a warm front. In this case, the front represents the leading edge of the approaching cold air. Because friction at the surface slows the rate of the front's advance, the air above pushes forward, causing the front to become steeper. Consequently, as it moves forward, it forces the warmer air mass to rise quite abruptly. If the warm air is unstable, this rapid uplift will often produce thunderstorms and heavy precipitation for a comparatively short time. In some cases, severe thunderstorms may develop as a squall line—a long line of massive clouds stretching for tens of kilometres (Figure 7.8).

Cold fronts normally move across a region at a faster rate than warm fronts. Thus, when both types are in close proximity, a cold front can overtake a warm front, which produces the *occluded stage* (Figure 7.5c). The colder air of the fast-moving cold front remains next to the ground, forcing the warm air ahead of it to rise along the **occluded front** (Figure 7.9). Either warm or cold occlusions develop, depending on the relative temperatures of the opposing air masses and the degree of modification they have undergone since they left their source regions. In warm occlusions, the warm air continues to rise fairly slowly over the cold air and the resulting clouds and precipitation is similar to that associated with a warm front. Cold occlusions develop where the trailing sector consists of colder denser air, which is actively forcing its way under the system. Uplift is more abrupt, producing conditions similar to those associated with cold fronts.

Eventually, the polar front is re-established in the *dissolving stage* (Figure 7.5d), but a pool of warm, moist air remains aloft in

7.8 Squall line This ominous cloud front is the leading edge of a severe squall line that will bring severe thundershowers and strong winds.

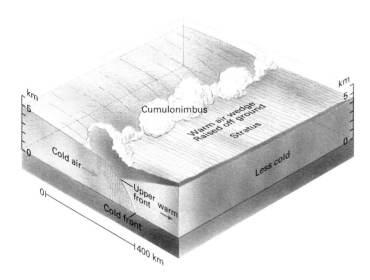

7.9 Occluded front In an occluded front, a cold front overtakes a warm front. The warm air is pushed upward, and it no longer contacts the ground. Abrupt lifting by the denser cold air produces precipitation. (Drawn by A.N. Strahler)

the atmosphere. Once the warm air mass is lifted completely free of the ground, the term TROWAL is sometimes used to indicate a **TRO**ugh of **W**arm air **AL**oft. As the moisture content of the pool of air lessens, precipitation dies out and the clouds gradually dissolve.

Keep in mind that a wave cyclone is quite a large feature—1,000 kilometres or more across. Also, the cyclone normally moves eastward, propelled by prevailing westerlies in the atmosphere. Therefore, the different stages of the system shown in Figure 7.5a-d develop progressively over the course of several days as it tracks eastward.

WEATHER CHANGES WITHIN A WAVE CYCLONE

Figure 7.10 shows two simplified weather maps of eastern Canada depicting conditions on successive days. The structure of the storm is defined by the isobars. Special line symbols represent the three kinds of fronts. Areas of precipitation are shown in grey.

The map on the left shows the cyclone in an open stage, similar to Figure 7.5b. The isobars show that the cyclone is a

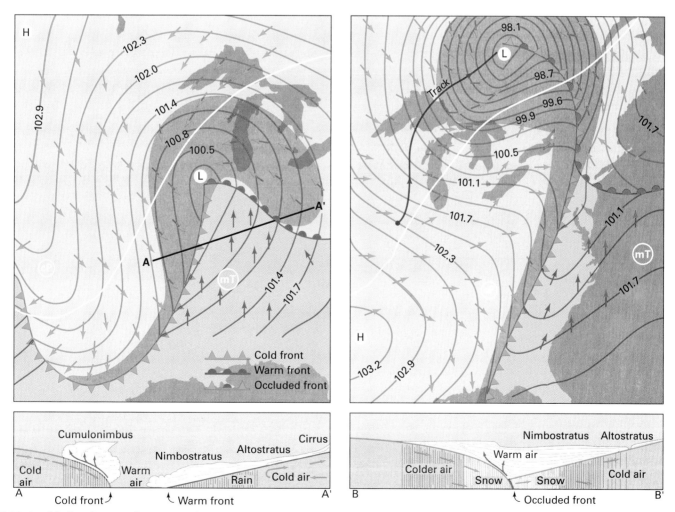

7.10 Simplified surface weather maps and cross-sections through a wave cyclone In the open stage (left), cold and warm fronts pivot around the centre of the cyclone. In the occluded stage (right), the cold front has overtaken the warm front, and a large pool of warm, moist air has been forced aloft.

low-pressure centre with inward-spiralling winds. The cold front is pushing south and east, supported by a flow of cold, dry continental polar air from the northwest filling in behind it. Note that the wind direction changes abruptly as the cold front passes. There is also a sharp drop in temperature behind the cold front as cP air fills in. The warm front is moving north and somewhat east, with warm, moist maritime tropical air following. The precipitation pattern includes a broad zone near the warm front and the central area of the cyclone. A thin band of precipitation extends down the length of the cold front. Cloudiness generally prevails over much of the cyclone.

A cross-section along the line A–A' (Figure 7.11) shows how the fronts and clouds are related. Along the warm front is a broad layer of stratus clouds. This extends ahead of the front and takes the form of a wedge with a thin leading edge of cirrus. As this wedge thickens, the sequence of clouds changes to altostratus, then to stratus, and finally to nimbostratus with steady rain.

Within the sector of warm air, the sky may partially clear with scattered cumulus. Along the cold front are cumulonimbus clouds associated with thunderstorms. These yield heavy rains but only along a narrow belt, and the skies clear quickly after the cold front passes. Note that the actual conditions that occur will depend on the position of the observer relative to the centre of the weather system and on the specific properties of the air masses involved.

Twenty-four hours later the cyclone has moved rapidly northeastward, its track shown by the red line. The centre has moved about 1,600 kilometres in 24 hours—a speed of just over 65 kilometres per hour. The cold front has overtaken the warm front, forming an occluded front in the central part of the disturbance. A high-pressure area, or tongue of cold polar air, has moved in to the area west and south of the cyclone, and the cold front has moved far south and east. Within the cold air tongue, the skies are clear, but where the warm air mass is lifted well off the ground, there is heavy precipitation.

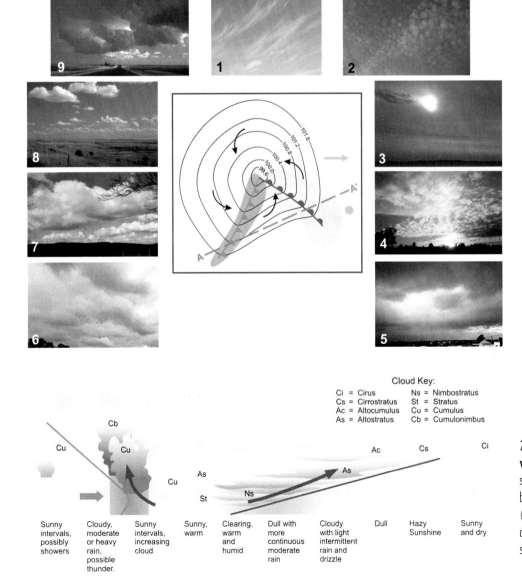

7.11 Representative cross-section of a well-developed midlatitude cyclone The sequence of clouds and weather conditions would be expected along, shown here as X-Y. cloud types: (1) cirrus, (2) cirrocumulus, (3) altocumulus, (4) altostratus, (5) nimbostratus, (6) stratus, (7) stratocumulus, (8) cumulus, (9) cumulonimubus

7.12 Daily world weather map A daily weather map of the world for a given day during July or August might look like this map, which is a composite of typical weather conditions. Note the trains of wave cyclones (lows) in various stages of development moving through the midlatitudes. (After M.A. Garbell)

CYCLONE TRACKS AND CYCLONE FAMILIES

Wave cyclones tend to form in certain areas and travel common paths until they dissolve. In the northern hemisphere, wave cyclones are heavily concentrated in the neighbourhood of the Aleutian and Icelandic lows. They commonly form in a succession, travelling in chains across the North Atlantic and North Pacific oceans (Figure 7.12). Many wave cyclones tend to form in groups that develop one after the other, until the opposing air masses have lost their individuality. The secondary wave cyclones typically form along the original trailing cold front. Thus, in the northern hemisphere, as the polar air pushes out from its source region, the eastward course that the successive low pressure centres follow is progressively displaced to the south. The sequence terminates in the formation of a broad zonal ridge of high pressure.

The western coast of North America commonly receives wave cyclones arising in the North Pacific Ocean, but they can also originate over the continent, especially in Alaska and northern Canada. Most of these tracks converge toward the northeast, often passing into the Gulf of St. Lawrence due to the general surface airflow directed by the Appalachian Mountains. Once in the North Atlantic, they tend to concentrate in the region of the Icelandic low, and then move into Europe.

In the southern hemisphere, storm tracks more often form along a single lane, following the parallels of latitude. The storm tracks are less variable because the extensive ocean surface circling the globe at these latitudes is broken only by the southern tip of South America.

The general eastward movement of wave cyclones is largely controlled by the paths of the Rossby waves and their relationship to the polar front jet streams. The curving path of the Rossby waves is a prime factor in wave cyclone development. Air in the atmosphere alternately converges or diverges in the frontal zones, and this is linked to surface conditions (Figure 7.13).

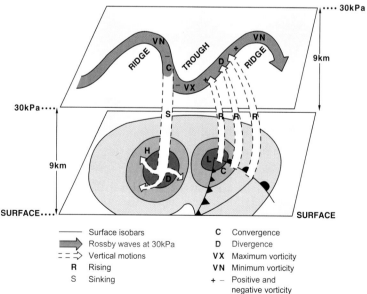

7.13 The meandering pattern of the Rossby waves in the atmosphere is linked to surface pressure conditions The fast moving air alternately converges and diverges as it bends toward and away from the equator, and this complements inflow and outflow of air in surface pressure systems.

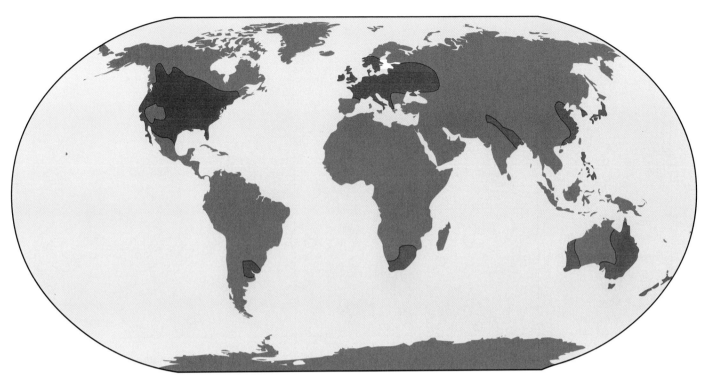

7.14 Tornado regions. Apart from North America, tornadoes occur quite frequently in western Europe, and Australia, and less commonly in South America, South Africa and the Far East.

However, mountain barriers and ocean surface temperatures also influence the routes followed by wave cyclones and their associated weather patterns. For example, the passage of a wave cyclone across the Rocky Mountains often fails to produce precipitation because orographic lifting removes the moisture on the windward side of the mountains. Similarly, unusually cold water in the north-central Pacific Ocean in winter tends to push storm tracks farther north across North America, whereas warmer surface water in this region tends to displace storm tracks well to the south of their normal position.

THE TORNADO

A **tornado** is a small but intense cyclonic vortex in which air spirals at tremendous speed. It is associated with thunderstorms spawned by fronts in midlatitudes. Tornadoes occur in several parts of the world (Figure 7.14), but nowhere are they as frequent, or as violent, as in North America. They also occur in Australia in substantial numbers and are occasionally reported from other midlatitude locations, such as Russia. In Canada, tornadoes are most frequently spotted in the Prairie Provinces, southern Ontario, and southern Quebec. Significant tornado damage has been reported in Windsor, Sudbury, and Barrie, Ontario. The best known tornadoes in Canada are the Regina tornado, which occurred on June 30, 1912; the Edmonton tornado (July 31, 1987); and more recently, the Pine Lake tornado. Pine Lake, located about 25 kilometres southeast of Red Deer, Alberta, was struck by a tornado at about 6 p.m. on July 14, 2000. The tornado, with winds estimated in excess of 300 kilometres per hour and accompanied by baseball-sized hail, cut a swath 800–1,500 metres wide for a distance of about 15 kilometres. The tornado caused 12 fatalities and about 100 injuries as it ploughed through the Pine Lake campground (Figure 7.15).

7.15 The Pine Lake Tornado A tornado of F3 magnitude (see Table 7.2) cut a swath more than 800 metres wide through the campground at Pine Lake, Alberta, on July 14, 2000, causing 12 fatalities and about 100 injuries.

7.16 Tornado Surrounding the funnel is a cloud of dust and debris carried into the air by the violent winds.

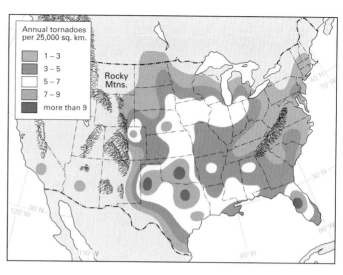

7.17 Frequency of occurrence of observed tornadoes in the United States and southern Canada The data shown on this map span a 30-year record, 1960–89. (Courtesy of Edward W. Ferguson, National Severe Storms Forecast Center, National Weather Service)

A tornado appears as a dark funnel cloud hanging from the base of a dense cumulonimbus cloud (Figure 7.16). At its lower end, the funnel may be 100 to 450 metres in diameter. The base of the funnel appears dark because of the density of condensing moisture, dust, and debris swept up by the wind. Wind speeds in a tornado exceed those recorded in any other type of storm. Estimates of wind speed run as high as 100 metres per second (about 360 kilometres per hour). As the tornado moves across the countryside, the funnel writhes and twists. Where it touches the ground, it can cause the complete destruction of almost anything in its path. The Fujita scale assesses the strength of a tornado (Table 7.2).

Tornadoes occur most commonly in the Mississippi valley, which has earned the name "tornado alley" (Figure 7.17). They are most common in the spring and summer but can occur in any month, with usually more than 1,000 occurring in any given year. In spring, cold, dry air from the high plains will override warm maritime air entering the central lowlands from Gulf of Mexico, so that directly above the warm surface air is a wedge of cold dry air at the 70–50 kilopascal level. Higher still, at the 30 kilopascal level, the polar front jet swings over the region, causing an area of upper air divergence that initiates surface convergence and rising air.

The presence of cold, dry air above warm, moist air creates an inherently unstable atmosphere, but initially a temperature inversion separates these different layers. However, the atmosphere becomes increasingly unstable as daytime warming continues and some lifting occurs at frontal zones. Eventually the inversion breaks and convective instability causes the moist surface air to rise rapidly. The strong wind shear that develops in severe thunderstorms provides the rotation that can lead to a tornado.

Table 7.2 The Fujita scale of tornado intensity

Scale	Estimated Wind Speed (km/h)	Typical Damage
F0	<120	**Light damage**—some damage to chimneys; branches broken off trees; shallow-rooted trees pushed over; sign boards damaged
F1	120–185	**Moderate damage**—peels surface off roofs; mobile homes pushed off foundations or overturned; moving automobiles blown off roads
F2	186–260	**Considerable damage**—roofs torn off frame houses; mobile homes demolished; boxcars overturned; large trees snapped or uprooted; light-object missiles generated; cars lifted off ground
F3	261–340	**Severe damage**—roofs and some walls torn off well-constructed houses; trains overturned; most trees in forest uprooted; heavy cars lifted off the ground and thrown
F4	341–430	**Devastating damage**—well-constructed houses levelled; structures with weak foundations blown away some distance; cars thrown; and large missiles generated
F5	431–530	**Incredible damage**—strong frame houses levelled off foundations and swept away; automobile-sized missiles fly through air in excess of 100 metres; trees debarked; incredible phenomena occur

TROPICAL AND EQUATORIAL WEATHER SYSTEMS

Weather systems of the tropical and equatorial zones show some basic differences from those of the midlatitudes. Upper-air winds are often weak, so air mass movement is slow and gradual. Air masses tend to be warm and moist, and so have quite similar characteristics. Thus, clearly defined fronts and large, intense wave cyclones are missing. On the other hand, strong convectional activity occurs because of the high moisture content of low-latitude maritime air masses. With these conditions, even slight convergence and uplifting can be enough to trigger precipitation.

EASTERLY WAVES AND WEAK EQUATORIAL LOWS

One of the simplest forms of tropical weather systems is an **easterly wave**, a slowly moving, convectively active trough of low pressure that is driven westward by the trade winds. In the Atlantic, the waves are generated by instability over northwestern Africa associated with the low latitude easterly jet. They are then directed westwards in the troposphere by the winds of the Azores high pressure cell. They first develop in April or May and form about every three or four days until October or November. They move at a rate of about 20–35 kilometres per hour. Approximately two easterly waves per week travel from Africa to North America during hurricane season. Easterly waves occur in latitudes 5° to 30° N and S over oceans, but not over the equator itself, where strong convective uplift generates the rising limb of the Hadley cells. Figure 7.18 shows the upper-troposphere isobars, wind patterns, and the zone of shower activity associated with an easterly wave. At the surface, a zone of weak low pressure underlies the axis of the wave. Surface airflow converges on the eastern, or trailing, side of the wave axis. This convergence causes the moist air to be lifted, producing scattered showers and thunderstorms. The rainy period may last a day or two as the wave passes.

Another related weather system is the *weak equatorial low*, a disturbance that forms near the centre of the equatorial trough. Moist equatorial air masses converge on the centre of the low, causing rainfall from many individual convectional storms. Several such weak lows are shown lying along the Intertropical Convergence Zone (ITCZ) on the world weather map (Figure 7.12). Because the map is for a day in July or August, the ITCZ is shifted well north of the equator. During this season, the rainy monsoon in progress in Southeast Asia is marked by a weak equatorial low in northern India.

POLAR OUTBREAKS

Another distinctive feature of low-latitude weather is the occasional penetration of powerful tongues of cold polar air from the midlatitudes into very low latitudes. These tongues are known as *polar outbreaks*. The leading edge of a polar outbreak is a cold front with squalls, which is followed by unusually cool, clear weather with strong, steady winds. The polar outbreak is best developed in the Americas. Outbreaks that move southward from the United States into the Caribbean Sea and Central America are called "northers" or "nortes," while those that move north from Patagonia into tropical South America are called "pamperos." One such outbreak is shown as a high pressure cell over South America on the world weather map (Figure 7.12). A severe polar outbreak may bring subfreezing temperatures to the highlands of South America, severely damaging coffee and other important crops.

TROPICAL CYCLONES

The most powerful and destructive type of cyclonic storm is the **tropical cyclone**, which is known as the *hurricane* in the western hemisphere, the *typhoon* in the western Pacific off the coast of Asia, and the *cyclone* in the Indian Ocean. The World Meteorological Organization classifies a tropical cyclone as a storm of tropical origin of small diameter with a minimum surface pressure below 95 kilopascals, with violent winds greater than 125 kilometres per hour, and accompanied by torrential rain. The storms revolve around a central "eye," which is about 30 to 50 kilometres in diameter and characterized by light winds and clear skies. The distinctive pattern of inward-spiralling bands of clouds and clear central eye make it easy to track tropical cyclones by satellite (Figure 7.19).

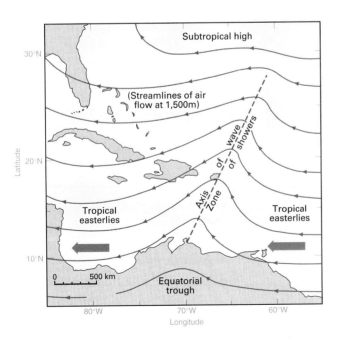

7.18 Easterly wave An upper-air easterly wave passing over the West Indies at an altitude of 1,500 metres (Data from H. Riehl, Tropical Meteorology, New York: McGraw-Hill)

temperatures are too low to ensure enough evaporation for the latent heat release required to maintain a sufficient supply of the energy for the storm. From 7° latitude to the equator, the Coriolis effect is too weak to initiate the storm's circular motion.

Once formed, the storms move westward through the trade-wind belt, often intensifying as they travel (Figure 7.20). In the northern hemisphere, they travel westward and northwestward through the trade-wind belts, and then turn northeast at about 30° to 35° N, where they are steered by westerly winds in the atmosphere. Their intensity lessens as they move over cooler ocean surfaces because their source of energy is reduced; they are further weakened by friction if they move over land. In the trade-wind belt, the cyclones travel 10 to 20 kilometres per hour; however, in the zone of the westerlies, their speed is more variable. Tropical cyclones can penetrate into the midlatitudes, bringing severe weather to the southern and eastern coasts of the United States. Although heavy rain and strong winds often persist as far north as the Maritime provinces, at these latitudes hurricanes are normally downgraded to tropical storms. Note that tropical cyclones never cross the equator.

An intense tropical cyclone is an almost circular storm of extremely low pressure. The storm's diameter may be 150 to 500 kilometres. Because of the strong pressure gradient, winds spiral inward at high speed, which often exceeds 30 to 50 metres per second (100 to 180 kilometres per hour). Barometric pressure in the storm centre commonly falls to 95 kilopascals or lower.

7.19 Hurricane Katrina in the Gulf of Mexico, August 29, 2005 Katrina was designated a Category 5 Hurricane (see Table 7.3), with sustained winds greater than 250 kilometres per hour. This image comes from the GOES-12 weather satellite.

This type of storm develops over oceans in 7° to 15° N and S latitudes, but not closer to the equator. High sea-surface temperatures, over 27°C, are required for tropical cyclones to form. Typically this condition is met in the western regions of tropical oceans in late summer. At latitudes higher than 15°, sea surface

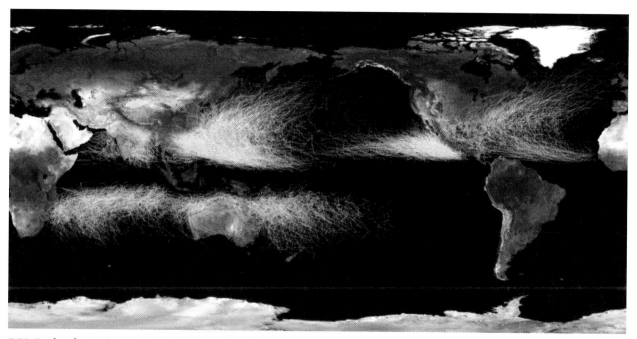

7.20 Tracks of typical cyclones Tropical cyclones always form over oceans. In the Atlantic, hurricanes originate off the west coast of Africa, and in the Caribbean Sea. Curiously, tropical cyclones do not form in the South Atlantic or southeast Pacific regions, and South America is never threatened by them. In the Indian Ocean, cyclones originate both north and south of the equator, and strike India, Pakistan, and Bangladesh, as well as the eastern coasts of Africa and Madagascar. Typhoons of the western Pacific also form both north and south of the equator, moving into northern Australia, Southeast Asia, China, and Japan.

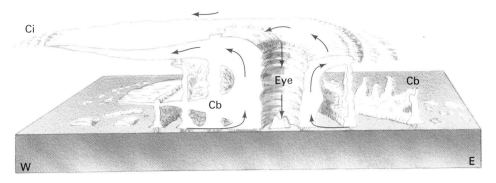

7.21 Anatomy of a hurricane In this schematic diagram, cumulonimbus (Cb) clouds in concentric rings rise through dense stratiform clouds. Cirrus clouds (Ci) fringe out ahead of the storm. The width of this diagram represents about 1,000 kilometres. (Redrawn from NOAA, National Weather Service)

Ascending air currents give rise to spirals of cumulonimbus clouds that converge on eye of the storm (Figure 7.21). This gives rise to very heavy rainfall, and as the precipitation forms, more energy is added to the storm through the release of latent heat. The cloud spirals are hundreds of kilometres long, but only a few kilometres wide. The distance between adjacent cloud spirals is 50–80 kilometres at the perimeter of the storm, and decreases toward the eye (Figure 7.22).

A characteristic feature of the tropical cyclone is its central eye, in which clear skies and calm winds prevail. The eye is a cloud-free vortex produced by the intense spiralling of the storm, and usually develops when sustained wind speeds exceed 120 kilometres per hour. The eye is formed through the combined effects of conservation of angular momentum and centrifugal force. Conservation of angular momentum causes objects to spin faster as they move toward the centre of the hurricane, so wind speeds increase. However, as speed increases, the air molecules are subject to an outward-directed centrifugal force. The centrifugal force increases with the degree of curvature and with faster rotation speeds. At about 120 kilometres per hour, the strong rotation of air balances the force of the inflow to the centre of the hurricane. This causes the air to ascend near the centre of the storm and form the eye wall.

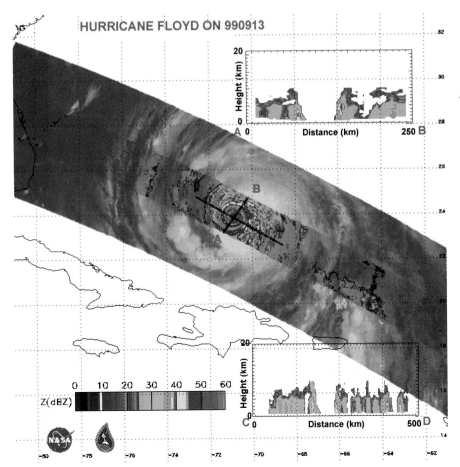

7.22 Tropical Rainfall Monitoring Mission (TRMM) data for Hurricane Floyd as it passed north and east of Cuba and the Greater Antilles in September 1999 The inward-spiralling cloud pattern of this tropical cyclone is clearly shown by the wide track of the visible and infrared scanner. The narrow track shows precipitation as detected by precipitation radar, and a passive microwave imager. The two cross-sections, A–B and C–D, show the structure of the storm. Precipitation is most intense in the yellow and red areas in the eye wall of the storm, but some of the outer precipitation bands visible in cross-section C–D also show intense precipitation. The TRMM satellite's orbit is strongly inclined so that it covers only the region from 35° N to 35° S. (NASA)

The eye wall consists of a ring of tall thunderstorms that produce heavy rains and usually the strongest winds. However, the strong rotation also creates a vacuum and causes some of the air flowing out the top of the eye wall to turn inward and sink to replace air lost from the centre. Hence, in the eye, air descends from high altitudes and is adiabatically warmed. As the eye passes over a site, calm prevails, and the sky clears. The eye of the storm passes over fairly quickly, after which the storm strikes with renewed ferocity, but with winds in the opposite direction.

The Simpson–Saffir scale rates the intensity of tropical cyclones (Table 7.3). This scale ranks storms based on the central pressure of the storm, mean wind speed, and height of accompanying **storm surge**, in which sea level is higher than normal. Category 1 storms are weak, while Category 5 storms are devastating. The top three hurricanes in the United States ranked in terms of intensity are the Great Labor Day Storm, which occurred on September 2, 1935, in Florida with a recorded minimum pressure of 89.2 kilopascals; Hurricane Katrina in August 2005, 90.4 kilopascals; and Hurricane Camille in August 1969, 90.9 kilopascals.

Tropical cyclones occur only during certain seasons. For hurricanes of the North Atlantic, the season runs from May through November, with maximum frequency in late summer or early autumn. In the southern hemisphere, the season is roughly the opposite. These periods follow the annual migrations of the ITCZ to the north and south with the seasons, and correspond to periods when ocean temperatures are warmest. In a typical year, there are about 60 hurricanes worldwide. About 16 are reported in the Atlantic, 18 in the eastern Pacific, and 25 in the western Pacific.

For convenience, weather forecasters tracking tropical cyclones give them names. They alternate male and female names in an alphabetical sequence, with the exception of ones beginning with Q, U, X, Y, and Z—those are reused every six years. Two sets of names are used—one for hurricanes of the Atlantic and one for typhoons of the Pacific. Names are reused, but the names of storms that cause significant damage or destruction are retired. The 2005 hurricane season was one of the worst on record with four Category 5 storms reported, including the infamous Katrina that caused considerable damage in New Orleans. A total of 28 tropical storms and hurricanes were reported in 2005. Following

on from Tammy, Vince, and Wilma, the Weather Service resorted to the Greek alphabet (Alpha, Beta, Gamma, Delta, Epsilon, and Zeta) for the six last storms of the season, which lasted from June 8 until January 7, 2006. The names Dennis, Katrina, Rita, Stan, and Wilma were retired following the 2005 season.

IMPACTS OF TROPICAL CYCLONES

Tropical cyclones can be tremendously destructive. Islands and coasts feel the full force of the high winds and flooding as tropical cyclones move onshore (Figure 7.23). Densely populated low-lying areas are particularly vulnerable to tropical cyclones. The deadliest tropical cyclone on record struck Bangladesh on November 13, 1970, with at least 500,000 deaths reported. In North America, the worst hurricane in terms of loss of life occurred on September 8, 1900, at Galveston, Texas, with an estimated 6,000–12,000 casualties. In terms of cost, Hurricane Katrina, in August 2005, caused an estimated US$200 billion in damage. (See *Eye on the Environment 7.1 • Hurricane Katrina.*) This compares with US$43.7 billion for Hurricane Andrew in August 1992 and Hurricane Charley at US$15 billion in August 2004. Hurricane Betsy, which occurred in

7.23 Bay St. Louis, Mississippi The lower photo, taken on August 31, 2005, reveals the devastation caused by Hurricane Katrina, which made landfall two days earlier. The resulting storm surge was estimated at more than six metres. The arrows provide comparative reference marks.

Table 7.3 Simpson–Saffir scale of tropical cyclone intensity

Category	Central Pressure kPa	Storm Surge m	Mean Wind Speed km/h
1 Weak	>98.0	1.2–1.7	119–153
2 Moderate	96.5–97.9	1.8–2.6	154–177
3 Strong	94.5–96.4	2.7–3.8	178–209
4 Very Strong	92.0–94.4	3.9–5.6	210–249
5 Devastating	<92.0	>5.6	>250

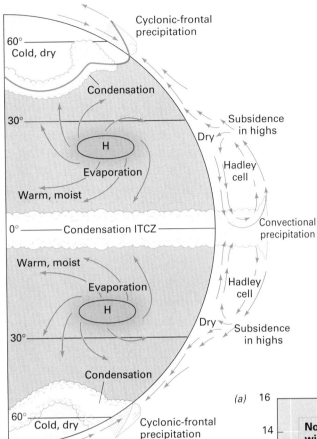

7.24 Global atmospheric transport of heat and moisture Major mechanisms of heat transport in the atmosphere include intertropical convergence, the Hadley cell circulation, and cyclonic-frontal precipitation in the midlatitudes.

September 1965, was the first to cause $1 billion in damage, and so is often referred to as Billion-Dollar Betsy. This would be equivalent to about $11 billion by 2006 standards.

POLEWARD TRANSPORT OF HEAT AND MOISTURE
ATMOSPHERIC HEAT AND MOISTURE TRANSPORT

The atmosphere is a fluid layer that transmits much of the Sun's energy to the surface of the Earth. Because solar energy does not warm the Earth's surface uniformly, the atmosphere and oceans exhibit a complex circulation pattern of air and water flows that acts to redistribute absorbed solar radiation more evenly through various convection loops. The process is referred to as *poleward heat transport*. In the atmospheric global convection system, warm, moist air rises along the ITCZ and gives rise to the **thermally direct** Hadley cells (Figure 7.24). The air then subsides and diverges in the subtropical high-pressure belts.

Figure 7.25 shows wind speeds associated the Hadley cell circulation for a typical day in December. Figure 7.25a shows merid-

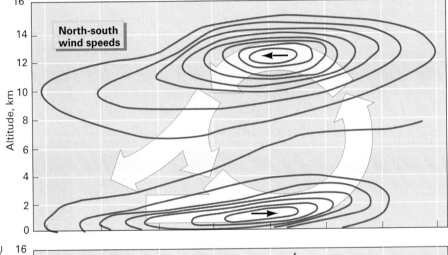

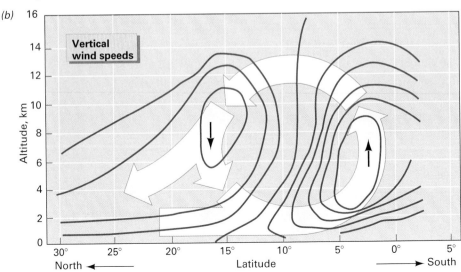

7.25 A cross-section of the northern hemisphere Hadley cell showing isotachs for a typical day in December (a) Horizontal component of velocity, ranging up to about four metres per second. (b) Vertical component, ranging up to about one centimetre (or .01 metres) per second. Note that the height of the cell is greatly exaggerated. (Copyright © A.N. Strahler)

ional (north-south) wind speeds, while 7.25b shows vertical (upward-downward) wind speeds. The horizontal component of the wind (7.25*a*) shows a series of nested ovals below an altitude of 4 kilometres, with a centre of strong horizontal motion near the surface. This flow is southward, as shown by the direction of the arrow. At about 10 to 15 kilometres, a second nest of isotachs (lines of equal velocity) indicates strong wind flow in a northerly direction. These two flows are the connecting legs of horizontal flow in the convection loop. The flows are strongest near 10° N latitude. The vertical component of the wind, shown in (*b*), also shows two centres of wind strength. Near the equator, motion is upward. At about 15° N latitude, the motion is downward.

Taken together, these four directional flows provide a single convection loop, indicated by the broad arrows. In this loop, heat is released by condensation in rising air near the equator and is carried by the air in motion to the upper part of the troposphere. Here, a portion of the heat is lost by direct radiation to space. As the air descends, some heat escapes toward the poles at lower altitudes. The global wind system is completed by the outflow of cold air from the polar regions. Because this airflow depends on the increase in density that results from loss of energy, the polar cell is also considered as thermally direct.

In the mid- and high latitudes, the Rossby wave mechanism also produces poleward heat transport (see Figure 5.25). Lobes of cold, dry polar or Arctic air (cP, cA, and cAA air masses) plunge toward the equator, while tongues of warmer, moister air (mT and mP air masses) flow toward the poles. At their margins, wave cyclones develop, releasing latent heat in precipitation, and providing a heat flow that warms the mid- and higher latitudes well beyond the capabilities of the Sun. The net effect is that heat gathered from tropical and equatorial zones is released in the zone of subsidence associated with the semi-permanent subtropical high pressure cells, where it can be conveyed into the midlatitudes by the motion of mT and cT air masses.

OCEANIC HEAT TRANSPORT

The motions induced by solar heating and influenced by the Earth's rotation also power the surface and deep currents in the oceans, which play a role in moving heat from one region of the globe to another, as well. Secondary forces contributing to the oceans' circulation arise from differences in density induced by climate. Evaporation in the subtropical high pressure cells increases salinity, whereas the diluting effect of heavy rainfall along the ITCZ reduces salinity. Chapter 6 described the thermohaline circulation of the ocean, in which the sinking of cold, salty water in the North Atlantic is linked to a slow-moving system of deep ocean currents, upwelling, and surface currents. As the surface currents flow northward, the water loses heat, which warms the overlying atmosphere. By the time the water reaches the north polar seas, it is chilled almost to its freezing point and sinks to the bottom of the Atlantic basin. This cold, Atlantic bottom water then heads slowly southward. Note that this circulation pattern, like the outflow of air from the polar regions, is also a type of sinking convection loop powered by a loss of heat and increase in density.

The ocean circulation acts like a heat pump in which sensible heat is acquired in tropical and equatorial regions and is moved northward into the North Atlantic, where it is transferred into the air. The amount of heat released is equivalent to about 35 percent of the total solar energy received by the Atlantic Ocean north of 40° latitude. This type of circulation does not occur in the Pacific or Indian Oceans.

CLOUD COVER, PRECIPITATION, AND GLOBAL WARMING

Global temperatures have been rising over the last 20 years. Satellite data show a rise in temperature of the global ocean surface of about 1°C over the past decade. Any rise in sea-surface temperature increases the evaporation rate, which raises the average atmospheric content of water vapour. Water in its vapour state is one of the greenhouse gases, and an increase in global water vapour in the atmosphere should enhance warming. In addition, water vapour can condense or sublimate to form clouds.

Clouds can have two different effects on the surface radiation balance. As large, white bodies, they can reflect a large proportion of incoming short-wave radiation back to space, thus acting to cool global temperatures. But cloud droplets and ice particles also absorb long-wave radiation from the ground and return part of that emission as counter-radiation. This absorption is an important part of the greenhouse effect, and it is much stronger for water as cloud droplets or ice particles than as water vapour. Thus, clouds also act to warm global temperatures by enhancing long-wave re-radiation from the atmosphere to the surface.

Whether more clouds will lead to an increase or decrease in global temperatures depends on cloud altitude, size, and droplet characteristics. Low, dense clouds tend to reflect solar radiation, but high clouds tend to trap outgoing long-wave radiation. Present satellite measurements indicate that the average flow of short-wave energy reflected by clouds back to space is about 50 watts per square metre, while the greenhouse warming effect of clouds amounts to about 30 watts per square metre. Thus, the net effect of clouds is a cooling of the planet by an energy flow of about 20 watts per square metre. However, computer models of global climate generally agree that increases in atmospheric water vapour will enhance long-wave warming more than short-wave cooling. Thus, increasing atmospheric water vapour will provide a positive feedback that enhances the greenhouse effect and so accentuates the surface warming.

EYE ON THE ENVIRONMENT | 7.1 Hurricane Katrina

More than 1,830 deaths have been attributed to Hurricane Katrina and the subsequent flooding it caused when it struck in August 2005. Damage along the low-lying Gulf Coast of the United States was reported as far as 160 kilometres from the storm's centre. Severe wind damage was noted as much as 150 kilometres inland. Unprecedented damage was inflicted on the city of New Orleans (Figure 1). Most of the damage was caused by the storm surge, which raised the sea level by an estimated six metres in some locations. About 80 percent of New Orleans was flooded for several weeks, because the surge breached several of the levees that protect the city from Lake Pontchartrain.

Hurricane Katrina formed in the Atlantic near the Bahamas on August 23, 2005, as a tropical depression associated with an easterly wave. The following morning the system was upgraded to a tropical storm and was named Katrina. Katrina continued to move westward and attained class 2 hurricane status as it reached the coast of Florida on the morning of August 25 (Figure 2).

Katrina weakened to a class 1 hurricane as it crossed Florida, but quickly regained its strength as it entered the Gulf of Mexico. On August 27, Katrina reached Category 3 intensity and continued to increase in size and intensity. By August 28, it was classed as a Category 5 hurricane, reaching peak strength in the early afternoon, with maximum sustained winds of 280 kilometres per hour and a minimum pressure of 90.2 kilopascals. Winds subsided during the night to 190 kilometres per hour, but increased to 205 kilometres per hour by about 6 a.m. the following day, when Katrina reached the Gulf Coast as a Category 3 hurricane.

Katrina maintained hurricane strength for about 240 kilometres inland across the state of Mississippi. It was downgraded to a tropical depression in Tennessee, when it was about 1,000 kilometres north of the Gulf of Mexico. Remnants of Katrina produced storms in parts of southern Ontario and Quebec on August 30. In Canada, the storm produced heavy rain and strong wind gusts. Port Colborne, Ontario received the heaviest rain in the region, with 102 millimetres falling that day. This was the first measurable rain in seven days. The storm passed quickly, with 2.8 millimetres of rain falling on August 31, followed by another seven-day dry spell. Some localized flooding and fallen trees were reported. The weather system finally dissipated over Labrador on August 31.

Hurricanes follow many paths as they approach North America, and they appear to move farther to the west during El Niño years. Although 2005 was not an El Niño year, it was unusually warm in many parts of the world, and the global annual temperature for combined land and ocean surfaces established a record that exceeded the previous record set in 1998 under the influence of an extremely strong El Niño (see table).

Figure 1 Destruction caused by Hurricane Katrina in New Orleans

The hottest years on record

Rank	Year
1	2005
1	1998
3	2002
4	2003
5	2004
6	2001
7	1997
8	1990
9	1995
10	1999
11	2000
12	1991
13	1987
14	1988
15	1994
16	1983
17	1996
18	1944
19	1989
20	1993

Perhaps for this reason, three Category 5 hurricanes formed in the Atlantic basin for the first time in a single season. Two of these, Katrina and Rita, passed over the oil-producing region in the Gulf of Mexico. The more westerly path of Hurricane Rita and the width of Hurricane Katrina together affected most of the 2,900 oil platforms in the region. The combined effect was that approximately 150 oil rigs were severely damaged; 36 rigs sank; and several floated free, having broken their moorings.

In preparation for Hurricane Katrina, daily production of oil was reduced by 1.4 million barrels a day—the equivalent to 95 percent of daily production—and natural gas production was reduced by about 35 percent. The impact of Katrina on the oil industry lasted several weeks. Twenty-one oil refineries, which produced almost 50 percent of the petroleum distillates in the United States, were still not functioning in mid-September. Four refineries suffered serious damage, and none of these was at full capacity at the end of 2005.

In addition to loss of production, there was a serious threat of oil pollution. At least 44 oil spills were reported, the largest of which involved about 4 million gallons. Of this total, 960,000 gallons were recovered, 2 million were contained, and 982,000 evaporated. More than 7 million gallons of oil escaped from industrial plants and storage depots, most of which was contained within earthen berms. A major concern was that the oil would have an impact on the Mississippi River. However, only a few minor oil slicks were reported, and these evaporated fairly quickly. There were no reports of major offshore spills into the Gulf of Mexico.

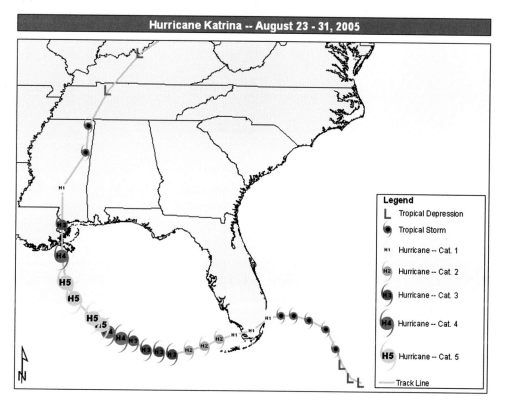

Figure 2 Tracking Hurricane Katrina

GEOGRAPHERS AT WORK

Air Pollution Climatology
by Dr. Ian McKendry, University of British Columbia

What proportion of air that you are breathing right now has travelled from other continents bringing with it pollution, dust particles and even microbes?

You might be surprised to know that air pollution is carried great distances and may significantly degrade the quality of the air that you breathe. This pollution may even contribute to glorious sunsets and more intense storms where you live.

Air pollution climatology is an important field of geography that focuses on the impacts of human activities on air quality. Air pollution climatologists are concerned with a range of problems, including the prediction of local smog episodes; the changing frequency of dust storms due to land use changes in arid regions of the world; the effect of air pollution on water quality in streams, lakes and oceans; and the impact of air pollution and dust on global climate. By examining the science behind such problems, air pollution climatologists seek to influence local, regional and international policies designed to control and reduce the harmful impacts of air pollution.

Air pollution climatology is the speciality of Ian McKendry, shown at work in the photo. His research focuses on the climatology of atmospheric particles (referred to as particulate matter). There are literally thousands of tiny particles in a cubic centimeter of air with most being too small to be seen easily. Particles are derived from many different sources including, windblown dust from the Earth's surface, salt particles from the ocean, and combustion products from forest fires, motor vehicles and industry. These tiny particles are important because they may be breathed deep into the lungs causing respiratory and cardiac problems. Recently Ian's work has ranged from the local to global scales. In the photograph, he is shown with a particle counting instrument mounted on a bicycle. In this work, he and his graduate student Amy Thai are investigating the levels of particulate matter pollution along the bicycle pathways in Vancouver British Columbia. At a much larger scale, Ian studies the climatology of dust transport across the Pacific Ocean to North America from the deserts of Africa and Asia. Recently, Ian and colleagues identified a case in which dust from a Saharan storm travelled 19,000 kilometres to British Columbia, Canada. Read more about Ian's work on our website.

Figure 7.26 summarizes some of the feedback loops in the global climate system. Increased concentrations of carbon dioxide and other greenhouse gases boost the greenhouse effect, increasing planetary temperature and raising atmospheric specific humidity. Since water vapour is itself a greenhouse gas, the increase in specific humidity forms a positive feedback and the change is amplified. Increased cloud cover also provides positive feedback, since liquid and solid water particles are more effective at greenhouse warming than water vapour alone. However, increased clouds provide negative feedback, as well, if they reflect solar radiation back to space, which would reduce the effect. Increased precipitation also could provide a negative feedback if it resulted in greater snow and ice cover, which would reflect more solar radiation back to space.

Systems with negative feedback loops will tend to be more stable, so that when changes occur, the system moderates them. Systems with positive feedback loops will tend to be more dynamic, since the structure of the system amplifies changes. It is difficult to determine at this time whether negative or positive feedback loops dominate the global climate system.

A Look Ahead

Annual cycles of temperature and precipitation in most regions are quite predictable, given the changes in wind patterns, air mass flows, and weather systems that occur with the seasons. These recurring annual cycles vary in response to global differences in energy. This creates spatially variable weather patterns and provides the basis for climate and climate classification. Chapters 8–10 discuss these topics.

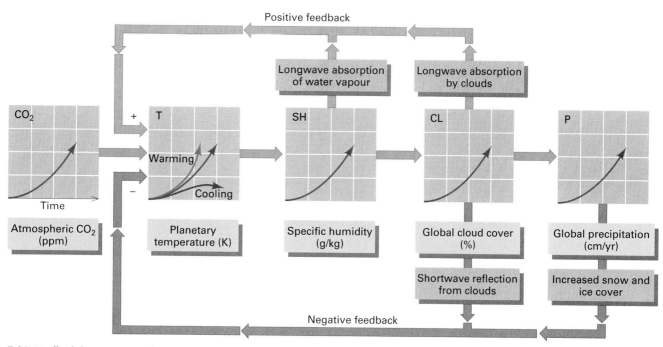

7.26 Feedback loops Positive and negative feedback loops in a coupled system of surface temperature and cloud cover.

CHAPTER SUMMARY

- A weather system is a recurring atmospheric circulation pattern with its associated weather. Weather systems include wave cyclones, travelling anticyclones, tornadoes, easterly waves, weak equatorial lows, and tropical cyclones.

- Air masses are distinguished by the latitudinal location and type of surface of their *source regions*. Air masses influencing North America include those of continental and maritime source regions, and of arctic, polar, and tropical latitudes.

- The boundaries between air masses are termed fronts. These include cold and warm fronts, where cold or warm air masses are advancing. In the occluded front, a cold front overtakes a warm front, pushing a pool of warm, moist air above the surface.

- *Travelling cyclones* include wave cyclones, tropical cyclones, and the tornado. The *travelling anticyclone* is typically a fair-weather system.

- Wave cyclones form in the midlatitudes at the boundary between cool, dry air masses and warm, moist air masses. In the wave cyclone, a vast inward-spiralling motion produces cold and warm fronts, and eventually an occluded front. Precipitation normally occurs with each type of front.

- Tornadoes are very small, intense cyclones that occur as a part of thunderstorm activity. Their high winds can be very destructive.

- Tropical weather systems include *easterly waves* and *weak equatorial lows*. Easterly waves occur when a weak low-pressure trough develops in the easterly wind circulation of the tropical

zones, producing convergence, uplift, and shower activity. Weak equatorial lows occur near the ITCZ. In these areas of low pressure, convergence triggers abundant convectional precipitation.

- Tropical cyclones can be powerful storms. They develop over warm tropical oceans and can intensify to become vast inward-spiralling systems of high winds with low central pressures. As they move onto land, they bring heavy surf and storm surges of very high waters.

- Global air and ocean circulation provides the mechanism for *poleward heat transport* by which excess heat moves from the equatorial and tropical regions toward the poles. In the atmosphere, the heat is carried primarily in the movement of warm, moist air poleward, which releases its latent heat when precipitation occurs. In the oceans, a global circulation moves warm surface water northward through the Atlantic Ocean. Heated in the equatorial and tropical regions, the surface water loses its heat to the air over the North Atlantic and sinks to the bottom. These heat flows help make northern and southern climates warmer than expected based on solar heating alone.

- Because global warming, produced by increasing carbon dioxide levels in the atmosphere, will increase the evaporation of surface water, atmospheric moisture levels will increase. This will enhance the greenhouse effect. But more clouds are likely to form, and this should cool the planet. Increased moisture could also reduce temperatures by increasing the amount and duration of snow cover.

KEY TERMS

air mass	easterly wave	storm surge	warm front
cold front	eye	thermally direct	wave cyclone
cyclonic precipitation	front	tornado	weather system
cyclonic storm	occluded front	tropical cyclone	

REVIEW QUESTIONS

1. Define air mass. What two features are used to classify air masses?
2. Compare the characteristics and source regions for mP and cT air mass types.
3. Describe a tornado. Where and under what conditions do tornadoes typically occur?
4. Identify three weather systems that bring rain in equatorial and tropical regions. Describe each system briefly.
5. Describe the structure of a tropical cyclone. What conditions are necessary for the development of a tropical cyclone? Give a typical path for the movement of a tropical cyclone in the northern hemisphere.
6. Why are tropical cyclones so dangerous?
7. How does the global circulation of the atmosphere and oceans provide poleward heat transport?
8. What is a feedback loop? Contrast the effects of negative and positive feedback loops on a system.
9. How does water, as vapour, clouds, and precipitation, influence global climate? How might water in these forms act to enhance or slow climatic warming?

EYE ON ENVIRONMENT 7.1   Hurricane Katrina

1. Where did Hurricane Katrina first form, and how did it develop and move?
2. What damage was sustained by South Florida and coastal Louisiana from Hurricane Katrina?
3. What were the maximum wind speeds of Katrina? What other phenomena were associated with this storm?
4. What effect does El Niño have on hurricanes?

VISUALIZATION EXERCISES

1. Identify three types of fronts and draw a cross-section through each. Show the air masses involved, the contacts between them, and the direction of air mass motion.
2. Sketch two weather maps, showing a wave cyclone in open and occluded stages. Include isobars and identify the centre of the cyclone as a low. Lightly shade areas where precipitation is likely to occur.
3. Sketch a cross-section of Hadley cell circulation, using arrows to indicate air flow direction. At what latitudes are northerly, southerly, upward, and downward flows strongest?

ESSAY QUESTIONS

1. Compare and contrast midlatitude and tropical weather systems. Be sure to include the following terms or concepts in your discussion: air mass, convectional precipitation, cyclonic precipitation, easterly wave, polar front, stable air, travelling anticyclone, tropical cyclone, unstable air, wave cyclone, and weak equatorial low.
2. Prepare a description of the annual weather patterns that your location experiences through the year. Refer to the general air mass patterns, as well as the types of weather systems that occur in each season.

Chapter Opener   Stormy skies at sunset. Low stratocumulus clouds (**A**) darken the sky in this passing weather system. Altocumulus cloud (**B**) aloft is illuminated by the setting sun, while near Earth's surface the red and orange hues (**C**) are beginning to colour the western horizon. Here, the sky is clearing as the weather system moves through.

Hurricane Katrina roars into Gulfport, Mississippi on August 30, 2005. Strong winds carry ocean spray (**A**) and surging water (**B**) onshore with sufficient force to knock over trees and power poles (**C**).

Chapter 8

EYE ON THE LANDSCAPE

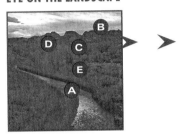

Desert rains in Canyon Lands National Park, Petrified Hollow, Utah, USA. **What else would the geographer see? . . . Answers at the end of the chapter.**

THE GLOBAL SCOPE OF CLIMATE

Keys to Climate
Temperature Regimes
Focus on Systems 8.1 • *Time Cycles of Climate*
Working It Out 8.2 • *Averaging in Time Cycles*
Global Precipitation
Seasonality of Precipitation
Climate Classification
Overview of the Climates
Dry and Moist Climates
A Closer Look
Special Supplement 8.3 • *The Köppen Climate System*

In its most general sense, **climate** is the average weather of a region. Climate descriptions could include any of the measures that express the state of the atmosphere, such as daily net radiation, barometric pressure, wind speed and direction, cloud cover, and precipitation type and intensity. However, most weather stations around the world don't regularly make observations as detailed as these. For this reason, comparisons of world climates generally rely on temperature and precipitation data, and are based on average values for each month, as well as seasonal variation during the year.

Temperature and precipitation strongly influence a region's natural vegetation—for example, forests generally occur in moist regions, and grasslands in dry regions. The natural vegetation cover is often a distinctive feature of a climatic region and typically influences the human use of the area. Temperature and precipitation are also important factors in the cultivation of crop plants. The development of soils, as well as the types of processes that shape landforms, are partly dependent on temperature and precipitation. For these reasons, the study of global climates is an important part of physical geography.

KEYS TO CLIMATE

Several of the principles discussed in earlier chapters are helpful in understanding the global scope of climate. First, the annual cycle of air temperature experienced at a weather station is primarily influenced by latitude and location on the continent:

- *Latitude*. Near the equator, temperatures are warmer and the annual temperature range is low. Toward the poles, temperatures are colder and the annual range is greater. These effects are produced by the annual cycle of insolation, which varies with latitude.
- *Coastal-continental location*. Coastal stations show a smaller annual range in temperature, while the range is larger at stations in continental interiors. This effect occurs because of different rates of heating and cooling on land surfaces and oceans.

Air temperature also has an important effect on precipitation:

- *Warm air can hold more moisture than cold air*. This means that colder regions generally have lower precipitation than warmer regions. Also, precipitation tends to be greater during the warmer months of the temperature cycle.

The primary driving force for weather is the flow of solar energy that the Earth and atmosphere receive. Energy flows vary according to the Earth's rotation on its axis and revolution around the Sun, and this is reflected in daily and seasonal cycles in temperature and precipitation. Other time cycles affect climate as well; *Focus on Systems 8.1 • Time Cycles of Climate* discusses these further.

TEMPERATURE REGIMES

Figure 8.1 shows typical patterns of mean monthly temperatures observed at stations around the globe. These patterns are referred to as *temperature regimes*—distinctive types of annual temperature cycles related to latitude and location. Each regime has been labelled according to its latitude zone: equatorial, tropical, midlatitude, and subarctic. Some labels also describe the location of the weather station in terms of its position on a landmass—"continental" or "west coast."

The equatorial regime (Douala, Cameroon, 4° N) is uniformly very warm. Temperatures are close to 27°C year-round. There are no pronounced temperature seasons because insolation is similar in all months. In contrast, the tropical continental regime (In Salah, Algeria, 27° N) shows a strong temperature cycle. Temperatures change from very hot when the Sun is high, near one solstice, to mild at the opposite solstice. The situation is quite different at Walvis Bay, Republic of Namibia (23° S), which

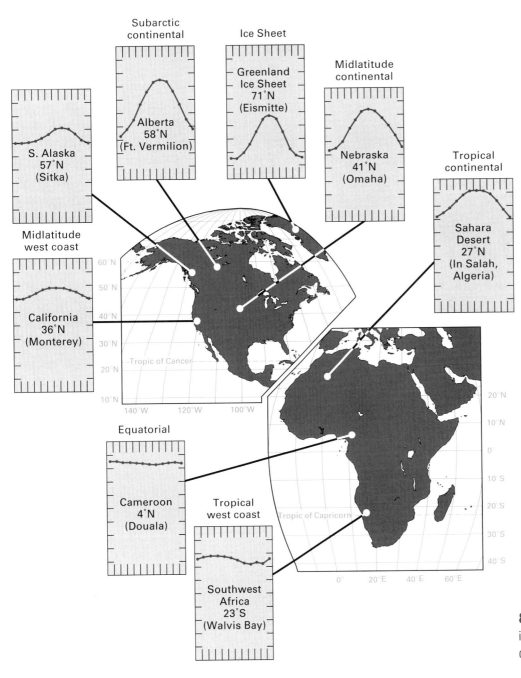

8.1 Temperature regimes Some important temperature regimes, represented by annual cycles of air temperature.

FOCUS ON SYSTEMS | 8.1   Time Cycles of Climate

One major class of time cycle evident in climate is the astronomical cycles related to the daily cycle of the Earth's rotation on its axis, and the annual cycle of the Earth's revolution around the Sun. In the midlatitudes, rotation results in a daily temperature cycle (a), which is produced by hourly changes in insolation and the surface energy balance. The Earth's revolution produces an annual cycle of daylight (b), which is determined by latitude and solar declination. Both curves are strongly related to the total daily energy flow received at the surface.

A standard **climograph** shows the annual cycles of temperature and precipitation derived from data averaged over many years (see Figure 8.7). Each of the red dots showing the mean air temperature of a given month is itself the average value of all of the monthly averages in the period of record. If the climographs of single years were plotted in sequence, as in (c), each year's curve would be different. The same would hold true for the precipitation bars on the climographs. Nevertheless, there is an annual rhythm in the climate data, even though the "beats" in this rhythm are far from uniform.

In many parts of the world annual climate data are available covering one or two centuries, and can be extrapolated over even longer time spans using surrogate data, such as annual tree rings. Here, the annual temperature value is reduced to a single number. The major time cycle for these data has an approximate length of 140 years. Superimposed is a shorter cycle of one decade. In Figure 4.24, which is a similar graph, annual values are shown by the yellow line, and the red line represents the five-year running average. The five-year running average helps to reveal significant cycles of moderate magnitude. The general, overall rise in values from left to right may be the start of a much larger cycle. The interpretation of this larger cycle is, of course, the subject of the ongoing debate about global warming.

Climate is always changing through time, showing rhythms of rise and fall of temperature and precipitation. The cycles observed in nature are nested in a hierarchy of sizes—smaller ones within larger ones. Scientists can plot the cycles of the past with some assurance, but prediction of trends to come is filled with uncertainty.

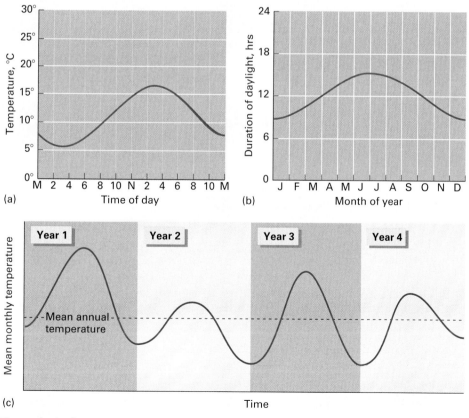

(a) Time of day

(b) Month of year

(c) Time

Time cycles in climate

WORKING IT OUT | 8.2   *Averaging in Time Cycles*

The weather at most stations can vary widely from day to day, or even from hour to hour. For the study of climate, however, analysis focuses on weather in a statistical, or average, sense. Typically, the averages for variables such as daily temperature and precipitation are calculated over decades.

Part (a) of the figure below shows monthly precipitation for a period of 20 years, from 1986 to 2005, as observed at St. John's, Newfoundland. At first inspection, the graph shows a rhythmic pattern to the precipitation, with sequences of wet months interspersed with drier months. But beyond that general observation, it is difficult to determine the shape of the typical annual cycle of precipitation. In fact, it is not easy to determine which month is, on the average, the wettest or the driest.

The typical monthly cycle of precipitation can be determined by calculating the *average*, or *mean*, of a number of observations—the sum of the observations divided by the number of observations

$$\overline{P} = \frac{1}{n}\sum_{i=1}^{n} P_i$$

where $\overline{P}$ (pronounced *P*-bar) is the mean precipitation for all observations in a particular month; P_i is the precipitation value associated with that month in the ith year; n is the number of years for which there are data for that particular month; and the subscript i denotes the number of the yearly observations, which ranges from $i = 1$ to n. The Greek letter (capital) sigma, Σ, is used to denote the sum, with the lower and upper notations showing that the sum is to be formed for each observation in the sequence from $i = 1$ to n.

The monthly averages for St. John's as listed in Environment Canada's Canadian Climate Normals, 1971–2000, are shown in part (b) of the figure to the right. Note that the monthly values appear to be much less variable than in the longer record. Maximum precipitation occurs in October, decreasing to a low in July. The average eliminates the year-to-year variability in the record. Part (c) shows the average for two five-year periods, 1986–1990 and 2001–2005. The

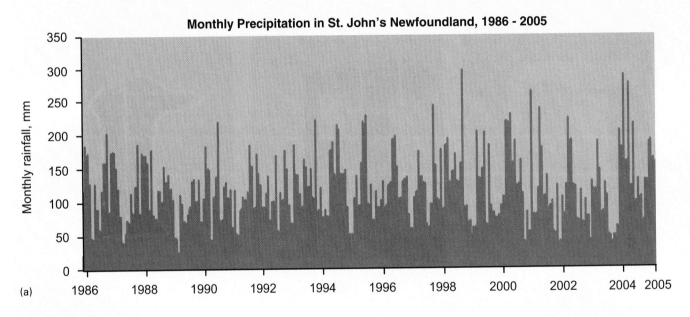

Monthly Precipitation in St. John's Newfoundland, 1986 - 2005

(a)

has about the same insolation cycle as In Salah. The tropical west coast regime at Walvis Bay has only a weak annual cycle and no extreme heat. The difference is due to the moderating effects of Walvis Bay's maritime location, an effect that is enhanced by the cold Benguela Current offshore. A similar moderating effect is shown in the two graphs for the midlatitude west-coast regime—Monterey, California (36° N) and Prince Rupert, B.C. (54° N).

In continental interiors, however, the annual temperature cycle remains strong. The midlatitude continental regime of Este-

van, Saskatchewan (49° N) and the subarctic continental regime of Fort Vermilion, Alberta (58° N) show annual variations in mean monthly temperature of about 35°C and 40°C respectively. The ice sheet regime of Greenland (Eismitte, 71° N) is characterized by severe cold all year.

Other regimes can be identified, and because they grade into one another, the list could be expanded indefinitely. However, what is important is that (1) annual variation in insolation, which is determined by latitude, provides the basic control of tempera-

period 1986–1990 is generally drier than the 2001–2005 period. This is the case for all months except February, April, and June, which received more precipitation during the earlier period.

In general, the longer the period of averaging, the closer the average will come to the true long-term average, and the smoother the cycle will be. Thus, averaging is a useful tool for revealing patterns and time cycles in data sequences.

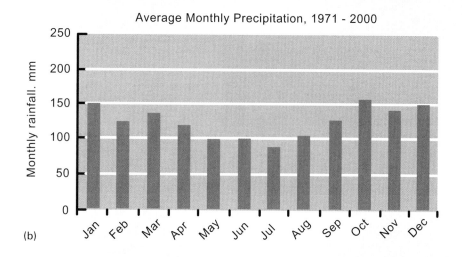

(b)

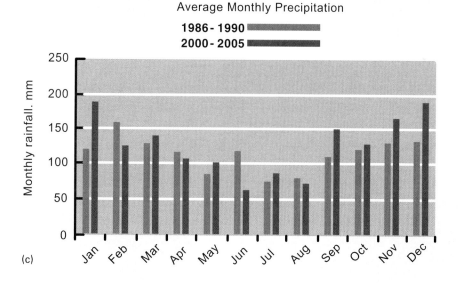

(c)

St. John's precipitation
Precipitation records and statistics for St. John's, Newfoundland, 48° N, moist continental climate (Data from Environment Canada)

ture patterns; and (2) the effect of location—maritime or continental—moderates that variation.

The temperature regimes in Figure 8.1 show the variation in mean monthly temperature through the year. Monthly temperature is the average of daily temperatures during the month, and daily temperature is the average of the daily maximum and minimum. Mean monthly temperature is averaged for a particular month over a long period of record, usually several decades. *Working It Out 8.2 • Averaging in Time Cycles* demonstrates the

power of averaging to reveal distinctive time cycles in data, such as the temperature regimes presented above.

GLOBAL PRECIPITATION

Global precipitation patterns are largely determined by air masses and their movements, which in turn are produced by global air circulation patterns. Figure 8.2 shows the general patterns expected for a hypothetical landmass that has most of the features of the Earth's continents. Five classes of annual

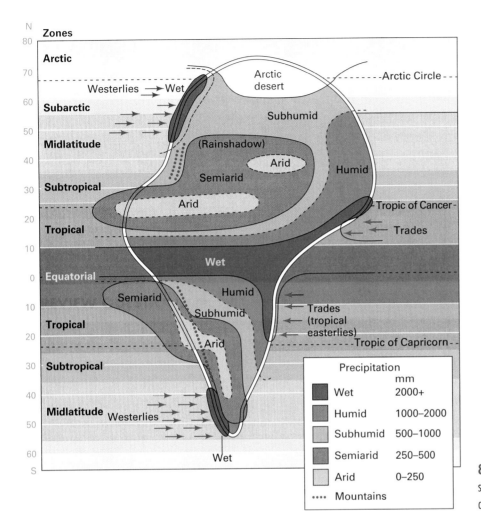

8.2 Precipitation on an idealized landmass A schematic diagram of the distribution of annual precipitation over a hypothetical landmass and adjoining oceans

precipitation—wet, humid, subhumid, semi-arid, and arid—are recognized on this landmass.

The equatorial zone shows a wet band stretching across the landmass. This band is produced by convectional precipitation in the weak equatorial lows that develop near the ITCZ (see Chapter 7). Note that the wet band widens and is extended away from the equator into the tropical zone along the eastern coast. This region is kept moist by the influence of trade winds, which move warm, moist mT air masses and tropical cyclones westward. Humid conditions continue along the east coasts into the midlatitude zones. In these regions, subtropical high-pressure cells tend to move mT air masses from the east, bringing them onshore in the summer, while in winter, wave cyclones bring cyclonic precipitation from the west.

Another important feature of the hypothetical landmass is the pattern of arid and semi-arid regions that stretches from tropical west coasts to subtropical and midlatitude continental interiors. In the tropical and subtropical latitudes, the arid pattern is produced by dry, subsiding air in persistent subtropical high-pressure cells. The aridity continues eastward and poleward into semi-arid continental interiors. These regions remain relatively

dry because they are far from regions that produce moist air masses. Rain shadow effects, created by coastal mountain barriers that block the flow of moist westerly winds from the ocean, may also contribute to dryness inland.

Another distinctive feature of the landmass is the pair of wet bands along the west coasts that extend from the midlatitudes into the subarctic zones. The heavy precipitation comes from the moist mP air masses that are carried by the prevailing westerlies in a succession of wave cyclones, and may be augmented by orographic uplift along coastal mountain ranges.

The high-latitude zone is depicted as Arctic desert, where precipitation is low because cold air can carry only a small amount of moisture. In addition, most of the Arctic Ocean is frozen for part of the year, which means that in winter, associated mP air masses are not readily distinguishable from their cP counterparts.

These general patterns can be seen in Figure 8.3, which shows mean annual precipitation for the Earth. The map uses **isohyets**, which are lines joining places with the same precipitation. Seven global precipitation regions can be discerned (Table 8.1) and are explained below. Note that the word "rainfall" is used for regions

Table 8.1 World precipitation regions

Name	Latitude Range	Continental Location	Prevailing Air Mass	Annual Precipitation Millimetres
1. Wet equatorial belt	10° N to 10° S	Interiors, coasts	mE	More than 2,000
2. Trade-wind coasts (wind-ward tropical coasts)	5–30° N and S	Narrow coastal zones	mT	More than 1,500
3. Tropical deserts	10–35° N and S	Interiors, west coasts	cT	Less than 250
4. Midlatitude deserts and steppes	30–50° N and S	Interiors	cT, cP	100–500
5. Moist subtropical regions	25–45° N and S	Interiors, coasts	mT (summer)	1,000–1,500
6. Midlatitude west coasts	35–65° N and S	West coasts	mP	More than 1,000
7. Arctic and polar deserts	60–90° N and S	Interiors, coasts	cP, cA	Less than 300

where all or most of the precipitation is rain; "precipitation" is used where snow is a significant part of the annual total.

1. **Wet equatorial belt.** This is a zone of heavy rainfall, more than 2,000 millimetres annually. It straddles the equator and includes the Amazon River Basin in South America, the Congo River Basin of equatorial Africa, much of the African coast from Nigeria west to Guinea, and the East Indies. Here the prevailing warm temperatures and high-moisture content of mE air masses favour abundant convectional rainfall. Thunderstorms are frequent year-round.

2. **Trade-wind coasts.** Narrow coastal belts of high rainfall, 1,500 to 2,000 millimetres, and locally even more, extend from near the equator to latitudes of about 25° to 30° N and S on the eastern sides of every continent or large island. Examples include the eastern coast of Brazil, Central America, Madagascar, and northeastern Australia. The rainfall on these coasts is supplied by moist mT air masses from warm oceans, brought over the land by the trade winds. As they encounter coastal hills and mountains, these air masses produce heavy orographic rainfall.

3. **Tropical deserts.** In striking contrast to the wet equatorial belt astride the equator are the two zones of vast tropical deserts lying approximately on the tropics of Cancer and Capricorn. These are hot, barren deserts, with less than 250 millimetres of rainfall annually and in many places less than 50 millimetres. They are located under the large, stationary subtropical cells of high pressure, in which the subsiding cT air mass is adiabatically warmed and dried. These desert climates extend from the west coast of the landmasses out over the oceans. Rain here is largely convectional and extremely unreliable.

4. **Midlatitude deserts and steppes.** Farther northward, in the interiors of Asia and North America between lat. 30° and lat. 50°N, are great deserts, as well as vast expanses of semi-arid grasslands known as **steppes**. Annual precipitation ranges from less than 100 millimetres in the driest areas to 500 millimetres in the moister steppes. The dryness results from the distance from ocean sources of moisture.

Located in regions of prevailing westerly winds, these arid lands typically lie in the rain shadows of coastal mountains and highlands. For example, the Cordilleran ranges of Alaska, British Columbia, Oregon, and Washington shield the interior of North America from moist mP air masses originating in the Pacific. When these air masses descend into the intermountain basins and interior plains, they are adiabatically warmed and dried and most of the moisture remains in the vapour state.

The mountains of Europe and the Scandinavian Peninsula are less obstructive, but because of the huge area of the Asian continent, much of the moisture in mP air masses from the North Atlantic is lost as it moves inland. In addition, any moisture present in the air in summer usually remains as vapour, because of high temperatures in the continental interior. In winter, the intense high pressure region that develops over Siberia further restricts airflow into the continent. Dry steppes give way to more arid conditions in Kazakhstan and eventually merge with the Gobi Desert in Mongolia, where dryness is accentuated by the blocking effect of the Himalayas on moist mT and mE air masses from the Indian Ocean.

The southern hemisphere has too little land in the midlatitudes to produce a true continental desert, but the dry steppes of Patagonia, lying on the lee side of the Andean chain, are roughly the counterpart of the North American deserts and steppes of Oregon and northern Nevada.

5. **Moist subtropical regions.** On the southeastern sides of the continents in lat. 25° to 45° N are the moist subtropical regions, with 1,000 to 1,500 millimetres of rainfall annually. In North America this area includes states such as Alabama, Georgia, and South Carolina, in Asia, areas such as the plains of the Yangtze and Yellow Rivers in Korea. Smaller areas are

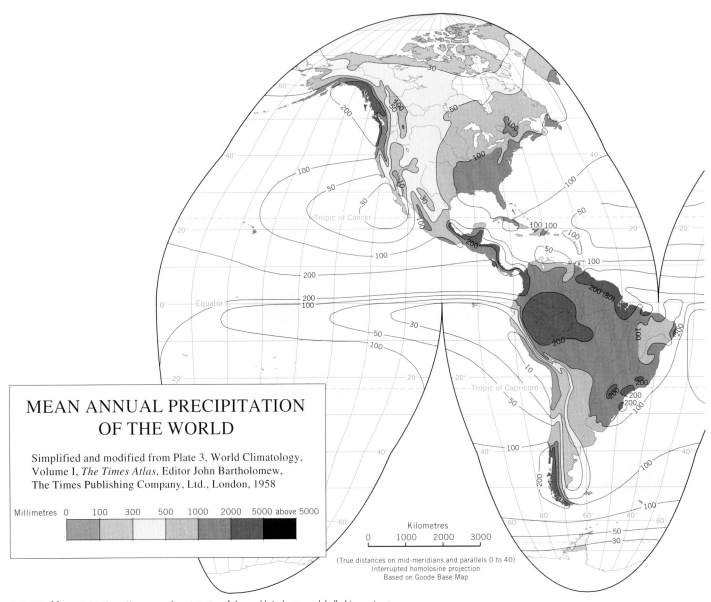

MEAN ANNUAL PRECIPITATION OF THE WORLD

Simplified and modified from Plate 3, World Climatology, Volume I, *The Times Atlas*, Editor John Bartholomew, The Times Publishing Company, Ltd., London, 1958

Millimetres 0   100   300   500   1000   2000   5000 above 5000

Kilometres
0   1000   2000   3000

(True distances on mid-meridians and parallels 0 to 40)
Interrupted homolosine projection
Based on Goode Base Map

8.3 World precipitation Mean annual precipitation of the world. Isohyets are labelled in centimetres.

found in the southern hemisphere in Uruguay, Argentina, and southeastern Australia. These regions are positioned on the moist western margins of the oceanic subtropical high-pressure centres. As a result, the lands receive moist mT air masses from the tropical ocean. Commonly, too, these areas receive heavy rains from tropical cyclones.

6. **Midlatitude west coasts.** Another distinctive wet location is on midlatitude west coasts of all continents and large islands lying between about 35° and 65° in the region of prevailing westerly winds. In these zones, abundant orographic precipitation occurs as a result of forced uplift of mP air masses. Where the coasts are mountainous, as in British Columbia and Alaska, southern Chile, Scotland, and Norway, the annual precipitation is more than 2,000 millimetres. The South Island of New Zealand expe-

riences a similar climate because of its situation in the zone of westerlies. During the Ice Age, this precipitation fed alpine glaciers that descended to the coast, carving the picturesque deep bays (fiords) that are so typically a part of this climate type.

7. **Arctic and polar deserts.** Northward of the 60th parallel, annual precipitation is generally less than 300 millimetres, except for the west-coast belts. Cold cP and cA air masses cannot hold much moisture, and consequently, they do not yield large amounts of precipitation. At the same time, however, the relative humidity is high and evaporation rates are low.

8. **Highland climates.** Highland climates are similar to the arctic and polar climates, but usually distinguished from these because they are influenced by high elevation rather than high latitude. Highland climates are also more variable than

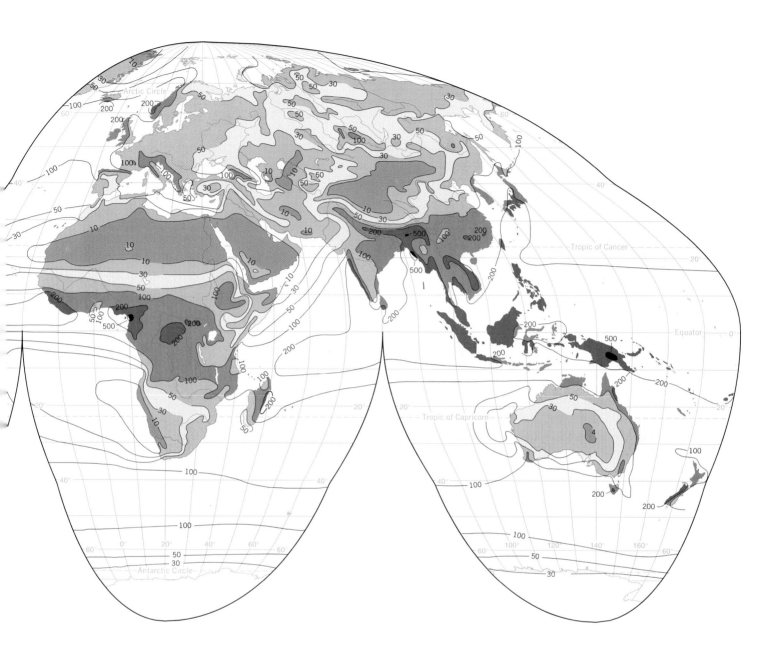

their high latitude counterparts, and conditions can change markedly over relatively small distances. Highland climates occur at all latitudes and in coastal and interior locations. They become colder at higher elevations, but have the same seasonal precipitation patterns as the regions that surround them.

SEASONALITY OF PRECIPITATION

Total annual precipitation is useful in establishing the character of a climate type, but it does not account for the seasonality of precipitation. The variation in monthly precipitation through the annual cycle is an important factor in describing climate. A pattern of alternating dry and wet seasons, instead of a uniform distribution of precipitation throughout the year, will affect the

natural vegetation, soils, crops, and human use of the land. It also makes a significant difference whether the wet season coincides with a season of higher or lower temperatures. If the warm season is also wet, growth of both native plants and crops will be enhanced. If the warm season is dry, the stress on growing plants increases and crops may require irrigation.

Monthly precipitation patterns can be described by three general types: (1) uniformly distributed precipitation; (2) a precipitation maximum during the high-Sun season (summer) when insolation is at its peak; and (3) a precipitation maximum during the low-Sun or cooler season (winter), when insolation is least. Note that the uniform precipitation pattern can include a wide range of conditions, from little or no precipitation in any month to abundant precipitation in all months.

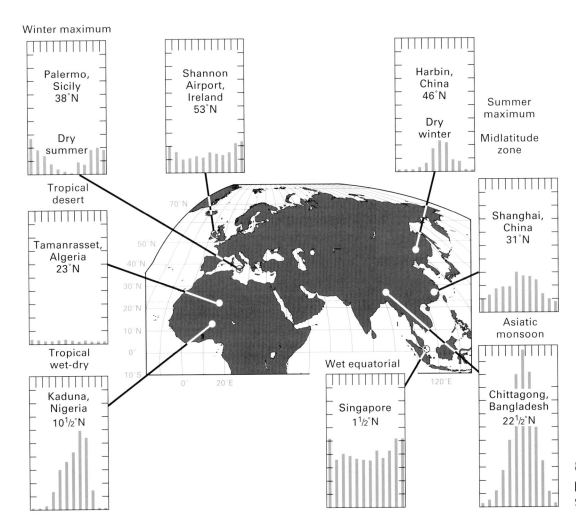

8.4 Seasonal precipitation patterns Eight precipitation types show various seasonal patterns.

The monthly precipitation diagrams shown in Figure 8.4 represent the main patterns found in various parts of the world. Two stations have precipitation distributed uniformly throughout the year: Singapore (1½° N), a wet station near the equator, and Tamanrasset, Algeria (23° N), a tropical desert station near the Tropic of Cancer. At Singapore, rainfall is abundant in all months, although some months are wetter than others. Tamanrasset has so little rain in any month that it scarcely shows on the graph.

Chittagong, Bangladesh (22½° N) and Kaduna, Nigeria (10½° N) both experience a wet summer season and a dry winter season. Chittagong is an Asian monsoon station, with a large amount of rainfall falling during the high Sun season. Kaduna, with about half the total annual precipitation, is a wet-dry tropical location. At both of these stations, wet season rainfall is associated with convective activity along the ITCZ.

A summer precipitation maximum also occurs at higher latitudes on the eastern sides of continents. Shanghai, China (31° N) shows this pattern in the subtropical zone. The same summer maximum persists into the midlatitudes. For example, Harbin, in eastern China (46° N), has a long, dry winter with a marked period of summer rain.

In contrast, Palermo, on the Italian island of Sicily (38° N), receives maximum precipitation in winter. This is an example of a Mediterranean climate, which is characterized by dry summers and moist winters. As the name suggests, this climate prevails in the lands surrounding the Mediterranean Sea, including southern Spain and France, Greece, Turkey, parts of Israel, and Morocco in North Africa. Southern and central California also experience a Mediterranean climate. In the southern hemisphere, this climate occurs in Chile, parts of South Africa, and southwestern Australia. In Mediterranean climates, summer drought is produced by subtropical high-pressure cells, which intensify and move poleward during the high-Sun season. The cT air that extends into these regions brings hot dry weather. In the low-Sun season, the subtropical high-pressure cells move toward the equator and weaken, and cyclonic storms bring precipitation to these regions.

The dry-summer, moist-winter cycle extends into coastal areas at higher midlatitudes. However, the difference between summer and winter precipitation becomes less and less marked, and the Mediterranean climate eventually gives way to the marine west coast type. The marine west coast climate is most developed in the Pacific Northwest, from Oregon to Alaska, but occurs as

GEOGRAPHERS AT WORK	**Tree-ring Records in Atlantic Canada**
	by Dr. Colin P. Laroque, *Mount Allison University*

Did you know we can learn what past climates were like, on a yearly basis, long before people collected instrumental weather data? For many areas of Canada, tree-ring records have provided a lot of information about past environments. Until recently, however, such information was scarce in Atlantic Canada.

In 2003, dendroclimatologist Colin Laroque established Atlantic Canada's first tree-ring laboratory, which dedicates its efforts to understanding past climates through tree-ring analysis. Colin and his team exploit the fact that trees incorporate information about their environment as variable widths of their annual growth rings. The researchers develop mathematical functions that relate known temperature and precipitation data to recent radial growth, and then carry these relationships backward

in time ("hindcast") for as long as the trees have been alive. The older the trees, the better, and Colin often uses timbers from old buildings to extend the record in this region of scarce old-growth forest.

Colin is finding that some tree species in this maritime climate reveal long-term Atlantic Ocean influences, including ties to the warm Gulf Stream current and the cold Labrador current. Colin's tree-ring data improve our models not only of past temperature and precipitation, but also of oceanic phenomena such as the North Atlantic Oscillation. The data ultimately contribute to stronger Global Circulation Models, which help us forecast future climate fluctuations in Canada and around the world. We can then predict the effects of these climate changes on trees—and, of course, on people.

Colin Laroque in an old-growth eastern hemlock forest in central Nova Scotia.

well in Britain, Ireland, and elsewhere in Western Europe; Shannon Airport in Ireland (53° N) is a characteristic location. Typically, less precipitation is received in summer than in winter. This is partly due to the blocking effects of subtropical high-pressure cells that extend poleward into these regions. The polar front also shifts further north in summer, which reduces the frequency of cyclonic storms passing through these regions. Also, cyclonic storms are less intense in summer, because the temperature and moisture contrasts between polar and arctic air masses and tropical air masses are weaker in summer.

CLIMATE CLASSIFICATION

Mean monthly values of air temperature and precipitation can describe the climate of a weather station and its nearby region quite accurately. Together, these two variables can produce an infinite set of conditions, which, for the purpose of global comparisons, must be presented in a more generalized way. Various methods have been devised to group climate information into distinctive climate types. In this way, distinctive regional characteristics can be described and maps compiled to show their distribution.

In 1918, Vladimir Köppen developed one of the earliest and best known approaches to organizing the vast amount of world climate information. It is an empirical system based on long-term temperature and precipitation records. The Köppen climate classification system uses a code of letters to group climates into classes according to predefined values of selected annual and seasonal temperature and precipitation. *A Closer Look 8.3 • The Köppen Climate System* at the end of this chapter presents a modified version of the Köppen system.

A second approach, proposed by C.W. Thornthwaite in 1946, is based on the concept of the water budget. In this scheme, precipitation inputs are balanced against water losses through evaporation, transpiration, runoff, and infiltration into the soil. Although Thornthwaite's approach was never popular as a method of climate classification, it is still widely used in agriculture to determine soil moisture levels and potential drought problems. Chapter 20 discusses the concept of the moisture budget in detail.

A more practical approach to climate classification—applied classification—defines climates according to some specified purpose. For example, the construction industry in Canada might divide the country according to high, medium, and low heating costs on the basis of mean winter temperatures; regional differences in roof design might be related to total snowfall loads; or summer temperature and humidity could be used to determine potential sales of air conditioners.

This textbook uses a genetic approach to climate classification, which is based on the factors that cause the climate at a given location or region. A comprehensive genetic classification system must incorporate all the factors that affect the common properties of the atmosphere, particularly temperature and precipitation. It groups regions according to the air masses and frontal zones that affect them. Air masses are classified according to the general latitude of their source regions and the surface—land or ocean—over which they develop. Latitude is an important determinant of air mass temperature and can impose pronounced seasonal variation as solar radiation varies over the year. Temperature will also influence the moisture content of an air mass; however, at any given latitude, this is primarily controlled by the distribution of land and ocean. Air mass characteristics therefore reflect the two most important climate variables—temperature and precipitation—and so provide a rational basis for classifying climates.

This method of classification follows quite logically from the previous discussions of global temperature and precipitation processes. It recognizes 13 distinctive climate types. Appendix 1 describes the procedure used to define these types. The procedure is based on an analysis of how air temperature and rainfall affect the amount of moisture held in the soil throughout the year. Therefore, there is some similarity with the Thornthwaite method.

Air masses acquire distinctive temperature and moisture properties based on the characteristics of their source regions. Adjacent air masses are separated by frontal zones across which the properties of the atmosphere change. However, the position of frontal zones varies with the seasons. For example, the polar-front zone generally lies across the midlatitudes of the United States in winter, but it moves northward to Canada during the summer (Figure 8.5). The seasonal movements of frontal zones therefore influence annual cycles of temperature and precipitation.

Figure 8.6 shows a general subdivision of air mass source regions into global bands, recognizing three broad climate groups: low-latitude (Group I), midlatitude (Group II), and high-latitude (Group III). They are described briefly as follows:

• *Group I: Low-Latitude Climates.* The region of low-latitude climates (Group I) is dominated by the source regions of continental tropical (cT), maritime tropical (mT), and maritime equatorial (mE) air masses. These source regions are related to the three most obvious atmospheric features that occur within their latitude band: the two subtropical high-pressure belts and the equatorial trough at the ITCZ. Air originating in the polar regions occasionally invades regions of low-latitude climates.

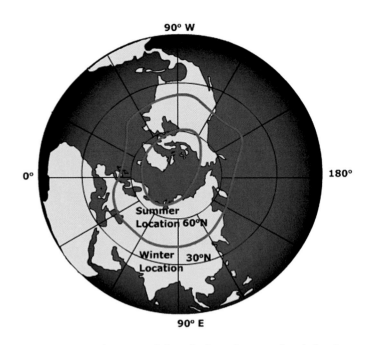

8.5 Mean seasonal position of the polar front In summer, the polar front lies at approximately 60–65° N; and in winter, it moves southward to about 35–40° N.

Easterly waves and tropical cyclones are important weather systems in this climate group.

• *Group II: Midlatitude Climates.* The region of midlatitude climates (Group II) lies in the polar-front zone—a zone of intense interaction between different air masses. Here tropical air masses moving toward the poles and polar air masses moving toward the equator are in conflict. Wave cyclones are normal features of this climate group.

• *Group III: High-Latitude Climates.* The region of high-latitude climates (Group III) is dominated by polar and Arctic (or Antarctic) air masses. In the Arctic belt of the 60th to 70th parallels, continental polar air masses meet Arctic air masses along an *Arctic-front zone*, creating a series of eastward-moving wave cyclones. In the southern hemisphere, there are no source regions in the subantarctic belt for continental polar air—only a great single oceanic source region for maritime polar (mP) air masses. The continent of Antarctica provides a single great source of the extremely cold, dry Antarctic air mass (cAA). These two air masses interact along the *Antarctic-front zone*.

Within each of these three climate groups are a number of climate types (or simply, climates)—four low-latitude climates (Group I), six midlatitude climates (Group II), and three high-latitude climates (Group III)—for a total of 13 climate types. The name of each climate describes its general nature and also suggests its global location.

A climograph can readily portray climate information (Figure 8.7). This example shows the annual cycles of monthly mean air temperature and monthly mean precipitation for

8.6 Climate groups and air mass source regions Using the concept of air mass source regions, five global bands associated with three major climate groups can be identified. Within each group is a set of distinctive climates with unique characteristics that are explained by the movements of air masses and frontal zones.

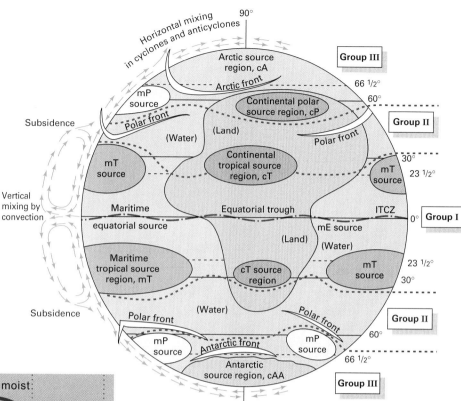

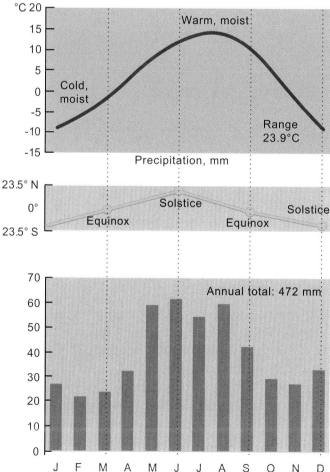

8.7 Representative climograph Banff, Alberta, at 51° N, is located in the Rocky Mountains at an elevation of 1,384 metres. The temperature range is 23.9°C and reflects the relatively high latitude of this location. Precipitation occurs in all months, and comes predominantly from cyclonic storms.

Banff, Alberta, located at 51° N. At the top of the climograph, the mean monthly temperature is plotted as a line graph. At the bottom, the mean monthly precipitation is shown as a bar graph. The annual range in temperature and the total annual precipitation are provided. Most climographs also display dominant weather features using picture symbols. For Banff, a dominant feature is the influence of cyclonic storms that bring precipitation in all months.

The world map of climates (Figure 8.8) shows the actual distribution of the 13 climate types derived from the genetic classification's three climate groups. In addition, highland climates are indicated for mountain areas. The map is based on data collected at a large number of weather stations. It is simplified because the climate boundaries are uncertain in many areas where stations are thinly distributed.

OVERVIEW OF THE CLIMATES

Each of the 13 climate types is briefly introduced here; Chapters 9 and 10 provide more complete descriptions.

Low-Latitude Climates (Group I)

- *Wet equatorial.* Warm to hot with abundant rainfall, this is the steamy climate of the Amazon and Congo basins and the East Indies.

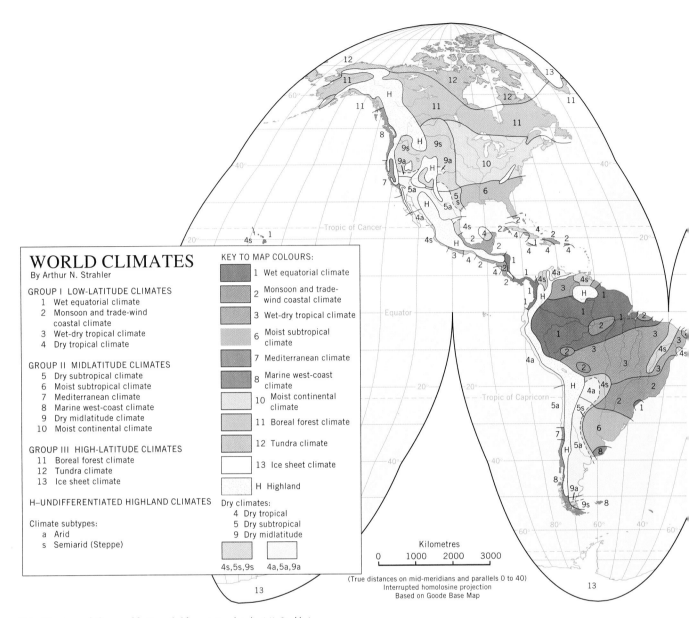

8.8 Climates of the world (Compiled from station data by A.N. Strahler)

• *Monsoon and trade-wind coastal.* This warm to hot climate has a very wet rainy season. It occurs in coastal regions that are influenced by trade winds or a monsoon circulation. The climates of Vietnam and Bangladesh are good examples.

• *Wet-dry tropical.* A warm to hot climate with distinct wet and dry seasons. The monsoon region of India falls into this type, as does much of the Sahel region of Africa.

• *Dry tropical.* The climate of the world's hottest deserts—extremely hot in the high-Sun season, a little cooler in the low-Sun season, with little or no rainfall. The Sahara desert, Saudi Arabia, and the central Australian desert are typical of this climate.

Midlatitude Climates (Group II)

• *Dry subtropical.* Another desert climate, but not quite as hot as the dry tropical climate, since it is found farther toward the poles. This type includes the hottest part of the American southwest desert.

• *Moist subtropical.* The climate of the southeastern regions of the United States and China—hot and humid summers, with mild winters and ample rainfall year-round.

• *Mediterranean.* Hot, dry summers and rainy winters mark this climate. Southern and central California, as well as the Mediterranean region—Spain, southern Italy, Greece, and the coastal regions of Lebanon and Israel—are prime examples.

• *Marine west-coast.* The climate of the Pacific Northwest—coastal Oregon, Washington, and British Columbia; Western Europe; and the South Island of New Zealand. Warm summers and cool winters, with more rainfall in winter, are characteristics of this climate.

• *Dry midlatitude.* This dry climate is found in midlatitude continental interiors. The steppes of central Asia and the Great Plains

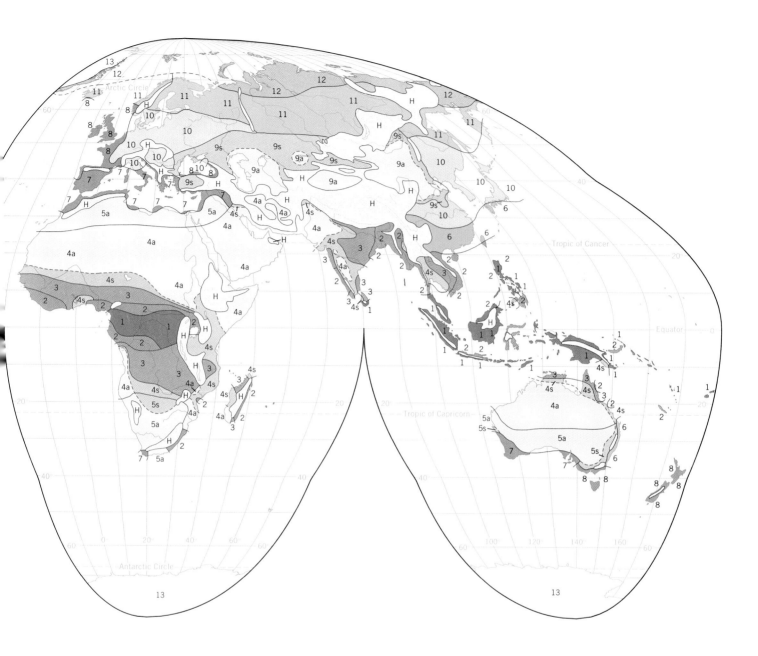

of North America are representative locales—warm to hot in summer, cold in winter, and with low annual precipitation.

- *Moist continental*. This is the climate of eastern Canada and most of the eastern United States—cold in winter, warm in summer, with ample precipitation through the year.

High-Latitude Climates (Group III)

- *Boreal forest*. Short, cool summers and long, bitterly cold winters characterize this snowy climate. Northern Canada, Siberia, and central Alaska are regions of boreal forest climate.
- *Tundra*. Although this climate has a long, severe winter, temperatures on the tundra are somewhat moderated by proximity to the Arctic Ocean. This is the climate of the coastal arctic regions of Canada, Alaska, Siberia, and Scandinavia.

- *Ice sheet*. The bitterly cold temperatures of this climate, restricted to Greenland and Antarctica, can drop below −50°C during the sunless winter months. Even during the 24-hour days of summer, temperatures remain well below freezing.

DRY AND MOIST CLIMATES

All but two of the 13 climate types introduced above are classified as either dry climates or moist climates. *Dry climates* are those in which total annual evaporation of moisture from the soil and transpiration from plants exceeds the annual precipitation. Generally speaking, dry climates do not support permanently flowing streams. The soil is dry much of the year, and the land surface is clothed with sparse plant cover—scattered grasses, shrubs, or cacti—or simply no plant cover. *Moist climates* are those with sufficient precipitation to maintain the soil in a moist condition through much of the year and to sustain the year-round flow of larger streams. Moist climates support forests and prairies.

Within the dry climates there is a wide range of aridity, from very dry deserts nearly devoid of plant life to moister regions that support a partial cover of grasses or shrubs. Two dry climate subtypes recognize this diversity: (1) *semi-arid* or steppe and (2) *arid*. The semi-arid subtype has sufficient precipitation to support a sparse cover of plants. The arid subtype ranges from extremely dry to transitional with semi-arid.

Two of the 13 climates show a seasonal alternation between a very wet season and a very dry season: the wet-dry tropical and Mediterranean climate types. The striking seasonal contrast gives both of them a special character, hence they are designated *wet-dry climates*. Table 8.2 summarizes the moist, dry, and wet-dry climates that occur within the three main climate groups.

A Look Ahead

This introduction to global climate has stressed the relationship between climate and the factors that influence annual cycles of temperature and precipitation. Temperature cycles may be uniform, seasonal (continental), or moderated by oceanic influences (marine). Precipitation may be uniform (ranging from scarce in all months to abundant in all months), may have a maximum at the time of high Sun, or may have a maximum at the time of low Sun. These temperature and precipitation cycles, in turn, are produced by the annual cycle of insolation, and by the global patterns of atmospheric circulation and air mass movements. The next two chapters examine the climates of the world in more detail.

A CLOSER LOOK   SPECIAL SUPPLEMENT 8.3   The Köppen Climate System

Air temperature and precipitation data have formed the basis for several climate classification systems. One of the most important of these is the Köppen climate system, devised in 1918. For several decades, this system, with various later revisions, was the most widely used climate classification method among geographers. Köppen was both a climatologist and plant geographer, so his main interest lay in finding climate boundaries that coincided approximately with boundaries between major vegetation types.

Under the Köppen system, each climate is defined according to assigned values of temperature and precipitation, calculated in terms of annual or monthly values. Any given station can be assigned to its particular climate group and subgroup solely on the basis of its temperature and precipitation records.

Note that mean annual temperature refers to the average of 12 monthly temperatures for the year. Mean annual precipitation refers to the average of the entire year's precipitation as observed over many years.

The Köppen system features a shorthand code of letters designating major climate groups, subgroups within the major groups, and further subdivisions to distinguish particular seasonal characteristics of temperature and precipitation. Five major climate groups are designated by capital letters as follows:

A *Tropical rainy climates*

The average temperature of every month is above 18°C. These climates have no winter season. Annual rainfall is large and exceeds annual evaporation.

B *Dry climates*

Evaporation exceeds precipitation, on average, throughout the year. There is no water surplus; hence, no permanent streams originate in B climate zones.

C *Mild, humid (mesothermal) climates*

The coldest month has an average temperature of under 18°C, but above –3°C; at least one month has an average temperature above 10°C. The C climates thus have both a summer and a winter.

D *Snowy-forest (microthermal) climates*

The coldest month has an average temperature of under –3°C. The average temperature of the warmest month is above 10°C. (Forest is not generally found where the warmest month is colder than 10°C.)

E *Polar climates*

The average temperature of the warmest month is below 10°C. These climates have no summer season.

Note that four of these five groups (A, C, D, and E) are defined by temperature averages, whereas one (B) is defined by precipitation-to-evaporation ratios. Groups A, C, and D have sufficient heat and precipitation for the growth of forest and woodland vegetation. Figure 1 shows the boundaries of the five major climate groups, and Figure 2 presents a world map of Köppen climates.

Subgroups within the five major groups are designated by a second letter according to the following code:

Table 8.2 Moist, dry, and wet-dry climate types

		Climate Type		
Climate Group		**Moist**	**Dry**	**Wet-Dry**
I:	Low-latitude climates	Wet equatorial Monsoon, and trade-wind coastal	Dry tropical (steppe; arid)	Wet-dry tropical
II:	Midlatitude climates	Moist subtropical Marine west-coast Moist continental	Dry subtropical (steppe; arid) Dry midlatitude (steppe; arid)	Mediterranean
III:	High-latitude climates	Boreal forest Tundra	Ice sheet	

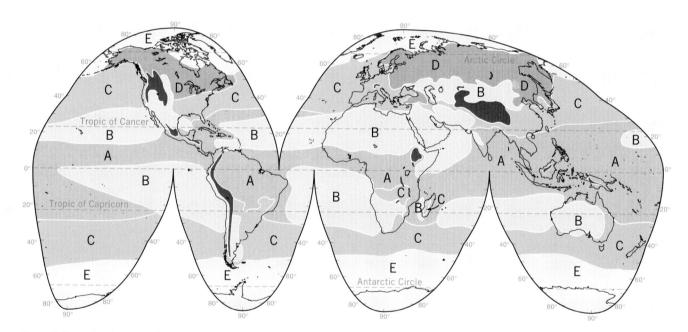

Figure 1 Generalized Köppen climate map Highly generalized world map of major climate regions according to the Köppen classification system. Highland areas are in black.

S *Semi-arid (steppe)*

W *Arid (desert)*

(The capital letters S and W are applied only to the dry B climates.)

f Moist, adequate precipitation in all months, no dry season. This modifier is applied to A, C, and D groups.

w Dry season in the winter of the respective hemisphere (low-Sun season).

s Dry season in the summer of the respective hemisphere (high-Sun season).

m Rainforest climate, despite short, dry season in monsoon type of precipitation cycle. (m applies only to A climates.)

From combinations of the two letter groups, 12 distinct climates emerge:

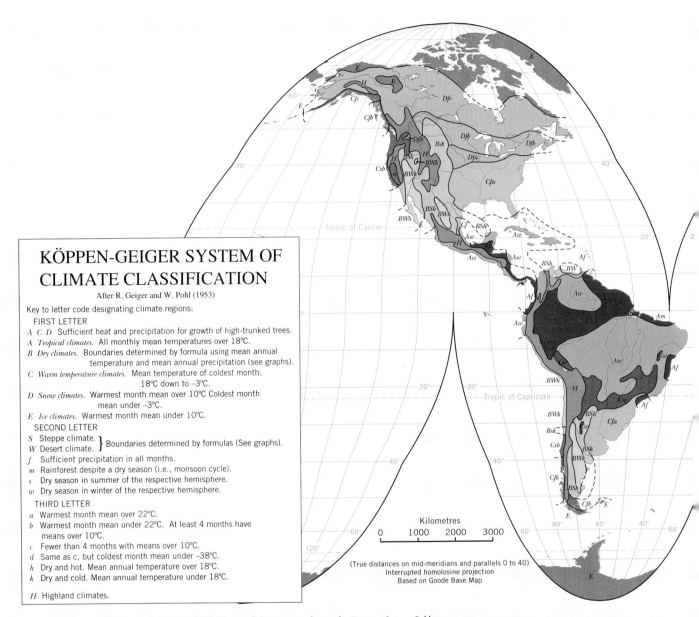

Figure 2 Köppen climates of the world World map of climates according to the Köppen–Geiger–Pohl system

A *Tropical rainforest climate*

The rainfall of the driest month is 60 millimetres or more.

Am *Monsoon variety of Af*

The rainfall of the driest month is less than 60 millimetres. The dry season is strongly developed.

Aw *Tropical savanna climate*

At least one month has rainfall less than 60 millimetres. The dry season is strongly developed.

Figure 3 shows the boundaries between Af, Am, and Aw climates, as determined by both annual rainfall and rainfall of the driest month.

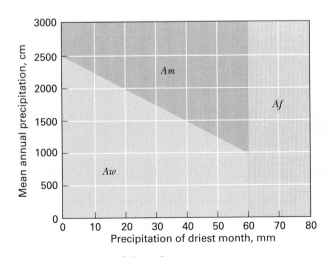

Figure 3 Boundaries of the A climates

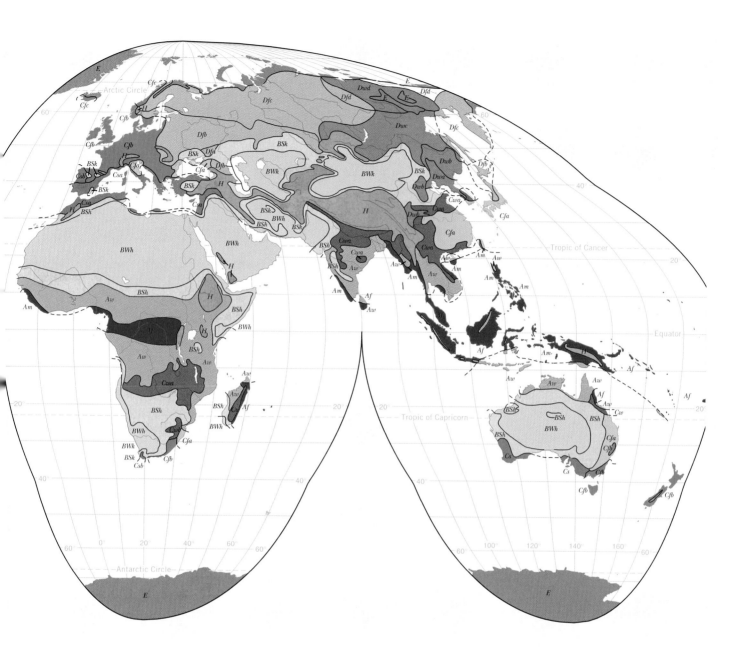

BS *Steppe climate*

A semi-arid climate characterized by grasslands, it occupies an intermediate position between the desert climate (BW) and the more humid climates of the A, C, and D groups. Boundaries are determined by formulas given in Figure 4.

BW Desert climate

Desert has an arid climate with annual precipitation of usually less than 400 millimetres. The boundary with the adjacent steppe climate (BS) is determined by formulas given in Figure 4.

C *Mild humid climate with no dry season*

Precipitation of the driest month averages more than 30 millimetres.

Cw *Mild humid climate with a dry winter*

The wettest month of summer has at least 10 times the precipitation of the driest month of winter. (Alternative definition: 70 percent or more of the mean annual precipitation falls in the warmer six months.)

Cs *Mild humid climate with a dry summer*

Precipitation of the driest month of summer is less than 30 millimetres. Precipitation is at least three times as much as the driest month of summer. (Alternative definition: 70 percent or more of the mean annual precipitation falls in the six months of winter.)

Df *Snowy-forest climate with a moist winter*

No dry season.

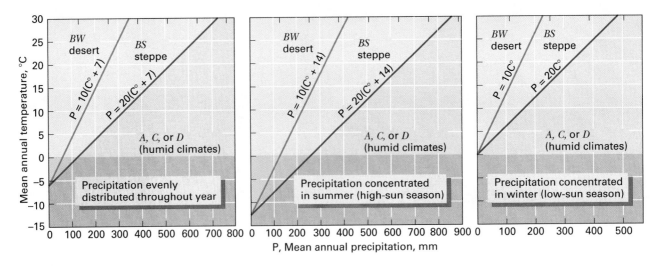

Figure 4 **Boundaries of the B climates**

CHAPTER SUMMARY

- Climate is the average weather pattern of a region. Because temperature and precipitation are measured at many stations worldwide, the combined annual patterns of monthly averages of temperature and precipitation can be used to assign climate types.
- *Temperature regimes* are typical patterns of annual variation in temperature. They depend on latitude, which determines the annual insolation cycle, and on location—continental or maritime—which enhances or moderates annual insolation.
- Global precipitation patterns are largely determined by air masses and their movements, which in turn are produced by global air circulation patterns. The main features of the global pattern of rainfall are the following:
  1. a wet equatorial belt produced by convectional precipitation around the ITCZ;
  2. trade-wind coasts that receive moist flows of mT air from trade winds as well as tropical cyclones;
  3. tropical deserts located under subtropical high-pressure cells;
  4. midlatitude deserts and steppes, which are dry because they are far from oceanic moisture sources;
  5. moist subtropical regions that receive westward flows of moist mT air in the summer and eastward-moving wave cyclones in winter;
  6. mid-latitude west coasts, which are subjected to eastward flows of mP air and occluded wave cyclones by prevailing westerly winds; and
  7. polar and Arctic deserts, where little precipitation falls because the air is too cold to hold much moisture.

- Annual patterns of precipitation can be described as uniform (ranging from abundant to scarce); high-Sun (summer) maximum; and low-Sun (winter) maximum.
- There are three groups of climate types, arranged by latitude. *Low-latitude climates (Group I)* are dominated by mE, mT, and cT air masses, and are largely related to the global circulation patterns that produce the ITCZ, trade winds, and subtropical high-pressure cells. They include
  1. Wet equatorial (warm to hot with abundant rainfall);
  2. Monsoon and trade-wind coastal (warm to hot with a very wet rainy season at high Sun);
  3. Wet-dry tropical (warm to hot with very distinct wet and dry seasons); and
  4. Dry tropical (extremely hot in the high-Sun season, a little cooler in the low-Sun season, with little or no rainfall).
- *Midlatitude climates (Group II)* lie in the polar-front zone and are strongly influenced by eastward-moving wave cyclones in which mT, mP, and cP air masses interact. They include
  1. Dry subtropical (a desert climate, not quite as hot as the dry tropical climate);
  2. Moist subtropical (hot and humid summers, mild winters, ample rainfall);
  3. Mediterranean (hot, dry summers, rainy winters);
  4. Marine west-coast (warm summers, cool winters, more precipitation in winter);
  5. Dry midlatitude (warm in summer, cold in winter, low annual precipitation); and

Dw *Snowy-forest climate with a dry winter*
No wet season.

ET *Tundra climate*
The mean temperature of the warmest month is above 0°C but below 10°C.

EF *Perpetual frost climate*
In this ice sheet climate, the mean monthly temperatures of all months are below 0°C.

To denote further variations in climate, Köppen added a third letter to the code. The meanings are as follows:

a With hot summer; warmest month is over 22°C; C and D climates.

b With warm summer; warmest month is below 22°C; C and D climates.

c With cool, short summer; less than four months are over 10°C; C and D climates.

d With very cold winter; coldest month is below −38°C; D climates only.

h Dry-hot; mean annual temperature is over 18°C; B climates only.

k Dry-cold; mean annual temperature is under 18°C; B climates only.

According to the complete Köppen climate code, BWk refers to a cool desert climate, and Dfc refers to a cold, snowy-forest climate with a cool, short summer.

6. Moist continental (warm in summer, cold in winter, ample precipitation through the year).
- *High-latitude climates (Group III)* are dominated by polar and Arctic (Antarctic) air masses. Wave cyclones mixing mP and cP air masses along the *arctic-front zone* provide precipitation in this region. Climates of high latitudes include
1. Boreal forest (short, cool summers, with very cold, snowy winters);
2. Tundra (very short or nonexistent summers with cold winters); and

3. Ice sheet (bitterly cold, even in summer).
- *Dry climates* are those in which precipitation is largely evaporated from soil surfaces and transpired by vegetation, so that permanent streams cannot be supported. Within dry climates, there are two subtypes: *arid* (driest) and *semi-arid* or steppe (a little wetter). In *moist climates*, precipitation exceeds evaporation and transpiration, providing for a sustained year-round stream flow. In *wet-dry climates*, strong wet and dry seasons alternate.

KEY TERMS

climate	climograph	isohyet	steppe

REVIEW QUESTIONS

1. Discuss the use of monthly records of average temperature and precipitation to characterize the climate of a region. Why are these measures useful?
2. Why are latitude and location (maritime or continental) important factors in determining the annual temperature cycle of a location?
3. How does air temperature, as a climatic variable, influence precipitation?
4. Describe three temperature regimes and explain how they are related to latitude and location.
5. Identify seven important features of the global precipitation map and describe the factors that produce each.

6. The seasonality of precipitation at a station can generally be described as following one of three patterns. Identify them, explaining how each pattern can arise and providing an example.
7. What are the important global circulation patterns and air masses that influence low-latitude (Group I) climates? What are their effects?
8. What air masses and circulation patterns influence the mid-latitude (Group II) climates and how?
9. Identify the air masses and frontal zones that are important in determining high-latitude climates (Group III) and explain their effects.

FOCUS ON SYSTEMS 8.1   Time Cycles of Climate

1. What are the two primary time cycles evident in climate data? How do they influence weather at a location?

2. Provide an example of another time cycle evident in temperature records of the past few centuries.

VISUALIZATION EXERCISES

1. Victoria, B.C. enjoys a midlatitude west-coast location. Sketch a climograph showing monthly temperature and precipitation.

2. Sketch a climograph for Sudbury, Ontario, noting that it has a midlatitude continental location.

ESSAY QUESTION

1. Sketch a hypothetical continent with a shape and features of your own choosing. It stretches from about 70° N to 40° S. Add some north-south mountain ranges, positioned where you like. Then select four different locations, describing and explaining the annual cycles of temperature and precipitation you would expect at each.

PROBLEMS   WORKING IT OUT 8.2 Averaging in Time Cycles

1. The table below shows monthly precipitation in millimetres at Dryden, Ontario, for 2001–2005. Find the average for each month of this five-year period.

	Jan	Feb	Mar	Apr	May	Jun	Jul	Aug	Sep	Oct	Nov	Dec
2001	15.5	16.4	11.7	80.3	124	126.9	106.6	23.1	39.9	85.3	25.7	24.4
2002	20.4	6.2	21.8	68.8	58.7	177.3	45.8	87	92.2	37.5	19.3	16.4
2003	18	15	13.6	24	68.7	80.2	79.2	116.6	204.8	45	16.9	32.7
2004	55.7	16.8	60.2	15	172.3	60.6	58.4	191.4	137.2	84.8	22.5	54.5
2005	39.5	13.4	24.4	45.6	86.1	154.1	95.3	93.8	68.8	43.2	62.4	29.3

2. Plot the monthly averages for 2001–2005, and then plot the 30-year monthly means. How do the two graphs compare?

Find answers at the back of the book

EYE ON THE LANDSCAPE |

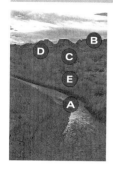

Chapter Opener   Desert rains in Canyon Lands National Park, Petrified Hollow, Utah, USA. Moisture is of paramount importance in arid climates, and following heavy storms, excess water will form streams (**A**) which help to mold the desert surface. Gullies and canyons (**B**) have been carved into the horizontally-bedded sandstone (see chapters 11 and 17) and contribute sediment to the alluvial fans (**C**) that accumulate at the base of the slopes. The plant cover consists primarily of sparely distributed shrubs on the well-drained slopes (**D**) to denser stands of small trees, shrubs, and herbs (**E**) where moisture is more abundant or persistent in the landscape.

Payton, Saskatchewan, snowy landscape with grain elevators. Sun angle is low during the Prairie winter and incoming radiation (**A**) is further reduced by clouds and ice crystals in the atmos-

phere. High winds can create bitter cold weather because of wind chill, and also create hazardous blowing snow (**B**) conditions. Snow will accumulate in drifts (**C**) where its movement is impeded by vegetation.

Chapter 9

EYE ON THE LANDSCAPE

> > > >

Todd River, Australia, near Alice Springs. The dry channel of the Todd River can at **Fly By: 23° 44' 50" S; 133° 52' 33" E**

What else would the geographer see? . . . Answers at the end of the chapter.

LOW-LATITUDE CLIMATES

Overview of the Low-Latitude Climates
The Wet Equatorial Climate (Köppen AF)
The Monsoon and Trade-Wind Coastal Climate (Köppen AF, AM)
The Low-Latitude Rainforest Environment
The Wet-Dry Tropical Climate (Köppen AW, CWA)
The Savanna Environment
Working It Out 9.1 • Cycles of Rainfall in the Low Latitudes
Eye on Global Change 9.2 • Drought and Land Degradation in the African Sahel
The Dry Tropical Climate (Köppen BWH, BSH)
The Tropical Desert Environment
Highland Climates of Low Latitudes

9.1 Low-latitude climographs These four climographs show the key features of the low-latitude climates.

OVERVIEW OF THE LOW-LATITUDE CLIMATES

The *low-latitude climates* lie for the most part between the tropics of Cancer and Capricorn. In terms of world latitude zones, the low-latitude climates occupy the entire equatorial zone (10° N to 10° S), most of the tropical zone (10–15° N and S), and part of the subtropical zone (25–35° N and S). The low-latitude climate region includes the equatorial trough of the intertropical convergence zone (ITCZ), the belt of tropical easterlies (northeast and southeast trades), and large portions of the oceanic subtropical high-pressure belt (Figure 8.6).

Figure 9.1 shows climographs for the four low-latitude climates. These climates range from extremely moist—the wet equatorial climate—to extremely dry—the dry tropical climate. They also vary strongly in the seasonality of their rainfall. In the wet equatorial climate, rainfall is abundant year-round. But in the

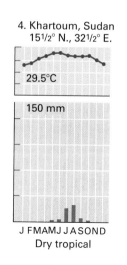

Low-Latitude Climates

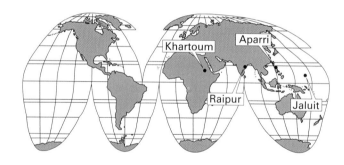

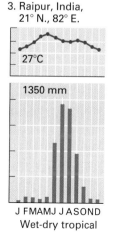

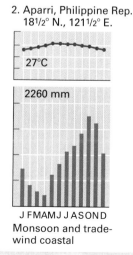

9.2 Heavy clouds associated with strong convective uplift along the ITCZ

wet-dry tropical climate, rainfall is abundant for only part of the year. During the remainder of the year little or no rain falls.

The seasonal distribution of precipitation at these latitudes is largely controlled by the annual movement of the Sun north and south of the equator and the corresponding shift in circulation that accompanies it. A zone of convective precipitation moves with the Sun as it travels toward its June solstice position at the Tropic of Cancer, culminating in a short wet season of two to three months at latitudes about 20°N. This wet season gets progressively shorter as the influence of the semi-permanent high pressure zone becomes more pronounced. Storms accompany the Sun as it moves toward its winter solstice position at the Tropic of Capricorn. Thus, near the equator, uplift along ITCZ provides precipitation year round. At latitudes 5° to 15° N and S, the effect

is two short wet seasons alternating with two short dry seasons, and at higher latitudes, the wet-dry tropical climate merges with the dry tropical climate, which receives very limited precipitation.

The seasonal temperature cycle also varies among these climates. In the wet equatorial climate, temperatures are nearly uniform throughout the year. In the dry tropical climate there is a strong annual temperature cycle. Table 9.1 summarizes these characteristics.

THE WET EQUATORIAL CLIMATE (KÖPPEN *AF*)

The **wet equatorial climate** is associated with the ITCZ, which is nearby for most of the year. The climate is dominated by warm, moist maritime equatorial (mE) and maritime tropical (mT) air masses that yield heavy convectional rainfall (Figure 9.2). Precipitation is plentiful in all months, and the annual total often exceeds 2,500 millimetres. However, there is usually a seasonal pattern to the rainfall, with a somewhat wetter period occurring during the time of year when the ITCZ migrates into the region. Remarkably uniform temperatures prevail throughout the year. Both mean monthly and mean annual temperatures are typically close to 27°C. In fact, the diurnal temperature range is normally greater than the annual range.

Figure 9.3 shows the world distribution of the wet equatorial climate. This climate is found in the latitude range 10° N to 10° S. The major regions where it occurs include the Amazon lowland

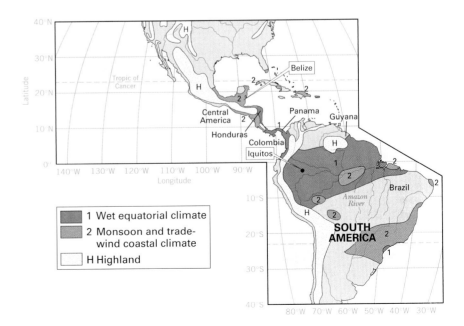

9.3 World map of wet equatorial and monsoon and trade-wind coastal climates (Based on Goode Base Map)

Table 9.1 Low-latitude climates

Climate	Temperature	Precipitation	Explanation
Wet equatorial	Uniform temperatures, mean near 27 °C.	Abundant rainfall, all months, from mT and mE air masses. Annual total may exceed 2,500 mm.	The ITCZ dominates this climate, with abundant convectional precipitation generated by convergence in weak equatorial lows. Rainfall is heaviest when the ITCZ is nearby.
Monsoon and trade-wind coastal	Temperatures show an annual cycle, with warmest temperatures in the high-Sun season.	Abundant rainfall, but with a strong seasonal pattern.	Trade-wind coastal: Rainfall from mE and mT air masses is heavy when the ITCZ is nearby, lighter when the ITCZ moves to the opposite hemisphere. Asian monsoon coasts: dry air flowing southwest in low-Sun season alternates with moist oceanic air flowing northeast, producing a seasonal rainfall pattern on west coasts.
Wet-dry tropical	Marked temperature cycle, with hottest temperatures before the rainy season.	Wet high-Sun season alternates with dry low-Sun season.	Subtropical high pressure moves into this climate in the low Sun-season, bringing very dry conditions. In the high-Sun season, the ITCZ approaches and rainfall occurs. Asian monsoon climate: alternation of dry continental air in low-Sun season, with moist oceanic air in high-Sun season bringing a strong pattern of dry and wet seasons.
Dry tropical	Strong temperature cycle, with intense hot temperatures during high-Sun season.	Low precipitation. Sometimes rainfall occurs when the ITCZ is near.	This climate is dominated by subtropical high pressure, which provides clear, stable air for much or all of the year. Insolation is intense during high-Sun period.

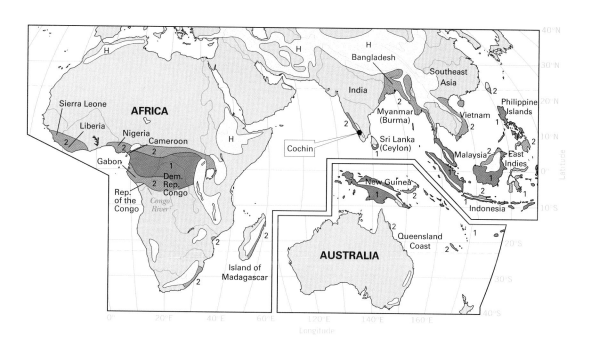

of South America, the Congo Basin of equatorial Africa, and the East Indies, from Sumatra to New Guinea.

Figure 9.4 is a climograph for Iquitos, Peru, a typical wet equatorial station situated close to the equator in the broad, low basin of the upper Amazon River (see Figure 9.3 for location). Notice the small annual range in temperature and the large annual rainfall total. Like the annual figures, monthly air temperatures are extremely uniform in the wet equatorial climate. Typically, mean monthly air temperature will range between 26° and 29°C for stations at low elevation in the equatorial zone.

THE MONSOON AND TRADE-WIND COASTAL CLIMATE (KÖPPEN *AF, AM*)

Like the wet equatorial climate, the **monsoon and trade-wind coastal climate** has abundant rainfall. But unlike the wet equatorial climate, the rainfall of the monsoon and trade-wind coastal climate always shows a strong seasonal pattern imposed by the migration of the ITCZ. In the high-Sun season, the ITCZ is nearby and monthly rainfall is greater. In the low-Sun season, when the ITCZ has migrated to the other hemisphere, subtropical high pressure dominates and monthly rainfall is less.

Figure 9.3 shows the global distribution of the monsoon and trade-wind coastal climate. The climate occurs in latitudes from 5° to 25° N and S.

As the name of the monsoon and trade-wind coastal climate suggests, two somewhat different situations produce this climate type. On trade-wind coasts, moisture-laden maritime tropical (mT) and maritime equatorial (mE) air masses produce rainfall. Trade winds move these air masses onshore onto narrow coastal zones. As the warm, moist air passes over coastal hills and mountains, the orographic effect produces convectional shower activity. Rainfall is also intensified by easterly waves, which are more frequent when the ITCZ is nearby. The east coasts of land masses experience this trade-wind effect because the trade winds blow from east to west. Trade-wind coasts are found along the east sides of Central and South America, the Caribbean Islands, Madagascar, Southeast Asia, the Philippines, and northeast Australia (see Figure 9.3).

The coastal precipitation effect also applies to the summer monsoon of Asia, when the monsoon circulation brings mT air onshore. However, the onshore monsoon winds blow from southwest to northeast, so the western coasts of land masses are exposed to this moist airflow. Western India and Myanmar are examples. Moist air also penetrates well inland in Bangladesh, providing the very heavy monsoon rains for which the region is well known.

The southwest monsoon season accounts for more than 75 percent of the annual rainfall over most of India. Figure 9.5 shows the typical progression of a monsoon over India. The dates for the onset and end on a southwest monsoon are determined by the respective increase and decrease in rainfall. The actual dates used are the middle dates of consecutive five-day periods, during which a there is noticeable rise or fall in precipitation.

One of the wettest places on the Earth is Cherrapunji, which lies at 1,370 metres in the foothills of the Himalayas in northeastern India. The average annual rainfall at Cherrapunji is 11,430 millimetres; however, in one 12-month period it received 22,987 millimetres, which still stands as a world record. It also holds the record for maximum rainfall in a single month—9,300 millimetres in July 1861. The heavy rain period is generally from May to September, when the rain-bearing monsoon airflow is forced to rise over the high terrain. Local agriculture is vulnerable to excessive soil erosion from the heavy rainfall. Also, because of the seasonality of the monsoon rainfall, Cherrapunji faces acute water shortages in some months.

The wettest place in the world is Mount Waialeale (1,598 metres) in Kauai, Hawaii, which receives more rain annually than Cherrapunji. Here, the average annual rainfall is 11,685 millimetres (Figure 9.6). The maximum annual rainfall recorded for Mount Waialeale was 16,916 millimetres in 1982.

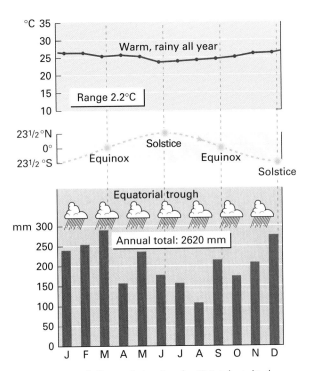

9.4 Wet equatorial climate Iquitos, Peru, lat. 3° S, is located in the upper Amazon lowland, close to the equator. Temperatures differ little from month to month, and there is abundant rainfall throughout the year.

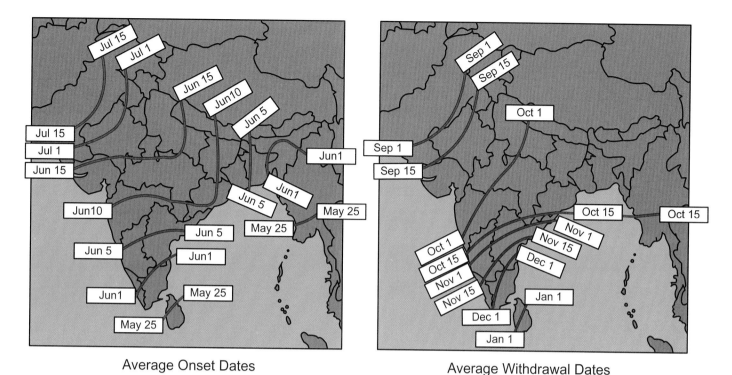

Average Onset Dates

Average Withdrawal Dates

9.5 Typical dates for the onset and withdrawal of the wet monsoon over India

In central and western Africa and southern Brazil, the monsoon pattern shifts the ITCZ over 20° of latitude or more (see Figure 5.7). Here, heavy rainfall occurs in the high-Sun season, when the ITCZ is nearby. Drier conditions prevail in the low-Sun season, when the ITCZ is more distant. Temperatures in the monsoon and trade-wind coastal climate, though warm throughout the year, also show an annual cycle. Warmest temperatures occur in the high-Sun season, just before arrival of the ITCZ brings clouds and rain. Minimum temperatures occur at the time of low Sun.

Figure 9.7 is a climograph for Belize City, in the Central American country of the same name (see Figure 9.3 for location). This east-coast city, located at lat. 17° N, is exposed to the tropical easterly trade winds. Rainfall is abundant from June through November, when the ITCZ is nearby. Easterly waves are common in this season, and occasionally, a tropical cyclone will bring torrential rainfall. Following the December solstice, rainfall is greatly reduced, with minimum values in March and April. At this time, the ITCZ lies farthest away, and the climate is dominated by subtropical high pressure. Air temperatures

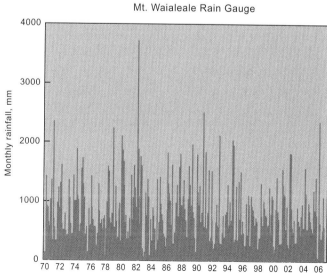

Mt. Waialeale Rain Gauge

9.6 Mount Waialeale, Hawaii; 36-year (1970–2006) monthly rainfall graph The rain gauge near the summit of Mt. Waialeale has been operating since 1910. The average yearly rainfall at the gauge between the years 1970-2006 is about 10,000 mm.

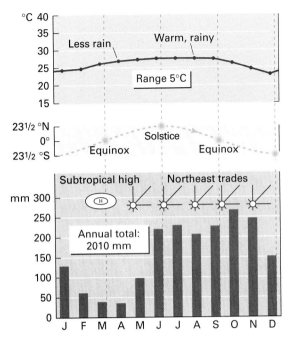

9.7 Trade-wind coastal climate This climograph for Belize City, a Central American east-coast city at lat. 17° N, shows a marked season of low rainfall following the low-Sun period. For the remainder of the year, precipitation is high, produced by warm, moist northeast trade winds.

show an annual range of 5° C with the maximum occurring in the high-Sun months.

The Asiatic monsoon shows a similar pattern, but there is an extreme peak of rainfall during the high-Sun period and a well-developed dry season of two or three months with little rainfall. The climograph for Cochin, India provides an example (Figure 9.8). Located at lat. 10° N on the west coast (see Figure 9.3 for location), Cochin receives the warm, moist southwest winds of the

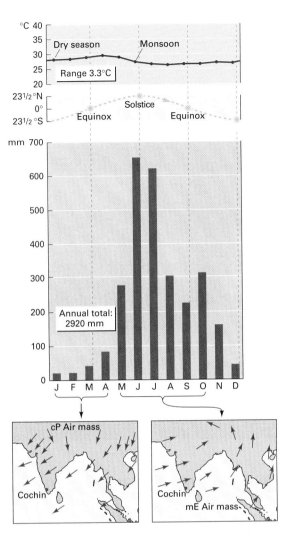

9.8 Monsoon coastal climate Cochin, India, on a windward west coast at lat. 10° N, shows an extreme peak of rainfall during the rainy monsoon, contrasting with a short dry season at the time of low Sun.

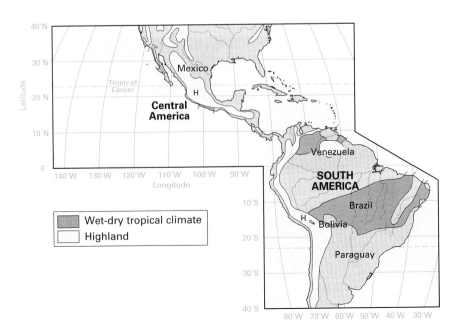

9.9 Rainforest of the western Amazon lowland, near Manaus, Brazil The river is a tributary of the Amazon. Note the many different types of trees of varying shapes.

summer monsoon. Monthly rainfall is extreme in both June and July. A strongly pronounced season of low rainfall occurs at the time of low Sun—December through March. Air temperatures show only a weak annual cycle, cooling a bit during the rains, so the annual range is small at this low latitude.

THE LOW-LATITUDE RAINFOREST ENVIRONMENT

The wet equatorial and monsoon and trade-wind coastal climates are quite uniform in temperature and have a high annual rainfall. These factors create the *low-latitude rainforest environment* (Figure 9.9). Here, abundant rainfall and prevailing warm soil temperatures promote decay and decomposition of rock to great depths, which has created a thick soil layer. The soil is typically

rich in iron oxides, which impart a deep red colour (see Chapter 20). This kind of soil has largely lost its ability to hold the nutrients plants need. However, many kinds of native plants have adapted to this soil. The most conspicuous are the trees of the evergreen rainforests, which may contain more than 3,000 different species in an area of only a few square kilometres. Rainforest species quickly recycle the essential plant nutrients that are released by the decay of fallen leaves and branches before they are leached from the impoverished soil. Chapter 23 discusses in detail the characteristics of the low-latitude rainforest environment.

THE WET-DRY TROPICAL CLIMATE (KÖPPEN *AW, CWA*)

In the monsoon and trade-wind coastal climate, the ITCZ moves into and away from the climate region to produce the seasonal cycle of rainfall and temperature. Farther from the equator, this cycle becomes stronger, and the monsoon and trade-wind coastal climate grades into the **wet-dry tropical climate**.

The wet-dry tropical climate is distinguished by a very dry season at low Sun that alternates with a very wet season at high Sun. During the low-Sun season, the ITCZ moves beyond the region and dry continental tropical (cT) air masses prevail. In the high-Sun season, when the ITCZ is nearby, moist maritime tropical (mT) and maritime equatorial (mE) air masses dominate. Cooler temperatures accompany the dry season but give way to a very hot period before the rains begin.

Figure 9.10 shows the global distribution of the wet-dry tropical climate, which is found at latitudes of 5° to 20° N and S in Africa and the Americas, and at 10° to 30° N in Asia. In Africa

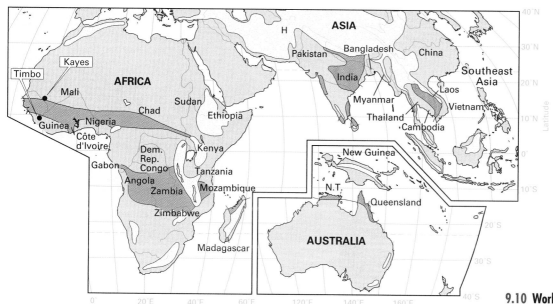

9.10 World map of the wet-dry tropical climate (Based on Goode Base Map)

and South America, the climate occupies broad bands on the poleward side of the wet equatorial and monsoon and trade-wind coastal climates. Because these regions are farther away from the ITCZ, less rainfall occurs during the rainy season, and subtropical high pressure can dominate more strongly during the low-Sun season. In central India and Indochina, mountain barriers protect the wet-dry tropical climate regions somewhat from the warm, moist mE and mT air flows associated with the trade winds and monsoons. These barriers create a rain shadow effect, which reduces rainfall during the rainy season and accentuates the dry season.

Figure 9.11 is a climograph for Timbo, Guinea, at lat. 10° N in West Africa (see Figure 9.10 for location). Here the rainy season begins just after the March equinox and reaches a peak in August, about two months following the June solstice. At this time, the ITCZ has migrated to its most northerly position, and moist mE air masses flow into the region from the ocean lying to the south. Monthly rainfall then decreases as the low-Sun season arrives and the ITCZ moves to the south. Three months—December through February—are practically rainless. During this season, subtropical high pressure dominates the climate, and stable, subsiding continental tropical (cT) air pervades the region. This air originates in the Sahara and brings with it the dry, dusty conditions of the *harmattan*—a dust-laden wind that blows from the east and northeast, carrying dust across western Africa and the eastern Atlantic.

Timbo's temperature cycle is closely linked to both the solar cycle and the precipitation pattern. In February and March, insolation increases, and air temperature rises sharply, resulting in a brief hot season. As soon as the rains set in, the effect of cloud cover and evaporation of rain causes the temperature to fall. By July, temperatures have resumed an even level.

A characteristic of the tropical wet-dry climate is its large year-to-year variability in precipitation. In this climate, rainfall occurs when the ITCZ migrates into the region. In some years, the ITCZ fails to migrate, greatly reducing rainfall. *Working It Out 9.1 • Cycles of Rainfall in the Low Latitudes* shows how to measure variation from year to year and looks at multi-year cycles at various tropical climate stations.

THE SAVANNA ENVIRONMENT

The wet-dry tropical climate is associated with the savanna environment. Here the native vegetation must survive alternating seasons of very dry and very wet weather. River channels that are not fed by nearby moist mountain regions are nearly or completely dry in the low-Sun dry season. In the rainy season, these river channels become filled with swiftly flowing, turbid water. Often, extensive tracts of land are flooded (Figure 9.12). However, the rains are not reliable, and agriculture without irrigation is hazardous at best. When there is no rain, a devastating famine can ensue. The Sahel region of Africa, discussed further in *Eye on Global Change 9.2 • Drought and Land Degradation in the African Sahel*, is a region well-known for such droughts and famines.

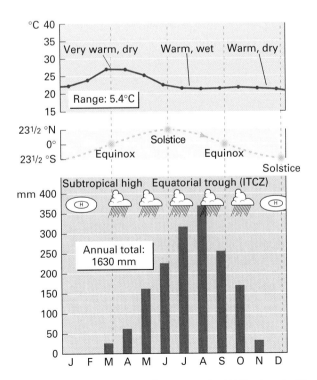

9.11 Wet-dry tropical climate Timbo, Guinea, at lat. 10° N, is located in West Africa. A long wet season during high Sun alternates with an almost rainless dry season during low Sun.

9.12 A flooded savanna Drainage channels are poorly developed in the relatively flat topography characteristic of savannas, and during the wet season the land can be flooded to a depth of several metres.

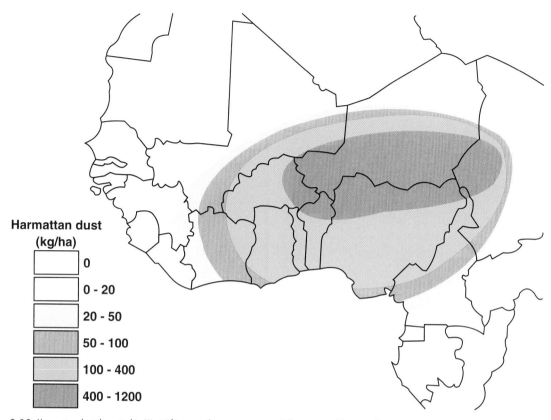

Harmattan dust (kg/ha)

☐	0
☐	0 - 20
☐	20 - 50
▨	50 - 100
▨	100 - 400
▨	400 - 1200

9.13 Harmattan dust deposited in West Africa contributes, on average, 3.8 grams per kilogram of nitrogen, 0.8 grams per kilogram of phosphorus, and 18.7 grams per kilogram of potassium to the soil.

Soils of the savanna environment are similar in their physical characteristics and fertility to those of the rainforest environment—that is, they are largely red soils of low fertility. However, substantial areas of the savanna environment have fertile soils developed and sustained by the slow accumulation of windblown dust from adjacent deserts (Figure 9.13). Equally important are highly fertile soils that occur along major through-flowing rivers. Annual flooding of these rivers leaves lowland deposits of fertile silt carried down from distant mountain ranges.

Most plants enter a dormant phase during the dry period, then come into leaf and bloom with the rains. For this reason, the native plant cover is described as rain-green vegetation (Figure 9.14). In

9.14 A comparison of the wet and dry season Wet and dry seasonal aspects of Mopane (*Colophospermum mopanc*) woodland in Zimbabwe

WORKING IT OUT | 9.1   Cycles of Rainfall in the Low Latitudes

The cycle of monthly precipitation shown on the climographs presented in Chapters 8–10 is basically controlled by the Sun's declination cycle, which rhythmically varies the insolation a location receives, depending on its latitude and the time of the year. This cycle produces an annual pattern of precipitation that is revealed by taking monthly averages over a long period. (See *Working It Out 8.2 • Averaging in Time Cycles*.)

Although this annual cycle is strong at most locations, there are also other cycles that influence rainfall over a sequence of years, rather than from month to month. This may give rise to several "wet" years followed by several "dry" years in a repeating sequence.

Rainfall cycles that span several decades are shown in the three graphs to the right. The graphs present the total rainfall for the wettest month at each location for a sequence of 46 years. The low-latitude stations that have been selected are (a) Padang, Sumatra, in the wet equatorial climate; (b) Mumbai, India, in the wet-dry tropical climate; and (c) Abbassia, near Cairo, Egypt, which is in the dry tropical climate. Notice the difference in vertical scale of the Abbassia record.

The data are presented in two ways, using bars and a superimposed continuous curved line that smoothes the rainfall totals. Many of the peak rainfalls appear to be followed by a period of low rainfall. This suggests that there is a rainfall cycle with a length, or period, of between two and three years. This cycle is particularly pronounced at Padang and Mumbai, especially in the early part of the sequence.

Another way to describe a cycle is by its *amplitude*. For a smooth wavelike curve, amplitude is the difference in height between a crest and the adjacent trough. That difference is expressed for these data in millimetres of rainfall. The varying amplitude of the curve is a measure of the *variability* of the precipitation. The difference in amplitude of the cycles at Padang and Mumbai suggests that rainfall at Mumbai has higher variability.

Variability can be measured by the *mean deviation* of all the values in the record. This requires the calculation of the deviation of each value. The deviation of an individual value is the difference between the value and the mean of all the values, taken without respect to sign. Thus, deviation is defined as

$$D_i = |P_i - \overline{P}|$$

where D_i is the deviation of P_i, the *i*th observation in the series; $\overline{P}$ is the mean of the series (see *Working It Out 8.2 • Averaging in Time Cycles*); and the two vertical bars | | indicate absolute value. (The absolute value is the value taken without respect to sign. It is always positive.) The mean deviation ($\overline{D}$) is then

$$\overline{D} = \frac{1}{n}\sum_{i=1}^{n} D_i = \frac{1}{n}\sum_{i=1}^{n} |P_i - \overline{P}|$$

where n is the number of observations.

Calculation of the mean and mean deviation for the three sets of observations yields the following results:

Station	Mean (mm)	Mean deviation (mm)
Padang	514	129
Mumbai	610	207
Abbassia	6.7	7.2

Mumbai has the largest mean value: 610 millimetres during its wettest month. Padang has a corresponding mean of 514 millimetres. Compared with these two stations, Abbassia is very dry, with a mean of less than 10 millimetres of rainfall in its wettest month. The mean deviations of these stations show that Mumbai has the largest year-to-year fluctuation. Padang is next, followed by Abbassia with a value of 7.2.

Note that if only mean deviation is considered, it appears that rainfall in the wettest month is constant at Abbassia. However, in many years, Abassia actually receives no rainfall during the wettest month, while in other years more than 30 millimetres falls. For the residents of this location, the variability is great, indeed!

Comparing mean deviation with the mean gives a measure of *relative variability*. Thus,

Station	Relative Variability
Padang	12.9 ÷ 51.4 = 0.25
Mumbai	20.7 ÷ 61.0 = 0.34
Abbassia	0.72 ÷ 0.67 = 1.07

In other words, the average deviation during the wettest months is 25 percent of the mean at Padang, 34 percent at Mumbai, and 107 percent at Abbassia. By this measure, Abbassia's rainfall is the most variable, while Padang's is the least variable.

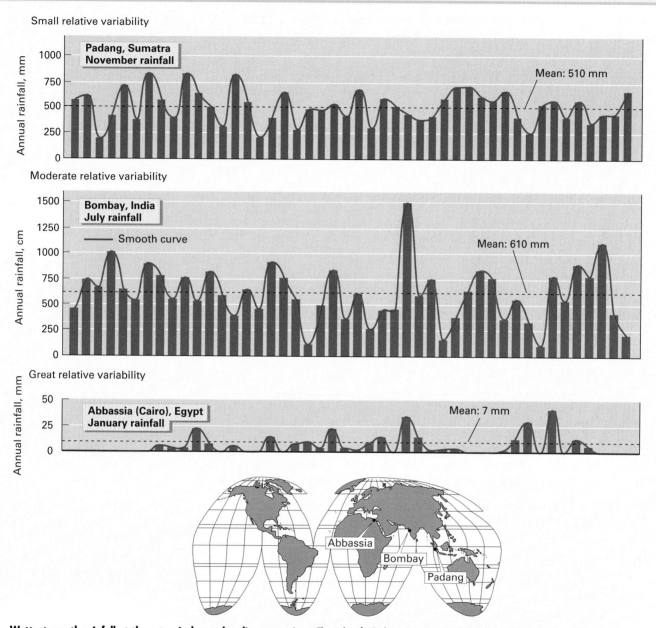

Small relative variability

Padang, Sumatra
November rainfall

Mean: 510 mm

Moderate relative variability

Bombay, India
July rainfall

—— Smooth curve

Mean: 610 mm

Great relative variability

Abbassia (Cairo), Egypt
January rainfall

Mean: 7 mm

Abbassia

Bombay

Padang

Wettest-month rainfall at three tropical wet-dry climate stations These data for Padang, Sumatra; Mumbai, India; and Abbassia, Egypt; show a tendency for rainfall in the wettest month to rise and fall in a two- to three-year cycle. Large variability is a characteristic of monthly rainfall in many low-latitude climates.

EYE ON GLOBAL CHANGE | 9.2 Drought and Land Degradation in the African Sahel

The wet-dry tropical climate is subject to years of devastating drought as well as years of abnormally high rainfall that can result in severe floods. Climate records show that two or three successive years of abnormally low rainfall (a drought) typically alternate with several successive years of average or higher than average rainfall. Such variability is a characteristic feature of the wet-dry tropical climate, and the native plants and animals inhabiting this region are well adapted to the rainfall regime.

West Africa's wet-dry tropical climate, including the adjacent semi-arid southern belt of the dry tropical climate to the north, provides a lesson on the human impact on a delicate ecological system. Figure 1 shows countries in this perilous belt, called the *Sahel* or *Sahelian zone*. From 1968 through 1974, all these countries were struck by a severe drought, and again from 1983–1985. Both nomadic cattle herders and grain farmers share this zone. During the drought, grain crops failed and foraging cattle could find no food to eat. In the worst stages of the Sahel drought, nomads were forced to sell their remaining cattle. Because the cattle were their sole means of subsistence, the nomads soon starved. Some 5 million cattle perished, and it has been estimated that 100,000 people died of starvation and disease in 1973 alone.

The Sahelian drought of 1968–1974 was an example of a phenomenon that, at that time, was called *desertification*—the permanent transformation of the land surface to resemble a desert, largely through human activities such as the destruction of grasses, shrubs, and trees by grazing animals and fuel wood harvesting. That term has now been abandoned in favour of *land degradation*. This degradation accelerates the effects of soil erosion, such as gullying of slopes and accumulations of sediment in stream channels (Figure 2). It also intensifies the removal of soil by wind.

Periodic droughts throughout past decades are well documented in the Sahel, as they are in other regions of the wet-dry

tropical climate. Figure 3 shows the percentage departure from the long-term mean of each year's rainfall in the western Sahel from 1901 through 1998. Note the wide year-to-year variation. Since about 1950, the duration of periods of continuous departures both above and below the mean seem to have increased substantially. The period of sustained high-rainfall years in the 1950s contrasts sharply with a series of severe drought episodes starting in 1971. To obtain an earlier record, scientists have examined fluctuations in the level of Lake Chad. In times of drought, the lake's shoreline retreats, while in times of abundant rainfall, it expands. The changes document periods of rainfall deficiency and excess—1820–40, below normal; 1870–95, above normal; 1895–1920, below normal.

The long-term precipitation record shows that droughts and wet periods are a normal phenomenon in the Sahel. His-

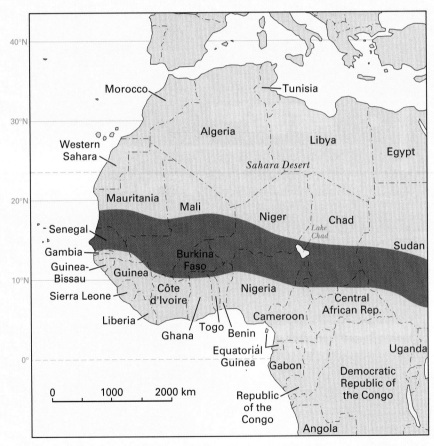

Figure 1 The African Sahel The Sahel, or Sahelian zone, shown in brown, lies south of the great Sahara desert of North Africa.

Figure 2 Sahelian drought At the height of the Sahelian drought, vast numbers of cattle had perished and even the goats were hard pressed to survive. Trampling of the dry ground prepared the region for devastating soil erosion by the rains that eventually ended the drought.

torically, the landscape was able to recover from the droughts during periods of abundant rainfall. Today, the pressures from the increased human and cattle populations keep the land degraded. As long as these populations remain high, land degradation will become a severe and permanent feature.

Climate models developed in the 1990s suggested that the Sahelian droughts could be explained by natural large-scale climate changes. However, in the early 2000s, these ideas were revised to incorporate the effect of air pollution on cloud cover. It has been suggested that an increase in abundance of sulphate aerosols from pollution results in smaller cloud droplets and increases the albedo of the cloud surfaces. This reduces convective rain activity in the tropics and limits the northward migration of the rains as the Sun moves to its July position at the Tropic of Cancer.

The effect of accelerated global warming in the coming decades on wet-dry tropical climate regions is difficult to predict. Some regions may experience greater swings between drought and surplus precipitation. At the same time, these regions may migrate toward the poles as a result of intensification of the Hadley cell circulation. Thus, the Sahel region may move northward into the desert zone of the Sahara. However, such changes are highly speculative. At this time, there is no scientific consensus on how climate change will affect the Sahel or other regions of the wet-dry tropical climate.

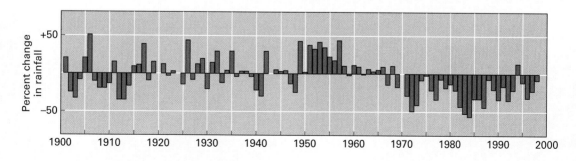

Figure 3 Rainfall in the Sahel Rainfall fluctuations for stations in the western Sahel, 1901–98, expressed as a percentage departure from the long-term mean. (Courtesy of Sharon E. Nicholson, Department of Meteorology, Florida State University, Tallahassee)

9.15 Fire in a wooded savanna in Australia

the dry season, the grasses turn to straw, and many of the tree species shed their leaves to cope with the drought. As the dry season progresses, there is increasing probability of fires (Figure 9.15).

Fires are especially common in Africa, where they are deliberately set to maintain farmland and grazing areas. The burning area shifts from north to south over the course of the year, as the seasons change (Figure 9.16). In January, when the rainy season comes to southern Africa, the fires burn extensively in the grasslands south of the Sahara Desert. By July, the fires have shifted to the southern part of the continent.

The dry season brings a severe struggle for existence to animals of the savanna lands. As streams and hollows dry up, the few muddy water holes must supply all drinking water. Danger of attack by carnivores greatly increases. In Africa, the herbivores must move across vast regions to find food and water. The annual migration begins at the start of the dry season with the wildebeest and zebra, followed in turn by gazelles and other antelope species. The sequence is determined by the type and quality of food available and the physiology of the animals. The zebra can feed on dry straw-like materials, whereas antelopes require more green and tender plants. However, even these well-adapted animals must endure unusual drought conditions in years when there is no rain (Figure 9.17).

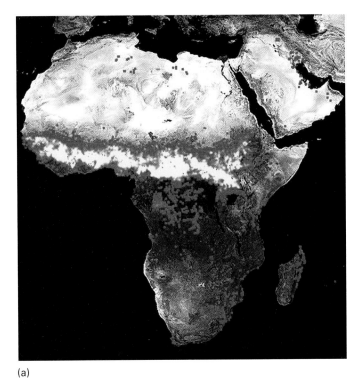

(a)

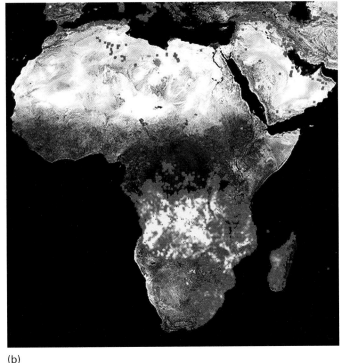

(b)

9.16 Seasonal fire patterns in Africa during January (a) and July (b) 2005. The images are compiled from fires detected by NASA's Moderate Resolution Imaging Spectroradiometer (MODIS) during consecutive 10-day periods. Red marks occasional fires in a location; yellow signifies frequent fires.

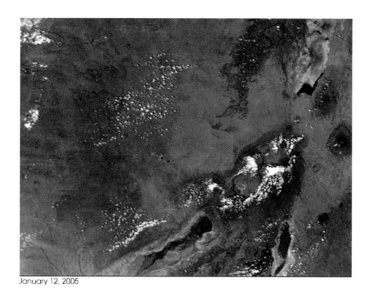 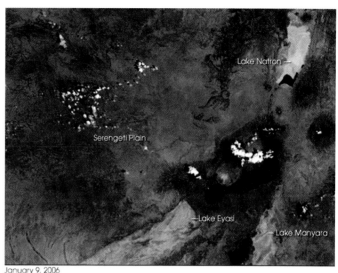

January 12, 2005

January 9, 2006

9.17 This pair of images from the Moderate Resolution Imaging Spectroradiometer (MODIS) on NASA's Terra satellite compares the landscape of the Serengeti in January 2005 (left) with January 2006 (right), when rainfall totals for the year were only 20 percent of normal in some areas.

THE DRY TROPICAL CLIMATE (KÖPPEN *BWH, BSH*)

The **dry tropical climate** is found in the centre and east sides of sub-tropical high-pressure cells. Here, strong subsidence and adiabatic warming inhibit condensation. Rainfall is rare and occurs only when unusual weather conditions move moist air into the region. Since skies are clear most of the time, the Sun heats the surface intensely, keeping air temperatures high. During the high-Sun period, heat is extreme. The record high air temperature for the Earth—58°C—was recorded in this climate type at Al Aziziyah, Libya. During the low-Sun period, temperatures are cooler. Given the dry air and lack of cloud cover, the daily temperature range is large.

The driest areas of the dry tropical climate are near the tropics of Cancer and Capricorn; rainfall increases toward the equator. Immediately adjacent to the dry tropical regions, precipitation comes in the form of a short rainy season at the time of the year when the ITCZ is nearby. As the wet period lengthens, the transitional wet-dry tropical climate occurs and the sparse, semi-arid vegetation is replaced by savanna.

Figure 9.18 shows the global distribution of the dry tropical climate. Nearly all of the areas lie in the latitude range 15° to 25° N and S. The largest region is the Sahara–Saudi Arabia–Iran–Thar desert belt of North Africa and southern Asia. This vast desert expanse includes some of the driest regions on the Earth. Another large region of dry tropical climate is the desert of central Australia. The west coast of South America, including portions of Ecuador, Peru, and Chile, is also very dry; however, temperatures there are moderated by a cool marine air layer associated with the Peruvian Current that blankets the coast.

Figure 9.19 is a climograph for a dry tropical station in the heart of the North African desert. Wadi Halfa, Sudan (see Figure 9.18 for location) lies at lat. 22° N, almost on the Tropic of Cancer. The temperature record shows a strong annual cycle with a very hot period at the time of high Sun, when three consecutive months average 32°C. Daytime maximum air temperatures are frequently between 43° and 48°C in the warmer months. There is a comparatively cool season at the time of low Sun, when the coolest month averages 16°C. Occasionally, frost can occur, especially in the adjacent highlands. The climograph does not show rainfall bars because precipitation averages less than 2.5 millimetres in all months. Over a 39-year period, the maximum rainfall recorded in a 24-hour period at Wadi Halfa was only 7.5 millimetres. In this environment, dew is an important source of moisture for plants and animals.

Although the world's dry climates consist largely of extremely *arid* deserts, there are, in addition, broad zones at the margins of the desert that are best described as *semi-arid*. These zones have a short wet season that supports the sparse growth of grasses on which animals (both wild and domestic) graze. Nomadic tribes and their herds of animals visit these areas during and after the brief moist period. In figures 8.8 (world climate map) and 9.18 (world map of dry climates), the two subdivisions of dry climates are identified by the letters *a* (arid) and *s* (semi-arid).

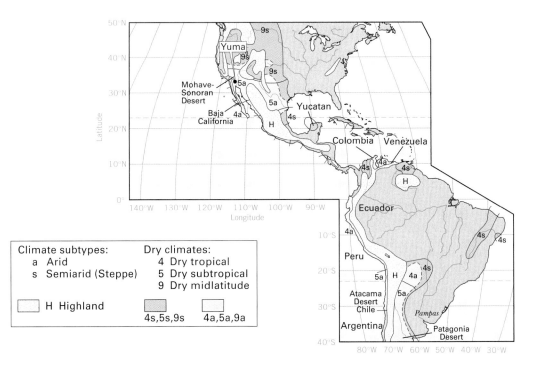

9.18 World map of the dry tropical, dry subtropical, and dry midlatitude climates The latter two climates are poleward and eastward extensions of the dry tropical climate with cooler temperatures.

Climate subtypes:
a   Arid
s   Semiarid (Steppe)

☐   H Highland

Dry climates:
4   Dry tropical
5   Dry subtropical
9   Dry midlatitude

▓ 4s,5s,9s   ☐ 4a,5a,9a

An example of the semi-arid dry tropical climate is that of Kayes, Mali (Figure 9.20). Located in the Sahel region of Africa, Kayes has a distinct rainy season that occurs when the ITCZ moves north in the high-Sun season. This precipitation pattern shows the semi-arid subtype as a transition between the dry tropical climate and the wet-dry tropical climate.

The Earth's desert landscapes are quite varied (Figure 9.21). In some areas of the arid desert, the surface consists of barren areas of drifting sand or sterile salt flats. However, in semi-arid regions, thorny trees and shrubs are often abundant, since the climate includes a small amount of regular rainfall.

An important variation of the dry tropical climate occurs in narrow coastal zones along the western edge of continents. These regions are strongly influenced by cold ocean currents and the upwelling of deep, cold water, which occurs just offshore. The cool water moderates coastal zone temperatures and reduces the seasonality of the temperature cycle. Figure 9.22 shows a climograph for Walvis Bay, a port city on the west coast of Namibia (southwest Africa), at lat. 23° S. Note the yearly cycle begins in July because this is a southern hemisphere station. The monthly temperatures are remarkably cool given its location nearly on the Tropic of Capricorn. The mean temperature of the warmest month is only 19°C, and for the coolest month, it drops to 14°C. This provides an annual range of only 5°C. Coastal fog is a persistent feature of this climate.

Another important location for this western coastal desert subtype is along the western coast of South America in Peru and Chile. Figure 9.18 shows a stretch of this coastline in northern Chile, where it is known as the Atacama Desert. Arica in northern Chile holds the record as the Earth's driest location, with an average annual precipitation of only 0.8 millimetres.

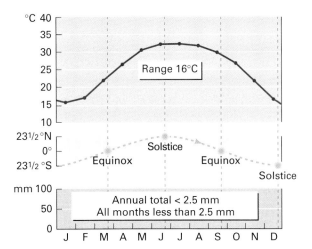

9.19 Dry tropical climate, arid desert Wadi Halfa is a city on the Nile River in Sudan at lat. 22° N, close to the Egyptian border. Too little rain falls to be shown on the graph. Air temperatures are very high during the high-Sun months.

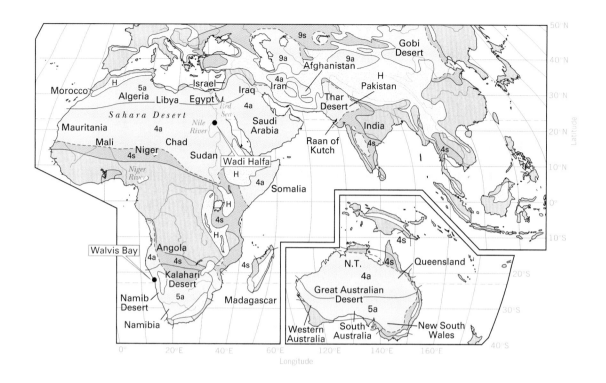

THE TROPICAL DESERT ENVIRONMENT

The tropical deserts and their bordering semi-arid zones compose a global environmental region sustained by subsiding air masses of the continental high-pressure cells. Because desert rainfall is so infrequent, river channels and the beds of smaller streams are dry most of the time. However, a sudden and intense downpour can cause local flooding of brief duration that transports large amounts of silt, sand, gravel, and boulders. These events are termed *flash floods* (see Chapter 16). Major river channels often end in flat-floored basins having no outlet. Here clay and silt are deposited and accumulate, along with layers of soluble salts. Shallow salt lakes occupy some of these basins.

Lake Eyre in Australia is the largest ephemeral lake in the world and for a long time was considered to be permanently dry. It actually comprises two lakes connected by the narrow

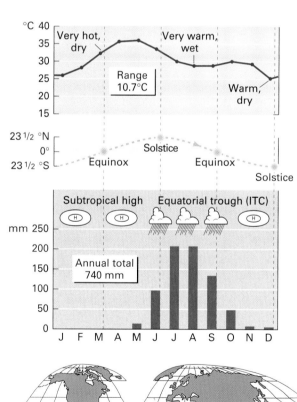

9.20 A semi-arid dry tropical climate Kayes, Mali, lat. 14° N, is located in western Africa, just south of the Sahara desert. Here, the climate is nearly rainless for seven months of the year. Rainfall occurs when the ITCZ moves northward and reaches the vicinity of Kayes in June or July. Temperatures are hottest in April and May, before the rainy season begins.

(a)

(b)

(c)

9.21 Desert Landscapes

(a) Sahara Desert This sandy plain in Algeria is dotted with date palms that tap ground water near the surface.

(b) Great Australian desert Red colours dominate this desert scene from the Rainbow Valley, south of Alice Springs, Australia.

(c) Baja Caifornia desert Rounded granite boulders and many plants adapted to dry desert conditions are visible in this scene from the Catavina desert of the northern Baja California Peninsula, Mexico.

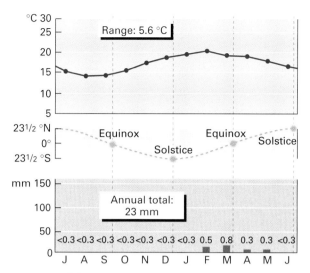

9.22 Dry tropical climate, western coastal desert subtype Walvis Bay, Namibia (southwest Africa) is a desert station on the west coast of Africa at lat. 23° S. Air temperatures are cool and remarkably uniform throughout the year.

Goyder Channel. At its lowest point, it is 15 metres below sea level. The drainage basin that supplies Lake Eyre is approximately a 1.2 million-kilometre-square area. However, annual average rainfall in the surrounding area is meager, although over the period of record has ranged from about 50 millimetres to 760 millimetres. Potential evaporation is about 3,500 millimetres per year. Consequently, rivers flow into the lake only after exceptional rainstorms. The first reported filling of Lake Eyre occurred in 1938 and it has filled and dried several times since. Because it has no outlet, its size and shape is sensitive to the amount of inflowing water it receives (Figure 9.23). The wet-dry phases have recently been linked to El Niño-Southern Oscillation (ENSO) events, with flooding occurring during strong, positive Southern Oscillations.

HIGHLAND CLIMATES OF LOW LATITUDES

Highland climates are cool to cold, usually moist, climates that occupy mountains and high plateaus. Generally, the higher the location, the colder and wetter is its climate. Temperatures are lower since air temperatures in the atmosphere normally decrease with altitude (see Chapter 4). Rainfall increases because orographic precipitation tends to be induced when air masses ascend to higher elevations (see Chapter 6). Highland climates are not usually included in the broad schemes of climate classification, since many small highland areas are simply not shown on a world map.

The character of a given highland area's climate is usually closely related to that of the climate of the surrounding lowland,

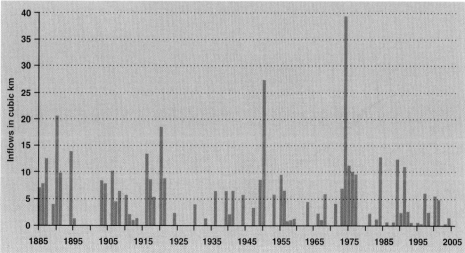

9.23 Lake Eyre, Australia Inflows of water occur periodically and have been linked to ENSO events.

particularly the annual temperature cycle and the occurrence of wet and dry seasons. The climographs for two Indian stations in close geographical proximity provide an example of this effect in the tropical zone (Figure 9.24). New Delhi lies in the Ganges lowland, and Simla is located at about 2,200 metres in the foothills of the Himalayas. When the hot-season temperature averages over 32°C in New Delhi, Simla is enjoying a pleasant 18°C. Notice, however, that the two temperature cycles are quite similar in appearance, with the minimum month being January for both. The annual rainfall cycles are also similar. New Delhi shows the typical rainfall pattern of Southeast Asia, with monsoon rains peaking in July and August. Simla has the same pattern, but the

amounts are larger in every month, and the monsoon peak is very strong. Simla's annual total rainfall is more than twice that of New Delhi.

Kilimanjaro, the highest mountain in Africa, and the tallest free-standing mountain in the world, provides some interesting information about the highland climate in the tropics. Kilimanjaro is situated at lat. 3°S and rises to 5,895 metres from the adjacent plains that are at about 2,000 metres above sea level (Figure 9.25). Kilimanjaro is influenced by the passage of the ITCZ, which brings two wet seasons. The highest precipitation occurs in March, April, and May, with a second peak from October to November. The southeastern slopes receive the highest precipitation with totals of about 1,800 millimetres at 2,400 metres elevation. At similar elevations on the north and west slopes, precipitation is about 1,000 millimetres. Desert-like conditions prevail at the summit, where annual precipitation, in the form of snow, is usually less than 100 millimetres. Temperatures at the base of the mountain average 25°–30° C, but can drop to −10° C on the summit. Nighttime frosts are usual at 3,000 metres year round.

Kilimanjaro is capped with glaciers, but recent measurements have shown that they have been shrinking very rapidly. Most people attribute this to global warming, but recent changes in atmospheric moisture patterns have been implicated. In 2003, researchers noted that the ice cover was about 20 percent of what it was in 1912, despite the fact that air temperatures at the summit have remained below freezing. Surprisingly, recession rates were much higher between 1912 and 1953 than those recorded between 1989 and 2003. This suggests that the glacier retreat on Kilimanjaro is an adjustment to climate shifts that occurred in the late nineteenth century.

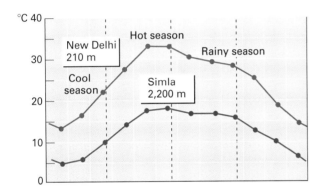

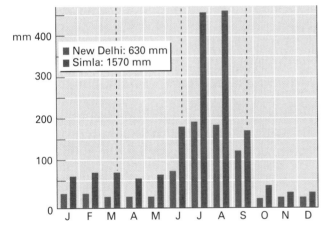

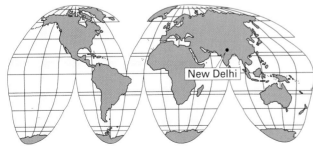

9.24 Climographs for New Delhi and Simla Both of these stations are in northern India but at different elevations. Simla is a welcome refuge from the intense heat of May and June in the Ganges Plain, where New Delhi is located. However, in July and August, Simla is much wetter.

9.25 The effect of altitude on tropical climates is dramatically illustrated by Mount Kilimanjaro in Tanzania. *Fly By: 03° 04' S; 137° 21' E*

A Look Ahead

Low-latitude climates are a study in contrasts, from the wettest to the driest of climates and from the hottest to the most uniform in temperature. They are dominated by two important features of atmospheric circulation: the ITCZ and the subtropical high-pressure cells. The next chapter discusses the climates of the mid- and high-latitudes. Although this group cannot claim the wettest or hottest of climates, it includes climates that are still very wet, very dry, and very hot, at least at certain times of the year. The polar-front boundary and the contrast of warm, moist air masses of the subtropics with colder, drier air masses of the polar regions is the most important factor in influencing the temperature and precipitation cycles of mid- and high-latitude climates.

CHAPTER SUMMARY

- *Low-latitude climates* are located largely between 30° N and S latitudes, and are controlled by the characteristics and annual movements of the ITCZ and the subtropical highs. These climates range from very moist to very dry and from quite uniform to very seasonal.
- In the wet equatorial climate, the ITCZ is always nearby, and so rainfall is abundant throughout the year. The annual temperature cycle is very weak; the daily range greatly exceeds the annual range.
- The monsoon and trade-wind coastal climate shows a dry period when the ITCZ has migrated toward the opposite tropic. With the return of the ITCZ, monsoon circulation and enhanced easterly waves provide increased rainfall. Temperatures are highest in the dry weather before the onset of the wet season.
- The wet equatorial and monsoon and trade-wind coastal climates provide the *low-latitude rainforest environment*.
- The wet-dry tropical climate has a very dry period at the time of low Sun and a wet season at the time of high Sun, when the ITCZ is near. Temperatures peak strongly just before the onset of the wet season. Associated with this climate type is the *savanna environment*.
- Extreme year-to-year variability in rainfall in the tropical wet-dry climate provides a constant threat of extreme drought, especially in the Sahel region of western and north-central Africa. The drought not only causes famine and disease, but also produces land degradation brought about by overgrazing and fuel wood harvesting.
- The dry tropical climate is dominated by subtropical high pressure and is often nearly rainless. Temperatures are very high during the high-Sun season. The *semi-arid* subtype of this climate has a short wet season and supports sparse grasslands; the *arid* subtype provides the tropical desert environment. In the desert, streams flow after heavy rain showers, which often produce *flash floods*.
- Highland climates of low latitudes normally show a similar seasonality to lowland climates of nearby regions, but are cooler and wetter.

KEY TERMS

dry tropical climate
highland climates

monsoon and trade-wind
    coastal climate

wet-dry tropical climate
wet equatorial climate

REVIEW QUESTIONS

1. Why is the annual temperature cycle of the wet equatorial climate so uniform?
2. The wet-dry tropical climate has two distinct seasons. What factors produce the dry season? The wet season?
3. Describe how seasonal climate patterns affect the savanna environment.
4. How do the arid (*a*) and semi-arid (*s*) subtypes of the dry climates differ?
5. How do low-latitude highland climates differ from their counterparts at low elevation?

EYE ON GLOBAL CHANGE 9.2 — Drought and Land Degradation in the African Sahel

1. What is meant by the term *land degradation?* Provide an example of a region in which land degradation has occurred.

2. Examine the Sahel rainfall graph carefully. Compare the pattern of rainfall fluctuations for the periods 1901–50 and 1951–2000. How do they differ?

VISUALIZATION EXERCISES

1. Sketch the temperature and rainfall cycles for a typical station in the monsoon and trade-wind coastal climate. What factors contribute to the seasonality of the two cycles?

ESSAY QUESTIONS

1. The ITCZ moves north and south with the seasons. Describe how this movement affects the four low-latitude climates.

2. Compare and contrast the low-latitude environments of Africa.

PROBLEMS — WORKING IT OUT 9.1 Cycles of Rainfall in the Low Latitudes

1. The data below are measurements of total precipitation for November for the years 1985–94 at San Juan, Puerto Rico. (November is the wettest month during this 10-year period.) Find the mean deviation of these data.

2. Calculate the relative variability by taking the ratio of the mean deviation to the mean. How does the result compare with those of the wettest months of Padang, Sumatra; Mumbai, India; and Abbassia, Egypt?

Find answers at the back of the book

Year	Precipitation (mm)
1985	115
1986	149
1987	190
1988	144
1989	126
1990	135
1991	156
1992	304
1993	110
1994	211

EYE ON THE LANDSCAPE |

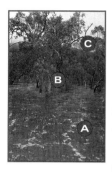

Chapter Opener Todd River, Australia, near Alice Springs. In this arid environment precipitation often comes as torrential rain and the otherwise dry river channels quickly fill with water and overflow their banks. The river carries a high sediment load (**A**) because materials are removed by sheet erosion (see chapter 16) from sparsely vegetated slopes (**B**). The eucalyptus trees (**C**) that line the river depend on floods for groundwater recharge and the nutrients in the alluvial deposits.

Wet and dry seasons in the Simen Mountains, northern Ethiopia. A noticeable transformation of the landscape occurs with the change in seasons. At the end of the rainy season most of the vegetation dies back (lower photo, see chapter 23). Only on some of the lower slopes (**A1**) and valley floors (**A2**) does sufficient moisture persist to keep the plants green. During the wet season (lower photo) water vapour in the atmosphere creates a more haze (**B**). Water has helped to carve the horizontally bedded sandstones (**C**) into a landscape of flat-topped mesas (**D**) and smaller buttes (**E**) (see chapter 17).

Chapter 10

EYE ON THE LANDSCAPE

> > > >

Alpine meadow in Mount Rainier National Park, Washington, USA. **What else would the geographer see? . . . Answers at the end of the chapter.**

Overview of the Midlatitude Climates

Working It Out 10.1 • *Standard Deviation and Coefficient of Variation*

The Dry Subtropical Climate

The Subtropical Desert Environment

The Moist Subtropical Climate

The Moist Subtropical Environment

The Mediterranean Climate

The Mediterranean Climate Environment

Focus on Systems 10.2 • *California Rainfall Cycles and El Niño*

The Marine West-Coast Climate

The Marine West-Coast Environment

Geographers at Work • *Dr. Trisalyn A. Nelson*

The Dry Midlatitude Climate

The Dry Midlatitude Environment

Eye on the Environment 10.3 • *Drought on the Prairies*

The Moist Continental Climate

The Moist Continental Forest and Prairie Environment

Overview of the High-Latitude Climates

The Boreal Forest Climate

The Boreal Forest Environment

The Tundra Climate

The Arctic Tundra Environment

Arctic Permafrost

The Ice Sheet Climate

The Ice Sheet Environment

Our Changing Climate

A Closer Look:

Eye on Global Change 10.4 • *Regional Impacts of Climate Change on North America*

This chapter continues with global climates, focusing on midlatitude and high-latitude regions. As seen in Chapter 9, the effects of latitude, continental location, and the movement of air masses and fronts can explain the temperature and precipitation cycles of the individual climates.

OVERVIEW OF THE MIDLATITUDE CLIMATES

The *midlatitude climates* extend from the midlatitudes into land areas in the subtropical zone. Along the western fringe of Europe, they also continue into the subarctic zone, reaching to the 60th parallel. Unlike the low-latitude climates, which are almost equally distributed between northern and southern hemispheres, nearly all of the midlatitude climate area is in the northern hemisphere. In the southern hemisphere, the land area in latitudes higher than the 40th parallel is so small that the climates are dominated by the great Southern Ocean and do not develop the continental characteristics of their counterparts in the northern hemisphere.

The midlatitude climates of the northern hemisphere lie in a broad zone of intense interaction between two groups of unlike air masses (see Figure 8.6). Maritime tropical (mT) air from the subtropical zone enters the midlatitudes, where it meets and conflicts with maritime polar (mP) and continental polar (cP) air along the discontinuous and shifting polar front.

In terms of prevailing pressure and wind systems, the midlatitude climates include the poleward halves of the great subtropical high-pressure systems and much of the belt of prevailing westerly winds (see Figure 5.16). As a result, weather systems, such as travelling low pressure systems and their associated fronts, characteristically move from west to east. This predominantly zonal airflow influences the distribution of climates across the North American and Eurasian continents.

The interaction between warm, moist air masses and cooler, drier air masses produces climates that are quite variable from day to day. *Working It Out 10.1* • *Standard Deviation and Coefficient of Variation* uses monthly precipitation from stations in two different midlatitude climates to show how variability is measured.

The midlatitude climates range from those with strong wet and dry seasons to those with precipitation that is more or less uniformly distributed through the year (Figure 10.1). Temperature

WORKING IT OUT | 10.1   *Standard Deviation and Coefficient of Variation*

Although the mean, or average, monthly precipitation or temperature is an important determinant of climate, the variability from one year to the next is also informative. In Working It Out 9.1 • Cycles of Rainfall in the Low Latitudes, the mean deviation was calculated as a measure of variation around the mean value of the data; the ratio of the mean deviation to the mean was used as a measure of relative variability.

A more common measure of variation is the *sample standard deviation*, which is defined for a sample of *n* precipitation values P_i, where i = 1, . . ., n, as

$$S_P = \sqrt{\frac{1}{n-1}\sum_{i=1}^{n}D_i^2}$$

where S_P is the sample standard deviation of *P*, and D_i is the deviation of the *i* th observation from the sample mean. To find the sample standard deviation, take each deviation and square it. Note that any deviations that are negative will become positive after squaring. Thus, there is no need to take the absolute value of the deviation, as was done in *Working It Out 9.1 • Cycles of Rainfall in the Low Latitudes*. Next, add the squared deviations, and divide that amount by *n* – 1—that is, one less than the number of samples. Finally, take the square root of the result.

The set of graphs on the left show an example of the sample standard deviation. Graphs (*a*) and (*b*) present monthly precipitation for two stations for a 10-year period from 1996 to 2005. Penticton, B.C. lies in the dry interior of the province and experiences a mean annual precipitation of 330 millimetres. Charlottetown, P.E.I., with a moist continental climate, has mean annual precipitation of 1,297 millimetres.

By comparing the two graphs, it appears that Charlottetown has greater variation, largely because many months are much wetter than those in Penticton. Graph (*c*) compares the monthly standard deviations for each sample. In all months, the value is larger for Charlottetown, which strengthens the impression gained from the comparison of graphs (*a*) and (*b*). The standard deviation for the sample of annual totals, shown in the last pair of bars, is larger than the monthly values for both locations since the annual total itself is always larger than those of individual months.

However, the standard deviation does not measure the relative variation—that is, the variation with respect to the mean. Graph (*d*) shows the *coefficient of variation* by month. The coefficient of variation for a sample is defined as the ratio of the standard deviation to the mean. That is,

$$CV_P = \frac{S_P}{\overline{P}}$$

where CV_P is the coefficient of variation for precipitation. Relative variation of monthly precipitation at Penticton is greater than at Charlottetown in all months. Note that the coefficients of variation for the annual totals (the last pair of bars in graph *d*) are smaller than those of the monthly averages, showing that the relative variation of the total—that is, for the entire year—is less.

The mean, standard deviation, and coefficient of variation of a sample are values that give basic information about the sample and its variability. They are referred to as *sample statistics* and are widely used in many branches of science.

cycles also are quite varied. The windward west coasts experience low annual ranges. In contrast, annual ranges are large in the continental interiors. Table 10.1 summarizes the important features of these climates.

THE DRY SUBTROPICAL CLIMATE (KÖPPEN *BWH, BWK, BSH, BSK*)

The **dry subtropical climate** is a poleward extension of the dry tropical climate, but because of the midlatitude location, the annual temperature range is greater. There is a distinct cool season at the time of low Sun, which is accentuated by invasions of cold continental polar (cP) air masses that also bring frontal storms to the region. As in the dry tropical climate, this climate type has both arid and semi-arid subtypes.

Figure 9.21 shows the global distribution of the dry subtropical climate. In North America, the Mojave and Sonoran deserts of the American southwest and northwest Mexico are representative. In South America, the dry subtropical climate extends as far south as Patagonia. A discontinuous band of dry subtropical climate stretches across North Africa, connecting with the Near East. It is also found in Southern Africa and southern Australia.

Figure 10.2, a climograph for Yuma, Arizona (lat. 33° N), illustrates the arid subtype of the dry subtropical climate. Monthly temperatures show a strong seasonal cycle with hot summers (mean 33°C) and cool winters (mean 13°C). Freezing temperatures can be expected at night in December and January. The annual range is 20°C. Annual precipitation, which totals about 80 millimetres, is small in all months but has peaks in

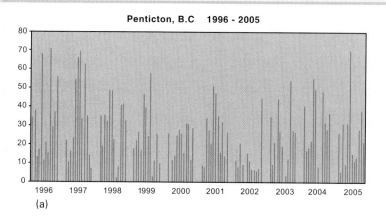

(a)

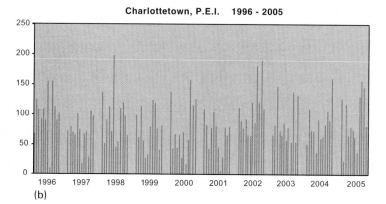

(b)

Precipitation data and statistics for Penticton, B.C., and Charlottetown, P.E.I.

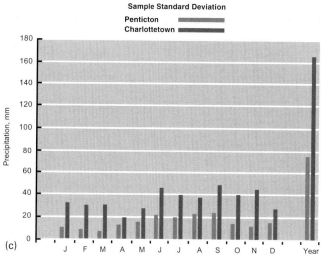

(c)

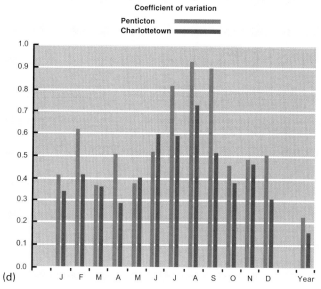

(d)

late winter and late summer. The August maximum is caused by the invasion of maritime tropical (mT) air masses, which bring thunderstorms to the region. Midlatitude depressions produce higher rainfalls from December through March. May and June are nearly rainless.

THE SUBTROPICAL DESERT ENVIRONMENT

The shift from a dry tropical to a dry subtropical climate type is gradual; however, the effect of the cooler, moister winters in midlatitudes becomes increasingly apparent in the landscapes. Although many soils in subtropical deserts are naturally saline, they may support an open cover of trees and shrubs (Figure 10.3). Deep-rooting species help to mitigate the problem of salinity by keeping the water table many metres below the surface.

Replacing native species with shallow-rooted crops can increase soil salinity through the process of *dryland salinization*. The rise in the water table that results when shallow-rooted crops are planted brings dissolved salts upward into the surface soils where it accumulates through evaporation. *Irrigation salinization* occurs when irrigation water soaks through the soil and causes the water-table to rise. Deterioration in the quality of the water supply can accentuate the problem. In the Shepparton district of Victoria, Australia, the water table has risen from a depth of about 30 metres prior to land clearance to 2 metres or less. Many farmers are now planting trees in the hopes of reversing the problem of salinization.

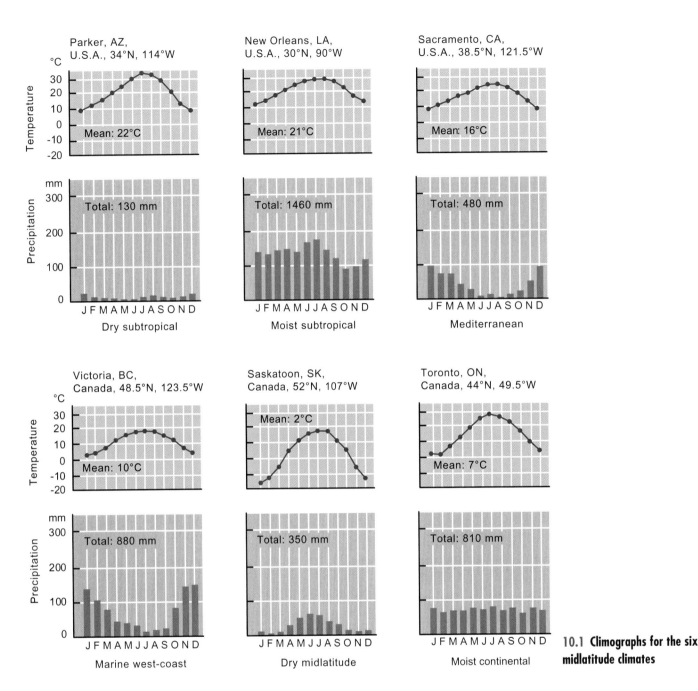

10.1 Climographs for the six midlatitude climates

THE MOIST SUBTROPICAL CLIMATE (KÖPPEN *CFA*)

Recall that circulation around the subtropical high-pressure cells provides a flow of warm, moist air onto the eastern side of continents (see Figure 5.12). This flow of maritime tropical (mT) air dominates the **moist subtropical climate**. Summer in this climate sees abundant rainfall, much of it convectional. Occasional tropical cyclones augment summer precipitation. In Southeast Asia, this climate is characterized by a strong monsoon effect, with much more rainfall in summer than in winter. Summer temperatures are warm, with persistent high humidity.

Winter precipitation in the moist subtropical climate is also plentiful, produced in midlatitude cyclones. Invasions of continental polar (cP) air masses are frequent in winter, bringing spells of below freezing temperatures. However, no winter month has a mean temperature below 0°C.

The moist subtropical climate is found on the eastern sides of continents in the latitude range 20° to 35° N and S (Figure 10.4). In the United States, this climate covers most of the southeast, from the Carolinas to east Texas. Another extensive area includes southern China, Taiwan, and the southernmost part of Japan. In the southern hemisphere, moist subtropical climates are found in Uruguay, Brazil, and Argentina, as well as in a narrow band along Australia's east coast.

Table 10.1 Midlatitude climates

Climate	Temperature	Precipitation	Explanation
Dry subtropical	Distinct cool or cold season at low-Sun period.	Precipitation is low in nearly all months.	This climate lies poleward of the subtropical high-pressure cells and is dominated by dry cT air most of the year. Rainfall occurs when moist mT air reaches the region, either in summer monsoon flows or in winter frontal movements.
Moist subtropical	Temperatures show a strong annual cycle, but with no winter month below freezing.	Abundant rainfall, cyclonic in winter and convectional in summer. Humidity generally high.	The flow of mT air from the west sides of subtropical high-pressure cells provides moist air most of the year. cP air may reach this region during the winter.
Mediterranean	Temperature range is moderate, with warm to hot summers and mild winters.	Unusual pattern of wet winter and dry summer. Overall, drier when nearer to subtropical high pressure.	The poleward migration of subtropical high-pressure cells moves clear, stable cT air into this region in the summer. In winter, cyclonic storms and polar frontal precipitation reach the area.
Marine west-coast	Temperature cycle is moderated by marine influence.	Abundant precipitation but with a winter maximum.	Moist mP air, moving inland from the ocean to the west dominates this climate most of the year. In the summer, subtropical high pressure reaches these regions, reducing precipitation.
Dry midlatitude	Strong temperature cycle with large annual range. Summers warm to hot, winters cold to very cold.	Precipitation is low in all months but usually shows a summer maximum.	This climate is dry because of its interior location, far from mP source regions. In winter, cP dominates. In summer, a local dry continental air mass develops.
Moist continental	Summers warm, winters cold with three months below freezing. Very large annual temperature range.	Ample precipitation, with a summer maximum.	This climate lies in the polar-front zone. In winter, cP air dominates, while mT invades frequently in summer. Precipitation is abundant, cyclonic in winter and convectional in summer.

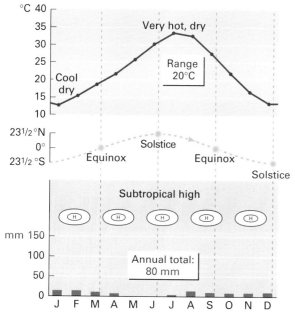

10.2 Dry subtropical climate Yuma, Arizona, lat. 33° N, has a strong seasonal temperature cycle. Compare with Wadi Halfa (Figure 9.22).

10.3 Growth in Saline Soils Myall Woodland comprised of *Acacia pendula* (Weeping Myall) in the Murray-Darling region of New South Wales, Australia.

Figure 10.5 is a climograph for Charleston, South Carolina, located on the eastern seaboard of the United States at lat. 33° N. In this region, a marked summer maximum of precipitation is typical. Total annual rainfall is abundant (1,200 millimetres), and ample precipitation falls in every month. The annual temperature cycle is strongly developed, with a large annual range of 17°C. Winters are mild, with the January mean temperature well above freezing.

THE MOIST SUBTROPICAL ENVIRONMENT

Winters are mild in regions with moist subtropical climates, but summers are hot and humid. The combination of warm summer temperatures and high rainfall can create uncomfortable conditions because it stresses the body's ability to cool itself. Several indices have been developed to describe seasonal conditions based on human perceptions. For example, cooling degree days and heating degree days are calculated from a base temperature of 18°C; that is, how much the mean air temperature on a given day is either higher than or lower than 18°C. Figure 10.6 shows these statistics for the southeastern United States. In July, cooling degree days exceed 450, indicating a high potential demand for energy for air conditioning. Conversely, the mild winter reduces potential demand for heating.

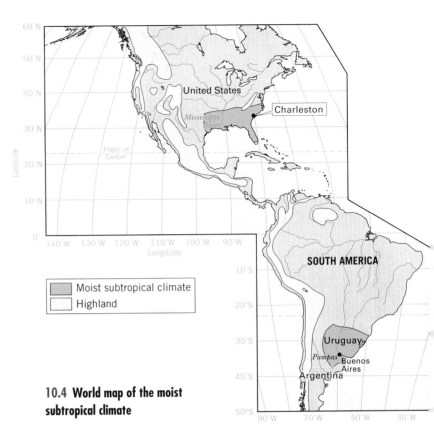

10.4 World map of the moist subtropical climate

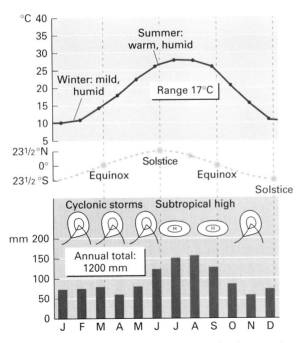

10.5 Moist subtropical climate Charleston, South Carolina, lat. 33° N, has a mild winter and a warm summer. There is ample precipitation in all months but a definite summer maximum.

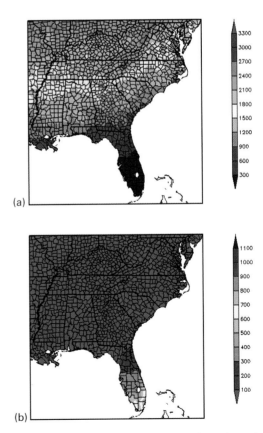

10.6 The calculation of temperature Heating (a) degree days and cooling (b) degree days (base 18°C) for the southeastern United States November-January, 2006-2007.

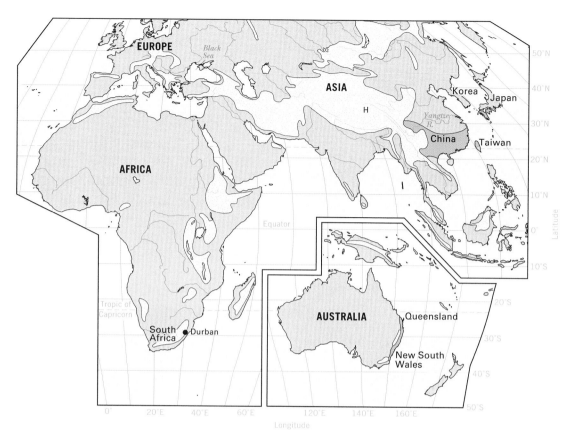

The comfort index (Figure 10.7) is derived from temperature and relative humidity readings, and is also subjectively based on people's apparent comfort or discomfort under given conditions. This index provides an "apparent temperature," that is, what the air temperature feels like when the effect of humidity is taken into account. Conditions that can lead to heat exhaustion or heatstroke are not uncommon in this region of the United States, and could become more frequent according to the Canadian and U.K. (Hadley) climate prediction models. Although the Hadley model predicts a less extreme trend in the summer heat index,

Relative Humidity (%)

°C	40	45	50	55	60	65	70	75	80	85	90	95	100
43	136												
42	130	137											
41	124	130	137										
40	119	124	131	137									
39	114	119	124	130	137								
38	109	114	118	124	129	136							
37	105	109	113	117	123	128	134						
35	101	104	106	112	116	121	126	132					
34	97	100	102	106	110	114	119	124	129	136			
33	94	96	99	101	105	108	112	116	121	126	131		
32	91	93	95	97	100	108	106	109	113	117	122	127	132
31	88	89	91	93	95	98	100	108	106	110	113	117	121
30	85	87	88	89	91	93	95	97	100	102	105	108	112
29	83	84	85	86	88	89	90	92	94	96	98	100	103
28	81	82	83	84	84	85	86	88	89	90	91	93	95
27	80	80	81	81	82	82	83	84	84	85	86	86	87

Air Temperature (°C)

Heat Index
(Apparent Temperature)

With Prolonged Exposure and/ or Physical Activity

Extreme Danger
Heat stroke or sunstroke highly likely

Danger
Sunstroke, muscle cramps, and/ or heat exhaustion likely

Extreme Caution
Sunstroke, muscle cramps, and/ or heat exhaustion possible

Fatigue possible

10.7 The comfort index of heat The heat index represents the apparent temperature based on prevailing air temperature and relative humidity conditions.

10.8 The effect of excessive runoffs Severe soil erosion in southwest China.

precipitation is expected to be higher than at present with a greater frequency of more intense weather events.

The abundant rainfall of the moist subtropical environment supplies rivers and streams through much of the year, and at times, flooding can occur from torrential rains associated with tropical cyclones. Excessive runoff in areas where the natural forest cover has been removed has led to severe soil degradation (Figure 10.8), and the heavy sediment loads carried by the rivers seriously impair water quality. Reforestation and other forms of erosion control are used extensively in an effort to combat the problem.

THE MEDITERRANEAN CLIMATE (KÖPPEN CSA, CSB)

The **Mediterranean climate** is located along the west coasts of continents in the latitude range 30° to 45° N and S (Figure 10.9). It is distinguished by its annual precipitation cycle, which has a wet winter and very dry summer. This is caused by the poleward movement of the subtropical high-pressure cells during the summer season, which brings dry continental tropical (cT) air into the region. In winter, moist mP air masses bring low pressure systems and ample precipitation.

In terms of total annual precipitation, the Mediterranean climate type varies from arid to humid, depending on location. Generally, the closer an area is to the tropics, the stronger the influence of subtropical high pressure will be, and thus the drier the climate. Annual precipitation is between 275 to 600 millimetres, with more than 65 percent of the total falling in the winter. Moisture recharge during the winter months must be enough to maintain the native vegetation through the summer drought.

The annual temperature range is generally moderate, with warm to hot summers and mild winters. Mean summer temperatures are about 25°C, but daytime maxima often exceed 35°C. In winter, temperatures average 10–12°C, but there is an occasional risk of frost in higher areas. Coastal zones between lat. 30° and 35° N and S typically have a smaller annual temperature range than elsewhere within this climate type.

More than half of the Mediterranean climate type is found in a discontinuous belt around the Mediterranean Sea from which it is named. It includes the countries of southern Europe, including parts of Portugal, Spain, France, and Italy, and extends through the Levant region of the eastern Mediterranean into North Africa. Elsewhere in the northern hemisphere, the Mediterranean climate type

10.9 World map of the Mediterranean and marine west-coast climates

occurs in central and southern California. Locations in the southern hemisphere include coastal Chile, the Cape Town region of South Africa, and parts of southern and western Australia.

Figure 10.10 is a climograph for Monterey, California, a Pacific coastal city at lat. 36° N. The annual temperature cycle is very weak. The small annual range and cool summer reflect the strong control of the cold California current and its cool, marine air layer. Fogs are frequent. Rainfall drops to nearly zero for four consecutive summer months. In California's Central Valley, the dry summers persist, but the annual temperature range is greater, so that daily high temperatures are similar to those of the adjacent dry climates.

THE MEDITERRANEAN CLIMATE ENVIRONMENT

The Mediterranean climate is attractive for human habitation, mainly because of year-round pleasant temperatures, especially where moderated by coastal influences. For many people, the mild winters are a welcome refuge from the severe winters of the mid-latitude continental interiors of Eurasia and North America. However, the low annual precipitation with dry summers makes fresh water scarce, requiring a large investment in technology to deliver enough water to meet the high demand.

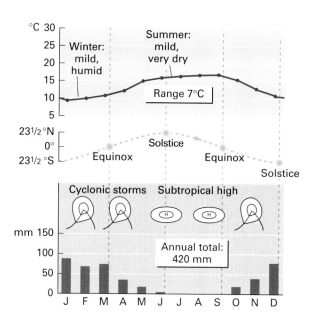

10.10 Mediterranean climate Monterey, California, lat. 36° N, has a weak annual temperature cycle because of its closeness to the Pacific Ocean. The summer is very dry.

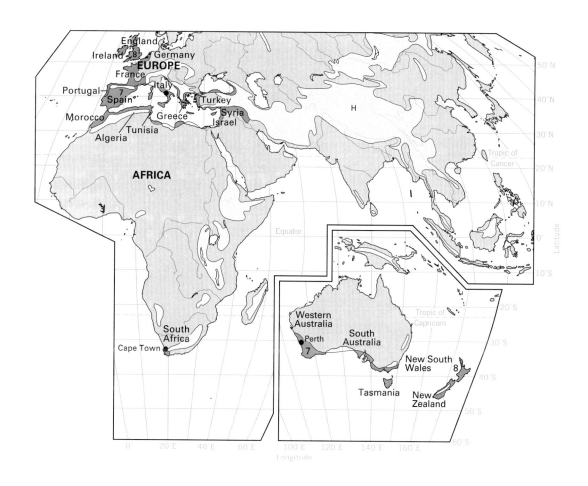

The problem is particularly acute in California, where massive engineering projects have been necessary to ensure reliable supplies. California's extensive water supply system includes reservoirs, groundwater basins, and interregional conveyance facilities, all of which are needed to reduce short-term water shortages. Snowpack in the Sierra Nevada watersheds is an important source of much of California's developed water supply.

In California, more than 90 percent of the water supply is used in agriculture, where evaporation during the hot summer months has resulted in salinity problems. In addition, many of the natural water courses are dry for long periods, altering the ecological balance of the stream ecosystems. A related concern is the high levels of chemicals carried into the streams by agricultural runoff.

Much of the irrigated land is used to grow citrus fruits and grapes for the expanding wine industry. Water supply is not the only factor that is critical for these operations. The risk of frost is also a concern, especially from cold air drainage into low-lying sites. To combat this problem, some farmers have installed wind machines to mix the lower air or to spray water onto their crops (Figure 10.11). Spraying water is effective during light frosts because latent heat released when the water freezes on the fruit is generally enough to prevent injury.

The persistent flow of stable air out of the high-pressure centres during summer brings several months of hot, dry weather, and fire is a frequent hazard at this time of the year. Wildfire is an integral part of all Mediterranean environments. Average fire frequency in most regions is 20–30 per year. The relatively dense canopy of live and dead fine stems and the volatile oils in the leaves of many of the native species make the vegetation very flammable. The fires burn fiercely and tend to spread rapidly, especially where they are fanned by hot, dry winds such as the Santa Ana (see Chapter 5). Fires as large as 10,000 hectares are frequently reported. The devastating fire in Portugal in 2003 destroyed more than 400,000 hectares of woodland. However, the native vegetation is well adapted to fire, and most plants will renew their growth from buried rootstocks or germinate from fire resistant seeds (Figure 10.12).

Another weather-related hazard in Mediterranean climate regions is excessive soil erosion that can occur when the rains return in winter. This can be especially bad when the vegetation cover has been burned off during the previous summer. However, overgrazing, especially by goats, has increased the probability of soil loss, especially in southern Europe. Large quantities of coarse mineral debris are swept down slopes and carried long distances by flooding streams (see Chapter 15).

Rainfall in the Mediterranean climate can be quite variable from year to year. Sometimes the weather patterns that provide winter precipitation fail to appear, leading to drought. In other years, precipitation may be much greater than normal. *Focus on Systems 10.2 • California Rainfall Cycles and El Niño* describes rainfall patterns in Santa Barbara over a 22-year period and the possible relation to El Niño cycles and volcanic eruptions.

THE MARINE WEST-COAST CLIMATE (KÖPPEN *CFB, CFC*)

The **marine west-coast climate** occurs in midlatitude west coasts where prevailing westerlies bring moist air onshore from the adjacent oceans. Precipitation comes mainly from low pressure systems. Where the coast is mountainous, the orographic effect causes a very large annual precipitation. Precipitation is plentiful in all months, but there is often a distinct winter maximum. In summer, subtropical high pressure extends into the region, reducing rainfall. The annual temperature range is relatively small for midlatitudes. The marine influence keeps winter temperatures mild, compared with inland locations at equivalent latitudes.

10.11 The risk of frost (a) Wind machines disperse frost pockets that may develop in areas that are susceptible to cold air drainage. (b) Latent heat released during freezing can prevent frost injury in sensitive crops.

10.12 Fire is a frequent threat in Mediterranean ecosystems. (a) Cork oak is well protected by its thick bark, which in this specimen has been recently stripped for commercial use. (b) New branches sprouting in fire-damaged Eucalyptus in Australia (c) Young coffeeberry shrub developing from undamaged tissues in the root of a burned plant in California.

The general latitude range of this climate is 35° to 60° N and S (Figure 10.10). Locations include the Pacific northwestern coast of North America, the British Isles, and parts of Western Europe. In the southern hemisphere it is associated with New Zealand, the southern part of Australia, and the Chilean coast south of 35° S.

Figure 10.13 is a climograph for Vancouver, British Columbia. The annual precipitation is high, and most of it falls during the winter months. Notice the greatly reduced rainfall in the summer months. The annual temperature range is 15°C, and mean temperatures in the winter months remain above freezing.

THE MARINE WEST-COAST ENVIRONMENT

In Europe, the native vegetation is broadleaf deciduous forest. Common species, such as oak, ash, and beech, provide a continuous and dense canopy in summer but shed their leaves completely in winter. Forest clearance for agriculture can be traced back more than 4,000 years. As well as being in great demand for houses, the trees also provided timber for the massive wooden warships of previous centuries. Wood was also used for charcoal in the iron industry prior to the discovery of coal. Consequently, except in rare situations, only scattered patches of forest remain.

GEOGRAPHERS AT WORK

The Domino Effect of Climate Change on our Forests
by Dr. Trisalyn A. Nelson, University of Victoria

Western Canada is currently experiencing the largest mountain pine beetle epidemic on record. As of 2006, the infestation was estimated to cover between 12 and 13 million hectares of forest. For comparison, this is equivalent to 12-13 million football fields or an area twice the size of New Brunswick. By 2013 it is anticipated that 80% of forests in British Columbia will be infested and the spread of beetles into Alberta is raising concerns about the potential for beetles to impact the boreal forest throughout Canada.

The mountain pine beetle epidemic is the result of fire suppression and climate change. Populations can only explode when there is a lot of food. For the mountain pine beetle the preferred food is mature pine and fire suppression has lead to more mature trees than naturally occur on the landscape. Climate change is also playing a role. The primary cause of mountain pine beetle mortality is cold temperatures. Either sustained cold spells or sudden cold snaps are required to reduce the mountain pine beetle population. Due to climate change, neither climatic condition is occurring.

If there has never been such a large epidemic, how do decision-makers effectively manage the forests? The Spatial Pattern Analysis and Research (SPAR) Laboratory, directed by Dr. Trisalyn Nelson, is conducting research to answer this questions. By studying the spatial and spatial-temporal patterns in trees infested by the mountain pine beetle they are generating a new understanding of how beetles disperse over landscapes, the type of environmental characteristics that host epidemic mountain pine beetle populations, and how the spatial pattern of beetle population dynamics

changes through time. Trisalyn and the SPAR lab are also assisting forest management by developing spatial analysis methods for detecting where and when beetle populations and forest conditions change. With climate change impacting many aspects of the natural environment, geographers need new tools for detecting and analyzing change in both space and time. SPAR lab is working to develop new spatial-temporal analysis tools to address this need.

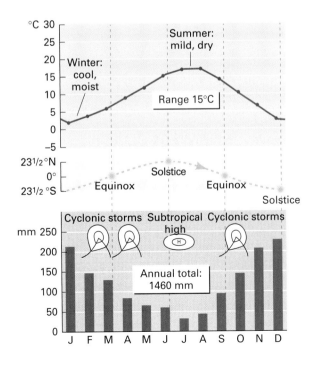

In western North America, southern Chile, and the South Island of New Zealand, much of the land is mountainous and has been heavily scoured by glaciers. Hence, the soils are poorly developed. Dense needleleaf forests of fir, cedar, hemlock, and spruce flourish in the wet mountainous coastal areas of British Columbia (Figure 10.14). They are important lumber trees and are considered one of the greatest timber resources in the world. The mountainous terrain is also suited to hydroelectric power, and many of the rivers have been

10.13 Marine west-coast climate Vancouver, British Columbia, lat. 49° N, has a large annual total precipitation but with greatly reduced amounts in the summer. The annual temperature range is small, and winters are mild for this latitude.

10.14 Needleleaf forest Douglas fir, hemlock, and cedar grow to great sizes in protected environments. Ferns and mosses provide a lush ground cover.

10.15 Minimizing the effect of clearcutting Riparian leave strips are designed to protect stream environments "as shown in this fly-by photo taken from a helicopter."

dammed in order to use the enormous water surpluses that run to the sea.

Much of the world's old growth forests have been destroyed, and the rainforest of the Pacific Northwest is one of the few regions that still contain relatively undisturbed areas. Clearcutting is the harvest method most widely used in British Columbia. Trees are removed from patches of forest some 50–60 hectares in size. Regulations are in place to minimize the environmental impact of clearcutting—for example, fish-bearing rivers and streams should be protected by riparian leave strips (Figure 10.15). However, such forest protection code regulations are sometimes difficult to enforce. Logging on steep slopes in this wet environment causes soil erosion problems, and slumping and landslides are common.

Water is an abundant natural resource in the marine west-coast environment. Many of the rivers are dammed for hydroelectricity production and to regulate the flow of the huge volumes of water that are returned to the Pacific Ocean each year. One of the most ambitious projects was the Kenny Dam (Figure 10.16). It was built in the 1950s to impound and reverse the flow of the Nechako River to supply electricity to the Alcan smelter in Kitimat, B.C. Despite many control structures, unusually heavy or prolonged precipitation events can result in flooding in many low-lying areas and cause disruption of roads and railways (Figure 10.17). In addition, high turbidity can lead to problems with water quality, as was the case in November 2006, when about a million residents in British Columbia's Lower Mainland region were instructed to boil their drinking water.

10.16 Regulating the flow of water for human benefit Many hydroelectricity projects have been constructed in British Columbia. At the Kenny Dam, built in the 1950s, water passes through a 16-kilometre tunnel under Mt. Dubose to the power generating facility that supplies the Alcan aluminum smelter at Kitimat.

FOCUS ON SYSTEMS | 10.2   California Rainfall Cycles and El Niño

California's coastal zone from San Francisco Bay to San Diego lies in a coastal subtype of the Mediterranean climate. It has a long, nearly rainless summer season, often cursed by disastrous wildfires. Relief from summer drought comes with winter storms that can bring heavy rainfalls, and with them, disastrous floods and mudflows. Substantial amounts of rain can fall in November or even earlier, but the months with the heaviest rainfall are December through March. In some years, heavy rainfall may continue well into April.

Graph (a) shows the year's total rainfall for Santa Barbara, a coastal city lying about midway between San Francisco and San Diego, for the period 1963–2005. The rain gauge is positioned at low elevation not far from the coast. Values are shown for the water year, beginning on October 1 and ending on September 30, in order to place all winter rainfall in the same recording period. The year-to-year relative variability is obviously great, with high peaks rising far above the mean value of 489 millimetres. The time cycle of these annual accumulations seems to be irregular in both its period and amplitude (see *Working It Out 9.1 • Cycles of Rainfall in the Low Latitudes*). Forecasting into the winter rain season from one year to the next is unreliable, as there is no obvious repeating period to rely on.

Graph (b) shows the recorded monthly values of the Southern Oscillation Index for the same period. The index is the difference between the monthly barometric pressure at Darwin, Australia,

and at the island of Tahiti. Note that the y-axis is reversed, so that negative values of the index (i.e., El Niños) correspond more closely with rainfall peaks. A persistent change in the upward (negative) direction signals El Niño; the downward (positive) direction signals La Niña.

Comparing the two graphs, we see that wet years 1978, 1983, 1995, and 1998 correspond nicely with El Niño values of the Southern Oscillation Index. However, El Niño years 1965–66, 1972, 1987, and 1991–93 are not associated with rainfall peaks. So the index does not seem to be a reliable predictor of high rainfall for Santa Barbara. Obvious La Niña events occur in 1971, 1974, 1975–76, 1989, and 1999–2001. Although these are not strongly related to individual low-rainfall years, annual rainfall during La Niña periods is always at or below the mean to some degree. From these comparisons, we can see that while there is some relationship between Santa Barbara rainfall and the Southern Oscillation Index, it is not all that reliable.

What other signals might predict a wet year for Santa Barbara? One possibility would be a rise in sea-surface temperature levels in the northern hemisphere of the eastern Pacific. Warmer sea-surface temperatures might add more moisture to maritime air masses that are drawn into occluded cyclones, which in turn provide much of the heavy coastal rainfall to southern California.

Graph (c) shows the Pacific Decadal Oscillation Index, which is constructed monthly from sea surface tem-

peratures in the North Pacific. The Pacific Decadal Oscillation has a time cycle of 20–30 years. In the graph, it is easy to pick out 1977 as a year when the index shifted from negative values (cool phase) to positive values (warm phase). Although there are a few negative peaks, the positive phase clearly continues until about 1999, when another cool phase seems to begin. Comparing this to Santa Barbara rainfall deviations, the four wettest years occurred within the warm phase. But there are many years within the warm phase of below-normal rainfall that also have strong positive index values. So again, the relationship between actual rainfall and the index is not very strong.

Another possible controlling factor is volcanic eruptions. Marked on graph (a) are the eruptions of El Chichon (1982) and Pinatubo (1991). Both of these events sent significant volumes of aerosols into the stratosphere, although the volume of Pinatubo aerosols was three times greater. (Recall from Chapter 5 that volcanic aerosols reflect more sunlight back to space, thus cooling the Earth's surface and lower atmosphere, and thus impacting global climate.) While El Chichon corresponds perfectly with high rainfall at Santa Barbara in the following winter, there is no similar response to Pinatubo.

Uncovering such connections between weather and climate phenomena in different parts of the world is a challenging and fascinating study that can be of real value to human populations affected by the vagaries of climate.

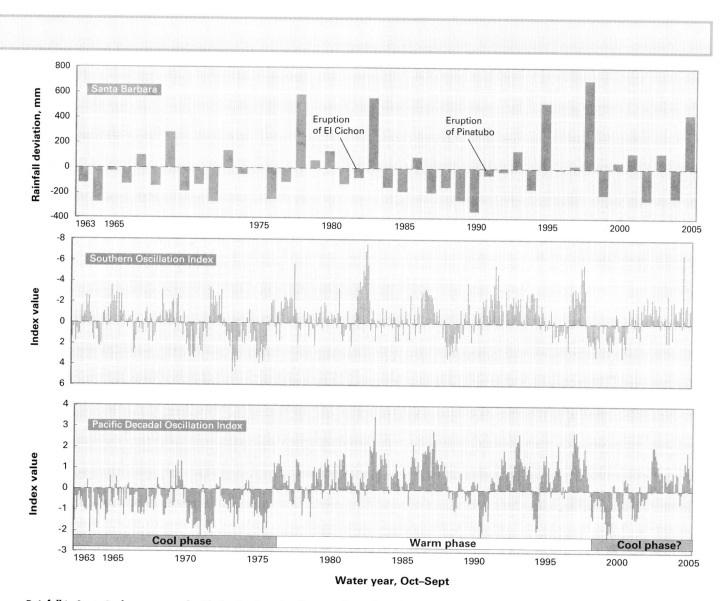

Rainfall in Santa Barbara compared with the Southern Oscillation Index and Pacific Decadal Oscillation Index   Note that the scale of the Southern Oscillation Index is reversed for easier comparison with the rainfall data. (Data from NOAA/NCDC/JISAO)

10.17 The destruction caused by flooding Heavy rains commonly cause flooding and infrastructure damage in the wet west-coast marine environment. Floodplain mapping has helped to minimize flood losses in susceptible areas.

THE DRY MIDLATITUDE CLIMATE (KÖPPEN *BWK, BSK*)

The **dry midlatitude climate**, at a latitude range of about 35 to 50° N, is limited almost exclusively to interior regions of North America and Eurasia. Typically the region lies in the rain shadow of mountain ranges. Maritime air masses are effectively blocked out much of the time, so that continental polar (cP) air masses dominate the climate in winter. In summer, a dry continental air mass of local origin is dominant. Summer rainfall is mostly convectional and is caused by occasional invasions of maritime air. The annual temperature cycle is strongly developed, with a large annual range. Summers are warm to hot, but winters are cold to very cold.

The largest expanse of the dry midlatitude climate is in Eurasia, and stretches from the southern republics of the former Soviet Union to the Gobi desert and northern China (Figure 9.18). In the central portions of this region lie true deserts of the arid climate subtype, with very low precipitation. In North America, the dry western interior regions, including the Great Basin, Columbia Plateau, and the Great Plains, are of the semi-arid subtype. In the southern hemisphere, a small area of dry midlatitude climate is found in southern Patagonia, near the tip of South America.

Figure 10.18 is a climograph for Lethbridge, Alberta, a semi-arid station located at lat. 50° N, just east of the Rocky Mountains. Total annual precipitation is 386 millimetres. Most of this comes in the form of convectional summer rainfall. Winter snowfall is light, averaging 130 centimetres over a 50-day season. The temperature cycle has a large annual range, with warm summers and cold winters. January, the coldest winter month, has a mean

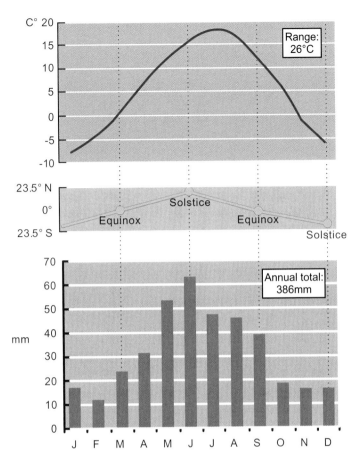

10.18 Dry midlatitude climate Lethbridge, Alberta lat., 50° N, shows a marked maximum of rainfall in the summer months. Winters are cold and dry. Figure 9.18 shows the location of this station.

temperature of –8°C, although the record low is –43°C, which occurred in 1950. The highest temperatures are in June, with a record of 39°C in 1973.

Annual precipitation in the dry midlatitude climate is quite variable and there may be large differences in precipitation from year to year. When dry conditions prevail for several years, the drought may be intense. This was the case in the mid-1930s when a series of drought years caused severe hardships in the central and western Great Plains of North America. During the drought, many centimetres of soil were removed from fields. Exceptionally intense dust storms transported the finer material out of the region, while the coarser silt and sand particles accumulated in drifts along fence lines and around buildings. The affected area became known as the Dust Bowl.

Drought is an ever-present threat in the dry midlatitude climates and its occurrence has serious implications for the region. *Eye on the Environment 10.3 • Drought on the Prairies* examines climate conditions during recent drought episodes in the Canadian Prairies.

THE DRY MIDLATITUDE ENVIRONMENT

The arid subtype of the dry midlatitude climate is restricted to the driest of midlatitude continental interiors. In North America, this cold desert environment supports a cover of sagebrush and associated low woody shrubs. Some of the drier valleys in the interior of British Columbia are of this subtype.

The steppe subtype of the dry midlatitude climate is found in large regions of interior North America and central Asia. In this subtype, the dominant natural vegetation comprises hardy perennial grasses that can tolerate the effects of low annual precipitation combined with a strongly continental thermal regime. The soils underlying these *short-grass prairies* are characteristically high in calcium, which is deposited when soil moisture evaporates during the hot, dry summers.

In North America, much of the grassland has been converted to cropland, which is mainly used in the production of spring wheat. The crops depend on water remaining in the soil from winter rains and snows and the precipitation that falls in late spring and early summer. In Russia, the rich wheat region of the Ukraine continues into a narrow zone far eastward across the steppes of Kazakhstan.

Soil water, not soil fertility, is key to wheat production over these vast steppe lands. In the western Great Plains, there has been a great increase in recent decades in the use of irrigation systems to pump groundwater to the surface and distribute it. This groundwater resource is rapidly being depleted and will ultimately be gone.

The fate of the Aral Sea is a dramatic example of the importance of water, in the dry midlatitude environment. Once the world's fourth largest lake, the Aral Sea has been shrinking for the last 40 years because of reduced inflows (Figure 10.19). The cause is both climatic and due to human activity. A series of dry years that occurred in the 1970s, particularly 1974–75, lowered discharge from the Amu Dar'ya and Syr Dar'ya rivers that drain into the Aral Sea. The period 1982–86 was also climatically dry. Long-term hydrological records since 1926 indicate that natural losses through evaporation and infiltration account for about 50 percent of the flow in these rivers. Annual discharge into the Aral Sea would normally average about 55 cubic kilometres.

By the 1960s, withdrawal of water for irrigation was about 40 cubic kilometres, which was in balance with inputs.

10.19 The Aral Sea has shrunk considerably in the past few decades and much of its former basin is now a dry, salty plain. The effect on native terrestrial and aquatic ecosystems has been dramatic; commercial fishing has been abandoned. *Fly By: 45° 05' N; 59° 24' E*

However, since that time, agricultural demands have increased, especially as a result of large-scale cotton production. The irrigated area had grown to nearly 6.5 million hectares by 1980, and demand for water increased to 132 cubic kilometres. This led to a progressive reduction in the area and volume of the Aral Sea and a corresponding increase in salinity. For example, in 1960, the Aral Sea covered 68,000 square kilometres and contained 1,090 cubic kilometres of water, with a salinity of 10 grams per litre; by 2000, the area was reduced to 23,400 square kilometres, with a volume of 162 cubic kilometres, and salinity rising to 38 grams per litre.

The environmental impact has been considerable. An estimated 43 million metric tonnes of salt are carried from the sea's dried bottom each year and deposited over an area of 150,000 to 200,000 square kilometres. Many native plant communities have disappeared. The forests, which once occupied the river deltas and shallow margins of the Aral Sea, were especially vulnerable to the three- to eight-metre drop in groundwater that has occurred. Loss of habitat has resulted in the elimination of all but 38 of the original 173 animal species that once occupied the area. Similarly, the aquatic ecosystems have been similarly devastated, with 20 of the 24 native fish species disappearing, together with the loss of commercial fishing.

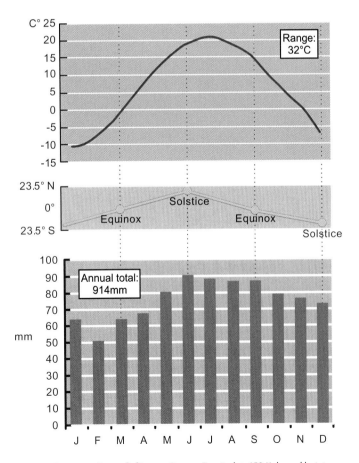

10.20 Moist continental climate Ottawa, Ontario, lat. 45° N, has cold winters and warm summers, making the annual temperature range very large.

THE MOIST CONTINENTAL CLIMATE (KÖPPEN *DFA, DFB, DWA, DWB*)

The **moist continental climate**, located in central and eastern parts of North America and Eurasia, lies in the polar-front zone. Seasonal temperature contrasts are strong, and day-to-day weather is highly variable. Ample precipitation throughout the year is increased in summer by invading maritime tropical (mT) air masses. Continental polar (cP) and continental arctic (cA) air masses dominate cold winters.

In eastern Asia—China, Korea, and Japan—the seasonal precipitation pattern shows more summer rainfall and a drier winter than in North America. This is an effect of the monsoon circulation, which moves moist maritime tropical (mT) air across the eastern side of the continent in summer and dry cP air southward through the region in winter. In Europe, the moist continental climate receives precipitation from mP air masses coming from the North Atlantic.

Ottawa, Ontario, lat. 45° N (Figure 10.20), provides an example of the moist continental climate. The annual temperature range is large. Summers are warm, with average temperatures in July of 21°C, but winters are cold, with four consecutive monthly means below freezing. Precipitation is ample in all months, with an annual total of 914 millimetres. The summer precipitation maximum is associated with invasions of maritime tropical (mT) air masses and thunderstorms that form along moving cold fronts and squall lines. Much of the winter precipitation is in the form of snow, which remains on the ground for long periods.

The moist continental climate is restricted to the northern hemisphere, occurring in latitudes 30° to 55° N in North America and Asia, and in latitudes 45° to 60° N in Europe (Figure 10.21). In Asia, it is found in northern China, Korea, and Japan. Most of central and eastern Europe has a moist continental climate, as does most of the eastern half of the United States, from Tennessee to the north, as well as the southernmost part of eastern Canada.

THE MOIST CONTINENTAL FOREST AND PRAIRIE ENVIRONMENT

When traced westward into the continental interior of North America and Asia, the moist continental climate becomes progressively drier. This gradient influences both soils and vegetation. For example, throughout much of southeastern Canada and the northeastern United States, the vegetation is mixed forest, with patches of conifers intermingled with broadleaf deciduous species. To the west, the forest gives way to *tall-grass prairie*, which used to be found in a belt running from Manitoba south into Nebraska. Farther west, in places like

Saskatchewan, the tall grass prairie is replaced by the medium and short grasses of the *mixed prairies*, in response to the transition to the dry midlatitude climate. Because of the availability of soil water through the warm summer growing season, the moist continental environment has an enormous potential for food production, and most regions have been under field crops or pasture for centuries.

Heavy snowfalls are not uncommon in these regions, especially in eastern North America, where warm moist air can move into the region along the Mississippi Valley and across the Great Lakes. One consequence of this is the *lake effect*, in which heavy snowfall accumulations develop on the eastern and northern shores of the Great Lakes (Figure 10.22). The amount and distribution of snowfall in the Great Lakes region from lake-effect snowfall depends on wind speed and direction, lake water temperatures, and the extent of ice cover. The lake effect contributes as much as 50 percent of the snowfall in some regions and is especially noticeable where prevailing winds are forced over topographic barriers.

In southeastern Ontario, snowfall accumulations of 300–400 centimetres occur along the shores of Lake Superior and Lake Huron. Where the land rises adjacent to the lakes, annual snowfall increases at a rate of about 60 centimetres per 100 metres of elevation. Major storms can drop up to 100 centimetres of snow in these areas, and the accompanying winds can cause heavy drifting.

Ice storms—or freezing rain—are another hazard in eastern Canada and New England. They typically develop when a warm air mass invades the region and overruns a shallow layer of cold air at the surface. Ice storms can cause considerable damage and inconvenience. They affect transportation especially, and interrupt power supplies when transmission lines break under the weight of the ice or trees collapse on them (Figure 10.23). Eastern Canada can expect about 15 ice storms each year, which typically last for only a few hours. Maintenance crews equipped with de-icing materials can usually deal with them quite quickly, and the power transmission infrastructure can usually cope with storms that deposit up to 50–60 millimetres of ice.

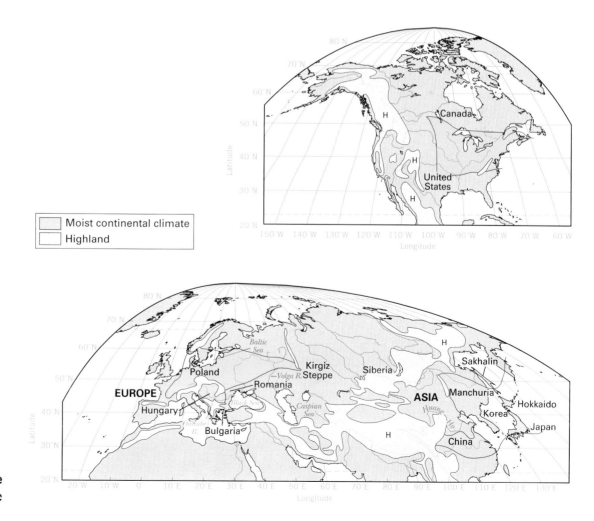

Moist continental climate
Highland

10.21 World map of the moist continental climate

EYE ON THE ENVIRONMENT | 10.3   Drought on the Prairies

Drought in the Canadian prairies is the most important factor that limits crop yield, and is a recurring threat to agriculture. The long-term precipitation record for Indian Head, Saskatchewan (see figure below), shows that precipitation was significantly lower than normal in the 1930s, as was the case throughout the much of the American Great Plains region. However, droughts of equal or greater magnitude have been recorded in the early 1960s, 1997, 2001, and 2003.

The winter of 2001–02 was warmer and drier than normal in the Prairies, resulting in limited snow accumulation,

so that low soil moisture conditions continued into the following growing season. The figure below shows precipitation for the agricultural year, which runs from September to the following August. The agricultural year includes the precipitation for soil moisture recharge after harvest, the winter precipitation available for soil moisture recharge, and growing season precipitation. The maximum extent of the drought is assessed on August 6, as rainfall after this date is too late to improve yields. More than 65 percent of the prairie cropland was affected by moderate, severe, or record

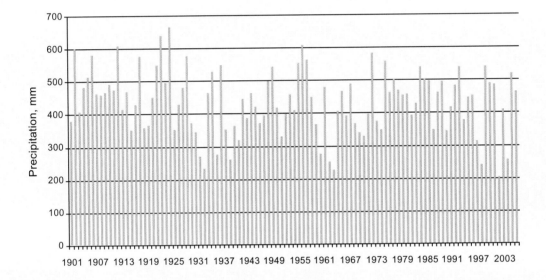

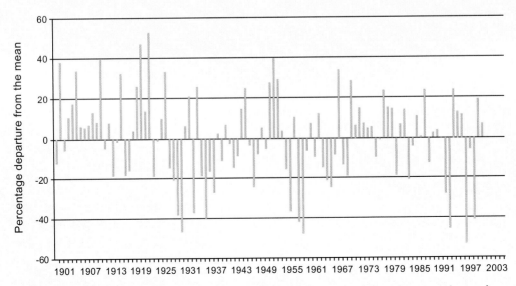

Figure 1 Annual precipication for Indian Head, Saskatchewan, for the period 1900–2005 (above) and percentage departure from the mean (below).

drought conditions, although some areas in southern Alberta and southwestern Saskatchewan actually received higher than average precipitation.

Annual crops such as wheat and barley are susceptible to drought at any time during their growth and development. However, forage grasses are more sensitive to moisture deficiencies in the fall and spring. Thus, drought conditions in one year have a pronounced carry over effect into the following year. Poor grass growth due to forage drought leads to reduced stocking rates during the summer and hay supply problems for winter feed. Figure 2 shows the extent of drought-affected pasture in 2002. Approximately 60 percent of prairie pastureland was severely affected.

Hydrologic drought caused by low precipitation and high water demand affects livestock operations. Water supplies declined in central Alberta, central Saskatchewan, and in some areas of western Manitoba during the summer of 2002, and at the peak of the drought, more than 70 percent of the farms were affected.

Although the cyclic occurrences of drought in the Great Plains cannot be prevented, improved farming practices, such as zero tillage, help to reduce its impact. Zero tillage has two general benefits —the stubble left from harvested grain crops forms an effective snow trap and by seeding directly into the stubble, there is less bare soil exposed to the wind.

Dust storms still occur on the prairies, although their frequency and magnitude are now reduced because of improved farming practices. In addition, the Prairie Shelterbelt Centre at Indian Head distributes trees to farmers to create wind breaks. In 2005, approximately 4 million trees were used to establish 220 kilometres of shelter belts in Saskatchewan, 121 kilometres in Manitoba, and 78 kilometres in Alberta. Since its inception in 1901, the centre has shipped nearly 600 million trees to almost 650,000 applicants and has established nearly 4,500 kilometres of shelter belts. An additional benefit is that the trees planted in 2005 will sequester an estimated 1.5 million tonnes of carbon dioxide by 2054.

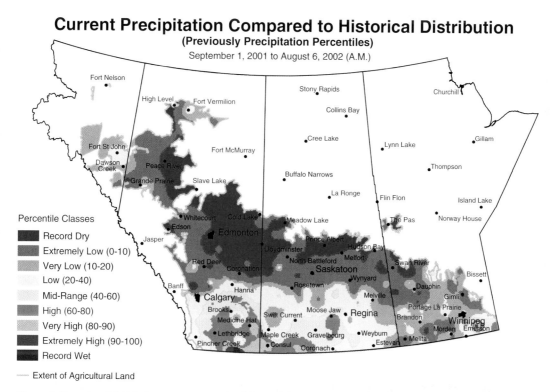

Current Precipitation Compared to Historical Distribution
(Previously Precipitation Percentiles)
September 1, 2001 to August 6, 2002 (A.M.)

Percentile Classes

- Record Dry
- Extremely Low (0-10)
- Very Low (10-20)
- Low (20-40)
- Mid-Range (40-60)
- High (60-80)
- Very High (80-90)
- Extremely High (90-100)
- Record Wet

— Extent of Agricultural Land

Figure 2 Drought of 2002 Precipitation compared with historical distribution for the 2002 agricultural year (September 1, 2001 to August 6, 2002).

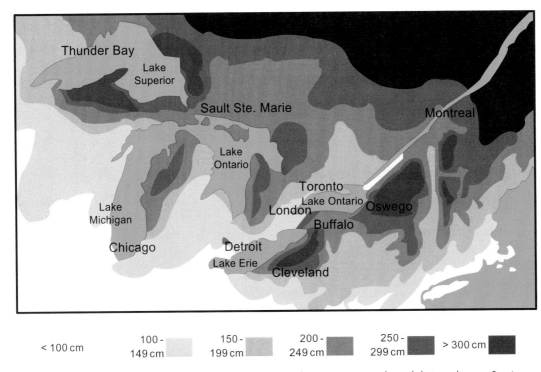

| < 100 cm | 100 - 149 cm | 150 - 199 cm | 200 - 249 cm | 250 - 299 cm | > 300 cm |

10.22 The lake effect caused by winds bringing moist air from the Great Lakes creates pronounced snow belts in southeastern Ontario.

In January 1998, a storm of unprecedented magnitude caused considerable damage and was directly linked to the deaths of at least 25 people. The storm lasted from January 4–10, and in that time deposited more than 100 millimetres of ice in some areas. It affected a relatively narrow band from Eastern Ontario to Nova Scotia and bordering areas in the northeastern United States. Millions of people were without power for days or weeks until the power grid was repaired. Damage was estimated at $4–6 billion.

In addition to the material damage caused by the large numbers of trees that collapsed under the weight of the ice, the ice storm impacted the agriculture sector since many of those trees were fruit trees or sugar maples. The Quebec maple sugar industry was devastated.

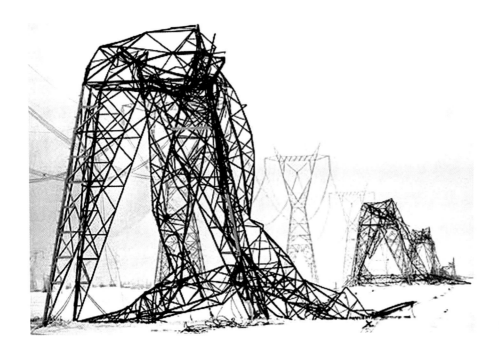

10.23 Ice storms are an expected feature of moist continental climates and can cause considerable damage.

OVERVIEW OF THE HIGH-LATITUDE CLIMATES

The *high-latitude climates* are mainly restricted to the northern hemisphere. They occupy the northern subarctic and arctic latitude zones. Only the ice sheet climate of the polar zones is present in both hemispheres. Table 10.2 provides an overview of the three high-latitude climates.

The dynamic wind pattern at high latitudes brings maritime polar (mP) air masses, formed over the northern oceans, into conflict with continental polar (cP) and continental arctic (cA) air masses on the continents. The Rossby wave system brings lobes of warmer, moister air toward the poles in exchange for colder, drier air moved toward the equator. The result of these processes is frequent wave cyclones produced along a discontinuous and constantly fluctuating arctic-front zone. In summer, tongues of maritime tropical air masses (mT) occasionally reach subarctic latitudes to interact with polar air masses and yield significant amounts of precipitation.

THE BOREAL FOREST CLIMATE (KÖPPEN *DFC, DFD, DWC, DWD*)

The **boreal forest climate** is a continental climate with long, bitterly cold winters and short, cool summers. It occupies the source region for cP air masses, which are cold, dry, and stable in the winter. Invasions of very cold cA air masses are common. The annual temperature range is greater than that of any other climate and is greatest in Siberia. Precipitation increases substantially in summer, when maritime air masses penetrate the continent with travelling cyclones, but the total annual precipitation is small. Although some areas of the boreal forest climate are moist, large areas in western Canada and Siberia have low annual precipitation.

In North America, the boreal forest climate stretches from central Alaska, across the Yukon and Northwest Territories, to Labrador on the Atlantic coast (Figure 10.24). In Europe and Asia, it reaches from the Scandinavian Peninsula eastward across Siberia to the Pacific. The latitude range for this climate type is from 50° to 70° N.

Figure 10.25 is a climograph for Fort Vermilion, Alberta, at lat. 58° N. Monthly mean air temperatures are below freezing for seven consecutive months. The summers are short and cool. Precipitation shows a marked annual cycle with a summer maximum, but the total annual precipitation is small. Winter snowfall remains over solidly frozen ground for about five months. Comparable data are also shown for Yakutsk, Siberia, at lat. 62° N. Here the mean annual temperature ranges between –42°C in January and 17°C in July.

THE BOREAL FOREST ENVIRONMENT

The Pleistocene ice sheets shaped much of the boreal forest climate region, and lakes and bogs are common features in the landscape. Most of the area is covered with coniferous forests that are exploited for pulp, paper, and lumber. The forests of Europe and Asia consist of closely related species, each of which is well adapted to the harsh environment and to periodic disturbance from fire. Each year several thousand fires burn in the coniferous forests of Canada, affecting about 1 percent of the forest area (Figure 10.26). Most fires are started by people, but about 40 percent are caused by lightning strikes.

Natural fires are typically preceded by rainless periods lasting at least one to two weeks, accompanied by high temperatures and low humidity. The probability of these conditions determines a fire climate. A fire climate is strongly seasonal at higher latitudes, and most fires occur in summer when long days and strong winds cause rapid drying. In the boreal forest, the average interval between successive fires is about 60 years.

Table 10.2 High-latitude climates

Climate	Temperature	Precipitation	Explanation
Boreal forest	Short, cool summers and long, bitterly cold winters. Greatest annual range of all climates.	Annual precipitation small, falling mostly in summer months.	This climate falls in the source region for cold, dry stable cP air masses. Travelling cyclones, more frequent in summer, bring precipitation from mP air.
Tundra	No true summer, but a short mild season. Otherwise, cold temperatures.	Annual precipitation small, falling mostly during mild season.	The coastal arctic fringes occupied by this climate are dominated by cP, mP, and cA air masses. The maritime influence keeps winter temperatures from falling to the extreme lows of interiors.
Ice sheet	All months below freezing, with lowest global temperatures on Earth experienced during antarctic winter.	Very low precipitation, but snow accumulates since temperatures are always below freezing.	Ice sheets are the source regions of cA and cAA air masses. Temperatures are intensely cold.

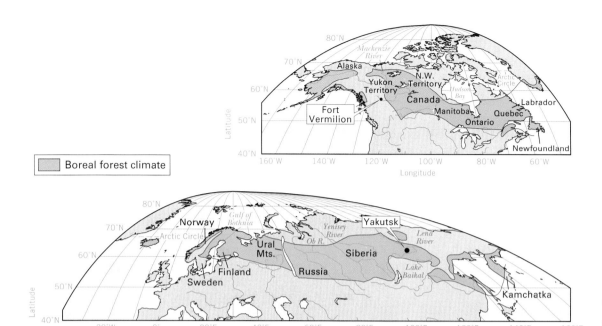

Boreal forest climate

10.24 World map of the boreal forest climate

In Canada, the danger of forest fires is incorporated into the Fire Weather Index. Several parameters determine the index, including temperature, rainfall, relative humidity, and wind speed. In addition, the calculations include the amount of litter on the forest floor and its moisture content. The Fire Weather Index is used to assess the fire risks in national parks.

Typical average values for the Fire Weather Index range between 2 and 10, but this can rise dramatically during individual fires (Figure 10.27). Index values decrease toward the east in response to the generally wetter climate conditions.

THE TUNDRA CLIMATE (KÖPPEN *ET*)

The **tundra climate** is dominated by polar (cP, mP) and arctic (cA) air masses. Winters are long and severe. A moderating influence of the nearby ocean water prevents winter temperatures from falling to the extreme lows found in the continental interior. There is a short mild season. Precipitation generally increases in the summer, when more moisture is available from the ice-free polar seas.

10.25 Boreal forest climate Extreme winter cold and a very large annual range in temperature characterize the climates of Fort Vermilion, Alberta, and Yakutsk, Siberia.

10.26 A blazing summer fire Fire is common in the boreal forest, especially during the long days of summer.

10.27 Fire ratings in Canada (a) Average values of the Canadian Fire Weather Index range from 2–10 across most of the boreal forest; (b) A fire rating of 14; (c) A fire rating of 34.

The world map of the tundra climate (Figure 10.29) shows tundra surrounding the Arctic Ocean and extending across the island region of northern Canada. It includes the Alaskan north slope, the Hudson Bay region, and the Greenland coast. In Eurasia, this climate type occupies the northernmost fringe of the Scandinavian Peninsula and Siberian coast. The Antarctic Peninsula (not shown in Figure 10.28) also experiences a tundra climate. The latitude range for this climate is 60° to 75° N and S, except for the northern coast of Greenland, where tundra occurs at latitudes greater than 80° N.

Figure 10.29 is a climograph for Baker Lake, located in Nunavut at lat. 64° N. A short mild period with temperatures slightly above freezing lasts for four months. The long winter is very cold. Total annual precipitation is 270 millimetres, with a pronounced late summer maximum. This is associated with the melting of the sea-ice cover and a warming of ocean water temperatures, which increases the moisture content of local air masses.

THE ARCTIC TUNDRA ENVIRONMENT

The term *tundra* describes both an environmental region and a major class of vegetation. An equivalent climatic environment—called *alpine tundra*—prevails in many global locations in high mountains above the timberline. The tundra environment is characterized by a steep temperature gradient close to the ground surface. Temperatures at the surface can be several degrees warmer than the air just a few centimetres above, and the short plants take advantage of this to complete their growth in the brief summer.

The tundra is a treeless zone that borders the open lichen woodland at the extreme range of the boreal forest. The transition zone is marked by stunted, deformed trees that have lost their branches due to abrasion from ice crystals that are carried in the bitter winds. In some places the forest-tundra boundary is fairly distinct. It coincides approximately with the 10° C isotherm of the warmest month and with the median position of the arctic front that separates cool arctic air masses from milder air originating in the Pacific.

ARCTIC PERMAFROST

Soils of the arctic tundra are poorly developed and consist of freshly broken mineral particles and varying amounts of partially decomposed plant matter. Peat bogs are numerous. Because soil water is solidly and permanently frozen not far below the surface, the summer thaw saturates the soil with water.

Perennially frozen ground, or **permafrost**, occurs throughout the tundra region and much of the adjacent boreal forest. Normally, a top layer of the ground will thaw each year during the mild season. This *active layer* of seasonal thaw is from 0.6 to 4 metres thick, depending on latitude, topography, and vegetation cover.

Continuous permafrost, which extends without gaps or interruptions under all surface features, coincides largely with the tundra climate, but also underlies much of the taiga in northern Canada and Siberia. *Discontinuous permafrost*, which occurs in patches separated by frost-free zones under lakes and rivers, is common in the boreal forests of North America and Europe and occurs sporadically even near the southern limits of the forest. Permafrost is sensitive to disturbance, so regulations have been developed to minimize the impact of mining and exploration activities in these regions. The impact of global warming is especially serious for coastal environments in this climate zone. Not

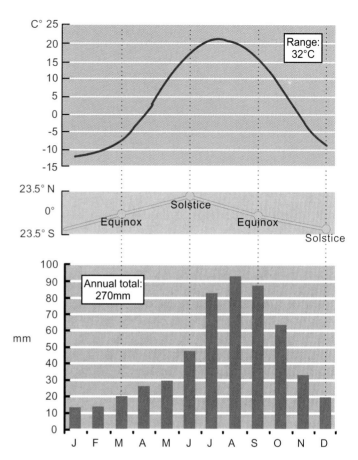

10.29 Tundra climate Baker Lake, Nunavut, 64° N.

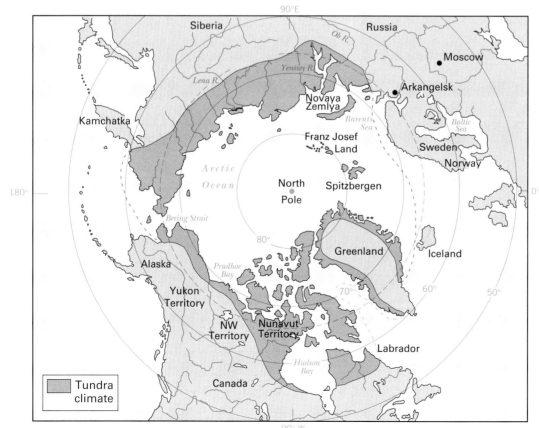

10.28 World map of the tundra climate

only will they be affected by rising sea level and loss of ice cover, but much of the coastline will be subject to slumping and erosion. Chapter 14 discusses permafrost in more detail.

THE ICE SHEET CLIMATE (KÖPPEN *EF*)

The **ice sheet climate** coincides with the source regions of arctic (A) and antarctic (AA) air masses, situated on the vast ice sheets of Greenland and Antarctica and over the polar sea ice of the Arctic Ocean. The mean annual temperature is much lower than in any other climate, and no monthly mean exceeds 0° C. Strong temperature inversions, caused by radiation loss from the surface, develop over the ice sheets. In Antarctica and Greenland, the high altitude of the ice sheet surface intensifies the cold. Strong cyclones with blizzard winds are frequent. Precipitation, almost all occurring as snow, is very low but accumulates because of the continuous cold.

Figure 10.30 shows temperature graphs for several ice sheet stations. Temperatures in the interior of Antarctica are far lower than those at any other place on Earth. The record low temperature of −88.3°C was recorded in 1958 at Vostok, located about 1,300 kilometres from the South Pole at an altitude of about 3,500 metres. Temperatures are considerably higher, month for month, at McMurdo Sound because it is located close to the Ross Sea and is at a low altitude.

THE ICE SHEET ENVIRONMENT

Because of low monthly mean temperatures throughout the year over the ice sheets, this environment is practically devoid of vegetation. Some algae and bacteria survive in the surface layers of the ice and in locally heated volcanic openings. The few species of animals found on the ice margins, such as polar bears and penguins, are associated with the marine habitat. Concern has been raised for both of these species as a result of significant changes in the ice conditions in recent years.

OUR CHANGING CLIMATE

In Chapter 8, climate was defined as a region's long-term average weather, based on mean monthly temperature and precipitation observed at weather stations. However, climate change has affected present weather conditions and will likely become more pronounced in the coming years. As noted in *A Closer Look: Eye on Global Change 4.4 • The IPCC Report of 2001*, recent human activity has raised global temperatures, which in turn have reduced global snow and ice cover and raised sea levels. Precipitation, enhanced by greater evaporation, has increased in mid-

and high-latitude regions, but has decreased in subtropical regions. The variability of weather has also increased, with more frequent extreme precipitation events.

For most of North America, temperatures of the twenty-first century are predicted to rise significantly, bringing warmer winters and hotter summers. Although precipitation will increase in many regions, warmer temperatures will cause more evapotranspiration, and the result will be more summer drought. Meanwhile, more frequent extremes of precipitation will enhance flooding and storm damage. *A Closer Look: Eye on Global Change 10.4 • Regional Impacts of Climate Change on North America* summarizes climate change predictions and impacts for nine regions in North America.

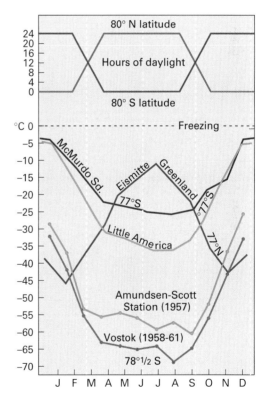

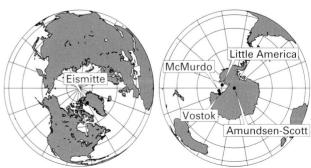

10.30 Ice sheet climate Temperature graphs for five ice sheet stations.

A CLOSER LOOK | EYE ON GLOBAL CHANGE 10.4 Regional Impacts of Climate Change on North America

By nearly all accounts, the Earth is getting warmer under the influence of human-induced releases of carbon dioxide into the atmosphere. Will the warming continue? What will be the effects on different regions? To answer these questions, we must rely on global mathematical climate models that, although imperfect, point the way to possible effects that have great implications for human society as well as natural ecosystems.

Eye on Global Change 4.4 • *The IPCC Report of 2001* described the changes in global climate we can expect for the balance of the twenty-first century. Recall that these predictions included the following:

• Global average temperatures rising 1.4–5.8°C, with temperatures of land areas, especially northern-latitude and winter temperatures, rising even more;

• Increased precipitation in northern mid- and high latitudes;

• Over land, more hot days and increased summer drought, coupled with increased precipitation in extreme events;

• Continued melting and retreat of ice caps and glaciers; and

• Continued rise in sea level by 9–88 centimetres.

Because the implications of these predictions are so far-reaching and have so many important consequences, Canada and the United States have completed studies predicting climate change and analyzing the effects by geographic region.[1] The table below presents these regional changes and their impacts under the headings of agriculture, forests and ecosystems, the water cycle, urban and human impact, transportation and infrastructure, coastal and marine environments, and natural hazards. From the table, it is clear that every region will suffer change and disruption, sometimes with far-reaching consequences.

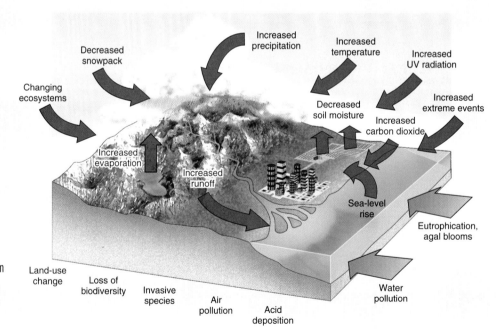

Multiple stresses of a changing climate Climate change will stress natural ecosystems and human activities in many ways. (After National Assessment Synthesis Team, U.S. Global Change Research Program)

---

[1] Canada has published a series of eight detailed reports on climate change within its regions as part of the Canada Country Study. They are summarized in two documents, *The National Summary for Policy Makers* and *Highlights for Canadians*, both issued in 1997 and available at http://www.ec.gc.ca/climate/ccs. For the United States, see National Assessment Synthesis Team, Climate Change Impacts on the United States. *The Potential Consequences of Climate Variability and Change*, U.S. Global Change Research Program, Washington, DC, 2000.

Impact of global climate change on regions of the United States and Canada

Region	Observed Climate Trends, Past Century	Future Climate Trends, Next Century	Agriculture	Forests and Ecosystems
Alaska, British Columbia, and Yukon	• Substantial warming—2°C since 1950s; especially interior in winter, where it has warmed by 4°C. • Growing season has lengthened 14 days since 1950s. • Precipitation has increased 30% since 1968.	• Warming to continue, 3–10°C. • Precipitation to increase 20–25%, with 10% decrease along south coast.	• Lengthened growing season will enhance productivity. • Summer drought on south coast may reduce productivity.	• Warming will increase forest productivity on the moist southern coast but reduce productivity in dry interior. • Increased disturbance, including insects, uprooted trees, and fire will reduce forest productivity. • Warming will enhance decay of soil organic matter, adding CO_2 to carbon cycle. • Potential shift of boreal forest northward into tundra zone.
Canadian Arctic	• Mackenzie River district has warmed by 1.5°C; arctic tundra by 0.5°C over last 100 years. • Arctic mountains and fiords of eastern arctic have cooled slightly.	• Future winter temperature increases of 5–7°C predicted over mainland and arctic islands. • Modest cooling in extreme eastern arctic. • Summer temperatures will increase up to 5°C on mainland, 1–2°C over marine areas. • Annual precipitation will increase up to 25%.	• Agricultural opportunities (e.g., irrigated wheat) will arise in central and upper Mackenzie River basin, but will be restricted by availability of suitable soils.	• The size of tundra and taiga/tundra ecosystems will decrease by as much as two thirds. • Freshwater species will migrate northward about 150 km per °C increase in temperature. • Seal, sea lion, and walrus populations will decline because of the recession of pack ice. • Muskoxen and high-arctic Peary caribou may become extinct.
Pacific Northwest	• Average annual temperature has increased 0.5–1.5°C over most of the region. • Annual precipitation has increased by 10%; 30–40% east of the Cascade Mountains. • Warm-dry and cool-wet years correlate with El Niño/Southern Oscillation and Pacific Decadal oscillation cycles.	• Annual temperatures will increase steadily, by 4–4.5°C by 2090. • Winter temperatures will rise by 4.5–6°C. • Precipitation will rise, possibly as much as 50% by 2090, with summer precipitation unchanged or slightly decreasing. • Extreme precipitation events will increase substantially.	• Wetter and warmer winters will shift the dryland farming cycle. • Summer droughts will impact irrigated agriculture in conflict with urban use.	• Salmon populations will decrease as a result of increased winter flooding, reduced summer and fall flows, and rising stream and estuary temperatures. • Conifer forest will be stressed by warm, dry summers, leading to pest infestations and fires. • Summer stress on seedlings will reduce conifer recruitment.
Canadian Prairies and Great Plains	• Temperatures have risen more than 1°C in past century; increases up to 3°C in Canadian Prairies and northern Great Plains	• Temperatures will increase and soil moisture decrease, even with increased rainfall. • More frequent heat events, with impact on people and livestock.	• Higher temperatures and lower soil moisture will decrease crop yields in Canadian Prairies. • Longer growing season may increase crop production at northern limits.	• Decreased severity of winters and changed habitats will alter distributions of big game, waterfowl, and game birds. • Some freshwater fish species may extend their ranges north.

Impact of global climate change on regions of the United States and Canada (continued)

Region	Observed Climate Trends, Past Century	Future Climate Trends, Next Century	Agriculture	Forests and Ecosystems
	• Annual precipitation decreased 10% in northern section, increased 10% in eastern section. • Texas has had more high-intensity rainfall events. • Snow season ends earlier in spring.	• Precipitation will increase 10–20% in northern area and decrease 10–20% in southern area and in Rocky Mountain rain shadow..	• Southern area and Rocky Mountain rain shadow will suffer reduced soil moisture and increased crop stress.	• Some seasonal and semi-permanent wetlands will dry up. • Some native species will be displaced by undesirable invasive species as climate changes.
Ontario and Midwestern states	• Northern portion has warmed by 2°C, southern portion has cooled by 0.5°C. • Precipitation has increased 10–20% in the twentieth century. • Substantial rise in number of heavy and very heavy precipitation events.	• Temperature will rise 3–6°C. • Average minimum temperature will rise as much as 0.5–1°C more than the maximum temperature. • Precipitation will increase 10–30%. • Precipitation increase will not keep pace with temperature increase, creating soil moisture deficits, reduced lake and river levels, more frequent drought.	• With longer growing season, double cropping will become more prevalent. • CO_2 fertilization will enhance crop yields. • Reduced soil moisture will reduce yields in some areas of the corn belt.	• Boreal forest will shift north and suffer more fire, insects, disease; transitional forests will shift north, as well. • Changes in water temperature will affect freshwater ecosystems, changing species compositions. • Runoff of excess nutrients and wastes into lakes and rivers will be enhanced by heavy precipitation events. • Decreasing lake levels will impact current wetlands. • Birds and wildlife populations will be impacted by changing habitats and food sources.
American West	• Temperatures have risen 1–3°C. • Precipitation has generally increased, some areas by greater than 50%; but Arizona has been drier. • Extreme precipitation events are more frequent. • Snow season has decreased by 16 days since 1951 in California and Nevada. • Exceptionally wet and dry periods.	• Temperature will increase 1.5–2°C by 2030s; 4.5–6°C by 2090s. • Precipitation will increase, especially winter precipitation over California. • Rockies will get drier. • More extreme wet and dry years. • Higher temperatures will increase heat stress, weeds, pathogens, etc. • Forage production will increase, and growing and grazing season will lengthen, with more precipitation and warmer temperatures.	• Increased precipitation will increase crop yields and decrease water demands. • Milder winter temperatures will lengthen the growing season and shift cropland areas northward.	• Grassland, woodlands, and forests will increase, with a loss of desert vegetation. • In the fragmented western landscape, species won't be able to migrate easily in response to warming; alpine ecosystems could become extinct. • Increases in fires during dry cycles and air pollution damage could reduce productivity.
South-eastern states	• Temperatures were warm during the 1920s–40s; cooled through the 1960s; warmed again starting in 1970s to levels of the 1920s and 1930s.	• Models predict warming from about 3–6°C, depending on model and region.0	• Models disagree on soil moisture trends; they could increase or decrease depending on amount of warming and change in precipitation.	• Models disagree on forest productivity; generally it will increase, but will decrease in some places for some ecosystems in some scenarios.

Region	Observed Climate Trends, Past Century	Future Climate Trends, Next Century	Agriculture	Forests and Ecosystems
	• Coastal regions warmed by 2°C. • Rainfall increased 20–30% or more over much of the region. • Strong El Niño and La Niña effects provide large seasonal and inter-annual variation in temperature and precipitation, with hurricanes more frequent in La Niñas.	• Models disagree on precipitation predictions, ranging from neutral or slight increase to 25% increase. • El Niño rainfall and La Niña droughts may intensify as atmospheric CO_2 increases. • Heat indexes will rise dramatically.	• Models predict different crop yield scenarios ranging from increases in interior regions to decreases for dryland crops. • Significant decreases in yields of corn, peanuts, and sorghum along the Gulf Coast.	• If soil moisture drops, pine forests will be replaced by savannas and grasslands.
Southern Quebec and North-eastern states	• Temperatures have increased as much as 2°C in last 100 years along coastal margins. • Precipitation has increased more than 20% over much of the region. • Precipitation extremes are increasing, drought decreasing. • The period between first and last dates of snow on the ground has decreased by seven days in the last 50 years.	• Temperatures will increase 3–5°C depending on area and model. • Winter minimum temperatures show greatest increases. • Precipitation projections range from neutral or modest increases to nearly 25%. • Winter snowfalls and periods of extreme cold will decrease. • Ice storms, rains over frozen ground, and rapid snow melting events will increase. • Heavy precipitation events will increase; hurricane frequency could increase.	• Temperature and rainfall increases may enhance crop productivity. • The earlier plants come into leaf, the longer the growing season will be. • More frequent extreme rainfalls and storms will affect local agriculture.	• Forest composition will change; northern hardwoods will move north as oak-hickory forests replace them; sugar maple will migrate out of New England. • Lobster populations will move northward. • A rise in sea level will reduce migratory bird habitat. • Trout populations will decrease in Pennsylvania. • Fall foliage will be muted as species composition changes with added warmth. • Milder winters will possibly increase insect-borne human and animal disease.
Canadian Atlantic Provinces	• Slight cooling experienced in past 50 years. • Higher frequency of extreme events.	• No strong evidence for significant warming. • Increased intensity and/or frequency of storms anticipated. • More temperature extremes, leading to milder winters, early extended thaws, late springs, early frosts.	• Increased frequency of storms could impact crop productivity.	• Fish habitat could be lost; distribution of fish species and migration patterns altered, with the interruption of life cycles. • Winter ranges of terrestrial birds could shift; seabirds could change in range, distribution, and breeding success. • Reduced snow cover could increase deer populations but reduce forest regeneration and species diversity. • Increased forest blowdowns and mortality with increased storms and insect outbreaks.

Impact of global climate change on regions of the United States and Canada

Region	Water Cycle	Urban/Human Impact	Transportation and Infrastructure	Coastal/Marine Enviroments	Natural Hazards
Alaska, British Columbia, and Yukon	• Warm temperatures and rainfall will enhance spring flooding. • Flood protection efforts on south coast rivers and streams may not be adequate.	• Summer drought along south coast and southern interior will place water supplies for urban areas in contention with agriculture. • Summer drought may impact urban air quality.	• Melting permafrost will damage roads, pipelines, and structures. • Sea-level rise will increase coastal flooding, placing docks and port facilities at risk.	• Sea-level rise could be as much as 30–50 cm. • Marine ecosystems of the Gulf of Alaska and Bering Sea already show large fluctuations with causes unknown; climate change effect is likely to be large and unpredictable.	• Retreat of glaciers will cause local flooding and enable landslides and mudflows, causing loss of life and infrastructure. • Permafrost will thaw in discontin-uous permafrost regions, causing erosion, landslides, sinking of ground surface, damage to forests, buildings, and infrastructure. • Sea ice will retreat, allowing coastal erosion and storm surges.
Canadian Arctic	• Increased evaporation in arctic regions from warmer atmos-phere and longer thaw period will decrease flows and levels of northward-flowing rivers. • The river ice season will decrease by one month by 2050; the ice sea-son for large lakes by two weeks.	• Subsistence of native peoples likely to be made more difficult as populations of mammals, fish, and sea birds fluctuate, and the lack of snow and ice makes hunting more difficult.	• Reduced sea ice will benefit offshore oil and gas operations. • Melting permafrost will negatively affect pipelines. • Shipping season will be extended, with easy transit through the northwest passage at times.	• The rise in sea level will flood coastal ecosys-tems, e.g., the Mackenzie Delta. • Reduced sea ice and increased water temperatures with unknown effects on marine biota.	• Melting of permafrost will cause land subsidence, forest loss, damage to structures. • Retreat of eastern mountain glaciers will produce soil instability with landslides and mudflows.
Pacific Northwest	• Warmer, wetter winters will increase flooding in rain-fed rivers and trade snow pack for runoff; snow pack will decrease by 75–125 cm. • Summer shortages will be more severe because of reduced snow pack and earlier melting; summer soil moisture will decrease 10–25%.	• The rise in sea level will require substantial investments to control coastal flooding, especially in southern Puget Sound, where subsidence is occurring. • Summer drought will impact urban areas through water supply, air quality.	• New investments will be needed for water resource management for winter floods and summer droughts.	• Increased frequency of severe storms will enhance storm surge flooding and coastal erosion, especially during El Niño events.	• Higher precipitation and more extreme events will increase soil saturation, mudflows, and mass movements. • Winter flooding will be enhanced, with large populations at risk.

Region	Water Cycle	Urban/Human Impact	Transportation and Infrastructure	Coastal/Marine Enviroments	Natural Hazards
American West	• Allocation conflicts and conflicting authorities may enhance vulnerability to drought. • Increased precipitation will increase water supplies, reduce demand, and ease competition among uses. • Greater runoff will increase hydropower production and ease some water quality problems. • In drier regions, expect reduced supply, power production, and more contention among users.	• The rise in sea level will impact coastal communities, with increased lowland flooding and cliff erosion. • Warmer temperatures will increase heat stress on urban populations and lower air quality in dry summers. • For tourism, more hiking and less skiing. • Less water available for recreation during dry years.	• Flood control will be required for enhanced winter rains. • New waterworks will be needed to provide supply during dry years.	• Rising sea levels will threaten coastal wetlands, e.g., San Francisco Bay, and reduce diversity.	• More extreme rainfall events will cause more mass movements of soil. • In wet years, floods will be more damaging.
Canadian Prairies and Great Plains	• Canadian hydro-electric power will compete with agriculture and urban uses of water. • More competition among agriculture, ecosystem, urban, industrial, and recreational users for water in American Great Plains. • Little hope for recharge of declining groundwater reserves in the southern Great Plains.	• The heat index will increase by as much as 25% in northern Texas, Oklahoma, and Kansas region, along with more frequent heat events. • The changing climate will impact small family farmers and ranchers harder than commercial agriculture.	• Reduced water levels during summer droughts will affect river transportation.		• River flooding from extreme events will be enhanced. • Sand Hills (dunes) may become mobile again if vegetation cover declines.
Ontario and Midwestern states	• Great Lakes levels will fall 0.3–1.5 metres; outflow to St. Lawrence Seaway will decrease 20–40%. • Increased national and international tension over lake and river waters. • Reduced river levels will reduce water quality.	• Reduced risk of life-threatening cold in winter but increased risk of life-threatening heat in summer. • Increased frequency of urban air pollution episodes. • River flooding will be more frequent.	• Reduced water levels on rivers and lakes will make water-based transportation more difficult and expensive and will require new docks and locks. • Reduced ice cover will lengthen lake/river shipping seasons.		• Flash flooding from more frequent heavy rains will wash sediment and agricultural pollutants into rivers and lakes. • Sustained heavy precipitation events may cause more catastrophic river flooding.

Impact of global climate change on regions of the United States and Canada (continued)

Region	Water Cycle	Urban/Human Impact	Transportation and Infrastructure	Coastal/Marine Enviroments	Natural Hazards
Southeastern states	• Stresses on water quality will be associated with intensive agricultural practices, urban development, coastal processes, and mining activities. • Higher temperatures that reduce dissolved oxygen, and contamination from agricultural runoff, untreated sewage, and chemical releases in catastrophic rainfall events will impact water quality.	• Hurricanes, floods, and heat waves will have major impacts on human population; climate change could enhance their frequency and severity. • Low air quality episodes from increased temperatures and pollution will accompany heat waves.	• Flooding from extreme events may create transportation difficulties on roads and rivers.	• Rising sea level, coupled with subsidence due to groundwater withdrawal, sediment compaction, wetland drainage, and levee construction, will cause continued loss of coastal wetlands. • Salt water intrusion of groundwater from rising sea level will continue to kill coastal forests. • Rising sea level increases the frequency of storm surge and flooding of low-lying areas by storms and causes retreat of barrier beaches.	• Coastal flooding in hurricanes can produce catastrophic loss of life.
Southern Quebec and Northeastern states	• Variations in water supply in Quebec may adversely affect power generation.	• Climate change will add significantly to stresses on major urban areas, such as electric power consumption, summer air quality, heat waves, etc. • Summer recreation season to increase, but winter ski season to decrease in length. • Loss of beachfront property and recreation areas from the rise in sea level.	• Low water levels in the St. Lawrence River will impact shipping and the marine environment. • The rise in sea level may require engineering structures for coastal protection.	• Estuarine water quality will fall as increased temperatures reduce dissolved oxygen and extreme rainfall events provide more polluted runoff and reduce salinity. • The rise in sea level will substantially increase loss of wetlands and marshes.	• The rise in sea level will enhance coastal flooding in hurricanes and northeasterly winds.
Canadian Atlantic provinces	• Less winter snow cover is predicted. • More extremes of excessive moisture and drought. • Declining runoff would reduce hydroelectric power generation.	• The rise in sea level will increase danger of flooding of urban areas.	• Changes in ice-free days would affect marine transportation and offshore oil and gas industry.	• The rise in sea level will affect coastlines from the Bay of Fundy to Newfoundland; it will also threaten Acadian dyke-land agriculture in Nova Scotia.	• The rise in sea level will enhance coastal flooding during severe storms.

The principal climate changes for North America forecasted by the global climate models in the Canadian and American reports include the following:

- *General air temperature increases, with more northerly locations showing greater increases than southerly locations.* These will lead to major shifts in ecosystems, as forests migrate farther north, grasslands and savannas expand, and tundra shrinks.
- *Warmer nighttime temperatures and warmer winter low temperatures.* These will increase growing seasons in many areas, as well as moderate extreme cold weather events. Frozen rivers, lakes, and sea ice in Canada will thaw sooner and freeze later, facilitating water-based transportation.
- *Higher summer temperatures and heat indexes in most regions.* Extreme heat events will become more frequent, severely stressing crops, livestock, human health, and water supplies. Heat and smog will lower the air quality in cities.
- *More precipitation in most regions.* Although precipitation will increase, rising temperatures will stimulate more evaporation, causing summer drought and water shortages in many regions.
- *More frequent extreme precipitation events.* These will lead to more flooding, both locally in thundershowers and squall lines, and regionally in hurricanes and extra-tropical cyclones.
- *Enhanced effects of El Niño and La Niña.* These cycles influence the frequency and magnitude of Pacific storms and Atlantic hurricanes, as well as wet and dry years in the southeast.

Considering specific climates, the changes and impacts can be highlighted as follows:

- *Mediterranean and Marine West-Coast Climates:* Pacific subtropical high pressure will move northward and intensify during the summer, increasing air temperatures and reducing precipitation. In the winter, the pressure gradient between Aleutian low pressure and Pacific subtropical high pressure will intensify, bringing more frequent and severe storms to the coast and inland regions. These effects will accentuate the wet-winter–dry-summer contrast in west-coast climates. Contention over water resources between urban and agricultural uses, as well as between political divisions, will be enhanced.
- *Boreal Forest and Tundra Climates:* Temperatures will moderate, particularly in winter, and the growing season will increase in length. Boreal forests will move northward, displacing tundra, and will in turn be replaced by mixed forests on their southern boundary. Fires, disease, and insect damage during prolonged summer water stress will reduce forest productivity. Extensive thawing of discontinuous permafrost will cause erosion, landslides and mudflows, sinking of the ground surface, and damage to forests and structures. Oxidation of organic matter in soils will release carbon dioxide, enhancing the greenhouse effect and exacerbating global warming.
- *Dry Subtropical and Dry Midlatitude Climates:* Winter precipitation in the arid southwest will increase, benefiting urban areas and agriculture. Deserts will retreat, and some will be replaced by grasslands and savannas. Rain shadow effects in the inter-mountain region and east of the Rockies will be enhanced, reducing soil moisture and stressing irrigation systems. While precipitation will increase in the northern regions, it will not keep pace with increased evaporation from rising temperatures, and summer soil moisture will decrease, impacting dry-land farming.
- *Moist Subtropical and Moist Continental Climates:* Although precipitation will increase in the Midwest, higher temperatures will reduce soil moisture, lowering water levels in lakes and rivers in summer. This will degrade water quality and affect river and lake transportation. In the southeastern states, warming is predicted, but models disagree on precipitation change. More frequent extreme precipitation events here will cause more frequent flooding. The rise in sea level will lead to increased damage to barrier beaches and structures from storm surges, as well as to saltwater encroachment on coastal groundwater tables. In the northeastern states, temperature increases will be more moderate, but still cause migration of forest species northward. The Canadian Maritime provinces will warm only slightly, if at all, but will be affected by the rise in sea level. In southern Ontario and Quebec, warming will be more intense, leading to species migration and ecosystem disruption.

These observations show that global warming and climate change will have major impacts on North Americans. To reduce the costs to society and the environment, we need to follow a two-pronged strategy. The first strategy is abatement. We need to reduce emissions of carbon dioxide and other greenhouse gases by reducing our dependence on fossil fuels. This will require both active energy conservation and shifting to alternative energy sources, such as solar, wind, and wave power. Water conservation will also be important in many regions, as water supplies dwindle and demand increases.

The second strategy is mitigation. Substantial investments in infrastructure will be needed, for example, in construction of new coastal and maritime structures where sea levels are rising or water levels are falling. In agriculture, new crop varieties and cropping practices will have to be developed. In nearly all economic endeavours, we will have to factor in the costs of dealing with a greater risk of extreme weather events and their impact.

Whatever the course of global warming, human society will be best served by anticipating and planning for a changing future, rather than by reacting to it as it occurs.

A Look Ahead

The Earth's climates are remarkably diverse, ranging from the hot, humid wet equatorial climate at the equator to the bitterly cold and dry ice sheet climate at the poles. Between these extremes are the other climates, each with distinctive features—such as dry summers or dry winters, or uniform or widely varying temperatures. The global environments associated with each climate type are also highly varied, and some are more hospitable to human habitation than are others. Climate exerts strong controls on vegetation and soils, especially at the global level; Chapters 20 to 23 discuss this in more detail.

CHAPTER SUMMARY

- *Midlatitude climates* are quite varied, since they lie in a broad zone of intense interaction between tropical and polar air masses.

- The dry subtropical climate is dominated by subtropical high pressure and resembles the dry tropical climate; however, it has a larger annual temperature range and a distinct cool season. Agriculture is possible with adequate water supply, but this has led to serious soil degradation through the processes of dryland salinization and irrigation salinization.

- Abundant rainfall with a summer maximum is a characteristic of the moist subtropical climate. The temperature cycle of this type includes cool winters with spells of below-freezing weather and warm, humid summers. This climate is often uncomfortably hot and humid in the summer. The natural vegetation cover in these regions is mainly broadleaf deciduous forest, but much of it has been cleared for agriculture. The heavy rainfall in these areas makes them prone to severe soil erosion.

- The Mediterranean climate is unique because its annual precipitation cycle has a wet winter and a dry summer. The temperature range is moderate, with warm to hot summers and mild winters. Water supply is a key factor in this climate where it is used mainly for agriculture. Some crops such as fruit are sensitive to frost and must be protected in areas where cold air drainage occurs.

- The marine west-coast climate, like the Mediterranean climate, has a winter precipitation maximum. The marine influence keeps temperatures mild with a narrow annual range. Forest is the native vegetation of the marine west-coast environment—dense needleleaf forest on the northern Pacific coast and broadleaf deciduous forest on the coast of western Europe. In Europe, this region has been under intensive cultivation for many centuries, and little forest remains. In North America, this climate zone is a source of many diverse forest products. Abundant precipitation is used to generate hydroelectricity, but flooding is common in low-lying areas and soil erosion is a hazard on steep slopes.

- The dry midlatitude climate has both *arid* and *semi-arid* subtypes. Both have warm to hot summers and cold to very cold winters. The semi-arid subtype typically has relatively fertile soils with a sparse cover of grasses. Wheat is the dominant crop, and cattle grazing is important. Because rainfall is highly variable, drought is a recurring event and has led to dust-bowl conditions in the past.

- The moist continental climate has ample precipitation with a summer maximum. Summers are warm and winters cold. Heavy snowfalls and ice storms commonly occur in this climate and can cause significant damage and disruption to day-to-day activities.

- *High-latitude climates* have low precipitation since air temperatures are low. The precipitation generally occurs during the short warm period.

- The boreal forest climate has long, bitterly cold winters. The boreal forest consists of needleleaf trees of pine, spruce, fir, and larch interspersed with patches of deciduous aspen, poplar, willow, and birch. Timber harvesting for pulp and lumber is the principal economic activity in the boreal forest environment. Long days and dry weather in summer lead to frequent fires.

- The tundra climate occupies arctic coastal fringes. Because of the marine influence, winter temperatures may not be as cold as those of the boreal forest climate. The tundra is a vegetation cover of scattered grasses, sedges, lichens, and dwarf shrubs, which take advantage of the warm microclimate close to the ground. Permafrost occurs throughout this climate region.

- The ice sheet climate is the coldest of all climates, with no monthly mean temperature above freezing.

- The global climate is changing. Temperatures are rising, and precipitation is increasing in mid- and high latitudes while decreasing in the subtropics. Extreme weather events are more frequent and more severe.

KEY TERMS

boreal forest climate
dry midlatitude climate
dry subtropical climate

ice sheet climate
marine west-coast climate
Mediterranean climate

moist continental climate
moist subtropical climate
permafrost

tundra climate

REVIEW QUESTIONS

1. What climate type is associated with the Mojave Desert? Describe some of the features of the desert environment.
2. Both the moist subtropical and moist continental climates are found on eastern sides of continents in the midlatitudes. What are the major factors that determine their temperature and precipitation cycles? How do these two climates differ?
3. What natural vegetation types are associated with the moist subtropical forest environment? Why are these areas sometimes uncomfortable in the summer?
4. Both the Mediterranean and marine west-coast climates are found on the west coasts of continents. Why do they experience more precipitation in winter than in summer? How do the two climates differ?
5. Identify the characteristic features of the natural vegetation of the Mediterranean climate environment. Describe human use of this environment for agriculture and the limitations the climate places on these uses.
6. Contrast the natural vegetation types and human use of the land in the marine west-coast environments of North America and Europe.

7. The dry midlatitude environment is one of great agricultural importance. Why is this? What natural hazards can occur and how do farmers attempt to minimize their impact?
8. The moist continental climate region is often subject to severe winter weather. Explain the nature, causes, and impacts of such events.
9. The subtropical high-pressure cells influence several climate types in the midlatitudes. Identify the climates and describe the effects of the subtropical high-pressure cells on them.
10. Both the boreal forest and tundra climate are climates of the northern regions; however, the tundra is found bordering the Arctic Ocean, and the boreal forest is located further inland. Compare these two climates taking into account coastal-continental effects.
11. The boreal forest environment includes vast areas of land that support conifer forests. Discuss their importance in terms of lumber and the natural hazards that might occur.
12. Distinguish between two types of permafrost.
13. What is the coldest climate on Earth? How is the annual temperature cycle of this climate related to the cycle of insolation?

FOCUS ON SYSTEMS 10.2 | California Rainfall Cycles and El Niño

1. What is the relationship between annual rainfall at Santa Barbara and the Southern Oscillation Index?

2. Is the relationship between annual rainfall and the Southern Oscillation Index reliable enough to predict high-rainfall years?

EYE ON THE ENVIRONMENT 10.3 | Drought on the Prairies

1. How does drought impact agriculture on the Canadian Prairies?

A CLOSER LOOK   EYE ON GLOBAL CHANGE 10.4   Regional Impacts of Climate Change on North America

1. Identify six major changes in the North American climate that are anticipated for the remainder of the twenty-first century and describe their general impact.
2. Summarize climate changes and effects on North America for the following climate pairs: Mediterranean and marine west-coast, boreal forest and tundra, dry subtropical and dry midlatitude, moist subtropical and moist continental.
3. How do climate change impacts on the water cycle vary from region to region in North America? Which region will be influenced most, in your opinion? Least?

VISUALIZATION EXERCISES

1. Sketch climographs for the Mediterranean climate and the dry midlatitude climate. What are the essential differences between them? Explain why they occur.
2. Imagine that South America is inverted so that the southern tip is now at about 10° N latitude and the northern end (Venezuela) is at about 55° S. The Andean mountain chain would still be on the west side. Sketch this continent and draw possible climate boundaries, using your knowledge of global air circulation patterns, frontal zones, and air mass movements.

ESSAY QUESTIONS

1. Discuss the role of the polar front and the air masses that come in conflict in the polar-front zone in the temperature and precipitation cycles of the midlatitude and high-latitude climates.
2. Compare and contrast the differing climate regions of Canada using Vancouver, Saskatoon, Toronto, and Yellowknife as examples.

PROBLEMS   WORKING IT OUT 10.1 Standard Deviation and Coefficient of Variation

1. The following data are monthly precipitation (in millimetres) for January and July at Gander, Newfoundland. For each month, find the sample standard deviation and the coefficient of variation. Which month has the largest standard deviation? Which has the highest coefficient of variation?

Year	January	July
1996	78	141
1997	118	52
1998	65	108
1999	109	79
2000	190	64
2001	53	97
2002	116	72
2003	75	95
2004	135	62
2005	150	118

Find answers at the back of the book

EYE ON THE LANDSCAPE |

Chapter Opener Alpine meadow in Mount Rainier National Park, Washington, USA. The red-coloured bracts of Indian paintbrush (**A**) and the purple lupines (**B**) add to the colour and diversity of perennial herbs in this alpine meadow. Strong radiation is received under clear sky conditions (**C**) at high elevations in temperate mountains and plants take advantage of this to complete growth and development before the onset of winter. Forest species, such as alpine fir, persist at tree line in small groups or islands (**D**) and in more exposed sites may develop a characteristic one-sided, flagged krummholtz form (**E**) (see chapter 23). Highly reflective ice persists throughout the year on glaciers (**F**) (see chapter 19). Small cap clouds (**G**) often develop on the high peaks as moist air rises over the mountains.

Snow melt in the mountains. A significant proportion of incoming solar energy is reflected by the ice and snow cover (**A**) which can persist until late-summer in some temperate mountains. Meltwater streams develop quickly with rising air temperature and help to remove the winter snowpack where it reached

the water's edge (**B**). Over the years the recurrent flow of water has smoothed the rock surfaces (**C**) in the channel through abrasive action of transported sediment. Small rapids occur where the stream bed is irregular (**D**) which helps to oxygenate the water. Talus slopes (**E**) common features in mountain climates where rock is loosened and dislodged by freeze-thaw action (see chapter 14), and the small glacial trough (**F**) attests to harsher conditions in previous (see chapter 19).

Part 3

CHAPTERS IN PART 3

11 Earth Materials and the Cycle of Rock Transformation

12 The Lithosphere and the Tectonic System

13 Volcanic and Tectonic Landforms

SYSTEMS AND CYCLES OF THE SOLID EARTH

Part 3 examines the systems and cycles of the solid Earth. Unlike surface energy and matter flows, such as heat and precipitation driven by solar radiation, the solid Earth systems are driven by the planet's internal energy. These two great systems of energy and matter flow also operate on very different time scales. The surface energy flow systems are influenced by the rhythms that arise from the daily rotation of the Earth on its axis and its annual revolution around the Sun. However, the cycles associated with the solid Earth systems are typically measured in thousands to millions of years. ■ Chapter 11 begins Part 3 by presenting the basic materials of the solid Earth—rocks and minerals—and some principles of their formation. The continuous production and conversion of one type of rock into another is described by the cycle of rock transformation, in which Earth materials are constantly created and recycled over geologic time. Forces deep within the planet drive part of the rock cycle. This same energy source causes the vast tectonic plates that form the Earth's crust to converge, collide, split, and separate. Because of this, the configuration of continents and ocean basins slowly changes. Chapter 12 discusses this slow, inexorable movement, called plate tectonics. The present pattern of plates and their motions explains the location of many geologic phenomena, such as earthquakes and volcanoes. These and other forms of geologic activity, such as folding and faulting, are the subject of Chapter 13. ■ Physical geographers are concerned with the solid Earth because the Earth's outer layer provides the continental surfaces that are carved into landforms by moving water, wind, and glacial ice. Landforms, in turn, influence the distribution of ecosystems and exert strong controls over human occupation of the lands. Landforms made by water, wind, gravity, and ice are the subject of Chapters 14 to 19.

Wave Rock, Western Australia. This granite outcrop has been undercut by weathering and erosion from flowing water. Its face is about 15 metres high and is stained by streaks of black algae.

Chapter 11

EYE ON THE LANDSCAPE

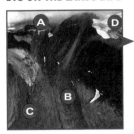

The Spectrum Range, Mt. Edziza
Provincial Park, British Columbia.
**What else would the geographer
see? . . . Answers at the end of the
chapter.**

EARTH MATERIALS AND THE CYCLE OF ROCK TRANSFORMATION

The Crust and its Composition
Rocks and Minerals
Igneous Rocks
Common Igneous Rocks
Intrusive and Extrusive Igneous Rocks
Chemical Alteration of Igneous Rocks
Sediments and Sedimentary Rocks
Clastic Sedimentary Rocks
Chemically Precipitated Sedimentary Rocks
Hydrocarbon Compounds in Sedimentary Rocks
Metamorphic Rocks
The Cycle of Rock Transformation
Focus on Systems 11.1 • Powering the Cycle of
   Rock Transformation

The study of the solid Earth begins by examining the minerals and rocks found at or near the Earth's surface. These Earth materials are constantly formed and reformed over geologic time in the rock cycle that ultimately depends on energy both from within the planet's interior and from the Sun.

THE CRUST AND ITS COMPOSITION

The outermost layer of the planet is the *Earth's crust*. It varies in thickness from about 8 to 80 kilometres and comprises the continents and ocean basins. The thickness of the crust is generally less than 10 kilometres under the oceans and averages about 40 kilometres in most continents (Figure 11.1). Approximately 10 percent of the crust exceeds 50 kilometres, with the greatest thickness reported in the Himalayas, where it is about 80 kilometres.

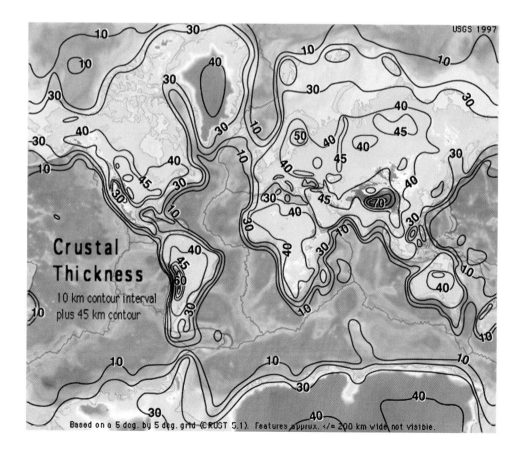

11.1 Thickness of Earth's crust The map uses a 10-kilometre contour interval (plus the 45-kilometre contour line) and was created directly from the 5 degree by 5 degree gridded crustal model CRUST 5.1.

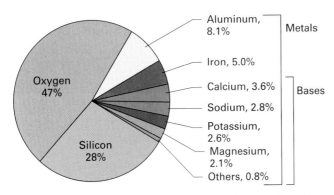

11.2 Crustal composition The eight most abundant elements in the Earth's crust, measured by percentage of weight.

11.3 Quartz crystals These large crystals of quartz form six-sided, translucent columns.

Figure 11.2 shows the eight most abundant elements of the Earth's crust in terms of percentage by weight. The predominant elements are oxygen (O) and silicon (Si), which together account for approximately 75 percent of the crust, by weight. Aluminum (Al) and iron (Fe), both of which are important to our industrial civilization, account for about 8 percent and 5 percent respectively. Four other metallic elements—calcium (Ca), sodium (Na), potassium (K), and magnesium (Mg), which occur in this order of abundance, from 3.6 to 2.1 percent—are essential nutrients for plant and animal life. Because of their role as nutrients, Ca, Na, K, and Mg are also important determiners of soil fertility (Chapter 21).

A few radioactive elements provide a nearly infinite source of energy that migrates slowly outward from the Earth's interior; the most notable are uranium (U), thorium (Th), and potassium (K). Chapter 12 discusses radiogenic heating and radioactive decay further. In addition to radioactive energy, a second source of internal energy is the residual heat generated by compression of cosmic debris when the Earth was created some 4.6 billion years ago. Both of these sources are important for powering the cycles of the solid Earth.

ROCKS AND MINERALS

The elements of the Earth's crust are usually bonded with other elements to form minerals. A **mineral** is a naturally occurring, inorganic substance that possesses a definite chemical composition and characteristic atomic structure. Most minerals have a crystalline structure; for example, quartz usually occurs as a clear, six-sided prism (Figure 11.3).

Minerals are combined into **rock**, which is essentially an assemblage of minerals in the solid state. Rock comes in many different compositions, physical characteristics, and ages. Most rock in the Earth's crust was formed many millions of years ago. But rock is also being formed currently as active volcanoes emit lava that solidifies on contact with the atmosphere or ocean. Rocks in

the Earth's crust fall into three major classes: igneous, sedimentary, and metamorphic.

Igneous rocks are solidified from mineral matter in a high-temperature molten state. The molten rock, or **magma**, cools and solidifies, either below the surface as intrusive rocks, or on the surface as extrusive rocks. Most of the more than 700 types of igneous rocks formed at some depth beneath the surface. Figure 11.4 shows the distributions of intrusive and extrusive igneous rocks in North America.

Sedimentary rocks are formed from layered accumulations of mineral particles, derived mostly by weathering and erosion of pre-existing rock. The various chemical, physical, and biological changes that occur as layers of sediment transform into rocks are collectively referred to as diagenesis. The resulting rock layers are called **strata** (Figure 11.5). Sedimentary rocks cover about 75 percent of the Earth's surface, but account for only about 5 percent of the total mass of the Earth's rocks. Thus, they form a comparatively thin layer over igneous and metamorphic rocks. Unlike these other rock types, sedimentary rocks often contain fossils, which makes them invaluable for interpreting Earth history. In fact certain classes of sedimentary rocks, such as chalk, comprise almost entirely the remains of organisms. Other sedimentary

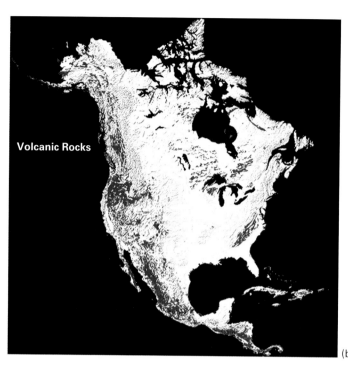

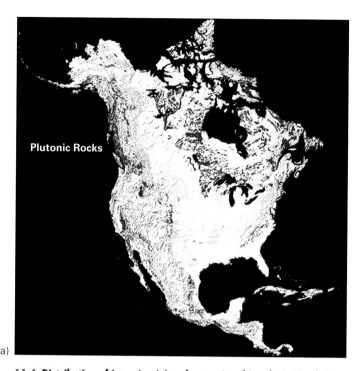

11.4 Distribution of intrusive (a) and extrusive (b) rocks in North America Although intrusive rocks are formed at a depth beneath the Earth's crust, they can be exposed at the surface as overlying rock layers that are eroded away over millions of years.

rocks, such as potash, originate as chemical precipitates through evaporation of mineral-rich waters. Figure 11.6 shows the distribution of sedimentary rocks in North America.

Metamorphic rocks are formed from igneous or sedimentary rocks that have been physically or chemically changed,

11.5 Sedimentary rocks These limestone formations in New Zealand are known locally as the Punakaiki or pancake rocks.

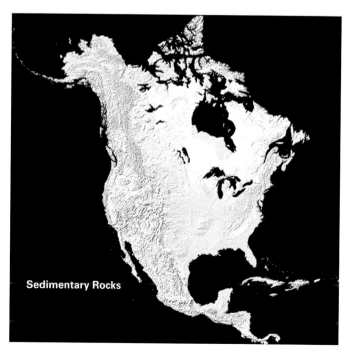

11.6 Distribution of sedimentary rocks in North America Sedimentary rocks are the most common of the rock types exposed at the Earth's surface. In North America, they are less common on the Canadian Shield and the western mountain ranges.

11.7 Distribution of metamorphic rocks in North America Metamorphic rocks are exposed throughout the Canadian Shield.

The three classes of rocks constantly transform from one to another in a continuous process through which the crustal minerals are recycled over the course of many millions of years. This is the **cycle of rock transformation,** or simply the **rock cycle.** In the process of magma formation, pre-existing rock of any class melts and later cools and solidifies to form igneous rock. In weathering and erosion, pre-existing rock breaks down and the fragments accumulate in layers that eventually become compacted, forming sedimentary rock. Heat and pressure, without melting, convert igneous and sedimentary rocks to metamorphic rock.

IGNEOUS ROCKS

The molten rock, or magma, cools and solidifies, with or without crystallization, either below the surface as intrusive (plutonic) rocks or on the surface as extrusive (volcanic) rocks. This magma is the result of partial melts of pre-existing rocks in either the Earth's mantle or crust. Typically, the melting is caused by one or more of the following processes: an increase in temperature, a decrease in pressure, or a change in composition.

Igneous rocks solidify from magma, a hot, molten liquid, formed from melting of Earth materials deep below the surface. The molten rock migrates upward from these magma chambers through fractures in older solid rock and eventually solidifies as igneous rock within or on top of the crust. Table 11.1 lists several common igneous rock types and their mineral crystal composition.

Most igneous rock consists of *silicate minerals*, chemical compounds that contain silicon and oxygen atoms combined with various proportions of other elements, particularly aluminum, iron, calcium, sodium, potassium, and magnesium. Figure 11.8 shows seven of the most common silicate minerals.

usually by the subsequent application of heat and pressure during episodes of significant crustal movement. The process of metamorphosis creates new minerals and rock structures. Certain minerals—for example, garnet—are considered metamorphic rocks, and can provide information about the temperatures and pressures under which specific metamorphic rocks form. The age of metamorphic rocks is often difficult to determine because the formation process disrupts the radiometric decay sequences. In North America, exposed metamorphic rocks occur most extensively in the Canadian Shield (Figure 11.7).

Table 11.1 Some common igneous rock types

Subclass	Rock Type	Composition
Intrusive (Cooling at depth, producing coarse crystal texture)	Granite	Felsic minerals, typically quartz, feldspars, and mica
	Diorite	Felsic minerals without quartz, usually including plagioclase feldspar and amphibole
	Gabbro	Mafic minerals, typically plagioclase feldspar, pyroxene, and olivine
	Peridotite	An ultramafic rock of pyroxene and olivine
Extrusive (Cooling at the surface, producing fine crystal texture)	Rhyolite	Chemically the same as granite
	Andesite	Chemically the same as diorite
	Basalt	Chemically the same as gabbro

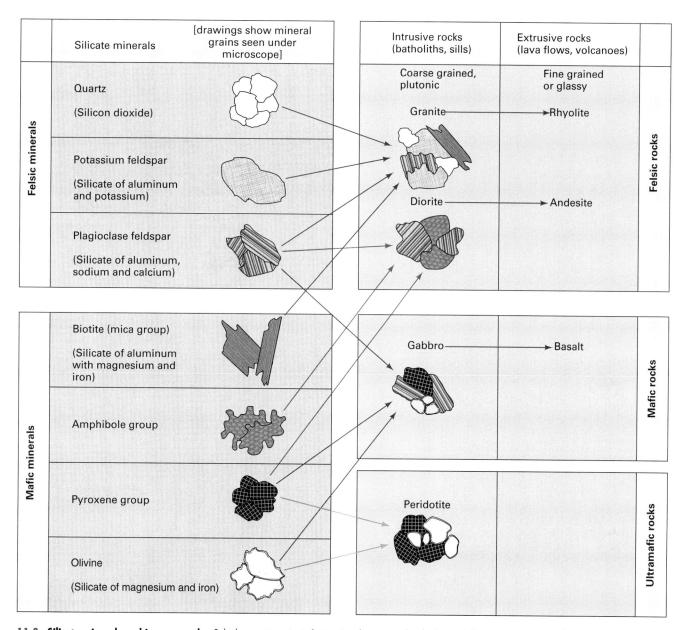

11.8 Silicate minerals and igneous rocks Only the most important silicate mineral groups are listed, along with four common igneous rock types. The patterns shown for mineral grains indicate their general appearance through a geological microscope, and do not necessarily indicate the relative amounts of the minerals within the rocks.

Quartz, which is silicon dioxide (SiO_2), is one of the most common minerals of all rock classes. It is quite hard and resists chemical breakdown. *Feldspars*, which are silicate-aluminum minerals, are also found in many rocks. One type, *potassium feldspar* ($KAlSi_3O_8$), contains potassium, whereas *plagioclase feldspar* ($NaAlSi_3O_8$) is rich in sodium. In anorthite ($CaAl_2Si_2O_8$), calcium is abundant. Quartz and feldspar form the **felsic** ("fel" for feldspar; "si" for silicate) mineral group. These minerals are light in colour (white, pink, or greyish) and lower in density than the other silicate minerals.

Other common silicate minerals include the *biotite, amphibole*, and *pyroxene* groups. All three contain aluminum, magne-

sium, iron, and potassium or calcium. These minerals are described as **mafic** ("ma" for magnesium; "f" from the chemical symbol Fe for iron). *Olivine*, comprising magnesium and iron silicates, is also included in this group, but differs from the others in that it lacks aluminum. All of these mafic minerals are dark in colour (usually black) and are denser than the felsic minerals.

COMMON IGNEOUS ROCKS

No single igneous rock is made up of all seven silicate minerals listed in Figure 11.8. Instead, a given rock variety contains three or four of those minerals as its major constituents. The mineral grains in igneous rocks are tightly interlocked, and the rock is normally very strong. Some of the most common igneous rocks are *granite, diorite, gabbro,* and *peridotite.*

Typical **granite** consists mainly of potassium feldspar, with lesser amounts of quartz and plagioclase feldspar, and some mica and amphibole. The actual composition of granite can be quite varied, with quartz, for example, ranging from between 20 and 60 percent, and feldspar between 10 and 65 percent, most of which is plagioclase. Because most of the volume of granite is of felsic minerals, granite is classified as a *felsic igneous rock.* Granite is a mixture of white, greyish or pinkish, and black grains, but the overall appearance is a light grey or pink colour (Figure 11.9a). *Diorite* lacks quartz and is mostly composed of plagioclase feldspar (60 percent) and secondary amounts of amphibole and pyroxene. Diorite is a light-coloured felsic rock that is only slightly denser than granite (Figure 11.9b).

Gabbro, in which the major mineral is pyroxene (60 percent), also contains a substantial amount of plagioclase feldspar (20 to 40 percent), and there may also be some olivine (0 to 20 percent). As the mafic minerals, pyroxene and olivine are dominant. Gabbro is classed as a *mafic igneous rock.* Gabbro is dark in colour and denser than the felsic rocks. *Peridotite* is dominated by olivine (60 percent). The rest is mostly pyroxene (40 percent). Peridotite is classed as an *ultramafic igneous rock,* even more dense than the mafic types.

These four common varieties of igneous rock show an increasing range of density, from felsic, through mafic, to ultramafic types. These density differences are reflected in the principal rock layers that comprise the solid Earth, with the least dense layer (mostly felsic and intermediate rocks) near the surface and the densest layer (ultramafic rocks) deeper in the Earth's mantle.

INTRUSIVE AND EXTRUSIVE IGNEOUS ROCKS

Magma that solidifies below the Earth's surface and remains surrounded by older, pre-existing rock is called **intrusive igneous rock**. The process of injection into existing rock is *intrusion.* Where magma reaches the surface, it emerges as **lava**, which solidifies to form **extrusive igneous rock**. The process of release at the surface is called *extrusion.*

Although both intrusive and extrusive rock can solidify from the same original body of magma, their outward appearances are quite different. Intrusive igneous rocks cool very slowly—over hundreds or thousands of years—and, as a result, develop large, readily seen, mineral crystals. Granite and diorite are good examples of *coarse-textured* intrusive igneous rocks.

Extrusive rock, which cools very rapidly, is *fine-textured,* and the individual crystals can only be seen through a microscope. Most lava solidifies as a dense, uniform rock with a dark, dull

(a)

(b)

11.9 Intrusive igneous rock samples (a) A coarse-grained granite. Dark grains are amphibole and biotite; light grains are feldspars; clear grains are quartz. (b) A coarse-grained diorite. Feldspar grains are light; amphibole and pyroxene grains are dark. Compare with granite in (a) — there are no clear quartz grains.

surface (Figure 11.10). Sometimes lava cools to form shiny *obsidian* or volcanic glass (Figure 11.11a). If the lava contains dissolved gases, it may solidify to form *scoria*, a rock with a frothy, bubble-filled texture (Figure 11.11b). The formation of extrusive igneous rock can be witnessed today in Hawaii and other regions where volcanic processes are active.

Intrusive and extrusive rocks of the same composition do not look alike, and so have different names. The distinctive chemical compositions of the intrusive rocks granite, diorite, and gabbro are also found in extrusive rocks (see Figure 11.8). Thus, **rhyolite** is the name for lava of the same composition as granite; **andesite** is lava with the composition of diorite; and **basalt** is lava with the composition of gabbro. Note that peridotite has no extrusive counterpart. Rhyolite and andesite are pale greyish or pink in colour, whereas basalt is black. Andesite and basalt are the two most common types of lava. Lava flows, along with particles of solidified lava expelled explosively during an eruption, typically accumulate as volcanoes (see Chapter 13).

A body of intrusive igneous rock is called a **pluton**. Granite typically accumulates in enormous plutons, called *batholiths* (Figure 11.12). As the hot fluid magma rises, it may melt and incorporate some of the older rock lying above it. A single batholith may extend down several kilometres and be many thousands of square kilometres in area.

Figure 11.12 shows other common forms of intrusions. One is a *sill*, a flat, layer-like body formed when magma forces its way between two pre-existing rock layers. A second kind of intrusion is the *dike*, a vertical, wall-like body formed when a rock fracture is forced open by magma. Dike rock tends to be fine-textured because it cools relatively close to the Earth's surface. In fact, these vertical fractures often conduct magma to the

(a)

(b)

11.11 Extrusive igneous rock samples (a) Obsidian, or volcanic glass. The smooth, glassy appearance is acquired when a gas-free lava cools very rapidly. This sample shows reddish-brown and black streaks caused by minor variations in composition. (b) A specimen of scoria, a form of lava containing many small holes and cavities produced by gas bubbles.

11.10 Fresh lava On Chain of Craters Road in Hawaii Volcanoes National Park, flow structures are clearly visible in this fresh black lava.

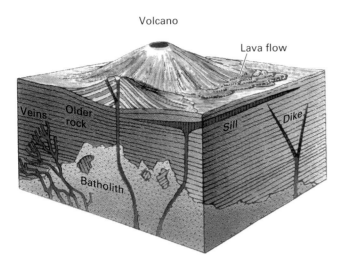

11.12 Volcanic rock formations This block diagram illustrates various forms of intrusive igneous rock plutons, as well as an extrusive lava flow.

surface (Figure 11.13). Magma entering small, irregular, branching fractures in the surrounding rock solidifies in a branching network of thin *veins*.

CHEMICAL ALTERATION OF IGNEOUS ROCKS

The minerals in igneous rocks are formed in magma chambers beneath the Earth's surface and begin to crystallize as the molten rock cools and solidifies. Extrusive igneous rocks are formed at the surface, but even deep-seated intrusive rocks can be exposed when the overlying crustal rocks erode away. The temperature and pressure at the surface is very different from the environment in which the rocks were formed. Also, they may be exposed to water and dissolved gases.

11.13 Exposure of a dike This dike radiates from the base of Shiprock, New Mexico (2,188 metres). This is one of a series of dikes that radiate from the base of this ancient volcano. *Fly By: 36° 41' N; 108° 50' W*

In this new environment, the minerals within an igneous rock may no longer be stable, and slow chemical changes begin to weaken their structure. These chemical processes cause **mineral alteration**, which is one form of *weathering*. (See Chapter 14 for related topics.) Weathering also includes physical forces, like frost action, that break up rock into small fragments and separate the component minerals, grain by grain. Fragmentation accelerates mineral alteration because it greatly increases the surface area that is exposed to chemically active solutions.

There are three main processes involved in chemical weathering: oxidation, hydrolysis, and dissolution. Many silicate minerals exposed at the surface undergo *oxidation* when they react with dissolved oxygen, particularly if they are enriched in iron. With oxidation, the silicate minerals are converted to *oxides*, in which silicon and the metallic elements, such as iron, bond completely with oxygen. Oxides are very stable. Quartz, with the composition silicon dioxide (SiO_2), is a common mineral oxide. It is long-lasting and is found abundantly in many types of rocks and sediments.

Although silicate minerals do not generally dissolve in water, some undergo chemical alteration through *hydrolysis*. This process is not merely a soaking or wetting of the mineral, but a true chemical change that produces a different mineral compound, in which water is added to the mineral structure itself. The products of hydrolysis are stable and long-lasting, as are the products of oxidation. **Clay minerals**, such as kaolinite, are a typical product of hydrolysis. A clay mineral has plastic properties when moist, because it consists of small, thin flakes that become lubricated by layers of water molecules. Clay minerals formed by mineral alteration are abundant in common types of sedimentary rocks.

When carbon dioxide dissolves in water, a weak acid—*carbonic acid*—is formed. Carbonic acid can dissolve certain igneous minerals, particularly potassium feldspar and biotite, by the process of dissolution. Generally dissolution occurs as a preliminary stage in the hydrolysis of igneous rocks. Dissolution is most effective when rocks are rich in calcium carbonate, and is the characteristic weathering process in sedimentary limestone.

SEDIMENTS AND SEDIMENTARY ROCKS

Sedimentary rocks are formed from mineral particles derived from pre-existing rock of any of the three rock classes, as well as from newly formed organic matter, and through chemical precipitation. For example, granite can weather to yield grains of quartz and particles of clay minerals derived from feldspars, thus contributing sand and clay to a sedimentary rock. Sedimentary rocks include rock types with a wide range of physical and chemical properties; Table 11.2 shows some of the most common kinds.

In the process of mineral alteration, solid rock is weakened, softened, and fragmented, yielding particles of many sizes and mineral compositions. When transported by a fluid medium—air, water, or glacial ice—these particles are known collectively as **sediment**. Dissolved mineral matter in solution also must be considered.

Streams and rivers carry sediment to lower land levels, where it can accumulate. The most favourable sites of sediment accumulation are shallow sea floors bordering continents. But sediments also accumulate in inland valleys, lakes, and marshes. Wind and glacial ice also transport sediment but not necessarily to lower elevations or to places suitable for accumulation. Over long time spans, the sediments can undergo physical or chemical changes, becoming compacted and hardened to form sedimentary rock.

There are three major classes of sediment. First is **clastic sediment**, which consists of inorganic rock and mineral fragments called *clasts*. Examples include the sand on a river bed or on an ocean beach. Second is **chemically precipitated sediment**, which consists of inorganic mineral compounds precipitated from a chemical solution. In the process of chemical precipitation, ions in solution combine to form solid mineral matter. A layer of rock salt, such as that found in dry lake beds in arid regions, is an example. Coral reefs can also be included in this class, as coral organisms precipitate calcareous exoskeletons that contribute directly to the growth of the reef. The third class is **organic sediment**, which consists of the tissues of plants and animals, accumulated and preserved after death. This might include a layer of peat in a bog, or the hard parts of organisms, such as sea shells.

Transported sediment is eventually deposited and gradually accumulates in approximately horizontal layers, called **strata**, or simply "beds." Individual strata are separated from those below and above by surfaces called stratification planes or bedding planes. Strata of widely different compositions can occur alternately, one above the other.

CLASTIC SEDIMENTARY ROCKS

Clastic sediments are derived from fragments of any of the rock groups and thus may include a wide range of minerals. Silicate minerals are the most important, both in original form and as altered by oxidation and hydrolysis. Quartz, feldspar, and clays usually dominate. Because quartz is abundant and resistant to alteration, it is the most important single component of the clastic sediments. Second in abundance are fragments of unaltered fine-grained parent rocks. Feldspar and mica are also commonly present. In addition, clay minerals are major constituents of the finest clastic sediments.

The range of particle sizes in a clastic sediment determines how easily and how far the particles are transported. Finer particles are more easily held suspended in the fluid, whereas coarser

Table 11.2 Some common sedimentary rock types

Subclass	Rock Type	Composition
Clastic (Composed of rock and/or mineral fragments)	Sandstone	Cemented sand grains
	Siltstone	Cemented silt particles
	Conglomerate	Sandstone containing pebbles of hard rock
	Mudstone	Silt and clay, with some sand
	Claystone	Clay
	Shale	Clay, breaking easily into flat flakes and plates
Chemically precipitated (Formed by chemical precipitation from sea water or salty inland lakes)	Limestone	Calcium carbonate, formed by precipitation on sea or lake floors
	Dolomite	Magnesium and calcium carbonates, similar to limestone
	Chert	Silica, a microcrystalline form of quartz
	Evaporites	Minerals formed by evaporation of salty solutions in shallow inland lakes or coastal lagoons
Organic (Formed from organic material)	Coal	Rock formed from peat or other organic deposits; may be burned as a mineral fuel
	Petroleum (mineral fuel)	Liquid hydrocarbon found in sedimentary deposits; not a true rock but a mineral fuel
	Natural gas (mineral fuel)	Gaseous hydrocarbon found in sedimentary deposits; not a true rock but a mineral fuel

particles tend to settle out. In this way, a separation of sizes, called *sorting*, occurs. Sorting determines the texture of the deposited sediment and of the sedimentary rock derived from it. The finest clay particles do not settle out unless they clump together. This process, called *flocculation*, normally occurs when river water carrying clay mixes with the saltwater of the ocean.

When sediments accumulate in thick sequences, the lower strata are subjected to the pressure produced by the weight of the overlying sediments. This compacts the sediments, squeezing out excess water. Cementation occurs as dissolved minerals re-crystallize where grains touch and in the spaces between mineral particles. Silicon dioxide (quartz, SiO_2) is slightly soluble in water, and so the cement is often a form of quartz, called *silica*, which lacks a true crystalline form. Calcium carbonate ($CaCO_3$) is another common cementing agent. Compaction and cementation harden the sediments into sedimentary rock.

The important varieties of clastic sedimentary rock are distinguished by the size of their particles. They include sandstone, conglomerate, mudstone, claystone, and shale. **Sandstone** is formed from fine to coarse sand and mostly consists of quartz (Figure 11.14). Very coarse-grained sedimentary rock containing numerous rounded pebbles of hard rock is called *conglomerate* (Figure 11.15).

Muddy sediments containing a high percentage of fine silt and clay together with some sand can harden to form *mudstone*. Compacted and hardened clay layers become *claystone*. Sedimentary rocks of mud composition are commonly layered in such a way that they easily break apart into small flakes and plates. The rock is then described as being *fissile* and is given the name **shale**. Shale, the most abundant of all sedimentary rocks, is composed largely of clay minerals. The compaction of the mud to form mudstone and shale results in a considerable loss of volume as water is driven out of the clay. One distinctive form of shale is oil shale. These are fine-grained sedimentary rocks containing relatively large amounts of organic matter from which significant amounts of hydrocarbons can be extracted by destructive distillation. Oil shale was deposited in a wide variety of environments including freshwater lakes, coastal swamps, and shallow marine areas. Most of the organic matter was derived from algae, although remains of land plants are sometimes present.

In 1909, a remarkable fossil bed was discovered in the Burgess Shale in Yoho National Park in the Canadian Rockies. The Burgess Shale fossils are exceptionally well-preserved and represent a unique diversity of Cambrian animal fossils dating from 545 to 525 million years ago (Figure 11.16). The sediment was deposited in a deep-water basin adjacent to an enormous algal reef that was hundreds of metres high. It is thought that many of the

11.14 Sandstone strata This photo of an eroded sandstone formation on the Colorado Plateau shows individual layers, or strata, that make up the rock. The layers were originally deposited within a filed of moving sand dunes.

fossilized organisms were living on mud banks close to the reef, but frequent slumping of the unstable sediment carried them into deeper waters, where they were subsequently preserved. Because of its importance, UNESCO. declared the Burgess Shale a world heritage site in 1981.

CHEMICALLY PRECIPITATED SEDIMENTARY ROCKS

Under favourable conditions, mineral compounds are deposited from the saline waters of the oceans or from lake water evaporation in dry climates. For example, surface deposits of sodium sulphate (Na_2SO_4 $10H_2O$) occurs in natural brines and crystal deposits in alkaline lakes in southern Saskatchewan (Figure 11.17). Sedimentary minerals and rocks deposited from concentrated solutions are called **evaporites**. The potash deposits mined in Saskatchewan are contained in the Prairie Evaporite Formation. Potash refers to potassium compounds and potassium-bearing materials, the most common being sylvite (potassium chloride). These deposits originated in the Devonian Period about 400 million years ago, and now lie at depths of more than 1,000 metres. Gypsum is another example of an evaporite, and gypsum

11.15 Conglomerate section This type of sedimentary rock is comprised of pebbles that have been rounded through transport in flowing water and subsequently cemented together finer sediment.

11.16 Burgess Shale trilobites Marine trilobites are one of many well-preserved groups of marine invertebrates found in the Burgess Shale in British Columbia. Other fossils include early crabs, sponges, and marine worms.

11.17 Chaplin Lake, Saskatchewan Evaporation from this shallow saline lake has left deposits of white sodium sulphate and other minerals. *Fly By: 50° 26' N; 106° 39' W*

11.18 Limestone The Canadian Rockies are mostly formed of limestones that were originally deposited in inland seas.

deposits in Nova Scotia also formed when a large water body evaporated as a result of warming and drying of the climate about 330 million years ago.

One of the most common sedimentary rocks formed by chemical precipitation is **limestone**, composed largely of the mineral *calcite*, which is calcium carbonate ($CaCO_3$). Marine limestone accumulated in thick layers in many ancient seas in past geologic eras (Figure 11.18). Limestone is often rich in fossils, but the fossil content is a minor part of most deposits. Most of the calcium begins as microscopic needle-like crystals of aragonite (a variety of calcite) produced by cyanobacteria (or blue-green algae), which accumulate on the seabed when organisms die and decompose. Over time, aragonite re-crystallizes to calcite and so is usually present only in younger limestones. A closely related rock, also formed by chemical precipitation, is *dolomite*, composed of calcium magnesium carbonate. Limestone and dolomite are grouped together as the *carbonate rocks*. They are relatively dense rocks with white, pale grey, or even black colour.

Sea water also yields sedimentary layers of silica in a hard, non-crystalline form called *chert*. Chert is a variety of sedimentary rock, but it also commonly occurs combined with limestone (cherty limestone). Chalk, a soft, white, porous form of limestone, is also a marine deposit and consists almost entirely of microscopic, single-celled algae called coccolithophores. These marine organisms are covered with ornate calcium carbonate plates and their remains are referred to as coccoliths. Massive chalk deposits, hundreds of metres thick, which accumulated in the Cretaceous Period (144–66 million years ago), are almost entirely composed of these calcareous coccoliths (Figure 11.19).

11.19 Chalk cliffs The Etretât cliffs of Normandy, France are composed of soft limestone, known as chalk.

HYDROCARBON COMPOUNDS IN SEDIMENTARY ROCKS

Hydrocarbon compounds (compounds of carbon and hydrogen) form a most important type of organic sediment. Although hydrocarbons occur both as solids (peat and coal) and as liquids and gases (petroleum and natural gas), only coal qualifies physically as a rock. At various times in the geologic past, plant remains accumulated on a large scale, accompanied by sinking of the area and burial of the compacted organic matter under thick layers of inorganic clastic sediments. *Coal* is the end result of this process (Figure 11.20). Individual coal seams are inter-bedded with shale, sandstone, and limestone strata.

Oil deposits and *natural gas* also are of organic origin, but they are classed as mineral fuels rather than as minerals. Oil and gas deposits commonly occupy open interconnected pores in a thick sedimentary rock layer; for example, a porous sandstone. The simplest arrangement of strata favourable to trapping these hydrocarbons is an upward arch (or anticline) of the type shown in Figure 11.21. Shale forms an impervious cap rock. A porous sandstone beneath the cap rock serves as a reservoir. Natural gas occupies the highest position, with the oil below it, and then water, in increasing order of density.

Yet another form of hydrocarbon fuels is *bitumen*, which is also called tar, asphalt, or pitch. In some locations, bitumen oc-

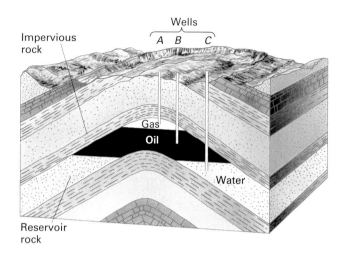

11.21 Trapping of oil and gas Idealized cross-section of an oil pool on a dome structure in sedimentary strata. Well *A* will draw gas, well *B* will draw oil, and well *C* will draw water. The cap rock is shale; the reservoir rock is sandstone. (Copyright © A.N. Strahler)

cupies pore spaces in layers of sand or porous sandstone. It remains immobile in the enclosing sand and will flow only when heated. Rock outcrops that contain bitumen often have black streaks where it has seeped out. Perhaps the best known of the great bituminous sand deposits are those at Fort McMurray, Alberta, where it is extracted by surface mining.

These **fossil fuels** have taken millions of years to accumulate. However, our industrial society is consuming them at a very rapid rate. These fuels are non-renewable resources. Once they are gone, there will be no more because the quantity produced by present geologic processes is scarcely measurable in comparison to the quantity stored through geologic time.

METAMORPHIC ROCKS

Any type of igneous or sedimentary rock may be altered by the tremendous pressures and high temperatures that accompany crustal deformation processes, such as mountain creation. The result is a rock so changed in texture and structure as to be reclassified as metamorphic rock. Mineral components of the parent rock are, in many cases, reconstituted into different mineral varieties. Re-crystallization of the original minerals can also occur. Five common metamorphic rocks are slate, schist, quartzite, marble, and gneiss (Table 11.3).

Slate is formed from shale that is heated and compressed by mountain-making forces. This fine-textured rock splits neatly into thin plates, which are commonly used in roofing. With increased heat and pressure, slate changes into **schist**. Schist exhibits foliation, which consists of thin, rough, irregularly curved planes in the rock (Figure 11.22). These are evidence of *shearing*—a stress that pushes the layers sideways, like a deck of

11.20 Coal deposits The Fording River Operation is located in the Elk Valley region of southeastern British Columbia, 29 kilometres northeast of the town of Elkford. The mine site comprises 20,304 hectares of coal lands, of which 3,592 hectares are currently being mined or are scheduled for mining.

Table 11.3 Some common metamorphic rock types

Rock Type	Description
Slate	Shale exposed to heat and pressure that splits into hard flat plates
Schist	Shale exposed to more intense heat and pressure that shows evidence of shearing
Quartzite	Sandstone that is "welded" by a silica cement into a very hard rock of solid quartz
Marble	Limestone exposed to heat and pressure, resulting in larger, more uniform crystals
Gneiss	Rock produced when clastic sedimentary rocks or intrusive igneous rocks are subjected to intense heat and pressure deep within the Earth

11.23 Banded gneiss This surf-washed rock surface at Pemaquid Point, Maine, exposes banded gneiss.

cards. Schist is set apart from slate by the coarse texture of the mineral grains, the abundance of mica, and occasionally the presence of scattered large crystals of newly formed minerals, such as garnet.

The metamorphic equivalent of conglomerate and sandstone is **quartzite**, which is formed by the fusion of quartz-rich sediments through the heat and pressure associated with tectonic activity. Limestone, after undergoing metamorphosis, becomes *marble*. During formation, calcite in the limestone is reconstituted into larger, more uniform crystals. Bedding planes are obscured, and masses of mineral impurities may be drawn out into swirling bands.

Finally, the metamorphic rock **gneiss** may form either from intrusive igneous rocks or from clastic sedimentary rocks that have been in close contact with intrusive magmas. A single description will not fit all gneisses because they vary considerably in appearance, mineral composition, and structure. One variety of gneiss is strongly banded into light and dark layers or lenses (Figure 11.23), which may have been bent into wavy folds. The segregation of light and dark minerals is indicative of intense metamorphosis, where temperatures as high as 600–700 °C have resulted in mineral migration. These bands have differing mineral compositions.

In 1999, the Japanese-built ASTER instrument (Advanced Spaceborne Thermal Emission and Reflection radiometer) was launched aboard NASA's Terra satellite platform. ASTER acquires multi-spectral images in three different waveband regions: visible and near-infrared, short-wave infrared, and thermal infrared. Minerals and rocks often have distinctive spectral signatures in near- and short-wave infrared domains, and images at these wavelengths are useful for geologic mapping.

A spectral region of particular interest is the thermal infrared, especially the wavelengths from 8 to 12 micrometres. Here the energy sensed is the radiant energy emitted by substances as a function of their temperature. However, some types of rocks and minerals emit more energy of a given wavelength at a given temperature than others. In this way, they, too, can be assigned spectral signatures in the thermal infrared region.

Figure 11.24 shows reflectance plotted against wavelength for two common minerals: quartz and calcite. Also shown are the spectral locations of the ASTER wavebands. In the left graph, both quartz and calcite are quite bright in ASTER's two visible bands (bands 1 and 2, green and red). They appear white to the eye, with calcite, which is slightly brighter in the red band, appearing to be a "warmer" white. In the short-wave infrared, however, calcite is significantly darker in the ASTER band 8 than in surrounding bands 7 and 9. The right graph shows spectral reflectance in the thermal infrared spectral region. Calcite is brightest in the ASTER band 13, while the darkest in the adjacent band 14. Quartz has a distinctive signature as well; it is about 40 percent brighter in bands 10–12 than in

11.22 Schist sample A specimen of foliated schist.

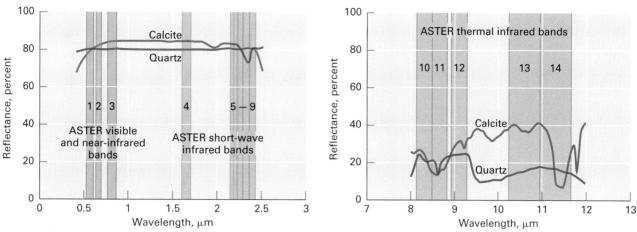

11.24 Reflectance spectra for quartz and calcite Wavebands in which ASTER acquires images are identified as numbered bands. Left, the visible and near-infrared spectral region; right, the thermal infrared spectral region.

13–14. Thus, both minerals also have unique spectral signatures in this spectral region.

Colour images in the three waveband regions were presented earlier in the third figure in *Geographer's Tools 3.5 Remote Sensing for Physical Geography*. In the infrared image (a), the different rock types appear in tones ranging from brown, grey, and yellow to blue. Image (b) displays the ASTER short-wave infrared bands 8, 6, and 4 as blue, green, and red. These bands distinguish carbonate, sulphate, and clay minerals. Limestone is yellow-green, while the clay mineral kaolinite appears in purplish tones. In (c), the ASTER bands 10, 12, and 13 are shown in blue, green, and red, respectively. Here, rocks rich in quartz appear in red tones, while carbonate rocks are green. Mafic volcanic rocks are visible in purple tones.

By comparing the three images, it is clear how the spectral information acquired by ASTER has the ability to make geologic mapping faster and easier. By using ASTER data, geologists can not only come to understand better a region's geologic history, but also identify the geologic settings in which valuable mineral ores may be found.

THE CYCLE OF ROCK TRANSFORMATION

The processes that form rocks, when taken together, constitute a single system that creates and recycles the Earth materials from one form to another over geologic time. The cycle of rock transformation—the rock cycle—describes this system (Figure 11.25). There are two environments: a surface environment of low pressures and

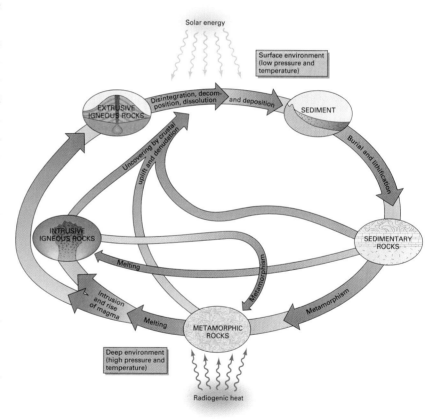

11.25 Cycle of rock transformation The three classes of rock are transformed into one another by weathering and erosion, melting, and exposure to heat and pressure, through a continuous process of rock formation and destruction.

temperatures, and a deep environment of high pressures and temperatures. The surface environment is the site of rock weathering and erosion, sediment transport, and sediment deposition. In this environment, igneous, sedimentary, and metamorphic rocks are exposed to air and water. Their minerals are altered chemically and

FOCUS ON SYSTEMS | 11.1   Powering the Cycle of Rock Transformation

In the cycle of rock transformation, the materials of the lithosphere are constantly being formed and transformed in both their physical and mineral composition. Deep within the crust, rocks and sediments are compressed, consolidated, baked, sheared, and sometimes melted. They eventually emerge at the surface either by extrusion, in the case of lava, by uplift, and by the stripping off of the overlying rock layers by erosion. Once at the surface, the rocks are physically and chemically weathered into sediment. The sediment then moves to low places where it accumulates and can be buried, thus completing the cycle.

What energy sources power the rock cycle? For the underground part of the cycle, the main power source is *radiogenic heat* that is slowly released by the radioactive decay of unstable isotopes. (See *Working It Out 12.1 • Radioactive Decay* for more details.) Isotopes of uranium (^{238}U, ^{235}U), thorium (^{232}Th), and potassium (^{40}K), along with the daughter products generated by their decay, are the source of nearly all of this heating.

Graph (a) to the right plots temperature against depth below the Earth's surface and shows that temperature increases rapidly at first then more slowly with depth. Most of the radiogenic heat is liberated within the uppermost 100 kilometres or so. This observation fits with chemical analyses, which show that isotopes of uranium, thorium, and potassium are most abundant in upper layers of continents. There is enough radiogenic heat to keep the Earth layers below the crust close to the melting point, and thus provide the power source for the heating required to form metamorphic and igneous rocks from pre-existing rocks.

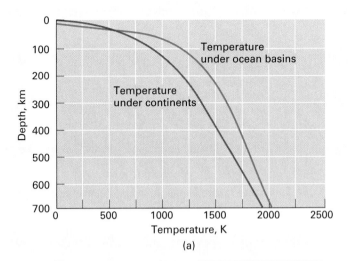

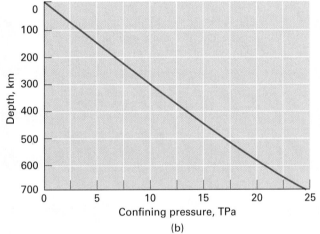

Change in temperature (a) and pressure (b) with depth from the Earth's surface (Copyright © A.N. Strahler)

broken free from the parent rock, yielding sediment. The sediment accumulates in basins, where deeply buried sediment layers are compressed and cemented into sedimentary rock.

Sedimentary rock forced down into the deep environment is heated by the slow radioactive decay of elements and subjected to high confining pressure. Here, it is transformed into metamorphic rock. Pockets of magma form in the deep environment and move upward, melting and incorporating surrounding rock as they rise.

Upon reaching a higher level, magma cools and solidifies, becoming intrusive igneous rock. Or it may emerge at the surface to form extrusive igneous rock. Either way, the cycle continues.

The rock cycle has been active since the Earth became solid and internally stable, continuously forming and reforming rocks of all three major classes. Not even the oldest igneous and metamorphic rocks found thus far are the "original" rocks of the Earth's crust, for they were recycled eons ago.

A secondary power source for the rock cycle is gravity. As sedimentary layers are buried more and more deeply, the pressure on the layers increases. Graph (b) plots pressure according to depth below the Earth's surface. This pressure brings mineral grains into very close contact. Water, often containing dissolved silica or calcium carbonate, is forced out of the sediments while leaving deposits of these minerals to cement the grains in place. Thus, the rock density increases.

Gravity and radiogenic heat are power sources that alter rock composition and structure. But what causes the motions of rocks, in which sediments and pre-existing rocks are buried deep within Earth and then later brought to the surface? These motions are part of the plate tectonic cycle, which is examined in Chapter 12.

The above-ground portion of the cycle of rock transformation is driven by several power sources, of which the Sun and gravity are the most important. Solar energy, and the unequal heating of the Earth's surface that it provides, is the primary energy source for the motions of the atmosphere and oceans. A lesser, but still important, power source for atmospheric and oceanic movements is the Earth's rotation, which transfers momentum to the fluid atmosphere and ocean. The tidal forces of the Sun and Moon also act to keep the ocean and atmosphere in motion.

Moving fluids—air, water, and glacial ice—transport rock particles, further reducing them to ever-finer sizes. In this way, the particles are also directly exposed to oxygen and water, allowing chemical alteration to occur. The force of gravity on the particles eventually brings them to sedimentary basins, where they accumulate, and the underground portion of the rock cycle begins. The figure below shows the flows of energy that power the cycle of rock transformation.

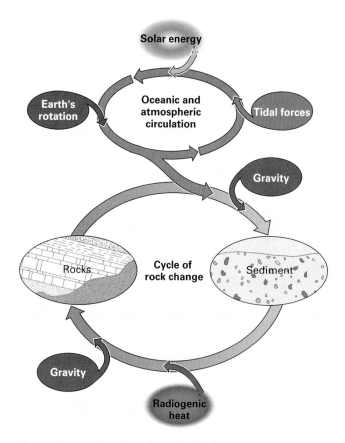

Flows of energy that drive the cycle of rock transformation

The loops in the rock cycle are powered by a number of sources, ranging from solar energy to internal, radiogenic heat. *Focus on Systems 11.1 • Powering the Cycle of Rock Transformation* describes these sources in more detail.

A Look Ahead

This chapter has described the minerals and rocks of the Earth's surface and the processes of their formation. These processes do not occur everywhere. Rather, there is a definite geographic pattern to the formation and destruction of rocks. This pattern is controlled by the way in which the solid Earth's brittle outer layer is fractured into great plates that split and separate and also converge and collide. The theory of plate tectonics provides the scheme for understanding the dynamics of the Earth's crust over millions of years of geologic time. Chapter 12 takes a closer look at plate tectonics.

CHAPTER SUMMARY

- The elements oxygen and silicon dominate the *Earth's crust*. Metallic elements, which include aluminum, iron, and the base elements, account for nearly all the remainder.
- Minerals are naturally occurring, inorganic substances. Each has an individual chemical composition and atomic structure. Minerals are combined into rock.
- *Silicate minerals* make up the bulk of igneous rocks. They contain silicon and oxygen together with some of the metallic elements.
- There are three broad classes of igneous rocks, depending on their mineral content. *Felsic* rocks contain mostly felsic minerals and are least dense; *mafic* rocks, containing mostly mafic minerals, are denser; and *ultramafic* rocks are most dense. Because felsic rocks are least dense, they are generally found in the upper layers of the Earth's crust. Mafic and ultramafic rocks are more abundant in the layers below.
- If magma erupts on the surface to cool rapidly as lava, the rocks formed are extrusive and have a fine crystal texture. If the magma cools slowly below the surface as a pluton, the rocks are intrusive and the crystals are larger. Granite (felsic, intrusive), andesite (felsic, extrusive), and basalt (mafic, extrusive) are three common igneous rock types.
- Most silicate minerals found in igneous rocks undergo mineral alteration when exposed to air and moisture at the Earth's surface. Mineral alteration occurs through *oxidation, hydrolysis,* or *solution*. Clay minerals are commonly produced by mineral alteration.

- Sedimentary rocks are formed in layers, or strata. Clastic sedimentary rocks are composed of fragments of rocks and minerals that usually accumulate on ocean floors. As the layers are buried more and more deeply, water is pressed out and particles are cemented together. Sandstone and shale are common examples.
- Chemical precipitation also produces sedimentary rocks, such as limestone. *Coal, petroleum,* and *natural gas* are hydrocarbon compounds occurring in sedimentary rocks as mineral fuels.
- Metamorphic rocks are formed when igneous or sedimentary rocks are exposed to heat and pressure. Shale is altered to *slate* or schist, sandstones become quartzite, and intrusive igneous rocks or clastic sediments are metamorphosed into gneiss.
- In the cycle of rock transformation, rocks are exposed at the Earth's surface, and their minerals are broken free and altered to form sediment. The sediment accumulates in basins, where the layers are compressed and cemented into sedimentary rock. Deep within the Earth, the heat of radioactive decay melts pre-existing rock into magma, which can move upward into the crust to form igneous rocks that cool at or below the surface. Rocks deep in the crust are exposed to heat and pressure, forming metamorphic rock. Mountain-building forces move deep igneous, sedimentary, and metamorphic rocks upward to the surface, providing new material for surface weathering and breakup, completing the cycle.

KEY TERMS

andesite	extrusive igneous rock	mafic	rhyolite
basalt	felsic	magma	rock
chemically precipitated sediment	fossil fuels	metamorphic rocks	sandstone
clastic sediment	gneiss	mineral	schist
clay mineral	granite	mineral alteration	sediment
cycle of rock transformation or rock cycle	igneous rocks	organic sediment	sedimentary rocks
	intrusive igneous rock	pluton	shale
evaporites	lava	quartz	strata
	limestone	quartzite	

REVIEW QUESTIONS

1. What is Earth's crust? What elements are most abundant in the crust?
2. Define the terms *mineral* and *rock*. Name the three major classes of rocks.
3. What are silicate minerals? Describe two groups of silicate minerals.
4. Name four types of igneous rocks and arrange them in order of increasing density.

5. How do igneous rocks differ when magma cools (a) at depth and (b) at the surface?

6. How are igneous rocks chemically weathered? Identify and describe three processes of chemical weathering.

7. What is sediment? Define and describe three types of sediments.

8. Describe two processes that produce sedimentary rocks, and identify at least three important varieties of clastic sedimentary rocks.

9. How are sedimentary rocks formed by chemical precipitation?

10. What types of sedimentary deposits consist of hydrocarbon compounds? How are they formed?

11. What are metamorphic rocks? Describe at least three types of metamorphic rocks, and the conditions under which each formed.

FOCUS ON SYSTEMS 11.1 | Powering the Cycle of Rock Transformation

1. How do pressure and temperature change with depth below the Earth's surface? Sketch a graph of temperature change with depth and explain its shape.

2. Identify and describe the sources of energy that power the cycle of rock change.

VISUALIZATION EXERCISES

1. Sketch a cross-section of the Earth showing the following features: batholith, sill, dike, veins, lava, and volcano.

2. Sketch the cycle of rock change and describe the processes that act within it to form igneous, sedimentary, and metamorphic rocks.

ESSAY QUESTION

1. Granite is exposed at the Earth's surface. Describe how mineral grains from this granite might be released, altered, and eventually become incorporated in a sedimentary rock. Summarize the processes that would incorporate the same grains in a metamorphic rock.

EYE ON THE LANDSCAPE |

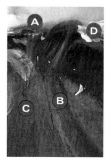

Chapter Opener   The Spectrum Range, Mt Edziza Provincial Park, British Columbia. The origin of this region is evident from the cones of stratovolcanoes (**A**) that can be seen in the background. The distinctive colours (**B**) are derived from the assortment of minerals present in the rock. The mountain summits are at about 2450 m and the resulting orographic Missingprecipitation, much oftext it in the form of snow, textis a characteristic Missingof this wet,text west-coast climate. Snowmelt in the warmer months runs quickly down the unvegetated slopes eroding channels (**C**), while persistent snow packs, small glaciers and ice falls are still present (**D**).

Sedimentary strata, Utah. The characteristic bedding of sedimentary rocks (**A**) is exposed in this cliff, which in desert areas will often erode away to leave remnants of a more extensive

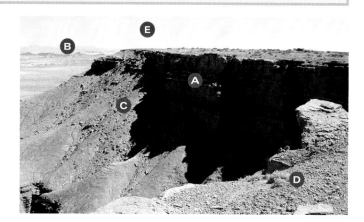

plateau, such as the mesas (**B**) which can be seen on the distant skyline. Rock which is loosened by weathering and erosion accumulates on the lower slopes (**C**). Plant cover (**D**) is sparse in this dry region, and only well-adapted species will survive the heat and aridity which develops under the clear desert sky (**E**).

Chapter 12

EYE ON THE LANDSCAPE

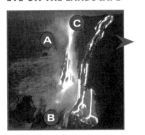

> > > >

The outpouring of basaltic lava in Hawaii.
What else would the geographer see? . . . Answers at the end of the chapter.

THE LITHOSPHERE AND THE TECTONIC SYSTEM

The Earth's Structure
The Earth's Interior
The Lithosphere and Asthenosphere
The Geologic Time Scale
*Working It Out 12.1 • Radioactive Decay and
    Radiometric Dating*
**Major Relief Features of the Earth's
Surface**
Relief Features of the Continents
Relief Features of the Ocean Basins
**Midoceanic Ridge and Ocean Basin
Features**
The Global System of Lithospheric Plates
Plate Tectonics
Plate Motions and Interactions
Tectonic Processes
Subduction Tectonics
Orogens and Collisions
Continental Rupture and New Ocean Basins
Continents of the Past

The outlines of the continents on a globe or map are so unique that they can be readily identified. These outlines are the result of forces within the Earth that have acted over millions of years. Various types of evidence suggest that during their long history, the continents had been close together, but were fractured and split apart. The theory describing the changing configuration of the continents through time is called *plate tectonics*.

Plate tectonic theory maintains that the Earth's outermost solid layer consists of huge, rigid plates that float on a layer of plastic rock. These plates are in slow, constant motion, powered by energy sources deep within the planet. When two plates come together, one may be forced under the other in a process called *subduction*. In this case, part of the subducted plate can melt, creating pockets of magma that move upward to the surface to create volcanic mountain chains and island arcs. Plates can also collide without subduction, forming mountain chains where rocks are fractured, folded, and compressed into complex structures. When crustal plates are split apart, rift valleys and ultimately ocean basins are created. Lava that moves upward to fill the gap forms submerged mountain chains such as the mid-Atlantic ridge. Thus, ocean basins expand with rifting and contract with subduction, while continents grow by collision and shrink by fracturing. Because plate motions are powered by forces within the Earth, they are independent of surface conditions.

Although internal tectonic processes operate independently, the processes that determine climate, vegetation, and soils are dependent on the major relief features and Earth materials that the tectonic setting provides. Thus, knowledge of plate tectonics is important to our understanding of the global patterns of the Earth's landscapes—including climate, soils, vegetation, and, ultimately, human activity.

THE EARTH'S STRUCTURE

From studies of earthquake waves, reflected from deep layers within the Earth, scientists have discovered that the Earth is far from uniform from its outer crust to its centre. It consists of a central core with several layers, or shells, surrounding it. The densest matter is at the centre, and each layer above it is progressively less dense. The Earth's structure dates back to its earliest history, when it was formed by accretion from a mass of gas and dust orbiting the Sun.

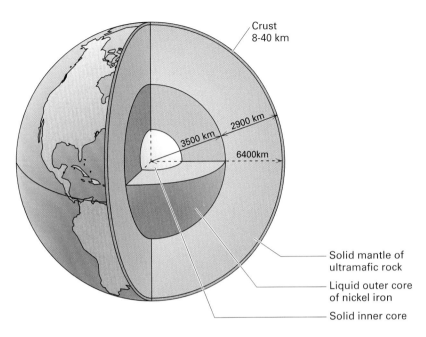

Crust
8-40 km

3500 km 2900 km

6400km

Solid mantle of
ultramafic rock

Liquid outer core
of nickel iron

Solid inner core

12.1 The Earth's interior This cutaway diagram of the Earth shows the inner core of iron, which is solid, surrounded by a liquid outer core. The mantle that surrounds the core is a thick inner layer of ultramafic rock. The crust is too thin to show at a correct scale. Thicknesses of layers are rounded to the nearest 100 units.

THE EARTH'S INTERIOR

The Earth as a whole is an almost spherical body approximately 6,400 kilometres in radius. The centre is occupied by the **core**, which is about 3,500 kilometres in radius (Figure 12.1a). Because earthquake waves suddenly change behaviour upon reaching the core, scientists have concluded that the outer core has the properties of a liquid. However, the innermost part of the core is in a solid state. Even though temperature increases toward the Earth's centre, the inner core remains solid because it is under very high pressure. Based on various geophysical data, it has long been inferred that the core consists mostly of iron, with some nickel. The core temperature is estimated at 3,000°C to 5,000°C.

Enclosing the metallic core is the **mantle**, a rock shell about 2,900 kilometres thick. Earthquake waves indicate that mantle rock is composed of mafic minerals similar to olivine (a silicate of magnesium and iron) and may resemble the ultramafic igneous rock peridotite. Temperatures in the mantle range from about 2,800°C near the core to about 1,800°C near the crust.

The outermost and thinnest of the Earth's layers is the **crust.** It is formed largely of igneous rock, but also contains substantial proportions of metamorphic rock and a comparatively thin upper layer of sedimentary rock. The base of the crust is sharply defined where it contacts the mantle. This contact is detected by the way in which earthquake waves abruptly change velocity at that level. The boundary surface between crust and mantle is called the *Moho*, a simplification of the name of the scientist Andrija Mohorovicic, who discovered it in 1909.

Earthquakes produce three basic types of waves, two of which travel deep within the Earth's interior, and the third travels near the surface (Figure 12.2). The faster of the two deep-seated wave types is called the primary or **P wave**. P waves propagate by alternately pushing and pulling the rock. These waves can travel through both solid and liquid rock material, as well as the water of the oceans. The slower waves are called the secondary or **S waves**. As an S wave develops, it shears the rock sideways at right angles to the direction it is travelling. S waves cannot propagate in liquid materials. The speed of P and S seismic waves depends on the density and properties of the rocks through which they pass. In most earthquakes, the P waves are felt first, followed by the shaking and twisting motion of the S waves, which are so damaging to buildings. **Surface waves** are similar to ripples of water that travel across a lake. Most surface waves can also travel across water. The centre of the earthquake is determined by the time of arrival of the different wave types at three locations on the surface.

Because seismographs do not detect S waves on the opposite side of the Earth from an earthquake's point of origin, it has been concluded that the core is liquid. The core's size has been determined by the distance at which S waves are detected. As P waves move through the Earth's interior, they refract or bend, with abrupt changes in direction occurring at the boundary between different layers. P waves entering the outer core are bent toward the Earth's centre so they reach a region opposite the earthquake's point of origin. Thus a **shadow zone** separates the P waves that pass through the mantle only from the P waves that pass through the mantle and the core. The fact that very weak P waves are felt in the shadow zone suggests that the inner core is solid.

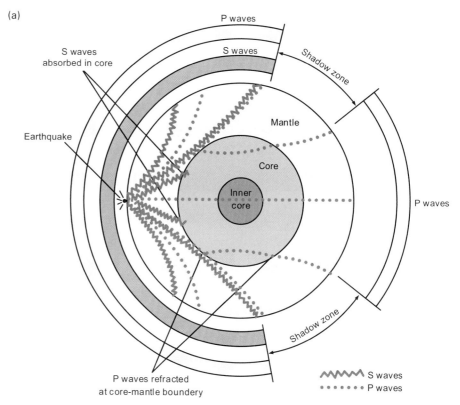

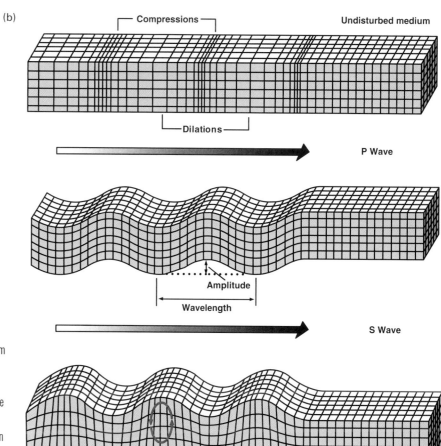

12.2 Major types of earthquake waves (a) This diagram shows how earthquake waves travel through the Earth. The existence of a shadow zone where neither P nor S waves arrive is evidence of a core. The inability of S waves to pass through the core suggests that at least the outer core is liquid. (b) P waves which have the highest velocity vibrate parallel to the direction of motion in a pulse-like manner; in S waves, movement is perpendicular to the direction of propagation and causes a twisting, heavy motion; surface (Rayleigh) waves travel in the same manner as waves in water, causing earth materials to follow in a circular path.

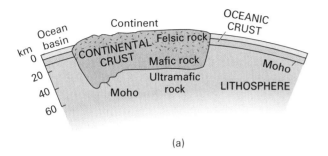

(a)

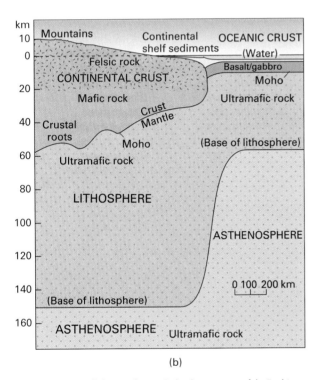

(b)

12.3 Outer layers of the Earth (a) Idealized cross-section of the Earth's crust and upper mantle. (b) Details of the crust and mantle at the edge of a continent, including the types of rocks found there. Also shown are the lithosphere and asthenosphere. (Copyright © A.N. Strahler)

The **continental crust** consists of a lower, denser zone of mafic composition, and an upper zone of less dense felsic rock (Figure 12.3). The average density of the continental crust is 2.7 grams per cubic centimetre (g/cm³). Because the felsic portion has a chemical composition similar to that of granite, it is commonly described as *granitic rock*, although much of it is actually metamorphic rock. There is no sharply defined separation between the felsic and mafic zones.

While the continental crust typically has two zones of different composition, the **oceanic crust** consists almost entirely of mafic rocks with the composition of basalt and gabbro. However, the basaltic material, with an average density of 2.9 grams per cubic centimetre, is typically found above the gabbro, which has a density of about 3.0 grams per cubic centime-

tre. Another key distinction between continental and oceanic crust is that the crust is much thicker beneath the continents (Figure 12.3). The crust is generally less than 10 kilometres thick in ocean basins, and averages about 40 kilometres under the continents (see Figure 11.1).

THE LITHOSPHERE AND ASTHENOSPHERE

The **lithosphere**, a zone of rigid, brittle rock, includes not only the crust, but also the cooler, upper part of the mantle. It ranges in thickness from 60 to 150 kilometres, and is thickest under the continents. Some tens of kilometres deep within the Earth, the brittle condition of the lithospheric rock gives way gradually to a plastic layer named the **asthenosphere**, where temperatures reach 1,400°C (Figure 12.3). The density of the asthenosphere increases from 3.7 to 5.5 grams per cubic centimetre with depth. However, at still greater depth in the mantle, the strength of the rock material increases again. Thus, the asthenosphere is a soft layer between the "hard" lithosphere above and a "strong" mantle rock layer below.

Because the asthenosphere is soft and plastic, the rigid lithosphere can easily move over it. The lithosphere consists of large **lithospheric plates**. A single plate can be as large as a continent and can move independently of the plates that surround it. Like great slabs of floating ice, lithospheric plates can separate from one another at one location, while elsewhere they may collide and push up mountain ranges. Along these collision zones, one plate moves under the edge of its neighbour.

THE GEOLOGIC TIME SCALE

The lithospheric plates have moved at various times in the geologic past. Table 12.1 lists the major geologic time divisions. All time before 570 million years (m.y.) ago is referred to as the *Precambrian Time*. Three *eras* of time follow: *Paleozoic*, *Mesozoic*, and *Cenozoic*. These eras are subdivided into *periods* and comprise the *Phanerozoic eon*.

The Cenozoic Era is particularly important in terms of the continental surfaces, because nearly all landscape features seen today have been produced in the past 65 million years. The Cenozoic Era is comparatively short in duration. It is subdivided into seven shorter time units called *epochs*.

Various techniques are used to establish the ages of rocks within the geologic time scale. One of the most important techniques is *radiometric dating*, which establishes the age of certain minerals within rocks using principles of radioactive decay. *Working It Out 12.1 • Radioactive Decay and Radiometric Dating* describes this process in more detail.

Table 12.1 Geologic time scale

Era	Period	Epoch	Duration (millions of years)	Age (millions of years)	Orogenies	Evolution of Life Forms
Cenozoic	Neogene	Holocene	(10,000 yr)			Human genus
		Pleistocene	2	2		
		Pliocene	3	5		
		Miocene	18			Hominoids
	Paleogene	Oligocene	11	23	Alpine ends	Whales
		Eocene	23	34	Cordilleran ends	Bats
		Paleocene	9	57	Alpine begins	Mammals
Mesozoic	Cretaceous		80	66		Flowering plants
	Jurassic		54	146		Dinosaurs (extinct)
	Triassic		51	200	Cordilleran begins	Turtles
Paleozoic	Permian		48	251		Frogs
	Carboniferous		60	299	Alleghany	Conifers; higher fishes
	Devonian		57	359	Caledonian/ Alleghany	Vascular plants; primitive fishes
	Silurian		28	416		
	Ordovician		44	444	Caledonian	
	Cambrian		54	488		Invertebrates
				542		

Precambrian Time extends to oldest known rocks, about 4 billion years ago.
Age of Earth as a planet: 4.6 to 4.7 billion years.
Age of universe: 17 to 18 billion years.

Source: International Commission on Stratigraphy, 2004.

MAJOR RELIEF FEATURES OF THE EARTH'S SURFACE

About 29 percent of the Earth's surface is land and 71 percent oceans. However, if the seas were to drain away, broad sloping areas lying close to the continental shores would be exposed. These continental shelves are covered by shallow water, less than 150 metres deep. Beyond the continental shelves, the ocean floor drops rapidly to depths of thousands of metres. Thus, it appears that the oceans have spread over the margins of the continents. If these submerged areas are considered part of the continents, land area would increase to 35 percent, and the ocean basin area would decrease to 65 percent. These revised figures represent the true relative proportions of continents and oceans.

RELIEF FEATURES OF THE CONTINENTS

Figure 12.4 shows some of the major relief features of continents and oceans, including mountain arcs, island arcs, ocean trenches, and the midoceanic ridge.

Broadly viewed, the continents consist of two basic subdivisions: active belts of mountain-building and inactive regions of old, stable rock. The mountain ranges in the active belts grow through one of two very different geologic processes. First is *volcanism*, the formation of massive accumulations of volcanic rock by extrusion of magma. Many lofty mountain ranges consist of chains of volcanoes built of extrusive igneous rocks.

The second mountain-building process is *tectonic activity*, the breaking and bending of the Earth's crust under internal forces. Tectonic activity usually occurs when the lithospheric plates

WORKING IT OUT | 12.1　　Radioactive Decay and Radiometric Dating

The Earth's interior is heated largely by the spontaneous decay of naturally occurring radioactive isotopes of certain elements. The number of positively charged particles, or *protons*, contained in an element's nucleus determines its properties. For example, the element uranium (U) has 92 protons. Nuclei of atoms also contain neutrons. The number of protons defines the atomic number of the element, and this combined with the number of neutrons determines the *atomic mass number*. Some elements are found in forms with different mass numbers. These forms are known as *isotopes*. For example, the most common isotope of uranium is ^{238}U, a form with an atomic mass of 238; another form is ^{235}U.

A key to understanding radioactivity is that certain isotopes are *unstable*, meaning that the composition of the isotope's nucleus can experience an irreversible change. This change process is known as *radioactive decay*. In some types of decay, protons may be lost or gained. This means that an atom of one element may be transformed into an atom of another. For example, the uranium isotope ^{238}U decays to form ^{234}Th, an isotope of the element thorium. This new isotope is known as a *daughter product*. Often the new isotope created will also be unstable, and will decay into yet another isotope of a different element. As this process continues, the result is a decay chain of daughter products that eventually ends in the formation of a stable isotope. For example, ^{238}U ultimately forms the stable lead isotope ^{206}Pb.

The significance of radioactive decay is that it provides an internal source of heat for the Earth—a source that accounts for the melting of solid rock to form magma, and thus the creation of igneous rocks. The decay of ^{238}U is only one of several radioactive decay chains that are important in heating the Earth from within. Other chains begin with ^{235}U, thorium-232 (^{232}Th), and potassium-40 (^{40}K).

The rate of an unstable isotope's decay is measured by its *half-life*—the time period in which the number of atoms of the isotope will be reduced by half. For example, the half-life of ^{238}U is 4.47 billion years, meaning that one gram of ^{238}U will shrink to half a gram after that length of time. In another 4.47 billion years, this half-gram will shrink to a quarter-gram, and so forth.

The graph on the next page shows an example of the decay of ^{40}K. This unstable potassium isotope has a half-life of 1.28 billion years. The *y*-axis shows the proportion of the original quantity remaining after the elapsed time, shown on the *x*-axis. This curve has a *negative exponential* form. When ^{40}K decays, one of two products is produced: calcium-40 (^{40}Ca) or, less frequently, argon-40 (^{40}Ar). Both ^{40}Ca and ^{40}Ar are stable isotopes and undergo no further decay.

Suppose that a quantity of ^{238}U is present in an igneous rock at the time it was formed early in the Earth's history. These atoms of ^{238}U will slowly transform themselves to ^{206}Pb at a rate that depends on the half-life of ^{238}U. The ratios of particular elemental isotopes are used in *radiometric dating* to

collide. Crustal masses raised by tectonic activity form mountains and plateaus. In some instances, volcanism and tectonic activity have combined to produce a mountain range, such as the Cascade Range in western North America. Tectonic activity can also lower crustal masses to form depressions that may be occupied by ocean embayments or inland seas. Chapter 13 discusses landforms produced by volcanic and tectonic activity.

Alpine Chains

Active mountain-building belts are narrow zones that are usually found along the margins of lithospheric plates. These belts, sometimes referred to as *alpine chains*, are characterized by high, rugged mountains, such as the Rocky Mountains in Canada, formed 70–40 million years ago, and the Alps of Central Europe. Active alpine chains are characterized by broadly curved lines on the world map (Figure 12.4). Each curved section is called a **mountain arc**.

The mountain arcs are linked in sequence to form two principal mountain belts. One is the *circum-Pacific belt* (shown in green), which rings the Pacific Ocean basin. In North and South America, this belt is largely on the continents and includes the Cordilleran ranges, such as the Coast and Cascade Mountains, and the Andes. In the western part of the Pacific basin, the mountain arcs lie well offshore from the continents and take the form of **island arcs**. Partly submerged, they join the Aleutians, Kurils, Japan, the Philippines, and other smaller islands. These island arcs are the result of volcanic activity. Between the larger islands, the arcs are represented by volcanoes rising above the sea as small, isolated islands.

The second chain of major mountain arcs forms the *Eurasian-Indonesian belt* (shown in blue in Figure 12.4). It starts in the west at the Atlas Mountains of North Africa and continues through the European Alps and the ranges of the Near East and Iran to join the

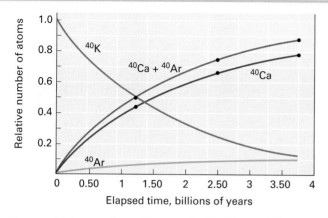

Exponential decay and growth curves for ^{40}K, ^{40}Ca, and ^{40}Ar (Copyright © A. N. Strahler.)

determine the time of formation or metamorphism of a rock using the following equation:

$$t = \frac{1}{k} \ln\left[\left(\frac{D}{M}\right) + 1\right]$$

where t = time of decay and k = a constant related to the half-life by the expression

$$k = 0.693/H$$

where H = the half-life, and 0.693 is the natural logarithm of 2. For the case of ^{238}U, $k = 0.693/4.77 = 0.145$.

Assume that a chemical analysis of a mineral shows that for every two atoms of ^{238}U, one atom of ^{206}Pb is present. The ratio of daughter to mother atoms is $D/M = 1/2 = 0.5$, and since $k = 0.693/4.47 = 0.145$ using units of billions of years (b.y.) then

$$t = \frac{1}{0.145} \ln (0.5 + 1)$$

$$= \left(\frac{0.405}{0.145}\right) = 2.61 \; b.y.$$

The practice of radioactive dating makes it possible to date rocks with certainty, based on careful analysis of the concentration of the radioactive isotopes in certain minerals. The oldest rocks so far discovered are ancient gneisses in northwestern Canada. They contain zircon crystals that are 3.96 billion years old. Ancient shield rocks from Antarctica have been dated at 3.93 billion years, and samples from Greenland at 3.80 billion years.

During the earliest eras of the Earth's history, continental crust was being vigorously formed and reformed, so rocks and minerals from these early times are long gone. However, some types of meteorites have radiometric ages of about 4.6 billion years, and these also have the same mixture of naturally occurring lead isotopes as the Earth's rocks. This latter fact leads geochemists to suspect strongly that these meteorites were formed from the same primordial matter as the Earth was, thus dating the formation of Earth at about 4.6 billion years ago.

Himalayas. The belt then continues through Southeast Asia into Indonesia, where it meets the circum-Pacific belt.

Continental Shields

Belts of recent and active mountain-building account for only a small portion of the continental crust. The remainder consists of comparatively inactive regions of much older rock. Within these stable regions are two types of crustal structure: continental shields and mountain roots. **Continental shields** are low-lying continental surfaces, beneath which are complex arrangements of ancient igneous and metamorphic rocks. Figure 12.5 shows the distributions of two classes of shields: exposed shields and covered shields. Some core areas of the shields are composed of rocks dating back to the Archean eon, 2.5 to 3.5 billion years ago. These ancient cores are exposed in some areas but covered in others.

Exposed shields include very old rocks, mostly from Precambrian Time, and have a complex geologic history. These regions occur extensively in Scandinavia, South America, Africa, Asia, India, Australia, and Canada. The exposed shields are typically regions of low hills and low plateaus, although in some areas large crustal blocks have been recently uplifted. Many thousands of metres of rock have eroded away from these shields throughout the past half-billion or more years.

The Canadian Shield is the greatest area of exposed Archaean rock in the world. It comprises mostly metamorphic base rocks from the Precambrian Time, many of which contain substantial mineral deposits, including nickel, copper, and gold. Diamonds are also mined from kimberlite in the Northwest Territories. Kimberlite is ultramafic igneous rock that has been brought to the surface from great depths in volcanic pipes and dikes. Despite being repeatedly uplifted and eroded, it has remained almost

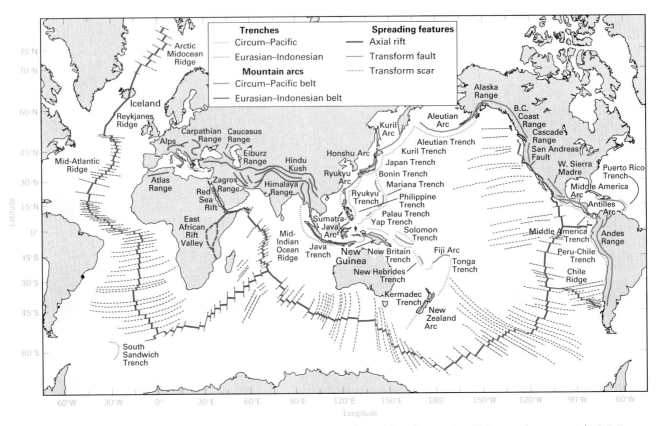

12.4 Tectonic features of the world Principal mountain arcs, island arcs, ocean trenches, and the midoceanic ridge. (Midoceanic ridge map copyright © A. N. Strahler)

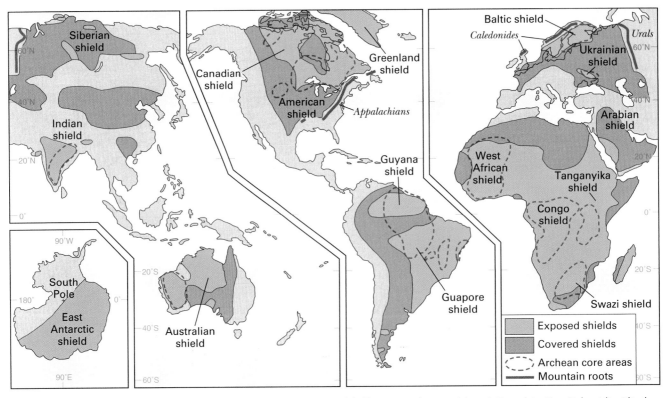

12.5 Continental shields Generalized world map of exposed and covered continental shields. Continental centres of the early Precambrian Time (Archean) lie within the areas encircled by a broken red line. Heavy brown lines show mountain roots of Caledonian and Hercynian orogenies. In North America, this is usually referred to as the Appalachian orogeny. (Based in part on data of R. E. Murphy, P. M. Hurley, and others. Copyright © A. N. Strahler)

12.6 Canadian Shield This outcrop of hard metamorphic rock has been shaped and smoothed by the action of ice and water.

entirely above sea level throughout its long history. The Canadian Shield was heavily glaciated during the late-Cenozoic Ice Age. The result is a low-relief, poorly drained, lake-studded area with a lot of exposed bedrock (Figure 12.6). Together with the covered shield areas to the west and south, the Canadian Shield comprises the North American craton, or ancient base of the continent.

Covered shields are areas of the continental shields that are covered by younger sedimentary layers, ranging in age from Paleozoic through Cenozoic eras. These strata accumulated at times when the shields were inundated by shallow seas. Marine sediments covered the ancient shield rocks in thicknesses ranging from hundreds to thousands of metres. These shield areas were then broadly arched upward to become land surfaces again. Erosion has since removed large sections of their sedimentary cover, but it still remains intact over vast areas. The shallow waters of Hudson Bay are presently receiving sediments carried by numerous rivers in what appears to be a renewed cycle of activity.

Ancient Mountain Roots

Remains of older mountain belts lie within the continental shields in many places. These *mountain roots* are mostly formed of Paleozoic and early Mesozoic sedimentary rocks that have been intensely bent and folded, and in some locations changed into metamorphic rocks—for example, slate, schist, and quartzite. Thousands of metres of overlying rocks have been removed from these old tectonic belts, so that only the lowest structures remain. Roots appear as chains of long, narrow ridges, rarely rising more than a thousand metres above sea level.

One important system of mountain roots was formed in the Paleozoic Era, about 400 million years ago. They are called the Caledonides, which form a highland belt across the northern British Isles and Scandinavia. They are also present in the Maritime Provinces and New England (Figure 12.5). The original collision between two lithospheric plates created high alpine mountain chains that have since been worn down to belts of subdued mountains and hills.

Three separate periods of mountain-building involving collisions of the North American continent with ancient landmasses took place between 420 and 280 million years ago and gave rise to the ancient Appalachian Mountains. The last phase, part of the Hercynian Orogeny, occurred when the combined continents of Europe and Africa, known as Gondwana, collided with the North American continent, then known as Laurentia. The tectonic processes that formed the Appalachians were also responsible for the formation of the Urals between Europe and Asia. These processes ended about 250 million years ago, when most of the smaller continents had come together to form the single supercontinent of Pangaea. The ancient Appalachian mountain range was eroded to an almost level plain, but a period of renewed uplift during the Cenozoic Era resulted in the general configuration seen today.

RELIEF FEATURES OF THE OCEAN BASINS

The relief features of ocean basins are quite different from those of the continents. Crustal rock of the ocean floors consists almost entirely of basalt, which is generally covered by a comparatively thin accumulation of sediments. Age determinations of the basalt and its overlying sediments show that the oceanic crust is, geologically, quite young. Much of the oceanic crust is less than 60 million years old, although some large areas have ages of about 65 to 135 million years. This is much younger than most of the continental crust, some of which is several billion years old.

MIDOCEANIC RIDGE AND OCEAN BASIN FEATURES

Figure 12.7 shows the important relief features of ocean basins, which are characterized by a central ridge structure that divides the basin approximately in half. This *midoceanic ridge* consists of submarine hills that rise gradually to a rugged central zone. In the centre of the ridge is a narrow, elongated depression known as the *axial rift*. The location and form of the axial rift suggest that the crust is being pulled apart along the line of the rift.

The midoceanic ridge and its principal branches can be traced through the world's ocean basins for a total distance of about 60,000 kilometres. Figure 12.4 shows the extent of the ridge. Beginning in the Arctic Ocean, it divides the Atlantic Ocean basin from Iceland to the South Atlantic. Turning east, it enters the Indian Ocean, where one branch extends into the Red Sea. The other branch continues east between Australia and Antarctica, and then swings across the South Pacific. Nearing South America, it turns north and reaches North America at the head of the Gulf of California.

On either side of the midoceanic ridge are broad, deep *abyssal plains* that belong to the *ocean basin floor* (Figure 12.7). Their average depth below sea level is about 5,000 metres. The level topography of the abyssal plains comprises fine sediment

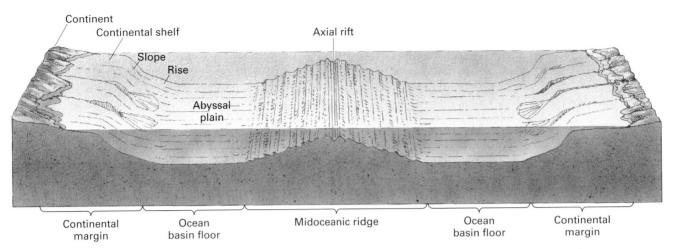

12.7 Ocean basins The main feature of the North and South Atlantic Ocean basins is the midoceanic ridge, a region where the crust is separating and moving apart.

that has settled slowly and evenly from ocean water above. Figure 12.8 shows many details of the ocean basins and their submarine landforms. The prominent central feature in the Atlantic Ocean is the Mid-Atlantic Ridge. The deep trenches of the western Pacific mark the positions of subduction arcs where oceanic crust is being forced down into the mantle.

Continental Margins

The *continental margin*, shown on the left and right sides of Figure 12.7, is the narrow zone in which oceanic lithosphere is in contact with continental lithosphere. As the continental margin is approached from the deep ocean, the ocean floor be-

gins to slope gradually upward, forming the *continental rise*. The floor then becomes much steeper on the *continental slope*. At the top of this slope is the edge of the *continental shelf*, a gently sloping platform with vertical relief of less than 20 metres. The average width of the continental shelves is about 80 kilometres, but this is quite variable. It is narrowest where a continent abuts the forward edge of an advancing oceanic plate, such as off the coast of Chile. It is about 120 to 160 kilometres wide along the eastern margin of North America. Some shallow seas overlie areas of continental shelf, such the North Sea and Persian Gulf. The continental rise, slope, and shelf together form the continental margin.

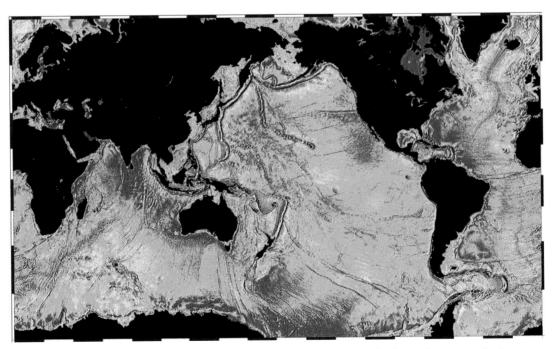

12.8 Undersea topography Portions of a map showing variations in the pull of gravity, which indicate undersea topography. Deeper regions are shown in tones of purple, blue, and green, while shallower regions are in tones of yellow and orange.

Figure 12.7 illustrates a symmetrical ocean floor model—a midoceanic ridge with ocean basin floors on either side. This model nicely fits the North Atlantic and South Atlantic Ocean basins. It also applies rather well to the Indian Ocean and Arctic Ocean basins. The margins of these symmetrical basins are described as *passive continental margins* because they have not experienced strong tectonic and volcanic activity during the last 50 million years. This is because the continental and oceanic lithospheres that join at a passive continental margin are part of the same lithospheric plate, and they move away from the axial rift as a single unit.

The margins of the Pacific Ocean basin are different from those of the Atlantic. Although the Pacific has a midoceanic ridge with ocean basin floors on either side, its margins are characterized by mountain arcs or island arcs with deep offshore *oceanic trenches*. Geologists refer to these trenched ocean-basin edges as *active continental margins*. Here oceanic crust is bent downward and forced under continental crust or, as in the case of island arcs such as Japan and Indonesia, under other oceanic crust. This creates trenches and induces volcanic activity. Trench floors can reach depths of nearly 11,000 metres, although most range from 7,000 to 10,000 metres.

THE GLOBAL SYSTEM OF LITHOSPHERIC PLATES

The Earth's crust comprises 19 distinctive lithospheric plates of various sizes (Figure 12.9). The Pacific plate occupies much of the Pacific Ocean basin, and consists almost entirely of oceanic lithosphere. Its relative motion is northwesterly; along most of the western and northern edge, there is a subduction boundary in which crustal plates are being forced down into the mantle. The eastern and southern edge is mostly a spreading boundary where material from the mantle rises toward the surface and forms an axial rift. A small area of continental lithosphere makes up the coastal portion of California and all of Baja California. An active transform fault (the San Andreas Fault) borders the California portion.

The American plate includes most of the continental lithosphere of North and South America, as well as the entire oceanic lithosphere lying west of the midoceanic ridge that divides the Atlantic Ocean. For the most part, the western edge of the American plate is a subduction boundary, where oceanic lithosphere is being forced beneath the continental lithosphere. The eastern edge is a spreading boundary. Most classifications recognize separate North American and South American plates.

The Eurasian plate is mostly continental lithosphere, but a belt of oceanic lithosphere fringes it on the west and north. The easternmost part of Asia, including the Kamchatka Peninsula, is part of the North American plate, but the nature of the boundary in this region is uncertain. The African plate has a central core of continental lithosphere nearly surrounded by oceanic lithosphere. The Red Sea, which separates the African Plate from the lesser Arabian plate, is an active rift zone.

The Austral-Indian plate is mostly oceanic lithosphere but contains two cores of continental lithosphere: Australia and India. Recent evidence shows that these two continental masses are moving independently and may actually be considered parts of separate plates. The Antarctic plate is almost completely enclosed by a spreading plate boundary. This means that the other plates are moving away from the pole. The Antarctica continent forms a central core of continental lithosphere completely surrounded by oceanic lithosphere.

Of the lesser plates, the Nazca and Cocos plates of the eastern Pacific are rather simple fragments of oceanic lithosphere bounded by the spreading Pacific midoceanic boundary to the west and by a subduction boundary to the east. The Philippine plate is noteworthy as having subduction boundaries on both east and west edges. Two small but distinct lesser plates—Caroline and Bismarck—lie to the southeast of the Philippine plate. The Arabian plate has two transform fault boundaries, and its relative motion is northeasterly. The Caribbean plate also has important transform fault boundaries. The tiny Juan de Fuca plate is steadily diminishing in size and will eventually disappear by subduction beneath the North American plate. Similarly, the Scotia plate is being consumed by the South American and Antarctic plates.

PLATE TECTONICS
PLATE MOTIONS AND INTERACTIONS

The motion of lithospheric plates and how they interact at their boundaries is collectively called **plate tectonics. Tectonic activity** describes all forms of breaking and bending of the lithosphere, including the crust. This type of deformation occurs on a large scale when lithospheric plates collide as they move about the Earth's surface. It is generally agreed that heat energy, derived from radioactive decay of unstable isotopes in the crust and mantle, moves the crustal plates. As the molten rock material rises beneath the oceanic lithosphere, the sea floor lifts, and the plate moves horizontally away from the spreading axis under the influence of gravity (Figure 12.10).

The oceanic plate's motion exerts a drag on the surrounding asthenosphere, which creates a horizontal flow current. At the same time, the plate undergoes cooling, becomes denser, and sinks steadily deeper. The process in which one plate is carried beneath another is called **subduction**. Subduction occurs because the oceanic plate is colder and denser than the asthenosphere through which it sinks. In the zone of subduction, the mantle

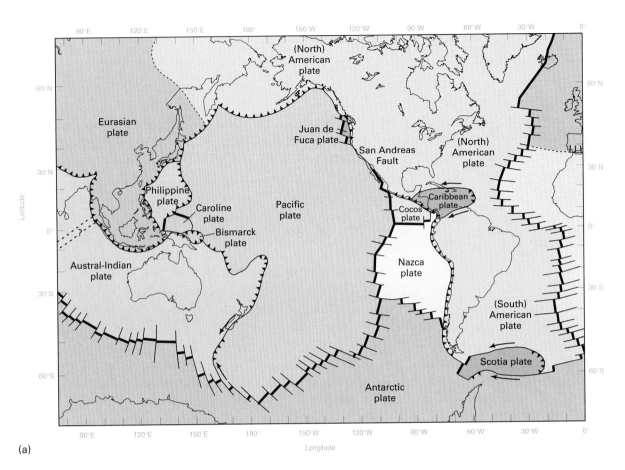

(a)

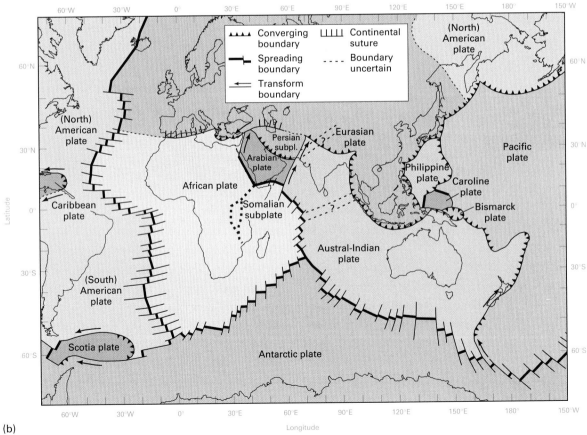

(b)

12.9 World map of lithospheric plates

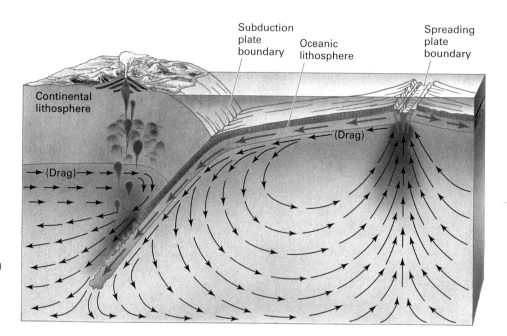

12.10 Mantle flow system and heat exchange A schematic diagram showing a system of mantle flow (blue arrows) that transports heat to the Earth's surface, where it is lost through volcanic activity associated with zones of subduction and midoceanic ridges. The crustal plates move away from the zone of upwelling magma (red arrows) and are eventually carried down to the mantle in the subduction zones. (Copyright © A. N. Strahler)

beneath the continental lithosphere becomes heated and the upper surface of the oceanic crust melts. The lower side of the descending slab is also heated and softened, and is incorporated into the mantle at great depth.

This convective system brings hot asthenosphere near the Earth's surface, where heat flows to the oceanic and atmospheric layers. It is a one-way energy system that depends on a radiogenic heat source, which is slowly being depleted. As time passes, the plate tectonic system will lose power and steadily slow its rates of motion. In the far distant future, the Earth's tectonic system will cease, and there will be no more earthquakes or volcanic eruptions.

Figure 12.11 shows three plates: X and Y are both made up of oceanic lithosphere, and Z represents a continental lithosphere. Both types of lithosphere can be thought of as "floating" on the soft asthenosphere. The separation of the lithospheric plates X and Y along the axis of a midoceanic ridge would be expected to

12.11 Schematic cross-sections showing some of the important elements of plate tectonics Diagram (a) is greatly exaggerated in vertical scale and emphasizes surface and crustal features. It shows only the uppermost 30 kilometres. Diagram (b) is drawn to true scale and shows conditions to a depth of 250 kilometres. Here the actual relationships between lithospheric plates can be examined, but surface features are too small to illustrate. (Copyright © A.N. Strahler)

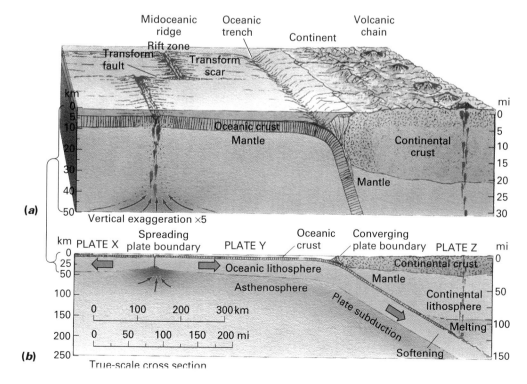

create a gaping crack in the crust, but magma continually rises from the mantle beneath to fill it. The magma appears as basaltic lava in the floor of the rift and quickly congeals. At greater depth under the rift, magma solidifies into *gabbro*, an intrusive rock of the same composition as basalt. Together, the basalt and gabbro continually form new oceanic crust. This type of boundary between plates is called a divergent or *spreading boundary*.

Where the oceanic lithosphere of plate Y collides with the continental lithosphere of plate Z, they form a *converging boundary*. Because the oceanic plate is comparatively thin and dense, in contrast to the thick, buoyant continental plate, the oceanic lithosphere bends down and plunges into the soft asthenosphere. The leading edge of the descending plate is cooler and therefore denser than the surrounding hot asthenosphere. As a result, the plate sinks under its own weight, once subduction has begun. The plate rock is gradually heated by the surrounding hot rock and thus eventually softens. The under portion, which is the same composition as mantle rock, is simply incorporated into the mantle as it softens.

However, the descending plate is covered by a thin veneer of sediment derived from oceanic and continental sources. This material can melt and become magma, which rises in bubble-like bodies to penetrate the thick continental lithosphere. Some magma cools and forms deep-seated plutons, while some reaches the surface and forms a chain of volcanoes parallel with the deep oceanic trench where subduction is occurring (Figure 12.12).

A third type of lithospheric plate boundary occurs where two lithospheric plates are in contact along a common boundary, but one plate merely slides past the other (Figure 12.13). This is a *transform boundary*; movement of the plates causes neither separation nor convergence. The fault plane along which motion occurs is a nearly vertical fracture extending down into the lithosphere. Transform boundaries are often found perpendicular to midoceanic ridges.

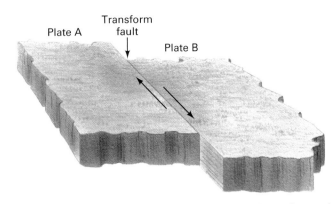

12.13 Transform boundary A transform boundary involves the horizontal motion of two adjacent lithospheric plates, one sliding past the other. (Copyright © A. N. Strahler)

TECTONIC PROCESSES

Generally speaking, prominent mountain masses and mountain chains (other than volcanic mountains) are elevated by one of two basic tectonic processes: *compression* and *extension* (Figure 12.14). Compressional tectonic activity acts at converging plate boundaries. The result is often an alpine mountain chain consisting of intensely deformed rock strata. The strata are tightly compressed into wavelike structures called *folds*. Extensional tectonic activity occurs where oceanic plates pull apart or where a continental plate undergoes breakup. As the crust thins, it fractures and is displaced, producing block mountains.

Alpine folds have a typical developmental sequence. Initially, the strata are folded into upwarps (anticlines) and downwarps

12.12 Cascade Mountain volcanoes This view, looking northeast across the summit of Mt. St. Helens *(Fly By: 46° 12' N; 122° 11' W)*, shows Mt. Rainier *(Fly By: 46° 51' N; 121° 45' W)* (4390 m) on the right at a distance of 80 km and Mt. Adams *(Fly By: 46° 12' N; 121° 29' W)* (3750 m) on the left about 50 km away.

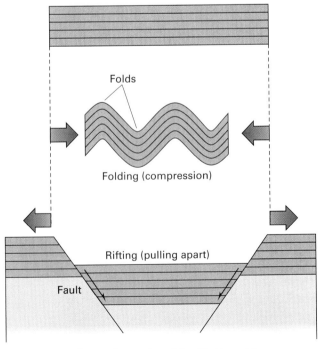

12.14 Two basic forms of tectonic activity Flat-lying rock layers may compress to form folds or pull apart to form faults by rifting.

(synclines). With further compression these folded strata may become *recumbent* as they are further overturned upon themselves (Figure 12.15). Accompanying the folding is a form of faulting in which slabs of rock are displaced along gently inclined fault surfaces. These are called *overthrust faults*. The rock slabs, called *thrust sheets*, are carried tens of kilometres across the underlying rock. In the European Alps, these thrust sheets are called *nappes*. Nappes may be thrust one over the other and build up to great heights. The entire deformed rock mass produced by compression is called an *orogen*, and the event that produced it is an *orogeny*.

SUBDUCTION TECTONICS

Converging plate boundaries, where subduction is taking place, are zones of intense tectonic and volcanic activity (Figure 12.16). The oceanic trench receives sediment from two sources. Deep-ocean sediment comprising fine clay and ooze is carried along on the moving oceanic plate. From the continent comes terrestrial sediment in the form of sand and mud brought by streams to the shore and then swept into deep water by currents. In the bottom of the trench, both types of sediment are intensely deformed and dragged down with the moving plate. The deformed sediment is then scraped off the plate and shaped into wedges that ride up, one over the other, on steep fault planes. The wedges accumulate at the plate boundary, forming an *accretionary prism* in which metamorphism takes place. This forms a new continental crust of metamorphic rock and builds the continental plate outward.

The accretionary prism is of relatively low density and tends to rise, forming a *tectonic crest*. The tectonic crest appears to be submerged in the figure, but in some cases it forms an island chain, or a *tectonic arc*, paralleling the coast. Between the tectonic crest and the mainland is a shallow trough, the *fore-arc trough*. This trough traps a great deal of terrestrial sediment, which accumulates in the basin-like structure. The bottom of the fore-arc trough continually subsides under the load of the added sediment. In some cases, the sea floor of the trough is flat and shallow, forming a type of continental shelf. Sediment carried across the shelf moves down the steep outer slope of the accretionary prism as *turbidity currents*. These dense tongue-like flows of suspended sediment water move rapidly downslope, like an underwater avalanche. The sediments eventually come to rest with the heaviest, coarsest grains settling first, grading to the finest, lightest grains above. Such materials when compacted and lithified are referred to as *turbidites*.

The descending lithospheric plate enters the asthenosphere where intense heating of the upper surface melts the oceanic crust, forming basaltic magma. As this magma rises, it melts and incorporates material from the base of the crust, and the change in chemical composition produces andesite magma. The rising andesite

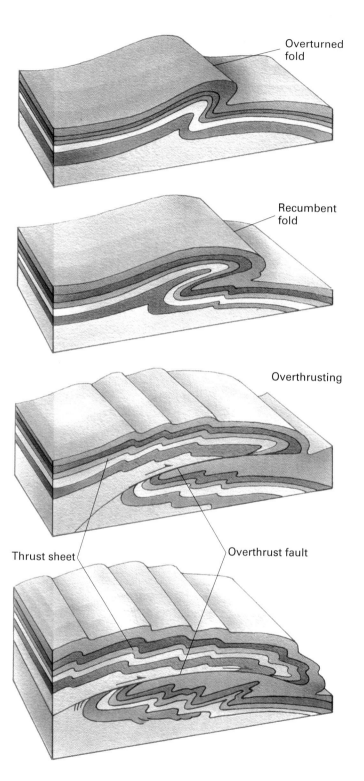

12.15 Folding in compressional tectonic activity These schematic diagrams show the development of a recumbent fold, broken by a low-angle overthrust fault to produce a thrust sheet, or nappe, in an alpine structure. (Based on diagrams by A. Heim, Geologie der Schweiz, vol. II-1, Tauschnitz, Leipzig, 1922)

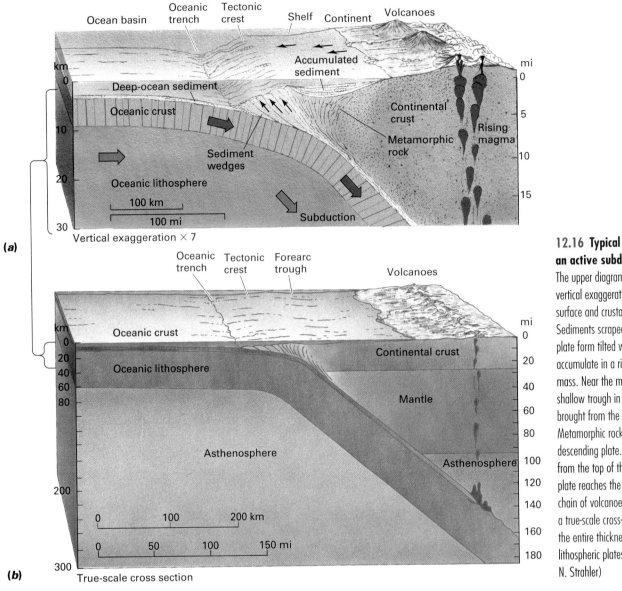

12.16 Typical features of an active subduction zone The upper diagram (a) uses great vertical exaggeration to show surface and crustal details. Sediments scraped off the moving plate form tilted wedges that accumulate in a rising tectonic mass. Near the mainland is a shallow trough in which sediment brought from the land accumulates. Metamorphic rock forms above the descending plate. Magma rising from the top of the descending plate reaches the surface to build a chain of volcanoes. Diagram (b) is a true-scale cross-section showing the entire thickness of the lithospheric plates. (Copyright © A. N. Strahler)

magma reaches the surface to form volcanoes, such as those in the Andes of South America, from which andesite gets its name.

OROGENS AND COLLISIONS

If two continental lithospheric plates converge along a subduction boundary, they must collide because the impacting masses are too thick and too buoyant to allow either plate to slip under the other. The result is an orogeny in which folding and faulting of the crustal rocks leads to the formation of nappes (Figure 12.17). A mass of metamorphic rock is formed between the joined continental plates, welding them together and terminating further tectonic activity along that collision zone. The collision zone is named a **continental suture**.

Continent–continent collisions occurred in the Cenozoic Era along a great tectonic line that marks the southern boundary of the Eurasian plate. The line begins with the Atlas Mountains of

North Africa, and runs across the Aegean Sea region into western Turkey. It appears again in the Zagros Mountains of Iran, where it can be followed discontinuously into the Himalayas. Each segment of this collision zone represents the collision of different northward-moving plates against the single and relatively immobile Eurasian plate. The European segment containing the Alps was formed when the African plate collided with the Eurasian plate. The Persian segment resulted from the collision of the Arabian plate with the Eurasian plate. The Himalayan segment represents the collision of the Indian portion of the Austral-Indian plate with the Eurasian plate.

Continent–continent collisions have occurred many times since the late Precambrian time. Several ancient sutures have been identified in the continental shields. For example, the Ural Mountains, which divide Europe from Asia, formed in this way toward the end of the Paleozoic Era.

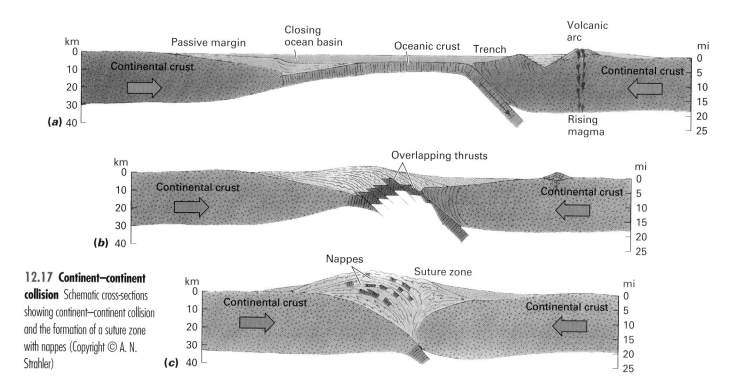

12.17 Continent–continent collision Schematic cross-sections showing continent–continent collision and the formation of a suture zone with nappes (Copyright © A. N. Strahler)

CONTINENTAL RUPTURE AND NEW OCEAN BASINS

Passive continental margins are formed when a single plate of continental lithosphere rifts apart. This process is called *continental rupture* (Figure 12.18) and starts when the crust is lifted and fractured, creating a region of upthrown block mountains and intervening basins. Eventually a long narrow *rift valley* appears (a). The widening rift valley is continually filled in with magma that solidifies to form new crust in the floor. Crustal blocks on either side slip down along a succession of steep faults, creating a mountainous landscape. The Rift Valley system of East Africa is a notable example of this stage of continental rupture. As separation continues, the rift valley widens and opens to the ocean. The ocean enters the narrow valley with a spreading plate boundary running down its centre (b). Rising magma in the central rift produces new oceanic crust and lithosphere. Widening of the basin can continue until a large ocean has formed and the continents are widely separated (c).

The youngest region of continental rupture is in the area of the Red Sea and the Gulf of Aden (Figure 12.19). Formation of the narrow Red Sea began about 30 million years ago and continues today at a rate of about one centimetre per year. As shown in the inset map, this region is a triple junction of three spreading boundaries created by the motion of the Arabian plate pulling away from the African plate. The appearance in 2005 of a 60-kilometre-long rift in the Afar Desert in Ethiopia is related to this.

During the ocean basin opening process, the spreading boundary develops a series of offset breaks that are connected by an active transform fault (see Figure 12.11). As spreading

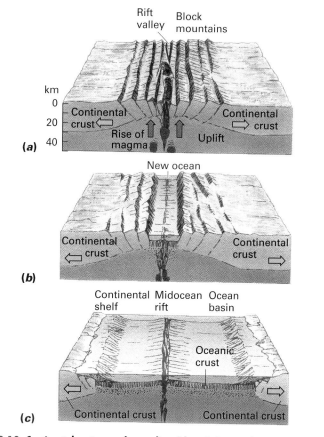

12.18 Continental rupture and spreading Schematic diagrams showing stages in continental rupture and the opening up of a new ocean basin. The vertical scale is greatly exaggerated to emphasize surface features. (a) The crust is uplifted and stretched apart, causing it to break into blocks that become tilted on faults. (b) A narrow ocean is formed, with a new oceanic crust underlying it. (c) The ocean basin widens, while the passive continental margins subside and receive sediments from the continents. (Copyright © A.N. Strahler)

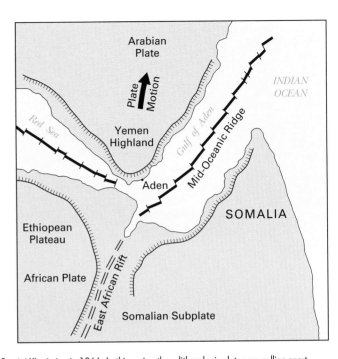

12.19 The Red Sea and Gulf of Aden This photograph was taken by astronauts on the Gemini XI mission in 1966. In this region three lithospheric plates are pulling apart.

continues, the offsets slide past each other and a scar-like feature forms on the ocean floor as an extension of the transform fault. These *transform scars* take the form of narrow ridges or cliff-like features and may extend for hundreds of kilometres. Transform scars are not, for the most part, associated with active faults, even though they are often labelled fracture zones on maps of the ocean floor.

CONTINENTS OF THE PAST

There is evidence to suggest the Earth's land surfaces existed as a single supercontinent, called **Pangaea**, about 200 million years ago. There is also evidence that an earlier supercontinent, *Rodinia*, was fully formed about 700 million years ago. Rodinia broke apart and its fragments were carried away in different directions. Later these ancient continents converged toward a common centre, where they collided and joined to form Pangaea. Assuming that the first supercontinent cycle began 3,000 million years ago (Middle Archean time), it is feasible that 6 to 10 such cycles could have occurred. This repeating cycle is called the Wilson Cycle. It can be considered in six stages (Figure 12.20).

- *Stage 1.* Embryonic ocean basin. The Red Sea separating the Arabian Peninsula from Africa is an active example (Figure 12.18b).
- *Stage 2.* Young ocean basin. The Labrador Basin, a branch of the North Atlantic lying between Labrador and Greenland, is an example of this stage.
- *Stage 3.* Old ocean basin (Figure 12.18c) This includes the entire vast expanse of the North and South Atlantic oceans and the Antarctic Ocean. Passive margin sedimentary wedges have become wide and thick.

- *Stage 4a.* The ocean basin begins to close as continental plates collide with it. New subduction boundaries begin to form.
- *Stage 4b.* Island arcs rise and grow into great volcanic island chains (Figure 12.16). These surround the Pacific plate; the Aleutian arc is an example.
- *Stage 5.* Closing continues. Formation of new subduction margins close to the continents is followed by arc–continent collisions. The Japanese Islands represent this stage.
- *Stage 6.* The ocean basin has finally closed with a collision orogen, forming a continental suture (Figure 12.17c). The Himalayan orogen is a recent example, with activity continuing today.

The hypothesis of a repeating cycle of supercontinents is now a fundamental theme in geology. Although modern plate tectonic theory is only a few decades old, the concept of a breakup of an early supercontinent into fragments that drifted apart dates back into the nineteenth century and beyond. Almost as soon as good navigational charts showing the continents became available, geographers became intrigued with the close correspondence in outline between the eastern coast of South America and the western coast of Africa. Credit for the first full-scale scientific hypothesis of the breakup of a single large continent belongs to Alfred Wegener, who offered geologic evidence, as early as 1915, that the continents had once been united and had drifted apart. He postulated that the supercontinent, Pangaea, existed about 300 million years ago in the Carboniferous Period.

Wegener noted that certain plant and animal fossils were found on several different continents (Figure 12.21a) and reasoned that it was impossible for these organisms to have travelled across the oceans. Similarly, broad belts of rocks of similar age and struc-

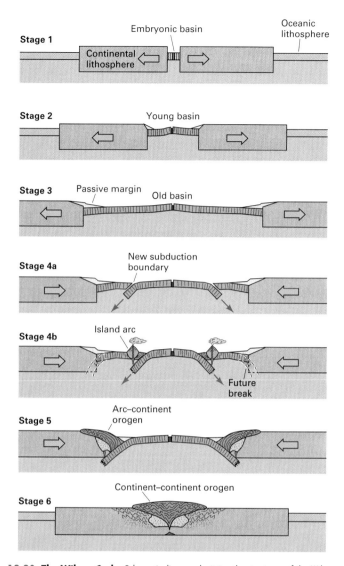

12.20 The Wilson Cycle Schematic diagram depicting the six stages of the Wilson Cycle. The diagrams are not to true scale. (From Plate Tectonics, copyright © Arthur N. Strahler, 1998. Used with permission)

tures occur in Africa and South America and their general location suggests they were formed when the continents were joined (Figure 12.21b). Other commonalities are evident in mountain chains that exhibit similar features in South America and South Africa; in North America, Africa, and Europe; and in Greenland and Europe (Figure 12.21c). Finally, there is evidence that a continental ice sheet covered parts of South America, southern Africa, India, and southern Australia in the Permian Period, about 300 million years ago (Figure 12.21d). Such a glacial episode is most likely if the oceans were much less extensive than they are today.

However, Wegener's explanation of the physical process that separated the continents was weak, and the theory of continental drift was strongly criticized on valid physical grounds. Wegener had proposed that a continental layer of less dense rock had moved like a great floating "raft" through a "sea" of denser oceanic crustal rock. Geologists could illustrate through established principles of physics that this proposed mechanism was impossible.

In the 1960s, however, seismologists showed beyond doubt that the thick lithospheric plates are in motion, both along the midoceanic ridge and beneath the deep offshore trenches of the continents. Geophysicists also used paleomagnetic data in crustal rock to conclude that the continents had moved great distances apart. Approximately 50 years later, Wegener's scenario was validated, but only by applying a mechanism to the process that was unknown at his time.

The continents are moving today. Data from orbiting satellites have shown that rates of separation, or convergence, between two plates are on the order of 5 to 10 centimetres per year (50 to 100 kilometres per million years). At that rate, global geography must have been very different in past geologic eras than it is today. Many continental riftings and plate collisions have taken place over the past two billion years. Single continents have fragmented into smaller ones, while at other times, small continents have merged to form large ones.

Modern reconstructions of the global arrangements of past continents have been available since the mid-1960s. Figure 12.22a shows stages in the positions and movements of the continents, starting in the Permian Period of the Mesozoic Era, about 250 million years ago. In the first map, Pangaea lies astride the equator and extends nearly to the two poles. Regions that are now North America and western Eurasia lie in the northern hemisphere, forming *Laurasia*. Regions that are now South America, Africa, Antarctica, Australia, New Zealand, Madagascar, and India lie south of the equator; together comprising *Gondwana*. Subsequent maps (b through d) show the breaking apart and dispersal of the Laurasia and Gondwana plates to yield their modern components and locations (e).

Note in particular that North America travelled from a low-latitude location into high latitudes, finally closing off the Arctic Ocean as a largely landlocked sea. This change may have been a major cause of the late-Cenozoic ice advances about two million years ago (see Chapter 19). Similarly, the Indian Peninsula started out from a near-subarctic southerly location in Permian times and moved northeast across the Thethys Sea to collide with Asia in the northern tropical savanna zone. In the future, the Atlantic Ocean basin will become much wider, and the westward motion of the Americas will cause a reduction in the width of the Pacific basin.

A Look Ahead

Plate tectonics provides a grand scheme for understanding the nature and distribution of the continents and oceans, which are the largest and most obvious features of the planet. Tectonic activity also accounts for other prominent features, such as volcanoes, mountain ranges, and rift valleys, and the earthquake activity that often accompanies them. The next chapter discusses the distribution, morphology, and forces that create these major structural features.

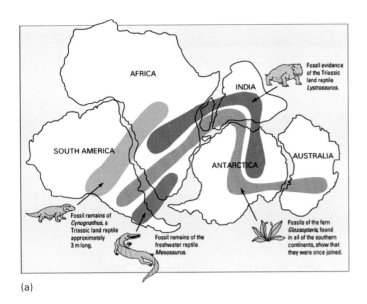

(a)

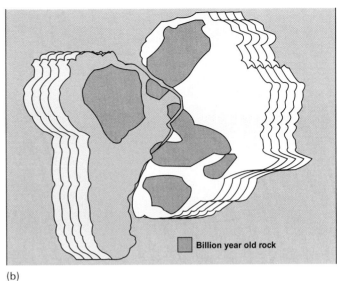

Billion year old rock

(b)

Matching Mountain Ranges
These mountain ranges exhibit
similar geology and age.

Applalachian Range

NORTH
AMERICA

Mauritania Atlas Range

AFRICA

SOUTH
AMERICA

Cape Fold Belt

Mountain Ranges

(c)

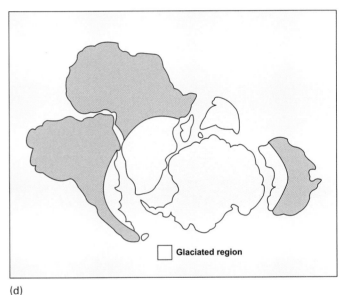

Glaciated region

(d)

12.21 Evidence for continental drift (a) The distribution of several fossil organisms suggests the continents were at one time joined. The distribution of present plant and animal groups provides additional evidence. (b) The distribution of age values obtained by potassium-argon and rubidium-strontium determinations appears to be almost identical for rocks in West Africa and those in potentially contiguous locations in South America. (c) Mountain chains that end at the edge of one continent can be traced to another continent. (d) The distribution of Permian glacial features suggests that the southern continents shared contemporaneous episodes of glaciation.

CHAPTER SUMMARY

- At the centre of the Earth lies its core: a dense mass of liquid iron and nickel that is solid at the very centre. Enclosing the metallic core is the mantle, composed of ultramafic rock. The outermost layer is the crust. Continental crust consists of two zones: a lighter zone of felsic rocks on top of a denser zone of mafic rocks. Oceanic crust consists only of denser, mafic rocks.

The activity of different types of earthquake waves provides details about the Earth's internal structure.

- The lithosphere, the outermost shell of rigid, brittle rock, includes the crust and an upper layer of the mantle. Below the lithosphere is the asthenosphere, a region of the mantle in which mantle rock is soft or plastic.

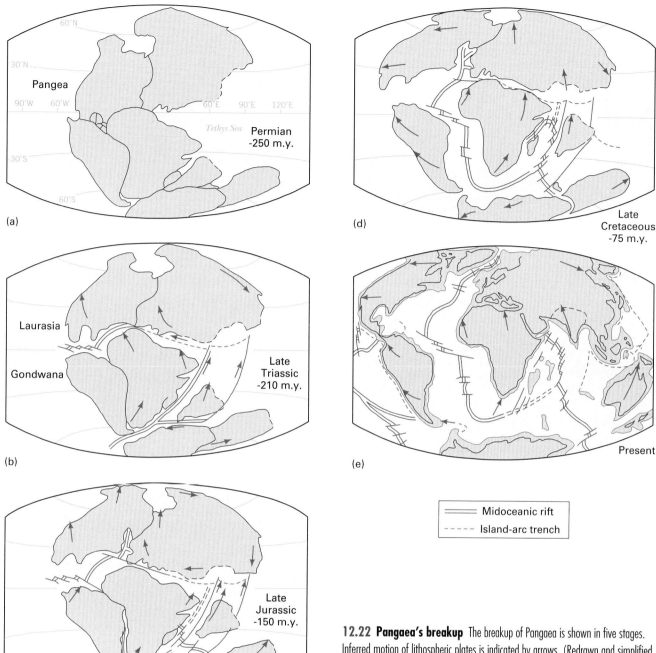

12.22 Pangaea's breakup The breakup of Pangaea is shown in five stages. Inferred motion of lithospheric plates is indicated by arrows. (Redrawn and simplified from maps by R. S. Dietz and J. C. Holden, *Jour. Geophysical Research*, vol. 75, pp. 4943–4951, Figures 2 to 6. Copyrighted by the American Geophysical Union. Used with permission)

- Geologists trace the history of the Earth through the geologic time scale. *Precambrian time* includes the Earth's earliest history. It is followed by three major divisions: the *Paleozoic, Mesozoic,* and *Cenozoic* eras.
- Continental masses consist of active belts of mountain-building and inactive regions of old, stable rock. Mountain-building occurs by *volcanism* and *tectonic activity. Alpine chains* include mountain arcs and island arcs. They occur in two principal mountain belts—the circum-Pacific and Eurasian-Indonesian belts.

- Continental shields are regions of low-lying igneous and metamorphic rocks. They may be *exposed* or *covered* by layers of sedimentary rocks. Ancient *mountain roots* lie within some shield regions.
- The ocean basins are marked by a *midoceanic ridge* with its central *axial rift.* This ridge occurs at the site of crustal spreading. Most of the *ocean basin floor* is *abyssal plain,* covered by fine sediment. As *passive continental margins* are approached, the *continental rise, slope,* and *shelf* are encountered. At *active continental margins,* deep *oceanic trenches* lie offshore.

- The two basic tectonic processes are extension and compression. Both processes can lead to the formation of mountains. *Extension* occurs in the splitting of plates, when the crust thins, fractures, and then pushes upward to produce block mountains. *Compression* occurs when lithospheric plates collide, shaping rock layers into *folds* that then break and move on top of one another along *overthrust* faults.

- *Continental lithosphere* includes the thicker, lighter continental crust and a rigid layer of mantle rock beneath. *Oceanic lithosphere* comprises the thinner, denser oceanic crust and rigid mantle below. The lithosphere is fractured and broken into a set of large and small lithospheric plates that move.

- Where plates move apart, a *spreading boundary* occurs. At *converging boundaries*, plates collide. At *transform boundaries*, plates move past one another on a *transform fault*. There are six major lithospheric plates: Pacific, American (North, South), Eurasian, African, Austral-Indian, and Antarctic.

- When oceanic lithosphere and continental lithosphere collide, the denser oceanic lithosphere plunges beneath the continental lithospheric plate in a process called subduction. A trench marks the site where the plate plunges down. Some subducted oceanic crust melts and rises to the surface, producing volcanoes. Under the severe compression that occurs with continent–continent collision, the two continental plates weld together in a zone of metamorphic rock named a continental suture.

- In *continental rupture*, extensional tectonic forces move a continental plate in opposite directions, creating a *rift valley*. Eventually, the rift valley widens and opens to the ocean, and new oceanic crust forms as the spreading continues.

- Plate movements are thought to be powered by *convection currents* in the plastic mantle rock of the asthenosphere.

- During the Permian Period, the continents were joined in a single, large supercontinent—Pangaea—that broke apart, leading eventually to the present arrangement of continents and ocean basins.

KEY TERMS

asthenosphere	crust	mountain arc	S wave
continental crust	island arcs	oceanic crust	shadow zone
continental shields	lithosphere	P wave	subduction
continental suture	lithospheric plates	Pangaea	surface wave
core	mantle	plate tectonics	tectonic activity

REVIEW QUESTIONS

1. Describe the Earth's inner structure, from the centre outward. What types of crust are present? How are they different?

2. How do geologists use the term *lithosphere*? What layer underlies the lithosphere, and what are its properties? Define the term *lithospheric plate*.

3. More recent geologic time is divided into three eras. Name them in order from oldest to youngest. How do geologists use the terms *period* and *epoch*? What age is applied to time before the earliest era?

4. What proportion of the Earth's surface is ocean? Land? Do these proportions reflect the true proportions of continents and oceans? If not, why not?

5. What are the two basic subdivisions of continental masses?

6. What term describes belts of active mountain-building? What are the two basic processes by which mountain belts are constructed? Provide examples of mountain arcs and island arcs.

7. What is a continental shield? How old are continental shields? What two types of shields are recognized?

8. Describe how compressional mountain-building produces folds, faults, overthrust faults, and thrust sheets (nappes).

9. What role does gravity play in the motion of lithospheric plates?

10. Name the six great lithospheric plates. Identify an example of a spreading boundary by general geographic location and the plates involved. Do the same for a converging boundary.

11. Describe the process of subduction as it occurs at a converging boundary of continental and oceanic lithospheric plates. How is the continental margin extended? How is subduction related to volcanic activity?

12. How does continental rupture produce passive continental margins? Describe the process of rupturing and its various stages.

13. What are transform faults? Where do they occur?

14. What is the Wilson Cycle of plate tectonics?

15. Using a sketch, describe the Cycle by which supercontinents are formed and reformed.

16. Provide a brief history of the idea of "drifting continents."

17. What was Wegener's theory on "continental drift?" Why was it opposed at the time?

18. Briefly summarize the reconstructed spatial history of the continents beginning with the Permian period.

VISUALIZATION EXERCISES

1. Sketch a cross-section of an ocean basin with passive continental margins. Label the following features: midoceanic ridge, axial rift, abyssal plain, continental rise, continental slope, and continental shelf.

2. Identify and describe two types of lithospheric plates. Sketch a cross-section showing a collision between the two types. Label the following features: oceanic crust, continental crust, mantle, oceanic trench, and rising magma. Indicate where subduction is occurring.

3. Sketch a continent–continent collision and describe the formation process of a continental suture. Provide a present-day example where a continental suture is forming, and give an example of an ancient continental suture.

ESSAY QUESTION

1. Suppose astronomers discover a new planet that, like the Earth, has continents and oceans. They dispatch a reconnaissance satellite to photograph the new planet. What features would you look for, and why, to detect past and present plate tectonic activity on the new planet?

PROBLEMS   WORKING IT OUT 12.1 Radioactive Decay and Radiometric Dating

1. The half-life of potassium-40 (^{40}K) is 1.28 billion years. What percent of a quantity of ^{40}K will remain after 1 billion years? 3 billion years?

2. The half-life of thorium-232 (^{232}Th) is 14.1 billion years. What percent of a quantity of pure ^{232}Th will remain after 5 billion years? 10 billion years? 15 billion years?

3. The unstable isotope carbon-14 (^{14}C) has a half-life of 5,730 years. How many years will it take for a pure sample of ^{14}C to reduce to 0.1 (10 percent)?

4. A geochemist analyzes a sample of zircon obtained from an igneous rock and observes that the ratio of ^{206}Pb to ^{238}U atoms is 0.448. Using the formula, what is the age of the sample?

5. The geochemist also determines that the atomic ratio of ^{207}Pb to ^{235}U is 9.56 for the same sample. For this decay process, the half-life is $H = 7.04 \times 10^8$ years. Calculate k and then determine the age from the D/M ratio of 9.56. Is it consistent with the age obtained from the ^{206}Pb/^{238}U ratio?

Find answers at the back of the book

EYE ON THE LANDSCAPE |

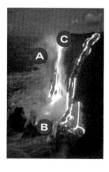

Chapter Opener Volcanic activity in Hawaii. The outpouring of basaltic lava (**A**) is associated with the movement an oceanic plate over a hot spot within the deeper aesthenosphere. The hot molten rock flows quite readily, but is quickly quenched in the sea water to form an apron at the base of the cliff (**B**). This addition of new material reduces the power of the waves (**C**) and slows the rate of shoreline erosion (see chapter 18)

Vestmannaeyjar, Iceland. This chain of islands is located to the south Iceland and was formed by periodic volcanic activity on the mid-Atlantic ridge. The most recent island, Surtsey, began to form in 1963. Once the eruption ceases, the protruding rock is

subject to the strong wave action of the Atlantic Ocean leading to the formation of steep cliffs (**A**). Volcanic soils are very fertile and vegetation quickly colonizes the land surface (**B**).

Chapter 13

EYE ON THE LANDSCAPE

> > > >

Deep-sea hydrothermal vent.
What else would the geographer see? . . . Answers at the end of the chapter.

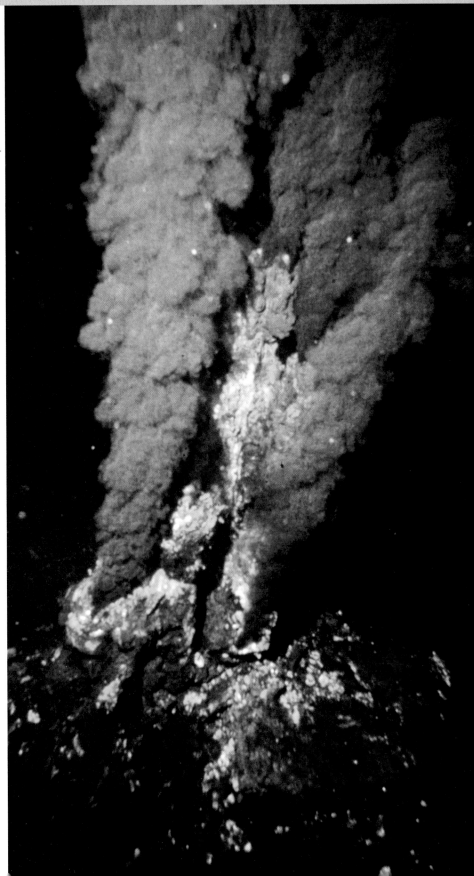

VOLCANIC AND TECTONIC LANDFORMS

Landforms
Volcanic Activity
Stratovolcanoes
Shield Volcanoes
Focus on Systems 13.1 • The Life Cycle of a Hotspot Volcano
Eye on the Environment 13.2 • Geothermal Energy Sources
Volcanic Activity across the Globe
Volcanic Eruptions as Environmental Hazards
Focus on Remote Sensing 13.3 • Remote Sensing of Volcanoes
Landforms of Tectonic Activity
Fold Belts
Faults and Fault Landforms
Earthquakes
Working It Out 13.4 • Earthquake Magnitude
Earthquakes and Plate Tectonics
Seismic Sea Waves

The two previous chapters on the Earth's materials and plate tectonics showed how forces within the Earth create major structural features. These same forces create landforms with distinctive characteristics on a regional scale and readily identifiable individual features in local landscapes. Stresses from compression generated by plate motions fold flat-lying rocks into wavelike forms, as seen in the Canadian Rockies. Faults mark zones in which rock layers break and move past one other, producing uplifted mountain blocks next to deep valleys. Similarly, upwelling magma spews forth explosively from vents and fissures, creating steep-sided volcanoes. Many of these tectonic processes are accompanied by strong earthquakes. This chapter provides an overview of the landscape features that are created by internal Earth forces.

LANDFORMS

Landforms are the natural physical features found on the Earth's surface, such as, mountain peaks, cliffs, canyons, plains, beaches, and sand dunes. Their creation, subsequent transformation, and ultimate removal from the landscape exemplify the dynamic nature of the planet. **Geomorphology** is the scientific study of the processes that shape landforms. The shapes of continental surfaces reflect the balance between two sets of forces. Internal Earth forces fold, fracture, and warp crustal materials through volcanic and tectonic processes. Additional processes and forces active at the Earth's surface, such as flowing water, wave action, glacial ice, and wind, modify the landscape by removing, transporting, and depositing mineral matter. Each geomorphological agent produces unique and distinctive landforms.

Landforms generally fall into two basic groups: initial landforms and sequential landforms. *Initial landforms* are directly produced by volcanic and tectonic activity. They include volcanoes and lava flows, as well as rift valleys and elevated mountain blocks in zones of recent crustal deformation. Radioactivity in the crustal and mantle rocks generates the energy for lifting molten rock and rigid crustal masses to produce the initial landform. Landforms shaped by processes and agents of denudation, such as river valleys and sand dunes, belong to the group of *sequential landforms*, which develop after the initial landforms have been created (Figure 13.1).

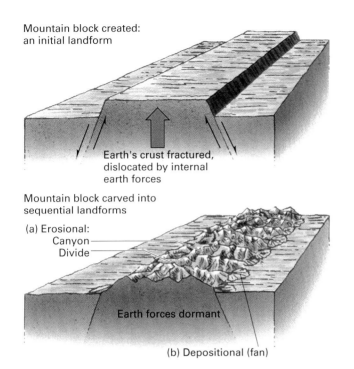

13.1 Initial and sequential landforms An initial landform is created, here by tectonic activity, then carved into sequential landforms. (Drawn by A.N. Strahler)

VOLCANIC ACTIVITY

Magma extruded by volcanic activity (*volcanism*) can form imposing mountain ranges comprising volcanic peaks and accumulated lava flows. A **volcano** is a conical or dome-shaped initial landform built of lava and ash emitted from a constricted vent in the Earth's surface. The magma rises in a narrow, pipe-like conduit from a magma reservoir lying beneath. Upon reaching the surface, magma may pour out in tongue-like lava flows, or it may be violently ejected in the form of solid fragments driven skyward under the pressure of confined gases. Ejected fragments, ranging in size from huge boulders to fine dust, are collectively called *tephra*. The size and shape of a volcano depends on the type of lava and the amount of tephra that it ejects during eruptions (Figure 13.2).

STRATOVOLCANOES

The nature of volcanic eruptions, whether explosive or subdued, depends on the type of magma. Felsic lava (rhyolite and andesite) has a high degree of viscosity; it is thick and sticky, and resists flow. Consequently, volcanoes of felsic composition typically have steep slopes, and lava does not usually flow far from the vent. When the volcano erupts, tephra falls on the area surrounding the crater and contributes to the structure of the cone. Included in the tephra are volcanic bombs, which are solidified masses of lava that can be the size of large boulders. They fall close to the crater.

13.2 Mount St. Helens This stratovolcano of the Cascade Mountain Range in southwestern Washington erupted violently on the morning of May 18, 1980, emitting a great cloud of condensed steam, heated gases, and ash from the summit crater. Within a few minutes, the plume had risen to a height of 20 kilometres. *Fly By: 46° 12' N; 122° 11' W*

The inter-layering of sluggish streams of felsic lava and eruptions of tephra produces **stratovolcanoes**, which are sometimes referred to as *composite volcanoes*, or composite cones, since they are formed from layers of ash and lava (Figure 13.3). Fine examples of stratovolcanoes include Mount Baker, Mount Rainier, and Mount Shasta in the Cascade Mountain Range; Mount Fuji in Japan; and Mount Mayon in the Philippines (Figure 13.4). Their tall, steep-sided cones usually become steeper toward the summit, where a bowl-shaped depression—the

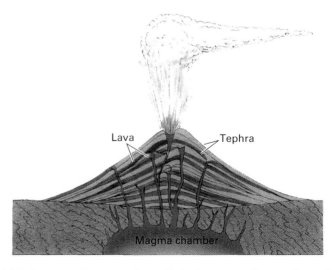

13.3 Anatomy of a stratovolcano Idealized cross-section of a stratovolcano with feeders from the magma chamber beneath. The steep-sided cone is built up from layers of lava and tephra. (Copyright © A. N. Strahler)

13.4 Mount Mayon Located in southeastern Luzon, the Philippines, Mount Mayon is often considered the world's most perfectly-shaped stratovolcanic cone. Its summit rises to an altitude of nearly 2,400 metres. Mount Mayon has erupted at least 40 times since 1616. The most recent eruption was in March 2000, causing the evacuation of a local population of about 55,000 people.

13.5 Nuée ardent A cloud of hot, dense volcanic ash, emitted by the Soufrière Hills volcano, courses down this narrow valley on the island of Montserrat in the Lesser Antilles. *Fly By: 16° 43' N; 62° 11' W*

crater—is located. The crater is the volcano's principal vent. Felsic lava holds large amounts of gas under high pressure, and usually produces explosive eruptions. Fine volcanic dust from these eruptions can rise high into the troposphere and stratosphere, travelling hundreds or thousands of kilometres before settling to the Earth's surface.

Another important form of emission from explosive stratovolcanoes is a cloud of white-hot gases and fine ash. This intensely hot cloud, called a *nuée ardente* or ash flow, travels rapidly down the flank of the volcanic cone, searing everything in its path. Mount Pelée, on the Caribbean island of Martinique, erupted in this way in 1902, destroying the city St. Pierre and killing all but two of its 30,000 inhabitants. A similar eruption of the Soufrière Hills volcano on Montserrat in 1997 left the southern two thirds of the island uninhabitable (Figure 13.5).

Calderas

One of the most catastrophic of natural phenomena is a volcanic explosion so violent that it destroys the entire central portion of the volcano. Vast quantities of ash and dust are emitted and fill the atmosphere for many hundreds of square kilometres around the volcano. Only a great central depression, named a **caldera**, remains after the explosion. Although some of the upper part of the volcano is blown outward in fragments, most of it settles back into the cavity left by the explosion.

Krakatoa, a volcanic island in Indonesia, exploded in 1883, leaving a huge caldera. Great seismic sea waves (*tsunamis*) generated by the explosion killed many thousands of people living in low coastal areas of Sumatra and Java. About 25 cubic kilometres of rock was blown out of the crater during the explosion. Vast

quantities of gas and fine particles of dust were carried into the stratosphere, where they contributed to the rosy glow of sunrises and sunsets seen around the world for several years afterward.

A classic example of a caldera produced in prehistoric times is Crater Lake, Oregon (Figure 13.6). The former volcano, named Mount Mazama, is estimated to have risen 1,200 metres higher than the present caldera rim. The great explosion and collapse occurred about 6,600 years ago. Ash layers associated with this eruption are reported in British Columbia, Alberta, and Saskatchewan, as well as the western United States, where they have been important stratigraphic tools in several archaeological studies.

13.6 Crater Lake Crater Lake, Oregon, is a water-filled caldera marking the remains of the summit of Mount Mazama, which exploded about 6,600 years ago. Wizard Island (centre foreground) built up on the floor of the caldera after the major explosive activity had ceased. It is an almost perfectly-shaped cone of cinders capping small lava flows. *Fly By: 42° 56' N; 122° 06' W*

Stratovolcanoes and Subduction ARCS

Most of the world's active stratovolcanoes lie within the circum-Pacific mountain belt. Andesitic magmas rise beneath volcanic arcs of active continental margins and island arcs. For example, the volcanic arc of Sumatra and Java lies over the subduction zone between the Australian plate and the Eurasian plate. Active subduction of the Pacific plate beneath the North American plate is associated with the Aleutian volcanic arc and the chain of volcanoes found in the Cascade Mountains of northern California, Oregon, and Washington. This merges with the Garibaldi volcanic belt of southern British Columbia, where the small Juan de Fuca plate is similarly being forced under the North American plate. Volcanic activity in Central America is linked with the subduction of the Cocos plate. In South America, the subduction zone between the Nazca and South American plates accounts for the stratovolcanoes that occur in various segments of the Andes Mountains.

SHIELD VOLCANOES

In contrast to thick, gassy felsic lava, mafic lava (basalt) is often highly fluid. It typically has a low viscosity and holds little gas. As a result, eruptions of basaltic lava are usually subdued and the lava can travel long distances, spreading out in thin layers (Figure 13.7). Large basaltic volcanoes are typically broad rounded domes with gentle slopes. They are called **shield volcanoes**. Hawaiian volcanoes are of this type. Smaller shield volcanoes are also found in Iceland and the Galapagos Islands, as well as in northern California and Oregon, where they are typically five to seven kilometres in diameter, with heights of 450 to 600 metres.

The shield volcanoes of the Hawaiian Islands are characterized by smooth, gently rising slopes that flatten near the top, producing a broad-topped mountain (Figure 13.8). Domes on the island of Hawaii rise to summit elevations of about 4,000 metres; however, because they have grown from the sea bed, their accumulated height is more than twice that. In width they range from 16 to 80 kilometres at sea level and up to 160 kilometres at their submerged base. Most of the lava flows from fissures (long, gaping cracks) on the flanks of the volcano.

Hawaiian lava domes have a wide, steep-sided central depression that may be three kilometres or more in diameter and several hundred metres deep. These large depressions are a type of collapsed caldera. Molten basalt is sometimes seen in the floors of deep pit craters that occur on the floor of the central depression or elsewhere over the surface of the lava dome.

Hotspots, Sea-Floor Spreading, and Shield Volcanoes

The chain of Hawaiian volcanoes was created by the motion of the Pacific plate over a *hotspot*—a plume of upwelling basaltic magma from very deep within the Earth's mantle. As the hot mantle rock rises, magma forms in bodies that melt their way through the lithosphere and reach the sea floor. Each major pulse of the plume sets off a cycle of volcano formation. However, the motion of the oceanic lithosphere eventually carries the volcano away from the location of the deep plume, and so it becomes extinct. Erosion processes wear the volcano away, and ultimately it becomes a low island. Continued attack by waves and slow settling of the island reduces it to a coral-covered platform. Eventually only a sunken seamount, or *guyot*, exists. *Focus on*

13.7 Lava flows from the Kilauea Volcano. Hot, fluid basaltic lava typically flows over existing surfaces which progressively adds to the size and height of shield volcanoes. *Fly By: 19° 25' N; 155° 17' W*

13.8 Basaltic shield volcanoes in Hawaii. At lower left is the now-cold Halemaumau pit crater, formed in the flow of the central depression of Kilauea volcano. In the distance, the snow-capped summit of Mauna Kea shows the characteristic form of a shield volcano; its elevation is over 4000 metres.

Systems 13.1 • The Life Cycle of a Hotspot Volcano describes this process in more detail.

A few shield volcanoes also occur along the midoceanic ridge, where sea-floor spreading is in progress. An outstanding example is Iceland, which is constructed entirely of basalt. Basaltic flows are superimposed on older basaltic rocks as dikes and sills formed by magma emerging from deep within the spreading rift. Mount Hekla, an active volcano on Iceland, is a shield volcano somewhat similar to those of Hawaii. Other islands consisting of shield volcanoes located along or close to the axis of the Mid-Atlantic Ridge are the Azores, Ascension, and Tristan da Cunha.

Where a mantle plume lies beneath a continental lithospheric plate, the hotspot may generate an enormous volume of basaltic lava that emerges from numerous vents and fissures and accumulates layer upon layer. The basalt may ultimately become thousands of metres thick and cover thousands of square kilometres. These accumulations are called *flood basalts.*

An important North American example is in the Columbia Plateau region of southeastern Washington, northeastern Oregon, and western Idaho. Here, basalts from the Cenozoic Era cover an area of about 130,000 square kilometres. Individual basalt flows are exposed along the walls of river gorges as cliffs in which vertical joint columns are conspicuous (Figure 13.9).

Small volcanoes, known as a *cinder cones* (Figure 13.10), are often found with shield volcanoes and basaltic flows. Cinder cones form when frothy basalt magma is ejected under high pressure from a narrow vent, producing tephra. The rain of tephra accumulates around the vent to form a roughly circular hill with a central crater. Cinder cones rarely grow to heights of more than a few hundred metres.

Although there is evidence of ancient volcanic activity throughout Canada, comparatively young volcanoes (i.e., less than 5 million years old) are found only in British Columbia and the Yukon. They are extensions of the Cascades Volcanic Belt that runs through Washington, Oregon, and California. Most are stratovolcanoes, some of which, such as Mount Nazko, have erupted in the past 7,000—8,000 years. Mount Meager, in the Garibaldi Volcanic Belt, erupted about 2,350 years ago and is considered the most recent explosive eruption in Canada, although several cinder cones have erupted in the past 500 years. The most recent eruption reported in Canada was the Iskut-Unuk River cinder cone event in 1904, located at lat. 56.58 N, long 130.55 W. Located in this area, Mount Edziza (2,787 metres) is a shield volcano similar to those in Hawaii. It began to form four million years ago. Although the last basalt flow on Mount Edziza occurred 10,000 years ago, about 30 cinder cones have since built up on its flanks.

Hot Springs and Geysers

Where hot rock material is near the Earth's surface, it can heat nearby groundwater to high temperatures. When the groundwater reaches the surface, it provides *hot springs* at temperatures not far below the boiling point of water (Figure 13.11). At some places, jet-like emissions of steam and hot water occur at intervals from small vents—producing *geysers* (Figure 13.12). Since the water that emerges from hot springs and geysers is largely groundwater that has been heated by contact with hot rock, this water is recycled surface water.

Hot springs and geysers are found wherever there are active or recently dormant volcanoes, for example, the North Island of New Zealand, Chile, Italy, Iceland, Japan, and California. In

13.9 Flood basalts Basaltic lava flows exposed in cliffs bordering the Columbia River in Washington. Each set of cliffs is a major lava flow. In cooling, vertical cracks form in tho lava, croating tall columns.

13.10 Cinder cone A cinder cone built up on the flank of Mount Edziza, a large shield volcano in northern British Columbia.

13.11 Mammoth Hot Springs Small terraces ringed by mineral deposits hold steaming pools of hot water as the spring cascades down the slope. This example of geothermal activity is from Yellowstone National Park, Wyoming. *Fly By: 44° 28' N; 110° 50' W*

13.12 Old Faithful Geyser An eruption of Old Faithful Geyser in Yellowstone National Park, Wyoming.

Canada, the best known locations are in Alberta, British Columbia, and the Yukon. In the 1880s, commercial development of hot springs began in Banff, which, because of its scenic mountain location, has become a world tourist attraction. British Columbia also has a significant number of commercial hot springs. Water emerges at a temperature of about 50°C in Canadian hot springs, so spas must mix it with cooler water before they can use it.

The heat from masses of lava close to the surface in areas of hot springs and geysers provides a source of energy for electric power generation, and numerous geothermal power stations are in operation. Geothermal facilities supply about 8,900 megawatts of electricity in 24 countries; this meets the needs of some 60 million people.

In Canada, drilling is underway at Mount Meager, to confirm its potential to become Canada's first commercial geothermal power station with a projected capacity of 100 to 200 megawatts. Temperatures as high as 225°C at depths of 800 metres have been recorded in test borings at this site. In the Yukon, geothermal energy is used to prevent water pipes from freezing. *Eye on the Environment 13.2 • Geothermal Energy Sources* provides more information about this resource.

VOLCANIC ACTIVITY ACROSS THE GLOBE

Figure 13.13 shows the locations of volcanoes that have been active within the last 12,000 years. Many volcanoes are located along subduction boundaries (see Figure 12.4). The "ring of fire" around the Pacific Rim is a prominent feature on the map. Other volcanoes, such as those in Iceland, are located on or near midoceanic rifts. Rifting has also produced volcanic regions in East Africa. Hotspot activity is also represented in Hawaii, the Society Island group, and MacDonald Island in the Pacific. *Focus on Remote Sensing 13.3 • Remote Sensing of Volcanoes* presents some volcano images taken using different techniques.

VOLCANIC ERUPTIONS AS ENVIRONMENTAL HAZARDS

Volcanic eruptions and lava flows are severe environmental hazards, often taking a heavy toll of plant and animal life and devastating human habitations. Complete loss of life and destruction of towns and cities are frequent in the history of peoples who live near active volcanoes. Loss occurs from the relentless advance of lava flows engulfing whole cities; from

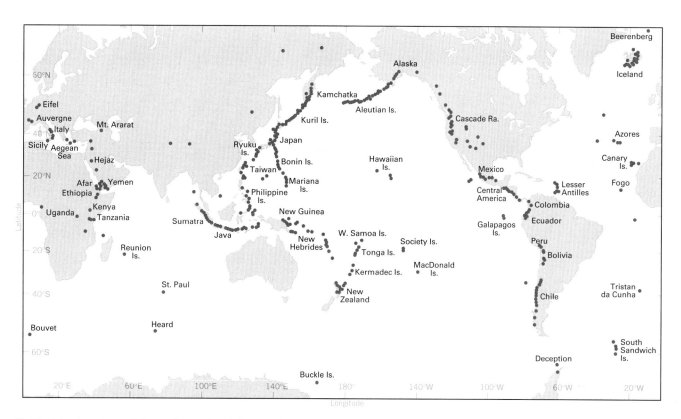

13.13 Volcanic activity of the Earth Dots show the locations of volcanoes known or believed to have erupted within the past 12,000 years. Each dot represents a single volcano or cluster of volcanoes. (From data of NOAA. Copyright © A. N. Strahler)

showers of ash, cinders, and volcanic bombs; from clouds of incandescent gases that descend the volcano slopes; and from violent earthquakes associated with volcanic activity. For habitations along low-lying coasts, there is the additional peril of great seismic sea waves, generated by eruptions of undersea or island volcanoes elsewhere.

In 1985, an explosive eruption of Nevado del Ruiz, a volcano in the Colombian Andes, caused the rapid melting of ice and snow at the summit. Mixing with volcanic ash, the water formed a variety of mudflow known as a *lahar*. Rushing down slopes at speeds of up to 145 kilometres per hour, the lahar was channelled into a valley on the lower slopes, where it engulfed a town and killed more than 20,000 people.

Scientific monitoring techniques are reducing the toll of death and destruction from volcanoes. By analyzing the gases emitted from the vent of an active volcano, as well as the minor earthquakes and local land tilting that precede a major eruption, scientists have successfully predicted periods of volcanic activity. Extensive monitoring of Mount Mayon and the Mexican volcano Popocatepetl to predict recent eruptions has led to evacuations that saved thousands of lives. However, not every volcano is well monitored or predictable.

Despite their potential for destructive activity, volcanoes are a valuable natural resource in terms of recreation and tourism.

British Columbia's Garibaldi Provincial Park preserves volcanic mountain vistas in a vast wilderness area of snow-covered peaks and swift rivers. Hawaii Volcanoes National Park recognizes the natural beauty of Mauna Loa and Kilauea—their breathtaking displays of molten lava are a living textbook of igneous processes. Similarly, Mount Rainier, Mount Lassen, and Crater Lake in the Cascade Mountain Range are national parks.

LANDFORMS OF TECTONIC ACTIVITY

Two basic forms of activity—compression and extension—are associated with the movement of the global lithospheric plates. Tectonic activity along converging plate boundaries is primarily compressional, with the severest compression occurring where continents collide. Compression also occurs in subduction zones, as the descending plate forces layers of sediment on the ocean floor against the overlying plate. In zones of rifting, the brittle continental crust is pulled apart and yields by faulting.

FOLD BELTS

Severe compressional stress caused by movement of continental plates can result in **folding** of sediments on the continental shelf or margin. The wavelike shapes imposed on the strata consist of

FOCUS ON SYSTEMS | 13.1   The Life Cycle of a Hotspot Volcano

The time cycles examined in earlier chapters are repetitive. That is, each cycle follows the previous one, in an endless procession. The astronomical cycles of Earth–Sun relationships, for example, generate repeating annual cycles of daily air temperature and monthly rainfall.

Here, attention focuses on a *life cycle*, which does not repeat, but rather consists of a continuous progression that has a number of stages from a beginning to an end. The higher organisms have life cycles, expressed in the various stages of growth, development, and aging from conception to death.

Non-organic life cycles also can be identified in nature. For example, the life of a raindrop starts with condensation of free gaseous water molecules, forming a tiny cloud droplet that in turn coalesces through collisions with other droplets to become a large drop. It ends its cycle by colliding with the surface of the Earth. Volcanoes have this kind of life cycle. A volcano can appear at a location where no other volcanic activity has taken place. The volcanic cone grows, then seemingly becomes dormant, and may end its life in a gigantic explosion that leaves a deep caldera.

The life cycle of a basaltic shield volcano that grows upward from the abyssal ocean floor is particularly interesting. This growth may take it from more than 5,000 metres below the sea surface to a height greater than 4,000 metres above the surface, thus building a huge mass of igneous rock on top of the sea floor. A unique feature of this type of volcano is that it rests on an oceanic lithosphere that is in steady horizontal motion over a soft asthenosphere.

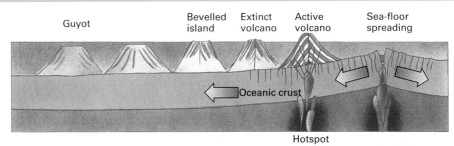

Volcanic chain A chain of volcanoes is formed by an oceanic plate moving over a hotspot. (Copyright © by A. N. Strahler)

The chain of Hawaiian volcanic domes is a good example of this type of volcano. However, this chain lies far away from the spreading boundary of the Pacific plate, which would be a natural source of magma. What, then, is the magma source for these islands? Based on geophysical evidence, geologists have concluded that there is a *hotspot* beneath the islands—a persistent source of heat providing pulses or bubbles of rising magma that built the Hawaiian chain, island by island, as the Pacific plate moved over the hotspot (see figure above).

The figure to the right shows the typical stages of the life cycle. In the first stage (1), magma rising from the hotspot on the ocean floor forms a low basaltic lava dome. After the dome reaches full height (2), a central caldera forms (3). A post-caldera stage (4) then follows in which large cinder cones fill the caldera and renew some of the original mass. Eventually, dormancy sets in and erosion processes lower the mountain's height, while waves cut back the coast (5). Fringing coral reefs that form on the dome become broader. The volcano is now fully extinct. No longer pushed upward by upwelling magma, the oceanic crust holding the volcano steadily subsides,

lowering its height further. Erosional bevelling finishes at the atoll stage (6), when a thick layer of reef corals and related lagoon sediments forms. In the final stage (7), continued crustal subsidence drowns the reef corals, and only a sunken island or seamount, called a *guyot*, remains.

It is now agreed that a *mantle plume*, arising far down in the asthenosphere, produced the hotspot that generated the entire chain of islands. As the hot mantle rock rises, magma forms in bodies that melt their way through the lithosphere to reach the sea floor. Each major pulse of the plume sets off a volcanic cycle. As the volcano moves away from the location of the deep plume, it undergoes the middle and late stages (5–7) of the cycle. This lithospheric motion has produced a long trail of sunken islands and guyots, shown on the map above (see also Figure 12.8). The Hawaiian trail, trending northwestward, is 2,400 kilometres long and includes a sharp bend to the north caused by a sudden change of direction of the Pacific plate. This distant leg consists of the Emperor Seamounts. Several other long trails of volcanic seamounts cross the Pacific Ocean basin. They, too, follow parallel paths that reveal the plate motion.

1. Deep marine stage

2. Shield-building stage

3. Caldera stage

4. Cinder cone stage

5. Erosional stage

6. Atoll stage

7. Guyot stage

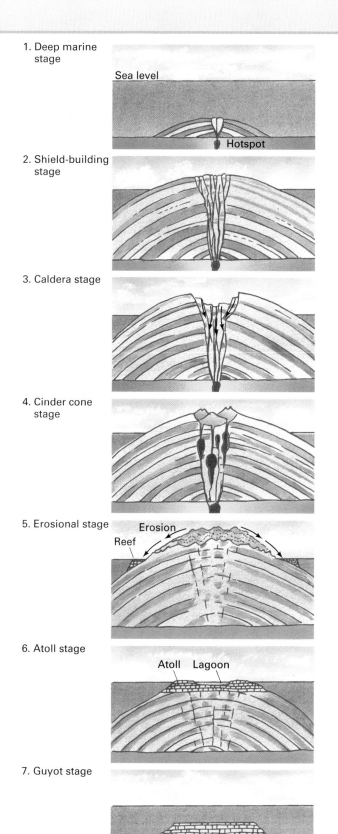

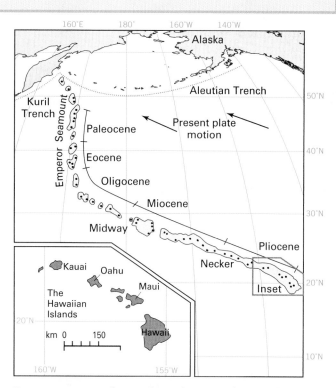

Hawaiian seamount chain in the northwest Pacific Ocean basin Dots are summits; the enclosed coloured area marks the base of the volcano at the ocean floor. (Copyright © by A. N. Strahler)

Life cycle of a typical Hawaiian volcano (Adapted with permission from Macdonald and Abbott, 1970, *Volcanoes in the Sea: The Geology of Hawaii*, Honolulu, University of Hawaii Press, p. 138, Figure 92)

EYE ON THE ENVIRONMENT | 13.2   Geothermal Energy Sources

Geothermal energy is energy in the form of sensible heat that originates within the Earth's crust and makes its way to the surface by conduction. Heat may be conducted upward through solid rock or carried up by circulating groundwater that is heated at depth before returning to the surface. Concentrated geothermal heat sources are usually associated with igneous activity, but deep zones of heated rock and groundwater that are not directly related to igneous activity also exist.

Observations made in deep mines and bore holes show that the rock temperature increases steadily with depth and attains very high values in the upper mantle. Heat within the Earth's crust and mantle is produced largely by radioactive decay, and the basic energy resource it provides can be regarded as limitless on a human scale.

It might seem simple enough to obtain all our energy needs by drilling deep holes at any desired location into the crust and letting the hot rock turn injected fresh water into steam, which could then be used to generate electricity. Unfortunately, at the depths usually required to furnish the needed heat intensity, crustal rock tends to close any cavity by rupture and slow flowage. Thus, geothermal locations occur where special conditions have caused hot rock and hot groundwater to lie within the range of conventional drilling methods.

Natural hot water and steam were the first type of geothermal energy source to be developed and at present account for nearly all production of geothermal electrical power. Wells are drilled to tap the hot water. When it reaches the surface, the water turns to steam under the reduced pressure of

Geothermal power plant This electricity-generating power plant at Wairakei, New Zealand, runs on steam produced by super-heated groundwater. Steam pipes in the foreground lead to the plant.

the atmosphere. The steam is fed into generating turbines to produce electricity, then condensed in large cooling towers (see photo above). The resulting hot water is usually released into surface stream flow, where it may create a thermal pollution problem. The larger steam fields have sufficient energy to generate at least 15 megawatts of electric power, and a few can generate 200 megawatts or more.

In certain areas, the intrusion of magma has been recent enough that the solidified igneous rock of a batholith is still very hot at depths of perhaps two to five kilometres. Rock in this zone may be as hot as 300°C and could supply an enormous quantity of heat energy. The planned development of this resource includes drilling into the hot zone and then shattering the sur-

rounding rock by hydrofracture—a method using water under pressure that is widely used in petroleum development. Surface water would then be pumped down one well into the fracture zone and heated water pumped up another well. Although some experiments have been conducted, this heat source has not yet been exploited in any practical way.

On a smaller scale, groundwater at 10–20°C can be used directly or with heat pumps to heat buildings. At present, more than 30,000 buildings in Canada are heated in this way. In Nova Scotia, water from the flooded Springhill coal mine is used for this purpose. Annual energy savings compared with conventional sources amount to about 600,000 kilowatt hours—and the process is non-polluting.

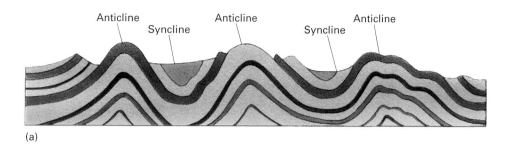

(a)

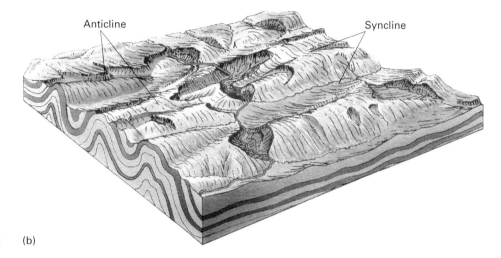

13.14 Anticlines and synclines Structural diagrams for the Jura Mountains in France and Switzerland show a variety of features that are common in the Canadian Rockies and other fold mountain areas: (a) typical cross-section and (b) landscape developed on folded strata. (From E. Raisz)

(b)

alternating arch-like upfolds, called **anticlines**, and trough-like downfolds, called **synclines** (Figure 13.14) Originally each mountain crest would have been associated with the axis of an anticline, with intervening valleys lying over the axis of a syncline. Some of the anticlinal arches may be partially removed by erosion processes creating step-like skylines in some areas. These features can be seen throughout the Canadian Rockies (Figure 13.15).

FAULTS AND FAULT LANDFORMS

A **fault** in the brittle rocks of the Earth's crust occurs when rocks suddenly yield to unequal stresses by fracturing. Faulting is accompanied by a displacement—a slipping motion—along the plane of breakage, or *fault plane*. Faults often extend great horizontal distances, and a fault line can sometimes be followed along the ground for many kilometres. Most major faults also extend down into the crust for at least several kilometres.

(a)

(b)

13.15 Folded strata (a) Complex folding in the Sullivan River area of the Rocky Mountains in British Columbia. (b) At 3,954 metres, Mount Robson in British Columbia is the highest peak in the Canadian Rockies. The compressed rock in the axis of an anticline is usually more resistant to denudation and many of the higher peaks in fold mountains are formed because of this.

Mount Vesuvius imaged by ASTER *Fly By: 40° 49' 20" N; 14° 25' 30" E*
(Image courtesy NASA/GSFC/MITI/ERSDAC/JAROS and U. S./Japan ASTER Science Team)

Mount Vesuvius erupted in A.D. 79, burying the nearby Roman city of Pompeii. In recent history, major eruptions of Mount Vesuvius were recorded in 1631, 1794, 1872, 1906, and 1944. The image of Mount Vesuvius was acquired by the Advanced Spaceborne Thermal Emission and Reflection Radiometer (ASTER) on September 26, 2000. Spatial resolution is 15 by 15 metres. Red, green, and blue colours in the image have been assigned to near-infrared, red, and green spectral bands, respectively. Vegetation appears bright red and urban areas in blue and green tones. The magnitude of development around the volcano shows that the impact of a major eruption would be catastrophic.

Mount Fuji lies about 100 kilometres southwest of Tokyo. Although it has the symmetry of a simple cone, it is actually a complex structure with two former volcanic cones buried within its outer form. Mount Fuji is considered an active volcano; it last erupted in 1707. This image (right) was acquired by the Shuttle Radar Topography Mission's interferometric synthetic aperture radar on February 21, 2000. This type of radar sends simultaneous pulses of radio waves toward the ground from two antennas spaced 60 metres apart. Very slight differences in the return signals can be related to the ground height. Vertical scale is doubled for visualization, so Mount Fuji and surrounding peaks appear twice as steep as they actually are.

Popocatepetl (5,470 metres), located only 65 kilometres from Mexico City, last erupted on December 18, 2000. This image was acquired by Landsat-7 on January 4, 1999, at a spatial resolution of 30 by 30 metres. Snow and ice flank the summit crater. Canyons carved into the volcano lead away from the summit. The lower slopes are thickly covered with vegetation, which appears green in this image.

Faulting occurs in sudden slippage movements that generate earthquakes. A single fault movement may result in slippage of as little as one centimetre or as much as 15 metres. Successive movements may occur many years or decades apart. Over long time spans, the accumulated displacements can amount to tens or hundreds of kilometres. In some places, clearly recognizable sedimentary rock layers are offset on opposite sides of a fault, allowing the accurate measurement of the total amount of displacement (Figure 13.16).

Normal Faults

One common type of fault associated with crustal rifting is the **normal fault** (Figure 13.17a). The plane of slippage, or fault plane, is steeply inclined. The crust on one side is raised, or upthrown, relative to the other, which is downthrown. A normal fault results in a steep, straight, cliff-like feature called a *fault*

13.16 Displaced strata Two normal faults are visible in this road cut in Canberra, Australia.

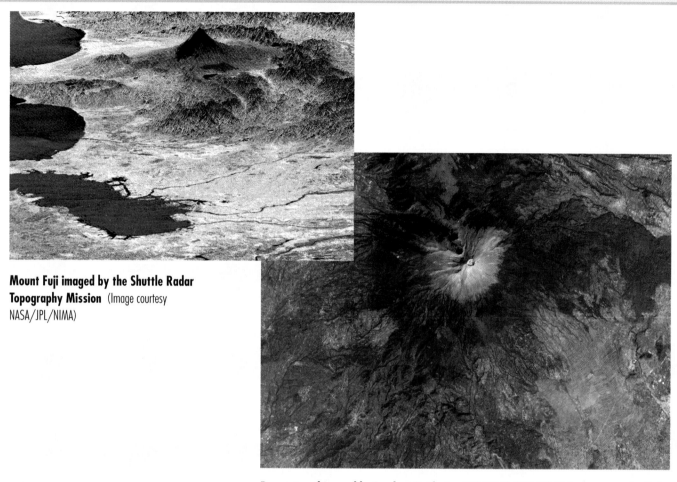

Mount Fuji imaged by the Shuttle Radar Topography Mission (Image courtesy NASA/JPL/NIMA)

Popocatepetl imaged by Landsat-7 *Fly By: 19° 01' N; 98° 37' W* (Image courtesy Ron Beck, EROS Data Center)

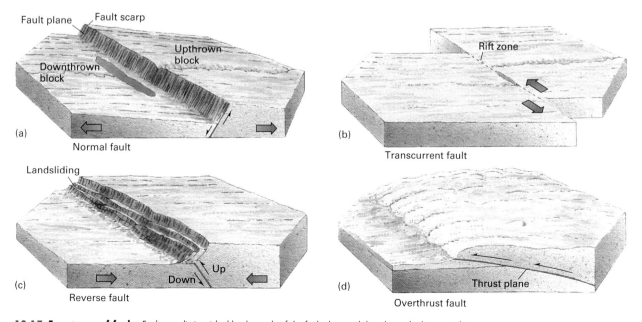

Fault plane — Fault scarp

Upthrown block

Downthrown block

(a) Normal fault

Rift zone

(b) Transcurrent fault

Landsliding

(c) Reverse fault — Up / Down

Thrust plane

(d) Overthrust fault

13.17 Four types of faults Faults are distinguished by the angle of the fault plane and the relative displacement that occurs.

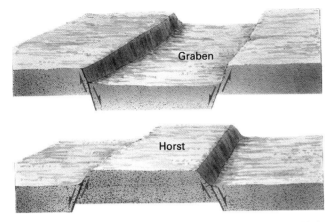

13.18 Initial landforms of normal faulting A graben is a downthrown block, often forming a long, narrow valley. A horst is an upthrown block, forming a plateau, mesa, or mountain. (A. N. Strahler)

scarp. Fault scarps range in height from a few metres to a few hundred metres. They are often many kilometres in length.

Normal faults are not usually isolated features. They commonly occur in multiple arrangements, often as intersecting sets of parallel faults, where land may drop down to form a *graben* or rise as a *horst* (Figure 13.18). The resulting topography is often a series trenches and plateaus. On a regional scale, *block mountains* are produced, as in the Basin and Range district of the western United States. The East African Rift Valley system, which extends some 3,000 kilometres from the Red Sea southward to the Zambezi River, illustrates the process on a continental scale (Figure 13.19).

The Rift Valley system consists of a number of graben-like troughs. Each is a separate rift valley ranging in width from about 30 to 60 kilometres, in which blocks have slipped down between neighbouring blocks as the land has spread apart. The floors of the rift valleys are above the elevation of most of the African continental surface. Major rivers and several long, deep lakes—Lake Malawi and Lake Turkana, for example—occupy some of the valley floors. The sides of the rift valleys typically consist of multiple fault steps (Figure 13.20), and sediments derived from the high plateaus fill the valley floors. Two stratovolcanoes, Mount Kilimanjaro (5,985 metres) and Mount Kenya (5,199 metres) have built up close to the Rift Valley.

Transcurrent Faults

A **transcurrent fault** (Figure 13.17b), also called a *strike-slip fault*, is formed when crustal blocks slide past each other horizontally along a fault plane that is more or less vertical. Normally, only a thin fault line is traceable across the surface. Because movement is lateral, no scarp, or only a very low one, is produced. Transcurrent faults are therefore similar to the major transform faults produced where lithospheric plates slide horizontally past one another.

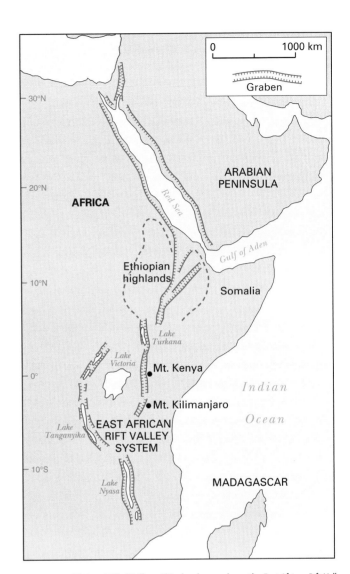

13.19 East African Rift Valley This sketch map shows the East African Rift Valley system and the Red Sea to the north.

13.20 The Rift Valley wall in Ethiopia Multiple fault scarps give the landscape a stepped appearance.

The San Andreas Fault is a good example of an active trans-current fault. It can be followed for a distance of about 1,000 kilometres from the Gulf of California to Cape Mendocino, on the Pacific coast, where it heads out to sea. The San Andreas Fault marks the boundary between the Pacific plate, which is moving toward the northwest, and the North American plate (Figure 13.21). Throughout the many kilometres of its length, the San Andreas Fault appears as a straight, narrow scar. In some places, this widens to a steep-sided valley, and elsewhere it becomes a low scarp.

Reverse and Overthrust Faults

In a *reverse fault*, the inclination of the fault plane is such that one side rides up over the other and a crustal shortening occurs (Figure 13.17c). Reverse faults produce fault scarps similar to those of normal faults, but the possibility of landslides is greater because the upthrust side tends to overhang the downthrust side. The *low-angle overthrust fault* (Figure 13.17d) involves predominantly horizontal movement, so that one slice of rock rides over the adjacent ground surface. The thrust slice may be up to 50 kilometres wide. Overthrust faulting is associated with the formation of nappes.

EARTHQUAKES

An **earthquake** is a motion of the ground surface, ranging from a faint tremor to a wild movement capable of shaking buildings apart (Figure 13.22).

Earthquakes can be produced by volcanic activity or when magma rises or recedes within a volcanic chamber, but most result from sudden slip movements along faults. Typically tectonic forces slowly bend the rock on both sides of the fault over many years. When a critical point is reached, the rocks on opposite sides of the fault move in different directions to relieve the strain. A large quantity of energy is instantaneously released in the form of seismic waves, which shake the ground. The waves move outward in widening circles from a point of sudden energy release, called the *focus*, and gradually lose energy as they travel outward in all directions. The **epicentre** is the point on the Earth's surface directly above the focus of an earthquake.

The *Richter scale* assesses the magnitude of earthquakes. The numbers on the Richter scale range from 0 to 9, but there is really no upper limit. For each whole unit increase (say, from 5.0 to 6.0),

13.21 The San Andreas Fault in southern California The fault is marked by a narrow trough. *Fly By: 35° 11' N; 119° 44' W*

13.22 Earthquake devastation This building did not survive the Mexico City earthquake of September 1985.

WORKING IT OUT | 13.4   Earthquake Magnitude

The magnitude of earthquakes is most commonly assessed by the *Richter scale*, devised in 1936 by the seismologist Charles F. Richter, and later modified in 1956. The **Richter magnitude scale** (M_L)—or simply the Richter scale—uses a single number to quantify the size of an earthquake. Earthquake magnitude is obtained by calculating the logarithm of the amplitude of the largest seismic wave detected on a seismograph. For each unit of increase in the Richter scale, the amplitude of the seismic wave increases by a factor of 10. The scale has neither a fixed maximum nor a minimum; however, several high-magnitude earthquakes have exceeded 8.9. Earthquakes of magnitude 2.0 are the smallest normally detected by human senses, but instruments can detect quakes as small as −3.0 on the scale. Note, magnitude −3.0 on a logarithmic scale means an earthquake is 0.0001 as strong as a magnitude 2 earthquake that might be detected by humans.

The **moment magnitude scale** (M_w), introduced in 1979, is based on the concept of the **seismic moment** (M_0) and provides a measure of the amount of energy released by an earthquake. The seismic moment combines the amount of movement on the fault with the area of the fault that ruptured and the type of the rock involved. The rigidity of the rock incorporates two components: stress (or force acting on the rock) and strain (deformation that occurs within the rock). Thus

Moment = Rock Rigidity × Fault Area × Slip Distance

$$M_0 = \mu A d$$

The moment magnitude is calculated as

$$M_W = \frac{2}{3}\left(\log_{10}\frac{M_0}{\text{N}\cdot\text{m}} - 9.1\right)$$

where M_0 is the seismic moment.

The constants in the equation are chosen so that estimates of moment magnitude roughly correspond with estimates based on the Richter magnitude scale. Thus, the relationship between the Richter scale and energy release is given by the following equation:

$$\log_{10} E = 4.8 + 1.5M_L$$

where E is the energy in joules and M_L is the Richter magnitude scale. By taking the exponent of both sides, the expression can be written as

$$E = 10^{(4.8+1.5M_L)} = 10^{4.8} \times 10^{1.5M_L}$$

Thus, if the Richter scale number increases by 1 unit, then the energy release will increase by a factor of $10^{1.5\times1}$, or 31.6. This means that for each unit increase in the Richter scale, say from 4.0 to 5.0, about 32 times more energy is released. For a two-unit increase, say from 4.0 to 6.0, the energy released will be $10^{1.5\times2} = 10^3 = 1,000$ times as large (see figure below). One advantage of the moment magnitude scale is that there is no value beyond which all large earthquakes are given the same magnitude. For this reason, moment magnitude is now the preferred scale for estimating the magnitude of large earthquakes. However, the Richter scale is generally used for earthquakes with a magnitude of less than 3.5.

While in North America, earthquakes are rated according to their magnitude, elsewhere in the world, the preference is still to use a scale of intensity. The *Rossi-Forel scale* devised in the latter part of the nineteenth century is generally regarded as the first of these descriptive scales. The 1873 version used

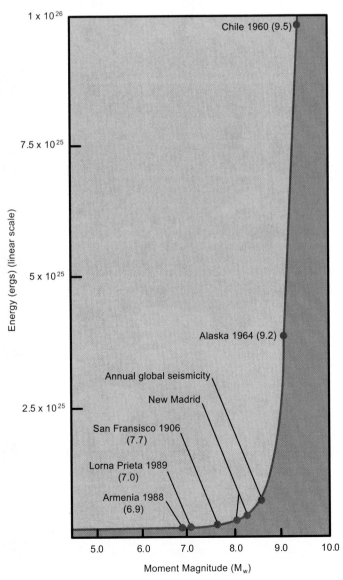

The moment magnitude scale and energy released by earthquakes

10 categories, and is interesting in that category X refers to "disturbance of the strata, fissures in the ground, rock falls from mountains." This scale was used until 1902 when the 10-point Mercalli scale was introduced. The Mercalli scale was changed to a 12-point scale and translated into English in 1931. It is now commonly referred to as the Modified Mercalli Scale. On this scale, the ground moves in waves or ripples, and large amounts of rock may move in a category XII earthquake.

A newer scale, the *Medvedev-Sponheuer-Karnik (MSK) scale*, came into widespread use in the 1970s and is still the standard reference in Russia, throughout the former Soviet Union, India, and Israel. The MSK scale incorporates natural landscape conditions, including landslides and tsunamis, into several of its categories, as shown in the table below. The *European Macroseismic Scale (EMS 98)*, the latest version of which was developed in 1998, is based largely on the principles the MSK scale uses.

The Medvedev-Sponheuer-Karnik (MSK) scale

1. Not perceptible	Not felt, registered only by seismographs. No effect on objects. No damage to buildings.
2. Hardly perceptible	Felt only by individuals at rest. No effect on objects. No damage to buildings.
3. Weak	Felt indoors by a few. Hanging objects swing slightly. No damage to buildings.
4. Largely observed	Felt indoors by many and felt outdoors by only very few. A few people are awakened. Moderate vibration. Observers feel a slight trembling or swaying of the building, room, bed, chair, etc. China, glasses, windows, and doors rattle. Hanging objects swing. Light furniture shakes visibly in a few cases. No damage to buildings.
5. Fairly strong	Felt indoors by most, outdoors by few. A few people are frightened and run outdoors. Many sleeping people awake. Observers feel a strong shaking or rocking of the whole building, room, or furniture. Hanging objects swing considerably. China and glasses clatter together. Doors and windows swing open or shut. In a few cases, window panes break. Liquids oscillate and may spill from fully filled containers. Animals indoors may become uneasy. Slight damage to a few poorly constructed buildings.
6. Strong	Felt by most indoors and by many outdoors. A few people lose their balance. Many people are frightened and run outdoors. Small objects may fall and furniture may be shifted. Dishes and glassware may break. Farm animals may be frightened. Visible damage to masonry structures, cracks in plaster. Isolated cracks on the ground.
7. Very strong	Most people are frightened and try to run outdoors. Furniture is shifted and may be overturned. Objects fall from shelves. Water splashes from containers. Serious damage to older buildings, masonry chimneys collapse. Small landslides.
8. Damaging	Many people find it difficult to stand, even outdoors. Furniture may be overturned. Waves may be seen on very soft ground. Older structures partially collapse or sustain considerable damage. Large cracks and fissures open up, rockfalls.
9. Destructive	General panic. People may be forcibly thrown to the ground. Waves are seen on soft ground. Substandard structures collapse. Substantial damage to well-constructed structures. Underground pipelines rupture. Ground fracturing, widespread landslides.
10. Devastating	Masonry buildings destroyed, infrastructure crippled. Massive landslides. Water bodies may be overtopped, causing flooding of the surrounding areas and the formation of new water bodies.
11. Catastrophic	Most buildings and structures collapse. Widespread ground disturbances, tsunamis.
12. Very catastrophic	All surface and underground structures completely destroyed. Landscape generally changed, rivers change paths, tsunamis.

The Japan Meteorological Agency seismic intensity (shindo) scale uses 10 categories, ranging from 0 to 7 (two levels are included in categories 5 and 6). It describes the intensity of an earthquake mainly in terms of the building damage incurred. Real-time earthquake reports in Japan are reported from a network of 180 seismographs and 600 seismic intensity meters and include information on how hard the ground shakes during the event.

(continued)

Japan Meteorological Agency seismic intensity (shindo) scale

Magnitude	Classification	Effects	Peak ground acceleration
7	Ruinous earthquake	In most buildings, wall tiles and windowpanes are damaged and fall. In some cases, reinforced concrete-block walls collapse.	Greater than 4 m/s²
6+	Violent earthquake	In many buildings, wall tiles and windowpanes are damaged and fall. Most un-reinforced concrete-block walls collapse.	3.15–4.00 m/s²
6−	Violent earthquake	In some buildings, wall tiles and windowpanes are damaged and fall.	2.50–3.15 m/s²
5+	Severe earthquake	In many cases, un-reinforced concrete-block walls collapse and tombstones overturn. Many automobiles stop due to difficulty in driving. Occasionally, poorly installed vending machines fall.	1.40–2.50 m/s²
5−	Severe earthquake	Most people try to escape from danger; some find it difficult to move.	0.80–1.40 m/s²
4	Strong earthquake	Many people are frightened. Some people try to escape from danger. Most sleeping people awake.	0.25–0.80 m/s²
3	Weak earthquake	Felt by most people in the building. Some people are frightened.	0.08–0.25 m/s²
2	Light earthquake	Felt by many people in the building. Some sleeping people awake.	0.025–0.08 m/s²
1	Slight earthquake	Felt by only some people in the building.	0.008–0.025 m/s²
0	Insensible	Imperceptible to people.	Less than 0.008 m/s²

Table 13.1 Comparative earthquake magnitudes on the Richter magnitude (M_L) and moment magnitude (M_W) scales

Earthquake	Richter Scale	Moment Magnitude Scale
New Madrid, MO, 1812	8.7	8.1
San Francisco, CA, 1906	8.3	9.2
Chile, 1960	8.5	9.6
Alaska, 1964	8.4	9.2
Northridge, CA, 1994	6.4	6.7

the amplitude of the earthquake wave increases by a factor of 10 and the quantity of energy released increases by a factor of 32. A value of 9.5 recorded in the Chilean earthquake of 1960 remains the highest to date. The Richter scale is based on the amount of ground shaking, as measured on a seismograph. It is one of three common measures of earthquake magnitude.

The oldest measure, the Mercalli scale devised in 1902, is based on the amount of damage caused by an earthquake and human response to it. The Mercalli scale is descriptive and originally included 10 categories; it was subsequently modified to a 12-point scale that ranges from I (most people do not notice, animals may be uneasy) to XII (all manmade structures are destroyed).

Table 13.2 Frequency of earthquakes of different moment magnitude (M_W) worldwide

Magnitude (M_W)	Number of Earthquakes per Year	Description
> 8.5	0.3	Great
8.0 - 8.4	1	
7.5 - 7.9	3	Major
7.0 - 7.4	15	
6.6 - 6.9	56	
6.0 - 6.5	210	Destructive
5.0 - 5.9	800	Damaging
4.0 - 4.9	6,200	Minor
3.0 - 3.9	49,000	
2.0 - 2.9	300,000	
0 - 1.9	700,000	

The moment magnitude scale, introduced in 1979, is based on the movement that occurs on a fault, not on how much the ground shakes during an earthquake. The moment magnitude, like the Richter scale, is logarithmic. However, the two scales cannot be directly compared, because they are based on different characteristics of an earthquake (Table 13.1). Table 13.2 shows the

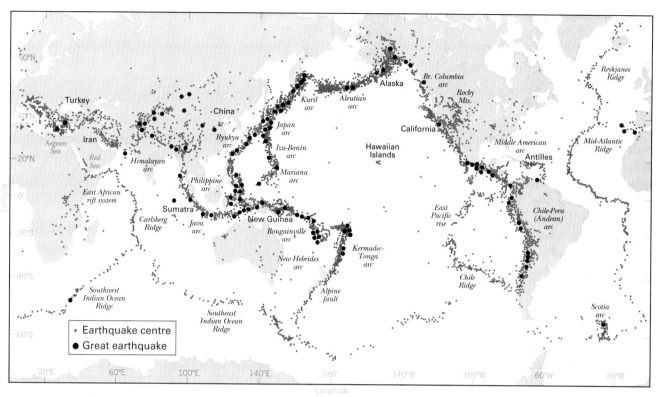

13.23 Earthquake locations This world map plots earthquake centre locations, highlighting the centres of great earthquakes. Centre locations of all earthquakes originating at depths of 0 to 100 kilometres during a seven-year period are shown by red dots. Each dot represents a single location or a cluster of centres. Black circles identify centres of earthquakes of Richter magnitude 8.0 or greater during an 80-year period. The map clearly shows the pattern of earthquakes occurring at subduction boundaries. (Compiled by A. N. Strahler from U.S. government data. Copyright © A. N. Strahler)

annual frequency of earthquakes of different magnitudes that occur worldwide. *Working It Out 13.4 • Earthquake Magnitude* provides more information on the Richter scale and other methods of assessing earthquake magnitude.

EARTHQUAKES AND PLATE TECTONICS

Most seismic activity occurs primarily near lithospheric plate boundaries. This is clearly revealed in Figure 13.23, which shows the location of all earthquake centres during a typical seven-year period. The greatest intensity of seismic activity is found along converging plate boundaries where oceanic plates are undergoing subduction. Strong pressures build up at the downward-slanting contact of the two plates. They are relieved by sudden fault slippages that generate earthquakes of large magnitude. This mechanism explains the great earthquakes experienced in Japan, Alaska, Central America, Chile, and other zones close to trenches and volcanic arcs of the Pacific Ocean basin.

Similar examples can be identified on the Pacific coast of Mexico and Central America, where the subduction boundary of the Cocos plate lies close to the shoreline. The great earthquake that devastated Mexico City in 1985 was centred in the deep trench offshore. Two great shocks in close succession, the first of magnitude 8.1 and the second of 7.5, damaged cities along the Pacific coast. Although Mexico City lies inland, about 300 kilometres from the earthquake epicentre, it experienced intense ground shaking of underlying saturated clay formations, with the resulting death toll of some 10,000 people (Figure 13.22).

Transcurrent faults on transform boundaries that cut through the continental lithosphere are also sites of intense seismic activity, with moderate to strong earthquakes. In North America, the best example is the San Andreas Fault, but similar activity is frequently reported along North Anatolian Fault in Turkey, where the Persian subplate is moving westward at its boundary with the European plate. In August 1999, a major earthquake with a magnitude of 7.4, centred near the city of Izmit in western Turkey, killed more than 15,000. A few months later, a quake of magnitude 7.2 occurred not far away on the same fault. Central and southeast Turkey shook again on related faults in 2002 and 2003. Further east, southwest Iran experienced a magnitude 6.6 earthquake in December 2003, and a magnitude 7.6 earthquake occurred in Kashmir in October 2005.

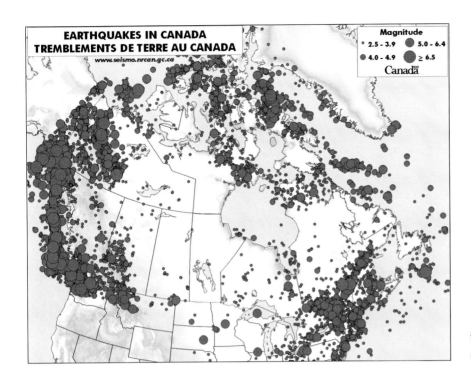

13.24 Earthquakes in Canada The strongest and most frequent earthquakes occur in western Canada, but they are also common in the St. Lawrence region and across the Arctic.

Spreading boundaries are the location of a third class of narrow zones of seismic activity related to lithospheric plates. Most of these boundaries are identified by the midoceanic ridge and its branches. For the most part, earthquakes in this class are moderate in intensity.

Earthquakes also occur at scattered locations over the continental plates, far from active plate boundaries. In many cases, no active fault is visible, and the geologic cause of the earthquake is uncertain. For example, significant earthquakes occur from time to time in southern Quebec and along the St. Lawrence River valley. The most recent of these is the Saguenay earthquake of 1988, with a magnitude of 5.9. However, in Canada the strongest earthquakes occur most frequently in British Columbia, particularly off Vancouver Island and the Queen Charlotte Islands, the Yukon, and the Northwest Territories (Figure 13.24).

Typically 30 to 50 earthquakes greater than magnitude 2.5 are reported in Canada each year. In recent years, the largest on record include an earthquake of magnitude 8.1 off the Queen Charlottes in 1949, and one of magnitude 7.4 in 1970 in the same region. Historically, the largest earthquake in Canada reportedly occurred in 1700, and is estimated to have had a magnitude of 9.0. It was associated with the undersea Cascadia thrust fault that runs in the Pacific Ocean from the vicinity of Vancouver Island to the waters off California.

SEISMIC SEA WAVES

An important environmental hazard often associated with a major earthquake centred on a subduction plate boundary is the *seismic sea wave*, or **tsunami**. A succession of these waves is often generated in the ocean by a sudden movement of the sea floor at a point near the earthquake source. The waves travel over the ocean in ever-widening circles, but they are not perceptible at sea in deep water (Figure 13.25).

When a tsunami arrives at a distant coastline, it causes a rise in water level. Normal wind-driven waves, superimposed on the heightened water level, strike places inland that are normally above their reach. Few have been as devastating as the Boxing Day (Asian) Tsunami of 2004. It was triggered by the undersea Suma-

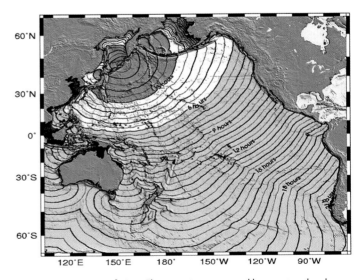

13.25 Tsunami travel time This tsunami was generated by a great earthquake with a moment magnitude of 8.3 on November 15, 2006, located near the Kuril Islands (46.6°N, 153.2°E, at a depth of 26.7 kilometres). Peak wave height was 1.5 metres, which was recorded in Hawaii.

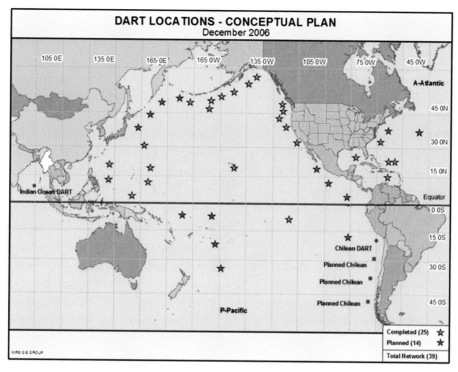

13.26 Planned location of DART buoys as of December 2006 The new system will expand monitoring capabilities throughout the entire Pacific and Caribbean basins, with additional buoys deployed in the Atlantic and one in the Indian Ocean.

tra-Andaman earthquake whose epicentre was off the west coast of Sumatra. The magnitude of the earthquake is officially rated at between 9.1 and 9.3, making it the second largest earthquake recorded on a seismograph. Seismic activity lasted for about 10 minutes and the vibrations were sufficiently powerful to cause the Earth to vibrate on its axis by more than a centimetre. Tsunamis with waves of up to 30 metres spread throughout the Indian Ocean, inundating coastal communities in places such as Indonesia, Sri Lanka, India, and Thailand. The estimated death toll exceeded 275,000 people.

The need to provide early warnings of tsunamis led to the establishment of the Pacific Tsunami Warning Center (PTWC) in Hawaii in 1949. A second centre at Palmer, Alaska was established in 1967 to serve the west coast of North America. Currently, the PTWC, comprising 26 countries, is responsible for monitoring seismological and tidal stations throughout the Pacific basin. The PTWC operational centre issues tsunami warnings and information about potentially tsunamogenic earthquakes to all nations in the Pacific. The global monitoring system is continually upgraded, and in 2005, plans were announced to expand the tsunami detection and warning system as part of the international Global Earth Observation System of Systems (GEOSS). Part of the plan calls for the deployment of Deep-ocean Assessment and

Reporting of Tsunami (DART) buoys for a fully operational tsunami warning system by mid-2007 (Figure 13.26).

A Look Ahead

The previous three chapters have surveyed the composition, structure, geologic activity, and initial landforms of the Earth's crust. The rocks and minerals that make up the Earth's crust and core are continuously recycled and transformed in a process that has occurred over some four billion years of geologic time. Powered by radiogenic heat stored in the Earth's interior, the cycle of rock transformation is part of the mechanism of plate tectonics that intricately links the geographical distribution of mountain ranges, ocean basins, and continents. Initial landforms, which result directly from volcanic and tectonic activity, occur primarily at the boundaries of spreading or colliding lithospheric plates.

Following this survey of the Earth's crust and the geologic processes that shape it, attention now turns to other landform-creating processes. Part 4 begins with the processes of weathering and mass wasting, which break up rock and move Earth materials downslope under the influence of gravity. Three chapters cover the importance of running water in shaping landforms, and Part 4 concludes with an examination of landforms created by wind, waves, and glacial ice.

CHAPTER SUMMARY

- Landforms are the surface features of the land, and geomorphology is the scientific study of landforms. *Initial landforms* are shaped by volcanic and tectonic activity, while *sequential landforms* are sculpted by agents of denudation, including running water, waves, wind, and glacial ice.
- Volcanoes are landforms created by the eruption of lava at the Earth's surface. Stratovolcanoes, formed by the emission of thick, gassy, felsic lavas, have steep slopes and tend to have explosive eruptions that can form *calderas*. Most active stratovolcanoes lie along the Pacific Rim, where subduction of oceanic lithospheric plates is occurring.
- At *hotspots*, rising mantle material provides mafic magma that erupts as basaltic lava. Because this lava is more fluid and contains little gas, it forms broadly rounded shield volcanoes. Hotspots occurring beneath the continental crust can also provide vast areas of *flood basalts*. Some shield volcanoes occur along the midoceanic ridge.
- The two forms of tectonic activity are compression and extension. Compression occurs at lithospheric plate collisions.

At first, the compression produces folding—*anticlines* (upfolds) and *synclines* (downfolds). If compression continues, folds may be overturned, and eventually overthrust faulting can occur.
- Extension occurs where lithospheric plates are spreading apart, generating normal faults. These can produce upthrown and downthrown blocks that are sometimes as large as mountain ranges or *rift valleys*. Transcurrent faults occur where two rock masses move horizontally past each other.
- Earthquakes occur when rock layers, bent by tectonic activity, suddenly fracture and move. The sudden motion at the fault produces earthquake waves that shake and move the ground surface in the adjacent region. Large earthquakes occurring near developed areas can cause great damage. Most severe earthquakes occur near plate collision boundaries.
- Seismic activity creates tsunamis, which can inundate coastal areas with waves of up to 30 metres. They are particularly devastating in the island nations of the Pacific and Indian oceans.

KEY TERMS

anticline	folding	Richter magnitude scale	transcurrent fault
caldera	geomorphology	seismic moment	tsunami
earthquake	landform	shield volcano	volcano
epicentre	moment magnitude scale	stratovolcano	
fault	normal fault	syncline	

REVIEW QUESTIONS

1. Distinguish between initial and sequential landforms. How do they represent the balance of power between internal Earth forces and the external forces of denudation agents?
2. What is a stratovolcano? What is its characteristic shape, and why does that shape occur? Where do stratovolcanoes generally occur and why?
3. What is a shield volcano? How is it distinguished from a stratovolcano? Where are shield volcanoes found, and why? Give an example of a shield volcano. How are flood basalts related to shield volcanoes?
4. How can volcanic eruptions become natural disasters? Be specific about the types of volcanic events that can devastate habitations and extinguish nearby populations.
5. Briefly describe the Rift Valley system of East Africa as an example of normal faulting.
6. How does a transcurrent fault differ from a normal fault? What landforms are expected along a transcurrent fault? How are transcurrent faults related to plate tectonic movements?
7. What is an earthquake, and how does it arise? How are the locations of earthquakes related to plate tectonics?
8. Describe a tsunami, including its origin and effects.

FOCUS ON SYSTEMS 13.1　The Life Cycle of a Hotspot Volcano

1. How does a life cycle contrast with a repeating time cycle?
2. Describe the stages in the life cycle of a basaltic shield volcano of the Hawaiian type.
3. What is a hotspot? What produces it? How is it related to the life cycle?

EYE ON THE ENVIRONMENT 13.2　Geothermal Energy Sources

1. What is the ultimate source of geothermal power? Where would you go, and why, to find a geothermal power source?
2. How is geothermal energy extracted? What environmental concerns arise in this process?

VISUALIZATION EXERCISES

1. Sketch a cross-section through a normal fault, labelling the fault plane, upthrown side, downthrown side, and fault scarp.
2. Sketch a cross section through a foreland fold belt showing rock layers in different colours or patterns. Label anticlines and synclines.

ESSAY QUESTIONS

1. Write a hypothetical news account of a volcanic eruption. Select a type of volcano—stratovolcano or shield—and a plausible location. Describe the eruption and its effects as it was witnessed by observers. Include any details you need, but be sure they are scientifically correct.
2. How are mountains formed? Provide the plate tectonic setting for mountain formation, and then describe how specific types of mountain landforms arise.
3. Discuss how magnitude and intensity are assessed and describe the advantages and disadvantages of each approach.

EYE ON THE LANDSCAPE |

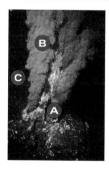

Chapter Opener Deep-sea hydrothermal vent. Hydrothermal vents, known Missingas black smokers, are associated mid-oceanic ridges. These chimney-like structures (**A**) are composed of sulphides which are deposited on volcanic rocks where plumes of hot, mineral-rich water (**B**) are vented onto the ocean floor. Despite the almost total lack of illumination (**C**) at these depths, hydrothermal vents support unique ecosystems that are dependent on the mineral deposits for their source of energy.

The Rapley Monocline, Utah. Sedimentary sandstones originally deposited in horizontal strata (**A**) have been sharply tilted by tectonic activity (**B**). Beyond the zone of dipping strata the rock retains the horizontal layering (**C**) typical of a monocline. The red colours in the weathered rock is from iron oxides (**D**). The river is carrying a heavy load of sediment eroded from the unvegetated surface. Deposition is occurring on the inner curve (**E**) of its me-

andering course with erosion causing undercutting on the outer banks (**F**).

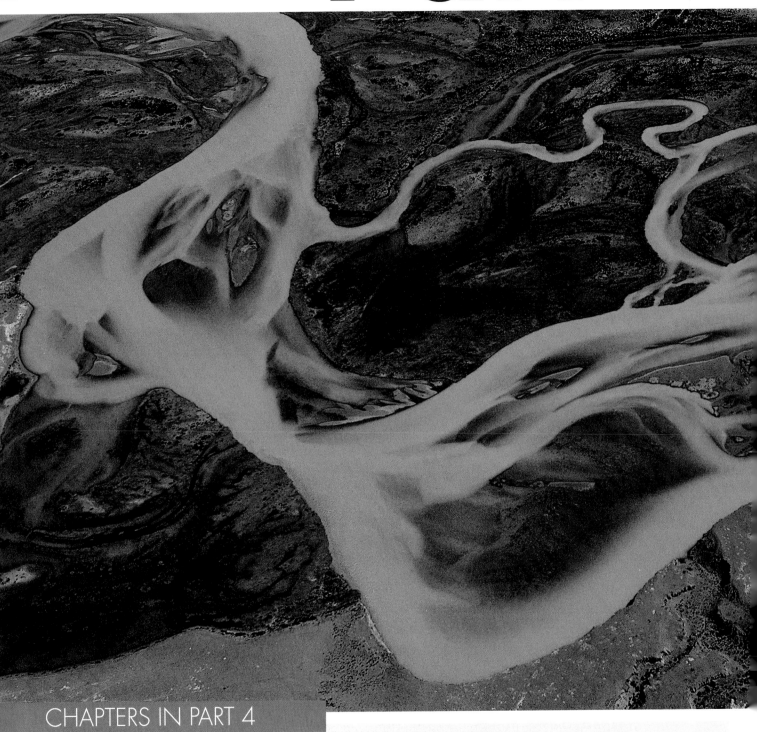

Part 4

CHAPTERS IN PART 4

14 Weathering and Mass Wasting
15 The Cycling of Water on the Continents
16 Fluvial Processes and Landforms
17 Landforms and Rock Structure
18 The Work of Wind and Waves

19 Glacier Systems and the Late-Cenozoic Ice Age

SYSTEMS OF LANDFORM EVOLUTION

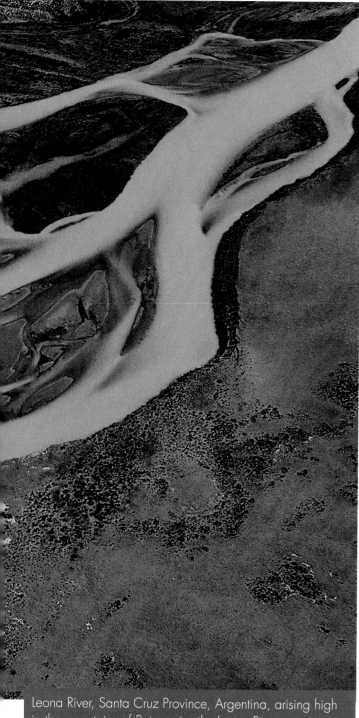

Leona River, Santa Cruz Province, Argentina, arising high in the mountains of Patagonia, the Leona River gets its milky-blue tint from the water of melting glaciers. The braided flow pattern is typical of rivers carrying large volumes of sediment.

Physical geography focuses on the life layer—the zone of interactions among the atmosphere, hydrosphere, lithosphere, and biosphere in which people live, breathe, and carry out their daily lives. After examining the matter and energy flow systems of the solid Earth in Part 3, the focus returns to the life layer to examine the matter and energy flow systems that shape the surface of the land. These systems operate largely on time scales that fall between those of the atmosphere and solid Earth—hundreds to thousands to a few million years. Systems of landform evolution are mostly powered by gravity—the constant attraction of the Earth's mass that moves solids (rock particles and soil) and fluids (water and glacial ice) downhill, creating landforms in the process. But gravity releases only potential energy that has been stored in positioning these solids and fluids above a base level, such as the ocean. The potential energy is ultimately derived from two sources: solar energy, which places liquid and solid water on the continents through precipitation, and the Earth's internal heat, which generates forces that uplift masses of rock and soil above the base level. ■ Chapter 14 begins Part 4 by examining the weathering processes that break up rock material, then turns to movements of masses of weathered material under gravity—landslides and earth flows, for example. Chapter 15 focuses on water at or near the land surface, in lakes, rivers, and as groundwater. Chapter 16 moves from the study of water as surface water and groundwater to consider its role as a landform-creating agent of erosion and deposition. Chapter 17 shows how fluvial action interacts with rock structures—folds, faults, and the like—to produce distinctive configurations of landforms related to those structures. Chapter 18 turns to two other landform-creating agents powered not by gravity, but by atmospheric motion: waves and wind. Chapter 19 concludes Part 4 by examining landforms created by glacial ice, both as high mountain glaciers and as vast continental ice sheets. Part 5 will complete the study of physical geography by focusing on the life layer from the viewpoint of the biosphere.

Chapter 14

EYE ON THE LANDSCAPE

Talus cones *Frost action has caused these cliffs to shed angular blocks of rock that accumulate in talus cones. In the foreground is the shore of Lake Louise in the Canadian Rockies.* **What else would the geographer see? . . . Answers at the end of the chapter.**

WEATHERING AND MASS WASTING

Physical Weathering
Frost Action
Salt-Crystal Growth
Unloading
Other Physical Weathering Processes
Chemical Weathering and its Landforms
Hydrolysis and Oxidation
Acid Action
Mass Wasting
Slopes
Soil Creep
Earth Flow
Environmental Impact of Earth Flows
Mudflow and Debris Flood
Landslide
Working It Out 14.1 • *The Power of Gravity*
Induced Mass Wasting
Induced Earth Flows
Scarification of the Land
Processes and Landforms of the Permafrost Regions
Permafrost
Permafrost Temperature
The Active Layer
Focus on Systems 14.2 • *Permafrost as an Energy Flow System*
Forms of Ground Ice
Thermokarst Lakes
Geographers at Work • *Konrad Gajewski, Ph.D.*
Forest Fires and Permafrost
Retrogressive Thaw Slumps
Patterned Ground and Solifluction
Alpine Tundra
Environmental Problems of Permafrost
Climate Change in the Arctic

Now that the study of the Earth's crust is complete—including its mineral composition, moving lithospheric plates, and tectonic and volcanic landforms—the focus can move to the shallow life layer itself. At this sensitive interface, the external processes of denudation carve sequential landforms from the rocks uplifted by the Earth's internal processes. The investigation of what happens to rock once exposed at the surface began in Chapter 11 with a description of mineral alteration of rock and the production of sediment, which is an essential part of the cycle of rock transformation. This chapter looks further at the softening and breakup of rock, called weathering, and how the resulting particles move downhill under the force of gravity, a process called *mass wasting*.

Weathering is the general term applied to the combined action of all processes that cause rock to disintegrate physically and decompose chemically due to exposure near the Earth's surface. There are two types of weathering. In *physical weathering*, rocks are fractured and broken apart, primarily by the growth of ice or salt crystals along rock planes and the penetration of minerals by watery solutions. In *chemical weathering*, rock minerals transform from those that were stable when the rocks were formed to those that are stable at surface temperatures and pressures. As seen in Chapter 11, these chemical processes include oxidation, hydrolysis, and acid solution. Weathering acts to produce **regolith**—a surface layer of weathered rock particles that lies above solid, unaltered rock. Weathering also leads to a number of distinctive landforms, which this chapter discusses.

This chapter also examines how gravity acts on rock fragments created by weathering to produce landforms. In this process, gravity induces the spontaneous downhill movement of soil, regolith, and rock fragments, but without the action of moving water, air, or ice. This downhill movement is called **mass wasting**. Movement of a mass of soil or weathered rock to lower levels takes place when the internal strength of the soil or weathered rock declines to a critical point where it can no longer resist the force of gravity. This failure of strength under the ever-present force of gravity takes many forms and scales; human activity causes or aggravates several forms of mass wasting.

This chapter concludes with a look at a suite of special landforms and geomorphic processes found in arctic lands. They are created primarily by the freezing and thawing of water, acting in concert with gravity.

PHYSICAL WEATHERING

Physical weathering, also known as *mechanical weathering*, produces regolith from massive rock by the action of forces strong enough to fracture the rock. The physical weathering processes this chapter discusses include frost action, salt-crystal growth, unloading, and wedging by plant roots.

FROST ACTION

One of the most important physical weathering processes in cold climates is *frost action*, the repeated growth and melting of ice crystals in the pore spaces of soil and rock fractures. In contrast to most other liquids, water expands when it freezes. The expansion of freezing water can fragment even extremely hard rocks, given many freeze and thaw cycles. Frost action and ice crystal growth produce a number of conspicuous effects and forms in all cold winter climates. Features caused by frost action and the buildup of ice below the surface are particularly visible in the tundra climate of arctic coasts and islands, and above the timberline in high mountains.

Almost everywhere, bedrock is cut through by systems of fractures called *joints*. These fractures are thought to occur as rocks exposed to heat and pressure cool and contract. Joints typically occur in parallel and intersecting planes, creating surfaces of weakness along which weathering can break rock into individual blocks. Since there is no relative movement of rock along the joints, they cannot be considered faults. Joints are important to the physical weathering of rocks because they admit water to the rock. This allows ice and salt-crystal growth to further fracture the rock, creating rock fragments and regolith.

In sedimentary rocks, the planes of stratification, or bedding planes, comprise another natural set of planes along which water can penetrate. Often these are cut at right angles by sets of joints. Comparatively weak stresses can separate joint blocks, while strong stresses are necessary to make fresh fractures through solid rock. The process of separating rock along joints and bedding planes is called *block separation* (Figure 14.1).

As chemical decomposition weakens coarse-grained igneous rock, water is able to penetrate the contact surfaces between mineral grains. Here, the water can freeze and exert forces strong enough to separate the grains. This form of breakup is called *granular disintegration* (Figure 14.1). The end product is a fine gravel or coarse sand in which each grain consists of a single mineral particle separated from the others along its original crystal or grain boundaries.

The effects of frost action can be seen in all climates that have a winter season with many alternations of freeze and thaw. Where bedrock is exposed on knolls and mountain summits, joint blocks are pried apart by water that freezes in joint cracks (Figure 14.2a).

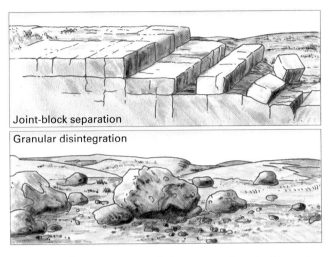

14.1 Bedrock disintegration Joint-block separation and granular disintegration are two common forms of bedrock disintegration. (Drawn by A. N. Strahler)

Under the most favourable conditions, such as on high mountain summits and in the arctic tundra, large angular rock fragments accumulate in a layer that completely blankets the bedrock beneath. The German word *felsenmeer* ("rock sea") describes such expanses of broken rock.

In high mountains, frost action on bare rock cliffs detaches rock fragments that fall to the cliff base. These loose fragments are known as *talus*. Where production of fragments is rapid, they accumulate to form *talus slopes*. Most cliffs are notched by narrow ravines that funnel the rock fragments into separate tracks. Each track, or chute, feeds a growing, cone-like talus body. Talus cones are arranged side by side along the base of the cliff. Fresh talus slopes are unstable: simply walking across the slope or dropping a large rock fragment from the cliff above will easily set off a sliding or rolling motion within the surface layer of fragments.

In fine-textured soils and sediments, composed largely of silt and clay, soil water freezes in horizontal layers or lens-shaped bodies. As these ice layers thicken, they heave the overlying soil layer upward. Prolonged frost heaving can produce minor irregularities and small mounds on the soil surface. A rock fragment at the surface can sometimes conduct heat from the soil to the cold night air and sky, causing perpendicular ice needles to grow beneath the fragment and lift it above the surface (Figure 14.2b). The same process acting on a rock fragment below the soil surface can eventually push the fragment to the surface.

Frost action is a dominant process in arctic and high mountain tundra environments, where it is a factor in the formation of a wide variety of unique landforms. The last section of this chapter investigates these landforms.

14.2 Frost Action (a) Ice crystal growth within the joint planes of rock can cause the rock to split apart. This split boulder, in the high country of the Sierra Nevada in California, is an example. (b) At night, water in the soil freezes at the surface, creating ice needles that can lift particles of soil or move larger stones.

(a)

(b)

SALT-CRYSTAL GROWTH

Closely related to the growth of ice crystals is the weathering process of rock disintegration by the growth of salt crystals in rock pores. This process, called *salt-crystal growth*, operates extensively in dry climates and is responsible for many of the niches, shallow caves, rock arches, and pits seen in sandstone formations. During long drought periods, *capillary action* moves groundwater to the surface of the rock. In this process, water is drawn into fine openings and passages in the rock by the same surface tension that gives a water droplet its rounded shape. As water evaporates from this porous outer zone of the sandstone, tiny crystals of salts, such as halite (sodium chloride), calcite (calcium carbonate), or gypsum (calcium sulphate), are left behind. Over time, the growth force of these crystals produces grain-by-grain breakup of the sandstone, which crumbles into sand and is swept away by wind and rain.

Zones of rock lying close to the base of a cliff are especially susceptible to breakup by salt-crystal growth because, at that location, the groundwater seeps downward and outward to reach the rock surface (Figure 14.3). In the southwestern United States, Native Americans occupied many of the deep niches or cave-like recesses formed in this way. Their cliff dwellings gave them protection from the elements and safety from armed attack.

Salt crystallization also damages masonry buildings, as well as concrete sidewalks and streets. Brick and concrete in contact with moist soil are highly susceptible to grain-by-grain disinte-

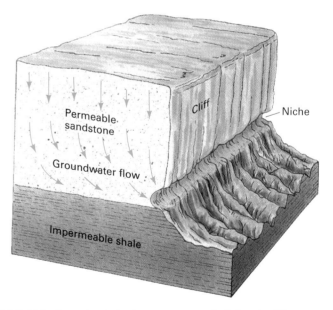

14.3 Niche formation In dry climates, water seeps slowly from the cliff base. Salt-crystal growth separates the grains of permeable sandstone, breaking them loose and creating a niche.

gration from salt crystallization. On damp basement floors and walls, the salt crystals can be seen as a soft, white, fibrous layer. The de-icing salts spread on streets and highways can be quite destructive. Sodium chloride (rock salt), widely used for this purpose, is particularly damaging to concrete pavements and walks, curbstones, and other exposed masonry structures.

Salt-crystal growth occurs naturally in arid and semi-arid regions. In humid climates, abundant rainfall dissolves salts and carries them downward to groundwater. (Chapter 15 describes how rainwater infiltrates soils and moves to the groundwater below.)

UNLOADING

A widespread process of rock disruption related to physical weathering results from *unloading*, also known as *exfoliation*, a process that relieves the confining pressure on underlying rock. Unloading occurs as erosion of overlying layers brings rock near the surface. Rock formed at great depth beneath the Earth's surface (particularly igneous and metamorphic rock) is in a slightly compressed state because of the confining pressure of overlying rock. As the rock above slowly wears away, the pressure decreases, and the underlying rock expands slightly in volume. This causes it to crack in layers that are more or less parallel to the surface, creating a type of jointing called *sheeting structure*. In massive rocks like granite or marble, thick curved layers or shells of rock break free in succession from the parent mass below.

If a sheeting structure forms over the top of a single large knob or hill of massive rock, it produces an *exfoliation dome*. Domes are among the largest of the landforms shaped primarily by weathering. In Yosemite Valley, California, where domes are spectacularly displayed, the individual rock sheets can be as thick as 15 metres.

OTHER PHYSICAL WEATHERING PROCESSES

Most rock-forming minerals expand when heated and contract when cooled. Where rock surfaces are exposed daily to the intense heat of the sun alternating with cool nights, the resulting expansion and contraction exert powerful disruptive forces on the rock. Although first-hand evidence is lacking, it seems likely that daily temperature changes can cause the breakup of a surface layer of rock already weakened by other agents of weathering.

Another mechanism of rock breakup is the growth of plant roots, which can wedge joint blocks apart, as seen in concrete sidewalk blocks uplifted and fractured by the growth of tree roots. This process is also active when roots grow between rock layers or joint blocks. Even fine rootlets in joint fractures can cause the loosening of small rock fragments and grains.

CHEMICAL WEATHERING AND ITS LANDFORMS

Chapter 11 investigated **chemical weathering** processes in its discussion of mineral alteration. The dominant processes of chemical change affecting silicate minerals are oxidation, hydrolysis, and carbonic acid action. Oxidation and hydrolysis change the chemical structure of minerals, turning them into new minerals that are typically softer and bulkier and therefore more susceptible to erosion and mass movement. Carbonic acid action dissolves minerals, washing them away in runoff. Chemical reactions proceed more rapidly in warmer temperatures, so chemical weathering is most effective in warm, moist climates.

HYDROLYSIS AND OXIDATION

Decomposition by hydrolysis and oxidation changes the minerals of strong rock into weaker forms that are rich in clay minerals and oxides. In warm, humid climates of the equatorial, tropical, and subtropical zones, thousands of years of hydrolysis and oxidation have resulted in the decay of igneous and metamorphic rocks to depths of up to 100 metres. The decayed rock material is soft, clay-rich, and erodes easily. To the construction engineer, deeply weathered rock is a major concern when building highways, dams, or other heavy structures. Although the weathered rock is soft and easy to move, its high clay content reduces its strength, and foundations built on the weathered rock can fail under heavy loads.

In dry climates, hydrolysis weathers exposed granite to produce many interesting boulder and pinnacle forms. Although rainfall is infrequent, water penetrates the granite along planes between crystals of quartz and feldspar. Chemical weathering of these surfaces then breaks individual crystal grains away from the main mass of rock, leaving rounded forms. The grain-by-grain breakup forms a fine desert gravel consisting largely of quartz and partially decomposed feldspar crystals.

ACID ACTION

Chemical weathering is also produced by *acid action*, largely that of *carbonic acid*. This weak acid forms when carbon dioxide dissolves in water. Rainwater, soil water, and stream water all normally contain dissolved carbon dioxide. Carbonic acid slowly dissolves some types of minerals. Carbonate sedimentary rocks, such as limestone and marble, are particularly susceptible to acid action. In this process, the mineral calcium carbonate dissolves and is carried away in solution in stream water.

Carbonic acid reaction with limestone produces many interesting surface forms, mostly of small dimensions. Outcrops of limestone typically show cupping (the formation of rounded cavities), rilling (the formation of surface valleys), grooving, and fluting in intricate designs (Figure 14.4). In a few places, the deep grooves and high wall-like rock fins are large enough to prevent people and animals from passing through. Carbonic acid in groundwater can dissolve limestone to produce underground caverns, as well as the distinctive landscapes that form when these caverns collapse. Chapter 15 will describe these landforms and landscapes.

wind, and the flow of glacial ice in landform-making processes. The remainder of this chapter considers the first of these landform agents—gravity. The following chapters will return to the other landform agents.

Everywhere on the Earth's surface, gravity pulls continuously downward on all materials. Bedrock is usually so strong and well supported that it remains fixed in place. However, when a mountain slope becomes too steep, bedrock masses can break free and fall or slide to new positions. When huge masses of bedrock are involved, the result can be catastrophic to towns and villages in the path of the slide. Such slides are a major environmental hazard in mountainous regions. They are one form of *mass wasting*, which, as defined earlier, is the spontaneous downhill movement of soil, regolith, and rock under the influence of gravity.

Because soil, regolith, and many forms of sediment are poorly held together, they are much more susceptible to movement under the force of gravity than hard, massive bedrock. On most slopes, at least a small amount of downhill movement is going on constantly. Although much of this motion is imperceptible, the regolith occasionally slides or flows rapidly.

The processes of mass wasting and the landforms they produce are extremely varied and tend to grade into each other. This chapter presents only a few of the most important forms of mass wasting (Figure 14.5). The top of the diagram provides three categories of information: the Earth material involved, the properties of the material, and the kind of motion that occurs. Boxes in

14.4 Solution features in limestone This outcrop of pure limestone in County Clare, Ireland, shows grooves and cavities formed by carbonic acid action.

In urban areas, air is commonly polluted by sulphur and nitrogen oxides. When these gases dissolve in rainwater, the result is acid precipitation. The acids rapidly dissolve limestone and chemically weather other types of building stones. The result can be damaging to stone sculptures, building decorations, and tombstones.

In the wet low-latitude climates, soil acids rapidly dissolve mafic rock, particularly basaltic lava. The effects of solution removal of basaltic lava are displayed in spectacular grooves, fins, and spires on the walls of deep alcoves in part of the Hawaiian Islands. The landforms produced are similar to those formed by carbonic acid action on massive limestone in the moist climates of the midlatitudes.

MASS WASTING

The discussion of weathering has described an array of processes that chemically alter rock and break it up into fragments. These rock fragments are subjected to gravity, running water, waves,

Kinds of earth materials:	Rock (dry)	Regolith, soil, alluvium, clays + water	Water + sediment
Physical properties:	Hard, brittle, solid	Plastic substance	Fluid
Kinds of motion:	Falling, rolling, sliding	Flowage within the mass	Fluid flow

Very slow	ROCK CREEP TALUS CREEP	SOIL CREEP		
		SOLIFLUCTION		
	LANDSLIDES:	EARTHFLOW (slump or flowage)		
	BEDROCK SLUMP · ROCKSLIDE	MUDFLOW		STREAM FLOW
Very fast	ROCKFALL	ALPINE DEBRIS AVALANCHE	DEBRIS FLOOD	

(Forms of mass movement)

14.5 Processes and forms of mass wasting

the body of the diagram contain the names of several forms of mass movements, arranged according to composition (left to right) and speed (top to bottom).

SLOPES

Mass wasting occurs on slopes. In physical geography, the term *slope* designates a small strip or patch of the land surface that is inclined from the horizontal. Thus, "mountain slopes," "hill slopes," or "valley-side slopes" describe some of a landscape's inclined ground surfaces. Slopes guide the downhill flow of surface water and fit together to form drainage systems within which surface-water flow converges into stream channels (Chapter 16). Nearly all natural surfaces slope to some degree. Few are perfectly horizontal or vertical.

Most slopes are mantled with regolith, which grades downward into solid, unaltered rock, known simply as **bedrock**. Regolith provides the source for **sediment**, which consists of rock and mineral particles that are transported and deposited by water, air, or even glacial ice. Both regolith and sediment comprise parent materials for the formation of *soil*, which is a surface layer of mineral and organic matter capable of supporting plant growth (see Chapter 20).

Figure 14.6 shows a typical hill slope that forms one wall of the valley of a small stream. Soil and regolith blanket the bedrock, except in a few places where the bedrock is particularly hard and projects in the form of *outcrops*. *Residual regolith* derives directly from the rock beneath and moves slowly down the slope toward the stream. Accumulations of regolith at the foot of a slope are called *colluvium*. Beneath the valley bottom are layers of regolith, called **alluvium**, which is sediment transported and deposited by the stream. The source of this sediment is regolith from hill slopes many kilometres upstream. All accumulations of sediment on the land surface, whether deposited by streams, waves and currents, wind, or glacial ice, are *transported regolith*, as opposed to residual regolith.

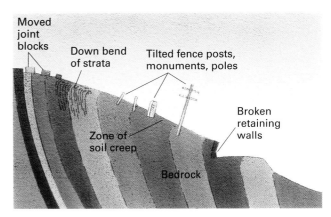

14.7 Indicators of soil creep The slow, downhill creep of soil and regolith appears in many ways on a hillside. (After C. F. S. Sharpe)

The thicknesses of soil and regolith are quite variable. Although the soil is rarely more than 1 or 2 metres thick, residual regolith on decayed and fragmented rock may extend down to depths of 5 to 100 metres, or more at some locations. In some places, there may be soil and regolith may be absent. The surface has been stripped down to the bedrock, which appears as an outcrop. In other places, following cultivation or forest fires, the fertile soil has partly or entirely eroded away, and severe erosion has exposed the regolith.

SOIL CREEP

On almost any soil-covered slope, soil and regolith move extremely slowly downhill, a process called **soil creep**. Figure 14.7 shows some of the indications of soil creep. Joint blocks of distinctive rock types are found far downslope from the outcrop. In some layered rocks such as shale or slate, edges of the strata seem to "bend" in a downhill direction. This is not true plastic bending but is the result of downhill creep of many rock pieces on small joint cracks. Creep causes fence posts and utility poles to lean downslope and even shift measurably out of line. Roadside retaining walls can buckle and break under the pressure of soil creep.

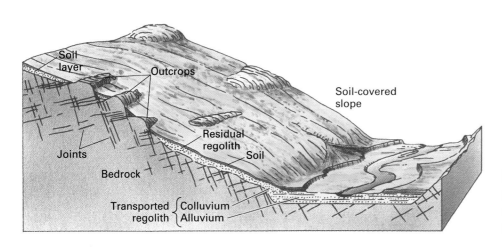

14.6 Soil, regolith, and outcrops on a hill slope Colluvium accumulates at the foot of the slope, while alluvium lies in the floor of an adjacent stream valley. (Drawn by A. N. Strahler)

Disturbance of the soil under the influence of gravity causes soil creep. Alternate drying and wetting of the soil, growth of ice needles and lenses, heating and cooling of the soil, trampling and burrowing by animals, and shaking by earthquakes all produce some disturbance of the soil and regolith. Because gravity exerts a downhill pull on every rearrangement, the particles gradually work their way downslope.

EARTH FLOW

In humid climate regions, a mass of water-saturated soil, regolith, or weak shale may move down a steep slope during a few hours in the form of an **earth flow** (Figure 14.8). At the top, the material slumps away, leaving a curved, wall-like scarp. Often the soil farther from the scarp moves more rapidly, leaving a series of steps as the soil mass slips downward. At the bottom, the sodden soil flows in a sluggish mass that piles up in ridges or lobes to form a bulging toe.

Shallow earth flows, affecting only the soil and regolith, are common on sod-covered and forested slopes that have been saturated by heavy rains. An earth flow may affect a few square metres, or it may cover an area of a few hectares. If the bedrock of a mountainous region is rich in clay (derived from shale or deeply weathered volcanic rocks), earth flows sometimes involve millions of tonnes of bedrock moving by plastic flowage like a great mass of thick mud.

Earth flows often block highways and railroad lines, usually during heavy rains. Generally, the flow rate is slow, so the flows are not a threat to life. However, damage to buildings, pavement, and utility lines is often severe where construction has taken place on unstable soil slopes.

ENVIRONMENTAL IMPACT OF EARTH FLOWS

One special form of earth flow has proved to be a major environmental hazard in parts of Norway and Sweden and along the St. Lawrence River and its tributaries in Quebec. This type of flow involves horizontally layered clays, sands, and silts accumulated during the ice age that form low, flat-topped terraces adjacent to rivers and lakes. Over a large area, which may be 600 to 900 metres across, a layer of silt and sand 6 to 12 metres thick begins to move toward the river, sliding on a layer of soft clay that has spontaneously turned into a near-liquid state. The moving mass also settles downward and breaks into step-like masses. Carrying along houses or farms, the layer ultimately reaches the river, into which it pours as a great disordered mass of mud. Figure 14.9 provides a block diagram of a classic example that occurred in 1898 near St. Thuribe, Quebec.

This type of earth flow is caused by *quick clays*—clays that spontaneously change from a solid condition to a near-liquid condition when subjected to a shock or disturbance. Quick clays are thought to have formed in the shallow waters of saltwater bays near the end of the ice age. When deposited, the thin plates of clay in these layers have a "house of cards" structure with a large proportion of water-filled space between plates. The salt water provides positively and negatively charged atoms that help to bind the structure and give it strength. But with the regional uplift that often occurs after ice sheets melt away, the quick clay is elevated above sea level, and fresh groundwater replaces the salt water. A mechanical shock, such as an Earth tremour, then causes the house of cards structure to collapse. Because such a large proportion of water (from 45 to 80 percent by volume) is present, the clay–water mixture behaves like a liquid.

A particularly spectacular example of this type of earth flow occurred in Nicolet, Quebec, in 1955. A clay layer beneath the town liquefied, carrying much of the town into the Nicolet River.

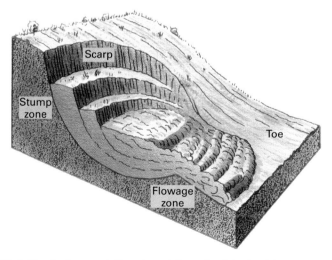

14.8 Sketch of an earth flow An earth flow with well-developed slump terraces in the upper part. The flow has produced a bulging toe. (After A. N. Strahler)

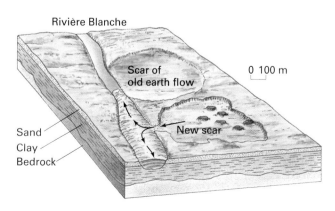

14.9 Earth flow at St. Thuribe This block diagram depicts the earth flow of 1898 near St. Thuribe, Quebec. (From C. F. S. Sharpe, *Landslides and Related Phenomena*, New York: Columbia University Press)

Fortunately, only three lives were lost, but the damage to buildings and a bridge ran into millions of dollars. A similar tragedy occurred in St. Jean Vianney in 1971, with 31 deaths, when nearly 8 million cubic metres of clay liquefied and flowed down a stream valley and into the Saguenay River.

MUDFLOW AND DEBRIS FLOOD

One of the most spectacular forms of mass wasting and a potentially serious environmental hazard is the **mudflow**. This stream of muddy fluid pours swiftly down canyons in mountainous regions (Figure 14.10). In deserts, where vegetation does not protect the mountain soils, local thunderstorms produce rain much more quickly than the soil can absorb it. As the water runs down the slopes, it forms a thin mud that flows to the canyon floors and then follows the stream courses. As it flows, it picks up additional sediment, becoming thicker and thicker until it is too thick to flow further. Great boulders are carried along, buoyed up in the mud, which engulfs and destroys roads, bridges, and houses in the canyon floor. Where the mudflow emerges from a canyon and spreads across an alluvial fan, severe property damage and even loss of life may be the result.

Mudflows are rapid events in which water, sediment, and debris follow slopes and river valleys to lower elevations. In contrast, earth flows are slower and are confined to collapsing slopes. Figure 14.5 shows this difference in the relative positions of these two types of mass wasting.

As explained in Chapter 13, mudflows that occur on the slopes of erupting volcanoes are called *lahars*. Heavy rains or melting snows turn freshly fallen volcanic ash and dust into mud that flows down the slopes of the volcano. Herculaneum, a city at the base of Mount Vesuvius, was destroyed by a mudflow during the eruption of A.D. 79. At the same time, the neighbouring city of Pompeii was buried under volcanic ash.

Mudflows vary in consistency, from a mixture similar to concrete emerging from a mixing truck to consistencies similar to that of turbid river floodwaters. The watery type of mudflow is called a *debris flood* or *debris flow* in the western United States, particularly in southern California, where it often occurs and with disastrous effects. The material carried in a debris flood ranges from fine particles to boulders to tree trunks and limbs. In mountainous regions on steep slopes, these flows are called *alpine debris avalanches*. The intense rainfall of hurricanes striking the eastern United States often causes debris avalanches in the hollows and valleys of the Blue Ridge and Smoky Mountains.

LANDSLIDE

A **landslide** is the rapid sliding of large masses of bedrock or regolith. Wherever mountain slopes are steep, there is a possibility of large, disastrous landslides (Figure 14.11). In Switzerland, Norway, or the Canadian Rockies, for example, villages built on the floors of steep-sided valleys have been destroyed by the sliding of millions of cubic metres of rock let loose without warning.

Landslides are triggered by earthquakes or sudden rock failures rather than by heavy rains or sheet floods. Thus, they are different from earth flows and mudflows, which are typically induced by heavy rains. Landslides can also result when excavation or river erosion makes the base of a slope too steep. Landslides can range from *rockslides* of jumbled bedrock fragments to *bedrock slumps* in which most of the bedrock remains more or less intact as it moves. The amazing speed with which rockslides travel down a mountainside is thought to be due to the presence of a layer of compressed air trapped between the slide and the ground surface. The air layer reduces frictional resistance and may prevent the rubble from touching the surface beneath.

Severe earthquakes in mountainous regions are a major cause of landslides. An example is the landslide of Santa Tecla, El Salvador, which killed hundreds of people on January 13, 2001 (Figure 14.12). The earthquake was centred off the southern coast of El Salvador and measured 7.6 on the Richter scale. It was felt as far away as Mexico City. In the shaking, a steep slope above the neighbourhood of Las Colinas collapsed, creating a wave of earth that swept across the ordered grid of houses and streets, burying hundreds of homes and their inhabitants. The same earthquake triggered other landslides in the region, with additional loss of life and property.

Aside from occasional local catastrophes, landslides have rather limited environmental influence because they occur only sporadically and usually in thinly populated mountainous regions. Small slides can, however, repeatedly block or break important mountain highways or railway lines.

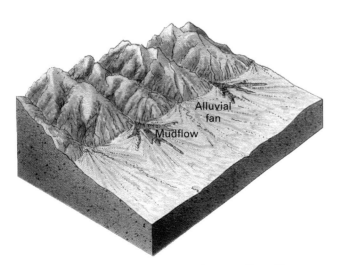

14.10 Mudflows in an arid environment Thin, stream-like mudflows issue occasionally from canyon mouths in arid regions. The mud spreads out on the fan slopes below in long, narrow tongues. (Drawn by A. N. Strahler)

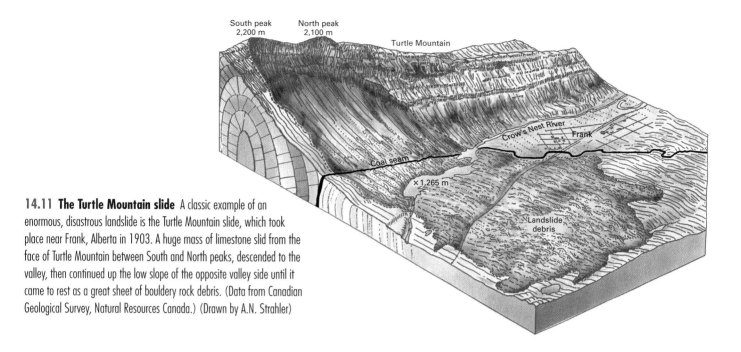

14.11 The Turtle Mountain slide A classic example of an enormous, disastrous landslide is the Turtle Mountain slide, which took place near Frank, Alberta in 1903. A huge mass of limestone slid from the face of Turtle Mountain between South and North peaks, descended to the valley, then continued up the low slope of the opposite valley side until it came to rest as a great sheet of bouldery rock debris. (Data from Canadian Geological Survey, Natural Resources Canada.) (Drawn by A.N. Strahler)

A large landslide releases a vast amount of gravitational energy. *Working It Out 14.1 • The Power of Gravity* documents this enormous power of gravity.

INDUCED MASS WASTING

Human activities can induce mass wasting in forms ranging from mudflows and earth flows to landslides. These activities include (1) piling up of waste soil and rock into unstable accumulations that can move spontaneously, and (2) removing the underlying support of natural masses of soil, regolith, and bedrock. Mass movements produced by human activities are called *induced mass wasting*.

In Los Angeles County, California, real estate development has been carried out on steep hillsides and mountainsides. Roads and home sites have been bulldozed out of the deep regolith, and the excavated regolith has been piled up to form nearby embankments. When saturated by heavy winter rains, these embankments can give way, producing earth flows, as well as mudflows and

14.12 Santa Tecla landslide A nearby earthquake of magnitude 7.6 triggered a landslide that carved a path of destruction through Santa Tecla, El Salvador, on January 13, 2001.

WORKING IT OUT | 14.1　The Power of Gravity

Directly or indirectly, gravity powers many of the processes that shape the landscape, from the erosion of running water in carving landforms (Chapter 16) to the processes of mass wasting—earth flows, mudflows, solifluction, and landslides—that are the subject of this chapter. By acting to move mineral matter and water downhill, gravity works over a span of time and is thus a source of power.

Let's briefly review gravitation, gravity, force, work, and power. *Gravitation* is the simultaneous attraction that occurs between two physical bodies. The strength of the attraction depends on the mass of each body and the distance between the two bodies. The Earth's attraction on an object near the Earth's surface is called *gravity*. It is an acceleration that acts on the mass of the object to produce a force according to the simple relationship

$$F = m \times g$$

where F is the force, m is the mass of the object, and g is the acceleration due to gravity (about 9.8 metres per second squared (m/s^2)). The mass is measured in kilograms, the acceleration is measured in metres per second squared, and the force is in newtons *(N)*.

Suppose now that gravity actually does some work; that is, it moves a mass through a distance. The work is measured as force multiplied by distance:

$$W = F \times d$$

where W is the work and d is the distance. The force is measured in newtons, the distance in metres, and the work done is measured in joules (J).

Power describes the rate at which work is accomplished: that is, work per unit of time, or

$$P = \frac{W}{T}$$

where P is the power, W is the work, and T is the time. Work is measured in joules (J), time is measured in seconds (s), and power is measured in watts (W). Recall from *Focus on Systems 3.1 • Forms of Energy* that work and energy have the same units and that energy is defined as the ability to do work. Thus, power is a measure of both a rate of energy flow and a rate at which work is done.

An example of the power of gravity is the Madison Slide—a catastrophic landslide that took place near Yellowstone National Park in 1959. A rock body with a volume of about 28 million cubic metres (28×10^6 m^3) moved a vertical distance downward of about 500 metres into the valley of the Madison River, attaining a velocity of about 150 kilometres per hour. Propelled by its own inertia, the flow then moved uphill on the far side of the river valley for a vertical distance of about 120 metres. In this slide, the potential energy of the position of the mass above the

---

debris floods that travel far down the canyon floors and spread out on the alluvial fan surfaces below, burying streets and houses in mud and boulders. Many debris floods in this area are also produced by heavy rains falling on mountain slopes stripped of vegetation by fire in the preceding dry season. Some of these fires are set carelessly or deliberately by humans.

INDUCED EARTH FLOWS

Examples of both large and small earth flows induced or aggravated by human activities are found in the Palos Verdes Hills of Los Angeles County. These movements occur in shale that tends to become plastic when exposed to water. The upper part of the earth flow subsides and slumps backward, as illustrated in Figure 14.8. The interior and lower parts of the mass slowly flow forward, producing a toe of flowage material.

The largest of the earth flows in the Palos Verdes Hills area was the Portuguese Bend "landslide," which affected an area of about 160 hectares. It was caused by the slippage of sedimentary

rock layers on an underlying layer of clay. The total movement over a three-year period was about 20 metres. Damage to residential and other structures totalled some US$10 million. Geologists attributed the earth flow to infiltration of water from septic tanks and irrigation water applied to lawns and gardens. A discharge of more than 115,000 litres of water per day from some 150 homes is believed to have sufficiently weakened the clay layer to start and sustain the flow.

SCARIFICATION OF THE LAND

Industrial societies now possess enormous explosives and machine power capable of moving great masses of regolith and bedrock from one place to another. This technology is used to extract mineral resources or move earth in the construction of highways, airfields, building foundations, dams, canals, and various other large structures. Both activities involve the removal of Earth materials, a process that destroys the pre-existing ecosystems and plant and animal habitats. When the materials are used to build

valley bottom was converted to kinetic energy during the slide, and the kinetic energy was dissipated as heat produced by friction during the sliding motion.

The slide moved material downhill for 500 metres, converting potential energy to kinetic energy and friction, then uphill another 120 metres. On the uphill leg, some kinetic energy converted back to potential energy as the mass moved upward against the pull of gravity. So the net release of potential energy was produced by a vertical distance change of $500 - 120 = 380$ metres.

How much energy was released? Or alternatively, how much work was done? The formulas to the left require knowledge of the mass, acceleration, and distance. Of these, only the mass is unknown. However, the volume of the landslide is 28 million cubic metres; what is missing is the density—the mass of the rock per unit of volume. Let's assume that the rock has the density of granite, which is about 2,700 kilograms per cubic metre (2.7×10^3 kg/m^3). Then mass can be calculated as follows:

$$m = 28 \times 10^6 \text{ m}^3 \times (2.7 \times 10^3 \text{ kg/m}^3)$$
$$= 7.56 \times 10^{10} \text{ kg}$$

Gravity will act on this mass with a force of

$$F = m \times g = 7.56 \times 10^{10} \text{ kg} \times (9.8 \text{ m/s}^2)$$
$$= 7.41 \times 10^{11} \text{ N}$$

This force is applied through a net distance of 380 metres, so the work done is

$$W = F \times d = 7.41 \times 10^{11} \text{ N} \times 380 \text{ m}$$
$$= 2.82 \times 10^{14} \text{ J}$$

The friction of the slide converted this work to heat.

The amount of work done is rather similar to the amount of electrical energy consumed in a day in the United States, which is about 2.33×10^{14} J/day. Thus, the energy released by the Madison Slide, if converted completely to electric power, could satisfy U.S. electrical needs for slightly more than one day.

To calculate the power expended by gravity in the slide, it is necessary to know how long the motion took. With some simple assumptions about the angles of the mountain slopes and the magnitude of friction in the earth flow, a reasonable guess for the time required is about 70 seconds. The power applied by gravity in the downward motion would then be

$$P = W/T = (2.82 \times 10^{14} \text{ J})/70 \text{ s} = 4.02 \times 10^{12} \text{ W}$$

Expressed in watts, the average rate of U.S. electric power consumption is about 2.70×10^9 W. The rate of energy release in the Madison Slide is therefore about 1,500 times greater than the rate at which the entire United States consumes electricity. This large number demonstrates the awesome power of gravity to shape and reshape the landscape.

up new land on adjacent surfaces, ecosystems and habitats are also destroyed—by burial. What distinguishes artificial forms of mass wasting from the natural forms is that machinery is used to raise Earth materials against the force of gravity. Explosives can produce disruptive forces many times more powerful than the natural forces of physical weathering.

Scarification is a general term for excavations and other land disturbances produced to extract mineral resources. The rock waste, known as *spoil* or *tailings*, accumulates at the site. Scarification takes the form of open-pit mines, strip mines, quarries for structural materials, sand and gravel pits, clay pits, phosphate pits, scars from hydraulic mining, and stream gravel deposits reworked by dredging.

Strip mining is a particularly destructive scarification activity. Strip mining of coal is practised where coal seams lie close to the surface or actually appear as outcroppings along hillsides (Figure 14.13). Earth-moving equipment removes the covering strata, or *overburden*, to expose the coal, which is lifted out by power

shovels. The piled-up overburden remains as spoil. When saturated by heavy rain and melting snow, the spoil can generate earth flows and mudflows that descend on houses, roads, and forest. Sediment from the spoil also clogs stream channels far down the valleys.

Because of these problems, strip mining is under strict control in most locations. Mine operators are not permitted to create hazardous spoil slopes, and they must restore spoil banks and ridges to a natural condition. In this process, called *reclamation*, they remove topsoil and save it as the mine is opened and worked. When spoil is returned to the mine cavity after extraction of the coal, mine operators cover it with the topsoil and plant suitable vegetation. With good management and proper planning, the resulting reclaimed land becomes usable again. But there are also many reclamation failures in which the land has been damaged beyond repair.

Scarification is on the increase. Driven by an ever-increasing human population, the demand for coal and industrial minerals

14.13 Contour strip mining This strip mine near Lynch, Kentucky follows an outcrop of coal along a hillside contour. A highway uses the winding bench at the base of the high rock wall.

used in manufacturing and construction is on the rise. At the same time, as the richer and more readily available mineral deposits are consumed, industry turns to poorer and less easily accessible grades of ore. As a result, the rate of scarification is further increased, making the success of reclamation efforts increasingly important.

PROCESSES AND LANDFORMS OF THE PERMAFROST REGIONS

The landscape of the treeless arctic and alpine tundra environment shows many distinctive effects of weathering and mass wasting in a climate dominated by a long, cold winter. During the short, warm season, surface thawing occurs, and the soil may be moist and vulnerable to mass wasting and water erosion. With the return of low temperatures, the freezing of soil water exerts a strong mechanical influence on the surface layer.

This intense frost action creates a distinctive set of landforms and landform-creating processes called the *periglacial system*. The adjective **periglacial** (*peri* meaning "near") describes an environment of intense frost action near the margins of active glaciers and ice sheets (which may actually be quite far from permanent ice and snow cover). The periglacial system is driven by the large annual range in temperature, which governs the flow of energy

into and out of the ground's surface layer. This cycle changes water from liquid to ice, with an increase in volume, and back to liquid again, with a decrease in volume. The expansion and contraction of water as it changes state, coupled with the pressure that ice crystals can exert as they grow, provides the mechanism for the movement of mineral particles individually and collectively and creates the distinctive processes that operate in the periglacial environment.

PERMAFROST

Ground that is perennially (year-round) below the freezing point of fresh water (0°C) is called **permafrost**. This includes mineral matter falling within the full range of particle sizes described in Chapter 11; that is, from clay up to boulders, bedrock, and ice. The term *permafrost* strictly refers only to ground temperature. Ice is commonly present in pore spaces in the permafrost or as free bodies or lenses, and is known as **ground ice**.

Permafrost prevails under most of the tundra climate region and in much of the boreal forest bordering the tundra. A distinctive feature of permafrost terrain is the shallow surface layer, called the *active layer*, which freezes and thaws each year. This layer ranges in thickness from about 15 centimetres to about 4 metres, depending on the latitude and the nature of the ground. Below the active layer is the *permafrost table*, the upper surface of the perennially frozen zone. Most lakes and rivers do not have permafrost beneath them, since the water at the bottom of these features is at or above 0°C throughout the year.

Figure 14.14 shows the distribution of permafrost in the northern hemisphere. Four zones appear on the map. The *continuous permafrost* zone, in which permafrost underlies more than 90 percent of the surface, coincides largely with the arctic tundra regions, but also includes a large part of the boreal forest in Siberia. *Discontinuous permafrost*, which occurs in patches separated by unfrozen ground, occupies much of the boreal forest of North America and Eurasia. A third zone—*subsea permafrost*—lies beneath sea level in the shallow continental shelf of the Arctic Ocean. It appears off the Asian coast and the coasts of Alaska, Yukon, and Northwest Territories in North America. Last is *alpine permafrost*, which is found at high elevations under frigid conditions in many parts of the world.

An examination of the permafrost map of the Northern Hemisphere shows some features that are related to climate. First is the general gradation from continuous permafrost in the north through discontinuous permafrost to no permafrost in the south, where the temperatures are generally higher. The zone of continuous permafrost dips southward in the interiors of both North America and Asia. As noted in the discussion of global climate, northern continental interiors become very cold in winter, since they are far from oceanic sources of moisture and heat. Another

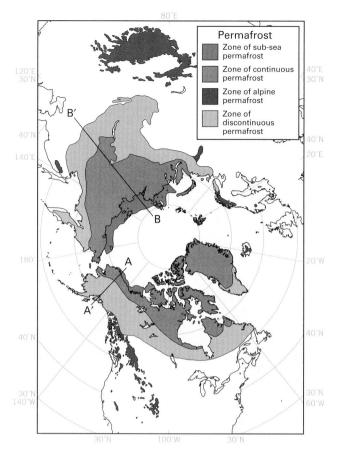

14.14 Permafrost map Distribution of permafrost in the northern hemisphere (Adapted from Troy L. Péwé, *Geotimes*, vol. 29, no. 2, p. 11, copyright the American Geologic Institute, 1984. Used with permission)

climate-related feature occurs in Europe. Scandinavia nearly lacks permafrost altogether, although it lies well within the latitude range of permafrost in North America and Asia. As described in Chapter 7, the thermohaline circulation of the Atlantic Ocean pumps large quantities of equatorial heat to high northern latitudes, and westerly winds carry the heat eastward to warm Europe's climate. Thus, warmer temperatures reduce the occurrence of permafrost in this region.

Figure 14.15 shows cross-sections of permafrost in Alaska and Asia that correspond with the permafrost map. Permafrost reaches depths of roughly 1,000 metres near the North Pole. Northern regions that remained uncovered by ice sheets during the last glacial advance show this greater thickness of permafrost. Without a cover of ice, the land here cooled to great depths during the frigid climate of the last ice age. Present-day subsea permafrost formed at the same time, when sea level was lower and the continental shelves were exposed to the cold atmosphere. Where ice sheets covered the land, continuous permafrost is shallower, reaching depths of 300 to 450 metres. Discontinuous permafrost is generally thinner.

Figure 14.16 shows a simplified map of permafrost in Canada. Continuous permafrost occupies the northernmost regions. As just noted, permafrost underlies more than 90 percent of the ground in continuous permafrost terrain. Even in very cold regions, some small areas of southern exposure or previous river channels may not be frozen. To the south, there are two classes of discontinuous permafrost: widespread discontinuous (50 to 90

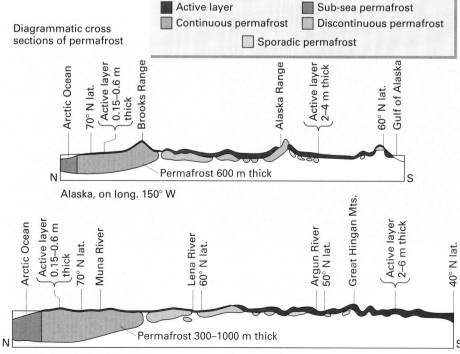

14.15 Permafrost map cross-sections
These north-south cross-sections of permafrost in Alaska and Asia are located on Figure 14.14. (From Robert F. Black, "Permafrost," Chapter 14 of P. D. Trask's *Applied Sedimentation*, copyright © by John Wiley & Sons. Reprinted with permission)

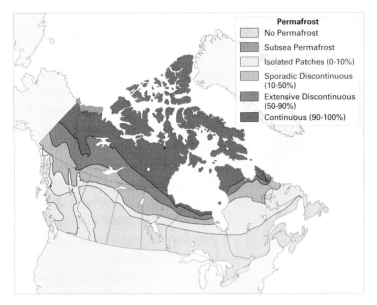

Permafrost
- No Permafrost
- Subsea Permafrost
- Isolated Patches (0-10%)
- Sporadic Discontinuous (10-50%)
- Extensive Discontinuous (50-90%)
- Continuous (90-100%)

14.16 Permafrost map of Canada The general distribution of permafrost shows continuous permafrost in the northernmost regions, with bands of widespread discontinuous permafrost, sporadic discontinuous permafrost, and isolated permafrost appearing progressively southward. Note the effect of elevation, which carries isolated permafrost southward in mountainous regions (alpine permafrost). (From National Atlas of Canada)

percent) and sporadic discontinuous (10 to 50 percent). A narrow band of isolated permafrost (0 to 10 percent) borders the southern rim of Canadian permafrost, with an arm extending southward along the Canadian Rockies to include alpine permafrost.

As in the global map, the general pattern in Canada corresponds with latitude, with continuous, extensive discontinuous, sporadic discontinuous, and isolated permafrost encountered in a southward progression. However, there are some deviations from this simple pattern. For example, the permafrost zones dip southward along the Western Cordillera. This is largely an effect of elevation, which lowers air temperatures. Continuous permafrost swings southward along the coast of Hudson Bay, compressing the zones in northern Ontario. This is due to persistent sea ice offshore for much of the summer, which lowers summer air temperatures near the coast.

PERMAFROST TEMPERATURE

At most locations, the average annual temperature of the upper 10 to 20 metres of the Earth's surface is close to the average in the air above. However, other factors affect the ground temperature in cold regions, and in most cases, mean annual air temperatures must be several degrees below 0°C before the mean ground temperature falls to 0°C and permafrost sets in. In fact, continuous permafrost is found only where mean annual air temperatures fall below −6 to −8°C. This effect is largely due to the thermal properties of water.

Water has the most effect on permafrost temperature in its snow state. As a blanket of ice crystals and air, snow has a low thermal conductivity, meaning that it acts as insulation to keep ground heat from flowing outward in winter. Thus, a deep cover of snow will keep the ground warmer. A 50-centimetre snow cover is enough to keep mean annual ground temperatures in mineral soils as much as 6°C higher than mean annual air temperatures.

The effects of snow cover on permafrost distribution are evident in the permafrost map of Canada. In general, the boundary between continuous and discontinuous permafrost coincides with the tree line that separates boreal forest from tundra. In the tundra zone, a thin cover of blowing snow leaves much of the ground bare or nearly bare, allowing soil heat to escape during the long arctic winter. Ground temperatures are lower and permafrost is nearly continuous. In the forest, a thicker blanket of snow slows winter cooling. Ground temperatures are warmer, and so unfrozen patches are more likely to occur in the forest zone. Thus, permafrost becomes discontinuous.

While snow keeps the ground warmer, other effects of water, namely its latent heat properties (see Chapter 6), tend to keep the ground colder. In the warm season, when ice in the active layer melts, large amounts of heat are taken up as ice turns to water. This heat comes from the atmosphere and is not available to raise ground temperatures. Evaporation of water from the surface of the soil, which occurs as long as the atmosphere is not saturated and the soil surface is moist, will also use heat that could otherwise warm the ground below.

What about the flow of heat through liquid water and solid ice? Ice conducts heat four times more easily than liquid water. Thus, in the depths of winter, heat escapes more readily from a surface layer of ground saturated with ice than it enters in summer through water-saturated soil. In this way, water in the soil acts like a one-way thermal filter to let heat out of the ground in the winter and block the entry of heat in the summer. The net effect of water's latent heat and thermal properties is that wet environments have colder ground temperatures than dry environments and are thus more likely to have permafrost below. The difference can be as much as 3°C.

The presence of vegetation also affects ground temperatures. A vegetation layer, such as a boreal forest stand, intercepts solar radiation in the warm season. This shade lowers ground temperatures. However, in most cases, the cooling effect of vegetation is less than the warming effect of snow cover.

Peat is the most important soil type influencing the occurrence of permafrost. Dry peat—a spongy material of tough, decomposing leaves and mosses, and fine branches of heath plants—has a low thermal conductivity and acts like an insulating blanket when it dries out in the summer. This effect reduces the flow of heat into the ground. In addition, peat holds a great deal of water, and as this evaporates lower down the soil profile in the large pores of the peat mass, latent heat is absorbed, further cooling the ground. Given these effects, it is no surprise to find persistent permafrost under thick peat deposits. The peatlands of northern Alberta are a classic example, where permafrost that formed during a cold period 200 to 400 years ago has persisted to today, even though the present climate does not favour the development of permafrost.

Permafrost is an example of an energy flow system in which the depth of the active layer and the base of the permafrost are determined by the balance of heat energy flowing into the ground from the surface and that upwelling from the Earth's interior. *Focus on Systems 14.2 • Permafrost as an Energy Flow System* describes this energy flow system in more detail.

Even in the areas of deepest continuous permafrost there are isolated pockets of ground that never fall below the freezing point. The Siberian word *talik* has been used to describe these pockets. Some taliks occur within permafrost as lenses of persistent unfrozen water. Others are large features underlying lakes and rivers. As noted earlier, water at the bottom of most lakes and flowing rivers, in spite of a winter ice cover, remains at or near 0°C, and thus the ground below remains unfrozen. The depth of the talik under a lake depends primarily on the lake's width. Under a small lake, the talik may be bowl-shaped and extend downward some tens of metres. Under a large lake, the unfrozen zone may extend below the base of nearby permafrost, thus creating a "hole" in the permafrost layer. Figure 14.17 sketches the taliks under three shallow lakes of increasing size.

THE ACTIVE LAYER

In all but the coldest of climates, air temperatures in the warm season exceed 0°C and a surface layer of frozen ground thaws. This is the *active layer*, which is defined as the layer between the ground surface and the top of permafrost that freezes and thaws each year. The active layer serves as a seasonal aquifer; that is, a water-rich layer of ground that provides plants with water during the warm season. Water stays within the active layer because it

cannot penetrate the permafrost. However, water can move laterally within the active layer, especially on a slope or to the top of the permafrost below. In this way, slight depressions accumulate water that cannot escape by sinking downwards. Thus, a summer walk across the tundra or through the boreal forest will often include walking through bogs.

The active layer contains boreal forest and tundra plant roots, and so is ecologically important. Photosynthesis, which requires the uptake of water from the soil, cannot begin until the root zone has thawed. Similarly, when the active layer freezes, photosynthesis shuts down for the winter. The active layer often contains an organic surface layer of plant remains, since in the cold and wet environment of the tundra and boreal forest, decay is slow and organic matter accumulates. The decay provides nutrients to plant roots, but also creates an acid soil to which many tundra and boreal forest plants have adjusted.

The thickness of the active layer depends on a number of factors, including the ground surface temperature, the thermal properties of the soil, and the temperature of the permafrost below. At the northern end of the continuous permafrost zone, the active layer can be quite shallow. Here, snow accumulation is limited and winter air temperatures are very low, so the ground surface becomes very cold. Summers are cool and short, so the thaw reaches a shallower depth. In the zone of discontinuous permafrost, where there is more snow and warmer air temperatures in winter, ground surface temperatures do not fall as low. As a result, a thicker active layer develops during the longer summers.

Soil water also has an effect. As noted in our discussion of permafrost temperatures, a high ice content in frozen soil requires more latent heat for melting, so summer thawing does not penetrate as deeply in wet soils. Dry soils, sands, gravels, and bedrock have relatively thick active layers since they all have little water.

Another important characteristic of the active layer is that its base tends to be rich in ice. When the active layer thaws, gravity causes water to accumulate at the bottom of the layer. As air temperatures drop at the end of the warm season, freezing moves downward from the ground surface, trapping water at the top of the permafrost. In many cases, freezing also proceeds from the bottom up as the latent heat of freezing is slowly conducted to the cold permafrost below. The result is the formation of lenses of nearly pure ice at the base of the active layer.

FORMS OF GROUND ICE

Ice is present in permafrost in a variety of forms. First, ice serves as a cement that binds particles of mineral matter into rock-like beds that are solid and tough as long as they are frozen. But there are also masses of nearly pure ice in the form of lenses, wedges, or

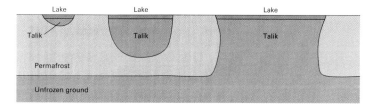

14.17 Taliks Taliks, or areas of unfrozen ground, underlie lakes and flowing rivers in permafrost terrains. As shown in this diagram, the taliks of larger lakes can completely penetrate the permafrost layer.

FOCUS ON SYSTEMS | 14.2   Permafrost as an Energy Flow System

The upper few hundred metres of the solid Earth receive heat from two directions—downward from the atmosphere (or ocean) above and upward from the Earth below. The flow of energy into and out of the ground at the surface depends on the surface energy balance. In winter, when the air is, on average, colder than the ground, heat energy will flow out of the ground to the atmosphere. In summer, when the air is warmer than the ground, heat energy will flow into the ground from the atmosphere. (*Focus on Systems 4.1 • The Surface Energy Balance Equation* described the surface energy balance in detail.)

The surface flow of heat energy into and out of the ground occurs mostly by conduction. Over the year, the greatest ground temperature variation occurs close to the surface. At increasing depth in the ground, the temperature varies less. The ground usually shows seasonal temperature fluctuations to depths of 10 to 20 metres. At 20 metres depth, the temperature range is small, perhaps about 0.1°C, and the ground remains near the mean annual surface temperature.

The figure to the right shows this effect for a mean annual surface temperature of −10°C. In the winter (left), air temperatures drop and heat flows out of the ground, as shown in the colour bar. In this example, the minimum surface temperature during the winter is −25°C. Summer conditions are shown on the right side of the graph. Here, the maximum surface temperature is 5°C and heat flows into the ground.

The two upper curves on the graph show the minimum and maximum annual temperatures reached at each depth. At the surface, the minimum and maximum values correspond to the annual values of −25°C and 5°C and the difference between the minimum and maximum values is large. With depth, the difference between the minimum and maximum values decreases as the curves converge. At about 14 metres annual variation ceases and the temperature remains near, although slightly above, the mean annual surface temperature of −10°C. Note that the minimum and maximum curves are not actually plots of temperature with depth at one time. Because the ground resists heating, changes in air tempera-

ture take time to penetrate the ground. The time lag increases with depth. In fact, the coldest temperature in the ground at 5 metres usually occurs about six months after the coldest surface conditions.

Note also that the uppermost layer of the ground warms above freezing. This is the active layer of permafrost that freezes and thaws each year.

In contrast to the seasonally varying flow of heat near the surface is the steady flow of heat upward from the Earth's interior. As seen in *Focus on Systems 11.1 • Powering the Cycle of Rock Transformation*, crustal temperature increases with increasing depth due to the constant flow of heat from the interior. This increase is shown by a straight slanting line on the lower part of graph. The temperature gradient varies somewhat, but it typically increases by about 3°C per 100 metres of depth. The increasing temperature eventually exceeds 0°C, which defines the base of the permafrost layer. In this example, this is reached at a depth of about 333 metres.

Suppose that the mean annual temperature is somewhat warmer—say −5°C instead of −10°C. In this case, the

veins. These masses of ice displace sediment and so add to the volume of frozen material.

Several different processes lead to the growth of ice in the ground. **Ice lenses** are more or less horizontal layers of ice that form as the active layer freezes back over at the end of the warm season. Freezing occurs as the *freezing front*—the location at which liquid water is changing to ice—moves downward into the soil and reaches the capillary fringe. Here, water is drawn upward from the saturated zone at the base of the active layer by capillary tension through tiny connected pores and pathways between mineral particles. As capillary water approaches the freezing front, it loses heat and becomes supercooled; that is, its temperature drops below 0°C. However, because the water is adsorbed on the surface of sediment grains, it does not form ice. When the supercooled water molecules reach the freezing front, they quickly join the molecular structure of the growing

ice mass, pushing aside any sediment grains, and forming a lens. As the freezing front descends into the active layer, the ice lenses usually form subparallel bands, separated or segregated by soil layers. This *segregated ice* is the most common form of ground ice in seasonally frozen ground and is primarily responsible for *frost heave* of soils; that is, the expansion of soils as they freeze.

Ice lenses also form from the bottom up as a freezing front ascends upward from cold permafrost below. Because water may exist in a supercooled state, ice growth can occur in the uppermost portion of the permafrost layer even though the temperature of the layer is at or slightly below 0°C. The result is that the top of the permafrost layer is often rich in segregated ice. This layer can also acquire segregated ice from the base of the active layer if the active layer thins in response to climate change or change in the surface cover.

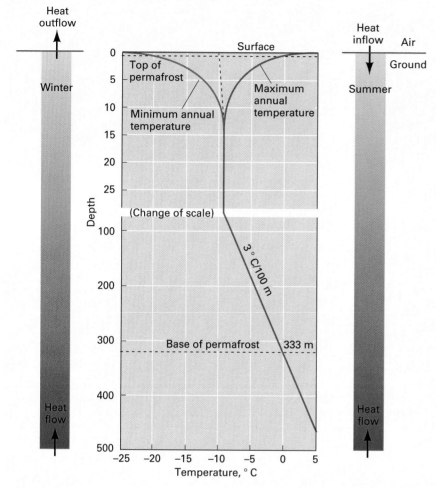

Maximum and minimum permafrost temperature profiles Mean annual maximum and minimum surface temperatures plotted against depth (Adapted with permission from A. H. Lachenbruch in Rhodes W. Fairbridge, Ed., *The Encyclopedia of Geomorphology* New York: Reinhold Publishing Corp.)

active layer at the surface would be thicker, as summer temperatures would be warmer and the soil would thaw more deeply. The bottom of the permafrost would also be not quite as deep. Thus, both the thickness of the active layer and the depth of permafrost are related to the mean annual surface temperature—the colder the temperature, the thinner the active layer and the deeper the permafrost base.

Permafrost is another example of an energy flow system in which inflows and outflows of energy tend to reach a balance, or equilibrium, over time. However, because heat flows so slowly through the ground, it takes a long time for permafrost to reach equilibrium when a climate change occurs—thousands of years in some cases. Permafrost bodies in some regions where mean annual surface temperature is now well above 0°C may be relics of the ice age, slowly wasting away, but still in place after 12,000 years of postglacial climate. Subsea permafrost is similarly in disequilibrium, since its great thicknesses are too deep to be sustained beneath ocean bottom temperatures above −2°C.

Lenses of segregated ice most commonly form in mineral deposits about the size of silt. The pore spaces, or *voids*, between silt particles are small enough to produce high capillary tension, while large enough to allow significant amounts of water to flow toward the freezing front. Clay, with its fine particle size, has a higher capillary tension than silt, but the spaces between clay particles are too small to permit significant water flow. Sand has larger voids than silt so water flow is no problem; however, the larger voids also reduce capillary tension.

Another common form of ground ice is the **ice wedge**. In many environments such as river flood plains, delta plains, and drained lake bottoms in the arctic environment, ice accumulates as vertical wedges in deep cracks in the sediment (Figure 14.18). Ice wedges originate in cracks that form when permafrost contracts during the extreme winter cold. As shown in Figure 14.19, surface meltwater enters the crack during the spring, moves downward into the permafrost, and freezes there. As the active layer thaws, the crack is covered with surface material, but the new ice is preserved in the permafrost. In the following winter, the crack may reopen, repeating the cycle. Repeated cracking and filling with ice over successive seasonal cycles causes the ice wedge to thicken. Ice wedges can grow as wide as three metres and as deep as five metres. The extra volume introduced into permafrost by the ice wedge displaces the ground upward to form low ridges on either side of the wedge, while a trough forms between the ridges, above the wedge.

Because the cracking relieves strain across a large horizontal surface, the cracks and wedges are typically interconnected in a network of surface troughs, called *ice-wedge polygons* (Figure 14.20). The ridges on each side of the wedges generally mark the edges of the polygons. These ridges are readily visible in the

14.18 Ice wedge Exposed by erosion of a coastal bluff near Tuktoyaktuk, Northwest Territories, this large wedge of ice pushed sediment aside and upward as it grew. Note the central thermal contraction crack, formed as a result of ground cooling in winter.

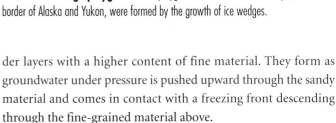

14.20 Ice-wedge polygons These polygons in the Alaskan north slope, near the border of Alaska and Yukon, were formed by the growth of ice wedges.

photo. Ice wedge polygons are typically 10 to 15 metres wide and are common in regions of continuous permafrost. The polygons are easy to see in lowlands; but on hills, downslope movement may fill the troughs, obscuring the polygonal outlines.

Massive icy beds are large ice lenses. They are layers of almost pure ice, up to 10 metres or more in thickness. In a typical situation, the icy beds occur on top of sandy or gravelly layers and un-

der layers with a higher content of fine material. They form as groundwater under pressure is pushed upward through the sandy material and comes in contact with a freezing front descending through the fine-grained material above.

A remarkable ice-formed feature of the arctic tundra is a conspicuous conical mound, called a *pingo* (Figure 14.21). The pingo has an ice core and slowly grows in height as more ice accumu-

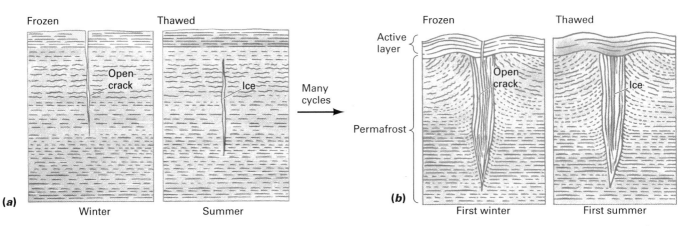

14.19 Formation of an ice wedge (a) In the beginning stage, an open crack appears during the winter and, in the spring, fills with meltwater that immediately freezes and expands. (b) After numerous seasonal cycles, the ice wedge has grown. (Adapted with permission from A. H. Lachenbruch in Rhodes W. Fairbridge, Ed., *The Encyclopedia of Geomorphology*, New York: Reinhold Publishing Corp.)

14.21 Ibyuk pingo Located on the Arctic coast near Tuktoyaktuk, Northwest Territories, this classic pingo began to form about 1,200 years ago in a drained lake bed. Since that time, coastline retreat has flooded the former lake basin, and it is now connected to Kugmallit Bay of the Beaufort Sea.

lates, forcing the overlying sediment upward. In extreme cases, pingos reach heights of 50 metres and have bases that are more than 600 metres in diameter.

Pingos typically form above sandy sediments on drained tundra lakes. Like massive icy beds, they depend on pressurized groundwater for their growth. Figure 14.22 shows one method of pingo formation. A shallow lake with an underlying sandy talik is drained,

perhaps by erosion of its outlet. Due to the high porosity of sand, the talik is rich in water. Without its insulating cover of water and ice, the lake bottom begins to freeze and permafrost penetrates downward. Recall that water, when frozen, has a slightly larger volume than it does as a liquid. Thus, as the water in the talik slowly converts to ice, it expands, placing pressure on the remaining water. The pressure pushes water upward at a weak point in the upper permafrost and forms an ice dome. As permafrost slowly engulfs the talik, the pingo grows. Pingos also form where water pressurized by other means underlies permafrost, such as at the foot of hill slopes.

THERMOKARST LAKES

Many areas of ice-rich permafrost are marked with shallow lakes formed by thawing of permafrost. These features are called *thermokarst lakes*. The word "karst," explained in Chapter 15, means a terrain of pits and hollows developed by uneven removal of limestone bedrock by solution. In the case of thermokarst, no solution occurs; the relief develops by melting of ground ice and subsequent settling of the ground. The majority of thermokarst lakes are less than 10 metres deep and rarely more than 2 kilometres wide.

Thermokarst lakes typically begin their formation after a chance event of thawing that occurs at a particular location. For example, a patch of trees in the boreal forest underlain by permafrost may be blown over in a storm. Without the thermal protection of the tree cover, the active layer deepens and the ice-rich upper portion of the permafrost melts. Losing its volume of ice lenses and layers, the ground settles, and a small water-filled depression forms. As the ground settles along its margins, the lake grows. Nearby trees lose their footing to the subsiding soil and tilt precariously ("drunken trees") or fall into the enlarging lake (Figure 14.23).

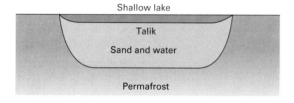

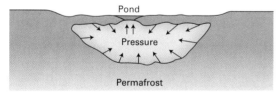

14.22 Pingo formation The draining of a shallow lake underlain by water-rich sandy sediments is the first step in the formation of a pingo. As the freezing of the lake bottom moves downward, volume expansion of ice places pressure on the unfrozen talik. The pressure causes water to well upward at a weak point or hole in the permafrost layer, where it freezes into a pingo. (After A. L. Washburn)

14.23 Thermokarst lake The radius of this thermokarst lake, in the boreal forest near Mayo, Yukon Territory, is expanding outward by about half a metre per year as the surrounding ice-rich permafrost thaws at its margins. The trees in the lake have gradually submerged as the ground ice melted and the ground surface subsided.

GEOGRAPHERS AT WORK

What Can Paleoclimatic Studies Show Us?
Konrad Gajewski, Ph. D., University of Ottawa

Konrad Gajewski is a climatologist whose research documents climate variability of the past in the Canadian Arctic. Paleoclimatic studies are based on the analysis of *proxy-climate* data, that is, a natural system that is affected in some way by changes in climate and which leaves a fossil record. A major source of paleoclimate data is lake sediments and the enclosed fossils, which are thus archives of past environments. The characteristics of the sediments and the enclosed fossils record conditions at the time of deposition.

Paleoclimate studies begin in the field, and the first step is collecting the appropriate data. Lake sediments are cored, and in the laboratory the characteristics of the sediments are measured and various microfossils are extracted. The cores are dated using radiocarbon or other methods.

One source of information about past climates is pollen grains that are produced by vascular plants. The vegetation is related to the climate and the pollen produced by the plants is also related to the climate. Since pollen grains are small and resistant to decay, they accumulate in lake sediments. Statistical methods are used to quantify the relation between pollen and plant abundance or pollen and climate directly, and these can be solved with fossil assemblages to provide quantitative *time series* of the climate for the past

The results of these studies show that the Arctic is already feeling the impacts of climate change. Paleoclimate studies have shown that climate changes can occur abruptly or slowly, they can be subtle or significant. The recent warming is unusual with respect to previous changes, showing the human impact on the climate.

Developing time series of past climates is only the first step in a paleoclimatology research program. Climatology is a spatial science, and the synthesis of many pollen diagrams is necessary.

Databases of pollen diagrams, as well as other paleoclimate proxy records such as rings, have been prepared and are used in a GIS to map the climate of the past 20,000 years.

Collecting a lake sediment core on the Boothia Peninsula, Nunavut.

A thermokarst lake's growth is slow at first as the shallow water freezes during winter. As thawing and subsidence proceed, however, the lake becomes deep enough to maintain a bottom temperature of 0°C or higher year round, and a talik begins to form. The expanding lake may join others formed nearby, for example, along a valley bottom, to create chains or strings of interconnected lakes.

What processes stop the growth of thermokarst lakes? First, a lake may run out of ice-rich permafrost if ground conditions change on its margins—for example, it may expand outward to a rim of bedrock or ice-poor permafrost. Many thermokarst lakes form along river flood plains, and they may be filled with sediment from a river flood. Aquatic plants or peat may invade the lake, creating a layer of roots and sediment at the edge that can freeze through in the winter, allowing the return of permafrost below. Lastly, thermokarst lakes may be drained by erosion of their outlet. This may occur suddenly, when a permafrost soil barrier or ice wedge constricting the outlet thaws. Once drained, permafrost can reform under the lake bottom and forest may grow again.

The formation of a thermokarst lake is dependent on subsidence due to the melting of ice-rich ground. Not all permafrost is rich enough in ice to permit the creation of a lake. In such cases, the active layer may temporarily deepen and subside following a disturbance, but will often return to a normal state as the disturbed surface recovers. As the following section describes, human activities can cause permafrost thawing and thus set off the formation of thermokarst terrain.

FOREST FIRES AND PERMAFROST

Forest fires are widespread agents of disturbance in the boreal forest. On average, 10,000 fires burn 2.5 million hectares of Canadian forest each year. Siberia also experiences large and frequent forest fires. Figure 14.24 shows the effects of a fire in the boreal forest.

14.24 Fire in the boreal forest These two views show the boreal forest near Minto, central Yukon Territory, (a) undisturbed and (b) after burning. The photos were taken on either side of the Klondike Highway, which served as a firebreak. Note the change in surface characteristics brought about by the fire.

(a)   (b)

Burning of the boreal forest normally raises ground temperatures by several degrees and results in a deepening of the active layer with accompanying thawing of the top of permafrost. The deepening of the active layer is not caused by the heat of the fire, but by effects such as drying of the ground, removal of shade, and destruction of insulating organic matter on top of the soil. Also, snow cover decreases, as there is less vegetation to trap winter snow. As seen in the discussion of permafrost temperatures, these factors tend to increase summer heat penetration or reduce winter heat loss, thus deepening the active layer. With the deepening of the active layer and thawing of ice-rich permafrost, land subsidence occurs on flat terrain and soils become wetter. In some cases, thermokarst terrain develops. In Siberia, many large areas of thermokarst lakes and depressions are thought to have been produced by huge fires that occur in warm, dry years. On slopes, melting of the top of permafrost can cause the formation of mud at the base of the active layer, so overlying sediment and burned forest vegetation together can form a downslope earth flow.

In most cases, the changes caused by forest fires are not permanent. The forest eventually regenerates and with the return of the forest cover and its insulating properties, the active layer becomes thinner and the permafrost table rises. Ice lenses grow, and the land surface rises.

RETROGRESSIVE THAW SLUMPS

An interesting and unique form of mass wasting in permafrost terrain is the *retrogressive thaw slump*, which may develop where erosion exposes the ice-rich upper layer of permafrost or massive icy beds (Figure 14.25). These slumps typically occur along eroding river banks, lake and ocean shores, and at road construction and mineral exploration sites. As the ground ice melts, the overlying sediments flow or fall down above the ice, providing a layer of mud at the foot of the slump.

Melting of the ground ice and growth of the slump can be rapid during the thaw season. The headwalls of thaw slumps have migrated at rates of as much as 16 metres per year and the slumps may reach lengths of over 650 metres from the point of initiation

14.25 Retrogressive thaw slumps These slumps are eroding ice-rich permafrost on Kendall Island, in the Mackenzie delta area of the western Arctic coast in Canada. Note the white ice wedges exposed at the heads of the slumps and the ice-wedge polygons on the nearby ground.

to the headwall. As such, they are the most rapidly developing geomorphic features in the North. The amount of material covering the ice face affects the thaw rate. For example, a thaw slump may stabilize when it encounters a layer of peat that falls down over the melting ice front and provides a cover of insulation. In most cases, sediment ultimately buries the melting ice front, stabilizing the slump. Coastal thaw slumps can be reinvigorated by the wave action of storms, which may re-expose ice-rich permafrost several years after the slump stabilized. Heavy rains that wash sediment from the ice face may also renew rapid retrogression.

PATTERNED GROUND AND SOLIFLUCTION

This chapter has described the effect of frost action on surface materials and noted that the growth of lenses and layers of ice in soil can move sediment to produce frost heaving. The growth of ice lenses usually causes stones to move upward or sideways within the soil profile. These effects, called *cryoturbation*, can produce considerable motion in the active layer. In arctic environments, cryoturbation can produce regular surface forms, such as circles, polygons, or nets of stones, or fields of low mounds of soil. These features are called **patterned ground**. The term "patterned ground" also includes ice-wedge polygons, described earlier. Table 14.1 summarizes several major categories of patterned ground. Circles, polygons, and nets are common features of permafrost terrain on flat or gentle slopes. Steps and stripes are found on steeper slopes and can also occur outside permafrost regions.

A common type of patterned ground in permafrost regions is the *mud hummock*, a low, rounded dome of soil found where

14.26 Earth hummock This earth hummock in boreal forest near Eagle Plains, northern Yukon Territory, has been excavated down to the frost table. The thaw depth in the centre of the hummock extends below the ground ice that is exposed at the sides.

fine-grained sediments overlie permafrost (Figure 14.26). Vegetation sometimes grows at the edges of the hummock. If vegetation covers the entire form, it is called an *earth hummock*. Figure 14.27 illustrates the formation of a mud hummock. As the hum-

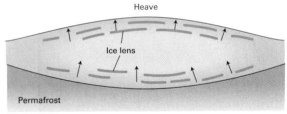

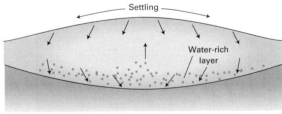

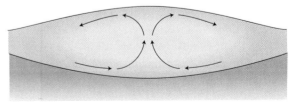

14.27 Formation of mud hummocks In the winter, the formation of ice lenses toward the top and bottom of the hummock causes an upward heaving motion. With the summer thaw, outward settling occurs at the top, and a water-rich layer of mud allows sediment movement inward and upward at the bottom of the hummock. (After J. R. Mackay)

Table 14.1 Forms of patterned ground

Circles	Circles of stones or rings of vegetation. Occurring singly or in groups. Typical size: 0.5–3 metres
Polygons	Unsorted polygons: Marked by furrows or cracks at edges. Variable in size. Example: ice-wedge polygons. Sorted polygons: marked by stones at edges and fine material in the centre. Typical size: usually less than 10 metres.
Hummocks	Earth hummocks (vegetated mounds) and mud hummocks (vegetated mounds with bare centres). Typical size: 2–4 metres
Steps	Terrace-like forms found on steep slopes. May be sorted, with a border of stones, or unsorted, with a vegetation-covered riser.
Stripes	Perpendicular to contours on steep slopes. May be sorted, with alternating coarse and fine debris, or unsorted, with vegetated troughs between stripes.

Source: Modified from A. L. Washburn, as summarized in M. A. Summerfield, *Global Geomorphology*, Longman-Wiley.

14.28 Stone circles Sorted circles of gravel abound on this low, wet plain underlain by permafrost in Broggerhalvoya, Spitsbergen, latitude 78° N. The circles in the foreground are 3 to 4 metres across; the gravel ridges are 20 to 30 centimetres high.

14.29 Solifluction Solifluction has created this landscape of soil mounds in the Richardson Mountains, Northwest Territories. Bulging masses of water-saturated regolith have slowly moved downslope, lubricated by a water-rich layer of sediment at the top of the permafrost.

mock freezes back in winter, ice lenses form in both the upper and lower parts of the active layer, causing the top of the hummock to heave upward and outward. As the summer thaw sets in, the heaved soil settles back, leaving a small mound. When the ice lenses at the permafrost table melt, a water-rich layer of soft mud forms. The pressure of overlying sediments moves the mud toward the bottom of the hummock floor, where it is pushed upward in the centre of the mound. The resulting circulation moves material inward along the bottom of the active layer, upward in the centre, and outward at the top.

In drier environments with coarser particles, a similar process yields stone circles. Here, rock fragments move to the edges of the form, where they accumulate. Figure 14.28 shows some impressive stone circles on Spitsbergen, an island in the Arctic Ocean. On slopes, the process of soil creep causes the polygons to elongate in the downslope direction and transform into parallel stone stripes.

A special type of earth flow characteristic of cold regions is **solifluction** (from the Latin words meaning "soil" and "to flow"). This occurs in late summer, when the ice-rich layer at the bottom of the saturated active layer melts to form a plastic mud. The active layer above then moves downslope as a single mass, typically at a rate of a few millimetres per year. Where the slope is covered with hummocks, they move individually and remain distinct. The process creates *solifluction terraces* and *solifluction lobes* that give tundra slopes a stepped appearance (Figure 14.29).

ALPINE TUNDRA

Most of the periglacial processes and forms of the low arctic tundra are also found in the alpine tundra of high mountains at mid- and high latitudes. The map in Figure 14.14 shows major regions

of *alpine permafrost* in the northern hemisphere. In alpine areas, many steep mountainsides have large exposures of bedrock and conspicuous talus slopes and cones formed of large angular blocks (see the chapter opening photo). Patterned ground and solifluction terraces occupy relatively gentle slopes where finer sediment tends to accumulate.

Alpine permafrost is restricted to elevations at which the mean annual air temperature is below freezing. The general elevation at which mean annual temperatures fall to 0°C will depend on latitude. In subtropical latitudes, it is necessary to reach elevations of 4,000 metres or higher, while at latitudes over 70°, permafrost may be encountered at sea level. The precise location also varies from continent to continent, just as the distribution of permafrost varies away from the moderating influence of the ocean on climate.

ENVIRONMENTAL PROBLEMS OF PERMAFROST

The permafrost environment is in many ways a delicate one in which small changes can have large impacts. Even a minor disturbance of the ground surface can change the thermal environment, leading to degradation of permafrost. The consequences of disturbance of permafrost terrain became evident in World War II, when military bases, airfields, and highways were quickly constructed in the North without considering the protection of the permafrost. Thawing led to rapid subsidence of the ground, tilting roads and runways, and swamping buildings as they cracked and settled irregularly.

Modern building practices in the North now place structures on insulated pilings with air space underneath, allowing cold winter air to maintain low soil temperatures. Water and sewage lines are not buried below ground, but are carried in above-ground conduits called utilidors. Roadways and runways crossing wet permafrost are built on thick pads of insulating gravel. Pipelines are carried on pilings over ice-rich permafrost and routed under streams through taliks. Structures such as dams are carefully engineered and often designed with a frozen core that is maintained by thermosiphons—devices that pump heat from the ground.

Another serious engineering problem of arctic regions is the behaviour of streams in winter. As the surfaces of streams and springs freeze over, the water beneath may burst out from below the ice, flow over the surface, and freeze into huge accumulations of ice. If this phenomenon occurs at a highway bridge or culvert, the roadway may wash out in spring when snowmelt occurs, the streams are full, and these structures remain blocked.

The lessons of superimposing technology on a highly sensitive natural environment were learned the hard way—by encountering undesirable and costly effects that were not anticipated. Continued economic development of arctic regions is likely to teach additional hard lessons.

CLIMATE CHANGE IN THE ARCTIC

As seen in Chapter 10 and in *A Closer Look: Eye on Global Change 10.4 • Regional Impacts of Climate Change on North America*, climate change is expected to bring warmer temperatures and increased precipitation to much of the polar regions. Substantial warming has already occurred, and predictions are that by the end of this century, winter temperatures will have increased by 3 to 10°C. Annual precipitation will increase by as much as 25 percent. What impact will these changes have on the arctic and boreal environment?

First, it seems certain that warming temperatures and increasing precipitation, especially in the form of snow, will deepen the active layer over broad areas of continuous permafrost. This is likely to be accompanied by extensive development of thermokarst terrain, where the ice-rich upper layers of permafrost melt and subside. This will result in mass movement on hill slopes, disrupting roads, pipelines, and other infrastructure. Sea-level rise along the shores of the Arctic Ocean will increase the number of retrogressive thaw slumps. The flow of northward-flowing rivers will decrease due to increased summer evaporation. Subsistence hunting by native peoples will become more difficult as populations of mammals, fish, and birds fluctuate and change their distribution. The winter, for overland and over-ice, travel will become shorter.

The boreal forest will migrate toward the North Pole, as will the border between continuous and discontinuous permafrost, as increased snowfall warms the ground under new forest. Discontinuous permafrost will decrease at the southern boundary, and the present zone of isolated permafrost will migrate northward. Warmer soil temperatures will increase the decay of soil organic matter, releasing carbon dioxide and methane, especially from peatlands, further boosting warming, while forest productivity may decline from summer drought stress, increased disease and insect damage, and more frequent fires. In short, major changes in the boreal environment seem to be in store.

A Look Ahead

This chapter examined two related topics: weathering and mass wasting. In the weathering process, rock near the surface breaks up into smaller fragments and often alters in chemical composition. In the mass wasting process, weathered rock and soil move downhill in slow or sudden mass movements. The landforms of mass wasting are produced by gravity acting directly on soil and regolith. Gravity also powers another landform-producing agent—running water—which the next three chapters discuss. The first deals with water in the hydrologic cycle, in soil, and in streams. The second deals specifically with how streams and rivers erode regolith and deposit sediment to create landforms. The third describes how stream erosion strips away rock layers of different resistance, providing large landforms that reveal underlying rock structures.

CHAPTER SUMMARY

- Weathering is the action of processes that cause rock near the surface to disintegrate and decompose into regolith. Mass wasting is the spontaneous downhill motion of soil, regolith, or rock under gravity.
- Physical weathering produces regolith from solid rock by breaking bedrock into pieces. *Frost action* breaks rock apart by the repeated growth and melting of ice crystals in rock fractures and *joints*, as well as between individual mineral grains. In mountainous regions of vigorous frost, fields of angular blocks accumulate as *felsenmeers*. Slopes of rock fragments form *talus cones*. In soils and sediments, needle ice and ice lenses push rock and soil fragments upward. *Salt-crystal growth* in dry climates breaks individual grains of rock free, and can damage brick and concrete. *Unloading* of the weight of overlying rock layers can cause some types of rock to expand and break loose into thick shells, producing *exfoliation domes*. Daily temperature cycles in arid environments are thought to cause rock breakup. Wedging by plant roots also forces rock masses apart.

- Chemical weathering results from mineral alteration. Igneous and metamorphic rocks can decay to great depths through *hydrolysis* and *oxidation*, producing a regolith that is often rich in clay minerals. *Carbonic acid* action dissolves limestone. In warm, humid environments, basaltic lavas can also show features of solution weathering produced by acid action.

- Mass wasting occurs on slopes that are mantled with regolith. Soil creep is a process of mass wasting in which regolith moves down slopes almost imperceptibly under the influence of gravity. In an earth flow, water-saturated soil or regolith slowly flows downhill. Quick clays, which are unstable and can liquefy when exposed to a shock, have produced earth flows in previously glaciated regions. A mudflow is much swifter than an earth flow. It follows stream courses, becoming thicker as it descends and picks up sediment. A watery mudflow with debris ranging from fine particles to boulders to tree trunks and limbs is called a *debris flow*. A landslide is a rapid sliding of large masses of bedrock, sometimes triggered by an earthquake.

- The scarification of land by human activities, such as mining, can heap up soil and regolith into unstable masses that produce earth flows or mudflows. The removal of supporting layers can also cause mass wasting by undermining the natural support of soil and regolith. These actions are called *induced mass wasting*.

- The tundra environment is dominated by the *periglacial system* of distinctive landforms and processes related to freezing and thawing of water in the *active layer* of permafrost terrain. Continuous permafrost and subsea permafrost occur at the highest latitudes, flanked by a band of discontinuous permafrost occurring within the boreal forest that serves as a transition to warmer regions. Permafrost is thickest toward the poles and under regions not covered by the ice sheets during the last ice age.

- A number of factors besides simple mean annual air temperature influence ground temperature. Snow insulates the ground from winter heat loss, so a thicker snow layer keeps the ground warmer. Water in the soil tends to lower ground temperatures, since the latent heat used in melting and evaporating water reduces soil warming during the annual thaw. Ice conducts heat more readily than water, so in winter a saturated, frozen active layer loses heat more rapidly than a dry active layer. Vegetation also tends to reduce ground temperatures. Ground doesn't freeze under lakes or rivers, so these water bodies are underlain by *taliks*—pockets of unfrozen ground surrounded by permafrost.

- The *active layer* overlies permafrost and thaws each year during the warm season. It contains the roots of plants as well as the water and nutrients needed to support them. The active layer is thinner in colder regions and where water is more abundant. Ice lenses often accumulate at the base of the active layer, on top of permafrost.

- *Segregated ice lenses* form as supercooled water migrates to a freezing front through capillary voids. Ice wedges form when water fills cracks in the active layer caused by shrinkage due to extreme winter cold. Ice wedges occur in systems of *ice-wedge polygons*. *Pingos* are distinctive domed hills of ice that form when shallow lakes are drained and permafrost invades the water-rich talik below, creating pressure as the water freezes and expands.

- *Thermokarst lakes* typically begin development when the surface cover is disrupted, triggering the melting of ice-rich permafrost. The ground settles to form a water-filled depression that expands by melting permafrost at its edges. The lake grows until it reaches dry permafrost or bedrock or is drained by erosion of its outlet.

- Forest fires disrupt the vegetation cover over permafrost, causing deepening of the active layer. If ice-rich permafrost melts, thermokarst terrain develops.

- *Retrogressive thaw slumps* occur where erosion exposes ground ice. Rapid melting leads to slumps that can move rapidly outward and cover large areas.

- Patterned ground includes circles, polygons, and nets that cover large areas of tundra and are the result of *cryoturbation*. *Hummocks* form when the formation and thawing of ice lenses at the top and bottom of the active layer lead to upwelling of sediment at the hummock centre with surface movement away and bottom movement toward the centre.

- Solifluction (soil flowage) occurs in late summer, when the bottom of the active layer thaws, releasing water that lubricates downhill movement of the active layer.

- Alpine tundra shows many of the same landforms and processes of arctic tundra, but is restricted to higher elevations where mean annual ground temperatures are below freezing. The elevation at which alpine tundra begins increases at lower latitudes.

- Disturbance of the surface layer in permafrost terrain by human activity can induce the formation of thermokarst. Construction of structures, roads, and pipelines must prevent thawing of ice-rich permafrost.

- Climate warming will thaw the surface layers of permafrost, producing extensive regions of thermokarst lakes and subsiding terrain. The boreal forest will migrate toward the pole, extending discontinuous permafrost further north, and at the same time, permafrost may disappear from its present southern boundary.

KEY TERMS

alluvium	ice lens	patterned ground	scarification
bedrock	ice wedge	periglacial	sediment
chemical weathering	landslide	permafrost	soil creep
earth flow	mass wasting	physical weathering	solifluction
ground ice	mudflow	regolith	weathering

REVIEW QUESTIONS

1. What does the term *weathering* mean? What are the types of weathering?
2. Define the terms *regolith, bedrock, sediment*, and *alluvium*.
3. How does frost action break up rock? Describe some landforms created by frost action and how they are formed.
4. How does salt-crystal growth break up rock? Give an example of a landform that arises from salt-crystal growth.
5. What is an exfoliation dome, and how does it form? Provide an example.
6. Name three types of chemical weathering. Describe how a chemical weathering process often alters limestone.
7. Define mass wasting and identify the processes it includes.
8. What is soil creep, and how does it arise?
9. What is an earth flow? What features distinguish it as a landform?
10. Contrast earth flows and mudflows, providing an example of each.
11. Define the term *landslide*. How does a landslide differ from an earth flow?
12. Define and describe induced mass wasting. Provide some examples.
13. Explain the term *scarification*. Provide an example of an activity that produces scarification.
14. What does the term *periglacial* mean?
15. Define and describe permafrost and some of its features, including ground ice, active layer, and permafrost table.
16. Identify the zones of permafrost and describe their general location, both globally and within Canada.
17. Identify the factors that affect the temperature of permafrost. Describe how and why each factor creates warmer or cooler ground temperatures.
18. What is a *talik*? How does the size of a talik under a lake depend on the width of the lake?
19. What is the *active layer*? Why is it important to plants? What factors influence the depth of the active layer and how?
20. How and when do ice lenses form in the active layer? Where are they found within the active layer and why?
21. Describe the formation of an ice wedge and the ridges found at the edges of ice-wedge polygons.
22. What is a *pingo*? How does it form?
23. Surface disturbance in permafrost terrain can trigger the formation of a thermokarst lake. Explain how this process occurs. What limits the growth of a thermokarst lake?
24. Explain the process that forms a retrogressive thaw slump. What limits the growth of the slump?
25. Identify five major categories of patterned ground.
26. What does *solifluction* mean? How and when does it occur?
27. Does alpine permafrost differ from arctic permafrost? How does the domain of alpine permafrost change with latitude?
28. What techniques are used to safely build structures, roads, and pipelines in permafrost terrains?
29. How will global climate warming affect boreal and arctic regions?

FOCUS ON SYSTEMS 14.2   Permafrost as an Energy Flow System

1. Contrast the energy flows reaching the ground through the surface from above and rising from below.
2. How does the profile of ground temperature with depth change from winter to summer in a permafrost region?
3. How is mean annual surface temperature related to the thickness of the active layer and the depth of permafrost?

VISUALIZATION EXERCISES

1. Define the terms *regolith, bedrock, sediment*, and *alluvium*. Sketch a cross-section through a part of the landscape showing these features and label them on the sketch.
2. Copy or trace Figure 14.5; then identify and plot on the diagram the mass movement associated with each of the following locations: Turtle Mountain, Palos Verdes Hills, Nicolet, Mount St. Helens, Madison River, and Herculaneum.
3. Sketch a profile of mean annual temperature through a section of active layer and permafrost, beginning in the atmosphere just above the ground surface. Sketch a second profile that would occur if the snow cover were twice as deep.
4. Draw a diagram of the process by which hummocks and stone circles form, demonstrating the role of ice lenses in the process.

ESSAY QUESTIONS

1. A landscape includes a range of lofty mountains elevated above a dry desert plain. Describe the processes of weathering and mass wasting that might be found on this landscape and identify their location.
2. Imagine yourself as the newly appointed director of public safety and disaster planning for your province. One of your first jobs is to identify locations where potential disasters, including those of mass wasting, threaten human populations. Where would you look for mass wasting hazards and why? In preparing your answer, you may want to consult maps of your province.
3. Compare and contrast the processes that produce ice lenses, ice wedges, massive icy beds, and pingos.
4. Identify the thermal properties of water, including latent heat associated with change of state, and the thermal conductivity of water, ice, and snow. Explain how these properties affect the temperature characteristics of permafrost and the active layer above it.

PROBLEMS   WORKING IT OUT 14.1 The Power of Gravity

1. The Madison Slide reached a velocity of at least 150 kilometres per hour. What was the kinetic energy of the slide at that point? Recall from *Focus on Systems 3.1 • Forms of Energy* that the formula for kinetic energy is $E = \frac{1}{2}mv^2$, where m is the mass (in kilograms) and v is the velocity (in metres per second). How does this value compare with the total energy released in the slide?

2. Physics predicts the velocity, v (in metres per second), of a falling body in the absence of friction to be $v = \sqrt{2gd}$, where g is the acceleration of gravity (in metres per second squared) and d is the distance of the fall from rest (in metres). What is the velocity of an object falling without friction for a distance of 500 metres? How does this compare with the velocity observed for the Madison Slide? Why are the two values different?

Find answers at the back of the book

EYE ON THE LANDSCAPE |

Chapter Opener Talus cones Individual rock strata (**A**) are readily visible in this photo. They are ancient sedimentary rocks that were thrust over and above younger rocks of the plains during an arc-continent collision. Strong and resistant to erosion, they now form the magnificent peaks of the northern Rockies in Alberta. At (**B**) are the remains of a small glacier, coated with grey blocks of talus. Chapter 12, on plate tectonics, describes overthrust faults and arc-continent collisions. Chapter 19 covers alpine glaciers.

Solution weathering, Kauai Kauai is the oldest of the Hawaiian Islands. Note the rounded dome shape of its outline (**A**), which is

characteristic of shield volcanoes (Chapter 13). The red-brown colours of soil, exposed on lower slopes (**B**), are those of iron oxides and indicate Oxisols (Chapter 20). Pocket beaches and an arch at (**C**) are products of wave action (Chapter 18).

Chapter 15

EYE ON THE LANDSCAPE

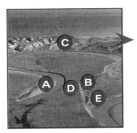

Waste water outflow from a desalination plant, Jahra region, Kuwait, into the Sea of Al-Doha. *Human activities require fresh water, and where it is scarce, desalination of sea water can produce potable water. However, the cost is high, both in energy and in the environmental impact of salt brine released to the ocean.* **What else would the geographer see? . . . Answers at the end of the chapter.**

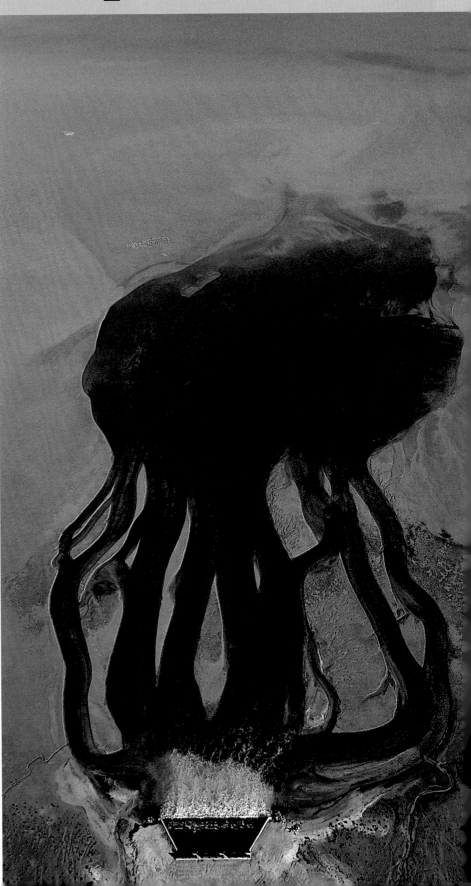

THE CYCLING OF WATER ON THE CONTINENTS

Groundwater
The Water Table Surface
Aquifers
Limestone Solution by Groundwater
Limestone Caverns
Karst Landscapes
Problems of Groundwater Management
Water Table Depletion
Contamination of Groundwater
Surface Water
Overland Flow and Stream Flow
Stream Discharge
Focus on Systems 15.1 • Stream Flow and its Measurement
Drainage Systems
Stream Flow
How Urbanization Affects Stream Flow
The Annual Flow Cycle of a Large River
River Floods
Flood Prediction
Working It Out 15.2 • Magnitude and Frequency of Flooding
The Red River Flood of 1997
Lakes
Origins of Lakes
Water Balance of a Lake
Eye on the Environment 15.3 • The Great Lakes
Saline Lakes and Salt Flats
Lake Temperature and Circulation
Seasonal Circulation in Lakes
Sediment in Lakes
Surface Water as a Natural Resource

Water is essential to life. Nearly all organisms require a constant flow of water or at least a water-rich environment to survive. Humans are no exception. Human activity depends on the fresh water that precipitation provides as part of the hydrologic cycle. This chapter focuses on two parts of the hydrologic cycle: water at the land surface and water that lies within the ground.

Recall from Chapter 6 that fresh water on the land surface and subsurface is only about 3 percent of the hydrosphere's total water. Most of this fresh water is locked into ice sheets and mountain glaciers. Groundwater accounts for a little more than half of 1 percent. Although this is a small fraction of global water, it is still many times larger than the amount of fresh water in lakes, streams, and rivers, which account for only 0.03 percent of the total water. Note also that groundwater can be found at almost every location on land that receives rainfall. In contrast, fresh surface water varies widely in abundance. In many dry regions, streams and rivers are nonexistent for most or all of the year.

Chapter 6 discussed atmospheric moisture and precipitation, describing the part of the hydrologic cycle in which water evaporates from ocean and land surfaces and then falls to the Earth as rain or snow. A portion of the precipitation returns directly to the atmosphere through evaporation. Another portion travels downward, moving through the soil under the force of gravity to become part of the underlying groundwater. Following underground flow paths, this subsurface water eventually emerges to become surface water, or it may emerge directly into the ocean. A third portion flows over the ground surface as runoff. As it travels, the water flow collects in streams, which eventually conduct the running water to the ocean.

This chapter traces the parts of the hydrologic cycle that include both the subsurface and surface pathways of water flow. The study of these flows is part of the science of *hydrology*, which is the study of water as a complex but unified Earth system.

Most soils are capable of absorbing the water from light or moderate rains by **infiltration**. In this process, water enters the small natural passageways between irregularly shaped soil particles, as well as the larger openings in the soil surface. These openings are formed by the borings of worms and animals, cracks produced by soil drying, cavities left from decay of plant roots, or spaces made by the growth and melting of frost crystals. A mat of decaying leaves and stems breaks the force of falling water drops and helps to keep these openings clear.

The precipitation that infiltrates the soil is temporarily held in the soil layer as soil water, occupying the *soil water belt*. Water within this belt can return to the surface and then to the atmosphere through a process that combines two components: direct evaporation from soil and transpiration by vegetation. As seen in Chapter 4, the term *evapotranspiration* describes the combination of these two forms of water vapour.

When rain falls too rapidly to pass through the soil, **runoff** occurs as a thin layer at the soil surface. This runoff is called **overland flow**. In periods of heavy, prolonged rain or rapid snowmelt, overland flow feeds directly into streams.

Overland flow also occurs when soil that is already saturated receives rainfall or snowmelt. Since soil openings and pores are already full, water cannot infiltrate the soil and drain down to deeper layers. Under these conditions, nearly all of the precipitation or snowmelt will run off.

Chapter 15 discusses how overland flow and stream flow erodes, transports, and deposits soil and sediment. These processes are important in shaping the landscape of modern river systems.

GROUNDWATER

Water from precipitation can continue to flow downward beyond the soil water belt. This slow downward flow under the influence of gravity is called *percolation*. Eventually, the percolating water reaches groundwater. **Groundwater** is the part of the subsurface water that fully saturates the pore spaces in bedrock, regolith, or soil, and so occupies the *saturated zone* (Figure 15.1). The **water table** marks the top of this zone. Above it is the *unsaturated zone* in which water does not fully saturate the pores. Here, water is held in thin films adhering to mineral surfaces. This zone also includes the soil water belt.

Groundwater moves slowly along deep flow paths, eventually seeping into streams, ponds, lakes, and marshes, where the land surface dips below the water table (Figure 15.2). Streams that flow throughout the year—perennial streams—derive much of their water from groundwater seepage.

THE WATER TABLE SURFACE

The position of the water table can be mapped in detail in areas where there are many wells (Figure 15.2). This is done by plotting the water heights in the wells and noting the change in elevation from one well to the other. The water table is highest under the highest areas of land surface—hilltops and divides. The water table declines in valleys, where it appears at the surface close to streams, lakes, or marshes.

The reason for this water table configuration is that water percolating down through the unsaturated zone tends to raise the water table, while seepage into lakes, streams, and marshes tends to draw off groundwater and lower the water table. These differences in water table level are built up and maintained because groundwater moves extremely slowly through the fine chinks and pores of bedrock and regolith. In periods of high precipitation, the water table rises under hilltops or divide areas. In periods of a water deficit, or during a drought, the water table falls (Figure 15.2).

Figure 15.2 also shows paths of groundwater flow. The direction of flow at any point depends on the direction of pressure at that point. Gravity always exerts a downward pressure, while

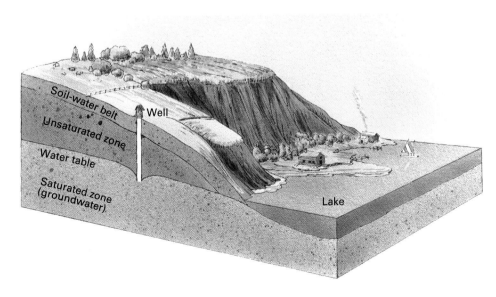

15.1 Zones of subsurface water Water in the soil water belt is available to plants. Water in the unsaturated zone percolates downward to the saturated zone of groundwater, where all pores and spaces are filled with water.

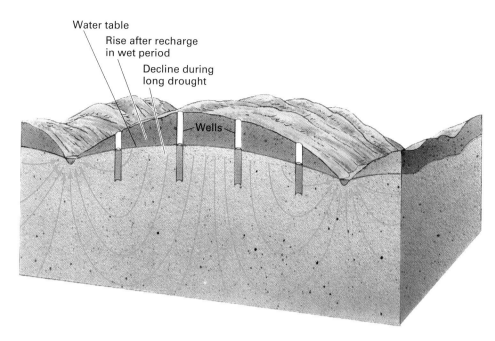

15.2 Water table surface The configuration of the water table surface conforms broadly with the land surface above it. It varies in response to prolonged wet and dry periods. Groundwater flow paths circulate water to deep levels in a very slow motion and eventually seep into streams.

the difference in the height of the water table between hilltop and streambed produces a sideways pressure. These downward and sideways pressures cause groundwater to flow in curving paths. Water that enters the hillside midway between the hilltop and the stream flows almost directly toward the stream. However, water that reaches the water table midway between the two streams flows almost straight down to great depths before curving and rising upward. Progress along these deep paths is very slow, while flow near the surface is much faster. The most rapid flow is close to the stream, where the arrows converge. Over time, the level of the water table tends to remain stable, and the flow of water released to streams and lakes balances the flow of water percolating down into the water table.

AQUIFERS

Sedimentary layers often exert a strong control over the storage and movement of groundwater. For example, clean, well-sorted sand—such as that found in beaches, dunes, or stream alluvium—can hold an amount of groundwater equal to about one third of its volume. A bed of sand or sandstone is thus often a good *aquifer,* that is, it contains abundant, freely flowing groundwater. In contrast, beds of clay and shale are relatively impermeable and hold little free water. They are known as *aquicludes.* In Figure 15.3, a shale bed is overlain by a bed of sandstone. Groundwater moves freely through the sandstone aquifer and across the top of the shale aquiclude, emerging in springs along the valley at the right. A small lens of shale accumulates a shallow *perched water table* above the main water table.

When an aquifer is situated between two aquicludes, groundwater in the aquifer may be under pressure and so will flow freely from a well that pierces the aquaclude. Figure 15.4 illustrates this self-flowing **artesian well**, created where a porous sandstone bed (aquifer) is sandwiched between two impervious rock layers (aquicludes). Precipitation on the hills where the sandstone is exposed provides water that saturates the sandstone layer, filling it to that elevation. Since the elevation of the well that taps the aquifer is below that of the range of hills feeding the aquifer, hydrostatic pressure causes water to rise in the well.

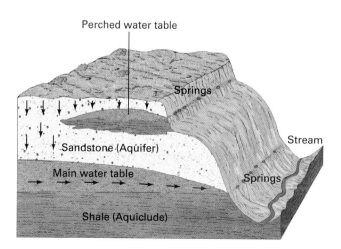

15.3 Groundwater in horizontal strata Water flows freely within the sandstone aquifer but very slowly in the shale aquiclude. A lens of shale creates a perched water table above the main water table. (Copyright © A. N. Strahler)

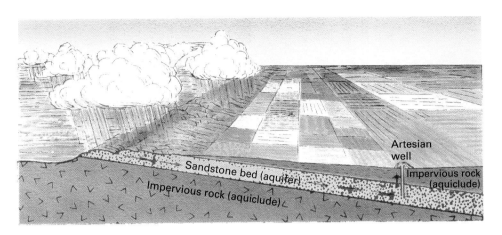

15.4 Artesian well A layer of sandstone between impervious layers provides a source of water under pressure that flows naturally to the surface. (Drawn by Erwin Raisz. Copyright © A. N. Strahler)

LIMESTONE SOLUTION BY GROUNDWATER

Chapter 13 discusses how, in moist climates, carbonic acid slowly dissolves limestone at the surface, producing lowland areas. The slow flow of groundwater in the saturated zone can also dissolve limestone below the surface, producing deep underground caverns. Some of these caverns collapse, causing the ground above to sink and a unique type of landscape to develop.

LIMESTONE CAVERNS

Limestone caverns are interconnected subterranean cavities in bedrock formed by the corrosive action of circulating groundwater on limestone (Figure 15.5). In stage 1, the action of carbonic acid is particularly concentrated in the saturated zone just below the water table. This dissolving process forms many kinds of underground landforms, such as tortuous tubes and tunnels, great open chambers, and tall chimneys. Subterranean streams flow in the lowest tunnels, carrying the products of solution that emerge along the banks of surface streams and rivers.

In stage 2, the stream valley has deepened, and the water table has been lowered to a new position. The cavern system previously formed is now in the unsaturated zone. Carbonate matter, known as *travertine*, is deposited on exposed rock surfaces in the caverns. Encrustations of travertine take many beautiful forms: stalactites (hanging rods), stalagmites (upward pointing rods), columns, and drip curtains (Figure 15.6).

KARST LANDSCAPES

Where limestone solution is very active, we find a landscape with many unique landforms. This is especially true along the Dalmatian coastal area of Croatia, where the landscape is called *karst*. Geographers apply this term to the topography of any limestone area where sinkholes are numerous and small surface streams are

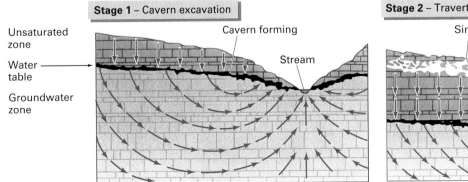

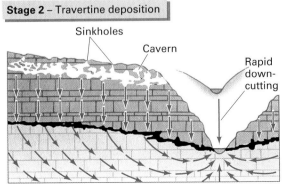

15.5 Cavern development Limestone dissolves at the top of the groundwater zone. When rapid erosion of streams lowers the water table, caverns result in the unsaturated zone. Water flow through the caverns results in travertine deposition. (Copyright © A. N. Strahler)

15.6 Inside a cavern Travertine deposits in the Papoose Room of Carlsbad Caverns in New Mexico. Deposits include stalactites (slender rods hanging from the ceiling), stalagmites (upward pointing rods), and sturdy columns, all formed when dripping groundwater evaporates, leaving travertine deposits of calcium carbonate behind.

15.7 Sinkholes Sinkholes in limestone are created by solution. This sinkhole is in Wood Buffalo National Park, Alberta.

PROBLEMS OF GROUNDWATER MANAGEMENT

Rapid withdrawal of groundwater has seriously affected the environment in many places. Increased urban populations and industrial developments require larger water supplies—needs that cannot always be met by building new surface water reservoirs. To

nonexistent. A *sinkhole* is a surface depression in a region of cavernous limestone (Figure 15.7). Some sinkholes are filled with soil washed from nearby hillsides, while others are steep-sided, deep holes. They develop where the limestone is more susceptible to weathering, or where an underground cavern near the surface has collapsed.

Figure 15.8 shows the development of a karst landscape. In an early stage, funnel-like sinkholes are numerous. Later, the caverns collapse, leaving open, flat-floored valleys. Examples of some important regions of karst or karst-like topography are the Mammoth Cave region of Kentucky, the Yucatan Peninsula, and parts of Cuba and Puerto Rico. In southern China and west Malaysia, the karst landscape is dominated by steep-sided, conical limestone hills or towers, 100 to 500 metres high (Figure 15.9). The towers are sometimes capped by beds of more resistant rock or impure limestone that dissolves more slowly. They are often riddled with caverns and passageways. In Canada, mountains in the west have numerous caves, while smaller examples exist in the limestone region of Ontario. The Castleguard Cave System in Alberta's Rocky Mountains is the largest, at more than 20 kilometres long.

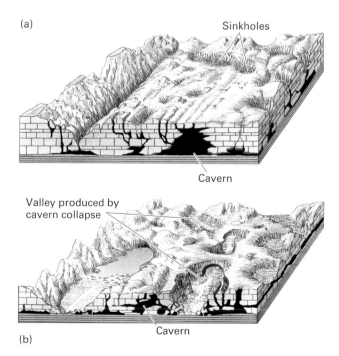

15.8 Evolution of a karst landscape (a) Rainfall enters the cavern system through sinkholes in the limestone. (b) Extensive collapse of caverns reveals surface streams flowing on shale beds beneath the limestone. Some parts of the flat-floored valleys can be cultivated. (Drawn by Erwin Raisz. Copyright © A. N. Strahler)

15.9 Tower karst White limestone is exposed in the nearly vertical sides of these towers near Guilin (Kweilin), Guanxi Province, southern China, at lat. 25° N.

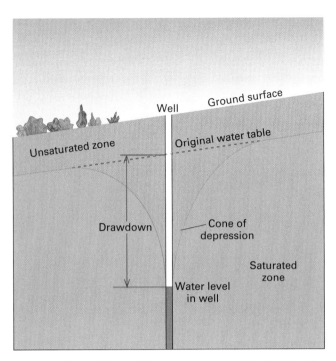

15.10 Drawdown and cone of depression in a pumped well As the well draws water, the water table is depressed in a cone with the well at its centre. (A.N. Strahler)

fill these needs, vast numbers of wells using powerful pumps draw huge volumes of groundwater to the surface, greatly altering nature's balance of groundwater discharge and recharge.

In dry climates, agriculture is often heavily dependent on irrigation water from pumped wells—especially since major river systems are likely to be already fully used for irrigation. Wells are also convenient water sources. They can be drilled within the limits of a given agricultural or industrial property and can provide immediate supplies of water without the need to build expensive canals or aqueducts.

In the past, the small well that supplied the domestic and livestock needs of a home or farm was actually dug by hand and sometimes lined with masonry. In contrast, a modern well supplying irrigation and industrial water is drilled by powerful machinery that can bore a hole 40 centimetres or more in diameter to depths of 300 metres or more. Drilled wells are lined with metal casings that exclude impure near-surface water and prevent the walls from caving in and clogging the tube. Near the lower end of the hole, in the groundwater zone, the casing is perforated to admit the water. The yield of a single drilled well ranges from as low as a few hundred litres per day in a domestic well to thousands of cubic metres per day in a large industrial or irrigation well.

WATER TABLE DEPLETION

As water is pumped from a well, the water level in the well drops. At the same time, the surrounding water table lowers in the shape of a downward-pointing cone, called the *cone of depression* (Figure 15.10). The difference in height between the cone tip and the origi-

nal water table is the *drawdown*. The cone of depression may extend out as far as 15 kilometres or more from a well where heavy pumping is taking place. Where many wells are in operation, their intersecting cones produce a general lowering of the water table.

Water table depletion often greatly exceeds recharge—the rate at which infiltrating water moves downward to the saturated zone. In dry regions, much of the groundwater for irrigation is drawn from wells driven into thick aquifers of sand and gravel. These deposits are often recharged by the seasonal flow of streams that originate from snowmelt high in adjacent mountains. Fanning out across the dry lowlands, the streams lose water, which sinks into the sands and gravels and eventually percolates to the water table below. The extraction of groundwater by pumping can greatly exceed this recharge by stream flow, lowering the water table. Deeper wells and more powerful pumps are then required. The result is exhaustion of a natural resource that is renewable only over very long time periods.

Another by-product of water table depletion is subsidence—a sinking of the land in response to the removal of water from underlying sediments. This problem has plagued a number of major cities that rely heavily on groundwater wells for their water supplies.

CONTAMINATION OF GROUNDWATER

Another major environmental problem related to groundwater withdrawal is the contamination of wells by pollutants that infiltrate the ground and reach the water table. The contamination of the groundwater supply in the community of Walkerton, Ontario illustrates this issue. On May 12, 2000, heavy rain washed a virulent strain of *E. coli* bacteria from cattle manure into the soil near the community and down to the groundwater and a shallow well that provides water for the town. Inadequate testing of the water drawn from the well failed to detect the highly toxic bacteria and the contaminated water was distributed in the town's water supply network.

The first cases of illness were reported on May 17, but an advisory warning residents that they must boil water to kill the bacteria before drinking and cooking with it wasn't issued until four days later. By then it was too late. The first death as a result of the contaminated water occurred on the following day, May 22. By May 31, six more people had died and more than 1,000 people in the small community fell seriously ill.

Eventually the water supply system was cleaned of the contamination. This involved, among other measures, a house-by-house disinfection of the water pipes. Clean, properly treated water flowed to the community again, but this tragedy was a wake-up call. By late July, the Ontario Ministry of the Environment had compiled a list of 131 municipalities in the province with "inadequate" water supply facilities. Before the end of August 2000, the provincial legislature had enacted new laws to ensure that drinking water is adequately treated throughout the province.

Thirty percent of Canadians—about nine million people—draw their water supply from groundwater. Efforts by federal and provincial agencies with jurisdictions in different regions continue to address what is an ongoing issue, especially in small, remote communities across the country.

There are other potential sources of groundwater contamination. Disposal of solid wastes poses a major environmental problem in developed countries because their advanced industrial economies provide an endless source of garbage. Traditionally, these waste products were trucked to the town dump and burned there in continuously smoldering fires that emitted foul smoke and gases. The partially consumed residual waste was then buried under earth.

In recent decades, a major effort has been made to improve solid-waste disposal methods. One method is high-temperature incineration, but this often leads to air pollution. Another is the sanitary landfill method in which waste is not allowed to burn.

Instead, layers of waste are continually buried, usually by sand or clay available on the landfill site. The waste is thus situated in the unsaturated zone. However, it can react with rainwater that infiltrates the ground surface. This water picks up a wide variety of chemical compounds from the waste and carries them down to the water table (Figure 15.11).

Once in the water table, the pollutants follow the groundwater flow paths. As the arrows in the figure indicate, the polluted water may flow toward a supply well that draws in groundwater from a large area. Once the polluted water has reached the well, the well water becomes unfit for human consumption. Contaminated water may also move toward a nearby valley, polluting the stream flowing there.

Another source of contamination in coastal wells is *saltwater intrusion*. Since fresh water is less dense than salt water, a layer of salt water from the ocean may lie under a coastal aquifer. When the aquifer is depleted, the level of salt water rises and eventually reaches the well from below, rendering the well unusable.

SURFACE WATER

So far, this chapter has examined how water moves below the land surface. Now, it turns to tracing the flow paths of surplus water that runs off the land surface and ultimately reaches the sea. Here, the primary concern is rivers and streams. (In general usage, "rivers" describe large watercourses, and "streams" smaller ones. However, the word "stream" is also a scientific term describing the channelled flow of any amount of surface water.)

OVERLAND FLOW AND STREAM FLOW

As seen earlier in this chapter, runoff that flows down the slopes of the land in broadly distributed sheets is overland flow. In contrast, *stream flow* is the flow of water along a narrow channel confined by lateral banks. Overland flow can take several forms. Where the soil or rock surface is smooth, the flow may be a continuous thin film, called *sheet flow*. Where the ground is rough or pitted, flow may take the form of a series of tiny rivulets connecting one water-filled hollow with another. On a grass-covered slope, overland flow is subdivided into countless tiny threads of

15.11 Movement of polluted groundwater
Polluted water, leached from a waste disposal site, moves toward a supply well and a stream. (Copyright © A N Strahler)

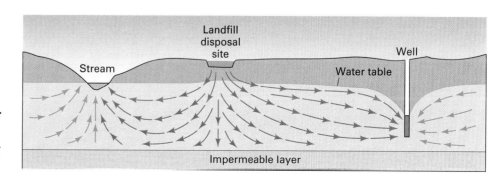

water, passing around the stems. Even in a heavy and prolonged rain, you might not notice overland flow on a sloping lawn. On heavily forested slopes, overland flow may pass entirely concealed beneath a thick mat of decaying leaves.

Overland flow eventually contributes to a stream, which is a much deeper, more concentrated form of runoff. A **stream** is a long, narrow body of flowing water that occupies a trench-like depression, or channel, and moves to lower levels under the force of gravity. The **channel** of a stream is a narrow trough. The forces of flowing water shape the trough to the most effective form for moving the water and sediment supplied to the stream. A channel may be so narrow that you can easily jump across it, or, in the case of the St. Lawrence River, as wide as two kilometres or more.

STREAM DISCHARGE

Stream flow at a given location is measured by its **discharge**, expressed as the volume of water in cubic metres passing a point in the river every second (see *Focus on Systems 15.1 • Stream Flow and its Measurement*). The discharge of streams and rivers changes from day to day, and records of daily discharges and floods of ma-

jor streams and rivers provide important information. They are used in planning the development and distribution of surface waters, as well as in designing flood-protection structures and generating models of how floods progress down a particular river system (see *Working It Out 15.2 • Magnitude and Frequency of Flooding*).

In Canada, the Water Survey of Canada monitors water levels in rivers and lakes using a network of about 2,500 gauges, many of which are operated in co-operation with provincial governments and other agencies. Figure 15.12 shows the relative discharge of major rivers in Canada. Canada's rivers drain to the three oceans that border the country: the Atlantic, the Pacific, and the Arctic (including Hudson Bay). The Arctic drainage basin is by far the largest at 7,444,000 square kilometres. The Atlantic (1,520,000 square kilometres) and the Pacific (1,009,000 square kilometres) are much smaller. A small portion of Canada's water from southern Alberta and Saskatchewan (21,000 square kilometres) flows to the Mississippi River and on to the Gulf of Mexico. The largest rivers in Canada are the St. Lawrence, which drains from the Great Lakes, and the Mackenzie River, which flows from Great Slave Lake to the Arctic Ocean. Both have an average annual

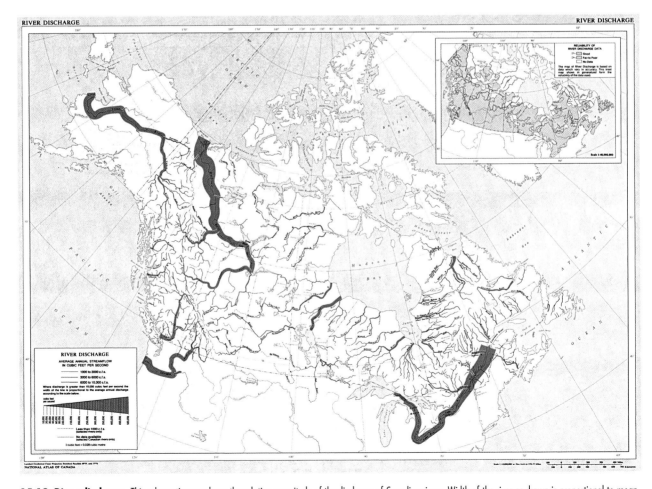

15.12 River discharge This schematic map shows the relative magnitude of the discharge of Canadian rivers. Width of the river as drawn is proportional to mean annual discharge. (From Natural Resources Canada)

discharge of less than 10,000 cubic metres per second. Hudson Bay, from its position as a small ocean in the middle of Canada, receives 30,900 cubic metres per second annually from the rivers flowing into it.

The discharge of most rivers increases downstream as a natural consequence of the way streams and rivers combine to deliver runoff and sediment to the oceans (Figure 15.12). The gradient also changes in a downstream direction. The general rule is the larger the cross-sectional area of the stream, the lower the gradient. Great rivers, such as the Mackenzie and Amazon, have gradients so low that they can be described as "flat." For example, the water surface of the lower Mississippi River falls in elevation about three centimetres for each kilometre downstream.

Rivers with headwaters in high mountains have characteristics that are especially desirable for using river flow for irrigation and for preventing floods. The higher ranges serve as snow storage areas, slowly releasing winter and spring precipitation in early or midsummer. As summer progresses, melting moves to successively higher levels. In this way, a continuous, sustained river flow is maintained. Among the snow-fed rivers of western Canada are the Saskatchewan, Yukon, Fraser, and Mackenzie.

DRAINAGE SYSTEMS

As runoff moves to lower levels and eventually to the sea, it becomes organized into a **drainage system**. The system consists of a branched network of stream channels, as well as the adjacent sloping ground surfaces that contribute overland flow to those channels. Between the channels on the crests of ridges are *drainage divides*, which mark the boundary between slopes that contribute water to different streams or drainage systems. The entire system is bounded by an outer drainage divide that outlines a more-or-less pear-shaped **drainage basin** or **watershed** (Figure 15.13).

Figure 15.13 shows a typical stream network within a drainage basin. Each fingertip tributary receives runoff from a small area of land surface surrounding the channel. This runoff flows downstream and merges with runoff from other small trib-

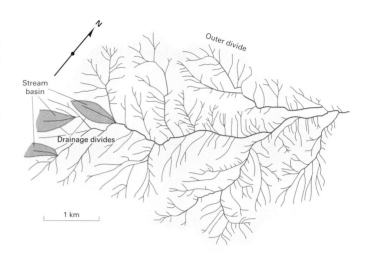

15.13 Channel network of a stream Smaller and larger streams merge in a network that carries runoff downstream. Each small tributary has its own small drainage basin, bounded by drainage divides. An outer drainage divide delineates the stream's watershed at any point on the stream. (Data from U.S. Geological Survey and Mark A. Melton)

utaries as they join the main stream. The drainage system thus provides a converging mechanism that funnels overland flow into streams and smaller streams into larger ones.

STREAM FLOW

A stream's discharge increases in response to a period of heavy rainfall or snowmelt. However, this response is delayed, as the movement of water into stream channels takes time. The length of delay depends on a number of factors, the most important of which is the size of the drainage basin feeding the stream. Larger drainage basins show a longer delay.

A simple graph, called a **hydrograph**, studies the relationship between stream discharge and precipitation by plotting the discharge of a stream with time at a particular stream gauge. Figure 15.14 is a hydrograph for a drainage basin of about 800 square kilometres, located in Ohio within the moist continental climate.

15.14 Sugar Creek hydrograph Four days of precipitation and stream flow at Sugar Creek, Ohio following a heavy rainstorm in August. (From Hoyt and Langbein, *Floods*, Princeton University Press. Used with permission)

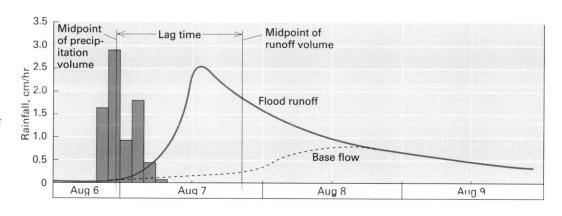

FOCUS ON SYSTEMS | 15.1   Stream Flow and its Measurement

To understand the flow of water in a river, we need to examine the geometry of flow in a stream channel (see the diagram at right). In (a), the velocity arrows show the flow pattern at the stream surface. Because of friction between the water and the banks, velocity is nearly zero at either bank and increases to a maximum near the stream's centre. In the vertical profile (b) along the centre line of the channel, velocity increases from the bottom of the channel to the water surface.

The discharge of a stream depends on the mean water velocity and cross-sectional area of the stream. For example, if a stream's mean velocity at a certain point is 2 metres per second, and if its cross-sectional area is 20 square kilometres, then in one second, 2 × 20 = 40 cubic metres flow past the point. This simple relationship is expressed by the equation

$$Q = A \times V$$

where Q is the discharge in cubic metres per second (m³/s), A is the cross-sectional area, and V is the mean velocity.

In any given vertical section through a river, the average velocity occurs at about six tenths of the depth from the surface. So to measure the discharge of a river, a geographer or hydrologist places a current meter at this depth in a number of equally spaced sections across the river (see the sketch of stream gauging on the opposite page). The current meter has a propeller or cups that rotate more quickly in higher velocity flow. Thus, the rate of rotation measures the velocity. The discharge in each section (width × depth × mean velocity) is added to the discharge in the other sections to obtain the total discharge, Q, of the river at that point.

The discharge of a river varies over time, which can be determined by carrying out repeated discharge measurements. An easier technique, however, is to measure the height of the water surface (called stage) rather than discharge. Stage is higher for greater discharges and lower for lesser discharges. A device to measure stage might consist of a ruler (called a staff gauge) permanently anchored in the river; however, an automated stage recorder is commonly used to obtain a continuous record of changes in stage. A stilling well is located in or beside the river with an intake to ensure the stage in the well is the same as the stage in the river. A float in the well is connected to a device that records the stage as the float rises and falls

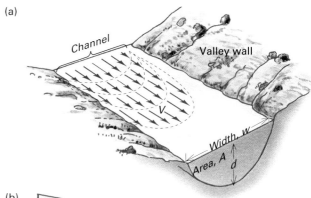

(a)

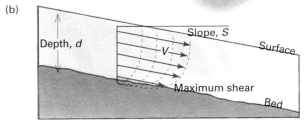

(b)

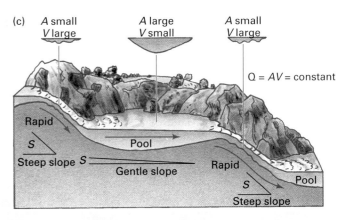

(c)

Characteristics of stream flow The flow velocity is greatest in the middle (a) and at the top (b) of the stream. (c) Mean velocity, cross-sectional area, and slope change in the pools and rapids of a stream section with uniform discharge (Copyright © A. N. Strahler)

with the water level. The relationship between stage and discharge is plotted from a number of measurements of both taken under different flow conditions. From this, the geographer can read the discharge for any measured stage and so plot a *hydrograph* of the variation of discharge or stage through time (see Figures 15.14 and 15.15).

In Canada, there are about 2,500 of these gauging stations on streams varying from small creeks to large rivers.

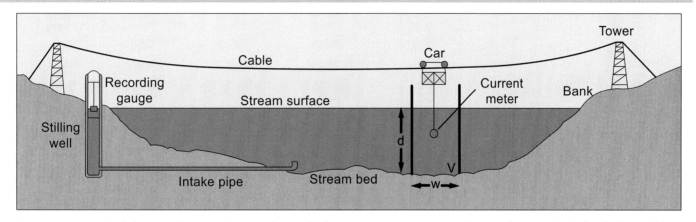

Gauging a stream The discharge in each increment of a river is determined by lowering a current meter to measure the mean velocity (V) and multiplying by the area of that increment—depth (d) x width (w). The total discharge is the sum of the discharges in each increment across the entire channel. For a large river, the operator must work from a cable car suspended above the river or from a bridge spanning the river. Small streams may be gauged by wading, while very large rivers like the St. Lawrence or the Mackenzie may be gauged from a boat. The photograph shows the current meter with a streamlined weight suspended below to stabilize it in the flow.

The Water Survey of Canada, in collaboration with provincial authorities, is responsible for their operation. Historical data from any of the stations, as well as "real time" data from many stations, are available from its website (www.wsc.ec.gc.ca). These data aid in understanding a river's previous behaviour (see *Working It Out 15.2 • Magnitude and Frequency of Flooding*). The stage recorder transmits its data to a regional office, which processes the data and quickly makes it available on the Internet. This is important to geographers and planners who monitor stream flow, for example, to predict the onset of a flood and so activate emergency measures in time to deal with the impact of the flooding.

Another important variable describing a stream is the downstream slope of the water surface *(S)*. Slope expresses the gradient or inclination as the difference in elevation between two points along the channel divided by the distance between them. For example, if points that are 1,000 metres apart along a stream are 100.2 and 100.0 metres above sea level, respectively, then the slope of the river in this region is (100.2 − 100.0)/1000 = 0.0002. There are no units in the expression of slope because the units of metres in the numerator cancel the units of metres in the denominator.

Slope is related to the stream's velocity. As a canoe moves along a stream, it will encounter pools, where the slope of the stream is low and the water moves slowly. Between these will be stretches of rapids or riffles, where the slope is steeper and the water moves more swiftly (see part (c) of the figure on the opposit page).

Suppose that no new tributaries enter the stream in the distance travelled. In this case, discharge (Q) is constant through this stretch. If the slope becomes steeper and the velocity (V) increases, the equation $Q = A \times V$ indicates that the cross-sectional area (A) of the stream must decrease, since Q is unchanged. In pools, the slope is low, velocity is low, and cross-sectional area is large. In the riffles, the situation is reversed—slope is steep, velocity is high, and cross-sectional area is small.

As a stream flows down its course, the potential energy due to the higher elevation of water in the stream's upper reaches converts into kinetic energy (the energy of the water in motion). Of course, gravitational force drives this process. This energy is required to overcome the friction between the water and the river bed, to generate turbulence in the water, and to erode and transport the sediment load of the river. Energy is also dissipated as heat (warming the water) and to a lesser extent as sound. The heat is eventually lost by conduction, radiation, or evaporation to the air above, and by conduction into the solid channel walls.

The smooth line on the graph represents the discharge of Sugar Creek, the main stream in the drainage basin, during a four-day period that included a rainstorm. A bar graph shows rainfall for the 12-hour storm, giving the number of centimetres of precipitation in each two-hour period. The average total rainfall over the watershed of Sugar Creek was about 15 centimetres. About half of this amount passed down the stream within three days. Some rainfall was held in the soil as soil water, some evaporated, and some infiltrated to the water table to be held in long-term storage as groundwater.

The rainfall and runoff graphs in Figure 15.14 show that prior to the onset of the storm, Sugar Creek was carrying a small discharge. This flow, which is supplied by the seepage of groundwater into the channel, is called *base flow*. After the heavy rainfall began, several hours elapsed before the stream gauge began to show a rise in discharge. This interval, called the *lag time*, indicates that the branching system of channels was acting as a temporary reservoir. The channels were at first receiving inflow more rapidly than they could pass down the channel system to the stream gauge.

Lag time is the difference between the time when half of the precipitation has occurred and the time when half the runoff has passed downstream. In the Sugar Creek example, the lag time was about 18 hours, with the peak flow reached almost 24 hours after the rainfall began. Note also that the stream's discharge rose much more abruptly than it fell. In general, the larger the watershed, the longer the lag time between peak rainfall and peak discharge, and the more gradual the rate of decline of discharge after the peak has passed. Another typical feature of a flood hydrograph is the slow but distinct rise in the amount of discharge contributed by base flow.

HOW URBANIZATION AFFECTS STREAM FLOW

The growth of cities and suburbs affects the flow of small streams in two ways. First, an increasing percentage of the surface becomes impervious to infiltration as it is covered by buildings, driveways, walks, pavements, and parking lots. In a closely built-up residential area with small lot sizes, the percentage of impervious surface may run as high as 80 percent.

As the proportion of impervious surface increases, overland flow from the urbanized area generally increases. This change acts to increase the frequency and height of flood peaks during heavy storms for small watersheds lying largely within the urbanized area. This also reduces recharge to the groundwater beneath, and this reduction, in turn, decreases the base flow contribution to channels in the same area. Thus, the full range of stream discharges, from low stages in dry periods to flood stages in wet periods, is made greater by urbanization.

A second change caused by urbanization is the introduction of storm sewers that quickly carry storm runoff from paved areas directly to stream channels for discharge. This shortens runoff travel time to channels, while also increasing the proportion of runoff by the expansion in impervious surfaces. The two changes together act to reduce the lag time of urban streams and increase their peak discharge levels. Many rapidly expanding suburban communities are finding that low-lying, formerly flood-free, residential areas now experience periodic flooding as a result of urbanization.

THE ANNUAL FLOW CYCLE OF A LARGE RIVER

In regions of wet climates, where the water table is high and normally intersects the important stream channels, the hydrographs of larger streams clearly show the effects of two sources of water: base flow and direct input, including overland flow. Figure 15.15 shows a hydrograph of the North Saskatchewan River at Prince Albert, Saskatchewan. This river drains a large watershed of 131,000 square kilometres on the Alberta and Saskatchewan Prairies. Its headwaters are high in the snowfields and glaciers of the Rocky Mountains to the west. Discharge during winter, from December to March, is low because almost all the precipitation falls as snow and is stored in the snowpack covering the drainage basin. At this time, base flow from groundwater contributes almost all the discharge to the river—about 100 cubic metres per second. Snowmelt on the Prairies in March and April provides the first discharge peak flows of the year. Contribution from the melt of the large snowpack in the headwaters is delayed until June due to the lag time along the 1,000 kilometres of river channel between the mountains and Prince Albert, and due to the later melt at higher elevations in the mountains. However, both factors contribute to groundwater storage and raise the base flow to about 300 cubic metres per second through the summer and into the autumn. Rainstorms, especially in August and September, cause secondary peaks in discharge and contribute to maintaining the base flow. In November, a cold snap at the onset of winter sets ice on the river and reduces the discharge. Water managers and farmers on the Prairies have recently become concerned that the shrinking of glaciers in the Rockies in a warming climate of the twenty-first century will result in a decline in discharge in large Prairie rivers, causing water shortages in this vital agricultural region.

RIVER FLOODS

The term **flood** is defined as the condition that exists when the discharge of a river cannot be accommodated within its normal channel. As a result, the water spreads over the adjoining ground, which is normally cropland or forest. Sometimes, however, homes, factories, and transportation corridors occupy the ground.

Most rivers of humid climates have a *flood plain*, a broad belt of low, flat ground bordering the channel on one or both sides that is

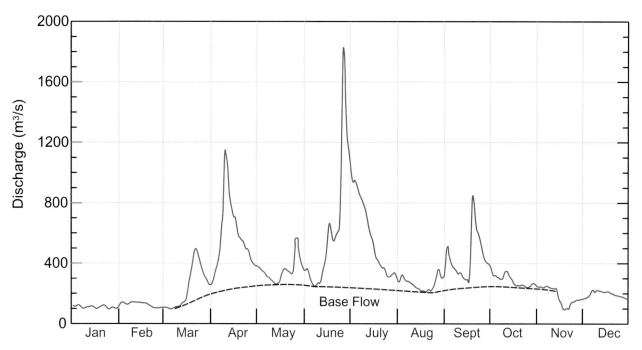

15.15 Hydrograph of the North Saskatchewan River at Prince Albert, Saskatchewan in 2005 The peaks in March and April are caused by snowmelt on the Prairies. The peak in June is associated with snowmelt in the Rocky Mountains. Other peaks are caused by rainfall in the drainage basin. (Data from the Water Survey of Canada)

flooded by stream waters about once a year (Figure 15.16). This flood usually occurs in the season when abundant surface runoff combines with the effects of a high water table to supply more runoff than the channel can carry. Annual inundation is considered a flood, even though its occurrence is expected and does not prevent

15.16 Flood plain This aerial view shows the flood plain of the Fraser River near Vancouver, British Columbia. The main river channel is in the foreground; Hatzic Lake, an oxbow, occupies the middle ground. Agricultural land on the flood plain is protected by dikes along the banks of the Fraser River.

the cultivation of crops after the water has subsided. Annual flooding does not interfere with the growth of dense forests, which are widely distributed over low, marshy flood plains in all humid regions of the world. The Water Survey of Canada, which provides a flood-warning service, designates a particular river stage at a given place as the *flood stage*. Above this critical level, inundation of the flood plain will occur. Higher discharges of water, the rare and disastrous floods that may occur as seldom as once in 30 or 50 years, inundate ground lying well above the flood plain.

Flash floods are characteristic of streams draining small watersheds with steep slopes. These streams have short lag times—perhaps only an hour or two—so when an intense rainfall event occurs, the stream quickly rises to a high level. The flood arrives as a swiftly moving wall of turbulent water, sweeping away buildings and vehicles in its path. In dry western watersheds, the flood water sweeps great quantities of coarse rock debris into the main channel, producing debris floods (see Chapter 13). In forested landscapes, flood water sweeps tree limbs and trunks, soil, rocks, and boulders downstream. Because flash floods often occur too quickly to warn affected populations, they can cause significant loss of life.

FLOOD PREDICTION

The magnitude of a flood is usually measured by the peak discharge or highest stage of a river during the period of flooding. As any river's flood history shows, large floods occur less frequently than smaller ones; that is, the greater the discharge or higher the stage, the less likely the flood. *Working It Out 15.2 • Magnitude and Frequency of Flooding* presents a graph of a river's flood frequency and shows how to determine the probability of various discharges given a record of peak annual discharges.

Another tool used to present the flood history of a river is the flood expectancy graph. Figure 15.17 shows flood expectancy graphs for two rivers. The key explains the meaning of the bar symbols. The graph for the Fraser River at Mission, British Co-

lumbia illustrates a large river responding to spring floods, with highest water levels in June and low levels in February and March. The red line indicates the stage above which flooding occurs. Note that daily stage values may significantly exceed the monthly averages shown on the graph, so would cause somewhat more frequent flooding for periods shorter than a month. However, because the flood plain in the lower reaches of the Fraser is heavily populated as part of the city of Vancouver and outlying municipalities, large dikes have been built to protect against floods of up to 15 metres.

WORKING IT OUT | 15.2   Magnitude and Frequency of Flooding

In nature, extreme events happen infrequently. For example, a region may experience many small earthquakes but only a few really large ones during a century. In other words, the greater the magnitude of an event, the less frequently it recurs.

The figure to the right illustrates this principle for the frequency of floods on the Lillooet River in British Columbia. Each dot on the graph plots the maximum average daily discharge of the river recorded in a particular year; that is, the peak flow, or largest flood, each year. The bottom horizontal scale indicates the probability that the given discharge will be equalled or exceeded in a given year. For example, a 20 percent discharge is interpreted to mean "a discharge of this magnitude (about 600 cubic metres per second) can be expected to be equalled or exceeded in 20 out of 100 years." The numbers on the top horizontal scale show the recurrence interval (or return period). This value is the probability percentage divided into 100. For example, the return period for the probability of 20 percent is 100/20 = 5 years. This means that, over a long period, a maximum annual flood of about 600 cubic metres per second can be expected once every five years. Keep in mind that the recurrence interval is only a way of expressing a probability. If a river experiences a "five-year flood" in one year, it does not mean that the next flood of the same magnitude will occur five years later. There is nothing to prevent two five-year floods from occurring in successive years. It's just not very likely.

Calculating the recurrence interval series is not difficult. First, assemble the maximum discharges for each year in a list. Next, reorder the values from greatest to least, assigning a rank of 1 to the largest value, 2 to the next largest, and so on. Then determine the recurrence interval for each flow from its rank, using the formula

$$I = (N + 1)/R$$

where I is the recurrence interval in years, N is the number of years in the record, and R is the rank of the value. Suppose that

a flow is the largest in a set of 44 years of observations. Its recurrence interval then is $I = (N + 1)/R = (44 + 1)/1 = 45/1 = 45$ years. Suppose that flow is the fifth largest value. Its recurrence interval would then be $I = (44 + 1)/5 = 45/5 = 9$ years.

Graphs such as these help geographers understand the changes in the hydrology of a river. Most of the largest floods (including those in 1984, 1991, and 2003 labelled on the graph) have occurred since 1980, perhaps in response to human-induced climate change. Urban planners and developers are also interested in flood frequency analysis. The Lillooet River valley is experiencing rapid development in part as a result of the 2010 Winter Olympics at nearby Whistler Mountain. Knowledge of the extent and severity of flooding is important in regulating the use of the flood plain.

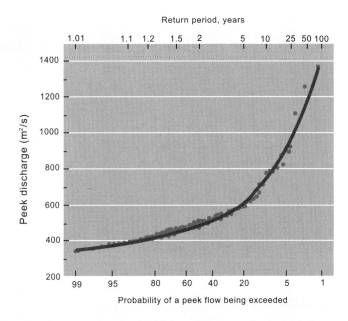

Flood frequency data for the Lillooet River at Pemberton, British Columbia Each dot is a measured maximum yearly discharge in an 88-year record. (Data from the Water Survey of Canada)

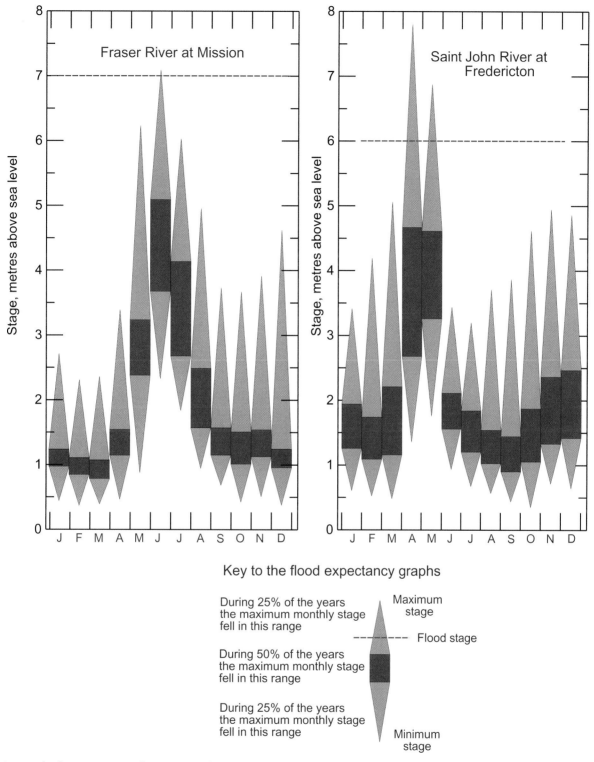

15.17 Flood expectancy graphs Maximum and minimum monthly stages of the Fraser River at Mission, British Columbia, and the Saint John River, at Fredericton, New Brunswick.

On the east coast, the Saint John River at Fredericton, New Brunswick has a similar annual stage pattern except that the spring snowmelt runoff peaks in April. The drainage basin of the Saint John River is less than one fifth the size of the Fraser, and the discharge is about one half. In British Columbia, runoff is delayed due to the later melt in the high mountains. A flood stage of 6 metres on the Saint John River represents the onset of overbank inundation. Damaging floods do not occur until the river's stage exceeds about 6.5 metres, which may happen in both April and May.

Canadian federal and provincial agencies coordinate flood forecasting from 10 regional centres across the country. When there is a threat of a flood, forecasters analyze precipitation patterns and the progress of high waters moving downstream. By examining the flood history of the rivers and streams concerned, they develop specific flood forecasts. They deliver these forecasts to communities within the associated district, which usually covers one or more large watersheds. Flood warnings are publicized by every possible means, and various agencies co-operate closely to plan the evacuation of threatened areas and the removal or protection of property.

THE RED RIVER FLOOD OF 1997

The flood on the Red River of southern Manitoba in 1997 was one of the most severe in recent Canadian history. The Red River flows northward from the United States, mainly Minnesota and North Dakota, through the city of Winnipeg and then on to Lake Winnipeg in central Manitoba. Its course is over the former lake bed of the glacial Lake Agassiz, the largest ice-marginal lake in North America created by the retreating continental glaciers (see marginal lakes and their basins, in Chapter 19). The broad, flat plain of this ancient lake floor provides an extensive area easily flooded by the rivers that flow across it. In addition, greater isostatic rebound of the Earth's crust in northern Manitoba has gradually raised the downstream portion of the river, reducing its slope and making it more difficult to transmit its water. (Isostatic rebound is a rising of the land following the retreat of a continental ice sheet and the loss of its weight pressing on the crust.)

The largest of at least seven major floods in recorded history occurred in 1826 (Figure 15.18a). A flood in 1950 did not reach the same level and discharge (Figure 15.18b); however, because the region was more heavily populated by then and Winnipeg had grown to a major Canadian city, the flood caused the largest evacuation of people in Canadian history. As a result of the 1950 flood, the 47-kilometre-long Red River Floodway (Figure 15.18c) was completed in 1968 as part of a flood protection scheme, which also included diversion of the Assiniboine River to Lake Manitoba. The purpose was to divert flood waters around the city of Winnipeg, and so protect the urban area from flooding.

The 1997 "flood of the century," with a return period of 110 years, severely tested these emergency measures (see *Working It Out 15.2 • Magnitude and Frequency of Flooding*). Four factors caused the flood. First, heavier than normal rains during the autumn of 1996 saturated the soil, leaving little opportunity for the ground to absorb the following spring's snowmelt. Second, snowfall in the winter of 1996–97 was greater than average, especially in the upper portions of the drainage basin. Fargo, North Dakota, was blanketed with 2.5 metres of snow. Third, a spring blizzard on April 5, shortly before the flood began, dumped another 50 centimetres of snow on the drainage basin. Fourth, the basin experienced one of the most rapid snowmelts on record, which proceeded from south to north. Thus, tributaries to the Red River experienced peak flows in sequence as the main flood proceeded north toward Winnipeg.

Flooding began on April 20 and the flood crested in Winnipeg on May 3. In anticipation of the flood, the Brunkild dike was enlarged to protect Winnipeg. Most of the previously constructed dikes around smaller communities held (Figure 15.18d). Nevertheless, 2,000 square kilometres of land was flooded and 22,000 people were evacuated from rural areas (Figure 15.18e). The Red River Floodway greatly reduced the impact of the flood in Winnipeg itself, although some 6,000 urban residents were also evacuated. Damage from the flood was estimated at $150 million, with another $200 million in agricultural losses.

LAKES

A **lake** is a body of standing water found within continental margins that is enclosed on all sides by land. The Caspian Sea, which contains 78,700 cubic kilometres of water, is by far the largest lake on the Earth. Other large lakes include Lake Baikal in central Asia (23,615 cubic kilometres) (Figure 15.19), Lake Tanganyika in the Great Rift Valley of Africa (18,940 cubic kilometres), and Lake Superior in central North America (12,251 cubic kilometres). Although these are immense water bodies, lakes range in size down to the smallest ponds containing only a few hundred cubic metres of water (Figure 15.20). Most lakes contain fresh water, although some are salty—some much saltier than the ocean.

Lakes are especially important in Canada, which contains more than half of the total surface area of lakes in the world. In fact, Canada has nearly half a million lakes with surface areas greater than one square kilometre, and they cover 8.4 percent of Canada's land area. In Canada, then, *lacustrine*, or lake-related, processes and landforms are extremely important in both the natural and human realms.

Lakes are dynamic environments. They respond to seasonal changes in climate and hydrology and in turn influence the land around them. Lakes modify climate by moderating surrounding air temperatures and adding water vapour to the local atmosphere. Lacustrine sediments, which settle year after year, are important repositories of environmental and paleoenvironmental information. Lakes are also important to humans' needs. The science of *limnology* studies the physical, chemical, and biological processes of lakes.

North America's Great Lakes are a unique resource of immense value to both Canada and the United States. *Eye on the Environment 15.3 • The Great Lakes* provides more information about these lakes' formation, their effects on local climate, and their degradation by water pollution.

(a)

(b)

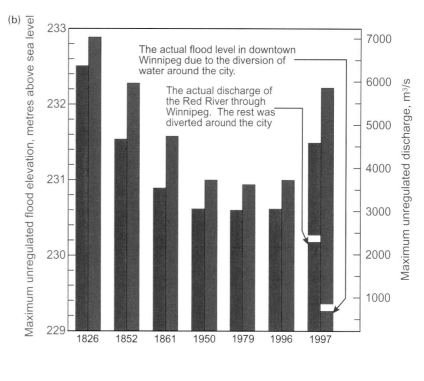

The actual flood level in downtown Winnipeg due to the diversion of water around the city.

The actual discharge of the Red River through Winnipeg. The rest was diverted around the city

(c)

(d)

(e)

15.18 (a) Map of southern Manitoba showing the extent of flooding of the Red River in 1826 and 1997. (b) This graph shows the maximum calculated flood levels (blue) and discharges (red) in unregulated floods of the Red River through downtown Winnipeg. Creation of the Red River Floodway in 1968 greatly reduced the impact of the floods. The actual values for the 1997 flood show that the diversion reduced one of the worst floods on record to a less significant event in the city. Normal water level is less than 224 metres above sea level. (c) The Red River Floodway (on the left) rejoins the Red River (entering from the right) north of Winnipeg. The purpose of the floodway is to divert large floods around the city and so protect the urban centre. (d) The community of St. Adolphe was protected by a dike during the 1997 flood of the Red River. The normal course of the Red River on the left of the photograph can be seen near the partially submerged trees. (e) A sand-bag dike protected this rural residence on the flood plain of the Red River. *Fly By: 49° 50' N; 96° 57' W*

15.19 Lake Baikal Lake Baikal, in southern Siberia, is shown here in an image acquired by astronauts aboard the space shuttle Endeavour. The lake is among the world's largest, with a length of 636 kilometres. It is also the deepest, at 1,741 metres, with its bottom 1,182 metres below sea level. *Fly By: 53° 26' N; 108° 26' E*

ORIGINS OF LAKES

The basins that contain lakes originate from many different processes acting on the Earth's surface. Tectonism creates rift valleys where the deepest lakes on the Earth are found. Meteorite impact and volcanic eruptions form craters that may contain lakes. Lakes are created on the flood plains of rivers by fluvial erosion and deposition as well as by the solution of limestone in karst regions. Landslides can block rivers to create lakes in river valleys, and rivers are often dammed artificially to create reservoirs for water supply, hydroelectric power generation, and a variety of other human uses.

In Canada, most lakes originated during the Ice Age when moving continental ice sheets eroded bedrock, carving depressions in areas of weak rock found within the Canadian Shield. Lakes were also formed in the sedimentary deposits left behind by the melting ice. (Chapter 19 provides a fuller account of lakes and landforms produced by glaciers and ice sheets.) Western Siberia is also rich in lakes created by glacial activity.

Although most lakes seem permanent in terms of human experience, they are normally short-lived features on the geologic time scale. Lakes that have stream outlets will gradually drain as their outlets erode to lower levels. Also, lake basins fill with sediment carried by inflowing streams and with organic matter

15.20 Freshwater pond This pond in northern Manitoba is among the smaller lakes in Canada. Vegetation is slowly growing inward at the edges, and eventually the pond will become a bog that supports wet forest. The small river in the background shows well-developed meanders and oxbows (see Chapter 16).

produced by plant and animals within the lake. As a group, the lakes of Canada are particularly young because the ice sheets that formed them retreated less than about 13,000 years ago. Even today, as much smaller alpine and arctic glaciers continue to retreat in response to a warming climate, new lakes are forming in their wake.

WATER BALANCE OF A LAKE

Lake levels vary; they usually rise in the spring and early summer and drop in the late summer and fall. Lake levels also rise after a period of heavy rainfall and fall after several weeks of dry weather. Thus, a lake is a repository of water that responds to changing inputs and outputs. It receives water as precipitation and inflow from streams and loses water from evaporation and outflow at its outlet. The lake's volume, and therefore the lake level, changes as these inputs and outputs vary.

The water balance of a lake can be expressed by the simple equation

$$\Delta V = P + I - E - O$$

where ΔV is the change in the volume of water in the lake, P is the precipitation that falls on the lake, I is the inflow of stream flow and groundwater to the lake, E is evaporation from the lake surface, and O is outflow from the lake (Figure 15.21). The time period over which the water balance is taken can vary. Often the quantities are taken on an annual basis and so are expressed in rates such as cubic kilometres per year. They can also be expressed as flow rates in metres per second, much like stream flow.

Changes in lake levels often follow cycles on different time scales. Annual cycles of precipitation and temperature modulate the lake's inputs and outputs and create regular seasonal variations in

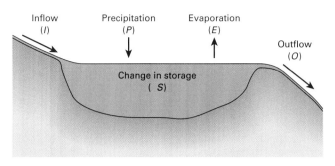

15.21 Water balance of a lake In a given time period, a lake receives water from inflow and precipitation and loses water from evaporation and outflow. The result is a change in volume.

lake levels. On a decadal time scale, sequences of cool and wet years followed by warm and dry years cause lake levels to rise and fall in response. Over hundreds to thousands of years, lakes can expand greatly or shrink and dry up in response to climate change.

Figure 15.22 plots monthly average lake levels of the Great Lakes. The annual cycle is obvious, with lowest lake levels in winter and highest levels in late spring and summer following the input of snowmelt. With warm summer temperatures, stream flow decreases and evaporation increases. Starting in July, mean water levels fall steadily through the remainder of the year.

The levels of Lake Ontario not only fluctuate seasonally, but also change significantly from year to year (Figure 15.23). For example, the 1930s was a period of warmer and dryer weather, and water levels in the Great Lakes were the lowest on record. The cooler and wetter conditions of the 1950s and 1970s resulted in higher than normal levels. Since the 1990s, levels have been declining in part due to management of water levels at dams on the St. Lawrence Seaway and perhaps in part as an early response to a warming climate.

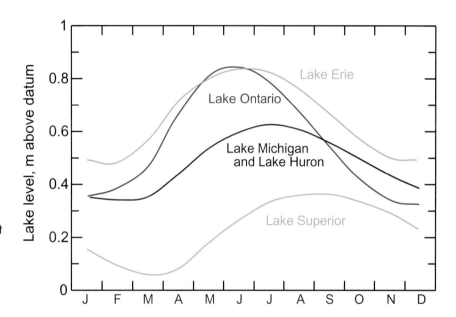

15.22 Mean monthly water levels of the five Great Lakes Water levels show a strong annual cycle. Data are mean monthly water levels over about 100 years and are shown with respect to the 1987 International Great Lakes Datum for each lake. (Data from the Canadian Hydrographic Survey)

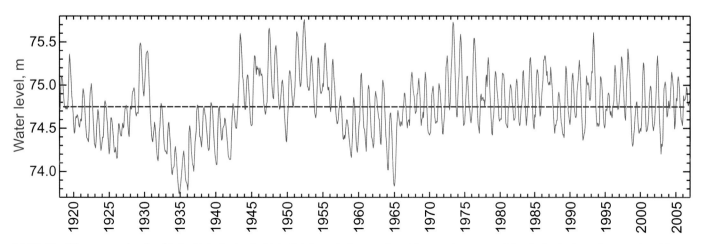

15.23 Monthly average levels of Lake Ontario Levels of Lake Ontario vary both seasonally and from year to year, as variations in climate change the parameters of the water balance. (Data from the Canadian Hydrographic Survey)

EYE ON THE ENVIRONMENT | 15.3   The Great Lakes

The Great Lakes—Superior, Huron, Michigan, Erie, and Ontario—along with their smaller bays and connecting lakes form a vast network of inland waters in the heart of North America. They contain 23,000 cubic kilometres of water—about 18 percent of all the fresh, surface water on the Earth. Only the polar ice caps and Lake Baikal in Siberia have a larger volume. Of the Great Lakes, Lake Superior is by far the largest. In fact, the volume of the other Great Lakes combined would not fill its basin.

The Great Lakes watershed contains a population of about 33 million people—22.8 million Americans and 9.2 million Canadians. The Great Lakes are especially important to Canada, since this population represents about 31 percent of the nation's total. The Great

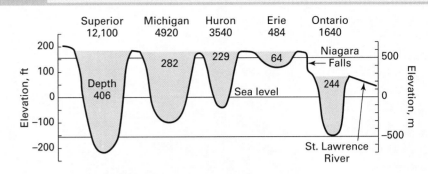

Figure 2 Volumes, elevations, and depths of the five Great Lakes Volumes are shown below the name of each lake in cubic kilometres (km³). Depths are shown in metres. (Modified from A. N. Strahler, The Earth Sciences. Used with permission)

Lakes basin is home to 45 percent of Canadian industry and 25 percent of Canadian agriculture. The lakes are an essential resource for drinking water, fishing, agriculture, manufacturing, transportation, and power generation.

The Great Lakes play an important role in moderating the climate of their region. Because water bodies heat and cool more slowly than adjacent land, autumn and winter temperatures near large lakes are warmer, while spring and summer temperatures are cooler. For the Great Lakes, the difference may be as much as 5 to 10°C, depending on the wind pattern over the lake.

The frost-free period near the lakes is up to 50 days longer each year. To the east of each of the lakes are snow belts where average snowfall ranges from 200 to 350 centimetres per year, approaching twice the amount that would occur otherwise. The extra snow is the result of evaporation from the relatively warm, free-water surface of the lake followed by condensation in cold, winter air.

Ice forms on the Great Lakes each winter. Although Lake Erie is the southernmost of the Great Lakes, it is normally more than 90 percent ice-covered in February because its volume is small and it has a limited ability to store heat. Lake Supe-

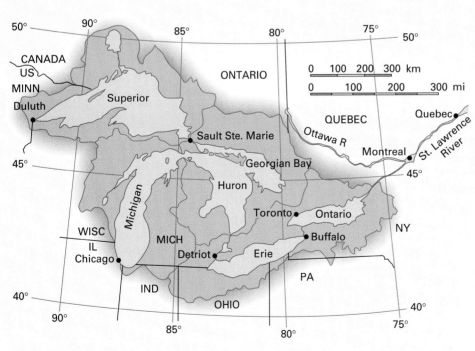

Figure 1 The Great Lakes and their watersheds The Great Lakes constitute a vast water resource lying astride the boundary between Canada and the United States. (©A.N. Strahler)

Lakes, it is normally more than 90 percent ice-covered in February because its volume is small and it has a limited ability to store heat. Lake Superior (at 75 percent), Lake Huron (70 percent), Lake Michigan (50 percent), and Lake Ontario (25 percent) all are less ice covered on average, although the amounts vary greatly from year to year as the climate fluctuates. For example, during the strong El Niño winter of 1983 (see *Eye on the Environment 5.4 • El Niño*), when temperatures in eastern North America were several degrees warmer than normal, ice cover ranged from 45 percent in Lake Huron to just over 10 percent in Lake Ontario. With global climate warming, the ice cover on the lakes will decrease, and by the last decades of this century ice may not form at all on some of them.

The sediments of the Great Lakes provide a record of the human settlement of their drainage basins. Analysis of sediment core samples shows that logging and clearing for agriculture caused accelerated erosion beginning in the nineteenth century. The input of pollutants associated with industrialization and urbanization increased sharply through the twentieth century.

By the 1960s and 1970s, excess nutrients, especially phosphorus from fertilizer, sewage, and detergents, had entered Lake Erie in amounts sufficient to cause a large increase in the lake's biological productivity. The large amount of biomass produced by this "fertilization" depleted the available oxygen dissolved in the water as it decayed. Suffocation threatened much of the aquatic life in the lake. (See *eutrophication* under "Pollution of Surface

Water" in this chapter for a fuller explanation of this process.) The Great Lakes Water Quality Agreement between Canada and the United States, first passed in 1972 and renewed in 1978, was the first step to remediate pollution of the lakes and was an important beginning in the recognition of their environmental degradation.

Pollution caused by a large number of substances from many sources continues to be a problem in the Great Lakes, even though vigilance has reduced the impact of many pollutants with the documentation of their presence and effects. Persistent organic pollutants are one group of particular concern. These include organic substances largely of industrial origin that are long lasting, highly mobile in the aquatic system, and toxic in very small amounts. Many of these compounds accumulate up the food chain as predators consume contaminated prey (see Figure 3). For example, the concentration of polychlorinated biphenyls (PCBs) in the algae of Lake Ontario has been measured at 0.02 parts per million (ppm). This very small amount is equivalent to about half a cubic millimetre in volume. However, this low concentra-

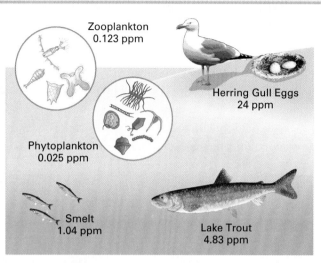

Figure 3 Concentration of PCBs in the Great Lakes food chain The degree of concentration of PCBs in each level of the Great Lakes aquatic food chain increases from aquatic plants to top predators. The eggs of fish-eating birds such as herring gulls have the highest levels. (From the U.S. Environmental Protection Agency)

tion has increased as it moved up the food chain; the eggs of herring gulls contain as much as 24 parts per million. This amount is still small, but it is 10 million times greater than the concentration in the lake water and is sufficient to cause deformities and high mortality in the chicks.

Human impact on the Great Lakes varies significantly from place to place. In 1987, the International Joint Commission identified 43 "Areas of Concern," 15 of which are in Canada. Most were located near major urban centres or sources of industrial pollution, especially where enclosed harbours or bays prevented dispersion of pollutants. Remedial action plans were proposed and implemented, and many of the sites have experienced much improvement. The Internet is an excellent source for up-to-date information on the Great Lakes and their water quality issues.

SALINE LAKES AND SALT FLATS

In moist climate regions with abundant precipitation, a lake fills until water escapes at the lowest point on its rim, forming an outlet. However, in dry regions, inflows are smaller, and if balanced by evaporation, no outlet is necessary. In this case, the water balance equation for the lake lacks the outflow term, and changes in level are determined by inflow and precipitation (gain) and evaporation (loss). If inflow and precipitation increase, the lake level rises. At the same time, the lake surface increases in area, allowing a greater evaporation rate, thus achieving a new balance. Similarly, if the region becomes more dry, reducing input and increasing evaporation, the water level falls.

Lakes without outlets commonly accumulate salts. Dissolved solids are brought into the lake by streams—usually streams that begin in distant highlands where a water surplus exists. Since evaporation removes only pure water, the salts remain behind and the salinity of the water slowly increases. *Salinity*, or degree of "saltiness," refers to the abundance of certain common ions in the water. Eventually, salinity levels reach a point where salts precipitate as solids (Figure 15.24). Notable examples of saline lakes include the Caspian Sea, Dead Sea, Aral Sea, and Great Salt Lake in Utah. In Canada, small saline lakes are found on the Prairies and in the interior of British Columbia.

In regions where climatic conditions consistently favour evaporation over input, the lake may disappear. A shallow basin covered with salt deposits, known as a *salt flat* or *dry lake*, remains. On rare occasions, these flats are covered by a shallow layer of water brought in by flooding streams originating in adjacent highlands. The Aral Sea has become saline and has greatly reduced in volume through human activity. The two major rivers that feed this vast central Asian water body have been largely diverted to irrigate agricultural lands. With its inflow greatly reduced, the lake has shrunk and become increasingly salty.

LAKE TEMPERATURE AND CIRCULATION

Lakes are not normally quiet bodies of still water. Wind drives water motion, producing waves that attack the lake shore and currents that mix the lake water. The level of mixing, however, depends on the temperature structure of the lake; that is, its thermal environment.

A lake receives most of its thermal energy from the Sun, and so at latitudes where there is a strong seasonal cycle of insulation, there is a strong seasonal cycle in the behaviour of lakes. As discussed in Chapter 2, short-wave solar radiation is more readily absorbed by water than by other natural surfaces. Recall that the term *albedo* expresses the proportion of solar radiation that a surface reflects back toward the sky. The albedo of water varies depending on the angle at which the Sun's rays strike the lake surface and the content of sediment in the water, but the average value for most lakes is less than 10 percent. Thus, water absorbs more than 90 percent of the solar radiation it receives.

A water body can absorb a substantial amount of solar energy. On a bright summer day, lake water in the midlatitudes of southern Canada can absorb as much as 600 watts per square metre (W/m^2) at noon directly from the Sun, as well as another 200 watts per square metre from solar radiation diffused through the atmosphere. The lake also receives long-wave radiation from the atmosphere, about 350 watts per square metre on a summer day and night.

The lake loses heat energy in several ways. First, the amount of long-wave energy the water surface radiates depends on its temperature. For example, a water surface at a temperature of 18°C loses about 400 watts per square metre in radiant energy flux. Cooling through latent heat flux by evaporation is also significant. Depending on water and air temperatures, it can be as much as 300 watts per square metre and is normally greatest in summer and autumn when the lake is warmest. The lake can also gain or lose heat by sensible heat transfer. The net result of these flows is that a strong annual insolation cycle creates a strong temperature cycle in a lake.

15.24 Salt encrustations These salt encrustations at the edge of Great Salt Lake, Utah, were formed when the lake level dropped during a dry period. *Fly By: 41° 13' N; 112° 44' W*

Chapter 3 noted that a water surface's temperature is cooler in the summer and warmer in the winter than the surrounding land surface. Two of the reasons for this effect were that water cools more intensely by evaporation from its free surface, and that water heats more slowly than typical earth materials such as rock or soil because of its high specific heat. *Specific heat* is the amount of heat energy in joules required to raise the temperature of one gram of material by one degree Celsius. The specific heat of water is 4.2 J/(g °C). For soil and rock, typical values of specific heat range from 0.8 to 2.5 J/(g °C). Thus, two to five times more energy is needed to raise the temperature of water than to raise the temperature of rock or soil to the same degree.

As noted in Chapter 3, solar energy penetrates water, so solar energy is absorbed and converted to heat below the surface. At a depth of 1 metre in clear water, more than half the solar radiation has been absorbed. By 10 metres, only about 10 percent remains. By 100 metres, only 3 percent is left. Within the solar spectrum, blue light penetrates most deeply. In very clear water, up to 60 percent of blue light remains at 60 metres depth. Red and near-infrared light is absorbed much closer to the surface—99 percent within 4 metres or less. This selective absorption of longer wavelengths gives clear water its characteristic blue-green colour.

Figure 15.25 illustrates the result of the absorption of solar energy near the lake surface along with the effect of mixing of the lake's upper waters by wind. Wind blowing across the surface exerts friction on the water that causes the water beneath it to move. Because water is a fluid, it has little resistance to movement, and so the velocity of water moved by wind is significant—about 10 percent of the wind velocity, on average.

The mixing is confined to the surface layer because of the density of water, which varies with temperature. At very close to 4°C, water attains its maximum density of 1,000 kilograms per cubic metre (kg/m³). From this point, density decreases both as temperature decreases, to 999.87 kilograms per cubic metre at 0°C, and as temperature increases, to 997.1 kilograms per cubic metre at 25°C. A difference of less than 3 kilograms per cubic metre between 4°C and 25°C in a substance already so heavy may not seem like much, but it is important. It means that warm water is light enough to "float" on top of cold water, and mixing between warm and cold water requires so much energy that it normally doesn't occur.

The effect of this density difference, coupled with solar heating in the uppermost portion of the lake, creates a warm, upper zone of less dense, lighter water called the *epilimnion*, and a cold, more dense, lower zone called the *hypolimnion* (Figure 15.25). Summer temperature in the epilimnion varies from about 25–30°C or so in midlatitude lakes, depending on the size of the lake and the time in the summer. Temperature in the hypolimnion is much lower and in many northern lakes remains near 4°C, the temperature of maximum density. These zones are separated by a zone in which the water temperature, and thus density, is changing rapidly with depth. This zone is called the *metalimnion*. The horizontal plane within the metalimnion where the temperature changes most rapidly with depth is called the *thermocline*. Because this layered structure is dependent on water density, it is very stable, and thus there is little water movement across the thermocline.

SEASONAL CIRCULATION IN LAKES

Because there is a large seasonal difference in solar energy in the midlatitudes and toward the poles, the temperature profile and resulting circulation in midlatitude lakes change through the year. Figure 15.26a shows the summer condition described above, in which wind-induced waves and currents create a strong circulation in the epilimnion. The hypolimnion below experiences a much weaker circulation that is driven by friction with currents at the bottom of the epilimnion. Very little energy or mass moves across the metalimnion.

In autumn, the temperature of the epilimnion falls as surface water cools, sinks, and mixes into the epilimnion. Eventually, the temperature of the epilimnion drops to that of the hypolimnion. Now the barrier caused by the density difference is gone, and the lake is able to circulate through all its depth (Figure 15.26b). This *overturn* of the lake distributes not only water, but also suspended

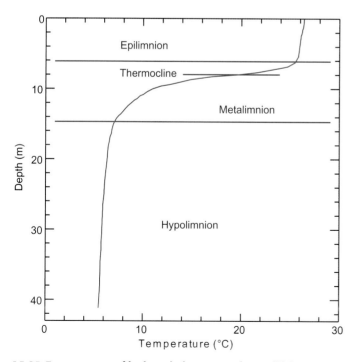

15.25 Temperature profile through the waters of a small lake in Ontario on August 5, 2006 *Absorption of solar energy warms the near-surface water. Mixing by wind redistributes the warm water downward. The result is that two distinct layers separated by a transition layer form in lakes at midlatitudes.*

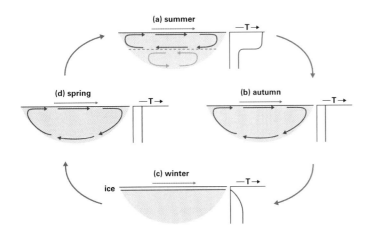

15.26 Water circulation and temperature in a midlatitude lake As the cycle of solar radiation varies through the year, the thermal structure and circulation within a lake changes. Overturn occurs in the spring and fall when the lake's temperature is uniform.

solids and dissolved gases throughout the water. For example, oxygen that respiring organisms (from bacteria to fish) in the hypolimnion have consumed all summer is replaced with oxygen from the atmosphere by an exchange across the air/water interface.

As winter sets in (Figure 15.26c), the upper part of the lake cools below 4°C. The near-surface water becomes slightly less dense, remains on top, and continues to cool, forming a layer of ice. The layer below the ice has a temperature of 0°C to 4°C. Because this layer is less dense than the water below it, mixing is inhibited, and a weak layered structure forms, which isolates upper and lower water masses. With the lake sheltered from wind-driven waves and currents by its ice cover, mixing ceases for the winter.

In spring, the ice melts, and the temperature throughout the lake becomes nearly uniform again. Another period of overturn occurs (Figure 15.26d), and water, suspended solids, and dissolved gases circulate freely throughout the lake. Lakes that experience overturn twice a year, in the spring and autumn, are referred to as *dimictic*. This term applies to most lakes in climates where the lake surface freezes, such as Canada and the northern United States. Lakes that are less than about 10 metres deep may not develop a hypolimnion, since substantial solar energy penetrates to that depth. In these lakes, mixing occurs throughout the year, and they are referred to as *polymictic*. In the Arctic, lakes may not become warm enough to establish a thermal structure in summer, and so spring and autumn mixing come together in one short period of mixing in the warm season when the lake ice melts. These lakes are *monomictic*.

SEDIMENT IN LAKES

Lakes receive water and sediment from inflowing streams and rivers, as well as from runoff down slopes adjacent to lake shores. When streams and rivers enter the lake, the velocity of the stream

water slows, depositing coarse-grained sediment—typically sand and gravel—to form a delta. However, fine-grained sediment—silt and clay—remains suspended in the stream water as it moves into the lake. While some suspended sediment can exit through a lake's outlet, most eventually settles out, creating layers of fine sediment that accumulate on the lake bottom. (Chapter 16 provides more details on how rivers carry sediment and build deltas.) In most lakes, an average of only about 0.5 to 1 millimetre of sediment accumulates each year, although in lakes where rivers deliver large loads of sediment, rates can exceed 1 centimetre per year.

Biological material produced in lakes is also deposited on the lake floor, where it only partially decomposes. The remains of the smallest organisms, such as *phytoplankton* (single-celled plants) and *zooplankton* (tiny animals) dominate this organic material.

Lakes can also make their own sediment. This occurs when chemical compounds dissolved in inflowing streams and groundwater precipitate and settle to the lake floor. This precipitation is common in regions underlain by limestone and other carbonate rocks. The limestone that makes up the bedrock is relatively soluble, and the dissolved load of calcium carbonate ($CaCO_3$) in streams and groundwater entering the lake is high. The solubility of calcium carbonate in water is inversely related to temperature. Cold water can carry more dissolved calcium carbonate than warm water. When carbonate-rich water in the epilimnion warms in the summer, the dissolved calcium carbonate will precipitate, forming tiny solid particles. Photosynthesis by plants in the water consumes carbon dioxide, as it does in land plants. This also facilitates the precipitation of calcium carbonate in the lake water. Several of the Great Lakes, including Lake Michigan and Lake Ontario, experience this *whiting* of their waters about every other year on average. The fine-grained precipitate is deposited around the margins of lakes in beds and banks of *marl*—soft, white, carbonate-rich mud. Some of the carbonate re-dissolves as the lake waters cool, but some accumulates on the lake floor.

The nature and composition of the sediment that accumulates on the lake bottom can reveal much about the environment of the lake and its surroundings. For example, the amount of inorganic sediment the lake receives is greater in wet years as inflowing streams and rivers increase their sediment loads. Temperatures and nutrient concentrations in the epilimnion affect the relative amounts of various species of phytoplankton and zooplankton and thus the types and amount of organic sediment they produce. Plant pollen falling into the lake is often preserved in lake-bottom sediments, and the mixture of pollen types can indicate the general climate of the region. In some cases, information on water temperature can be gleaned from the atomic composition of precipitated carbonate or silica. This method uses the ratio of two oxygen isotopes found in the precipitate, which is sensitive to temperature.

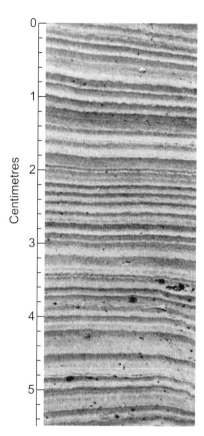

15.27 Lacustrine sedimentary record This photograph illustrates a portion of a core of sediment from Ape Lake in the Coast Mountains of British Columbia. Each light–dark couplet is the sediment deposited in one year, called a varve. The dark sediment is clay deposited in the quiet, ice-covered lake during winter when inflow was small. The light sediment is silt and fine sand deposited in summer when inflowing streams were more active. Notice that the thickness and appearance of each varve is different, illustrating the variability in the climate and hydrology that determines sediment delivery to the lake.

Thus, the sediments deposited on the lake floor provide a record of the environmental history of the lake and its drainage basin. In some cases, even seasonal patterns can be seen in this record. For example, if the rivers flowing into the lake introduce a pulse of clastic sediment every year during spring melt, or if a plankton bloom occurs every summer, then annual layers, called *varves*, may be preserved in the sediments (Figure 15.27). Like annual tree rings, the number of varves can determine the age of the lake. Each varve can provide information on the events that happened during that year, and comparison with the other varves can determine how the lake's environment changed over time.

Sediments can slump and flow over other sediments. Waves and currents can also erode and redistribute sediment throughout the lake. Organisms can burrow into the sediments, releasing old material to settle out again, while leaving a hole to be filled by new material. Thus, it takes a thorough and careful understanding of the deposition processes to read a lake's history.

15.28 Lake sediment corer This long probe, shown here on the deck of a research vessel, is lowered into the mud of the lake bottom, collecting a core of sediment that can capture the history of the lake.

Many lakes have sedimentary records that go back for tens of thousands of years—in some cases, more than a million years. By studying cores of lake sediment recovered from the lake floor (Figure 15.28), limnologists can reconstruct the environmental history of the region over these long time scales. In the glaciated terrains of Canada and the northern United States, lake sediments reveal where and when the climate warmed and how forests migrated northward following the retreat of the continental ice sheets. Observed over the last 10,000 years or so of the present interglacial period, these records also establish the natural variations in climate that occurred before the presence of humans began large-scale modification of the Earth's surface and atmosphere.

SURFACE WATER AS A NATURAL RESOURCE

Fresh surface water is a basic natural resource essential to human agricultural and industrial activities. Runoff held in reservoirs behind dams provides water supplies for great urban centres, such as New York City and Vancouver. When diverted from large rivers, it provides irrigation water for highly productive lowlands in dry regions, such as the Sacramento and San Joaquin valleys of California and the Nile Valley of Egypt. Runoff is also used in hydroelectric power generation, where the gradient of a river is steep, or routes of inland navigation, where the gradient is gentle.

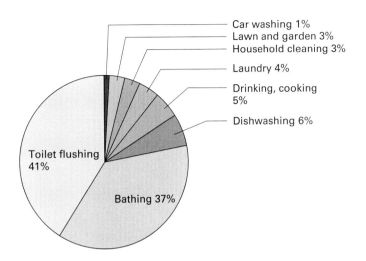

15.29 Domestic water use This chart shows how an average home in Canada uses water. (Environment Canada)

Our heavily industrialized society requires enormous supplies of fresh water for its sustained operation. Urban dwellers consume water in their homes at rates of 150 to 400 litres per person per day (Figure 15.29). Large quantities of water are used for cooling in air conditioning units and power plants. Much of this water is obtained from surface water supplies, the demand for which increases daily.

Unlike groundwater, which represents a large water storage body, fresh surface water in the liquid state is stored only in small quantities. (An exception is the Great Lakes system.) Recall from Chapter 6 that the global quantity of available groundwater is about 20 times larger than that stored in freshwater lakes, and the water held in streams is only about one one-hundredth of that in lakes. Because of small natural storage capacities, surface water can be drawn only at a rate comparable with its annual renewal through precipitation. Dams provide useful storage capacity for runoff that would otherwise escape to the sea, but once the reservoir is full, water use must be scaled to match the natural supply rate averaged over the year. Development of surface water supplies brings on many environmental changes, both physical and biological, and these must be taken into account when planning for future water developments.

A LOOK AHEAD

This chapter focused on water, including the precipitation that runs off the land and flows into the sea and the drainage system of a stream that conducts this flow of runoff. The smallest streams catch runoff from slopes, carrying the runoff into larger streams. The larger streams, in turn, receive runoff from their side slopes and pass their flow on to still larger streams. However, this cannot happen unless the gradients of slopes and streams allow the water to keep flowing downhill. This means that the landscape has been shaped and organized into landforms that are an essential part of the drainage system. The shaping of landforms within the drainage system occurs as running water erodes the landscape, which is the subject of the next chapter.

CHAPTER SUMMARY

• The fresh water on land surfaces accounts for only a small fraction of the Earth's water. Since it is produced by precipitation over land, it depends on the continued operation of the hydrologic cycle for its existence. The soil layer plays a key role in determining the fate of precipitation by diverting it in three ways: to the atmosphere through *evapotranspiration*, to groundwater through *percolation*, and to streams and rivers as runoff.

• Groundwater occupies the pore spaces in rock and regolith. The water table marks the upper surface of the *saturated zone* of groundwater, where pores are completely full of water. Groundwater moves in slow paths deep underground, recharging rivers, streams, ponds, and lakes by seeping upward and thus contributing to runoff. Solution of limestone by groundwater can produce *limestone caverns* and generate *karst* landscapes.

• Wells draw down the water table and, in some regions, lower the water table more quickly than it can be recharged. Groundwater contamination can occur when precipitation percolates through contaminated soils or waste materials. Landfills and dumps are common sources of groundwater contaminants.

• Runoff includes *overland flow*, moving as a sheet across the land surface, and *stream flow*, which is confined to a channel in streams and rivers. Rivers and streams are organized into a drainage system that moves runoff from slopes into channels and from smaller channels into larger ones. The discharge of a stream measures the flow rate of water moving past a given location. Discharge increases downstream as tributary streams add more runoff.

• The hydrograph plots the discharge of a stream at a location through time. Since it takes time for water to move down slopes and into progressively larger stream channels, a *lag time* separates peak discharge from peak precipitation. In larger streams, the lag time will be longer and the discharge's rate of decline after the peak will be more gradual. Because urbanization typically involves covering ground surfaces with impervious materials, urban streams exhibit shorter lag times and higher peak discharges. Annual hydrographs of streams from wet regions show an annual cycle of *base flow* on which discharge peaks related to individual rainfall episodes are superimposed.

- Floods occur when river discharge increases and its usual channel can no longer contain the flow. Water spreads over the *flood plain*, inundating low fields and forests adjacent to the channel. When discharge is high, flood waters can rise beyond normal levels to inundate nearby areas of development, causing damage and sometimes taking lives. *Flash floods* occur in small, steep watersheds and can be highly destructive.

- Lakes are standing bodies of water, ranging in size from the Caspian Sea to small ponds. Canada has nearly half a million lakes that cover 8.4 percent of its land area. Lakes are important as sources of water, and for transportation, recreation, fisheries, irrigation, electric power generation, and other uses.

- Lakes are formed by tectonism, volcanism, river erosion and deposition, solution, mass wasting, glacial action, and other processes. They disappear when erosion of their outlets drains them, or when they fill with sediment or dry up.

- The water balance of a lake is determined by inflow from streams and groundwater and precipitation, and outflow from evaporation and loss at its outlet. Change in the water balance through the year normally creates a seasonal cycle of lake levels, with high water in the spring and early summer and low water in the late summer, fall, and winter. Over decades, sequences of wet and dry years cause lake levels to rise and fall in response.

- Where lakes occur in inland basins without outlets, they are often saline. When climate changes, such lakes can dry up, creating *salt flats* or *dry lakes*.

- Midlatitude lakes experience a strong annual cycle of thermal heating. Coupled with the density change of water with temperature, warm-season heating creates a warm *epilimnion* atop a cold *hypolimnion* separated by a transitional *metalimnion*. In *dimictic* lakes, wind-driven mixing occurs twice a year as lake water temperature becomes uniform just before and after the surface water freezes over. *Monomictic* lake water mixes once per year during a short warm-season thaw. *Polymictic* lake water is shallow and mixes throughout the year.

- Lakes accumulate sediment from several sources. Inflowing rivers provide suspended sediment that settles to the lake bottom. *Phytoplankton* and *zooplankton* contribute organic sediment. Precipitation of calcium carbonate can create a *whiting* of lake water and leave *marl* in lake beds. The sediments on a lake bottom provide a record of the lake's history and prior environments.

- Groundwater and surface water resources are essential for human activities. Human civilization is dependent on abundant supplies of fresh water for many uses. But because the fresh water of the continents is such a small part of the global water pool, utilization of water resources takes careful planning and management.

KEY TERMS

artesian well	drainage system	infiltration	stream
channel	flood	lake	water table
discharge	groundwater	overland flow	watershed
drainage basin	hydrograph	runoff	

REVIEW QUESTIONS

1. What happens to precipitation falling on soil? What processes are involved?
2. How and under what conditions does precipitation reach groundwater?
3. How do caverns form in limestone? Describe the key features of a karst landscape.
4. How do wells affect the water table? What happens when pumping exceeds recharge?
5. How is groundwater contaminated? Describe how a well might become contaminated.
6. Define discharge (of a stream) and the two quantities that determine it. How does discharge vary in a downstream direction? How does gradient vary in a downstream direction?
7. What is a drainage system? How are slopes and streams arranged in a drainage basin?

8. Define the term *flood*. What is the flood plain? What factors are used in forecasting floods?

9. What factors combined to make the Red River flood of 1997 so large? What kinds of engineering structures may be constructed to reduce the size and effect of floods?

10. Define a lake. Name several of the Earth's largest lakes. Why are lakes especially important in Canada?

11. How are lakes formed? Identify typical processes of lake formation. What processes cause lakes to disappear?

12. Write an equation for the water balance of a lake and define the terms involved. How does the water balance influence the level of a lake?

13. Under what circumstances does a lake become saline? How does the water balance equation for a saline lake differ from that of a freshwater lake?

14. Explain how solar energy penetrates and warms a lake. What portion of the lake is most intensely heated?

15. Describe how the density of water depends on temperature. At what temperature is water most dense? How does water temperature and therefore density affect mixing in a lake?

16. Explain the formation of a layered structure in a midlatitude lake in the warm season, identifying the epilimnion, metalimnion, hypolimnion, and thermocline.

17. At what times of the year and why does the water of a midlatitude lake mix completely? Why do some lakes only mix once per year? Why are some lakes always well mixed?

18. Identify the sources of sediment that accumulates at the bottom of a lake.

19. How can the sediments found on the bottom of a lake be used to reconstruct the history of a lake and its environment?

20. How is surface water used as a natural resource?

FOCUS ON SYSTEMS 15.1 Stream Flow and its Measurement

1. How are the cross-sectional area and velocity of a stream related to the flow?

2. What two types of flow occur in a stream, and how do they differ?

3. How does the energy dissipated by stream flow differ in pools and rapids?

EYE ON THE ENVIRONMENT 15.3 The Great Lakes

1. Why are the Great Lakes important to us?

2. How did the Great Lakes form? Have the levels and outlines of the lakes changed in the past? Explain.

3. How do the Great Lakes affect local climate?

4. What concerns about environmental quality have been raised for the Great Lakes, particularly Lakes Erie and Ontario?

VISUALIZATION EXERCISES

1. Sketch a cross-section through the land surface showing the position of the water table and indicating flow directions of subsurface water motion with arrows. Include the flow paths of groundwater. Be sure to provide a stream in your diagram. Label the saturated and unsaturated zones.

2. Why does water rise in an artesian well? Illustrate with a sketched cross-sectional diagram showing the aquifer, aquicludes, and the well.

3. Sketch the annual cycle of water level of a midlatitude lake in a continental climate, such as Lake Ontario.

ESSAY QUESTIONS

1. A thundershower causes heavy rain to fall in a small region near the headwaters of a major river system. Describe the flow paths of that water as it returns to the atmosphere and ocean. What human activities influence the flows? In what ways?

2. Using Figure 15.26 as a model, construct seasonal circulation diagrams and temperature profiles for dimictic, monomictic, and polymictic lakes.

3. Imagine yourself a recently elected mayor of a small city located on the banks of a large river. What issues might you be concerned with that involve the river? In developing your answer, choose and specify some characteristics for this city—such as its population, its industries, its sewage systems, and the present uses of the river for water supply or recreation.

PROBLEMS	WORKING IT OUT 15.2 Magnitude and Frequency of Flooding

1. The data below are highest mean daily flows in each year from 1970 to 2004 in the Grand River at Galt (now part of Cambridge), Ontario. Rank the data from greatest to least and determine the recurrence interval for each flow using the formula. If two values are equal, use the average of the two ranks in determining the recurrence interval for the flow. What are the magnitudes of floods with recurrence intervals closest to 1, 2, 5, 10, and 30 years? (*Hint:* Many word processing programs and spreadsheets have the ability to sort data and thus can be used to order the list quickly and easily.)

Year	Discharge (m³/s)	Year	Discharge (m³/s)	Year	Discharge (m³/s)
1970	229	1982	479	1994	235
1971	355	1983	265	1995	251
1972	733	1984	224	1996	380
1973	432	1985	637	1997	570
1974	856	1986	565	1998	358
1975	561	1987	430	1999	96
1976	501	1988	255	2000	530
1977	446	1989	256	2001	431
1978	429	1990	318	2002	289
1979	698	1991	532	2003	247
1980	425	1992	470	2004	447
1981	368	1993	477		

(Water Survey of Canada)

2. Note that the percent probability that a flow will be equalled or exceeded is simply $100/I$, where I is the recurrence interval. What is the percent probability that a flood flow in any one year will exceed 250 cubic metres per second? 500 cubic metres per second? What flow will be equalled or exceeded in 25 percent of all years?

Find answers at the back of the book

EYE ON THE LANDSCAPE	

Chapter Opener   Waste water outflow, desalinization plant. Notice how the brine plume remains distinct after entering the ocean (**A**). The brine is considerably saltier, and therefore denser, than the ocean water, so it tends to flow for some distance under the ocean water which appears lighter blue in colour. The water cascades over the spillway (**B**) and flows in a turbulent manner (**C**) in the main channels that lead away from the structure. Flow is slower in the lateral channels (**D**) that curve to the left and right. The exposed channel forms (**E**) indicate that discharge can be higher than shown in the photograph The gradient of the sandy coastal plain (**F**) is quite gentle and in places the discharge channels have developed a braided pattern (**G**). The concentrated discharge has caused down-cutting of the channels (**H**) in this sparsely vegetated (**I**) arid landscape..

Fish ladder. Hydro dams form impassable barriers (**A**) to fish and significantly alter the flow regime of rivers by creating deep pools

(**B**) or shallows (**C**) which constantly fluctuate as power demands are met. Migrating fish can by-pass these structures along fish ladders (**D**) which replicate natural stream flows and gradients. The V-shaped profile (**E**) is characteristic of a naturally eroded river valley.

Chapter 16

EYE ON THE LANDSCAPE

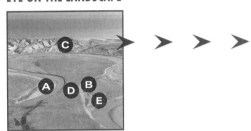

Horton River Northwest Territories, Canada.
What else would the geographer see? . . . Answers at the end of the chapter.

FLUVIAL PROCESSES AND LANDFORMS

Fluvial Processes and Landforms
Erosional and Depositional Landforms
Slope Erosion
Accelerated Erosion
Rilling and Gullying
The Work of Streams
Stream Erosion
Stream Transportation
Capacity of a Stream to Transport Load
Stream Deposition
*Working It Out 16.1 • River Discharge and
    Suspended Sediment*
Aggradation and Flood Plain Development
Stream Gradation
Landscape Evolution of a Graded Stream
*Focus on Systems 16.2 • Stream Networks as
    Trees*
Waterfalls
Dams
Entrenched Meanders
Fluvial Processes in an Arid Climate
Alluvial Fans

Running water has sculpted most of the world's land surface. Moving as a sheet across a land surface, water picks up particles and carries them down slopes into a stream channel. When rainfall is heavy, streams and rivers swell, lifting large volumes of rock debris and transporting it downstream. In this way, running water erodes mountains and hills, carves valleys, and deposits sediment. This chapter describes the work of running water and the landforms it creates.

Running water is one of four fluid agents that erode, transport, and deposit mineral and organic matter. The other three are waves, glacial ice, and wind. Recall that *denudation* is the total action of all processes that wear away exposed rock and transport the resulting sediments to the sea or closed inland basins. Denudation is an overall lowering of the land surface. If left unchecked, it would eventually reduce a continent to a nearly featureless, sea-level surface. However, tectonic activity continually elevates continental crust well above the ocean basins. The result is that running water, waves, glacial ice, and wind have always had material available to mold into a diversity of landforms.

FLUVIAL PROCESSES AND LANDFORMS

Landforms shaped by running water are described as **fluvial landforms**. They are shaped by the **fluvial processes** of overland flow and stream flow (see Chapter 15). Weathering and mass wasting (Chapter 14) operate in concert with overland flow, providing the rock and mineral fragments that are carried into stream systems.

Fluvial landforms and fluvial processes dominate continental land surfaces across the world. Throughout geologic history, glacial ice has been present only in mid- and high latitudes and in high mountains. Landforms made by wind action are mostly found in arid regions, although even there fluvial action is important. The action of waves and currents is restricted to the narrow contact zone between oceans and continents.

EROSIONAL AND DEPOSITIONAL LANDFORMS

All agents of denudation perform the geological activities of erosion, transportation, and deposition. When an initial landform such as an uplifted crustal block is created, it is immediately subjected to

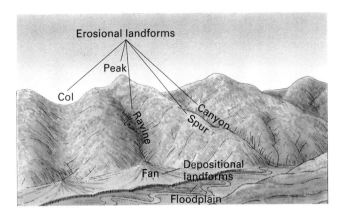

16.1 Erosional and depositional landforms (A.N. Strahler)

denudation and particularly to the force of running water. Valleys are formed where fluvial agents erode away rock. Between the valleys are ridges, hills, or mountain summits, representing the remaining parts of the crustal block that running water has not yet carved. These sequential landforms, shaped by progressive removal of the bedrock mass, are *erosional landforms*. Fragments of soil, regolith, and bedrock are transported and deposited elsewhere to make an entirely different set of surface features—*depositional landforms*. Typical landforms created by fluvial processes include ravines, canyons, peaks, spurs, and cols. Fans, built of rock fragments below the mouth of a ravine, are also depositional landforms, as are the flood plains that form on river valley floors (Figure 16.1).

SLOPE EROSION

Fluvial action starts with overland flow, which consists of a thin film of water or tiny rivulets that move across the ground. In humid climates, **overland flow** begins when precipitation exceeds the storage capacity of the surface (Figure 16.2). Some precipitation, whether rain or snow, will be retained by the vegetation and

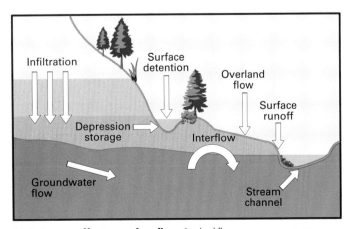

16.2 Factors affecting surface flow Overland flow commences as soon as infiltration, surface detention, and the water retention capacity of plant surfaces are exceeded. Local and regional topography and underlying geology control the ability of the water to become organized into channels and drainage systems.

be returned to the atmosphere through evaporation or sublimation. Excess rain and melting snow eventually reaches the ground, where some will infiltrate into the soil, and some will be retained in small depressions and behind obstructions such as little twigs and leaves. Once the storage capacities of the soil and surface depressions are full, the water will begin to move downslope in a process called *infiltration excess overland flow*.

Infiltration excess overland flow is different from *Hortonian overland flow*, or unsaturated overland flow, which typically occurs in arid and semi-arid regions under intense rainfall, especially where surface sealing may have limited soil infiltration capacity. Hortonian overland flow is also applicable to impervious surfaces, such as exposed bedrock or paved surfaces in urban environments.

As it moves across the surface, overland flow will entrain small particles of mineral matter, beginning the first stage in fluvial erosion (Figure 16.3). At this stage, the water is not concentrated into a well-defined channel, and movement of particles occurs as **sheet erosion**. The effects of sheet erosion are often hard to distinguish, especially in vegetated areas, because soil loss is so gradual. Early signs of sheet erosion include bare areas and puddles of water forming as soon as rain falls. Although only a thin layer of soil is being removed, the cumulative effect over time causes significant degradation.

Runoff occurs whenever excess water on a slope cannot be absorbed into the soil or trapped on the surface. In Canada, runoff is generally highest in the spring when the subsoil is still frozen and the overlying layers are quickly saturated by snowmelt. As the water moves across the surface, it exerts a tractive force on the loose soil particles, which become entrained in the flow. The size of the particles that can be transported depends on the gradient of the surface, the volume of water that moves across it, and the degree to which the particles are bound by plant roots or held down by a mat of leaves. Soil particles carried in runoff typically range from about 0.001 to 1.0 millime-

16.3 Overland flow The transport of soil particles by overland flow across this saturated ground is the initial stage in fluvial erosion.

tre in diameter. Under normal rainfall intensities, larger particles tend to settle out after moving only short distances, but finer materials that are suspended in the water can travel long distances. In addition to solid materials, the water will carry dissolved mineral matter in the form of ions produced by acid reactions or direct solution.

A steeper slope increases water's erosive power because water moves faster across the surface. Slope length is also important since a longer slope allows more runoff to accumulate and so increases the degree of scouring. The ability of soils to resist erosion is a measure of soil erodibility, which is based on the physical characteristics of each soil. Generally, soils with faster infiltration rates, higher levels of organic matter, and improved structure have a greater resistance to erosion. Soils with sand, sandy loam, and loam textures tend to be less erodible than silt, very fine sand, and certain clay-textured soils. Properties that determine erodibility, such as soil aggregation and shear strength, show systematic seasonal variation and are strongly affected by climatic factors, including rainfall distribution and frost action.

Shear strength changes significantly over shorter time periods, especially in response to soil moisture conditions. In fine-textured soils, cohesive forces that bind the particles mainly determine shear strength. Cohesion is an intermolecular attractive force that acts between particles; it is very significant for clay-size particles. Cohesion is virtually non-existent in sands; these coarser grains are held by friction. In moist soil, water films create a negative pore water pressure; that is, the water films produce suction, which increases cohesion by drawing particles together. In saturated soil, positive pore water pressures can develop, which push particles apart. As well as reducing cohesion, these pressures reduce friction by reducing the number and area of points of contact between grains.

This ongoing removal of soil is part of the natural geological process of denudation and occurs everywhere that precipitation falls on land. Under stable, natural conditions in a humid climate, the erosion rate is slow enough that a soil with distinct horizons forms and is maintained. Under equilibrium conditions, some soil is removed each year, but is replaced by material freshly weathered from the underlying regolith. Soil scientists refer to this activity as the *geologic norm.*

ACCELERATED EROSION

In contrast, the rate of soil erosion may be enormously speeded up by human activities or natural events of unusual magnitude to produce a state of *accelerated erosion.* The soil is removed much faster than it can be formed, progressively exposing the uppermost soil horizons (Figure 16.4). Accelerated erosion is often initiated by the removal of the plant cover by fire, logging, or other disturbances. No foliage remains to intercept rain, and the pro-

16.4 Accelerated soil erosion On steep slopes or where the vegetation cover has been disturbed, soil can be removed at a faster rate than it forms.

tection of a cover of fallen leaves and stems is removed. Consequently, the raindrops fall directly on the mineral soil.

The impact of falling raindrops on bare soil lifts and moves soil particles. This process is called *splash erosion* (Figure 16.5), which can disturb many tonnes of soil during a single rainstorm. Particle movement by rain splashes is usually greatest during short-duration, high-intensity rainstorms. Runoff water can easily remove lighter aggregate materials, such as silt and fine sand, that are detached in this way. On a sloping ground surface, splash erosion slowly shifts the soil downhill. More important, rain splashes can break up soil aggregates. Finer clays and silts can seal up the pores in the soil, making it much less able to absorb water. Reduced infiltration, in turn, increases the depth of overland flow from a given amount of rain, increasing the rate of soil erosion.

An effect of vegetation removal is a reduction in the resistance of the ground surface to the force of erosion under overland flow. For example, on a grassy slope, overland flow causes little soil erosion, because the energy of the moving water is dissipated in friction with the grass stems. On a heavily forested slope, the surface layer of leaves, twigs, roots, and fallen tree trunks counters the force of overland flow. Without such a cover, the eroding force is applied directly to the bare soil surface, easily dislodging grains and sweeping them downslope.

With an increased depth of overland flow and no vegetation cover to absorb the eroding force, soil particles are easily picked up and moved downslope. Eventually, they reach the base of the slope, where the gradient lessens as it merges with the valley floor. Here, the particles come to rest and accumulate in a thickening layer called **colluvium**. Because this deposit is built by overland flow, it has a sheet-like distribution and may be inconspicuous, except where it collects around features such as tree trunks.

16.5 Soil erosion by rain splash A large raindrop (above) lands on a wet soil surface, producing a miniature crater (below). Grains of clay and silt are thrown into the air, and the soil surface is disturbed.

If not deposited as colluvium, the sediment carried by overland flow eventually reaches a stream and is transported downstream, where it may accumulate as **alluvium** in layers on the valley floor. Alluvium refers to any stream-laid sediment deposit;

it could include particle sizes ranging from large pebbles and gravel to sand, silt, and clay. Coarse alluvium chokes the channels of small streams and can cause the water to flood broadly over the valley floors. The finest particles are distributed widely by flood waters, and, in this way, extensive alluvial deposits are gradually built up in increments. Typically, flood-plain soils are fertile and highly developed for agriculture.

The quantity of sediment that overland flow removes from a unit area of ground surface in a given unit of time is called *sediment yield*. Figure 16.6 gives annual average sediment yield and runoff by overland flow from several types of upland surfaces in northern Mississippi. Notice that both surface runoff and sediment yield decrease greatly with the increased effectiveness of protective vegetation cover. Sediment yield from cultivated land undergoing accelerated erosion is more than 10 times greater than that from pasture and about 1,000 times greater than that from pine plantation land.

In drier areas, such as midlatitude steppes, the sparse, short-grass cover is normally strong enough to slow the pace of erosion. However, the natural equilibrium is highly sensitive and easily upset. Depletion of the plant cover by fires or by grazing of domesticated animals can easily set off rapid erosion. These sensitive, marginal environments require cautious use because they lack the potential to recover rapidly from accelerated erosion once it has begun.

RILLING AND GULLYING

Accelerated soil erosion is a constant problem in cultivated regions with a substantial water surplus. Where land slopes are steep, runoff from torrential rains can cut more pronounced channels in the surface through the process of *rill erosion*. In this way, a series of closely spaced channels can be scored into the soil and regolith in a short time (Figure 16.7a). If the rills are not destroyed by soil tillage, they may join together into larger channels. These deepen rapidly and soon become *gullies*—steep-walled, trenches whose upper ends grow progressively upward (Figure

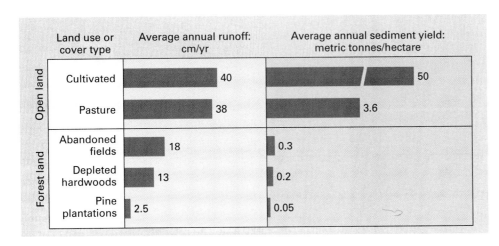

	Land use or cover type	Average annual runoff: cm/yr	Average annual sediment yield: metric tonnes/hectare
Open land	Cultivated	40	50
Open land	Pasture	38	3.6
Forest land	Abandoned fields	18	0.3
Forest land	Depleted hardwoods	13	0.2
Forest land	Pine plantations	2.5	0.05

16.6 Runoff and sediment yield This graph shows that both runoff and sediment yield are much greater for open land than for land covered by shrubs and forest. (Data from S. J. Ursic, U.S.D.A.)

16.7 Rills and gullies (a) Rills can develop quickly on soil-covered slopes when the vegetation cover is disturbed. They can be a problem in agricultural areas, if heavy rains occur before the crop is well developed. (b) If left unchecked, more runoff progressively concentrates in the rills and leads to more pronounced erosion and the formation of deep gullies.

16.7b). A rugged, barren topography results when accelerated soil erosion is allowed to proceed unchecked.

Erosion due to a high rate by overland flow occurs naturally in some semi-arid and arid lands where the process creates *badland topography*. Easily eroded clay formations underlie badlands. Erosion rates are too fast to permit plants to take hold, so no soil can develop. This develops a maze of small stream channels with steep gradients.

The landscape of Alberta's Dinosaur Provincial Park provides a striking example of badland topography (Figure 16.8). Here, the exposed horizontal strata date from the late Cretaceous period. The sand and mud deposits were left by rivers that flowed some 75 million years ago. The present landscape dates from end of the Pleistocene Epoch (10,000–15,000 years ago) when meltwater streams issuing from the retreating ice sheets began to carve the distinctive landscape. Final sculpting of the many hills and hoodoos is attributed to the effects of flash floods in this semi-arid region.

16.8 Badlands Gullies cut into soft sedimentary rock have created the dissected topography of Dinosaur Provincial Park in Alberta.

THE WORK OF STREAMS

The geomorphological work of streams consists of three related activities: erosion, transportation, and deposition. (Note that "stream" in this context refers to any channelled water flow, and the word is used interchangeably with "river"). **Stream erosion** is the progressive removal of mineral material from the floor and sides of the channel, whether bedrock or regolith. **Stream transportation** consists of movement of the eroded particles dragged over the stream bed, suspended in the body of the stream, or held in solution as ions. **Stream deposition** is the accumulation of transported particles on the stream bed and flood plain, or on the floor of a lake, where it may be temporarily held until it is carried to the oceans. Erosion cannot occur without some transportation taking place, and the transported particles must eventually come to rest. Thus, erosion, transportation, and deposition are simply three phases of a single activity.

STREAM EROSION

Streams erode in various ways, depending on the force of the flowing water, the nature of the channel materials, and the tools with which the current is armed. The force of flowing water is derived from its mass and acceleration due to gravity (9.81 metres per second squared (m s^{-2})). By extension, the work that a stream performs is expressed as force × distance. Thus, a stream performs a certain amount of work (expressed in joules) to move a pebble a given distance. The ability to do work is a measure of the stream's energy. The water molecules in a stream acquire potential energy, due to the height of the land surface above sea level. Potential energy is dissipated as the water flows downhill, but the river system gains kinetic energy as it flows. Kinetic energy dislodges and transports materials in the stream channel or converts to heat generated by friction between the water molecules.

The force of the flowing water not only sets up a dragging action on the bed and banks, but also causes particles to strike against these surfaces. Dragging and impact can easily erode alluvial materials, such as gravel, sand, silt, and clay. This form of erosion, called *hydraulic action*, can excavate enormous quantities of channel material in a short time. The undermining of the banks causes large masses of material to slump into the river, where the particles quickly separate and become part of the river's load. This process of bank caving is an important source of sediment during high river stages and floods.

Erosion of alluvial channels occurs when the flowing water carries sand and other materials. This begins when the water's force exceeds the threshold shear stress of the channel bed materials. The threshold shear stress of mineral particles can be approximated from their immersed weight.

The force of flowing water is greatly affected by friction between the water and the channel bed and sides, and by channel shape. Friction increases near a river bed where rock particles of various sizes create a rough surface. Irregularities in the bed will create turbulence, which also reduces velocity.

The average flow velocity in a stream channel can be calculated from Manning's equation, an empirical formula that uses channel gradient and the roughness of the sand, gravel, and cobble surfaces over which the water flows. It is written as:

$$V = \frac{1.49}{n} R^{2/3} S^{1/2}$$

where V is the velocity in metres per second, R is the hydraulic radius in metres (hydraulic radius is calculated by dividing the cross-sectional area of the channel by the wetted perimeter), S is the slope of energy line in metres per metre, and n is the coefficient of roughness, specifically known as Manning's n.

Manning's roughness coefficient is a function of friction along the channel bed. If the flow is deep relative to the bed material size, Manning's n can be estimated by

$$n = 0.048 \, D_{50}^{1/6}$$

where D_{50} (in metres) is the median size of the bed materials (i.e., half of the material is larger than this diameter).

Typical values for n in mountain streams with beds of cobbles with large boulders range from 0.050 to 0.070, while those with gravel, cobbles, and few boulders range from 0.04 to 0.05, compared with 0.028–0.035 for rivers greater than 30 metres wide.

Mineral particles of any size that are carried by a swift current will strike against the rock floor and walls of a channel. This process of mechanical wear of channel materials is called *abrasion* and is the principal method of deepening and widening a river channel in bedrock. The transported fragments also become chipped, broken, and rounded, and are reduced in size through *attrition*. The rolling of cobbles and boulders over the stream bed further crushes and grinds the smaller fragments to produce a

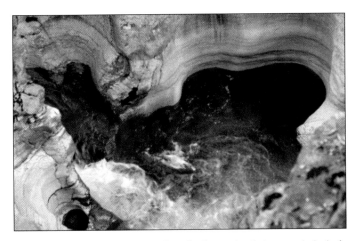

16.9 Potholes These potholes have been deeply incised in the limestone bedrock of Maligne Canyon, near Jasper, Alberta.

wide assortment of grain sizes, which are then sorted by the flowing water. As a result, channel materials are generally coarser in the upstream sections of river and get progressively finer downstream.

Abrasion is the principal means of fluvial erosion in bedrock that is too strongly consolidated to be affected by simple hydraulic action. Potholes are distinctive erosional features formed by abrasion. They occur when a shallow depression in the bedrock of a stream bed acquires stones, which are spun around and around by the flowing water, carving deep cylindrical depressions (Figure 16.9).

In addition to the mechanical processes of scour and abrasion, flowing water can also remove some types of bedrock, such as limestone, through chemical solution. This process is called *corrosion* and arises when the rock over which the river flows slowly dissolves. Effects of corrosion are conspicuous in limestone channels that develop cupped and fluted surfaces.

STREAM TRANSPORTATION

Stream load refers to the materials a stream carries, which it does in three ways, depending on particle size and the energy of the flowing water (Figure 16.10). Finer particles, such as clay and silt, are carried in *suspension*; the turbulent motion of the flowing water holds them within the water column. This fraction of the transported matter is the *suspended load* and normally accounts for the bulk of the material transported. Sand, gravel, and larger particles move as *bed load* by rolling, bouncing, or sliding along the channel floor. *Dissolved matter* is transported invisibly in the form of chemical ions. All streams acquire some dissolved ions through mineral alteration of the channel surfaces or of the particles in their loads (Figure 16.11). A fourth way streams can carry some materials is by *flotation*. Although mainly restricted to buoyant organic matter such as wood, mineral particles can be rafted on ice in this way.

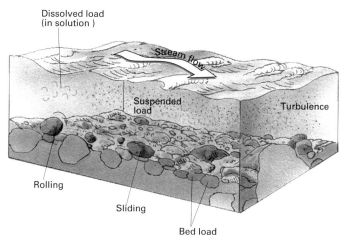

16.10 Sediment transport Streams transport materials in three ways. Coarse particles move along the channel bed as bed load, finer sediment may be carried in the water column as suspended load, and materials carried in ionic form compose the dissolved load.

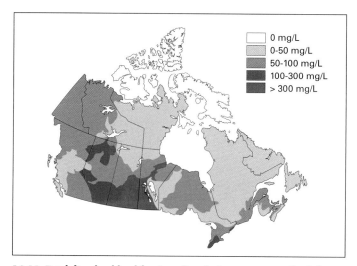

16.11 Total dissolved load for Canadian drainage systems The chalky nature of glacial tills in western Canada contributes to the high dissolved load carried by rivers draining from the Prairie provinces. Dissolved loads are less in rivers associated with the Canadian Shield. The above key shows total dissolved loads in mg/L.

Particles begin to move once the force of flowing water exceeds the critical shear stress of the bed materials. The force necessary for entrainment is called *fluid drag*, which is determined by the water's density and velocity. Fluid drag exerts a horizontal force that pushes against the particles, as well as a vertical component due to turbulence that lifts the particles. When these forces overcome friction and cohesion between particles, the channel sediment will begin to move. Once the particle is lifted, it is subject to gravity and so will have a tendency to settle back to the bed. Sediment transport in rivers is essentially determined by the same forces that influence wind patterns. The principal difference is that water exerts a greater force because it is about 9,000 times denser than air. Hence rivers can move larger calibre particles.

Fine sands about 1 millimetre in diameter tend to move most readily and can be entrained at flow velocities as low as 20 centimetres per second (Figure 16.12). Critical flow velocities increase for finer sediments, such as clays, because cohesion has bound the particles together into larger aggregates. Similarly, high flow rates are required to transport coarser sands and gravels because of their higher mass.

A river carries as much as 90 percent of its load in suspension and the largest river systems discharge considerable amounts of sediment into the oceans. For example, the annual sediment load of the Amazon River is calculated at 1,200 million tonnes, India's Brahmaputra-Ganga system at 1,100 million tonnes, and the Mississippi River at 159 million tonnes. Another river with a high suspended sediment load is the Huang He, or Yellow River, in China, which is so named because of the vast amount of yellow silt it the carries in suspension. The Huang He's drainage basin has an area of about 865,000 square kilometres and, based on the amount of sediment that the river transports, the annual denudation rate is estimated at about 3,000 tonnes per square kilometre, most of which is removed from the easily eroded loess plateau through which it flows.

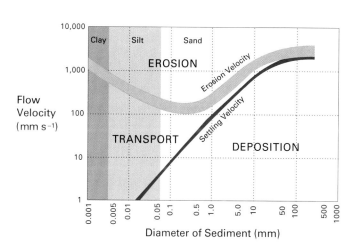

16.12 Relationship between flow velocity and particle transport in streams Higher velocities are required to entrain and transport larger calibre materials because of their density. Transport of fine silts and clays also requires a relatively high velocity because they tend to bind together through cohesion. Flowing water most readily moves sands in this range.

16.13 Average annual suspended sediment load (in tonnes per year) for selected rivers in Canada The heavy sediment loads characteristic of western rivers can be attributed to high rainfall and significant areas of unconsolidated materials within the drainage basins. In eastern Canada, most rivers flow over resistant bedrock.

In Canada, the river with the highest average annual suspended sediment load is the Mackenzie, at 100 million tonnes. Much of the sediment the Mackenzie River carries is derived from unconsolidated materials left by the ice sheets or from sediment supplied from tributaries arising in the wetter climates of western mountains. In contrast, the rivers of the drier Prairie provinces carry much lower sediment loads, while those in eastern Canada transport relatively little sediment because they flow mainly over bedrock that is more resistant to erosion (Figure 16.13).

CAPACITY OF A STREAM TO TRANSPORT LOAD

The maximum solid load of debris that a stream can carry at a given discharge is a measure of the *stream capacity*. This load is usually measured in units of metric tonnes per day passing downstream at a given location. Total solid load includes both bed load and suspended load. A stream's capacity to carry suspended load increases as its velocity increases, because the swifter the current, the more intense the turbulence. The capacity to move bed load also increases with velocity. In fact, the capacity to move bed load increases according to the third to fourth power of the velocity. In other words, when a stream's velocity is doubled in times of flood, its ability to transport bed load is increased from eight to sixteen times. Thus, most of the conspicuous changes in the channel of a stream occur during a flood. *Working It Out 16.1 • River Discharge and Suspended Sediment* shows how the suspended sediment load of a stream increases with discharge.

STREAM DEPOSITION

A stream delicately adjusts to the supply of water and rock waste from upstream sources. Where a stream flows in a channel of hard bedrock, the channel's form cannot deepen quickly in response to rising waters and may not change much even during extreme flood events. However, a stream flowing in a channel that is cut into thick layers of silt, sand, and gravel is very sensitive to changes in flow rates and sediment load. When water flow increases, alluvial channels may widen and deepen. As the flow slackens or more sediment is available, perhaps because of slumping along the banks of the channel upstream, the stream will deposit material on the bed.

When bed load increases, exceeding the transporting capacity of the stream, the sediment accumulates in the form of bars of sand, gravel, and pebbles. These deposits raise the elevation of the stream bed, a process called **aggradation**. The development of a variety of channel forms reflects this type of adjustment to flow regime and sediment load. There are three general channel patterns: straight, meandering, and braided. These develop in response to the amount and calibre of the sediment supply, the gradient of the stream, and the stability of the channel as determined by its resistance to erosion. Because these parameters are all highly variable, many different forms of river channels have developed.

An example of a straight channel is a fast-flowing mountain stream in a bedrock channel, with alternating sections of rapids and deeper pools (Figure 16.14a). Even where a stream's banks are relatively straight, the channel along which the water flows typically swerves around gravel bars. The course of a meandering channel consists of a series of broad, sweeping curves (Figure 16.14b). The spacing between adjacent meanders increases with the size of the river and is approximately 12 times the channel width. The shapes and positions of meanders change over time as the outer bank erodes and deposition builds up the inner bank. In a braided channel, the water flows in numerous channels that separate and reunite (Figure 16.14c). The geometry of a braided channel changes continually as varying flow rates deposit and scour sediment.

AGGRADATION AND FLOOD PLAIN DEVELOPMENT

Sediment introduced at any point in a stream will be gradually spread along its whole length and slowly build up the land surface through aggradation. In addition to sediment acquired by stream erosion, overland flow carries material to streams in the normal process of slope degradation. Higher amounts may be delivered to a stream following disturbance of the plant cover, for example, in a forest fire or through logging or poor farming techniques.

WORKING IT OUT | 16.1   River Discharge and Suspended Sediment

A river carries more sediment during a flood than during periods of normal flow. This happens for a number of reasons.

- First, as the discharge increases, more water is available to carry sediment downstream.
- Second, as discharge increases, so does the velocity of the stream flow; and as the velocity of the water flow increases, so does the intensity of the turbulence that keeps suspended sediment in motion. Thus, more sediment can be carried in a cubic metre of swiftly moving water.
- Third, as the water rises and increases in velocity, the force of the stream flow on the bed increases. The stream then erodes sediment in its bed, scouring and deepening the channel, a process that also increases the sediment available for transportation.
- Fourth, the heavy precipitation that causes a flood generates overland flow that drags more sediment into upstream river channels. Thus, more sediment is available for transport.

The graph to the right plots discharge against the suspended sediment load for the Oldman River at Lethbridge, Alberta. The discharge is measured in cubic metres per second (m³/s) and the suspended sediment load in metric tonnes per day (t/d). The points are daily measurements collected from April to October 1990. The graph shows that as discharge increases, so does suspended sediment. For example, when the discharge increases from 10 to 100 cubic metres per second, the suspended sediment load increases from about 8 to 800 tonnes per day. Thus, when discharge increases by a factor of 10, sediment load increases by a factor of about 100.

This type of increase follows a power function of the form

$$y = ax^b$$

here $b > 1$. This gives a straight line on a log-log plot.

In the case of the relationship between suspended sediment load and discharge for the Oldman River at Lethbridge,

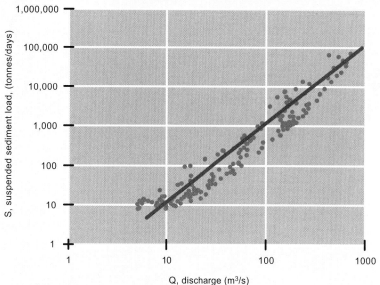

Sediment load and discharge The increase in suspended sediment load with an increase in discharge is plotted on logarithmic scales. The dots are daily observations for April–October 1990, for the Oldman River at Lethbridge, Alberta. (Data from the Water Survey of Canada)

the measurements come close to fitting a line with the following equation:

$$S = 0.015Q^{2.30}$$

where S is the suspended sediment load (tonnes per day) and Q is the discharge (cubic metres per second). This can also be written in log-log form, as a straight-line function:

$$log\ S = log(8.72) + 3.87\ log\ Q = 0.94 + 3.87\ log\ Q$$

The rapid increase in suspended load with increasing discharge is characteristic of nearly all rivers. Thus, most rivers carry out transport work in times of high flow. Much more sediment can be carried downstream during a single small flood than over many months of normal river flow. In addition, a large flood may move vast volumes of sediment that has been accumulating in the channel and flood plain, waiting for a rare and extreme event.

(a)

(b)

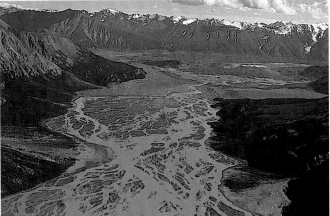

(c)

16.14 Characteristic channel forms (a) Many mountain streams have straight channels with sections of pools and rapids as the water flows over the coarse bed materials. (b) The sweeping curves of a meandering channel usually develop best on a flood plain. *Fly By: 53° 14' N; 105° 05' W* (c) Braided streams consist of numerous channels that divide and reunite in response to varying rates of sediment transport and deposition. *Fly By: 35° 33' N; 75° 35' 02" E.*

16.15 North Saskatchewan River near its source at the mountains of Alberta. The milky blue colour of the water is due to fine suspended sediment that is released in the glacier meltwater.

Other causes of aggradation are related to major changes in global climate, such as the effects glaciation. Many rivers in Canada are supplied by glacier meltwater that is heavily laden with sediment (Figure 16.15).

Accumulated alluvium has filled many valleys to depths of several tens of metres (Figure 16.16a). If the source of bed load is cut off or greatly diminished and the volume of water is unchanged, less of the stream's energy is needed for sediment transport. This increases the energy of the flowing water and results in channel scour. Where the stream gradient is gentle, this renewed degradation typically results in lateral cutting and subsequent growth of meanders (Figure 16.16b). This process gradually excavates the valley alluvium until the channel encounters hard bedrock. This essentially prevents the removal of the remaining alluvium and creates step-like **alluvial terraces** at the sides of the valley (Figure 16.17).

(a)

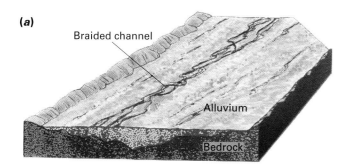

(b)

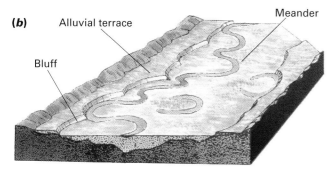

(c)

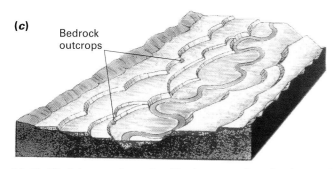

16.16 Alluvial terrace formation Alluvial terraces can form when the removal of alluvial fill in a valley is prevented by the greatly reduced rate of lateral erosion when the river encounters bedrock. (Drawn by A. N. Strahler)

Rivers can create a sequence of terraces when the course of the river moves back across the valley where the alluvial fill is deeper (Figure 16.16c). As each terrace is cut deeper into the original sediments, the river leaves a relic flood plain at a higher elevation. Each flat terrace surface is separated from the adjacent terrace by a steep slope, with the lowest terrace dropping down to the active channel. If a river cuts evenly into the alluvial sediments, terraces form at the same elevation on both sides of the valley. These are called *paired terraces. Unpaired terraces* occur when the stream encounters material on one side of the valley that does not erode easily. This leaves a terrace on one side with no corresponding terrace on the resistant side.

Meanders are dynamic features of the flood plain landscape; they change their shape and position over time as a result of different rates of erosion and deposition on their inner and outer banks (Figure 16.18). They are associated with **alluvial rivers** that transport sediment in channels that are free to migrate across the

16.17 Alluvial terrace Multiple terraces have formed along the Thompson River near Ashcroft, British Columbia.

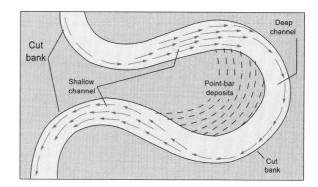

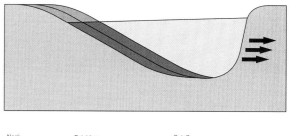

16.18 Meander bend Idealized map and cross-profile of a meander bend of a large alluvial river. Arrows show the position of the swiftest current. (Copyright © A. N. Strahler)

16.19 Floodplain features This river in northern Saskatchewan meanders across a broad flood plain. Some of the oxbow lakes that mark the former course of the river are filling with organic debris.

16.21 Anabranching river The divided channels in this anabranching river are semi-permanent features that are separated by well-vegetated levees and sandbars.

flood plain as they are subjected to scour and deposition. On the inside of each bend, alluvium accumulates as a long, curving deposit of sediment, called a *point bar*. On the outside of the bend, the bank is undercut and collapses. In this way, the configuration of the meander changes and eventually a narrow neck develops, which is cut through. This shortens the river course and leaves a meander loop to form a *cutoff*. This is quickly followed by deposition of silt and sand across the ends of the abandoned channel, producing an *oxbow lake*. The oxbow lake gradually fills in with fine sediment brought in during high floods and with organic matter produced by aquatic plants. Eventually, oxbow lakes convert into swamps, but they retain their identity indefinitely (Figure 16.19).

In time, the meandering river widens the **flood plain**, so that broad level areas lie on both sides of the river channel (Figure 16.20). A river flowing across a flood plain has a low channel gradient and, in unregulated systems, will characteristically experience overbank flooding every few years. During floods, water spreads from the

main channel over adjacent flood plain deposits. In this way, the flood plain is slowly built up by alluvium that the river deposits. Floodwaters also bring an infusion of dissolved mineral nutrients, which help retain the high natural fertility of flood plain soils. As the current slackens, the coarser particles are deposited in a zone adjacent to the channel. The result is an accumulation of higher land on either side of the channel known as a **natural levee**.

The flood plain extends to a point where land elevation precludes flooding, and its boundary is usually marked by a line of *bluffs*. Between the levees and the bluffs is lower ground, called the *backswamp*. Rivers will often divide into several channels and rejoin as they cross a flood plain to produce a pattern that is similar to that of braided streams. These are referred to as anabranching rivers and are distinguished from braided streams by the relative permanence of their channels (Figure 16.21). Each channel in an anabranching river is generally bordered by a well-defined natural levee, which characteristically is stabilized by trees and shrubs. The islands that form are quite extensive, so adjacent channels are more widely separated than in braided streams.

STREAM GRADATION

Most major stream systems have experienced thousands of years of runoff, erosion, and deposition. Over time, the gradients of stream segments tend to adjust so that they carry only the average load of sediment that they receive from slopes and inflowing channels. If more sediment accumulates each year in the stream channel than can be carried away, the surface of the channel will build up and the slope of the stream down to its mouth will increase. But a steeper gradient increases stream velocity, allowing the stream to carry more sediment. Eventually, the slope will reach a point at which the stream carries away only the sediment that it receives. If sediment flow to the stream decreases, the stream will

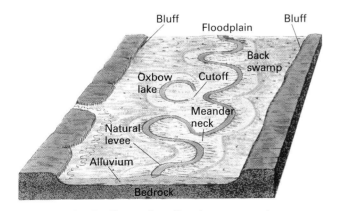

16.20 Flood plain landforms of an alluvial river As meanders migrate downriver, they create a variety of landforms, including oxbow lakes, cutoffs, and natural levees. (Drawn by A. N. Strahler)

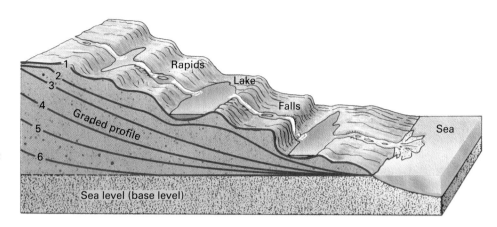

16.22 Stream gradation Schematic diagram of gradation of a stream. Originally, the channel consists of a succession of lakes, falls, and rapids. (Copyright © A. N. Strahler)

gradually erode its channel. This will reduce the gradient and also reduce the stream's ability to transport sediment. In time, a new equilibrium will be established between channel slope, flow velocity, and sediment load. Since every channel segment experiences this process, eventually the whole stream will tend toward a coordinated network that carries only the sediment load contributed by the drainage basin. A stream in this equilibrium condition is called a **graded stream**.

Figure 16.22 shows how a graded stream might develop on a landscape that has been uplifted, perhaps by faulting. A series of *stream profiles* (labelled 1–6) plots the elevation of the stream with distance from the sea. At first, the stream is not graded, with large variations in the slope profile. Water accumulates in shallow de-

pressions to form lakes, which overflow from higher to lower levels. Rapids and falls are abundant. As time passes, fluvial action slowly erodes the landscape. Each stream segment seeks its own equilibrium slope, and the stream profile smoothes out, as shown in profile 3. At this stage, the profile is now graded. In time, this *graded profile* is steadily lowered as the landscape is further eroded (curves 4 through 6).

LANDSCAPE EVOLUTION OF A GRADED STREAM

Figure 16.23 illustrates the stream gradation process over a landscape in a series of block diagrams. In (a), *waterfalls* and *rapids* occur in portions of the channel with steep gradients. Flow

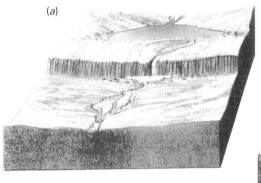

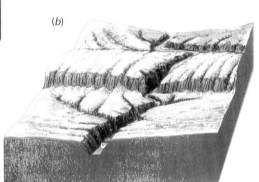

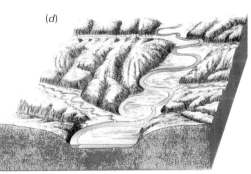

16.23 Evolution of a graded stream and its valley (a) A stream is established on a land surface dominated by landforms of recent tectonic activity. (b) Gradation is in progress. The lakes and marshes are drained. The gorge is deepening, and the tributary valleys are extending. (c) The graded profile is attained. Flood plain development is beginning, and the widening of the valley is in progress. (d) The flood plain has widened to accommodate meanders. Flood plains now extend up tributary valleys. (Drawn by E. Raisz; copyright © A. N. Strahler)

16.25 Point bar The deposition of sediment as a point bar on the inner bank of the meander shows how the river has migrated laterally as it cuts into the steep valley side.

16.24 Deeply incised canyon This section of the Thompson River in British Columbia has cut a deep canyon into the limestone bedrock. Rapids indicate that the channel is not graded.

velocity at these points is greatly increased, abrasion of bedrock is intense, and the falls are cut back. At the same time, the ponded stretches of the stream are first filled by sediment and are later lowered in level as the outlets are cut down. In time, the lakes disappear and the falls transform into rapids.

In the early stages of gradation, the capacity of the stream exceeds the load supplied to it, so little or no alluvium accumulates in the channels. Abrasion continues to deepen the major channels, resulting in steep-walled *gorges* or *canyons* (Figure 16.24). Weathering and mass wasting of canyon walls contribute an increasing supply of rock material. Also, overland flow supplies debris to develop branches of the stream network.

The erosion of rapids reduces the gradient of a slope angle to more closely approximate the average gradient of that section of the stream (Figure 16.23b). At the same time, branches of the main stream are extended into higher parts of the original land mass. These carve out many small drainage basins. Thus, the original tectonic landscape gradually transforms into a fluvial landform system.

As the landscape continues to change, the channel gradients decrease and the capacity for streams to move their bed load also decreases. Simultaneously, the load supplied to the stream from the entire area upstream increases. So the load supply eventually matches the stream's capacity to transport it. Thus, the major streams achieve the graded profiles (Figure 16.23c). The first indication that a stream has attained a graded condition is the beginning of flood plain development. This is initiated when the river

begins to migrate laterally, cutting into the side slopes flanking its channel as a sinuous course develops. Point bars form as material is deposited on the inside of each bend, which progressively widens to produce a crescent-shaped area of low ground (Figure 16.25).

As lateral cutting continues, the width of the flood plain increases, and the channel develops the sweeping meandering course (Figure 16.23d). In this way, the flood plain widens into a continuous belt of flat land in the valley floor. Flood plain development reduces the frequency with which the river attacks and undermines the adjacent valley wall. Weathering, mass wasting, and overland flow can then act to reduce the steepness of the valley side slopes (Figure 16.26). As a result, in a humid climate, the valley's gorge-like aspect gradually disappears and eventually gives way to an open valley with soil-covered slopes protected by vegetation. The time required for the stream to reach a graded condition and erode a broad valley is usually many millions of years. However, similar landforms may develop over a much shorter period when other factors influence river channel development. For example, the South Saskatchewan River near Saskatoon meanders across a broad flood plain and exhibits many of the features associated with the later stages of fluvial landscape development. The general topographic setting here was primarily created by continental ice sheets, which retreated from this part of Canada only about 12,000 years ago.

The gradient and channel form of each portion of a stream network adjust to carry the sediment load it receives. However, these properties vary from the small streams at the headwaters of the network to the broad river channels found farther downstream. One way to study the properties of streams is to organize them by their general status in the drainage basin. This process progressively unites the channel segments that compose a branching river network from its headwaters to its mouth. *Focus on Sys-*

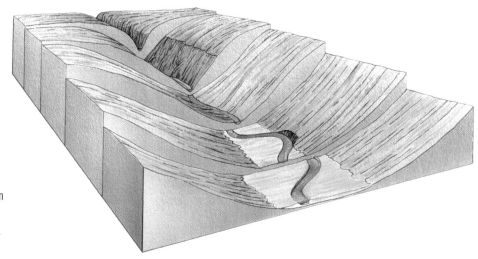

16.26 Evolution of side slopes Following stream gradation, the valley walls become gentler in slope, and soil and weathered rock cover the bedrock. (From W. M. Davis; copyright © A. N. Strahler)

16.27 Victoria Falls Located on the Zambezi River at the border of Zimbabwe and Zambia in southern Africa, Victoria Falls formed where a fault created a zone of weakness in the bedrock. *Fly By: 17° 55' 30" S; 25° 51' 25" E*

tems 16.2 • *Stream Networks as Trees* describes this system of **stream ordering** further.

WATERFALLS

The stream gradation process smoothes the profile of a stream by draining lakes and removing falls and rapids (see Figure 16.22). Thus, large waterfalls on major rivers are comparatively rare. Faulting and dislocation of large crustal blocks have caused spectacular waterfalls on several east African rivers, for example, Victoria Falls (Figure 16.27). Here, the shattered rocks along a fault have created a zone of weakness that the Zambezi River has eroded.

Another class of large waterfalls involves new river channels resulting from past glacial activity. Erosion and deposition by large moving ice sheets greatly disrupted drainage patterns in Canada and other northern regions, creating lakes and shifting river courses to new locations. Niagara Falls, which consists of the larger Horseshoe Falls (about 800 metres wide) and the American Falls (about 325 metres wide), is a good example (Figure 16.28).

16.28 Niagara Falls Niagara Falls formed where the river passes over the eroded edge of a massive limestone layer. *Fly By: 43° 04' 41" N; 79° 04' 32" W*

FOCUS ON SYSTEMS | 16.2 — Stream Networks as Trees

In mathematics, a branching system of lines and points is called a "tree," and it is part of the discipline called *topology*. A map of a drainage system, such as presented in Figure 1, shows the tree-like form of a typical channel network. It consists of two kinds of information. First is the "skeleton," a set of connected line segments that represent the stream channel system. A second set of connected lines consists of all the drainage divides—those topographic crests that direct the overland flow onto adjacent slopes. When the divides are mapped, they delineate individual drainage basins. Thus, a network of basins can be considered a topological set.

Topology recognizes curving and twisting lines and also assumes that both lines and points have no breadth or thickness; they are drawn that way only for ease of representation. The basis of topology is shown in Figure 2. What is usually called a "point" is a *node*; a "line" is an *arc* (or *edge*). Nodes are positioned at both ends of an arc. Arcs can form a closed loop containing a *region*. An assemblage of arcs connected by nodes is called a *network*. All these forms can lie in a space of any dimension, although here only topologies in real spaces of either two dimensions (planes) or three dimensions (volumes) are considered.

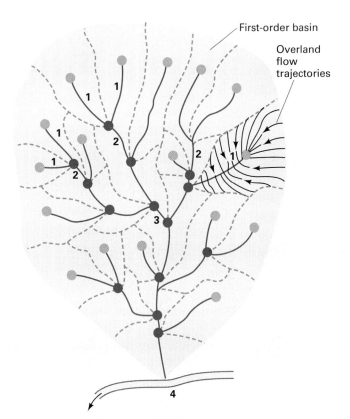

Figure 1 Schematic map of a third-order drainage basin showing both the stream channel system and the drainage basin network (Copyright © 1996 by A. N. Strahler)

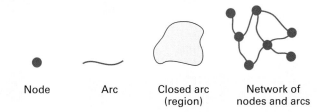

Figure 2 Basic forms used in topology

Figure 3 is a formal topological diagram of a type of network called a *rooted tree*. It consists solely of nodes and arcs and recognizes two types of nodes: *inner nodes* and *outer nodes*. Each inner node connects with three arcs, and each outer node connects with only one inner node. A third kind of node, called the *root*, serves as a starting point for the construction. This topological system serves well as a model, not only of a stream system, but also of several kinds of biological systems, among them the branching patterns of higher plants (trees, shrubs, veins of leaves), and even the respiratory pathways of the human lung. In these natural systems, a gaseous or fluid agent (air, water, sap) flows into or out of the system.

The formation of Niagara Falls began about 12,500 years ago when water from the early Lake Erie drained across the Niagara Escarpment at a point near Lewiston, New York, some 12 kilometres from the present falls. The course of the Niagara River makes a right angle bend at the Whirlpool, where it joins an older drainage route that was filled with glacial materials deposited from the former ice sheet.

The Niagara Escarpment is formed from a gently inclined layer of hard Silurian (approximately 400 million years old) limestone, beneath which lies more easily eroded shale. The Niagara River has gradually eroded the edge of the limestone layer, to produce the steep gorge marked by Niagara Falls at its head. The height of the falls is currently 52 metres, although the river has a clear drop of only 21 metres before reaching a mass of rock that collapsed in 1954. Niagara Falls is receding at a rate of about 30 centimetres per year. This is slower than might occur naturally because as much as 75 percent of the water is diverted for hydroelectric power generation.

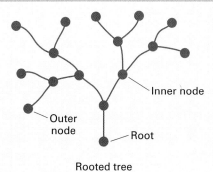

Figure 3 Topological diagram of a rooted tree

In Figure 1, the rooted tree represents a map of a natural stream system. It shows not only the network, but also the area of ground surface that encloses each stream arc. This surface slopes down from its perimeter—a drainage divide—to meet each stream channel and thus provides the channel with its water and sediment. In topology, it is a *region*, consisting of an area bounded by a closed loop of nodes and arcs. In this case, the loop bounding the region is composed of the arcs and nodes of the drainage divide network. All of these features—stream tree, drainage divide network, and regions that contribute flow to the channels—lie above a reference level and constitute a three-dimensional system. In the figure, they are projected upon a flat base, so the map is described as *planimetric*.

Geomorphologists use a convenient ordering system for the arcs and nodes within a tree. A single arc with nodes at each end is called a *stream segment*. The outermost (terminal) segments are designated as *first-order segments* (or *first-order streams*). Where two first-order segments join, they generate a *second-order segment*, and so forth, until they reach the single root segment. That terminal node probably joins an existing stream channel of the same or higher order,

but it could also terminate at the shoreline of a lake or ocean. For each stream segment, there is an enclosing drainage basin of the same order. The map in Figure 1 illustrates an example of a *first-order basin*, showing the overland trajectories of surface runoff.

Consider a landscape that has reached a stable condition in which watersheds have occupied all the upland surface. (See blocks a, b, and c of Figure 16.23.) Given time, the streams in the flow network will become graded. This means that each watershed is nicely adjusted in its form and slope so that all of the precipitation that falls on it passes along to the outlet. In this state, each stream channel segment has the channel form and gradient necessary to denude its drainage area and to transport erosional debris to the next segment.

When two segments of the first order join to form one second-order segment, some adjustments must be made. Below this junction, the discharge will (on average) have doubled. At the next downstream junction, that of two second-order segments, discharge will have doubled again. A basic assumption in hydraulic geometry is that the larger the channel cross-section, the more efficient the stream. Efficiency increases because the proportion of energy expended through friction with the bed and banks decreases. In compensation, the stream's gradient has diminished, an adjustment that operates by a natural feedback process. In this way, the graded profile of a stream is developed and maintained.

The idea of connected networks distributed in real space is an important one in many subfields of geography. In the case of stream channel networks, the topological ordering of stream segments has provided a useful and informative way to study streams and the way that fluvial action erodes landscapes.

DAMS

Because most large rivers with a steep gradient do not have falls, dams are necessary to create the vertical drop required to spin the turbines of electric power generators. Hydroelectric power is comparatively inexpensive, non-polluting, and renewable. Most large dams serve a second purpose of providing fresh water for urban use and irrigation. Some 45,000 dams have been constructed worldwide, with a combined surface area of 0.66×10^{12} square metres.

This compares with a surface area of 1.5×10^{12} square metres for the world's freshwater lakes. The Rogun Dam on the Vakhsh River in Tajikistan is world's highest at 335 metres. Canada's highest dam, the Mica Dam located 135 kilometres north of Revelstoke in British Columbia (243 metres), is the twelfth highest in the world (Figure 16.29); it was commissioned in 1973.

The gross reservoir capacity of the Mica Dam is 24,670 million cubic metres, and the lake that has formed behind it extends some 130 kilometres, with a surface area of about 115 square kilometres.

16.29 Mica Dam *Canada's tallest dam, the Mica Dam near Revelstoke, British Columbia, holds back 24,670 million cubic metres of water and has created a lake with a surface area of about 115 square kilometres.*

16.30 The Three Gorges area of China *The projected depth of the reservoir will be 135 metres. As well as flooding an area of high scenic and ecological significance, the reservoir will displace some 1.3 million people.*

The Caniapiscau Reservoir with a surface area of 4,318 square kilometres is the largest in Canada. Located on the the Caniapiscau River in Quebec, its waters flow into the James Bay hydroelectric complex. This is about half the area of the world's largest reservoir, Lake Volta in Ghana, which covers 8,500 kilometres and is 520 kilometres long. The Three Gorges reservoir in China is projected to be the world's biggest hydroelectric facility; it will produce about eight times more power than that generated at Niagara Falls. The reservoir began filling in 2003, and will occupy the present position of the scenic Three Gorges area (Figure 16.30). The projected depth of the reservoir is 135 metres; it will contain 39 billion cubic metres of water covering an area of 1,084 square kilometres.

The loss of scenic and recreational resources behind dams, as river valleys, rapids, and waterfalls are drowned, is of major concern. Dam construction also destroys ecosystems adapted to the river environment. In addition, deposition of sediment behind a dam rapidly reduces the lake water's holding capacity, and within a century or so, it may be almost completely filled. This reduces the lake's ability to provide consistent supplies of water and hydroelectric power. Thus, it is not surprising that new dam projects can meet with stiff opposition from concerned local citizens' groups and national environmental organizations.

ENTRENCHED MEANDERS

When a broadly meandering river is uplifted by rapid tectonic activity, the river's gradient increases, so that it cuts downward into the bedrock. This forms a steep-walled inner gorge. On either side lies the former flood plain, now a flat terrace high above river level. Runoff rapidly strips off any river deposits left on the ter-

races, and floods no longer reach the terraces to restore eroded sediment.

Uplift may also lead to the formation of *entrenched meanders* (Figure 16.31). Although entrenched meanders are not free to shift in the same way that meanders do on alluvial flood plains, they can enlarge slowly and produce cutoffs. The high, rounded hills created in this way are separated from the valley wall by the deep abandoned river channel and the shortened river course. Under unusual circumstances, where the bedrock includes a strong, massive sandstone formation, meander cutoff leaves a *natural bridge* formed by the narrow meander neck (Figure 16.32).

16.31 Entrenched meanders *The Goosenecks of the San Juan River in Utah are deeply entrenched river meanders in horizontal sedimentary rock layers. The canyon, carved from sandstone and limestone, is about 370 metres deep.* ***Fly By: 37° 12' N; 109° 59' W***

16.32 Natural bridge This natural bridge in Utah was formed when the neck of an entrenched meander was pierced by the river that formerly flowed in this dry region.

FLUVIAL PROCESSES IN AN ARID CLIMATE

Desert regions look different from humid regions both in terms of vegetation and landforms. Vegetation is sparse or absent, and land surfaces are mantled with coarse mineral material—sand, gravel, rock fragments, or bedrock itself. Although deserts have low precipitation, running water actually forms many landforms in desert regions. A specific location in a dry desert may experience heavy rain only once in several years and, at these times, the stream channels carry water and work as agents of erosion, transportation, and deposition. Fluvial processes are especially effective in shaping desert landforms because the sparse vegetation cover offers little or no protection to the land surface. On barren slopes, the flow of water carries large quantities of rock debris into the streams. In a few minutes, a dry channel transforms into a raging flood of muddy water heavily charged with rock fragments (Figure 16.33).

16.33 Flash flood A flash flood created by a thunderstorm has filled this desert channel in the Tucson Mountains of Arizona with fast flowing, turbid waters. A distant thunderstorm produced the runoff.

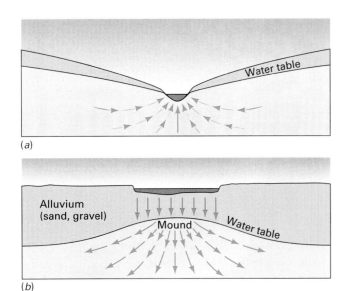

(a)

(b)

16.34 Groundwater and stream flow In humid regions (a), a stream channel receives ground water through seepage. In arid regions (b), stream water seeps out of the channel and into the water table below. (Copyright © A. N. Strahler)

16.35 Alluvial fans Extensive alluvial fans extend onto the floor of Death Valley in California. The canyons from which they originate have carved deeply into a great uplifted fault block.

An important contrast between regions of arid and humid climates is the way in which the water enters and leaves a stream channel. In a humid region, with a high water table sloping toward a stream channel, groundwater moves steadily toward the channel and seeps into the stream bed, producing permanent streams (Figure 16.34a). In arid regions, the water table normally lies far below the channel floor, and a stream flowing across a gravel or sand plain will lose water by seepage (Figure 16.34b). Loss of discharge by seepage and evaporation rapidly depletes the flow of streams in alluvium-filled valleys of arid regions. As a result, aggradation occurs and braided channels are common. Streams of desert regions are often short and end in alluvial deposits on the floors of shallow, dry lakes.

ALLUVIAL FANS

One common landform built by braided, aggrading streams is the **alluvial fan**, a low cone of alluvial sands and gravels (Figure 16.35). The apex, or central point of the fan, lies at the mouth of a canyon or ravine. The fan is built out on an adjacent plain. Alluvial fans are of many sizes, and some desert fans are many kilometres wide.

Fans are built by streams carrying heavy loads of coarse rock waste from a mountain or upland region. The braided channel shifts constantly, but its position is fixed at the canyon mouth. The lower part of the channel, below the apex, sweeps back and forth. This activity accounts for the semicircular fan form and the downward slope that radiates away from the apex.

In regions of internal drainage, where there is no outflow to the sea, streams respond only to local base level determined by their elevation relative to the floor of the enclosed basin. Each arid basin becomes a closed system for the transport of sediment. Only the hydrologic system is open, with water entering as precipitation and leaving as evaporation. This leads to the accumulation of fine sediment and precipitated salts and produces a level salt flat or **playa**. In some playas, shallow water forms a salt lake.

A Look Ahead

Running water, the primary agent of denudation, does not act equally on all types of rocks. Some rocks are more resistant to erosion. As a result, fluvial erosion can create unique and interesting landscapes that reveal rock structures of various kinds. Chapter 17 discusses the effect of rock structure on landform development.

CHAPTER SUMMARY

- This chapter has covered the landforms and land-forming processes of running water, one of the four active agents of *denudation*. Like other agents, running water erodes, transports, and deposits rock material, forming both *erosional* and *depositional* landforms.
- The work of running water begins on slopes, producing colluvium, where overland flow moves soil particles downslope, and alluvium, when the particles enter stream channels and are later deposited. In most natural landscapes, *soil erosion* and soil formation rates are more or less equal, a condition known as the *geologic norm*. *Badlands* are an exception in which natural erosion rates are very high.
- The work of streams includes erosion, transportation, and deposition. Where stream channels are carved into soft materials, large amounts of sediment can be obtained by *hydraulic action*. Where stream channels flow on bedrock, channels are deepened only by the abrasion of bed and banks by large and small mineral particles. Both the *suspended load* and *bed load* of rivers increase greatly as velocity increases. Velocity, in turn, depends on gradient.
- There are three basic channel forms. Straight channels tend to develop on steep terrain in a stream's headwaters; meandering channels develop sweeping curves where the channel gradient is very low; braided streams with multiple channels are common in streams that have heavy sediment loads.
- When provided with a sudden inflow of rock material, for example by glacial action, streams build up their beds by aggradation. When that inflow ceases, streams resume downcutting, leaving behind alluvial terraces.
- Large rivers with low gradients that move large quantities of sediment are called alluvial rivers. The meandering of these rivers forms *cutoff* meanders, *oxbow lakes*, and other typical landforms. Alluvial rivers are sites of intense human activity. Their fertile flood plains yield agricultural crops and provide easy transportation paths. When a region containing a meandering alluvial river is uplifted, *entrenched meanders* can result.
- Over time, streams become graded—their gradients are adjusted to move the average amount of water and sediment supplied to them by slopes. Lakes and *waterfalls*, created by tectonic, volcanic, or glacial activity, are short-lived events, geologically speaking, that give way to a smooth, *graded* stream *profile*. Grade is maintained as landscapes erode toward base level.
- Although rainfall is scarce in deserts, running water is very effective there in producing fluvial landforms. Desert streams, subject to flash flooding, build alluvial fans at the mouths of canyons. Water sinks into the fan deposits, creating local ground water reservoirs. Fine sediments and salts carried by streams accumulate in playas, from which water evaporates, leaving sediment and salt behind.

KEY TERMS

aggradation	colluvium	meander	stream deposition
alluvial fan	flood plain	natural levee	stream erosion
alluvial river	fluvial landforms	overland flow	stream load
alluvial terrace	fluvial processes	playa	stream ordering
alluvium	graded stream	sheet erosion	stream transportation

REVIEW QUESTIONS

1. List and briefly identify the four flowing substances that serve as agents of denudation.
2. Describe the process of slope erosion. What is meant by the geologic norm?
3. Contrast the two terms *colluvium* and *alluvium*. Where on a landscape would you look to find each one?
4. What special conditions are required for badlands to form?

5. When and how does sheet erosion occur? How does it lead to rill erosion and gullying?

6. In what ways do streams erode their bed and banks?

7. What is stream load? Identify its three components. In what form do large rivers carry most of their load?

8. How is velocity related to the ability of a stream to move sediment downstream?

9. How does stream degradation produce alluvial terraces?

10. Why is fluvial action so effective in arid climates, considering that rainfall is scarce? How do streams in arid climates differ from streams in moist climates?

11. Describe how playas and alluvial fans form an arid desert landscape.

FOCUS ON SYSTEMS 16.2   Stream Networks as Trees

1. Identify the components of a rooted tree and use them to sketch an example of a stream network.

2. How do geomorphologists order the arcs and nodes of a stream as a rooted tree?

3. How does the efficiency of a stream change with stream order? How does this explain the gradual reduction in slope of a graded stream in a downstream direction?

VISUALIZATION EXERCISES

1. Compare erosional and depositional landforms. Sketch an example of each type.

2. What is a graded stream? Sketch the profile of a graded stream and compare it with the profile of a stream draining a recently uplifted set of landforms.

3. Sketch the flood plain of a graded, meandering river. Identify key landforms on the sketch. How do they form?

ESSAY QUESTIONS

1. The North Saskatchewan River rises in the Rocky Mountains, crosses the foothills, flows through the agricultural regions of Alberta and Saskatchewan, and finally reaches the sea. Describe the fluvial processes and landforms you might expect to find on a journey along the river from its headwaters to Hudson Bay.

2. What would be the effects of climate change on a fluvial system? Choose either the effects of cooler temperatures and higher precipitation, or warmer temperatures and lower precipitation.

PROBLEMS   WORKING IT OUT 16.1 River Discharge and Suspended Sediment

1. What suspended load would you project for the Oldman River at Lethbridge when the discharge is 50 cubic metres per second? 500 cubic metres per second? If the discharge of the Oldman River doubles, what is the effect on its suspended sediment load?

2. A hydrologist measures discharge and sediment load for a river and plots the points as logarithms on ordinary graph paper. A line is fitted to the points with the following equation:

$$\log S = 2.50 + 2.12 \log Q$$

Rewrite this equation as a power function of the form $S = aQ^b$.

Find answers at the back of the book

EYE ON THE LANDSCAPE |

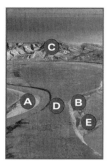

Chapter Opener   Horton River Northwest Territories, Canada. Differential flow velocities result in deposition of transported sediment in the form of point bars (**A**) on the inner banks of a meander and removal of materials on the outer banks (**B**). Gullying (**C**) on the adjacent highland will also contribute sediment to the river. The intervening meander neck (**D**) eventually may be breached to form a cut-off. The presence of water in the secondary channel (**E**) suggests this channel is still active in spring when flow increases due to snowmelt throughout the drainage basin, although the flood stage has now passed.

Terraces bordering the Tasman River, New Zealand. The active channels of the Tasmine River occupy the floor of a much broader valley that was cut earlier into the bedrock (**A**) probably during deglaciation when considerably more water flowed along its course. Sediment deposited in the channel as a level floodplain has been progressively eroded to form a series of terraces (**B**). At

times of high flow the present river actively erodes its banks causing slumping (**C**). This contributes to the high sediment load which is deposited as sand and gravel bars (**D**) during periods of low flow. The main channel (**E**) meanders across the full width of the active floodplain, with secondary channels (**F**) separating and rejoining to create a braided pattern, some of which have become disconnected and persist as stagnant backwaters (**G**).

Chapter 17

EYE ON THE LANDSCAPE

What else would the geographer see? . . . Answers at the end of the chapter.

LANDFORMS AND ROCK STRUCTURE

Rock Structure as a Landform Control
Strike and Dip
Focus on Remote Sensing 17.1 • Remote Sensing of Rock Structures
Landforms of Horizontal Strata and Coastal Plains
Arid Regions
Drainage Patterns on Horizontal Strata
Coastal Plains
Working It Out 17.2 • Properties of Stream Networks
Landforms of Folded Rock Layers
Sedimentary Domes
Fold Belts
Landforms Developed on Other Land-Mass Types
Erosion Forms on Fault Structures
Exposed Batholiths
Deeply Eroded Volcanoes
Impact Structures
The Geographic Cycle of Land-Mass Denudation

Denudation wears down all rock masses exposed at the land surface. However, different types of rocks wear down at different rates. Some erode easily, while others are highly resistant to erosion. Generally, weak rocks will underlie valleys, and strong rocks will underlie hills, ridges, and uplands. This means that the pattern of landforms on a landscape can reveal the sequence of rock layers underneath it. The sequence, in turn, is determined by the type of rock structure present—for example, a series of folds, or perhaps a dome, in a set of rock layers. Thus, there is often a direct relationship between landforms and rock structure.

Repeated episodes of uplift followed by periods of denudation can bring to the surface rocks and rock structures that formed deep within the crust. In this way, even ancient mountain roots may eventually appear at the surface and be subjected to denudation. The subsequent removal of overlying rock layers can also expose batholiths and other igneous rock structures produced by magmas that cool at depth.

Rock structure controls not only the locations of uplands and lowlands, but also the placement of streams and the shapes and heights of the intervening divides. A distinctive assemblage of landforms and stream patterns develops on each of the major types of crustal structures. Recall that landforms are classified as either initial or sequential in origin. Initial landforms are produced directly by volcanic activity, folding, and faulting. Denudation soon converts these initial landforms into sequential landforms, the shape, size, and arrangement of which are controlled by the underlying rocks and their structure. This chapter discusses the sequential landforms that arise through erosion of rock structures.

ROCK STRUCTURE AS A LANDFORM CONTROL

As denudation progresses, landscape features develop according to patterns of bedrock composition and structure. Figure 17.1 shows five layers of sedimentary rock, together with a mass of much older igneous rock. The sediments were originally deposited in horizontal layers on top of the igneous rock, but folding and tilting has occurred in an ancient orogeny. The diagram shows typical landforms associated with these rock types.

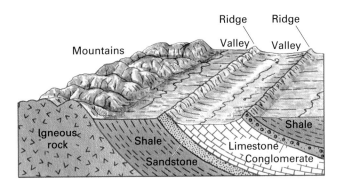

17.1 Landforms and rock types Landforms evolve through the slow erosional removal of weaker rock, leaving the more resistant rock standing as ridges or mountains. The cross-section shows conventional symbols used by geologists for each rock type. (Drawn by A. N. Strahler)

Shale is a weak rock that erodes easily and forms low valley floors. Limestone, subjected to solution by carbonic acid in rainwater and surface water, also forms valleys in humid climates. In arid climates, however, limestone is a resistant rock and usually stands high to form ridges and cliffs. Sandstone and conglomerate are typically resistant to denudation and form ridges or uplands. As a group, the igneous rocks are also resistant—they typically form uplands or mountains. Although metamorphic rocks in general are more resistant to denudation, the resistance of individual metamorphic rock types—marble or schist, for example—varies.

Figure 17.2 illustrates the important rock structures and their accompanying landforms. The resistant beds form ridges or belts of hills, while weaker rocks form lowlands or valleys. The upper four drawings show landforms and rock structures of sedimentary origin. The structures include horizontal layers, gently sloping layers, and layers that are folded or warped into domes. The remaining drawings show fault blocks, plutons, metamorphic belts, and eroded volcanoes.

The broad features of landscapes controlled by the rock structures shown in Figure 17.2 are particularly visible from orbiting spacecraft. *Focus on Remote Sensing 17.1 • Remote Sensing of Rock Structures* provides some satellite images in which rock structures are clearly visible.

STRIKE AND DIP

Most rock layers are not flat but are tilted at an angle. This is readily seen in bedding layers of sedimentary rock strata (Figure 17.3). The tilt and orientation of a natural rock plane are measured against a horizontal plane. Figure 17.4 shows a water surface resting against tilted sedimentary strata. The water surface can be used as the horizontal plane. The angle formed between the rock plane

17.2 Landforms associated with various rock structures (Drawn by A. N. Strahler)

FOCUS ON REMOTE SENSING | 17.1   **Remote Sensing of Rock Structures**

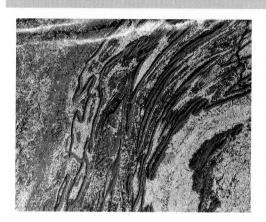

Figure 1 Ridge-and-valley landscape (Courtesy Earth Satellite Corporation)

With its 30-by-30-metre spatial resolution and 175-by-175-kilometre image size, the Landsat satellite provides a perspective from space that captures a large region in a single view. This scale is ideal for revealing the relationship between landforms and rock structure.

Figure 1 shows the ridge-and-valley country of south-central Pennsylvania in a colour-infrared presentation. The land surface has zigzag ridges formed by bands of hard quartzite. The strata were crumpled into folds during a continental collision that took place more than 200 million years ago to produce the Appalachian Mountains. In this September image, the red colour depicts natural

vegetation, while the blue colours identify mainly agricultural fields. Condensation from passing aircraft formed the cloud bands.

Kauai (Figure 2) is the oldest and most heavily dissected of the shield volcanoes that make up the island chain of Hawaii. The radial drainage pattern of streams and ridge crests leading away from the central summit is a primary feature of this image. The intense red colours of the northern and western slopes identify lush vegetation, watered by orographic rainfall from the prevailing northeast trade winds. Note the cumulus clouds that have appeared on the lower slopes and ridge crests where the orographic effect is strong. On the west side of the island is an arid region of rain shadow, with bright green tones identifying the Waimea Canyon. In this colour infrared image, the bright green corresponds to red rocks.

The Brandberg Massif (Figure 3), a dome-shaped plateau in Namibia, covers an area of approximately 650 square kilometres. This granitic pluton is about 120 million years old and rises 1,900 metres above the surrounding arid plains, with its summit at 2,573 metres above sea level. The steep slopes are devoid of soil and much of the surface consists of relatively unbroken granite sheets. Because of its height, rainfall is somewhat more frequent than elsewhere in the region, and during occasional storms, runoff will flow in the ravines that radiate from the centre.

The Manicouagan Impact Structure (Figure 4) in Quebec originated some 214 million years ago from the impact of an asteroid about 5 kilometres in diameter. The original diameter of the crater was

Figure 3 This image of the Brandberg Massif in Namibia was acquired by Landsat 7's Enhanced Thematic Mapper plus (ETM+) sensor.

about 100 kilometres, but scouring by subsequent glacial activity has greatly altered its form. The central area is of igneous and metamorphic rocks. Evidence of impact comes from shattercones—conical fracture surfaces ranging in size from a few centimetres to over a metre—and the presence of glass-like maskelynite produced from altered feldspars. The unusual annular lake was formed following construction of a hydroelectric dam.

Figure 2 Kauai, Hawaii (NASA/JPL)

Figure 4 This natural-colour image of the Manicouagan Impact Structure region in Quebec was acquired on June 1, 2001, by the MISR (Multi-Imaging SpectroRadiometer) nadir (vertical-viewing) sensor aboard NASA's Terra platform.

17.3 Tilted Rock strata

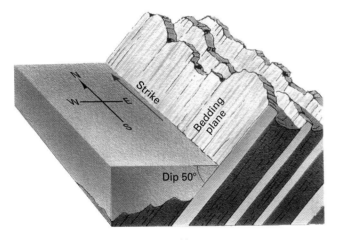

17.4 Strike and dip (Drawn by A. N. Strahler)

LANDFORMS OF HORIZONTAL STRATA AND COASTAL PLAINS

Extensive areas of the ancient continental shields are covered by thick sequences of horizontal sedimentary rock layers that were deposited in shallow inland seas at various times over the past 600 million years, since the Precambrian Era. Uplifted with little disturbance other than minor crustal warping or faulting, these areas became continental surfaces underlain with horizontal strata.

ARID REGIONS

In arid climates, where vegetation is sparse and the action of overland flow especially effective, sharply defined landforms develop on horizontal sedimentary strata (Figure 17.5). The normal sequence of landforms is a sheer rock wall, a *cliff*, which develops at the edge of a resistant rock layer. At the base of the cliff is an inclined slope, which flattens out into a plain beyond.

Erosion strips away successive rock layers, leaving behind a broad platform capped by hard rock layers. This platform is usually called a **plateau**. Cliffs retreat as near-vertical surfaces because storm runoff and channel erosion rapidly washes away the weak clay or shale formations exposed at the cliff base. When undermined, the rock in the upper cliff face repeatedly breaks away along vertical fractures.

Cliff retreat produces a **mesa**, essentially a small plateau bordered on all sides by cliffs. Mesas represent the remnants of a formerly extensive layer of resistant rock. As the retreat of the surrounding cliffs reduces the size of a mesa, it maintains its flat top. Eventually, it becomes a small steep-sided hill known as a **butte** (Figure 17.6). The resistant rock layers crop out at different levels, typically producing a series of steps. Further erosion may produce a single, tall column before the landform is totally consumed.

and the horizontal plane is called the **dip**. The amount of dip is stated in degrees, ranging from 0 for a horizontal rock plane to 90 for a vertical rock plane. In this example, the dip is 50°. The line of intersection between the inclined rock plane and the horizontal plane is oriented according to the points of a compass. This compass direction gives the **strike** of the rock, which by convention is always measured relative to north. For the example shown in Figure 17.4, the strike is along a north-south line and would be written as 360°. Of course, in reality there is rarely a convenient water surface available, hence strike is defined as the compass direction, relative to north, of the line formed by the intersection of a rock layer with an imaginary horizontal plane.

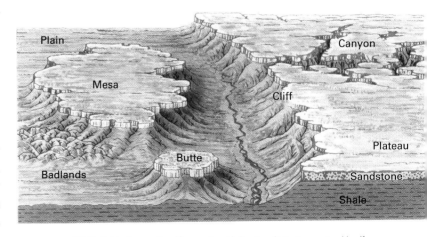

17.5 Arid climate landforms In arid climates, distinctive erosional landforms develop in horizontal strata. (Drawn by A. N. Strahler)

17.6 Monument Valley Mesas and buttes have developed in the shale, siltstone, and sandstone in Monument Valley on the borders of Utah and Arizona. The area is part of the extensive Colorado Plateau. The predominant red colour comes from iron oxide in the weathered rocks. *Fly By: 37° 03' N; 110° 06' W*

DRAINAGE PATTERNS ON HORIZONTAL STRATA

The stream **drainage pattern** that develops on a land mass is often related to the underlying rock type and structure (Figure 17.7). Regions of horizontal strata show broadly branching stream networks formed into a *dendritic drainage pattern* (Figure 17.8). This type of drainage pattern has a number of interesting properties, two of which are described in *Working It Out 17.2 • Properties of Stream Networks.*

COASTAL PLAINS

Coastal plains are found along passive continental margins that are largely free of tectonic activity. They are underlain by nearly horizontal strata that slope gently toward the ocean. Figure 17.9a shows a coastal zone that has recently emerged from beneath the sea. Formerly, this zone was a shallow continental shelf that accumulated successive layers of sediment brought from the land and distributed by currents. On the newly formed land surface, streams flow directly seaward, down the gentle slope. A stream

17.7 Drainage patterns Distinctive stream drainage patterns develop in response to local geologic factors. The resulting channel patterns are classified on the basis of their form and density. (a) Dendritic drainage, (b) Dome drainage, (c) Trellis drainage, (d) Radial drainage

(a)

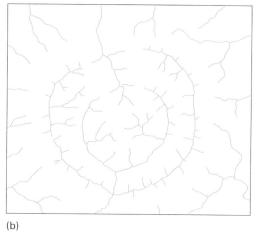
(b)

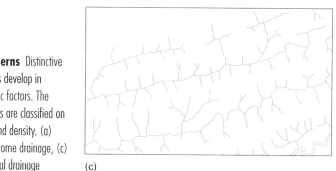
(c)

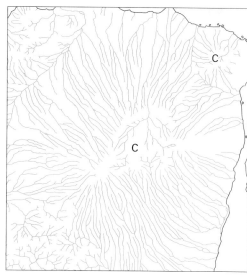
(d)

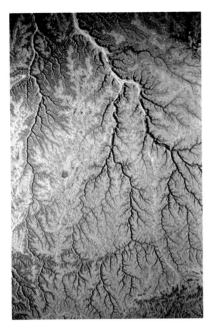

17.8 Dendritic drainage This branching dendritic drainage pattern is developed on the tertiary shale and limestone of the Hadhramut Plateau in Yemen. *Fly By: 15° 25' N; 48° 28' E*

that follows the initial gradient of a new land surface in this way is called a *consequent stream*. Consequent streams form on many kinds of initial landforms, such as volcanoes, fault blocks, or beds of drained lakes.

In an advanced stage of coastal plain development, a new series of streams and topographic features form (Figure 17.9b). Where more easily eroded strata (usually clay or shale) are exposed, denudation is rapid, making *lowlands*. Between them rise broad belts of hills called **cuestas**. Cuestas are commonly underlain by layers of sand, sandstone, limestone, or chalk that dip away from the

cuesta at a low angle. The lowland lying between the area of older rock—the oldland—and the first cuesta is called the *inner lowland*. Streams that develop along the trend of lowlands parallel with the shoreline are known as *subsequent streams*. They take their position along any belt or zone of weak rock and therefore closely follow the pattern of rock exposure. Subsequent streams occur in many regions and are commonly associated with folds, domes, and faults.

The drainage lines on a fully dissected coastal plain combine to form a *trellis drainage pattern* (Figure 17.7c). In this pattern, the tributaries of a main stream are arranged at right angles. The subsequent streams are at about right angles to the consequent streams.

LANDFORMS OF FOLDED ROCK LAYERS

Areas underlain by gently upwarped or downwarped rock layers form domes and basins. More intense folding produces the wavelike strata of anticlines and synclines.

SEDIMENTARY DOMES

A distinctive land mass type is the **sedimentary dome**, a circular or oval structure in which strata have been forced upward at its centre. Sedimentary domes can occur where igneous intrusions at great depth have caused the overlying sediments to bow upwards. In other cases, upthrusting on deep faults may have been the cause. Figure 17.10 illustrates a sedimentary dome's erosion features. Strata are first removed from the summit region of the dome, exposing older strata beneath. Eroded edges of steeply dipping strata form sharp-crested sawtooth ridges called *hogbacks*.

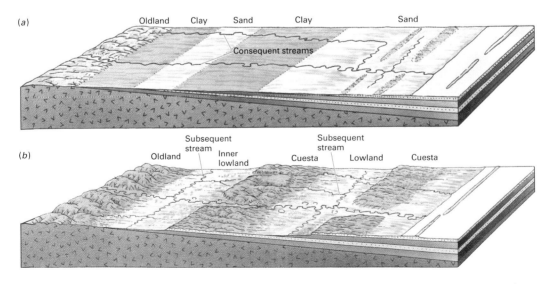

17.9 Development of a broad coastal plain (a) Early stage—the plain recently emerged (b) Advanced stage—cuestas and lowlands developed (Drawn by A. N. Strahler)

17.10 Dome erosion Erosion of sedimentary strata from the summit of a dome structure near Sinclair, Wyoming. The strata are partially eroded, forming an encircling hogback ridge. The massive triangular slabs forming the sawtooth pattern are often referred to as flatirons.

17.12 Water gap The Big Horn River has cut though the Sheep Mountain anticline in the Big Horn Basin, Utah, to expose the up-arched sedimentary rock strata.

The stream network on a deeply eroded sedimentary dome shows dominant subsequent streams forming a circular system, called an *annular drainage pattern* (Figure 17.7b). The shorter tributaries make a radial arrangement. The total pattern resembles a trellis pattern bent into a circular form.

FOLD BELTS

Figure 17.11 shows the distinctive landforms resulting from fluvial denudation of anticlines and synclines in folded strata. Deep erosion of simple, open folds produces a **ridge-and-valley landscape.** Weaker rocks such as shale and limestone erode away, leaving hard strata, such as sandstone or quartzite, to stand in bold relief as long, narrow ridges. Note that anticlines are not always ridges. If a resistant rock type at the centre of the anticline erodes

through to reveal softer rocks underneath, an *anticlinal valley* may form (Figure 17.11a). A *synclinal mountain* is also possible. It occurs when a resistant rock type is exposed at the centre of a syncline, and the rock stands up as a ridge (Figure 17.11b). The geometry of the folds may also contribute to the differential resistance of the rock mass; strata in a syncline tend to be compressed, while in an anticline, extensional forces may weaken and fracture the rock.

On eroded folds, the stream network is distorted into a trellis drainage pattern (Figure 17.7c) characteristically composed of long, parallel subsequent streams occupying the narrow valleys associated with the weaker rock. In a few places, a larger stream may cut across a ridge in a deep, steep-walled *water gap* (Figure 17.12).

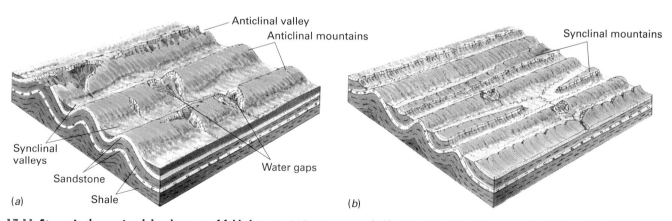

17.11 Stages in the erosional development of folded strata (a) Erosion exposes a highly resistant layer of sandstone or quartzite, which controls much of the ridge-and-valley landscape. (b) Continued erosion partly removes the resistant formation but reveals another below it. (Drawn by A. N. Strahler)

WORKING IT OUT | 17.2   Properties of Stream Networks

The erosion and deposition of streams that are connected into networks produce fluvial landforms. For this reason, the properties of stream networks have always been of interest to geomorphologists, who study many of these properties with reference to stream order. Chapter 16 noted that the un-branched headwater segments of a drainage network are first-order streams. When two first-order streams join, the downstream segment is of second order. The junction of sec-ond-order segments form third-order stream segments, and so forth.

In a given drainage basin, there are many more first-order segments than second-order; similarly, there are more sec-ond-order segments than third-order, and so on. Thus, the number of segments decreases with increasing stream order. First-order stream segments are on average shorter than sec-ond-order segments; third-order segments are longer still, and so forth. Thus, a higher-order stream segment will have to flow for a longer distance before joining another segment of the same order.

The table (below) provides data from the Allegheny River in Pennsylvania for an area upstream from a particular point on a seventh-order segment of the network. To provide the in-formation, detailed topographic maps were carefully studied over a large area. Each stream segment was identified, and its length measured and tabulated by stream order.

Allegheny River drainage basin characteristics

Stream Order	Number of Segments	Bifurcation Ratio	Mean Segment Length (km)	Cumulative Mean Segment Length (km)	Length Ratio
u	N_u	R_b	L_u	L_u^*	R_L
1	5,966		0.15	0.15	
		3.9			3.2
2	1,529		0.48	0.63	
		4.0			2.7
3	378		1.29	1.9	
		5.7			3.1
4	68		4.0	5.9	
		5.3			2.8
5	13		11.3	17.2	
		4.3			2.8
6	3		32.2	49.4	
		3.0			
7	1		13+		
		$\overline{R_b} = 4.37$			$\overline{R_L} = 2.92$

Looking first at the number of segments, note that the value drops quickly as order increases. However, the ratio of the number of segments of one order to the next is roughly constant. This ratio is called the *bifurcation ratio* and is shown in the third column. For example, the ratio of first- to second-order segment numbers is 5,966/1,529 = 3.9, meaning that each second-order segment branches to form 3.9 first-order segments, on average. In this drainage basin, the bifurcation ratio varies from 3.0 to 5.7; the average value is 4.37. In alge-braic notation, the bifurcation ratio, R_b, is

$$R_b = N_u/N_{u+1}$$

where N_u is the number of segments of stream order u and N_{u+1} is the number of segments of stream order $u + 1$.

Part (a) of the figure on the opposite page plots the number of segments against stream order for the Al-legheny River. This graph is *semi-logarithmic* (*semi-log*): that is, the x-axis is linear, while the y-axis is logarithmic. Note that each point falls close to a straight line. This line follows an equation of the form

$$N_u = R_b^{k-u} \quad (1)$$

where k is the order of the largest stream in the network be-ing considered—7 in the case of the Allegheny River drainage network in the table. Thus, the equation for the Allegheny is

$$N_u = 4.37^{(7-u)}$$

using the average bifurcation ratio, 4.37, for R_b. With this for-mula, the number of segments expected for stream orders 1 to 7 will be 6,964, 1,594, 365, 83, 19, 4, and 1 respectively.

The figure also shows data for a fifth-order stream net-work in the Big Badlands region of South Dakota. For this stream network, the average bifurcation ratio is 3.47. The Big Badlands region differs considerably from the Allegheny River environment in that a given area of easily eroded badland ter-rain has many more stream channels. However, the lines plot-ted for Big Badlands and the Allegheny River are close to parallel, indicating that they have similar average bifurcation ratios. In fact, bifurcation ratios for all natural stream networks tend to range between 3 and 5, no matter what type of sub-strate is subjected to fluvial action. This observation was first made by the hydraulic engineer Robert E. Horton, and thus

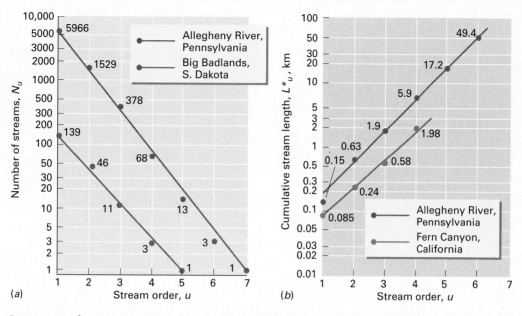

Stream network properties Numbers of segments (a) and mean segment length (b) plotted against stream order (Data from M. E. Morisawa, K. G. Smith, and J. C. Maxwell)

equation (1) above is sometimes referred to as Horton's law of stream numbers.

The average length of stream segments is expected to increase with stream order. The table confirms this, with mean lengths for the Allegheny ranging from 0.15 kilometres for first-order segments to 32.2 kilometres for sixth-order segments. As in the case of the bifurcation ratio for stream segment numbers, length ratio R_L for mean length of segments from two adjacent orders can be defined as

$$R_L = L_{u+1}/L_u$$

where L_u is the mean length of segments of order u. As the table shows, values of R_L are rather similar to values of R_b. For the Allegheny data, the average length ratio is $R_L = 2.92$.

Like the number of stream segments, the progression of lengths also closely follows a straight line on a semi-log graph, provided that it is plotted as the cumulative mean length. This value, shown in the fifth column of the table, is derived from

$$L_u^* = \sum_{i=1}^{u} L_u$$

where L_u^* is the cumulative mean length of segments of order u, and i is a whole number counting from 1 to u.

The equation for the straight line that connects values of Lu^* on a semi-log graph is

$$L_u^* = L_1 R_L^{(u-1)} \text{ (2)}$$

where L_1 is the mean length of first-order segments. For the case of the Allegheny, the table shows that $L_1 = 0.15$ kilometres. Using the average length ratio $R_L = 2.92$, this gives

$$L_u^* = 0.15 \times [2.02^{(u-1)}]$$

Part (b) of the figure plots the cumulative mean lengths for the first six orders of the Allegheny stream network. The points lie close to a straight line. Also shown are data from a fourth-order stream network at Fern Canyon, California. The fact that both lines are nearly parallel shows that cumulative mean length tends to increase by the same factor despite differences in climate, substrate, and rock type. Horton also derived equation (2), which is called Horton's law of stream lengths.

17.14 Plunging folds Folded strata with crests that plunge downward give rise to zigzag ridges following erosion, as illustrated by Sheep Mountain near Lovell, Wyoming. The ridge in the centre of the photograph is a plunging anticline; the valley in the left foreground is a plunging syncline. *Fly By: 44° 39' N; 108° 12' W*

17.13 Eroded anticline The axis of the Salt Anticline, in the Paradox Basin region in southeastern Utah and southwestern Colorado, lies more or less parallel to the surface.

The folds illustrated in Figure 17.11 are continuous and even-crested. They produce ridges and valleys that are approximately parallel and continue for great distances (Figure 17.13). In some fold regions, however, the axis of the folds is not parallel to the surface, so they rise or descend to form *plunging anticlines* and *plunging synclines*. Plunging folds give rise to a zigzag line of ridges (Figure 17.14), as can be seen in *Focus on Remote Sensing 17.1 • Remote Sensing of Rock Structures.*

LANDFORMS DEVELOPED ON OTHER LAND-MASS TYPES
EROSION FORMS ON FAULT STRUCTURES

Active normal faulting produces a sharp surface break—a *fault scarp*—that can create a rock cliff hundreds of metres high (Figure 17.15). Erosion quickly modifies a fault scarp, but because the fault plane extends down into the bedrock, its effect on erosional landforms persists for long spans of geologic time. Thus, a fault scarp produced by fault movement is distinct from a *fault-line scarp*, which is produced by differential erosion of rocks weakened by faulting, or where faulting brings softer rocks in contact with harder rocks. Because the fault plane is a zone of weak rock that has been crushed during faulting, it is often occupied by a subsequent stream.

EXPOSED BATHOLITHS

Batholiths, huge intrusions of igneous rock that are formed deep below the Earth's surface, are sometimes uncovered by erosion of the overlying rock materials. Because batholiths are typically com-

posed of resistant rock, they erode into hilly or mountainous uplands. Batholiths of granitic composition are a major component in the mosaic of ancient shield rocks. For example, they are found throughout the Canadian Shield.

One of the largest single batholiths in North America, the Coast Range batholith, encompasses an area of 182,500 square kilometres that extends from northwestern Washington State to the Yukon. The bedrock is mainly granite that was intruded in

17.15 Erosion of an uplifted fault block The near-vertical face of Mount Yamnuska in Alberta marks a fault scarp along which the rock has been uplifted.

17.16 Exfoliation dome Sheets of granite fracture by expanding when removal of overlying rock reduces pressure. The loosened slabs fall away to form an apron around the base of the dome.

17.17 Spheroidal weathering of an exhumed, granite intrusion Through a combination of mechanical and chemical processes, this jointed granite mass has eroded into a collection of rounded boulders. This formation, located in Australia's Northern Territories, is known as Devil's Marbles.

the Middle Jurassic and early Tertiary periods, 170 to 45 million years ago. Between the intrusions are zones of metamorphosed, folded, and faulted volcanic and sedimentary rock. The granitic rocks are thought to be the roots of deeply eroded volcanoes. The presence of younger volcanoes, such as Mount Baker in northwestern Washington and Mount Garibaldi in British Columbia, reflect the volcanic origin of the landscape. Batholiths are typically associated with rich mineral deposits. For example, the Guichon Creek batholith near Kamloops, although only about 1,000 square kilometres, is the principal copper ore reserve in British Columbia.

Small bodies of granite, representing domelike projections of batholiths that lie below, are often found surrounded by ancient metamorphic rocks into which the granite was intruded. These prominent features are known as **monadnocks** or **inselbergs**. Sugar Loaf Mountain in Rio de Janeiro is an example. A monadnock develops because it is more resistant to erosion than the bedrock of the surrounding area. The distinctive rounded form develops as the outermost layers of rock break apart and slide off to expose the new layer within (Figure 17.16).

The rounded shape is best maintained where the exposed rock lacks fractures, allowing weathering processes to act uniformly on its surface. Rock breakup is mainly due to the reduction in pressure that occurs as overlying rock is removed. This causes the granite to expand and fracture into large sheets that eventually fall away. These distinctively rounded features are called exfoliation domes. Small intrusions may be reduced to a collection of granite boulders known as a **kopje**. They are especially common in arid regions, such as Southern Africa and Australia (Figure 17.17). Similar rock outcrops formed by weathering of hill summits in Britain are called *tors*.

DEEPLY ERODED VOLCANOES

Volcanoes are divided into two general types: stratovolcanoes and shield volcanoes (see Chapter 13). Stratovolcanoes are typically constructed by explosive eruptions resulting in layers of viscous andesite lava flows and tephra, while shield volcanoes are built primarily of thin, runny flows of basalt. Thus, these two types of volcanoes produce different types of landforms as they erode away.

The distinctive initial landform produced by a stratovolcano is the volcanic cone. Once formed, the cone is typically dissected by stream action. The system of streams on a dissected cone forms a *radial drainage pattern* (Figure 17.7d). This is a consequent stream network because it is mainly determined by the slope of the initial land surface. When the volcanic episode ends and the volcano becomes dormant or extinct, a lake will usually occupy the caldera. These features are well illustrated by Mount Ruapehu, an active stratovolcano on the North Island of New Zealand (Figure 17.18).

17.18 Mt. Ruapehu, New Zealand This image is made from topography data collected by the Shuttle Radar Topography Mission aboard the space shuttle Endeavour, launched on February 11, 2000, and imagery collected by the Landsat satellite on October 23, 2002.

17.19 Volcanic neck Shiprock in New Mexico is a volcanic neck enclosed by a weak shale formation. The peak rises about 520 metres above the surrounding plain. In the foreground and to the left, wall-like dikes extend far out from the central peak (see also Figure 11.13).

Mount Ruapehu (2,797 metres) is located at the southern end of the Taupo Volcanic Zone (TVZ), an active crustal subsidence zone that is prone to faulting. Water from Ruapehu has created a classic radial drainage pattern. A crater lake has formed at the summit, which contains 8–10 million cubic metres of acidic water. Deposits of erupted ash and rock can at times cause the lake to fill above the solid rim of the caldera. During subsequent eruptions, water expelled from the lake can breach these temporary dams and mix with rock debris to form devastating mudflows or **lahars**.

Water-saturated lahars move great quantities of debris over long distances in a short time period. Debris accumulations can be several hundred metres thick, so they destroy or smother anything in their path. Lahars have been common throughout Ruapehu's history. The most active periods of lahar deposition occurred about 75,000 to 65,000 years ago and 23,000 to 14,000 years ago, at a time when cooler climatic conditions favoured glacial activity, increasing the supply of rock debris. Ten major eruptions have occurred since 1861, and in that same period, there is evidence of at least 60 lahars. Figure 17.18 shows the deeply carved path of a lahar.

With continued erosion, an extinct stratovolcano may be gradually reduced to a remnant *volcanic neck*, formed of lava that solidified in the pipe of the volcano. Radiating from it are wall-like dikes created from magma that filled radial fractures around the volcano's base. Ship Rock, New Mexico is a classic example of a volcanic neck with radial dikes (Figure 17.19). Ship rock rises 520 metres above desert plain. It is mainly composed of volcanic breccia that formed about 1,000 metres below the surface. Six dikes radiate from neck dikes. They are composed mainly of alkali feldspar; biotite and phlogopite (potassium magnesium aluminum silicate hydroxide) micas; and diopside, a member of the pyroxene group (see Chapter 11). Radiometric dating indicates that the dikes formed about 27 million years ago.

A similar landform of volcanic origin is Devils Tower, Wyoming, which rises 382 metres above the sedimentary rocks that surround it (Figure 17.20). This steep-sided igneous feature may also be an erosional remnant of a volcanic neck, although its exact origin is subject to debate. Like Ship Rock, it is made of magma that solidified at a shallow depth below the surface; it is dated at 40 million years old. The resistant rock is fine-grained phonolite, an extrusive rock that is rich in potash feldspar. The distinctive, predominantly hexagonal columns formed as the rock cooled and contracted.

Shield volcanoes, such as those in Hawaii, show erosion features that are quite different from those of stratovolcanoes. Radial consequent streams cut deep canyons into the flanks of an extinct shield volcano, and these open out into deep, steep-walled amphitheaters, such as Waimea Canyon on the island of Kauai (Figure 17.21). Over time, the original surface of the shield volcano is entirely obliterated, leaving a rugged mountain mass composed of sharp-crested divides and deep canyons (see *Focus on Remote Sensing 17.1 • Remote Sensing of Rock Structures*).

IMPACT STRUCTURES

An interesting group of landscape features are those that have originated from the impact of meteors or other extraterrestrial bodies. One of the best examples is the Barringer Meteorite Crater, also known as Meteor Crater, near Flagstaff, Arizona (Figure 17.22). The crater is 1.2 kilometres in diameter and 200 metres deep. It was formed by the impact of a nickel-iron meteorite approximately 49,000 years ago. The typical impact speed of a

17.20 Devils Tower, Wyoming This igneous remnant comprises fine-grained rock that cooled and contracted to give a distinctive columnar structure. *Fly By: 44° 5' 24" N; 104° 42' 55" W*

17.21 Advanced erosion of a shield volcano Waimea Canyon, over 760 metres deep, has eroded into the flank of an extinct shield volcano on Kauai, Hawaii. Gently sloping layers of basaltic lava flows are exposed in the canyon walls.

17.22 Barringer Meteorite Crater near Flagstaff, Arizona The nickel-iron meteorite that created this crater is estimated to have been 50 metres in diameter and was travelling at about 13 kilometres per second. Its impact weight was about 150,000 tonnes. Some fragments of the meteorite have been recovered, but most of it appears to have vapourized on impact.

meteorite, about 20 kilometres per second, produces a crater that is 10 to 20 times larger than the meteorite's diameter. Most craters are bowl-shaped depressions around which the ejected material is deposited. Rocks below the crater are significantly altered and usually consist of a shallow layer of breccia made up of coarse, angular fragments of the original rocks. The impact has fractured the bedrock at greater depths. Metamorphism through impact melting is also evident.

About 200 impact structures have been identified on the Earth, 29 of which are in Canada (Figure 17.23). The largest on record is at Sudbury, Ontario, with an approximate diameter of 250 kilometres. This impact is dated at 1,850 million years ago. The Manicouagan impact site (see *Focus on Remote Sensing 17.1 • Remote Sensing of Rock Structures*) is about 100 kilometres in diameter and formed about 214 million years ago. Another large crater formed near Chicxulub, Mexico, about 65 million years ago, following the impact of a meteorite that is estimated to have been about 10 kilometres in diameter. At the time of its formation, the crater measured nearly 100 kilometres in diameter and was 14 kilometres deep. However, subsequent collapse of the perimeter wall increased the crater's diameter to more than 150 kilometres. It is possible that this impact event may have caused a significant change in the Earth's climate and is perhaps linked to the extinction of the dinosaurs.

THE GEOGRAPHIC CYCLE OF LAND-MASS DENUDATION

The Earth's landscapes range from mountain regions of steep slopes and rugged peaks to regions of gentle hills and valleys, to extensive, nearly flat plains. Given that flowing water is the prin-

cipal agent of land-mass denudation, one way to view these landscapes is as stages in a cycle that begins with rapid uplift and follows with erosion by streams. This cycle, called the **geographic cycle**, was first described by William Morris Davis. The landscape in the cycle's initial phase, called the *youthful stage*, consists of many drainage basins and their branching stream networks. The region is rugged, with steep slopes and high, narrow crests (Figure 17.24a).

After initial uplift, the main streams draining the region establish a graded condition. They transport rock debris out of each drainage basin at the same average rate as the debris is being contributed from the land surfaces within the basin. Eventually, the debris movement lowers the land surface, and the average elevation steadily declines. This decline is accompanied by a reduction

17.23 Earth impact sites The distribution of known impact sites comes from a program that began in 1955 at the Dominion Observatory in Ottawa. The Planetary and Space Science Centre at the University of New Brunswick currently manages it.

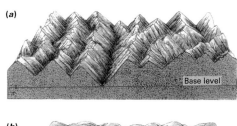

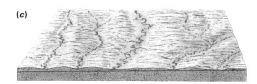

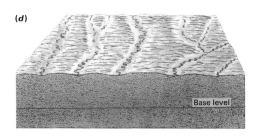

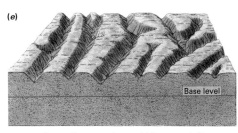

17.24 The geographic cycle (a) In the youthful stage, relief is great, slopes are steep, and the erosion rate is rapid. (b) In the mature stage, relief is greatly reduced, slopes are gentle, and the erosion rate is slow. Soils are thick over the broadly rounded hill summits. (c) In old age, after many millions of years of fluvial denudation, a peneplain forms. Slopes are very gentle, and the landscape is an undulating plain. Flood plains are broad, and the stream gradients are extremely low. The entire land surface lies close to base level. (d) The peneplain is uplifted. (e) Streams trench a new system of deep valleys in the phase of land-mass rejuvenation. (Drawn by A. N. Strahler)

in the average gradients of all streams. As the sharp peaks and gorges of the youthful stage give way to rounded hills and broad valleys, the geographic cycle enters the *mature stage* (b).

As time passes, the gradients of streams and valley-side slopes of the drainage basins gradually become lower. In theory, the ultimate goal of the denudation process is to reduce the land mass to a featureless plain at sea level. In this process, sea-level, which is considered to be at a lower elevation than the land mass, represents the lowest level, or *base level*, of fluvial denudation. But because the rate of denudation becomes progressively slower, the land surface approaches base-level at an increasingly slower pace. In reality, the ultimate goal can never be reached. Instead, after the passage of some millions of years, the land surface is reduced to a gently rolling surface of low elevation, called a **peneplain** (c). With the evolution of the peneplain, the landscape has reached *old age*.

Production of a peneplain requires a high degree of crustal and sea-level stability for a period of many millions of years. One region that has been cited as a possible example of a contemporary peneplain is the Amazon-Orinoco Basin of South America. This vast region is a stable continental shield of ancient rock with very low relief.

Inevitably, crustal deformation will occur, and this may uplift the peneplain (Figure 17.24d). The base level is now far below the land surface. Streams begin to cut into the land mass and carve deep, steep-walled valleys (e). This process is called *rejuvenation*. With the passage of many millions of years, the landscape will be carved into the rugged stage shown in (a), and the cycle will repeat.

While Davis's idealized geographic cycle is useful for understanding landscape evolution over very long periods, it does little to explain the diversity of the features observed in real landscapes. Most geomorphologists now approach landforms and landscapes from the viewpoint of *equilibrium*. This approach explains a landform as the product of forces acting upon it, including both forces of uplift and denudation, with the characteristics of the rock material playing an important role. Thus, we find steep slopes and high relief where the underlying rock is strong and highly resistant to erosion. Even a "youthful" landscape may be in an enduring equilibrium state in which hill slopes and stream gradients remain steep in order to maintain a graded condition while eroding resistant rock, such as granite.

Another problem with Davis's geographic cycle is that it applies only where the land surface is stable over long periods. However, crustal movements are frequent on the geologic time scale, and few regions of the land surface remain untouched by tectonic forces. Recall also that continental lithosphere floats on a soft asthenosphere. As erosion strips layer upon layer of rock from a land mass, the land mass becomes lighter and is buoyed upward. The process of crustal rise in response to unloading is known as *isostatic compensation*. Thus, a more appropriate model of landscape development is one in which tectonic processes operate continuously and erosional processes adjust accordingly, rather than as a sudden event followed by denudation.

A Look Ahead

The landscapes of most regions on the Earth's land surface are produced by fluvial processes acting on different rock types. Hence, running water is by far the most important agent of denudation. The three remaining agents of denudation are waves, wind, and glacial ice, which the final two chapters of Part 4 discuss.

CHAPTER SUMMARY

- This chapter has added a new dimension to the realm of landforms produced by fluvial denudation—the variety and complexity introduced by differences in rock composition and crustal structure. Streams carving up a land mass are controlled to a high degree by the structure of the rock on which they act. In this way, distinctive drainage patterns evolve.

- In arid regions of horizontal strata, resistant rock layers produce vertical *cliffs* separated by gentler slopes on less resistant rocks. The drainage pattern is *dendritic*. Where strata are gently dipping, as on a coastal plain, the more resistant rock layers stand out as cuestas, interspersed with lowland valleys on weaker rocks. *Consequent streams* cut across the cuestas toward the ocean, and *subsequent streams* follow valleys. This forms a *trellis drainage pattern*.

- Where rock layers are arched upward into a dome, erosion produces a circular arrangement of rock layers outward from the dome's centre. Resistant strata form *hogbacks*, and weaker rocks form lowlands. The drainage pattern is *annular*.

- In fold belts, the sequence of *synclines* and *anticlines* brings a linear pattern of rock layers to the surface. Resistant strata form ridges, and weaker strata form valleys. Like the coastal plain, the drainage pattern will be trellised.

- Faulting provides an initial surface along which rock layers move—the *fault scarp*. This feature can persist as a fault-line scarp long after the initial scarp is gone.

- Exposed batholiths are often composed of uniform, resistant igneous rock. They erode to form a dendritic drainage pattern. Monadnocks of intrusive igneous rock stand up above a plain of weaker rocks.

- Stratovolcanoes produce lava flows that initially follow valleys but are highly resistant to erosion. At the last stages of erosion, all that remains of stratovolcanoes are necks and dikes.

- Meteor craters that originated from the impact of extraterrestrial meteorites and asteroids are found throughout the world. Dust clouds resulting from such impacts might be associated with climate perturbations and changes in the Earth's plant and animal life.

- Denudation of the Earth's landscape is primarily by flowing water. The geographic cycle of land-mass denudation was an early concept that suggested landform development begins with tectonic uplift followed by a sequence of features that are produced mainly by fluvial erosion. The cycle progresses through youth, maturity, to old age, by which time the land surface has levelled to a peneplain. The cycle is renewed when the land is once again uplifted.

- The geographic cycle has been superceded by the equilibrium approach, in which landforms are viewed as the product of uplift and denudation—continuous processes that act on rocks of varying resistance to erosion.

KEY TERMS

butte	drainage pattern	lahar	plateau
coastal plain	geographic cycle	mesa	ridge-and-valley landscape
cuesta	inselberg	monadnock	sedimentary dome
dip	kopje	peneplain	strike

REVIEW QUESTIONS

1. Why is there often a direct relationship between landforms and rock structure? How are geologic structures that formed deep within the Earth exposed at the surface?

2. Which of the following types of rocks—shale, limestone, sandstone, conglomerate, and igneous rocks—tends to form lowlands? Uplands?

3. How are the tilt and orientation of a natural rock plane measured?

4. How are coastal plains formed? Identify and describe two landforms found on coastal plains. What drainage pattern is typical of coastal plains?

5. What types of landforms are associated with sedimentary domes? How are they formed? What type of drainage pattern would a dome have and why? Provide an example of an eroded sedimentary dome and describe it briefly.

6. How does a ridge-and-valley landscape arise? Explain the formation of the ridges and valleys. What type of drainage pattern is found on this landscape?

7. What type of landform(s) might be associated with a fault? Why?

8. Do shield volcanoes erode differently from stratovolcanoes? How?

9. Describe the evolution of a fluvial landscape according to the geographic cycle. What is meant by rejuvination?

10. What is the equilibrium approach to landforms? How does it differ from interpretation using the geographic cycle?

VISUALIZATION EXERCISES

1. Identify and sketch a typical landform of flat-lying rock layers found in an arid region.

2. Sketch a ridge-and-valley landscape, identifying the following features: syncline, anticline, synclinal valley, anticlinal mountain, anticlinal valley, water gap.

ESSAY QUESTION

1. Imagine the following sequence of sedimentary strata: sandstone, shale, limestone, and shale. What landforms would you expect to develop in this structure if the sequence of beds is (a) flat-lying in an arid landscape; (b) slightly tilted as in a coastal plain; (c) folded into a syncline and an anticline in a fold belt; and (d) fractured and displaced by a normal fault? Use sketches in your answer.

2. What does the cycle of land-mass denudation mean? Describe the processes and stages that were originally incorporated into this cycle. How do modern views differ from those proposed earlier?

PROBLEMS WORKING IT OUT 17.2 Properties of Stream Networks

1. A sixth-order stream network has an average bifurcation ratio of 3.5. How many stream segments would you expect for streams of orders 1–6?

2. A fourth-order stream network has an average mean segment length ratio of 3.2, and the mean length of first-order stream segment is 0.12 kilometres. What are the expected cumulative mean lengths for stream segments of orders 2–4? Note

that the (non-cumulative) mean length for order u is simply the difference between the cumulative length at order u and the cumulative length at order u-l. That is, $L_u = L_u^* - L_{u-1}^*$. Using this relationship, derive the expected (non-cumulative) mean lengths for segments of orders 2–4.

Find answers at the back of the book

EYE ON THE LANDSCAPE |

Chapter Opener Entrenched meanders on the San Juan River, Utah. The horizontally laid sandstone strata (**A**) are the dominant structural element in this region. However, gradual uplift of the plateau, accompanied by erosion by the river has resulted in lowering the main channel to form entrenched (or incised) meanders. Some lateral erosion occurs on the outer bends of the meanders which maintains the near-vertical gorge-like wall. Deposition occurs on the inner bank (**B**) and as channel bars (**C**) at times of reduced flow. Differential rates of weathering and erosion caused by varying resistance of the rock strata create step-like features (**D**) often seen in dry climates. The importance of flowing water is seen in the dissected slopes and small alluvial fans (**E**) which have accumulated at the base of the slopes.

Monument Valley, Utah. The extensive plateau that formerly occupied this region has been dissected into a series of mesas (**A**)

and buttes (**B**). The eroded sediment has mostly been carried away by water and wind, but a thin sandy soil supports a sparse cover mainly of coarse grasses and sagebrush (**C**). Rockfalls (**D**) accumulate at the base of the cliffs, but the general absence of an apron of debris suggests these mass-wasting episodes occur infrequently.

Chapter 18

EYE ON THE LANDSCAPE

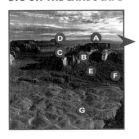

Arid landforms—a flat-topped mesa.
**What else would the geographer
see? . . . Answers at the end of the
chapter.**

THE WORK OF WIND AND WAVES

The Work of Wind in Terrestrial Environments

Transport by Wind

Dust Storms

Erosion by Wind

Depositional Landforms Associated with Wind

Sand Dunes

Types of Sand Dunes

Coastal Foredunes

Loess

The Work of Wind in Coastal Environments

Wind and Waves

The Work of Waves

Marine Erosion

Marine Transport and Deposition

Littoral Drift and Shore Protection

The Tides

The Physics of Tides

Tide Cycles and Amplitudes

Tidal Currents

Types of Coastlines

Coastlines of Submergence

Emergent Coastlines

Neutral Coastlines

A Closer Look:

Eye on Global Change 18.1 • *Global Change and Coastal Environments*

Organic Coastlines

Fault Shorelines and Marine Terraces

Rising Sea Level

Geographers at Work • *Dr. Marilyne Jollineau*

The rotation of the Earth on its axis combined with unequal solar heating produces a system of global winds in which air moves as a fluid across the Earth's surface. This motion of air produces a frictional drag that can transport surface materials. When winds blow across areas not protected by vegetation, they can pick up fine particles and carry them long distances. Winds can also move coarser particles, such as sand, forming dunes. Similarly, when winds blow over broad expanses of water, they generate waves. Waves then expend their energy at coastlines, eroding rock, moving sediment, and creating distinctive landforms. Another agent that acts in coastal regions is the tide—the rhythmic rise and fall in sea level generated by the gravitational attraction of the Sun and Moon. The movement of tides is especially important in estuaries.

THE WORK OF WIND IN TERRESTRIAL ENVIRONMENTS

The geomorphic work of wind in terrestrial environments is collectively referred to as **aeolian** processes, and includes erosion, transportion, and deposition of Earth materials. The ability of winds to carry out geomorphological work is related to the density of air, the velocity at which it moves, and the size and quantity of mineral particles that are being transported. Wind is generally not powerful enough to remove material from the surface of moist soils, or from soils protected by vegetation. However, it is an effective geomorphic agent in arid and semi-arid regions where vegetation is sparse and there is an abundant supply of unconsolidated surface materials. Wind is also effective in coastal environments, where beaches provide abundant supplies of loose sand.

Wind velocities increase rapidly with height above the surface. Close to the ground, there is a thin zone in which air movement is negligible. This zone of little or no wind is called the *laminar sublayer* and its thickness depends on the roughness of the surface. For soils, the depth of the laminar sublayer is about 0.03 of the average diameter of mineral grains. For example, the zone would be 1 millimetre thick on a surface covered with pebbles with an average diameter of 30 millimetres.

Air flow is quite turbulent, and in the lowest few metres of the atmosphere air typically rises at a velocity that is approximately one fifth of the average wind speed. However, in the laminar sublayer, laminar flow replaces turbulent flow and air flow is parallel to the surface. Hence, the thickness of the stationary air layer at the ground surface, coupled with the average velocity of upward motion in overlying air, affects wind's ability to transport sediment.

In order to move Earth materials, winds must exceed a threshold velocity, which varies according to the size of the mineral particles (Figure 18.1). **Threshold velocity** is the velocity required to entrain a particle of a given diameter and increases with the square root of the particle size. Thus larger, heavier particles require a higher wind velocity to move them. However, clay particles tend to become cohesively bonded into larger aggregates. As a result, the threshold velocity required to entrain clay is similar to that of sands. For this reason, the optimum particle size for wind erosion is about 0.08 millimetres, which is the size of very fine sand and silt.

The ability of wind to erode is termed **wind erosivity**. Erosivity is expressed mathematically as

$$E = V^3 \rho$$

where E is erosivity, V is wind velocity, and ρ is air density (1.2 kilogram per cubic metre).

Wind erosivity is an exponential function of wind velocity. Hence, if the wind velocity doubles, from say 5 to 10 metres per second, erosivity increases 8-fold; an increase in wind speed from 5 to 15 metres per second would result in a 27-fold increase in erosivity. Thus, the action of wind in removing and transporting sediment is facilitated by strong, regular prevailing winds.

Sand grains, up to 2 millimetres in diameter, are normally the largest particles that can be kept in suspension by strong winds, of say 14 metres per second (50 kilometres per hour). Particles of this size will quickly return to the surface as the wind speed drops.

The maximum rate at which a wind-transported particle settles through the air is called the terminal fall velocity, which can be calculated from Stoke's Law as

$$w = \frac{2(\rho_p - \rho_a)gr^2}{9\mu}$$

where w is terminal fall velocity, ρ_p is density of particle, ρ_a is density of air, g is acceleration due to gravity, r is radius of the particle, and μ is viscosity of the air.

The terminal fall velocity for clay is less than 0.0001 metres per second; for silt, it ranges from 0.001 to about 0.08 metres per second; and for coarse sand, it can be as high as 10 metres per second. Hence, the larger the particle, the greater the wind speed required to keep it moving above the ground surface.

TRANSPORT BY WIND

Wind transports Earth materials in different ways depending on their size and mass. The smallest and lightest particles are held in **suspension** in the atmosphere. Materials carried in suspension are typically less than 0.2 millimetres in diameter. This is the maximum size that can be picked up under normal wind conditions near the Earth's surface. Once aloft, wind can transport these materials great distances from their source, before precipitation or coagulation remove them from the atmosphere. Severe windstorms can keep larger particles aloft in turbulent eddies. Most of the silty loess deposits found extensively in North America, Europe, and China originated from wind-transported debris left by the ice sheets of the late-Cenozoic Ice Age.

When wind reaches a critical velocity of about 4.5 metres per second, larger particles such as sand become entrained in the air flow and begin to roll through **traction**. Strong winds can move small pebbles in this way. Traction accounts for about 20 to 25 percent of material moved by wind. Wind alone cannot lift sand grains from the surface, but some are thrown into the air through collision. Once the sand particles are airborne, they follow curving paths determined by the horizontal wind velocity and the force of gravity. This process, termed **saltation**, accounts for 75 to 80 percent of sediment transport by wind.

Saltation usually lifts sand grains a centimetre or so above the surface, and moves them downwind a few centimetres at a time, in a series of little jumps (Figure 18.2). The rate at which the particles move is about 30 to 50 percent of the wind's speed. When a particle strikes the surface, the force of impact is transferred to another particle, which may in turn be lifted into the air or moved forward through **creep**.

Wind blowing across loose sand creates a rippled surface of crests and troughs that are perpendicular to the wind direction (Figure 18.3). The distance between adjacent crests corresponds to the average distance travelled by airborne grains during saltation.

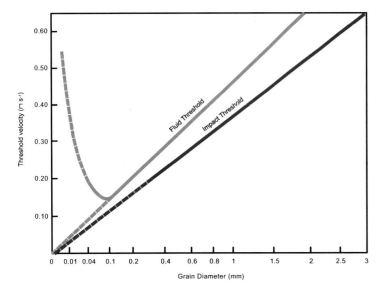

18.1 Geomorphic work of wind Threshold velocities for wind erosion, transportation, and deposition for surface materials of different diameters.

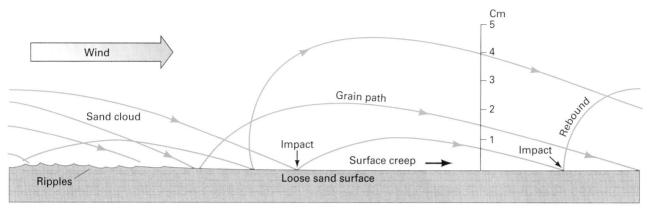

18.2 Saltation Sand particles travel in a series of long leaps. (After R. A. Bagnold, *The Physics of Blown Sand and Desert Dunes*, London: Methuen, 1941)

18.3 Sand ripples The rippled surfaces of sand deposits form at right angles to the prevailing winds; the strength of the wind determines the scale of these features.

Sorting occurs in these microtopographic features, with the coarsest materials collecting at the crests. This is opposite to the sorting that occurs in dune formation, where the coarsest materials generally settle in the troughs.

DUST STORMS

Even in severe sandstorms, sand particles generally move within a metre of the surface and rarely higher than two metres. However, strong, turbulent winds blowing over barren surfaces can lift finer materials hundreds of metres, forming a dense **dust storm** (Figure 18.4). In semi-arid grasslands, a dust storm is generated where cultivation or grazing has stripped ground surfaces of protective vegetation cover. The bouncing motion of soil particles under strong winds disturbs the soil surface, and with each impact, fine dust is released and carried aloft by turbulence.

18.4 Dust storm A cloud of fine dust sweeps across this savanna plain in eastern Kenya.

18.6 Blowout in a sand dune The dune surface is susceptible to deflation as soon as the protective vegetation cover is removed.

18.5 Atmospheric dust transport Dust from the Sahara region of North Africa is carried westward over the Azores and across the Atlantic Ocean.

A severe dust storm approaches as a dense cloud extending from the surface to heights of several thousand metres. Light intensity within dust clouds can decrease considerably, and visibility can diminish to a few metres in the fine, choking dust. Winds blowing out of deserts, such as the Sahara, are often heavily laden with fine particles (Figure 18.5). As noted in Chapter 9, the harmattan is a hot, dusty wind that affects West Africa as prevailing wind direction changes with the seasonal movement of the ITCZ.

EROSION BY WIND

Dust picked up by the wind is removed from the surface through **deflation**; this is the fundamental process of aeolian erosion. As well as lifting particles into the air, deflation also moves loose material along the ground. In dry climates, much of the ground surface is susceptible to deflation because of the sparse vegetation cover. In the same way, materials are removed from dry river beds, beaches, and areas of recently formed glacial deposits.

Induced deflation is a frequent occurrence when short-grass prairie in a semi-arid region is cultivated without irrigation.

Plowing disturbs the natural soil surface and grass cover, and in drought years, when vegetation dies out, the unprotected soil is easily eroded by wind. Much of the Great Plains region of North America has suffered such deflation, experiencing dust storms generated by turbulent winds.

Human activities in very dry regions contribute measurably to the creation of high dust clouds. For example, in the Thar Desert of northwest India and Pakistan, the continued trampling of fine-textured soils by grazing animals and humans produces a semi-permanent dust cloud that extends to a height of nine kilometres.

A typical landform produced by deflation is a shallow depression called a *deflation hollow*. These subtle topographic features may range from a few metres to a kilometre or more in diameter, and are usually only a few metres deep. In some areas, such as sand dunes, a small depression can form a **blowout** where the plant cover has been broken or disturbed (Figure 18.6).

Wind deflation is selective. The finest particles, those of clay and silt, are lifted and raised into the air, but sand grains travel within a metre or two of the ground. Gravel fragments and rounded pebbles can be rolled or pushed over flat ground by strong winds, but they do not travel far. They become easily lodged in hollows or between other large grains. Deflation will remove fine particles, leaving coarser, heavier materials behind as *lag deposits*. In this way, rock fragments ranging in size from pebbles to small boulders are progressively concentrated on the surface, and eventually form what is known as **desert pavement**.

Desert pavement acts as an armour that effectively protects the finer particles from deflation (Figure 18.7), although the stony surface is easily disturbed. Almost half of the Earth's desert surfaces are stony deflation zones. A surface of this kind in the Sahara is called a *reg*. In Australia, these landforms are known as *gibber plains*. The Gobi Desert of central Asia gets its name from

18.7 Desert pavement A desert pavement formed of closely fitted rock fragments is produced through the progressive removal of the finer particles by deflation.

18.8 Ventifacts Rocks etched by wind-blown particles in the Dry Valleys of McMurdo Sound, Antarctica.

the rounded pebbles, or *gobi*, that are found there. The term **hammada** describes a desert surface from which wind has removed most of the regolith, leaving only bedrock surfaces scattered with large rocks.

These stony and rocky landscapes are in marked contrast to the extensive sand seas or *ergs* that form in the Sahara and other desert environments. Sand sheets are level to gently undulating areas covered with sand grains that are too large to be moved by saltation under the prevailing wind conditions. They account for approximately 40 percent of aeolian depositional surfaces. One of the largest, the Selima Sand Sheet in southern Egypt and northern Sudan, occupies 60,000 square kilometres.

A dark, blackish-purple shiny stain, called *desert varnish*, is often found on the surfaces of desert rocks that are no longer etched by sand grains. Desert varnish chemically consists mainly of manganese, iron oxides, and clay minerals. It was previously thought that the shiny coating was formed by colonies of bacteria that preferentially concentrated manganese in their tissues. However, recent studies suggest that silica dissolved from other minerals is responsible for the glaze that forms over the rocks.

Particles moved by the wind are effective at wearing away exposed rock surfaces, and wind **abrasion** has produced several features in arid environments. On the smallest scale, rock surfaces are pitted and grooved by particles driven against them by the wind. Rocks that have been cut, and sometimes polished, by the abrasive action of wind are called *ventifacts* (Figure 18.8). They are generally formed from hard, fine-grained rocks. Smaller stones often have flattened surfaces or facets. The facets are cut in sequence, and as the stone becomes undercut, it topples over and exposes another surface to the wind. Often they develop three curved facets, giving them a somewhat triangular appearance.

Stones that acquire this form are called *dreikanters*. At a larger scale are undercut *pedestal rocks* (Figure 18.9). *Yardangs*, tens of metres high and several kilometres long, are formed where wind abrasion has sculpted outcropping bedrock.

DEPOSITIONAL LANDFORMS ASSOCIATED WITH WIND

Wind-transported particles remain suspended until the wind velocity drops; until the particles collide and stick together, which increases their mass; or until precipitation washes them from the atmosphere. In arid areas, deposition is initiated primarily by small obstructions, such as plants and stones, which increase surface roughness and reduce wind velocity. Small mounds of sand will form as particles and settle out on the lee side of such

18.9 Pedestal rocks Pedestal, or mushroom, rocks are undercut by the abrasive action of wind-blown sand. The upper part of the rock is above the zone of abrasion, and its rough appearance contrasts with the smoother rock at the base.

18.10 Dune formation With a sufficient supply of sand, a small mound that forms in the shelter of a desert shrub can grow into a large dune.

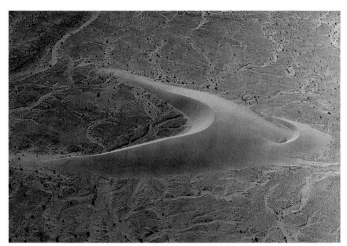

18.11 Migrating barchan dunes This aerial view shows a large barchan dune moving from right to left. At its apex is a smaller barchan dune that in time will merge with it. *Fly By: 21° 56' N; 55° 15' E*

obstructions (Figure 18.10). Once formed, the mound will further disrupt air flow and, with a sufficient supply of sand, it may grow into a large dune.

SAND DUNES

A **sand dune** is any mound of loose sand molded by the wind. Dunes form where there is a source of sand; for example, a sandstone formation that weathers easily to release individual grains, or perhaps a beach supplied with abundant sand from a nearby river mouth. Dune sand is most commonly composed of quartz, which is extremely hard and largely immune to chemical decay. Dunes must be free of a vegetation cover in order to form and move, and active dunes constantly change shape depending on wind strength and sand supply. They become inactive when stabilized by vegetation or when wind patterns or sand sources change. Active and stabilized sand dunes cover about 26,000 square kilometres in Canada. Most of these are found in the Prairie provinces; about 45 percent of the total occurs in Alberta, 36 percent in Saskatchewan, and 10 percent in Manitoba. The dune areas in the Prairie provinces originated from the exposed deposits left behind by the continental ice sheets.

TYPES OF SAND DUNES

A common type of sand dune is the isolated **barchan**, or *crescentic*, **dune**. Barchan dunes usually rest on a flat, pebble-covered ground surface and tend to be arranged in chains extending downwind from the sand source. The life of a barchan dune may begin as a sand drift on the lee side of some obstacle, such as a small hill, rock, or shrub. Once a sufficient mass of sand has formed, it begins to move downwind, taking on a distinctive crescent shape with the points directed downwind (Figure 18.11). On the upwind side of the dune, the sand slope is gentle and

smoothly rounded. On the downwind side, within the crescent, is the steep *slip face*.

Wind-blown sand moves up the gentle upwind side of the dune by saltation or creep to form a high, sharp crest (Figure 18.12). From this crest, the grains move on to the steep slip face of the dune, where they come to rest. The slip face maintains a more or less constant angle from the horizontal, which is known as the *angle of repose*. The accumulation of individual wind-carried sand grains makes the slip face increasingly steeper until it becomes unstable and the outermost layer of sand on the face slips down the dune slope, restoring the angle of repose. For loose sand, this angle is about 35°.

The rate at which sand moves over a dune surface increases exponentially with wind velocity (Figure 18.13). Sand transport is negligible at wind velocities below 20 kilometres per hour, and even at 50 kilometres per hour, it amounts to less than 500 kilograms per day. Although the downwind progress of dunes is imperceptible, over the course of a year, a barchan dune can migrate as much as 25 metres.

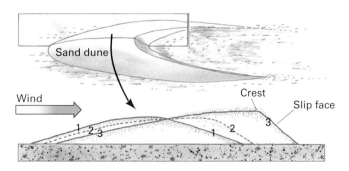

18.12 Growth of a dune and development of a slip face The gradient on the upwind slope is normally 10–12°, compared with about 35° on the steep downwind slope. (After R. A. Bagnold)

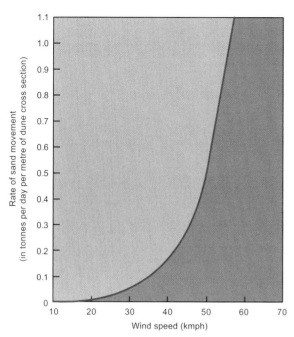

18.13 Rate of sand transport at different wind velocities Except under strong winds, the rate of sand transport is low, hence dunes progress downwind very slowly.

Where sand is so abundant that it completely covers the solid ground, dunes take the form of wave-like ridges separated by trough-like furrows. These dunes are called **transverse dunes** because, like ocean waves, their crests trend at right angles to the direction of the dominant wind (Figure 18.14). The entire area may be called a *sand sea*. The sand ridges have sharp crests and are asymmetrical; the gentle slope is on the windward side and the steep slip face on the lee side. Deep depressions lie between the dune ridges. Sand seas require enormous quantities of sand, supplied by material weathered from sandstone formations or from sands in nearby alluvial plains. Transverse dune belts also form adjacent to beaches that supply abundant sand and have strong onshore winds.

Another group of dunes belongs to a family called **parabolic dunes**, in which the curve of the dune crest is bowed outward in the downwind direction. This curvature is the opposite of the barchan dune. A common type of parabolic dune is the *coastal blowout dune*, which forms adjacent to beaches. Here, large supplies of sand are available, and prevailing winds blow the sand toward land. Deflation forms a saucer-shaped depression, heaping the sand in a curving ridge. On the land side is a steep slip face that advances over the lower ground and buries the vegetation.

On semi-arid plains where winds are strong, groups of parabolic blowout dunes develop to the lee of shallow deflation hollows. Sand is caught by low bushes and accumulates on a broad, low ridge (Figure 18.15). These dunes have no steep slip faces and may remain relatively immobile. In some cases, the dune ridge migrates downwind, drawing the dune into a long, narrow form with parallel sides, called a **hairpin dune**.

Longitudinal dunes (or *sief dunes*) form where sand supply is more limited. The long, narrow ridges are parallel to the direction of the prevailing wind. These dune ridges may be many kilometres long and cover vast areas of tropical and subtropical deserts in Africa and Australia (Figure 18.16).

Some dunes acquire more elaborate forms. For example, the radially symmetrical *star dunes* are pyramidal sand mounds with slip faces on three or more sides that radiate from the high centre of the mound (Figure 18.17). Star dunes tend to accumulate in areas with multidirectional wind regimes and grow in height rather than move laterally. They have been used for centuries as landmarks in the otherwise featureless landscape of the Sahara and Arabian Deserts.

18.14 Transverse dunes in the Rub' al Khali (Empty Quarter) of the Arabian Peninsula In this largely uninhabited region, prevailing winds from the north have molded the abundant supply of sand into transverse dunes. *Fly By: 19° 14' N; 53° 04' E*

18.15 Active parabolic dune and hairpin dunes, Dongara, Western Australia The active dune lobe is 700 metres wide. An extensive deflation basin has formed upwind. Acacia trees growing on the margins of the lobe trap the sand as the dune migrates downwind, forming the trailing ridges, which in this location have been stabilized as hairpin dunes.

18.16 Longitudinal dunes These parallel dune ridges are in the Empty Quarter of the Arabian Peninsula. *Fly By: 19° 17' N; 48° 20' E*

18.17 Star dunes in Saudi Arabia's Empty Quarter Radially symmetrical star dunes can attain heights of about 500 metres and are the tallest form of dunes. *Fly By: 19° 17' N; 48° 20' E*

COASTAL FOREDUNES

A narrow belt of dunes in the form of irregularly shaped hills and depressions typically develops on the land side of sand beaches. These **coastal foredunes** normally bear a cover of beach grass and a few other species of plants capable of survival in the severe environment (Figure 18.18). The sparse plant cover traps sand moving landward from the adjacent beach. As a result, the foredune ridge is built up several metres above high-tide level.

Foredunes form a protective barrier for the tidal lands that develop behind them. In a severe storm, waves may cut away the upper part of the beach, and although waves may then attack the foredune barrier, they will not usually breach it. Between storms, the beach is rebuilt, and, in due time, wind action restores the dune ridge.

Where the dune vegetation is disturbed, a blowout, or hole in the dune, will rapidly develop and extend as a trench across the dune ridge. Storm waves may then funnel through the gap, spreading sand onto the tidal marsh or lagoon behind the dune.

If badly eroded, the gap can become a new tidal inlet for ocean water.

LOESS

In several large midlatitude areas of the world, the surface is covered by deposits of wind-transported silt that has settled out from dust storms over many thousands of years. This material, known as **loess**, generally has a uniform yellowish colour and lacks any visible layering (Figure 18.19). Loess tends to break away along vertical cliffs wherever it is exposed, such as by the cutting of a stream. It is also easily eroded by running water and is subject to rapid gullying when the protective vegetation cover is removed.

Loess is important to world agriculture and forms the parent matter of the rich black soils particularly suited to the cultivation of grains. The thickest deposits of loess are in northern China, where layers over 30 metres thick are common and a maximum thickness of 100 metres has been measured. Extensive loess

18.18 Coastal foredunes, Queen's County, Prince Edward Island Beach grass thriving on coastal foredunes has trapped drifting sand to produce a dune ridge.

18.19 Wind-transported silt This thick layer of loess in New Zealand was deposited during the late-Cenozoic Ice Age. Loess has excellent cohesion and often forms vertical faces as it wastes away.

deposits are also found in Central Europe, Argentina, and the United States. In the United States, thick loess deposits lie in the Missouri–Mississippi Valley. Large areas of the prairie region of Indiana, Illinois, Iowa, Missouri, Nebraska, and Kansas are underlain by loess ranging in thickness from 1 to 30 metres. Extensive deposits also occur in Tennessee and Mississippi in areas bordering the lower Mississippi River flood plain, and in the Palouse region of northeast Washington and western Idaho.

The American and European loess deposits are directly related to the continental glaciers of the late-Cenozoic era. At the time when the ice covered much of North America and Europe, a generally dry winter climate prevailed in the land bordering the ice sheets. Strong winds blew southward and eastward over the bare ground, picking up silt from the flood plains of braided streams that discharged the meltwater from the ice.

THE WORK OF WIND IN COASTAL ENVIRONMENTS

WIND AND WAVES

A wave is a way of transmitting energy from one place to another. The energy in ocean waves is derived from the wind. The size and strength of an ocean wave is determined by wind speed and direction, and by the **fetch** or size of the expanse of water over which the wind acts. A longer fetch and greater wind speed produces more powerful waves.

The condition of the sea is the basis for the descriptive Beaufort scale, originally devised in 1805. As currently used, the Beaufort scale is somewhat modified, with the addition of wind speeds and wave heights (Table 18.1).

Table 18.1 The Beaufort Scale is an empirical scale that is based on the observable effects that winds of different strength have on the ocean surface (on land the effects on vegetation and other features is used)

Force	Wind (km/h)	World Meteorological Office (WMO) Classification	Appearance of Wind Effects	
			On Water	**On Land**
0	Less than 2	Calm	Sea surface smooth and mirror-like	Calm, smoke rises vertically
1	2–6	Light Air	Scaly ripples, no foam crests	Smoke drift indicates wind direction, wind vanes are still
2	7–11	Light Breeze	Small wavelets, crests glassy, no breaking	Wind felt on face, leaves rustle, vanes begin to move
3	12–19	Gentle Breeze	Large wavelets, crests begin to break, scattered whitecaps	Leaves and small twigs constantly move, light flags are extended
4	20–30	Moderate Breeze	Small waves, 0.3–1.2 metres, become longer, numerous whitecaps	Dust, leaves, and loose paper are lifted, small tree branches move
5	31–39	Fresh Breeze	Moderate waves, 1.2–2.4 metres, take a longer form, many whitecaps, some spray	Small trees in leaf begin to sway
6	40–50	Strong Breeze	Larger waves, 2.4–4.0 metres, whitecaps common, more spray	Larger tree branches move, whistling in wires
7	51–61	Near Gale	Sea heaps up, waves 4.0–6.1 metres, white foam streaks off breakers	Whole trees move, resistance felt walking against wind
8	62–74	Gale	Moderately high (4.0–6.1-metre) waves of greater length, edges of crests begin to break into spindrift, foam blows in streaks	Trees swaying, some resistance when walking into wind
9	75–87	Strong Gale	High waves (6.1 metres), sea begins to roll, dense streaks of foam, spray may reduce visibility	Slight structural damage occurs, slate blows off roofs
10	88–102	Storm	Very high waves (6.1–9.1 metres) with overhanging crests, sea white with densely blown foam, heavy rolling, lowered visibility	Rarely experienced on land, trees broken or uprooted, considerable structural damage
11	103–117	Violent Storm	Exceptionally high (9.1–13.7-metre) waves, foam patches cover sea, visibility more reduced	
12	117+	Hurricane	Air filled with foam, waves over 13.7 metres, sea completely white with driving spray, visibility greatly reduced	

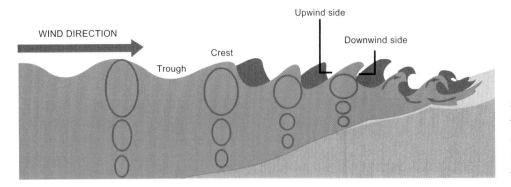

WIND DIRECTION

Trough

Crest

Upwind side

Downwind side

18.20 Deep-water waves appear to carry water forward as they pass. In fact, the water molecules move in a circular motion and maintain their general position. Closer to shore, drag on the sea bed causes waves to steepen, fall forward, and rush up the beach.

When the wind blows across an initially calm surface, local pressure differences generated by small eddies begin to disturb the water surface and generate small ripples, or *cat's paws*. As the ripples grow, the wind begins to exert a shear stress on the windward side, and also creates a pressure differential across the growing waves, which furthers their development.

Waves are classified into two types: *wind* (or *gravity*) *waves* and *swells*. Wind waves are short-lived waves created by local wind conditions. When the wind drops, the waves subside and disappear. These short, choppy waves have a chaotic appearance and tend to move in many directions. Usually, smaller waves are superimposed on the larger ones. While wind waves tend to be locally generated, swells come from the deep sea, and are produced by strong, persistent winds blowing across long stretches of water. Swells do not diminish when the wind dies, and can travel thousands of kilometres from their origin. For this reason, sizeable waves can still be seen, even when there is little or no wind. In a swell, the wave crests have a long period and may take 10–15 seconds to pass. Swells are not steep and appear to be slow-moving because of their lengths. Wind waves tend to be relatively steep and typically last for only 4 to 6 seconds. They appear to be fast-moving, because their crests are closely spaced; however, this is not the case (Table 18.2).

In the open sea, waves travel through water, but do not push the water forward as they pass. As a wave arrives, it lifts the water molecules, which then move forward, down, and back so that each particle completes a circle. The circling motion near the surface sets off smaller circling motions below it (Figure 18.20).

THE WORK OF WAVES

Throughout this section the term **shoreline** refers to the shifting line of contact between water and land. The broader term **coastline** includes the shallow water zone in which waves perform their work, as well as beaches and cliffs shaped by waves, and coastal dunes. The coastlines of large lakes exhibit many features and processes similar to those of marine coastlines. In North America, the Great Lakes provide many examples of coastal processes.

The most important agent shaping coastal landforms is wave action. Waves travel across the deep ocean with little loss of energy, but when they reach shallow water, the drag of the sea floor slows and steepens them. However, the top of the wave maintains its forward velocity and eventually falls down, creating a *breaker*, which surges forward up the beach. The total energy of a wave is proportional to twice the square of its height. Since wave height is limited by the available kinetic energy in the wind, wave energy, which includes both potential and kinetic energy, is roughly proportional to wind speed. This energy is expended primarily in the constant churning of mineral particles as waves break on the shore.

MARINE EROSION

Marine erosion is caused in four general ways. The force of the waves alone and of air trapped and compressed by the weight of the falling water can be sufficient to dislodge Earth materials through hydraulic action. Abrasion occurs when the waves pick up rock fragments and propel them against the shoreline. The rock

Table 18.2 General characteristics of waves (Wave period is the time between successive peaks, wave length is the distance between successive peaks, apparent speed is the rate at which each peak passes, and depth is the depth to which the effects of the wave can be felt. These data are for pure wave forms, which is rarely the case at sea).

Wave Period (seconds)	Wave Length (metres)	Apparent Speed* (kilometres per hour)	Depth (metres)
2	6.1	11.1	3.0
3	14.0	16.7	7.0
4	25.0	22.2	12.5
5	39.0	27.8	19.5
6	56.1	33.3	28.0
7	76.5	38.9	38.1
8	100.0	44.4	50.0
9	126.5	50.0	63.1
10	156.1	55.6	78.0
11	189.0	61.1	94.5
12	224.6	66.7	112.4
13	263.7	72.2	132.0
14	306.0	77.8	153.0
15	351.1	85.2	175.6

* Note, water moves in a circular motion in a wave, but does not move forward.

18.23 Marine erosion In a typical sequence, a headland is first undercut from both sides, creating an arch, and then a stack forms when the arch collapses.

18.21 Retreating shoreline, Rhode Island This marine scarp of Pleistocene sediments is being rapidly eroded, threatening the historic Southeast Lighthouse.

fragments are gradually rounded and reduced in size through attrition as they grind against each other. Finally, the weakly acidic nature of sea water can dissolve certain rocks, such as limestone and chalk, by solution (or corrosion).

Where the coastline comprises unconsolidated material, the force of the forward-moving water alone can easily dislodge it. Under these conditions, the shoreline may recede rapidly, to form a steep bank, or *marine scarp* (Figure 18.21). Pronounced erosion occurs where deep water offshore allows waves to break at the foot of rocky headland with little loss of energy in friction with the sea bed. The impact of the waves causes the headland to be undercut at its base. A **wave-cut notch** marks the line of the most intense wave

erosion (Figure 18.22), and as it increases in size, the unsupported rock above collapses to form a *marine cliff*. The collapsed rock provides temporary protection from the waves, but they eventually wear it down and remove it, and the process continues. In this way, the cliff erodes toward the shore, maintaining its form as it retreats.

More resistant material may form a headland that extends farther out to sea. The softer rock on either side of the headland erodes to form bays where sediment will accumulate in a pocket beach. As the headland becomes more exposed to the waves, the erosion rate increases. However, the approaching waves are also influenced by the configuration of the seabed, which causes them to bend around the headland by refraction. In this way, waves will attack the headland from both sides, creating enlarged *sea caves*. The waves may pierce the headland to form a *marine arch*. Further erosion weakens the arch, which collapses, leaving a *stack* (Figure 18.23). Then, eventually, wave action topples and levels the stack.

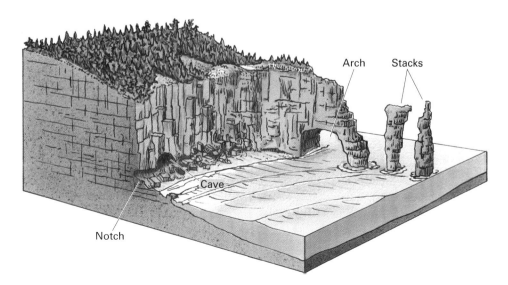

18.22 Landforms of sea cliffs Wave action undercuts the marine cliff, maintaining its form. (Drawn by E. Raisz)

18.24 A wave-cut abrasion platform exposed at low tide As the wave-cut platform gets wider, the energy of the approaching waves is dissipated in the shallow water and eventually the base of the cliff is protected from further erosion. Sub-aerial processes, such as slumping, then reduce the angle of the cliff.

As the sea cliff retreats, continued wave abrasion forms an *abrasion platform* (Figure 18.24). Abrasion continues to erode and widen this almost level rock pavement. However, because the water is shallow on the abrasion platform, much of the wave energy is dissipated as friction and the cliff's rate of retreat slows. The form of the cliff is then increasingly influenced by sub-aerial processes, such as slumping, which slowly reduce its height and abruptness. If a beach is present, it is little more than a thin layer of gravel and cobblestones on top of the abrasion platform.

MARINE TRANSPORT AND DEPOSITION

Much of the material that is deposited in coastal environments is actually supplied by rivers that discharge their sediment loads into the sea. For example, the Fraser River carries about 20 million tonnes of suspended sediment annually, the Mackenzie River about 100 million tonnes, and the St. Lawrence about 6.5 million tonnes. In comparison, the average annual sediment load delivered to the Gulf of Mexico by the Mississippi River is more than 200 million tonnes. To this is added the rock debris derived directly from marine erosion. Once in the sea, waves and currents transport the finer materials, until they are temporarily deposited as a **beach**, sand bar, or similar feature. Although many beaches are formed of particles of fine to coarse quartz sand, some beaches are composed of rounded pebbles or cobbles. Still others are formed from fragments of volcanic rock, as in Hawaii, or even from shells and broken coral.

The alternate landward and seaward flow of water generated by breaking waves moves beach materials laterally on the beach face as *beach drift* (Figure 18.25a). When a breaker collapses, a foamy, turbulent sheet of water washes up the beach slope. This *swash* is a powerful surge that causes materials to move toward land. The water returns to the sea as the *backwash*. The direction

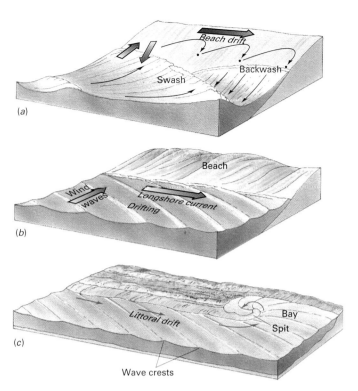

18.25 Coastal sediment transport (a) Swash and backwash move particles along the beach in beach drift. (b) Waves set up a longshore current that moves particles by longshore drift. (c) Littoral drift, produced by these two processes, creates a sandspit.

that waves advance toward a shore is mainly determined by where they originated relative to the coastline. Typically waves approach the shore obliquely. After the wave has spent its energy, the backwash flows perpendicularly down the beach slope, and particles come to rest down the beach from their initial position. Over time, the particles travel long distances along the shore, thus the wave movement becomes a significant form of sediment transport.

Beaches form where sand is in abundant supply, and although they may appear to be permanent features, the sand is continually moving along the beach face. In addition, more powerful waves will gouge the sand and carry it offshore, where they deposit it as a sand bar and carry it back and forth into the deeper water. Later smaller constructional waves bring it back to the beach. In this way, a beach may retain a fairly stable configuration over many years. As well, with this back and forth movement, sediment is also carried along in the shallow water offshore.

As waves approach a shoreline at an angle, a current is set up parallel to the shore. This is known as a *longshore current* (Figure 18.25b). When wave and wind conditions are favourable, this current is capable of carrying sand along the sea bed in a process called *longshore drift*. Beach drift and longshore drift act together to move particles in the same direction. The combined process is called **littoral drift** (Figure 18.25c)

18.26 Sandspit This sandspit is growing in a direction toward the observer. Several examples occur around Cape Cod, Massachusetts. *Fly By: 41° 36' N; 70° 00' W*

Littoral drift is a constructive process that shapes shorelines in two ways. Materials are moved along the beach in the direction of the prevailing winds. Where a bay exists at the end of a relatively straight stretch of coastline, the sand is carried out into open water as a long finger, or *sandspit* (Figure 18.26). As the sandspit grows, it can form a barrier, or *bar*, across the mouth of the bay. Beaches typically disappear at headlands and promontories, where wave action is intense and littoral drift quickly moves sediment away. Waves are refracted around headlands, causing sediment to accumulate as crescent-shaped pocket beaches (Figure 18.27).

Coastal erosion, deposition, and littoral drift are part of a larger system involving the movement of sediment brought to the sea by rivers and ultimately deposited on the ocean floor. At the mouth of a river, a beach will broaden in response to an influx of sediment and remain wider for some distance along the coast as the extra sediment is carried in the littoral current. Beaches even-

18.27 Pocket beaches On an embayed coast, sediment is carried from eroding headlands into the bays, where pocket beaches are formed. Examples on Vancouver Island, British Columbia can be seen at *49° 11' N; 126° 01' W.*

tually end where a submarine canyon interrupts the flow of sand in the littoral current. Offshore submarine canyons cross the continental shelf and extend out to great depths. Sediment is carried into these canyons, where it forms undersea fans and cones along the continental rise. Many submarine canyons were cut by streams during the late-Cenozoic Ice Age, when sea level was 100 metres or more below present levels.

This process is particularly important along the Pacific coast of North America. On the Atlantic coast, the continental shelf is wider and there are fewer submarine canyons. A notable exception is the Hudson Canyon at the southwest end of Long Island, New York. Consequently, east coast beaches tend to be wider, and barrier islands commonly run parallel to the coast. On the east coast, sand is lost primarily during hurricanes and other storms. Sand from the barrier islands is either pushed inland or transported far offshore where it settles on the sea bed.

The flow of sediment from rivers is an important part of this so-called coastal sediment cell. Many coastal communities have experienced beach recession because the rivers that feed their coastlines have been dammed, trapping sediment upstream and keeping it out of the flow system. With less sediment flowing through the coastal system as a whole, less is available for storage on the beach, and cliffs are more rapidly cut back.

LITTORAL DRIFT AND SHORE PROTECTION

When sand arrives at a particular section of the beach more rapidly than it is carried away, the beach widens through the process of **progradation**. A beach will prograde if wave action decreases, depriving littoral drift of its energy. The reverse process, **retrogradation**, occurs when sand is removed more rapidly than it is brought in, narrowing the beach. Along stretches of shoreline affected by retrogradation, the beach may be seriously depleted or even disappear entirely.

In some circumstances, structures can be installed to cause progradation. For example, *groynes*—barriers of large rocks, concrete, or wood, built at right angles to the shoreline—installed at close intervals along the beach will trap sediment moving along the shore as littoral drift (Figure 18.28).

THE TIDES
THE PHYSICS OF TIDES

An ocean tide refers to the cyclical rise and fall of the ocean's surface as a result of the gravitational potential of the Moon and Sun. Tide propagation and amplitude are influenced by the rotation of the Earth and the shapes and depths of the ocean basins.

According to Isaac Newton's law of gravity, the gravitational force exerted by a celestial body such as the Moon or Sun is

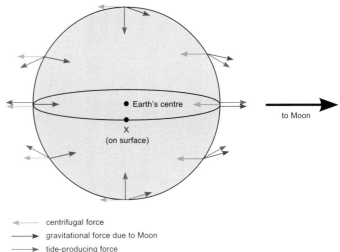

18.28 Groynes constructed across a beach Groynes are strong barriers, usually constructed from wood or stones, that are designed to reduce longshore transport of sand along exposed beaches.

proportional to its mass but inversely proportional to the square of the distance. The equation is

$$F = G\frac{m_1 m_2}{r^2}$$

where F is the magnitude of the gravitational force (N), G is the gravitational constant (6.67×10^{-11} N m^2 kg^{-2}), m_1 is the mass of the first object (i.e., the Earth) in kilograms, m_2 is the mass of second object (i.e., the Moon) in kilograms, and r is the distance between objects in metres.

The mass of the Sun is 1.98892×10^{30} kilograms, which is considerably greater than the Moon's mass of 7.36×10^{22} kilograms. The mean distance between the Sun and the Earth is 149.5 million kilometres, compared with 385,000 kilometres between the Moon and the Earth. Hence, the gravitational force of the Sun on the Earth is about 178 times that of the Moon. However, tides are not produced by the absolute pull of gravity exerted by the Sun and Moon, but by the differences in the gravitational fields, or *gravity gradient*, produced by these two bodies across the Earth's surface.

The gravity gradient changes at a rate that is inversely proportional to the cube of the distance. Because the Moon is closer to the Earth, its gravitational force field varies much more strongly than that of the Sun. Consequently, the gravitational component of the Sun's **tide-generating force** is about 46 percent of that of the Moon. Given that the Earth's diameter is 12,680 kilometres, a point on the Earth's surface nearest the Moon is approximately 378,660 kilometres away, compared with 391,340 kilometres for a point on the opposite side of the Earth. This means that the gravitational force of the Moon is about 3.5 percent greater on the side of the Earth nearer the Moon.

In addition to gravity, the tide-generating force includes a centrifugal component caused by the revolution of the Earth

18.29 The tide-generating force The centrifugal force due to the Earth's revolution around the Sun is the same for all locations, whereas the gravitational force exerted by the Moon is directed toward the centre of the Moon. The tide-generating force is the result of these two components.

around the Sun. The centrifugal force experienced by all points on the Earth is the same in magnitude and direction, whereas the gravitational force exerted by the Moon always points to the centre of the Moon (Figure 18.29). The net effect is that on the side of the Earth toward the Moon, the gravitational force exceeds the centrifugal component, and on the side of the Earth away from the Moon, the centrifugal component exceeds the gravitational force. The tide-generating force acts predominantly in the vertical direction, but also contains a horizontal component termed the *tractive force*.

The Earth's rotation on its axis does not alter the net effect of the Earth's revolution on the gravitational and centrifugal forces. The only effect that rotation has on the tides is that it moves the entire tide-generating force field around the Earth each day. This relates to another important concept associated with tides—that of the **geoid**. The geoid refers to the shape that the Earth would assume if it were entirely covered with water and responding to the forces acting upon it. In this hypothetical situation, sea level would represent the height of the water created by the resultant forces created by revolution, gravitational attraction, and rotation. The geoid is deformed as the forces acting upon it change.

At the point on the Earth directly beneath the Moon (the sub-lunar point), the lunar gravitational force is greatest, and consequently, the ocean surface is pulled and rises. On the side facing away from the Moon (the antipodal point), the Moon's gravity is at its weakest, which effectively allows the water to move away from the Earth. Assuming that the Earth is completely covered with water and responded instantly to the changing forces acting upon it, this would create what is called the **equilibrium tide**.

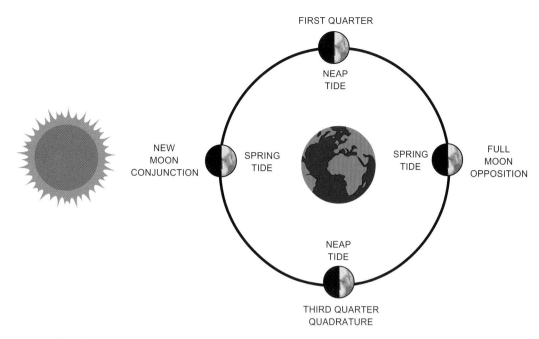

18.30 Effect of the Sun on tides At new Moon and full Moon, the Sun's gravitational force is added to that of the Moon, creating spring tides. Neap tides occur at the first and third quarters when the gravitational force of the Sun counteracts that of the Moon.

Because the equilibrium tide has two bulges, one on each side of the Earth, there would be two high tides and two low tides each lunar day. This is called a **semidiurnal lunar tide** and has a period of 12.42 hours.

TIDE CYCLES AND AMPLITUDES

The Moon revolves around the Earth once every 27.3 days in the same direction as the Earth rotates. Because of this, it takes 24 hours and 50 minutes for a point on the Earth's surface to return to the same position relative to the Moon. Thus, the tidal cycle tends to be almost an hour later on each successive day. The Sun's role in tide generation is most apparent over the course of a month. As the Moon revolves around the Earth, it is sometimes on the same side of the Earth as the Sun, and at other times, it is on the opposite side of the Earth. Between these extremes, the relative positions of the Sun, Earth, and Moon form a right angle (Figure 18.30).

At the time of new Moon, when Sun and Moon are in *conjunction*, and at full moon (*opposition*), the tide-generating forces of the Sun and Moon are acting in the same direction. The net effect is to produce **spring tides**, which have the greatest range between high and low water. About a week after the new moon or full moon, at the time of the first and third quarters, the Moon, the Earth, and the Sun are in *quadrature*. At this time, the peaks of the lunar tides tend to coincide with the troughs of the solar tides and so counteract each other. This produces **neap tides**, in which the tidal range between high and low water is at its lowest

(Figure 18.31). Additional variation in the tidal ranges at specific locations arises over the course of the year as the Sun migrates 23.5 degrees north and south of the equator. Because the Moon axis is inclined at 5 degrees, the sub-lunar point can move by an angular distance of 23.5 degrees, plus or minus 5 degrees north and south of the equator during the year. But note that this cycle of variation from 18.5 degrees to 28.5 degrees takes 18.6 years to complete.

Ideally the gravitational and centrifugal forces acting on the geoid would generate two high tides and two low tides each day. In reality, this is complicated by several factors, including the shape

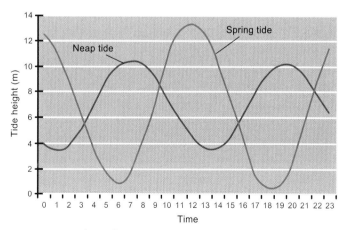

18.31 Spring tides and neap tides The effect of the relative positions of the Sun and Moon during each lunar month is dramatically illustrated by the tides at Hopewell Rocks on the Bay of Fundy, Now Brunswick.

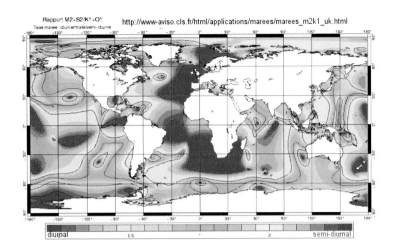

18.32 Global tide patterns Depth and configuration of the ocean basins are important determinants of tidal regimes.

flooded and exposed by the tides, salt-tolerant plants grow. The mudflats slowly build up to approximately the level of high tide, to become a *salt marsh* and eventually form a thick layer of peat at the surface. Tidal currents maintain their flow through the salt marsh by means of a complex network of winding tidal streams.

TYPES OF COASTLINES
COASTLINES OF SUBMERGENCE

A number of distinctive coastlines have developed as a result of the various tectonic and geomorphological processes that have occurred worldwide. One group of coastlines has formed by **submergence**. One of the most common of these is the *ria coast*, a deeply embayed coast resulting from submergence of valleys previously carved by rivers (Figure 18.33a). The attack

and distribution of the ocean basin, friction, the Coriolis effect, and inertia. As a result, tidal regimes are classified into three general types. *Diurnal tides* have one high tide and one low tide in each tidal day. *Semidiurnal tides* have two high and two low tides each tidal day. *Mixed tides* have two high tides and two low tides each tidal day, but the tidal ranges during these events are not equal.

Diurnal tides occur in the Gulf of Mexico, Southeast Asia, and the southern parts of the Pacific Ocean (Figure 18.32). Semidiurnal tides are common in northern Canada, Europe, southern Africa, and New Zealand. Mixed tides occur on the west coast of Canada and the United States, the Caribbean, the Mediterranean Sea, the Persian Gulf, southern Australia, and the East Indies.

TIDAL CURRENTS

In bays and estuaries, the changing tide sets in motion currents of water known as *tidal currents*. When the tide begins to fall, an *ebb current* sets in and carries water seaward. This flow ceases about the time when the tide is at its lowest point. A landward current, the *flood current*, starts as the tide begins to rise. Ebb and flood currents perform several important functions along a shoreline. First, the currents that flow in and out of bays through narrow inlets are very swift and can scour the inlet. This keeps the inlet open, despite the tendency of shore-drifting processes to close the inlet with sand.

Second, tidal currents carry large amounts of fine silt and clay in suspension. This silt and clay come from streams or from bottom mud that has been agitated by storm waves. The fine sediments settle on the floors of some bays and estuaries and gradually accumulate as tidal mud flats. Although the mud flats are alternately

(a)

(b)

18.33 Coastlines of submergence (a) Ria coastline is a submerged river system characterized by broad, branching, rather sinuous and shallow embayments, as shown above in Sydney, Australia. (b) Fjords are steep-sided and deep-submerged glacial troughs. Excellent examples can be seen around the southern tip of Greenland: *60° 31′ N; 44° 18′ W.*

of waves forms cliffs on the exposed seaward sides of headlands, and sediment accumulates in the form of beaches at the heads of bays.

The *fjord coast* is deeply indented by steep-walled fjord that form when glacial troughs are submerged (Figure 18.33b). Many fjords exceed 1,000 metres in depth, although characteristically they are deeper in the middle and at the head. This is thought to illustrate the greater erosive power of the ice in these landward sections, possibly because the glaciers were floated off of the underlying bedrock as they moved out to the sea. Many fjords have a shallow *threshold* at their mouths that limits water circulation. Fjords can extend many kilometres inland; the longest is the Scoresby Sund on Greenland, stretching for 350 kilometres (it is also the deepest at 1,500 metres). The second largest is the Sognefjord in Norway at 204 kilometres. Because sediment rapidly sinks into the deep water, beaches are rare in fjords.

EMERGENT COASTLINES

Coastlines of **emergence** form when underwater landforms are exposed by a drop in sea level or when the Earth's crust is elevated. The *barrier-island coast* is a recently emerged coastal plain. Here, the offshore slope is very gentle, and sand accumulates a short distance from the coast through wave action (Figure 18.34a). Behind the barrier island, there is normally a

lagoon, a broad expanse of shallow water that is filled in places with tidal deposits. A characteristic feature of barrier islands is the presence of gaps, known as *tidal inlets*. Strong currents flow alternately landward and seaward through these gaps as the tide rises and falls. In heavy storms, the barrier island may be breached by new inlets. Along rocky coasts, raised beaches may be visible where former wave-cut platforms become exposed well above sea level (Figure 18.34b).

NEUTRAL COASTLINES

Neutral coastlines develop when new land is built out into the ocean. **Delta** coasts form where sediment-laden rivers enter the sea. A rapid reduction in velocity of the current as it pushes out into the standing water causes deposition. The coarser sand and silt particles settle first, while the fine clays continue out into deeper water. When the fresh water comes in contact with salt water, the finest clay particles coagulate and settle out of the water. Deltaic deposits exhibit distinctive layering. Sediment initially settles into the shallow sea water to form bottomset beds. With time, these beds are covered by steeply inclined foreset beds, which form when the river loses its power to transport its load as flow is no longer contained in a channel. The topset beds form a superficial cover of horizontal layers as the shoreline continues to prograde. Typically, the river channel divides and subdivides into lesser channels called *distributaries*.

(a)

(b)

18.34 Emergent coastlines (a) Barrier islands form offshore from sands and gravels deposited by longshore currents. *(Fly By: 34° 39' N; 76° 31' W)* (b) The level terrace is a raised beach formed from a former wave-cut platform that ran back to the old cliff line. The angle of the cliffs has decreased by sub-aerial denudation, and a low cliff has developed along the active shoreline.

A CLOSER LOOK Eye on Global Change 18.1 Global Change and Coastal Environments

Global climate change over the remainder of the twenty-first century will have major impacts on coastal environments.[1] The changes include increases in sea surface temperature and sea level, decreases in sea-ice cover, and changes in salinity, wave climate, and ocean circulation.

According to the 2001 report of the Intergovernmental Panel on Climate Change, the global heat content of the ocean has been increasing since at least the late 1950s. Between 1950 and 1993, the change in sea-surface temperature was approximately 0.3°C, about half that of the average land-surface temperature. Sea level rose between 10 and 20 centimetres. Sea-ice cover of the northern hemisphere in the spring and summer has decreased by 10 to 15 percent since the 1950s. Since 1970, El Niño episodes, which affect ocean circulation, as well as the intensity of severe storm tracks, have become more frequent, more intense, and more persistent than during the previous 100 years.

What changes are in store? Between 1990 and 2100, global average surface temperature is predicted to increase between 1.4 and 5.8°C, and if the present pattern persists, average sea-surface temperatures will account for one third of that elevation. Sea level will rise by between 9 and 88 centimetres, depending on the temperature scenario. Snow and ice cover will continue to decrease, and mountain glaciers and icecaps will continue their dramatic retreat. Tropical cyclone peak wind and peak precipitation intensities will increase, and El Niño extremes of flood and drought will be exaggerated.

Coastal Erosion

These predicted changes do not bode well for coastal environments. Most coastal erosion occurs in severe storms, when high winds generate large waves and push water up onto the land in storm surges. Global warming will increase the frequency of high winds and heavy precipitation events, thus amplifying the effects of severe storms. More frequent and longer El Niño events will increase the severity and frequency of Pacific storms on the North American coast. During

Coastal erosion Wave action has undermined the bluff beneath these two buildings, which collapsed onto the beach below.

opposing La Niña events, Atlantic hurricane frequency and intensity will increase, with higher risk of damage to structures and coastal populations.

How will the rise in sea level affect coastlines? About 70 percent of sandy shorelines have retreated over the past 100 years or so, 10 percent have advanced, while the remainder have remained stable. Sea-level rise enhances coastal erosion in storms, but in the long term, its effect is to push beaches, salt marshes, and estuaries landward. In an unaltered landscape, the migration of these features landward and seaward with the rise and fall of sea level over thousands of years is a natural process without significant ecological impact. However, it is a serious problem when the rise is rapid. Estuaries become shallower and more saline, beaches disappear and are replaced by sea walls, and salt marshes are drained to reduce inland flooding. In this way, the most productive areas of the coast are squeezed between a rising ocean and a retreating shoreline.

Sea-level rise will not be uniform. Modifying factors of waves, currents, tides, and offshore topography can magnify the rise, depending on the location. For example, some models predict doubled rates of sea-level rise for the North American Pacific coast and the western Arctic shoreline.

Subsidence and Sea-Level Rise

Land subsidence is a contributing factor to the impact of sea-level rise. In an unaltered environment, rivers bring fine sediment to the coastline that settles in estuaries and is also carried by waves, currents, and tides into salt marshes and

[1] *See* Climate Change 2001: The Scientific Basis, Contribution of Working Group I to the Third Assessment Report of the Intergovernmental Panel on Climate Change, *IPCC, Cambridge University Press, 2001*; and Climate Change 2001: Impacts, Adaptation, and Vulnerability, Contribution of Working Group II to the Third Assessment Report of the Intergovernmental Panel on climate Change, *IPCC, Cambridge University Press, 2001, which are used as the basis for this discussion.*

Mississippi Delta marshland   Rising sea level and increasing subsidence have endangered marshlands of many river deltas. Shown here is a marshland of the Mississippi Delta in Louisiana.

A new inlet   Storm waves from Hurricane Isabel breached the North Carolina barrier beach to create a new inlet, as shown in this aerial photo from September 2003. Notice also the widespread destruction of the shoreline in the foreground, with streaks of sand carried far inland by wind and wave action. As the sea level rises, ocean waves will attack barrier beaches with increasing frequency and severity.

mangrove swamps. As the fine sediment accumulates, it slowly compacts, forming rich, dense layers of silt and clay mixed with fine organic particles. However, many coastlines are now fed by rivers that have been dammed at multiple points along their courses; this reduces the amount of fine sediment brought to the coast. Without a constant inflow of new sediment, coastal wetlands slowly sink as the older sediments that support them become compacted. This subsidence increases the effects of sea-level rise. Changes in river flows also affect estuaries as a result of changes in the frequency of severe precipitation events and summer droughts.

Delta coasts are especially sensitive to sediment starvation and subsidence. Here, subsidence rates can reach two centimetres per year. The Mississippi River has lost about half of its natural sediment load, and sediment transport by such rivers as the Nile and Indus has been reduced by 95 percent. Extracting ground water from deltas also increases subsidence. The Chao Phraya delta near Bangkok and the old deltas of the Huang and Changjing rivers in China are important examples.

According to recent estimates, sea-level rise and subsidence could cause the loss of as much as 22 percent of the world's coastal wetlands by the 2080s. Coupled with losses directly related to human activity, coastal wetlands could decrease by 30 percent. This reduction would have major effects on commercially important fish and shellfish populations, as well as other organisms in the marine food chain.

Coral Reefs

Coral reefs, like coastal wetlands, perform important ecological functions, such as harbouring marine fish and nursing their progeny. Coral reefs rank among the most diverse and productive of communities, and some contain more species than rainforests. They also serve as protective barriers to coastlines, reducing the effects of storm waves and surges. However, more than half of the total area of living coral reefs is considered to be threatened by human activities ranging from water pollution to coral mining.

It appears that simple sea-level rise will not be a factor because healthy coral reefs are able to grow upward at a rate equal to or greater than projected sea-level rise. However, the increase in sea-surface temperature that will accompany global warming is of major concern. Many coral reefs appear to be at or near their upper temperature limits. When stressed, for example, by a rise in temperature, many corals respond by "bleaching." In this process, they expel the algae that live symbiotically inside their structures, leaving the coral without colour. However, the algae are necessary for the coral's continued survival. The bleaching may be temporary if the stress subsides, but if permanent, the corals die. Major episodes of coral bleaching have been associated with strong El Niño events, where water temperatures increased by at least 1°C. During the very strong El Niño of 1997–98, for example, the Indian Ocean experienced a major coral bleaching event that was especially prominent on Australia's Great Barrier Reef.

Another concern is the effect of increased atmospheric carbon dioxide levels on corals. With higher concentrations of atmospheric carbon dioxide, more carbon dioxide dissolves in sea water, causing the water to become slightly less alkaline. This in turn shifts the solubility of calcium carbonate, making

it less available to the corals to use in building their skeletal structure. Whether this will create yet another source of stress for coral reefs is still the subject of study.

High-Latitude Coasts

Many pristine stretches of arctic shoreline are threatened by global warming. In these regions, the shoreline is sealed off for much of the year by sea ice. The shoreline is also buttressed by permanent ground ice that bonds unconsolidated sediment into a rocklike mass. This frozen ground is resistant to wave action and thus helps to protect the shoreline from erosion. In some areas, massive ice beds underlie major portions of the coastline.

With global warming, the shoreline is less protected by these mechanisms. Sea ice melts earlier and returns later, in-

---

GEOGRAPHERS AT WORK

Eye in the Sky: Mapping and Monitoring Wetland Ecosystems Using Remote Sensing
by Dr. Marilyne Jollineau, Brock University

Wetlands cover approximately 6% (8 million km²) of the Earth's surface and are extremely diverse as a result of differences in climate, topography, hydrology, geology, vegetation, water chemistry, soils, and other factors, such as human disturbance. Despite this diversity, wetlands are generally distinguished by the presence of water, moist soil conditions, and hydrophytic (i.e., water-tolerant) vegetation.

Historically, adjectives that were used to describe wetlands included: dark, mysterious and unattractive. Perceived as wastelands, wetlands were sources of mosquitoes and infectious diseases. Seen as a hindrance to more 'productive' land uses, many wetlands (i.e., more than 50%) in North America were drained and filled for agriculture and urban development. Gradually, perceptions about wetlands have changed as wetland functions (e.g., irreplaceable habitat, flood protection and nutrient cycling) and values (e.g., recreation, tourism and education) have become better understood.

Today, the realization that wetlands are critically linked to major environmental issues, such as climate change, has made wetland protection an important management issue. Climate change is expected to contribute to wetland degradation and loss; thus, wetland protection has become even more important. The notion that many wetland functions and values are not adequately covered by their terrestrial and aquatic counterparts has also made their protection important. In order to effectively manage wetland ecosystems, it is imperative to know their location. Within this context, wetland mapping and monitoring is essential for proper wetland management. Wetland mapping and monitoring is the specialty of Dr. Marilyne Jollineau, shown in the photo on the right. Dr. Jollineau analyses data acquired by airborne and spaceborne remote-sensing devices that measure electromagnetic energy reflected from wetland environs. Using these data, she is able to extract information about wetlands (e.g., wet-

land vegetation type, composition, extent, distribution and condition) and their changes, using image-analysis techniques. This information is useful to wetland managers responsible for protecting wetlands and responding to ecosystem changes as they emerge. This information can also be used by wetland scientists and managers for a range of other activities; from assessing the impacts of climate change on wetland ecosystems to measuring the progress toward the objectives of different wetland management programs, and their effectiveness, over time.

creasing the season of wave action. Greater expanses of open sea allow larger waves to build. Ground ice thaws to a greater depth, releasing more surface sediment and allowing waves to scour the shore more deeply. Near-shore sediments, stored in cliffs or bluffs, also thaw and are released to shoreline processes. Global warming is expected to be especially severe at high altitudes. Rapid coastal recession under wave attack is already reported for many ice-rich coasts along the Beaufort Sea.

It is apparent that global climate change will have major impacts on coastal environments, with broad implications for ecosystems and natural resources. It will take careful management to reduce these impacts on both human and natural systems.

Arctic shoreline A pristine arctic shoreline, such as this one on the Beaufort Sea, will be subjected to rapid change as the global climate warms. Thawing of ground ice will release beach sediments and the early retreat of sheltering sea ice will expose the shoreline to enhanced summer wave attack.

Deltas can have a variety of outlines. The Nile Delta (Figure 18.35a) is triangular in shape. The Mississippi Delta has long branching fingers that grow far out into the Gulf of Mexico. This form is called a bird's-foot delta (Figure 18.35b). Sediment discharge from the Mississippi amounts to about 1 million metric tonnes per day. Delta growth is often rapid, ranging from 3 metres per year for the Nile to 60 metres per year for the Mississippi.

A second form of neutral shoreline is the *volcano coast* that builds up through deposition of ash and lava. In most cases, the volcano is partly constructed below water level. Low cliffs occur when wave action erodes fresh deposits. Beaches are typically narrow, steep, and composed of fine particles of the dark extrusive rock.

ORGANIC COASTLINES

Coral-reef coasts are unique in that new land is constructed by organisms. Corals secrete rocklike deposits of mineral carbonate. As coral colonies die, new ones are built on top of them, accumulating as limestone **coral reefs**. Coral fragments are torn free by waves, and the pulverized fragments accumulate as sandy beaches.

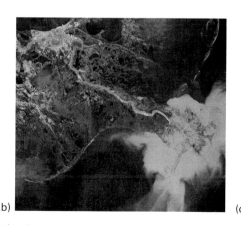

(a)   (b)   (c)

18.35 River deltas on neutral coastlines (a) The Nile Delta is a classic triangular shape. (b) The Mississippi Delta is an example of a bird's-foot delta, in which the distributary channels build levees as they extend into open water. (c) A volcanic coast builds into the sea from accumulations of lava or ash from successive eruptions.

Coral-reef coasts occur in warm, tropical, and equatorial waters between lat. 30° N and 25° S. For vigorous coral growth, the sea water must remain warm (above 20° C), well-aerated, and free of suspended sediment. For this reason, corals thrive where they are exposed to the open sea. There are three distinctive types of coral reefs: fringing reefs, barrier reefs, and atolls. *Fringing reefs* are built as platforms attached to shore. *Barrier reefs* lie offshore and are separated from the mainland by a lagoon (Figure 18.36). Narrow gaps occur at intervals in barrier reefs, through which water returns to the open sea. **Atolls** are more or less circular coral reefs enclosing a lagoon but have no island inside. Most are built on a foundation of volcanic rock that has subsided below sea level. (See *Focus on Systems 13.1 • The Life Cycle of a Hotspot Volcano.*)

FAULT SHORELINES AND MARINE TERRACES

Raised shorelines are formed through tectonic uplift as a result of faulting, or by isostatic rebound following the melting of ice sheets. If cliffs were present along the former coastline, the abrasion platforms can develop terrace-like features, known as *marine terraces*, or raised beaches when raised above the current sea level (see Figure 18.34b). Because it is no longer eroded at its base, the steep cliff line is gradually reduced by flowing water and other sub-aerial processes, such as slumping. Conversely, down-faulted zones can be flooded to produce a *fault coast.*

RISING SEA LEVEL

Global warming will result in a rise of sea level estimated at 9 to 88 centimetres between now and 2100. Some of the rise is due to the simple thermal expansion of the upper layers of the ocean, which will grow warmer. The melting of glaciers and snowpacks as air temperatures rise is contributing the remainder. Sea-level rise will have effects ranging from displacement of estuaries to enhanced coastal erosion. Depending on the amount of rise, some low-lying islands will disappear along with their inhabitants. *A Closer Look: Eye on Global Change 18.1 • Global Change and Coastal Environments* documents the causes and effects of sea-level rise more fully.

A Look Ahead

This chapter has described the processes and landforms associated with wind action, either directly or through wind-driven ocean waves. The power source for wind action is the Sun. The Sun is ultimately responsible for powering the last active landform-making agent of erosion, transportation, and deposition—glacial ice. By evaporating water from the oceans and returning that water to the lands as snow, solar power creates the bodies of solid ice that form as mountain glaciers and ice sheets. Compared with wind and water, glacial ice moves much more slowly. Glacial ice moves sediment forward like a vast conveyor belt, depositing it at the ice margin, where the glacier melts. By plowing its way over the landscape, glacial ice also shapes the local terrain—removing loose rock from hillsides and plastering sediments underneath its mass. This slow but steady action is very different from that of water, wind, and waves, and produces another set of distinctive landforms. The next chapter discusses this movement of glacial ice.

18.36 Fringing reefs Coral reefs fringe the Island of Moorea, Society Islands, in the South Pacific. The island is a deeply dissected volcano with a history of submergence. Tahiti lies in the background. A good example can be seen at *18° 52' S; 159° 47' W.*

CHAPTER SUMMARY

- This chapter has described the landforms of wind and waves, both of which are indirectly powered by the Earth's rotation and the unequal heating of its surface by the Sun.
- Wind is a landform-creating agent that acts by moving sediment. Deflation occurs when wind removes mineral particles—especially silt and fine sand, which can be carried long distances. Deflation creates blowouts in semi-arid regions and lowers playa surfaces in deserts. In arid regions, deflation produces dust storms.
- Sand dunes form when a source, such as a sandstone outcrop or a beach, provides abundant sand that can be moved by wind action. Barchan dunes are arranged individually or in chains leading away from the sand source. Transverse dunes form a sand sea of wave-like forms arranged perpendicular to the wind direction. Parabolic dunes are arc-shaped—coastal blowout dunes are an example. Longitudinal dunes parallel the wind direction and cover vast desert areas. Coastal foredunes are stabilized by dune grass and help protect the coast against storm wave action.
- Loess is a surface deposit of fine, wind-transported silt. It can be quite thick and typically forms vertical banks. Loess is easily eroded by water and wind. In eastern Asia, the silt forming the loess was transported by winds from extensive interior deserts located to the north and west. In Europe and North America, the silt was derived from fresh glacial deposits during the Pleistocene Epoch.
- Waves act at the shoreline—the boundary between water and land. Waves expend their energy as breakers, which erode hard rock into marine cliffs and create marine scarps in softer materials.
- Beaches, usually formed of sand, are shaped by the swash and backwash of waves, which continually work and rework beach sediment. Wave action produces littoral drift, which moves sediment parallel to the beach. This sediment accumulates in bars and sandspits, which further extend the beach. Depending on the nature of longshore currents and the availability of sediment, shorelines can experience progradation or retrogradation.
- Tidal forces cause sea level to rise and fall rhythmically, and this change of level produces tidal currents in bays and estuaries. Tidal flows redistribute fine sediments within bays and estuaries, which can accumulate with the help of vegetation to form salt marshes.
- Coastlines of submergence result when coastal lands sink below sea level or sea level rises rapidly. Ria and fiord coasts are examples. Coastlines of emergence include barrier-island coasts and delta coasts. Coral reef coasts occur in regions of warm tropical and equatorial waters. Along some coasts, rapid uplift has occurred, creating raised shorelines and marine terraces.
- Global sea level is predicted to rise sharply in the twenty-first century due to both volume expansion of warmer sea water and increased melting of glaciers and snowpacks. Future rises may be very costly to human society as estuaries are displaced, islands are submerged, and coastal zones are subjected to frequent flooding.

KEY TERMS

abrasion, aeolian, atoll, barchan dune, beach, blowout, coastal foredune, coastline, coral reef, creep, deflation, delta, desert pavement, dust storm, emergence, equilibrium tide, fetch, geoid, hairpin dune, hammada, littoral drift, loess, longitudinal dune, neap tide, parabolic dune, progradation, retrogradation, saltation, sand dune, semidiurnal lunar tide, shoreline, spring tide, submergence, suspension, threshold velocity, tide-generating force, traction, transverse dune, wave-cut notch, wind erosivity

REVIEW QUESTIONS

1. What is the energy source for wind and wave action?
2. What is deflation, and what landforms does it produce? What role does the dust storm play in deflation?
3. How do sand dunes form? Describe and compare barchan dunes, transverse dunes, star dunes, coastal blowout dunes, parabolic dunes, and longitudinal dunes.
4. What is the role of coastal dunes in beach preservation? How are coastal dunes influenced by human activity? What problems can result?
5. Define the term loess. What is the source of loess, and how are loess deposits formed?
6. How are waves formed and what determines their size and strength?
7. What landforms can be found in areas where bedrock meets the sea?
8. What is littoral drift, and how is it produced by wave action?
9. Identify progradation and retrogradation. How can human activity influence retrogradation?
10. What causes the tides? Why are there several different types of tidal regimes?
11. What key features identify a coastline of submergence? Identify and compare the two types of coastlines of submergence.
12. Under what conditions do barrier-island coasts form? What are the typical features of this type of coastline? Provide and sketch an example of a barrier-island coast.
13. What conditions are necessary for the development of coral reefs? Identify three types of coral-reef coastlines.
14. How are marine terraces formed?

A CLOSER LOOK

Eye on Global Change 18.1   Global Change and Coastal Environments

1. Review the observed and predicted changes in global climate that will affect coastal environments.
2. Identify the global change factors that will affect coastal erosion and describe their impacts.
3. How have human activities induced land subsidence in wetlands? Why are the wetlands of delta coasts particularly at risk?
4. What impact will global climate change have on coral reefs?
5. Why is global warming expected to cause coastal recession of arctic shorelines?

VISUALIZATION EXERCISES

1. Draw a map, illustrating winds coming from the north, at the top of the page. Then sketch the shapes of the following types of dunes: barchan, transverse, parabolic, and longitudinal.
2. Describe the features of delta coasts and their formation. Sketch and compare the shapes of the Mississippi and Nile deltas.

ESSAY QUESTIONS

1. Wind action moves sand close to the ground in a bouncing motion, whereas it lifts and carries silt and clay longer distances. Compare landforms and deposits that result from wind transportation of sand with those that result from wind transportation of silt and fine clay particles.
2. Consult an atlas to identify a good example of each of the following types of coastlines: ria coast, fjord coast, barrier-island coast, delta coast, coral-reef coast, and fault coast. For each example, provide a brief description of the key features you used to identify the coastline type.

EYE ON THE LANDSCAPE |

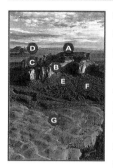

Chapter Opener   Arid landforms. The flat topped mesa (**A**) is comprised of horizontally-bedded sandstones overlying more massive sandstones (**B**) which form a sheer cliff. Where weathering and erosion have enlarged the joints and other zones of weakness the bedrock has acquired a dissected, and more rugged form (**C**). Over time the weaker materials will be removed leaving isolated buttes (**D**). Coarse rock debris forms a talus slope (**E**) at the base of the cliff which over time is further broken down and sorted and the surface etched into gullies (**F**) by flowing water. The finer materials have been transported and deposited by wind to form dunes (**G**).

Coastal features near Port Cambell, Australia. A dramatic coastline has developed by the constant wave action on the horizontally bedded limestone (**A**). Weaker zones in the bedrock have been eroded to form narrow inlets (**B**) which are extended by undercutting and cliff falls. As the inlets extend landward, the power of the waves is progressively reduced by friction on the shallowing seabed, and a narrow beach (**C**) is formed. The more resistant rock remains as narrow promontories (**D**) or may be separated from the mainland to form a small island (**E**). A wave cut platform or marine bench (**F**) has formed a protective apron at the base of the promontory (**F**) which helps to reduce the rate of recession. A secondary zone of weakness has been pierced by the pounding waves to form a search (**G**). This may eventually collapse to leave a stack; the rock (**H**) is likely the remnants of a stack.

Chapter 19

Mt. Angel glacier, Columbia Icefields, British Columbia.
What else would the geographer see? . . . Answers at the end of the chapter.

GLACIER SYSTEMS AND THE LATE-CENOZOIC ICE AGE

Glaciers

Alpine Glaciers

Focus on Systems 19.1 • A Glacier as a Flow System of Matter and Energy

Focus on Remote Sensing 19.2 • Remote Sensing of Glaciers

Landforms Made by Alpine Glaciers

Glacial Troughs and Fjords

Eye on Global Change 19.3 • Glacier Retreat— A Global Overview

Ice Sheets of the Present

Sea Ice and Icebergs

The Late-Cenozoic Ice Age

Glaciation during the Ice Age

Landforms Made by Ice Sheets

Erosion by Ice Sheets

Development of the Great Lakes

Deposits Left by Ice Sheets

Moraines

Outwash and Eskers

Drumlins and Till Plains

Marginal Lakes and Their Deposits

Geographers at Work • Dr. Shawn Marshall

Investigating the Late-Cenozoic Ice Age

Causes of the Late-Cenozoic Ice Age

Eye on Global Change 19.4 • Ice Sheets and Global Warming

Causes of Glaciation Cycles

Holocene Environments

Much of northern North America and Eurasia was covered by massive sheets of glacial ice as recently as one million years ago. As a result, glacial ice has played a dominant role in shaping landforms of large areas in midlatitude and sub-arctic zones. Glacial ice still exists today in two great accumulations—the Greenland and Antarctic ice sheets—and in many smaller masses throughout the Arctic and in high mountains.

The glacial ice sheets of Greenland and Antarctica strongly influence the radiation and heat balance of the globe. They reflect much of the solar radiation they receive, and their permanently cold surface air temperatures help drive the system of meridional heat transport. In addition, these enormous ice accumulations represent stored water in the solid state and so are important components of the global water balance. When the volume of glacial ice increases, such as occurred during the Late-Cenozoic Ice Age, the sea level must fall. Conversely, when ice sheets melt away, the sea level rises. Today's coastal environments evolved during the rising sea level that followed the melting of the last ice sheets of the Ice Age.

GLACIERS

Where a great thickness of ice exists, the pressure on the ice at the bottom makes the ice lose its rigidity and become plastic. This allows the ice mass to respond to gravity, slowly spreading out over a larger area or moving downhill. On steep mountain slopes, the ice can also move by sliding. Movement is the key characteristic of a *glacier*, defined as any large natural accumulation of land ice affected by present or past motion.

Glacial ice accumulates when the average winter snowfall exceeds the amount of snow lost in summer by ablation. The term **ablation** means the loss of snow and ice by sublimation, melting, and evaporation. When winter snowfall exceeds summer ablation, a layer of snow is added each year to what has already accumulated. As the snow compacts by surface melting and refreezing, it turns into a granular ice and then overlying layers compress it into hard crystalline ice. When the ice mass is so thick that the lower layers become plastic, outward or downhill flow starts, and the ice mass is now an active glacier.

Glacial ice forms where temperatures are low and snowfall is high. These conditions can occur both at high elevations and at high latitudes. In mountains, glacial ice can form even in tropical and equatorial zones if the elevation is high enough to keep average annual temperatures below freezing. In the Andes, the snowline at the equator is at an elevation of 4,600 metres compared with 5,000 metres on Mount Kilimanjaro and Mount Kenya in Africa. In the Himalayas, the snowline is generally about 5,200 metres. The higher snowline is attributed to warm summer conditions associated with the continental location of this Asian massif. Snowline in the European Alps and the Canadian Rockies is at about 2,600 metres, but drops to about 1,000 metres in the Yukon. Orographic precipitation encourages the growth of glacial ice. In high mountains, glaciers flow from high-elevation collecting grounds down to lower elevations, where temperatures are warmer. Here the ice disappears by ablation.

Typically, mountain glaciers are long and narrow because they occupy former stream valleys. This distinctive type is called an **alpine glacier** (Figure 19.1a). Small glaciers that persist in mountain basins are called *cirque glaciers* (Figure 19.1b). Others, called *hanging glaciers*, end as icefalls when they no longer extend down to the main valley (Figure 19.1c). A regionally extensive ice accumulation in a mountain region that feeds several glaciers is called an icefield. The Columbia Icefield, covering some 325 square kilometres of mountainous terrain, straddles the border of Alberta and British Columbia and can be accessed from the Icefields Parkway that runs between Banff and Jasper, Alberta (Figure 19.1d).

(a)

(b)

(c)

(d)

19.1 Canadian glacier types (a) Salmon Glacier, British Columbia is confined between the walls of the valley it has carved. (b) A cirque glacier on Bylot Island, Nunavut (c) Angel Glacier, a hanging glacier on Mount Edith Cavell in Jasper National Park, ends in an icefall. (d) The Columbia Icefield covers 325 square kilometres of mountainous terrain between Alberta and British Columbia.

19.2 Glacial abrasion At the smallest scale, glacial abrasion can polish a bedrock surface until it is quite smooth. Scratches or striations can develop where debris is dragged over the surface, and crescentic chattermarks may occur in hard rock types from the impact of boulders under the ice.

In arctic and polar regions, prevailing temperatures are low enough that snow can accumulate over broad areas, eventually forming a vast layer of glacial ice several thousand metres thick. This extensive type of ice mass is called an **ice sheet**. The largest are in Greenland and Antarctica, but smaller examples occur on Ellesmere Island, Baffin Island, and elsewhere in the Canadian Arctic. Many of the glaciers that lead away from the polar ice sheets terminate in the ocean, where they break into icebergs. These glaciers that end in the ocean or in a lake are called **tidewater glaciers**.

Glacial ice normally contains abundant rock fragments ranging from huge angular boulders to pulverized rock flour. Some of this material has eroded from the rock floor on which the ice moves. In alpine glaciers, rock debris also derives from material that slides or falls from valley walls onto the ice.

Glaciers are capable of eroding and depositing great quantities of sediment. The predominant glacial erosion process is *abrasion*, caused when the glacier scrapes and grinds rock fragments against bedrock. Abrasion rates increase with the speed of ice movement, as this determines the amount of abrasive fragments that pass over the bedrock and the rate at which glaciers carry eroded materials. Larger fragments of hard rock are the most effective abrasives.

Erosion also occurs because of the great pressure that develops beneath the ice mass, which is strong enough to cause failure in weak rocks such as shale. The manner in which rocks are moved by glaciers is also a factor. Rocks that are rolled along under the ice are more effective at weakening bedrock than rocks that move by sliding. Once the rock is weakened, moving ice can

lift out blocks of bedrock by *plucking*. Together, abrasion and plucking act to smooth the bedrock over which a glacier moves (Figure 19.2). The glacier eventually deposits the rock debris it has obtained at its terminus, where the ice melts. Both erosion and deposition result in distinctive glacial landforms.

ALPINE GLACIERS

Figure 19.3 illustrates a number of alpine glacier features. A glacier occupies a sloping valley between steep rock walls. Snow collects at the upper end in a bowl-shaped depression, called the **cirque**. The upper end lies in a **zone of accumulation**. Layers of snow in the process of compaction and recrystallization are called *firn*.

The smooth firn field is slightly bowl-shaped. The flow of glacial ice beneath the firn carries the ice down the valley out of the cirque. The rate of the ice flow accelerates where the valley floor becomes steeper, and deep crevasses develop as an icefall. The lower part of the glacier lies in the **zone of ablation**. The equilibrium line marks the boundary between the accumulation zone and the ablation zone. In the ablation zone, the rate of ice wastage is rapid, and as the ice thins, it loses its plasticity. At its *terminus*, the glacier carries abundant rock debris, which is released by melting.

Although the uppermost layer of a glacier is brittle, the ice beneath behaves as a plastic substance and flows slowly. Glaciers flow downhill in response to gravitational forces generated by the accumulated weight of snow and ice. Within a single glacier, flow rates are generally highest at the **equilibrium line**, where the ice

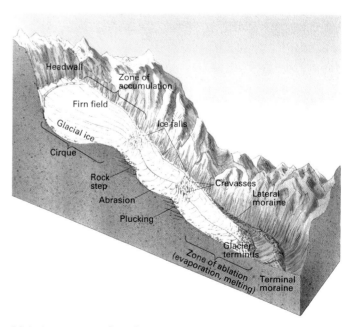

19.3 Cross-section of an alpine glacier Ice accumulates in the glacial cirque, then flows downhill, abrading and plucking the bedrock. Glacial debris accumulates at the glacier terminus. (After A.N. Strahler)

FOCUS ON SYSTEMS | 19.1   A Glacier as a Flow System of Matter and Energy

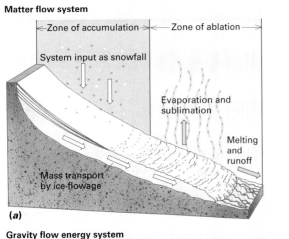

Matter flow system

Gravity flow energy system

Thermal flow energy system

Matter and energy flow in an alpine glacier system (Copyright © A.N. Strahler)

An alpine glacier provides a good example of a system of coupled matter and energy flow. Figure a shows the pathways of a matter flow system. In the zone of accumulation, snowfall provides input to the glacier as the snow becomes compacted and recrystallized to solid ice. During warm summer periods, some snow is lost to melting, evaporation, and sublimation, but the annual balance is on the side of accumulation.

When the glacier reaches lower elevations, it loses more water. The balance shifts to a net loss of ice over the year, and the glacier enters the zone of ablation. With further descent, the loss rate accelerates with increasing mean annual temperature and the glacier's cross-section shrinks rapidly. No more ice remains at the terminus.

Over years, the glacier flow system reaches a state of equilibrium in which the excess water in the form

is thickest. Glacier movement tends to be slow in the upper parts of the accumulation zone where ice is colder and more rigid and near the snout where the ice thins rapidly. Changes in gradient of the underlying rock surfaces also influence glacier movement. Where the gradient is steeper, the glacier will accelerate and become thinner; this is called extending flow. Areas of extending flow are usually characterized by large tensional cracks called crevasses. Conversely, compressive flow occurs in response to a reduction in gradient. Here, the ice thickens and pressure ridges deform the surface.

Flow rate also changes vertically within the ice, and is most rapid near the surface in the midline of the glacier, where friction with the valley floor and walls decreases. Movement occurs through internal deformation of ice under compressive stress; it is caused by slippage within and between ice crystals. Alpine glaciers also move by basal sliding. In this process, the ice moves downhill, lubricated by meltwater and mud at its base. The thin layer of water that separates the ice from the underlying bedrock comes from pressure melting the ice against obstacles. As the ice passes the rock obstruction, it refreezes through the process of *regelation*, and in this way, releases latent heat that contributes to melting of adjacent ice.

Where glaciers encounter weakly consolidated materials, movement also occurs through bed deformation. Bed deformation is effective in regions where underlying materials are saturated with water, which further reduces their strength, allowing the glacier to move across them with little resistance. The absence of water at the base of glaciers in very cold, polar regions limits the effectiveness of basal sliding and bed deformation; movement here is mainly by internal deformation.

Rates of ice flow vary between glaciers and generally range from 3 to 300 metres per year. The highest rates, exceeding 12

of ice in the zone of accumulation is balanced by the loss of water in the zone of ablation. Within the glacier, there is a continuous flow of ice, and each year, newly formed ice moves from the zone of accumulation to the zone of ablation. For the most part, the glacier's shape and appearance remain unchanged, although there may be some small variation from year to year. However, this balance is easily upset by changes in the average annual rates of accumulation or ablation, causing the glacier's terminus to move forward or retreat.

If temperatures cool and snowfall increases, more ice will form in the zone of accumulation. The equilibrium line—the boundary between the zone of accumulation and zone of ablation—will move to a lower elevation. Thus, the glacier will take longer to melt at lower elevations, so the terminus will extend farther down the valley. Eventually, the glacier will reach a new state of equilibrium as a larger, thicker glacier terminat-

ing farther from its source. If conditions were to become warmer or drier, the changes would reverse, producing a new equilibrium state in which a thinner glacier terminates at a higher elevation.

As the glacier flows downhill, it transports rock debris, which grinds away the underlying bedrock to create a valley with a distinctive U-shape. The glacier also receives rock fragments from side slopes. This debris, ranging from large blocks to fine rock flour, is transported to lower elevations and shaped into different types of depositional landforms. The glacier's transport of debris can be treated as a separate open flow subsystem powered by gravity (Figure b).

Another power source for this subsystem is solar energy. Solar energy provides water at high elevations through precipitation, thus contributing potential energy as an input. Gravity powers the downhill motion of the ice. In this motion, potential energy is converted to kinetic energy, and the ki-

netic energy is dissipated as heat generated by friction. Gravity also powers the landform-making processes of glacial erosion, transportation, and deposition.

Glacial ice exports "cold" from higher elevations to lower ones as a thermal flow subsystem (Figure c). When ice descends to low elevations, its surface is subjected to short-wave heat flows from solar radiation. This input provides a positive net radiation balance during the day. Most of the energy flow melts the ice, leaving less energy available to warm the overlying air through sensible heat transfer. Also, less energy remains to sublimate the snow and evaporate the meltwater, thus reducing the flow of latent heat to the atmosphere. This makes the local climate colder.

These three interlinked systems demonstrate the complexity of energy and matter flow systems. Analysis of energy and matter budgets of systems thus requires careful study.

---

kilometres per year, are currently recorded on the glaciers of the West Greenland ice sheet. Flow rate is dependent on several factors. A thicker glacier that generates more pressure by its own weight will tend to flow more rapidly than a thinner glacier. For this reason, the faster-moving glaciers tend to occur in areas of high snow accumulation. Temperature is also important. Warmer ice not only deforms more easily than cold ice, it also usually generates more meltwater at its base. Hence, temperate glaciers typically flow more rapidly than those in polar regions. In extremely cold regions, ice can be frozen onto the bedrock, and basal flow ceases.

Glaciers with steep ice-surface gradients flow faster than those with gentler gradients, since the shear stress generated in response to gravity is proportionately greater. The nature of the underlying surface further affects the rate of movement. Easily deformable rock, such as shale, favours basal sliding, which increases rate of move-

ment. Faster movement also occurs across impermeable rock surfaces where meltwater is retained at the base of the ice.

Flow rates are generally relatively constant, but some glaciers experience episodes of very rapid movement. A period of cool snowy weather can produce a *kinematic wave* or bulge in the glacier surface due to an increase in accumulation. Flow increases in response to a greater mass of ice. Glacial *surges* likely develop in response to an increase in meltwater beneath the ice, which enhances basal sliding. A surging glacier may travel down a valley at speeds of more than 60 metres per day for several months. Most glaciers do not experience surging. In one dramatic example, the snout of the Bruarjokull glacier in Iceland surged forward 45 kilometres at a rate of 5 metres per hour.

An alpine glacier is a good example of a flow system of matter and energy. Matter, in the form of ice and rock debris, flows downhill under the power of gravity. Potential energy is converted

Because glaciers are often found in inaccessible terrain or in extremely cold environments, they are difficult to survey and monitor. Satellite remote sensing therefore provides an invaluable tool for studying both continental and alpine glaciers.

Some of the world's most spectacular alpine glaciers are found in South America, along the crest of the Andes in Chile and Argentina, where mountain peaks reach as high as 3,700 metres. Cerro San Lorenzo (Figure 1) is a glacial horn. Leading away from the horn to the south is a long, sharp ridge, or arête. To the left of the peak is a cirque, now only partly filled with glacial ice. These features were carved during the Late-Cenozoic Era, when the alpine glaciers were larger and filled the now-empty glacial troughs behind the ridge to the right.

Northwest of Cerro San Lorenzo lies the San Quentin glacier, shown in Figure 2. This ASTER colour-infrared image was acquired at 15-metre spatial resolution and shows the glacier in fine detail. The snout of the glacier ends in a shallow lake of sediment-laden water. Note the low, semicircular ridge a short distance from the lake at the extreme left of the image—this is a terminal moraine, marking the limit of the ice tongue in the recent past. The intense red colour indicates a thick vegetation cover.

The world's largest ice cap covers nearly all of Antarctica. At its margin are outlet glaciers, where ice flow into the ocean is quite rapid. Figure 3 shows the Lambert Glacier, one of the largest and longest of Antarctica's outlet glaciers. The image was acquired by Canada's Radarsat radar imager. Because this type

Figure 1 Andean alpine glacial features H, horn (San Lorenzo Peak); A, arête; C, cirque; T, glacial trough (Image acquired by astronauts aboard the International Space Station in December 2000, courtesy of NASA)

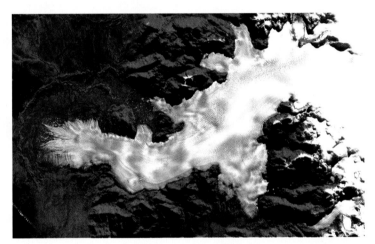

Figure 2 San Quentin Glacier, Chile The San Quentin Glacier as imaged by ASTER (Image acquired on May 2, 2000, courtesy of NASA/GSFC/MITI/ERSDAC/JAROS and U.S./Japan ASTER Science Team)

into kinetic energy as the ice moves and is then dissipated as heat through friction. *Focus on Systems 19.1 • A Glacier as a Flow System of Matter and Energy* provides more details.

LANDFORMS MADE BY ALPINE GLACIERS

Mountains are eroded and shaped by glaciers, and as the ice melts away, the resultant landforms are exposed (Figure 19.4). The initial landscape (a) is a region sculpted entirely by weathering, mass wasting, and fluvial processes. The highlands have a smooth, rounded appearance and are covered with a thick layer of regolith and soil.

A climatic change results in the accumulation of snow in the heads of the higher valleys and cirques carved by the grinding motion of the ice (b). The cirques are deepened by intense frost shattering of the bedrock near the masses of compacted snow. Glaciers fill the valleys and are integrated into a system of tributaries that feed the main glacier. The cirques grow steadily larger. Their rough, steep walls soon replace the smooth, rounded slopes of the original highland mass. Where two cirque walls intersect from opposite sides, a jagged, knifelike ridge, or *arête* (Figure 19.5a), is created. When three or more cirques grow together, a sharp-pointed peak, or *horn* (Figure 19.5b), forms.

of radar system can see through the clouds, it is ideal for mapping a large area in a short time period. It also allows the measurement of the velocity of glacier flow by comparing paired images acquired at different times (in this case, 24 days apart) using a technique called radar interferometry. The image uses colour to show the velocity. Brown tones indicate little or no motion, showing both exposed mountains and stationary ice. Green, blue, and red tones indicate increasing velocity. Velocity and direction are also shown by the arrows superimposed on the image. Glacial flow is most rapid at the left, where the flow is channelled through a narrow valley, and at the right, where the glacier spreads out and thins to feed the Amery Ice Shelf.

Figure 2 Lambert Glacier, Antarctica The Lambert Glacier as imaged by Radarsat during the 2000 Antarctic Mapping Mission. (Image courtesy of the Canadian Space Agency/NASA/Ohio State University, Jet Propulsion Laboratory, Alaska SAR Facility)

Debris carried in, on, or under the ice is eventually deposited as ridges or piles of rock in various configurations in the glaciated landscape. A **lateral moraine** is a debris ridge formed along the edge of the ice adjacent to the trough wall (Figure 19.6a).

Where two ice streams join, marginal debris is dragged as a narrow band on the ice surface. This midstream feature is called a *medial moraine* (Figure 19.6a). At the terminus of a glacier, rock debris accumulates in a **terminal moraine**, which forms an embankment that curves across the valley floor (Figure 19.6b). Even though the terminus of a glacier may be stationary, the ice still delivers debris to the snout. As the ice melts away, the debris is released and accumulates as a ridge across the valley. A terminal moraine marks the farthest advance of the glacier. When a glacier retreats, *recessional moraines* may be deposited where the glacier snout remained stationary long enough to produce a mound of material.

GLACIAL TROUGHS AND FJORDS

Glacier flow constantly deepens and widens the land surface it passes over. After the ice has disappeared, a deep, steep-walled **glacial trough** remains. The trough typically has a U-shaped

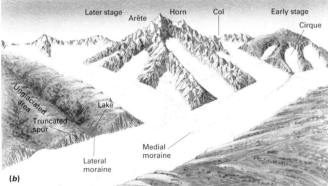

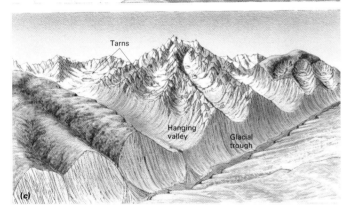

19.4 Landforms produced by alpine glaciers (a) Before glaciation sets in, the region has smoothly rounded divides and narrow, V-shaped stream valleys. (b) After glaciation has been in progress for thousands of years, new erosional forms develop. (c) The disappearance of the ice exposes a system of glacial troughs. (Drawn by A.N. Strahler)

(a)

(a)

(b)

19.5 Glacial erosion The arête or knife-edge ridge (a) and pyramid peak or horn (b) are distinctive erosional features of alpine glaciation.

19.6 Glacial moraines (a) Debris carried at the margin of a glacier forms lateral moraines. When two ice streams converge, their lateral moraines form a medial moraine. *(Fly By: 57° 23' N; 131° 22' W)* (b) Terminal moraines deposited at the snout of a glacier mark the farthest advance of the ice. *(Fly By: 61° 12' 37" N; 139° 31' 06" W)*

(b)

19.7 Glacial valleys and glacial lakes (a) Glacial troughs have distinctive U-shaped cross-profiles. (b) Hanging valleys develop where less powerful tributary glaciers join a more deeply eroded trough. (c) Tarns are small, rounded lakes that occupy the floors of cirques. (d) Long ribbon or finger lakes form in the floor of glacial troughs. Numerous examples can be found near *60° 24' N; 131° 16' W.*

cross-profile (Figure 19.7a). U-shaped troughs carved by tributary glaciers generally have a smaller cross-section and are less deeply eroded. Because the floors of the tributary troughs lie high above the level of the main trough, they are called *hanging valleys*. Streams later occupy these valleys, providing scenic waterfalls that cascade over their lips (Figure 19.7b). In the smaller troughs and cirques, the bedrock is unevenly excavated, so their floors contain basins and steps. Small lakes, called *tarns*, occupy the rock basins (Figure 19.7c). Major troughs sometimes hold large, elongated *ribbon lakes*, also known as finger lakes (Figure 19.7d), but the majority have deep alluvial fills. Aggrading streams that issue from the receding ice fronts are heavily laden

with rock fragments, and deposits of alluvium extend far down the valley.

In coastal areas, deeply excavated glacial troughs are typically flooded with sea water as the ice recedes, creating **fjords**. Fjords are found mainly along mountainous coasts between lat. 50° and 70° N and S. On these coasts, glaciers were nourished by heavy orographic snowfall, associated with the marine west-coast climate. Excellent examples of fjords are found in British Columbia, where the coastline is comparable to that of Norway. Today, fjords are opening up along the Alaskan coast, where some glaciers are melting back rapidly and ocean waters are filling their troughs

EYE ON GLOBAL CHANGE | 19.3   Glacier Retreat—A Global Overview

Glaciologists use the term *mass balance* to describe the difference between the amount of ice that is added to a glacier and the ice that a glacier loses over a period. Since the end of the Little Ice Age in 1850, glaciers worldwide have generally been in retreat, and a significant reduction in glacier area has been recorded in the last 20–30 years. This increasingly negative trend in glacier mass balance (Figure 1) is associated with climate changes linked to variations in both temperature and snowfall. Monitoring in the Cascade Mountains of Washington has concluded that glacier retreat is associated with a 25 percent decline in winter snowpack since 1946 and a 0.7°C increase in summer temperatures during the same period.

North America

For the past two decades, laser technology has measured the elevation of many glaciers, and data indicate glacial thinning is occurring throughout western North America. In Glacier National Park, which encompasses 4,080 square kilometres of mountainous terrain in Alberta, British Columbia, and northwestern Montana, all of the mountain glaciers have receded dramatically since 1850, the date recognized as the end of the Little Ice Age in this region (Figure 2). For example, the Illecillewaet Glacier in British Columbia has re-

treated two kilometres since 1887. Computer analysis suggests that if atmospheric warming continues at the present rate, all glaciers in Glacier National Park will be eliminated by 2030, and even under present conditions, they are predicted to disappear by 2100. Although most glaciers in North America are currently receding, the Taku Glacier in Alaska has advanced about 6.5 kilometres since 1890, mainly because calving at its front ceased when the fjord it flows into refilled with ice in 1948.

Europe

The glaciers of Europe are in a general state of retreat. In Switzerland, a 20 percent reduction in glacier area has been recorded since 1980. The Aletsch Glacier, the largest in Switzerland, has retreated 2,600 metres since 1880, of which 800 metres has occurred in the past 25 years. Only about 5 percent of the glaciers in the Swiss and Italian Alps are currently advancing. This has fallen dramatically since about 1980 when nearly 90 percent of the glaciers in the Ital-

ian Alps and more than 60 percent of the glaciers in the Swiss Alps were reportedly advancing. However, the current situation is not very different from conditions in the 1950s.

In Scandinavia, the majority of glaciers also are retreating. Norway has experienced a significant decrease since 2000, probably as a result of several consecutive years of little winter precipitation and record warm summers in 2002 and 2003. In Iceland, some glaciers have reportedly receded by up to two kilometres in the past 30 years.

Africa

In Africa, glaciers are found only on Mount Kilimanjaro (5,895 metres), Mount Kenya (5,199 metres), and the

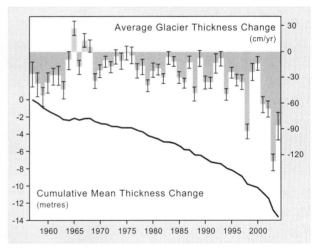

Figure 1 Glacier mass balance The volume of glacier ice has steadily declined worldwide since the late 1960s.

ICE SHEETS OF THE PRESENT

The enormous ice sheets of Antarctica and Greenland represent accumulations of ice that are thousands of metres thick. The Greenland Ice Sheet has an area of 1.7 million square kilometres and occupies close to 90 percent of the entire island of Greenland (Figure 19.8a). Its surface has the form of a broad, smooth dome. Only a narrow, mountainous coastal strip of land is exposed. The Antarc-

tic Ice Sheet covers 13 million square kilometres (Figure 19.8b). It is thicker than the Greenland Ice Sheet, with maximum accumulations of 4,000 metres. Both ice sheets are developed on large, elevated land masses in high latitudes.

At some locations, long tongues extend from ice sheets as *outlet glaciers* that reach the sea at the heads of fjords. From the floating edge of the glacier, huge masses of ice break off and drift

and the Gangotri Glacier in India, which has retreated 30 metres per year since 2000.

Andes

In the northern Andes, most glaciers are less than 1 square kilometre in size. Measurements from the Chacaltaya Glacier in Bolivia indicate that it lost 65 percent of its volume and 40 percent of its thickness between 1992 and 1998; it is predicted to disappear by 2015. Further south in Peru, glaciers cover a 725-kilometre area. From 1977 to 1983, general glacial recession was reported at 7 percent in this region. Recent studies on the Quelccaya Ice Cap, the largest in Peru, has shown significant retreat and thinning, with the rate of retreat on one glacier measured at 155 metres per year from 1995 to 1998. The glaciers in Patagonia at the southern end of South America are reportedly receding at a faster rate than any other region in the world. Here, ice has retreated by as much as 1 kilometre since the early 1990s.

New Zealand

In New Zealand, mountain glaciers have shown accelerated rates of recession since 1920, and in the past three decades, pro-glacial lakes have appeared behind the terminal moraines of several glaciers.

(a)

(b)

Figure 2 Grinnell Glacier, Glacier National Park, Montana These photographs taken in 1981 (a) and 2005 (b) show the retreat of the glacier.

Rwenzori Mountains, which rises to 5,109 metres on the borders of Uganda and Zaire. The ice cover on the summit of Kilimanjaro has decreased dramatically since 1912. Little ice was reported in 2005, and it is predicted that the summit will be ice-free by 2025. The Rwenzori Mountains also shows evidence of glacier retreat. However, because these mountains are supplied by moisture from the Congo region, retreat will likely be slower than on the isolated peaks of Mount Kilimanjaro and Mount Kenya.

Himalayas

The formation and growth of pro-glacial lakes at the termini of many glaciers in the Himalayas indicate glacier melting and retreat. Studies in China have concluded that 50 percent of the Himalayan glaciers were retreating in 1970, but this increased to 95 percent after 1990. Regional examples include the Khumbu Glacier, the main route up Mount Everest, which has retreated five kilometres since 1953; the Rongbuk Glacier on the north side of Mount Everest, which is retreating at a rate of 20 metres per year;

out to open sea with tidal currents to become icebergs. An important glacial feature of Antarctica is the presence of great plates of floating glacial ice, called *ice shelves* (Figure 19.8b). Ice shelves are fed by the ice sheet, but they also accumulate new ice through the compaction of snow. No ice sheet exists near the North Pole, which is positioned in the Arctic Ocean. Here, ice occurs as various forms of floating sea ice.

SEA ICE AND ICEBERGS

Free-floating ice on the sea surface is of two general types: sea ice and icebergs. Sea ice floats on the surface of the ocean. It is formed by direct freezing of ocean water, which due to salinity occurs at about minus 1.8°C. Sea ice that has frozen along coasts is referred to as fast ice. Pack ice consists of extensive areas of freely floating sea ice that is detached from land. Wind and currents

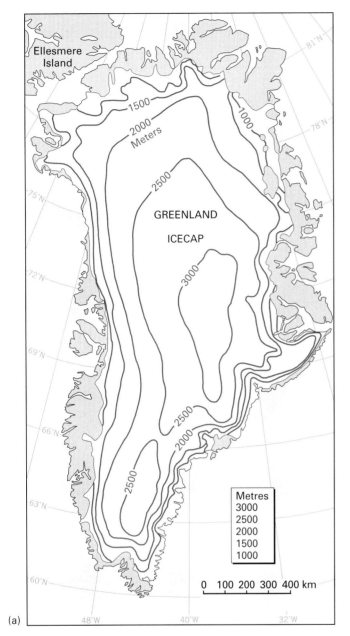

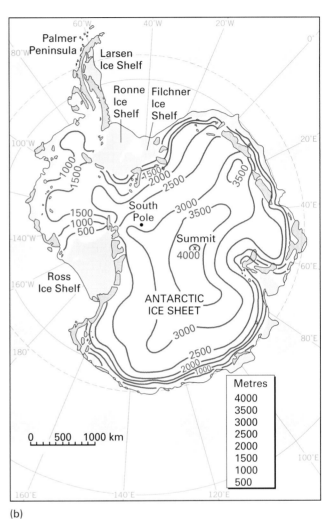

19.8 Greenland and Antarctica (a) The surface of the Greenland Ice Sheet rises to over 3,000 metres. (b) At its highest point, the Antarctic ice sheet rises to over 4,000 metres. (Based on data from R. F. Flint, *Glacial and Pleistocene Geology*, London: John Wiley & Sons, 1957)

break pack ice up into smaller patches called ice floes (Figure 19.9). The narrow strips of open water between these floes are known as *leads*. Where winds forcibly bring ice floes together, the ice margins buckle and turn upward into pressure ridges.

In contrast, icebergs are bodies of land ice that have calved from the ocean termini of glaciers and ice shelves. Because they are composed of ice originating from glaciers, icebergs are not considered sea ice. They are slightly less dense than sea water, so they float low in the water. About 83 percent of the bulk of an iceberg is submerged. Most of the icebergs that drift along Canada's east coast originate from Baffin Island and Greenland. A major difference between sea ice and icebergs is thickness. Sea

ice does not exceed five metres in thickness, while icebergs may be hundreds of metres thick. Over the past 30 years, sea ice has decreased significantly in the Arctic, but has increased in the Antarctic.

Sea ice forms and melts with the polar seasons and plays an important role in regulating climate. Relatively warm ocean water is insulated from the cold polar atmosphere except where cracks in the ice allow exchange of heat and water vapour. This affects regional cloud cover and precipitation. In the Arctic, some sea ice persists year after year. In late winter, Arctic sea ice typically covers about 15 million square kilometres, but decreases to about 8 million square kilometres by the end of summer (Figure

19.9 Ice floes A Canadian Coast Guard icebreaker in warmer weather.

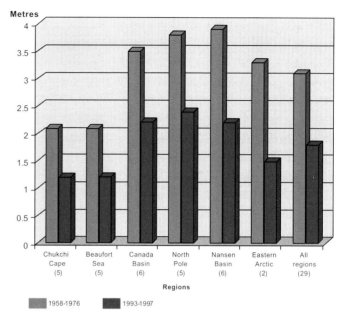

19.11 Decrease in Arctic sea ice thickness for the period 1958–1997
Average reduction in ice drift is about 40 percent and is most pronounced in the eastern Arctic. (Graph derived from Geophysical Research Letters, vol. 26)

19.10). Antarctic sea ice mainly develops during the southern hemisphere in winter, and at its maximum extent covers about 20 million square kilometres of the Southern Ocean; by summer's end it is reduced to about 4 million square kilometres.

Records from the Arctic indicate that the extent of sea ice has been declining since at least the early 1950s. Passive microwave and radar instruments carried aboard RADARSAT and other satellites now obtain precise measurements of ice. Dramatic reductions in regional ice cover have been recorded in recent years.

The present long-term decline in end-of-summer sea ice is now estimated at approximately 8 percent per decade. The reduction in ice area is particularly noticeable off the coasts of Alaska and Siberia. Sea ice thickness likewise has declined by about 1.3 metres over the past 40 to 50 years (Figure 19.11).

THE LATE-CENOZOIC ICE AGE

The present ice sheets in Greenland and Antarctica provide an impression of the landscape that would have existed over much of Canada at the peak of the **Late-Cenozoic Ice Age** (or simply the **Ice Age**) that began in late Pliocene time, perhaps 2.5 to 3.0 million years ago. The Ice Age is associated with the last three epochs of the Cenozoic Era: the Pliocene, Pleistocene, and Holocene. Previously, geologists associated the Ice Age with the Pleistocene Epoch, which began about 1.6 million years ago. However, new evidence obtained from deep-sea sediments shows that the Ice Age began earlier than this.

The period during which continental ice sheets grow and spread outward over vast areas is known as a **glaciation**. Glaciation is associated with a general cooling of average air temperatures over the regions where the ice sheets originate. At the same time, ample snowfall must persist over the growth areas to allow the ice masses to build in volume.

When the climate warms or snowfall decreases, ice sheets become thinner and begin to shrink. Eventually, the ice sheets may melt completely. This period is called a *deglaciation*. Following a deglaciation, but before the next glaciation, is a period in which a

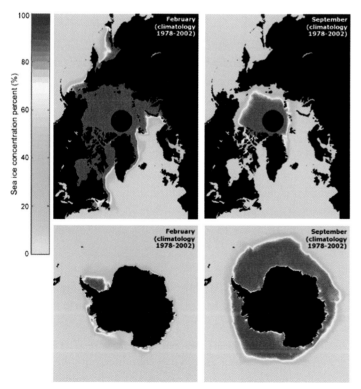

19.10 Seasonal variation in Arctic and Antarctic sea ice, 1978–2002 (National Snow and Ice Data Center, University of Colorado)

Table 19.1 Chronology of the recent glacial and interglacial episodes

Time (yrs BP)	North America		Alps		Northern Europe	
<12,000			Interglacial			
12,000–110,000	Glaciation Wisconsinan	Interglacial	Glaciation Würm	Interglacial	Glaciation Weichselian	Interglacial
		Sangamon		Riss-Würm		Eemian
130,000–200,000	Illinoisan		Riss		Saalian	
		Yarmouth		Mindel-Riss		Holsteinian
300,000–455,000	Kansan (includes 4 lesser glaciations)		Mindel		Elsterian	
		Aftonian		Günz-Mindel		Cromerian
620,000–680,000	Nebraskan		Günz		Menapian	

mild climate prevails—an **interglaciation**. The last interglaciation began about 140,000 years ago and ended between 120,000 and 110,000 years ago. A succession of alternating glaciations and interglaciations, spanning a period of 1 to 10 million years or more, constitutes an *ice age*.

The last four major ice advances of the Cenozoic Ice Age are known in North America as the Nebraskan, Kansan, Illinoisan, and Wisconsinan glaciations (Table 19.1). Different names are applied to contemporaneous episodes in Europe and other parts of the world. It is generally thought that the Earth is currently experiencing another interglaciation that started about 18,000 years ago. In the preceding *Wisconsinan Glaciation*, ice sheets covered much of North America, northern Europe, and northern Asia (Figure 19.12). In addition, extensive ice covers developed in the Andes Mountains in South America, the South Island of New Zealand, and less extensively in Australia.

GLACIATION DURING THE ICE AGE

During the Wisconsinan Glaciation, most of Canada was engulfed by the vast Laurentide Ice Sheet that was centred in the general vicinity of what is now Hudson Bay. Ice spread into the United States, covering most of the land lying north of the Missouri and Ohio rivers, and extending eastward into New Eng-

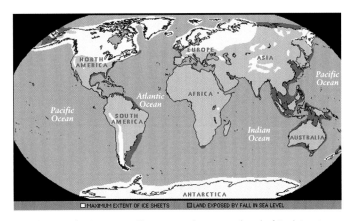

19.12 During the Wisconsinan Glaciation, ice sheets covered much of North America, northern Europe, and northern Asia.

land. Alpine glaciers of the western mountains coalesced into a single ice sheet that spread to the Pacific shores and met the Laurentide sheet lying to the east. A small area south of Lake Superior escaped inundation by the Wisconsinan ice. Known as the Driftless Area, it was apparently bypassed by the glacial lobes that moved on either side.

At the maximum spread of the Wisconsinan ice sheets, sea level was as much as 200 metres lower than today, exposing large areas of continental shelf throughout the world. The lower sea level explains why the ice sheets extended far out into what is now open ocean. The continental ice sheets covered vast areas beneath ice several kilometres thick. This great mass of ice depressed the Earth's crust by hundreds of metres at some locations. With deglaciation, the ice disappeared and the crust began to move upward, quickly at first, then more slowly. This crustal movement is known as *isostatic rebound*. Upward movement is still occurring today in some locations.

Important evidence of isostatic rebound is the position of ancient shorelines, now raised above sea level by the crustal motion. Sea level rose rapidly during the last deglaciation. The crust, still depressed, was flooded with ocean waters, providing wave action at the shoreline that created beaches. As the crust rebounded upward, the beaches were elevated, leaving them hundreds of metres above present levels. By dating these ancient shorelines, geologists have established the amount and rate of uplift that occurred in response to the unloading of ice. Many elevated shorelines are found today in the coastal belts of Hudson Bay (Figure 19.13).

Two of the largest *epicontinental seas* that extend into continental land masses are Hudson Bay and the Baltic Sea. Both are centred in areas of active isostatic rebound (Figure 19.14). In Hudson Bay, maximum uplift is presently about 100 metres. In the Baltic region, the land has risen by as much as 275 metres. The crust in these regions is expected to continue to rise, and eventually these bodies of water will be much smaller or disappear entirely. Gravity measurements predict another 100-metres rise in the Hudson Bay region.

19.13 Raised beaches on Devon Island Isostatic rebound has created a series of raised beaches in this high Arctic site following the melting of the Laurentide Ice sheet.

The upward motion of the crust in isostatic rebound is most rapid immediately following the melting of the ice; the rate decreases over time (Figure 19.15). North Bay, located significantly south of the centre of the ice sheet, shows the earliest rebound. Note the steeper rebound for James Bay, which is located near the centre of the ice sheet, and was depressed most by the ice.

LANDFORMS MADE BY ICE SHEETS

EROSION BY ICE SHEETS

Like alpine glaciers, ice sheets are highly effective eroding agents. The slowly moving ice scrapes and grinds away a lot of solid bedrock, leaving behind smoothly rounded rock masses. These show countless grooves and scratches trending in the general

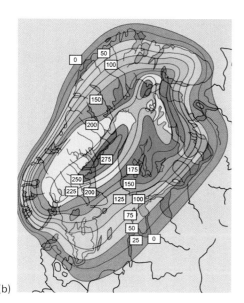

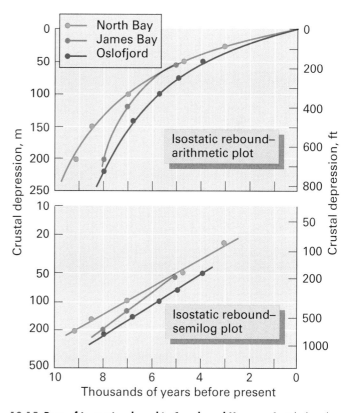

19.15 Rate of isostatic rebound in Canada and Norway Crustal rebound of more than 200 metres has occurred at North Bay, James Bay, and Oslo, Norway. Note that both axes on the graph are numerically reversed. Time increases from left to right, and upward crustal movement is shown by a higher position on the graph. (Data from R. F. Flint, W. Ferrand, and W. A. Heiskanen)

19.14 Isostatic rebound (a) Hudson Bay and (b) the Baltic Sea are located in areas that were greatly depressed by the weight of continental ice sheets (contours are in metres).

19.16 Glacial abraded grooves and striations Rock fragments dragged across bedrock cause abrasions ranging in scale from fine scratches that effectively polish the rock to deep gouges and grooves.

direction of ice movement (Figure 19.16). Sometimes the ice polishes the rock to a smooth, shining surface. The evidence of ice abrasion is common throughout glaciated regions of North America. More pronounced shaping of solid bedrock by the moving ice can form a *roche moutonnée* (Figure 19.17). The side of the rock that the ice approached is usually smoothly rounded, while glacial plucking on the downstream side leaves the rock irregular and blocky.

Ice sheet activity created many hundreds of lake basins. These range in size from the numerous small lakes scattered across the Canadian Shield (Figure 19.18) to Great Bear Lake, Northwest Territories, which has a surface area of 31,153 square kilometres, making it the world's eighth largest lake. Drainage is generally poor on the Canadian Shield as the surface consists of rocky, ice-smoothed hills with an average relief of only 30 metres. Between the lake-filled basins, the land is often boggy muskeg where the water table is near the surface. Ice sheets excavated linear depressions where the bedrock was weak and the flow of ice was channelled by the presence of a valley. The Finger Lakes of western New York State were produced in this way.

DEVELOPMENT OF THE GREAT LAKES

It is generally accepted that the Great Lakes did not exist before the Ice Age, and although a system of large lakes developed during earlier glacial and interglacial episodes, the configuration of the present Great Lakes is the product of the post Wisconsinan deglaciation that began about 14,800 years ago.

The Great Lakes formed in a low interior basin of old sedimentary rocks. During the Ice Age, ice sheets advanced over this basin, scouring and lowering the surface by as much as 500 metres below the surrounding terrain. The last continental glacier began its retreat about 18,000 years ago. At this time, drainage was southward to the Gulf of Mexico. By 14,000 years ago, the southern portions of the Great Lakes were beginning to emerge, and by 12,500 years ago, vast lakes formed in the ice-free portions of the basin (Figure 19.19a). Dammed by the ice mass to the north, these *pro-glacial lakes* reached levels many tens of metres higher

19.17 A glacially abraded roche moutonnée Glacial action abrades the rock into a smooth form as it rides over the rock summit, then plucks bedrock blocks from the lee side, producing a steep, rocky slope.

19.18 Lakes on the Canadian Shield Numerous lakes fill glacially scoured bedrock depressions throughout the Canadian Shield.

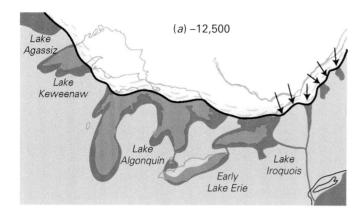

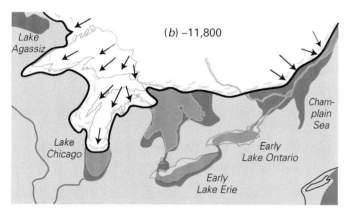

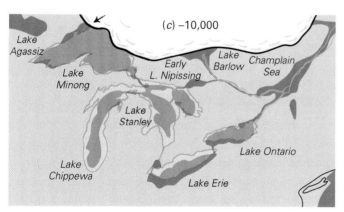

19.19 Emergence of the Great Lakes Three stages in the evolution of the Great Lakes as the continental ice sheets retreated at the end of the last glacial period.

However, isostatic rebound soon raised this pathway, causing the lakes to rise and shifting the outflow to the present route from southern Lake Huron through lakes Erie and Ontario to the St. Lawrence River.

DEPOSITS LEFT BY ICE SHEETS

The term **glacial drift** includes all varieties of rock debris deposited as a result of continental glaciation. There are two major types of drift. **Stratified drift** consists of layers of clays, silts, sands, or gravels. These materials are deposited by meltwater streams or in bodies of water adjacent to the ice. **Till** is an *unstratified* mixture of rock fragments, ranging in size from clay to boulders, that is deposited directly from the ice without subsequent water transport. As the glacial ice melts in a stagnant marginal zone, the rock fragments it holds are lowered to the underlying ground surface where they form a layer of debris (Figure 19.20). This *ablation till* shows no sorting and often consists of a mixture of sand and silt, with many angular pebbles and boulders. Beneath this residual layer, there may be a basal layer of dense *lodgement till*, consisting of clay-rich debris previously dragged forward beneath the moving ice. Where till forms a thin, more or less even cover, it is called a *ground moraine*.

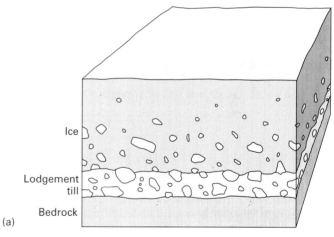

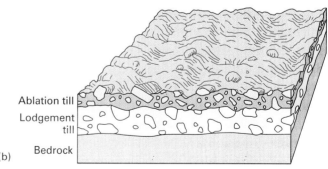

19.20 Glacial till (a) As ice passes over the ground, sediment and coarse rock fragments are pressed together to form a layer of lodgement till. (b) When the overlying ice stagnates and melts, the rock debris it contains is left as the residual deposit of ablation till. (Copyright © A.N. Strahler)

than at present and, in many areas, they covered the present outlines of the Great Lakes (Figure 19.19b).

At that time, drainage patterns were very different, with the western lakes draining to the Mississippi through the ancestral St. Croix and Desplaines rivers, and the eastern lakes draining to the Hudson River through the Mohawk Valley. Following a short readvance (b), the retreat of the ice sheet continued, and water levels dropped as an outlet to the northeast was uncovered through Lake Nipissing and the Ottawa River (Figure 19.19c).

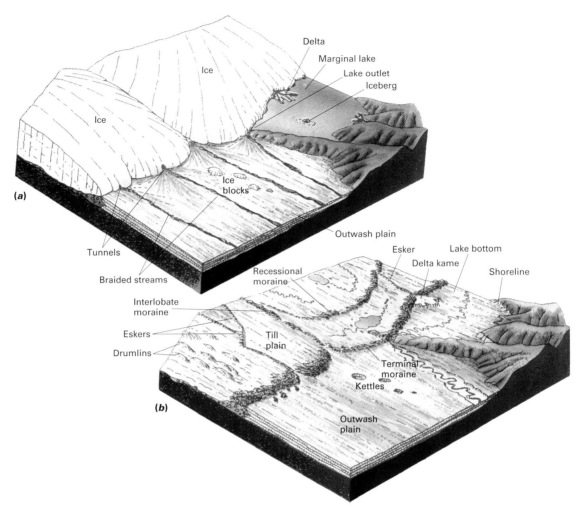

19.21 Marginal landforms of continental glaciers (a) With the ice front stabilized and the ice in a wasting, stagnant condition, meltwater creates various depositional features. (b) The ice has wasted completely away, exposing a variety of new landforms made under the ice. (Drawn by A.N. Strahler)

The thickness of glacial drift over the parts of North America that were formerly covered by late-Cenozoic ice sheets is quite variable and depends on the amount of debris in the ice, the pre-existing topography, as well as the length of time that the ice margin stayed at a given location. For example, extensive areas of the Northwest Territories are overlain by till that is generally less than 10 metres thick.

The form and composition of deposits left by ice sheets depends on the prevailing conditions at the time of the ice sheet's existence. Figure 19.21a shows a region partly covered by an ice sheet with a stationary front edge. This occurs when the rate of ice ablation balances the amount of ice brought forward by the ice sheet. Although the positions of the ice fronts fluctuated many times, there were long periods when they were essentially stable, which allowed thick deposits of drift to accumulate.

MORAINES

The ice sheet's transport work is much the same as that of an alpine glacier, but on a vaster scale. The ice carries material forward and dumps it at the margin where it will pile up if not removed. The ice sheet deposits rock fragments that moved within it in the ablation zone at the ice margin. Glacial till that accumulates at the margin forms an irregular heap of rubble—the *terminal moraine*. After the ice has disappeared (Figure 19.21b), the moraine appears as a belt of knobby hills. Small lakes may occupy depressions within this undulating landscape of mounds and ridges; this irregular landscape is often referred to as *knob-and-kettle* topography.

Terminal moraines form great curving patterns. The outward downstream curvature indicates that the ice advanced as a series of great *ice lobes*, each with a curved front (Figure 19.21a). Where two lobes come together, the moraines curve back and fuse together into a single *interlobate moraine* pointing upstream. During the recessional phase, the ice sheet begins to shrink because conditions no longer favour ice accumulation. If the ice front pauses for some time at various positions during its retreat, morainal belts similar to the terminal moraine form. As with alpine glaciers, these belts are called recessional moraines

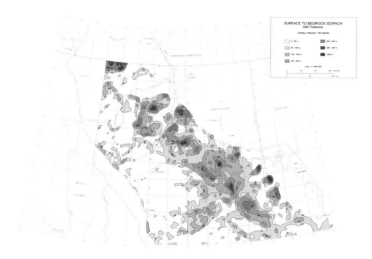

19.22 Glacial drift thickness in the Canadian Prairies Glacial drift, up to 300 metres thick in same places, has created a comparatively level surface across much of the Canadian Prairies.

(Figure 19.21b). They run roughly parallel with the terminal moraine but are often thin and discontinuous.

Glacial till covers areas between the morainal ridges. This cover is often inconspicuous since it does not form a prominent landscape feature. A layer of till can obscure, or entirely bury, the landscape that existed before glaciation. Where thick and smoothly distributed, the layer forms a level *till plain*. Plains of this origin are widespread throughout the central lowlands of the United States and southern Canada. In parts of Alberta and Saskatchewan, drift thickness varies from 300 metres in a few pre-glacial valleys to zero on some of the higher rock outcrops (Figure 19.22). Till has mostly filled bedrock lows, such as pre-glacial river channels, producing the subdued relief that characterizes the Canadian Prairies. The numerous *sloughs* found across Alberta, Saskatchewan, and Manitoba are mostly *kettle holes* that formed as ice lenses in or under the glacial drift eventually melted away, leaving surface depressions that later filled with water (Figure 19.23a).

(a)

(b)

(c)

19.23 Landforms of Continental Ice Sheets (a) **Prairie sloughs** Ponds and small lakes that formed in depressions in the undulating moraine deposits are a characteristic feature of the prairie pothole region. *(Fly By: 53° 04' N; 104° 55' W)* (b) **Esker** An esker is a curving ridge of sand and gravel, marking the bed of a river of meltwater flowing underneath a continental ice sheet near its margin. (c) **Drumlin** This small drumlin, located in Peterborough, Ontario, shows a tapered form, indicating that the ice moved from left to right. *(Fly By: 48° 17' N; 78° 12' W)*

19.24 Sub-glacial stream This sediment-laden stream emerges from a tunnel at the terminus of the Fox glacier in New Zealand. *(Fly By: 43° 39' S; 170° 12' E)*

OUTWASH AND ESKERS

Figure 19.21 shows a smooth, sloping plain lying in front of the ice margin. This is the *outwash plain*, formed of stratified drift left by braided streams issuing from the ice. An outwash plain is an example of a *pro-glacial deposit*, since it developed beyond the ice margin. Outwash plains are built of layer upon layer of sands and gravels.

In some cases, large streams that emerge from tunnels in the ice carry meltwater (Figure 19.24). The streams form when the ice front stops moving for many kilometres back from the margin. After the ice has gone, the position of a former ice tunnel is marked by a long, sinuous ridge of sediment. These features are called **eskers** and represent the deposits of sand and gravel on the tunnel floors (Figure 19.23b). Eskers can be many kilometres long. One of the longest in Canada is the Thelon esker, which runs west from Duwant Lake in the Northwest Territories for about 800 kilometres. An esker is an example of an *ice-contact deposit*, a landform that develops on or under an ice sheet.

DRUMLINS AND TILL PLAINS

Another common feature formed from glacial till is the **drumlin**, a smoothly rounded, oval hill resembling the bowl of an inverted spoon (Figure 19.23c). Drumlins invariably lie in a zone behind the terminal moraine. They often occur in groups or swarms and may number in the hundreds—the term "basket of eggs topography" is occasionally used to describe drumlin fields. The long axis of each drumlin parallels the direction of ice movement, with the steeper, broader end facing the oncoming ice. Several factors influence the ratio of length to width of drumlins. Length probably increases in response to ice velocity and the length of time the flowing ice acted upon the drumlin. Length likely decreases when the sediment strength increases, making the drumlin less likely to become deformed.

Drumlins seem to have formed under moving ice by a plastering action in which layer upon layer of clay boulders were spread and shaped into their final forms. However, the origin of drumlins is not well understood, and several theories have been put forward to explain them. Some believe that drumlins began as pre-existing features that were reshaped by erosion during a readvance of the ice sheet. A second theory proposes that till was deposited and molded behind an obstacle or in a cavity under the ice. Another idea suggests that drumlins were sculpted into streamlined forms by catastrophic meltwater discharge events beneath the ice. A widely accepted hypothesis involves the flow of deformable sediment onto and around cores of less-deformable material. Pore water pressure greatly influenced the process of deformation and deposition. Deposition occurs where water can drain from the deposits, but prevention of this deformed the sediments.

MARGINAL LAKES AND THEIR DEPOSITS

When ice sheets advanced toward higher ground, ice blocked valleys that may have opened out northward. Under these conditions, marginal glacial lakes formed along the ice front (see Figure 19.21a). Streams of meltwater from the ice were able to build *glacial deltas* into these marginal lakes. Once they filled to capacity, glacial lakes began to drain along the lowest available channel, which was progressively deepened, allowing more water to drain from the lake. Because glacial lakes were continuously supplied with enormous volumes of meltwater, the process typically produced wide valleys called spillways or overflow channels (Figure 19.25). Overflow channels appear as prominent valleys, usually several kilometres across. The streams that now flow in them are no longer supplied with vast amounts of meltwater, and because they are much smaller than the valley they occupy, they are referred to as underfit streams.

19.25 Glacial spillway The broad Qu'Appelle Valley in Saskatchewan was cut by meltwater draining from a pro-glacial lake. The valley contained a river that was much larger than the one that presently occupies it. *(Fly By: 40° 34' N; 103° 23' W)*

After the ice has disappeared and the lake has dried, several distinctive features remain in the landscape. Exposed lake beds remain as *glaciolacustrine plains* underlain by soils that are predominantly clay; these plains often contain extensive areas of marshland. For example, the level terrain near Winnipeg, Manitoba, is a legacy of former glacial Lake Agassiz (Figure 19.26). Lake Agassiz covered much of Manitoba, as well as parts of Ontario, Saskatchewan, Minnesota, and North Dakota. It formed about 12,000 years ago from glacial meltwater that was prevented from flowing northward by remnants of the Laurentide Ice Sheet. Old shorelines indicate that the size and depth of Lake Agassiz varied as the climate periodically warmed and cooled.

In many locations, the sediments deposited in former glacial lakes contain **varves** (see Figure 15.27). Varved sediments comprise alternate layers of clay and silt, usually only one to two centimetres thick. The layers reflect seasonal deposition in the lakes at the ice margin. When temperatures warm in the high sun season, meltwater streams begin to flow and bring sediment into the ice-

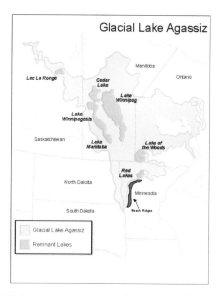

19.26 Glacial Lake Agassiz The largest of the ice marginal lakes. Evidence of glacial Lake Agassiz occurs over an area of approximately 950,000 square kilometres, although the lake did not cover this entire area at one time.

GEOGRAPHERS AT WORK

The Mysterious Disappearance of the Glaciers
by Dr. Shawn Marshall, University of Calgary

Mountain glaciers and polar icefields have been in a general state of retreat since the Little Ice Age, with this retreat accelerating in recent decades as the Earth has warmed. Many small glaciers, such as those in Montana and Spain, have disappeared from the maps. Larger glaciers and icefields in most of the world's mountain ranges are well out of equilibrium with current climate and will continue to melt back, transforming the alpine landscape and altering the climate and hydrology of the world's mountain and polar regions. On regional scales, this has implications for water resources[1], alpine ecology, and tourism/recreation in mountain areas. Globally, sea level rise has been accelerating as a result of warming-induced retreat of mountain glaciers and parts of the Greenland Ice Sheet. Greenland harbours a volume of ice equivalent to about 7 m of global sea level, and there is evi-

dence that much of this ice sheet melted away during the last major heat wave that the world experienced, the Eemian interglacial period[2]. During this time, about 125,000 years ago, orbital variations produced Arctic temperatures that were 3-4°C warmer than present. This is similar to the warming that is expected in the Arctic by the end of this century.

While it is clear that the world's glaciers and icefields will continue to disappear in a warmer climate, it is uncertain how rapidly this may occur. How much sea level rise can we expect this century? What are the expected impacts on water resources and mountain environments? Dr. Shawn Marshall's group at the University of Calgary is studying glacier–climate processes in the Canadian Rockies, the Canadian Arctic, Iceland, and Greenland to help quantify how glaciers will respond to different climate scenarios. His group uses a blend of

field and modelling studies to examine the exchanges of mass and energy that drive glacier advance and retreat. Process models being developed and calibrated through the field studies are being applied to continental scales to simulate the possible fate of the icefields which—until this century—have helped to shape and define the landscape in western and northern Canada.

[1]Barnett, T.P, J.C. Adam, and D.P. Lettenmaier, 2005. Potential impacts of a warming climate on water availability in snow-dominated regions. Nature, 438, 303-309.
[2]Cuffey, K.M. and S.J. Marshall, 2000. Sea level rise from Greenland Ice Sheet retreat in the last interglacial period. Nature, 404, 591-594.

EYE ON GLOBAL CHANGE | 19.4   Ice Sheets and Global Warming

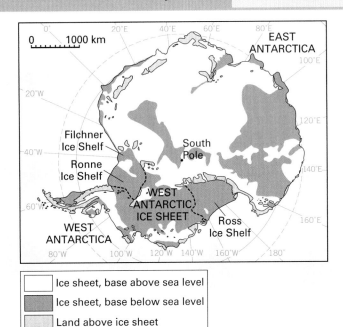

West Antarctic Ice Sheet A map of Antarctica showing the West Antarctic Ice Sheet. (Adapted with permission from C.R. Bentley, *Science*, Vol. 275, p. 1077. Copyright © American Association for the Advancement of Science)

The Antarctic Ice Sheet holds 91 percent of the world's ice, and if it were to melt entirely, mean sea level would rise by about 40 metres. The melting of the Greenland Ice Sheet, which holds most of the remaining volume of land ice, would release additional water. Although global warming will accelerate glacial melting and thinning of the ice at the edges of ice sheets, increased precipitation, held by warmer air over the ice sheets, should produce a net growth in ice sheet thickness. Thus, unless the warming is much greater than anticipated, it seems most unlikely that global warming will melt these ice sheets completely. However, other factors may be at work.

In 1993 and 1998, NASA researchers measured the height of the southern half of the Greenland Ice Sheet using an aircraft-mounted laser altimeter. They discovered that marginal thinning had reduced the volume of the ice sheet by about 42 cubic kilometres in that five-year period. The thinning was primarily on the eastern side of the ice sheet. On the western side, melting was more in balance with new snow accumulation. The loss of this ice volume is not large enough to have much effect on global sea level. However, a substantial flow of fresh meltwater into the ocean could influence the thermohaline circulation of the northern Atlantic (see Chapter 6), triggering an ocean cycle of several centuries' duration that would bring very cold winter weather to Europe.

Recently, attention has focused on the West Antarctic Ice Sheet, shown in the map to the right. Much of this vast expanse of Antarctic ice is "grounded" on a bedrock base that is well below sea level. Attached to the grounded ice sheet are the Ross, Ronne, and Filchner ice shelves, under which is sea water. They are also grounded at some points where the shelves enclose islands and overlay higher parts of the undersea topography. At present, the grounded ice shelves hold back the flow of the main part of the ice sheet.

Geophysicists regard this as an unstable situation. A rapid melting or deterioration of the ice shelves would release the back pressure on the main part of the ice sheet, allowing it to advance and thin rapidly. As the ice sheet thins, reduced pressure at its base would allow sea water to enter, and soon most of the ice would be afloat. The added bulk would then raise sea level by as much as six metres.

A key feature of the West Antarctic Ice Sheet is the presence of a number

free lake. Turbulence in the lake caused by wind and water circulation permits the coarser sediments to settle to the lake bed. When streams stop flowing in the low sun season, the finer clays are able to settle through the still waters beneath the ice cover on the lake.

INVESTIGATING THE LATE-CENOZOIC ICE AGE

The study of Ice Age glacial history progressed significantly in the 1960s when research into paleomagnetism was combined with core samples taken from the deep ocean floor. Paleomagnetism is a technique used to measure the absolute age of certain types of sediments. The Earth's magnetic field experienced many sudden reversals of polarity in the Cenozoic Era. The absolute ages of these reversals has been firmly established, and further study of the composition and chemistry of the layers within the core samples has provided a record of ancient temperature cycles in the air and ocean.

Deep-sea cores reveal a long history of alternating glaciations and interglaciations going back at least 2 million years and possi-

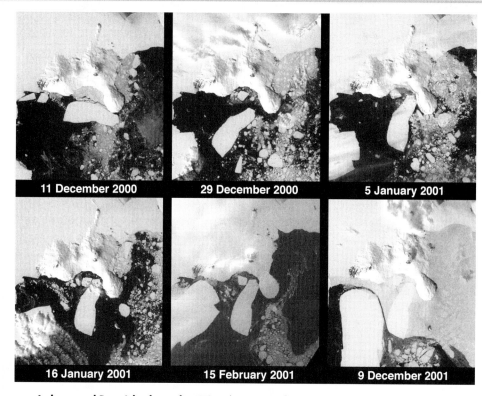

Icebergs and Ross Island seen by MISR This sequence of images, acquired by NASA's MISR instrument, shows the motion of huge icebergs near Ross Island, Antarctica, at the edge of the Ross Ice Shelf.

creasing flow will offset the effects of slowing Ross Shelf ice streams is not known.

Either way, new evidence suggests that there has been at least one collapse of the West Antarctic Ice Sheet during an interglacial period. Sediments extracted from below the ice sheet show the presence of marine organisms that indicate freely circulating ocean water at some time within the last two million years. Moreover, temperatures at the time of the collapse are believed to be not much greater than those today.

Meanwhile, the climatic warming of the past few decades seems to be causing the ice on the shelves to thin and fracture more easily, freeing huge icebergs. The photo below shows six satellite images of Ross Island, at the western edge of the Ross Ice Shelf, acquired in 2000 and 2001 by the MISR instrument. The island is about 75 kilometres long and includes Mount Erebus (leftmost peak), a volcano with an elevation of 3,743 metres. The first three images show a vast iceberg, designated C-16, as it rotates a quarter-turn counter-clockwise over the course of a month and then stops. Slowly moving in from the left is another huge iceberg, designated B-15A. In the last image, sea ice from the previous winter has expanded to surround C-16, and B-15A presses against it.

of ice streams within the sheet. Ice flow in these streams is much more rapid than in the surrounding ice mass, probably because geothermal activity provides enough heat to melt the ice at the base. This creates a liquid bottom layer that lubricates the ice motion. The ice streams blend gradually into the ice shelves, and represent their main supply source.

The flow of the ice streams is essential to maintaining the ice shelves. If these streams slowed, the ice shelf might retreat. This could release the unstable western ice sheet, producing catastrophic flooding. In 2002, it was reported that the ice streams feeding the Ross Ice Shelf were getting thicker and flowing more slowly. If the slowing of the ice streams continues at the present rate for another 70 to 80 years, they would stop altogether, denying the West Antarctic Ice Sheet an influx of ice. However, ice flow into other parts of the West Antarctic Ice Sheet seems to be accelerating. Whether this in-

bly 3 million years before present. The cores show that more than 30 glaciations occurred in late-Cenozoic time, spaced at intervals of about 90,000 years.

CAUSES OF THE LATE-CENOZOIC ICE AGE

Three principal causes are postulated for the Ice Age cycle of glaciations and interglaciations. First is a change in the relative positions of continents on the Earth's surface as a result of plate

tectonic activity. Second is an increase in the number and severity of volcanic eruptions. Third is a reduction in the Sun's energy output.

In Permian period, only the northern tip of Eurasia projected into the polar zone (see Chapter 12). As the Atlantic Basin opened up, North America moved westward and toward the North Pole to a position opposite Eurasia, with Greenland located between them. These plate motions brought an enormous land-mass area into a high latitude and surrounded the polar ocean with land. Because

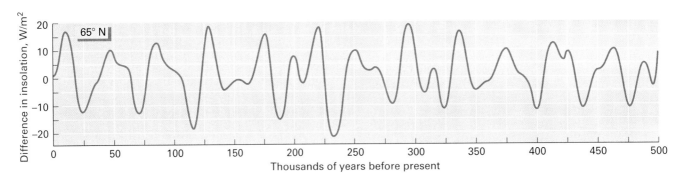

19.27 The Milankovitch Curve Is named after Milutin Milankovitch, the astronomer who first calculated it in 1938. The vertical axis shows fluctuations in summer daily insolation at lat. 65°N for the last 500,000 years. These are calculated from mathematical models of the change in the distance between the Earth and the Sun and the change in axial tilt over time. The zero value represents the present value. (Based on calculations by A. D. Vernekar, 1968. Copyright © A.N. Strahler)

the flow of warm ocean currents into the polar ocean was greatly reduced or totally cut off, ice sheets grew. The polar ocean was ice-covered much of the time, and average air temperatures in high latitudes decreased enough to allow ice sheets to grow on the surrounding continents. In addition, Antarctica moved southward during the breakup of Pangaea to a position over the South Pole, where it was ideally situated to develop a large ice sheet. The uplift of the Himalayan Plateau might also have modified weather patterns sufficiently to trigger the Ice Age.

The second theory suggests volcanic activity increased in late-Cenozoic time. Volcanic eruptions produce dust veils that linger in the stratosphere and reduce the intensity of solar radiation reaching the ground (see Chapter 3). Temporary cooling of near-surface air temperatures follows these eruptions. Although the geologic record shows periods of high levels of volcanic activity in the Miocene and Pliocene epochs, their role in initiating the Ice Age has not been convincingly demonstrated on the basis of current evidence.

Another possible cause of the Ice Age is a slow decrease in the Sun's energy output over the last several million years, perhaps as part of a long-term cycle of energy fluctuation. There is not enough data yet to identify this mechanism. However, satellites that probe the Sun's atmosphere and monitor its changing surface continue to acquire information.

CAUSES OF GLACIATION CYCLES

Several timing and triggering theories have been proposed for the glacial cycles of the Ice Age. The most widely accepted of these is the **astronomical hypothesis**, which is based on well-established motions of the Earth in its elliptical orbit around the Sun (see Chapter 3). Perihelion, the point in the orbit where the Earth is nearest the Sun, presently occurs about December 5; aphelion, when the Earth is farthest from the Sun, occurs about July 5. Astronomers have observed that the orbit slowly rotates

on a 108,000-year cycle, thus shifting the absolute time of perihelion and aphelion by a small amount each year. In addition, the orbit's shape varies on a cycle of 92,000 years, becoming more and less elliptical. This changes the distance between the Earth and the Sun and therefore the amount of solar energy the Earth receives at each point of the annual cycle.

The Earth's axis of rotation also experiences cyclic motions. The tilt angle of the axis varies from about 22 to 24 degrees on a 40,000-year cycle. The axis also "wobbles" on a 26,000-year cycle, moving in a slow circular motion. This change in the direction of the axis in space is called **precession**.

As a result of these cycles in axial rotation and solar revolution, the annual insolation experienced at each latitude changes from year to year. Figure 19.27 shows summer insolation received at lat. 65° N for the last 500,000 years, as calculated from these cycles; the graph is called the Milankovitch curve. The dominant cycle of the curve has a period of about 40,000 years; however, every second or third peak seems to be higher. Peaks at about 12,000, 130,000, 220,000, 285,000, and 380,000 years ago have been associated with rapid melting of ice sheets and the onset of deglaciations, as revealed by other dating methods involving ancient ice cores and deep lake sediment cores. It is now generally agreed that cyclic changes in insolation, operating through complex interactions among the atmosphere, ocean, and continental surfaces, are the primary cause of the glaciation cycles during the Ice Age.

HOLOCENE ENVIRONMENTS

The elapsed time span of about 10,000 years since the end of the Wisconsinan Glaciation is called the *Holocene Epoch*. It began with a rapid warming of ocean surface temperatures. Continental climate zones then quickly shifted toward the poles, and plants soon re-established in glaciated areas.

Three major climatic periods occurred during the Holocene Epoch leading up to the last 2,000 years. These periods are inferred from studies of changes in vegetation cover types observed in fossil pollen and spores preserved in glacial bogs. The earliest of the three is known as the Boreal stage, which was characterized by boreal forest vegetation in midlatitude regions. A general warming followed until the Atlantic stage, with temperatures somewhat warmer than they are today, was reached about 8,000 years ago. Next came a period of below-normal temperatures, the Sub-boreal stage, from 5,000 to 2,000 years before present (BP).

Through the availability of historical records and more detailed evidence, the climate of the past 2,000 years can be described in a finer detail. A warm period occurred between 1,000 to 800 years BP. The Little Ice Age followed from 550 to 150 years BP, during which valley glaciers made new advances and extended to lower elevations.

Within the last century, global temperatures have slowly increased and are projected to increase more rapidly for at least the next 100 years. This has profound implications for existing glaciers and the continental ice sheets of Greenland and Antarctica. *Eye on Global Change 19.4 • Ice Sheets and Global Warming* discusses the potential impact on these ice sheets.

A Look Ahead

The preceding group of chapters has reviewed landform-making processes that operate on the surface of the continents. Human influence on landforms is felt most strongly on surfaces of fluvial denudation due to changes caused by agriculture and urbanization. Landforms shaped by wind and by waves and currents are also highly sensitive to changes induced by human activity. Only continental ice sheets maintain their integrity and are to date largely undisturbed. However, they are increasingly sensitive to climate changes induced by industrial activity.

CHAPTER SUMMARY

- Glaciers form when snow accumulates to a great depth, creating a mass of ice that is plastic in its lower layers and flows outward or downhill from a centre in response to gravity. As they move, glaciers can deeply erode bedrock by abrasion and plucking. The eroded fragments, incorporated into the flowing ice, leave depositional landforms when the ice melts.

- Alpine glaciers develop in cirques in high mountain locations. They flow down valleys on steep slopes, picking up rock debris and depositing it in lateral and terminal moraines. Through erosion, glaciers carve distinctive U-shaped glacial troughs in mountainous regions. Glacial troughs will become fjords if later submerged by a rising sea level.

- Ice sheets are huge plates of ice that cover vast areas. They are present today in Greenland and Antarctica. The Antarctic Ice Sheet includes *ice shelves*—great plates of floating glacial ice. Icebergs form when glacial ice flowing into an ocean breaks into great chunks and floats free. *Sea ice*, which is much thinner and more continuous, is formed by direct freezing of ocean water and accumulation of snow.

- An ice age includes alternating periods of glaciation, *deglaciation*, and interglaciation. During the past 2 to 3 million years, the Earth has experienced the Late-Cenozoic Ice Age. During this ice age, continental ice sheets expanded and melted as many as 30 times. The most recent glaciation was the *Wisconsinan Glaciation*,

in which ice sheets covered much of North America. Contemporaneous expansion of ice occurred in Europe, as well as parts of northern Asia and southern South America.

- Moving ice sheets create many types of landforms. Bedrock is grooved and scratched. Where rocks are weak, long valleys can be excavated to depths of hundreds of metres. The melting of glacial ice deposits glacial drift, which may be stratified by water flow or deposited directly as till. Moraines accumulate at ice edges. *Outwash plains* are built up by meltwater streams. Tunnels within the ice leave streambed deposits, or eskers. Till may spread out smoothly and thickly under an ice sheet, leaving a *till plain*. This may be studded with elongated till mounds, called drumlins. Many small lakes have developed on the undulating till plains, especially in parts of the Canadian Prairies. Meltwater streams built *glacial deltas* into lakes formed at the ice margin and line lake bottoms with clay and silt. These features remain after the lakes have drained. Level *glaciolacustrine plains* are common features throughout the Canadian Prairies.

- Several factors have been proposed to explain the cause of present ice age glaciations and interglaciations. These factors include ongoing change in the global position of continents, an increase in volcanic activity, and a reduction in the Sun's energy output. Individual cycles of glaciation seem strongly related to cyclic changes in the distance between the Earth and the Sun and axial tilt.

KEY TERMS

ablation	esker	interglaciation	tidewater glacier
alpine glacier	fjord	Late-Cenozoic Ice Age	till
astronomical hypothesis	glacial drift	lateral moraine	varves
cirque	glacial trough	precession	zone of ablation
drumlin	glaciation	stratified drift	zone of accumulation
equilibrium line	ice sheet	terminal moraine	

REVIEW QUESTIONS

1. How does a glacier form? What factors are important? Why does a glacier move?
2. Distinguish between alpine glaciers and ice sheets.
3. What is a glacial trough and how does it form? What is its basic shape?
4. Where are ice sheets present today? How thick are they?
5. Contrast sea ice and icebergs, including the processes by which they form.
6. Identify the Late-Cenozoic Ice Age. When did it begin? What was the last glaciation in this cycle? When did it end?
7. What areas were covered with ice sheets by the last glaciation? How was sea level affected?
8. What are moraines? How do they form? What types of moraines are there?
9. Identify the landforms and deposits associated with stream action at or near the front of an ice sheet.
10. Identify the landforms and deposits associated with deposition underneath a moving ice sheet.
11. Identify the landforms and deposits associated with lakes that form at ice sheet margins.
12. What cycles are known to affect the amount of solar radiation received by the Earth's polar regions?
13. What is the Milankovitch curve? What does it show about warm and cold periods during the last 500,000 years?
14. How have environments changed during the Holocene Epoch? What periods are recognized, and what are their characteristics?

FOCUS ON SYSTEMS 19.1   A Glacier as a Flow System of Matter and Energy

1. Describe the matter flow system of an alpine glacier. How can a glacier achieve a steady state?
2. What two energy flow subsystems can be recognized in the flow of an alpine glacier? Describe each.

EYE ON GLOBAL CHANGE 19.4   Ice Sheets and Global Warming

1. What process seems to be offsetting the melting rate of ice sheets as the climate warms?
2. Why is the West Antarctic Ice Sheet considered unstable? What is the role of ice streams in maintaining the present size of the ice sheet?
3. What change has recently been observed in the flow of ice streams into the Ross Ice Shelf? What are its implications?

VISUALIZATION EXERCISES

1. What are some typical features of an alpine glacier? Sketch a cross-section along the length of an alpine glacier and label it.
2. Refer to Figure 19.8b, which shows the Antarctic continent and its ice cap. Identify the Ross, Filchner, and Larsen ice shelves. Use the scale to measure the approximate area of each in square kilometres. Consulting an atlas, identify a province in Canada that is nearest in area to each ice shelf.

ESSAY QUESTIONS

1. Imagine that you are planning a car trip to the Canadian Rockies from Winnipeg. What glacial landforms might you expect to find in the mountains, and how do they differ from those you would encounter as you travel there?

2. At some time during the latter part of the Pliocene Epoch, the Earth entered an ice age. Describe the nature of this ice age and the cycles that occurred within it. What are the proposed explanations for the cause of an ice age and its cycles? What cycles have been observed since the last ice sheets retreated?

EYE ON THE LANDSCAPE |

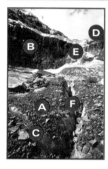

Chapter Opener   Mt. Angel glacier, Columbia Icefields, BC. Angular rock fragments (**A**) loosened by freeze-thaw activity from exposed limestone (**B**), or eroded by glacial action now mantle the ice. This superficial cover regulates melting by insulating the underlying ice (**C**). The upper section of the glacier (**D**) has thinned considerably and down-valley movement is limited to a small ice-fall (**E**). Meltwater channels (**F**) are dissecting the ice surface and will promote further melting through direct exposure to incoming radiation,

Alpine glacier, Alaska, USA. Fresh snow (**A**) covers the surface of the glacier upstream, but towards its terminus (**B**) temperatures are conducive to melting, and the darker colour is caused by rock debris exposed on the ice. Medial moraines (**C**) appear as dark stripes where rock materials are concentrated on the surface of the glacier. The trim-line (**D**) on the lower valley wall suggests that

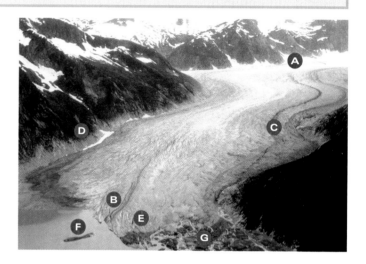

recent ablation has lowered the surface of the glacier by several metres. Crevasses (**E**) have formed near the snout where melting contributes water and sediment to the proglacial lake (**F**), with mounds of transported debris (**G**) accumulating elsewhere as the glacial recedes.

Part 5

CHAPTERS IN PART 5

20 Soil Systems
21 Systems and Cycles of the Biosphere
22 Biogeographic Processes
23 The Earth's Terrestrial Biomes

SYSTEMS AND CYCLES OF SOILS AND THE BIOSPHERE

Sagebrush dominates the landscape in this dry intermontane environment with occasional stands of aspen and ponderosa pine where soil conditions are more favourable.

P art 5 focuses directly on the life layer, where biological and physical processes interact. Soils are formed by the physical processes that weather and transport rock material, as well as the biological processes that result in the decay of organic matter. These processes are strongly influenced by temperature and the abundance of water at the surface; hence, they are functions of climate. Climate also describes the availability of sunlight, water, and heat, which affect photosynthesis and thus determine plant growth. Moreover, individual landforms present distinct habitats for the development of local soil types, as well as local biotic communities. ■ Soils and the biosphere provide a logical conclusion to a study of physical geography because these topics relate to many of the systems, cycles, and processes that earlier chapters introduced. Chapter 20 provides an overview of the key characteristics of soils, the important processes of soil formation, the classification of soils, and the global distribution of soil types. Chapter 21 describes ecosystems, focusing on energy and the cycling of major nutrient elements by ecosystems and their environments. Chapter 22 discusses the biogeographic processes that govern the distribution of plants in space and time. Finally, Chapter 23 examines the nature of biomes and their global distribution patterns, especially as they relate to climate types.

Chapter 20

EYE ON THE LANDSCAPE

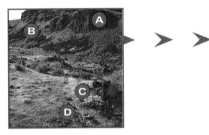

Horsethief Butte State Park, Washington. What else would the geographer see? . . . Answers at the end of the chapter.

SOIL SYSTEMS

The Nature of the Soil
Soil Colour and Texture
Soil Structure
Soil Minerals
Soil Colloids
Soil Acidity and Alkalinity
Soil Moisture
The Soil Water Balance
A Simple Soil Water Budget
Working It Out 20.1 • Calculating a Simple Soil Water Budget
Soil Development
Soil Horizons
Soil-Forming Processes
Soil Temperature
Surface Configuration
Biological Processes
The Global Scope of Soils
Soils of Canada
Soils of the World

This chapter describes the processes that form soils and give them their distinctive characteristics, soil classification, and global distribution of soil types. **Soil** is the uppermost layer of the land surface that plants use and depend on for nutrients, water, and physical support. The many different types of soils reflect the high variability in factors and processes that influence soil development. It is generally agreed that soils are a function of five main factors: parent material, topography, climate, biological factors, and time.

For most soils, the bulk of the material present is disintegrated rock. The type of rock from which the soil is derived constitutes the parent material, which has an important influence on the soil's physical and chemical properties. For example, soils formed from granites are often rich in clay. They may be water retentive and quite fertile because of the various minerals released from the weathered feldspars. Sandstone, on the other hand, typically forms sandy soils that drain quite rapidly and are comparatively low in nutrients.

Surface erosion by overland flow and runoff is rapid where slopes are steep and less water infiltrates compared with the erosion on gentle slopes. As a result, soils generally are thinner on steeper slopes. Poor drainage in low-lying areas also tends to slow soil development and may lead to anaerobic conditions. This can affect soil chemistry and produce bluish-green colours due to the presence of iron sulphide. In areas of undulating terrain, it is common to find soils becoming progressively wetter downslope. This results in a hydrologic sequence and is characteristic of areas of hummocky glacial terrain, such as parts of the Canadian Prairies. Gentle slopes, therefore, are considered ideal for soil development, as drainage is good and the erosion rate can be compensated by the addition of weathered parent material.

Climate, measured by precipitation and temperature, is an important determinant of soil properties also. Precipitation controls the downward movement of nutrients and other chemical compounds in soils. Temperature acts to control the rate of decay of organic matter that falls to the soil from the plant cover or is provided to the soil by the death of roots. When conditions are warm, decay organisms work efficiently, readily consuming organic matter. Thus, organic matter is generally low in soils of the tropical and equatorial zones. Under cooler conditions, decay proceeds more slowly, and organic matter is more abundant in the soil. In desert areas, vegetation growth is slow or absent, so organic matter will be low, regardless of temperature conditions.

Some of North America's richest soils develop in the continent's interior under a cover of thick grass sod. The deep roots of the grass, in a cycle of growth and decay, deposit nutrients and organic matter throughout the soil layer. In northern regions, conifer forests provide a surface layer of decaying needles that keep the soil quite acidic. This acidity allows nutrients to be washed below the root depth, resulting in low fertility.

Other biotic factors play a role in soil development. Microscopic organisms, such as bacteria, are not only important for decomposition, they also play a role in soil chemistry, such as nitrogen cycling. Worms and other large organisms constantly rework the soil and help to mix soil components, maintain soil structure, and improve drainage and aeration.

The characteristics and properties of soils take time to develop. For example, a fresh deposit of mineral matter, like the clean, sorted sand of a dune, may require hundreds to thousands of years to acquire the structure and properties of a sandy soil.

THE NATURE OF THE SOIL

Soil, as the term is used in soil science (or *pedology*), is a natural surface layer that contains living matter and can support plants. The soil consists of matter in all three states: solid, liquid, and gas. It includes both *mineral matter* and *organic matter*. Mineral matter is largely derived from rock material, whereas organic matter is of biological origin and may be living or dead. Living matter in the soil consists of not only plant roots, but also many kinds of organisms, including micro-organisms.

Soil scientists use the term *humus* to describe finely divided, partially decomposed organic matter in soils. Some humus rests on the soil surface, and some is mixed through the soil. Rainfall that percolates through the soil gradually carries the finest humus particles downward to lower soil layers. When abundant, humus can give the soil a brown or black colour.

The soil atmosphere comprises oxygen and other gases that diffuse from the air, as well as gases, such as carbon dioxide and methane, that are derived from respiration and decomposition processes. Soil water will generally contain dissolved substances, such as nutrients. The solid, liquid, and gaseous matter in soil is continuously interacting through chemical and physical processes, making soil a dynamic layer.

Although it is commonly assumed that soils occur everywhere, large expanses of continents possess a surface layer that cannot be called soil. For example, dunes of moving sand, bare rock surfaces of deserts and high mountains, and surfaces of fresh lava near active volcanoes have no soil layer.

Soil characteristics are developed over a long time period through a combination of many processes acting together. Physical processes break down rock fragments into smaller and smaller pieces. Chemical processes alter the composition of the original rock, producing new minerals. Taken together, these physical and chemical changes are referred to as *weathering* (see Chapter 14) and are part of the process by which soils develop their properties and characteristics.

In most soils, the inorganic material present consists of fine mineral particles. The term **parent material** describes all forms of mineral matter that are suitable for transformation into soil. Parent material normally is derived from the underlying *bedrock* (Figure 20.1), although in some cases, materials transported by streams or other geomorphological agents may be deposited in sufficient thickness to act as parent material. Over time, weathering processes weaken, disintegrate, and break bedrock apart, forming a layer of *regolith*, or residual mineral matter.

SOIL COLOUR AND TEXTURE

The most obvious feature of a soil is its colour. Soils of the Prairies have a black or dark brown colour because they contain abundant humus. The presence of iron-containing oxides creates red or yellow colours, while greenish-blue colours can indicate very moist conditions. The Munsell colour system is used to standardize the assessment of soil colour. The system is based on three components—hue (a specific colour), value (lightness and darkness), and chroma (colour intensity)—that are arranged in a book of colour chips

20.1 A cross-section through the land surface In this cross-section, vegetation and forest litter lie on top of the soil. Below is regolith, produced by the breakup of the underlying bedrock.

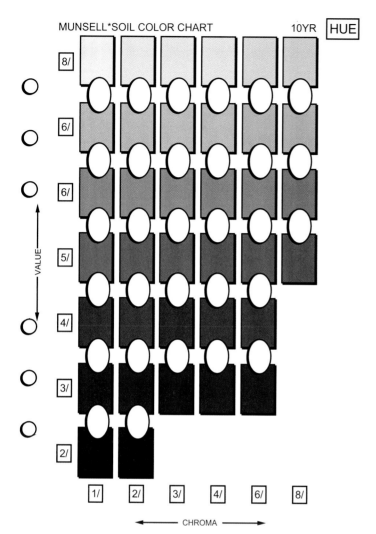

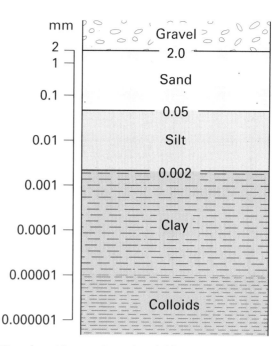

20.3 Mineral particle sizes Size grades, which have names like sand, silt, and clay, refer to mineral particles within a specific size range.

20.2 A sample page from the *Munsell Soil Color Book* Three components are used to assess soil colour. Hue refers to a specific colour and is represented by the pages in the book, in this case 10YR; value, or brightness, of that colour decreases from top to bottom on the page; and chroma, or intensity, changes across the page. Thus, a soil that matches the colour chip in the fourth row of the sixth column is assigned the Munsell notation 10 YR 5/8, and is described as yellowish brown.

(Figure 20.2). A sample of soil is visually matched to a colour chip and assigned the corresponding Munsell notation.

In some areas, soil colour may be inherited from the parent mineral material, but more generally, soil colour is generated by soil-forming processes. For example, a white surface layer in soils of dry climates often indicates the presence of mineral salts brought upward by evaporation. A pale, ash-grey layer near the top of soils of the boreal forest climate results when organic matter and nutrients are washed downward, leaving only pure, light-coloured mineral matter behind.

The mineral matter of the soil consists of individual particles of various sizes. The proportion of particles that fall into each of three size grades—sand, silt, and clay —determines **soil texture**. Figure 20.3 shows the diameter range of each of these grades. The finest of all soil particles are called **colloids**. Gravel and other particles larger than two millimetres in diameter are removed prior to assessing soil texture.

Soil texture is described by a series of names that emphasize the dominant particle size, whether sand, silt, or clay. With practice, the relative proportions of each size class can be determined by hand—sand feels gritty, silt is slippery, and clay is sticky—but texture usually is determined in a laboratory using methods that are based on the settling rate of different sized particles. Figure 20.4 gives examples of five soil textures with typical percentage compositions. A *loam* is a mixture containing a substantial proportion of each of the three grades. Loams are classified as sandy, silty, or clay-rich when one of these grades is dominant.

Texture largely determines the ability of the soil to retain water. Coarse-textured (sandy) soils have many small passages between the mineral grains that quickly conduct water to deeper layers. If the soil is a fine clay, the spaces between the particles are much smaller, and water will penetrate more slowly. Finer soils also tend to retain more water.

SOIL STRUCTURE

Whereas soil texture refers to the sizes of the individual particles that make up the soil, *soil structure* refers to the way in which soil grains are grouped together into larger masses, called peds, that

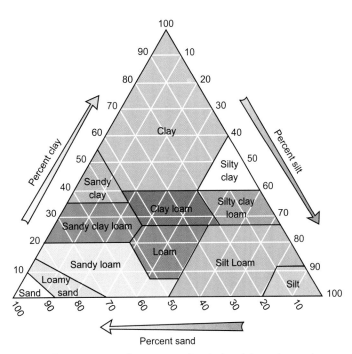

20.4 Soil textures Once the proportions of sand, silt, and clay are known, the textural classes can be read from a graph, as shown for five different soils.

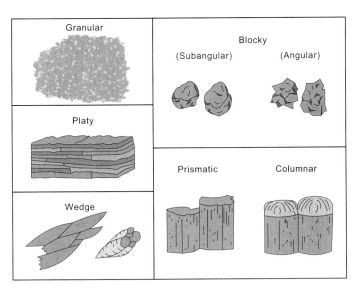

20.5 Types of soil structure Soil structure refers to how soil particles are combined to form crumbs or larger units. Structure, like texture, is important for air and water movement within a soil.

are bound together by soil colloids. Small peds, roughly shaped like spheres, give the soil a granular or crumb-like structure (Figure 20.5). Larger peds provide an angular, blocky structure. Peds form as a result of various processes including seasonal climatic changes, which cause wetting and drying, and freezing and thawing. Soils with a well-developed granular or blocky structure are easy to cultivate. Soils with high clay content can lack peds and are sticky and heavy when wet; when dry, they may be too hard to manipulate easily.

SOIL MINERALS

Soil scientists recognize two classes of minerals abundant in soils: primary minerals and secondary minerals. The *primary minerals* are compounds present in unaltered rock. They are mostly silicate minerals, with varying proportions of aluminum, calcium, sodium, iron, and magnesium. Primary minerals account for a large fraction of the solid matter of many kinds of soils, but they play no important role in sustaining plant or animal life.

When primary minerals are exposed to air and water at or near the Earth's surface, their chemical composition slowly changes. This process is part of *mineral alteration* (see Chapter 11). The primary minerals change into **secondary minerals**. In terms of soils, the most important secondary minerals are the *clay minerals*. They form the majority of fine mineral particles in soils and are essential to soil development and fertility because of their ability to hold base ions.

Mineral oxides are secondary minerals that are derived by chemical oxidation. In soils, the most important are the *sesquioxides* of aluminum and iron. Aluminum sesquioxide (Al_2O_3) is formed from the combination of three atoms of oxygen with two atoms of aluminum. The mineral *bauxite* is a combination of aluminum sesquioxide and water molecules bound together. It occurs as hard, rocklike lumps in layers below the soil surface. In areas of unusually high concentration, bauxite is mined as an aluminum ore. When combined with water molecules, iron sesquioxide (Fe_2O_3) forms *limonite*, a yellowish to reddish mineral that imparts reddish and brown colours in soils. It is also familiar as rust.

SOIL COLLOIDS

Soil colloids consist of particles smaller than 0.000,01 millimetres. Mineral colloids are usually very fine particles of clay minerals that are thin and plate-like. When well-mixed in water, these very small particles remain suspended indefinitely, giving the water a murky appearance. Soil colloids are also derived from organic material that is resistant to decay.

The nature of the clay minerals in a soil determines its *base status*. Clay minerals that can hold abundant base ions give soils a *high base status* and generally will be highly fertile. If the clay minerals hold a smaller supply of bases, the resulting *low base status* will result in low fertility. Humus colloids also have a high capacity to hold bases, so the presence of humus is usually associated with potentially high soil fertility.

Soil colloids attract soil nutrients that are dissolved in soil water. Colloid surfaces tend to be negatively charged because of their molecular structure, and thus attract and hold positively charged

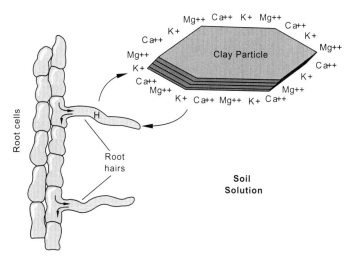

20.6 Nutrient exchange Plant nutrients can move from the adsorption sites on colloids into the soil water solution where they are available for root uptake. Colloidal particles have negative surface charges that attract and hold positively charged ions.

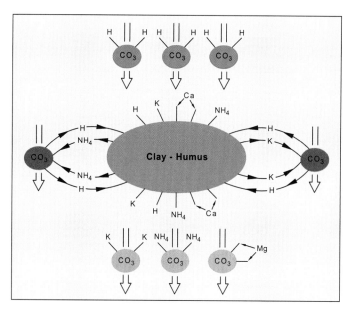

20.7 Soil leaching Carbon dioxide is a soluble gas that combines with water in the atmosphere to form very weak carbonic acid. As rainwater percolates through the soil, nutrient ions attached to the soil colloids are replaced by hydrogen ions. In this example, potassium, ammonium, and magnesium are removed as carbonates. Over time, the exchange of nutrients for hydrogen increases soil acidity and reduces soil fertility.

ions (cations) derived from the dissociation of chemical compounds. One important group of cations is the **bases**, which include plant nutrients such as calcium (Ca^{++}), magnesium (Mg^{++}), and potassium (K^+). Although colloids retain these ions in the soil, they also release them to plants (Figure 20.6).

SOIL ACIDITY AND ALKALINITY

Generally, cations are held tightly enough on adsorption sites to restrict their loss through leaching. However, any cation present in the soil solution can be exchanged with other cations attached to the colloids. Also, any cation in solution can potentially be removed from the soil by percolating water. The removal of plant nutrients from a soil through leaching is primarily associated with the abundance of hydrogen ions in the soil solution. Hydrogen (H^+), unlike

the bases, is not considered a plant nutrient. The presence of hydrogen ions, and to a lesser degree, aluminum (Al^{+++}), makes the soil solution acidic, which can affect soil fertility and plant growth when these ions displace nutrient bases from the soil colloids. Once displaced, the bases can be gradually washed out of the soil; this, in turn, increases its acidity (Figure 20.7).

The degree of a soil's acidity or alkalinity is designated by its pH value. A pH value of 7 represents a neutral state. Values lower than 7 are in the acid range, values higher than 7 are in the alkaline range. The availability of plant nutrients is related to soil pH (Table 20.1).

Table 20.1 Soil acidity and alkalinity

pH	4.0 4.5	5.0	5.5	6.0 6.5	6.7 7.0	8.0	9.0	10.0	11.0
Acidity	Very strongly acid	Strongly acid	Moderately acid	Slightly acid	Neutral	Weakly alkaline	Alkaline	Strongly alkaline	Excessively alkaline
Lime requirements	Lime needed except for crops requiring acid soil		Lime needed for all but acid-tolerant crops	Lime generally not required	No lime needed				
Occurrence	Rare	Frequent	Very common in cultivated soils of humid climates			Common in subhumid and arid climates			Limited areas in desert

Source: Based on data from C. E. Millar, L. M. Turk, and H. D. Foth, *Fundamentals of Soil Science*, New York: John Wiley & Sons, 1972.

Availability of most nutrients is greatest in the general pH range of 6.5 to 7.5, although for some, such as iron, availability increases in more acidic soils; for others, such as molybdenum, availability increases under more alkaline conditions. High soil acidity is typical of cold, humid climates where organic acids are added to the soil from partially decomposed plant matter. In dry climates, soils are typically alkaline, often because of high sodium concentrations.

SOIL MOISTURE

Besides providing nutrients for plant growth, the soil layer serves as a reservoir for the moisture that plants require. The soil receives water from rain and melting snow. Some of the water can run off the soil surface and flow into brooks, streams, and rivers, eventually reaching the sea. Water that sinks into the soil can return to the atmosphere as water vapour, either by evaporation or transpiration from plant leaves, which together are referred to as *evapotranspiration*. In addition, some water can flow completely through the soil layer to recharge supplies of groundwater at depths below the reach of plant roots.

Precipitation that infiltrates and moistens the soil results in *soil water recharge*. Once the soil layer's maximum capacity to hold water is reached, water movement continues downward. If no further water enters the soil for a time, the excess continues to drain; however, some water will cling to the soil particles through *capillary tension*. It remains there until it evaporates or is absorbed by plant roots.

When a soil has been saturated by water and then drains freely under gravity until no more water moves downward, the soil is said to be at *field capacity*. For most soils, field capacity is achieved within two or three days. Field capacity depends largely on the soil's texture (Figure 20.8). Fine-textured clay soils hold more water than coarse-textured sandy soils, because fine particles have a much larger surface area per unit of volume.

Wilting point is an agricultural term that approximates the water storage level below which plants can experience moisture stress (Figure 20.8). Wilting point also depends on soil texture, because fine particles hold water more tightly, making it difficult for plants to extract. Thus, plants can wilt in fine-textured soils even though more soil water is present than in coarse-textured soils. The difference between the field capacity of a soil and its wilting point is the maximum available water capacity; this is greatest in loamy soils.

THE SOIL WATER BALANCE

Water in the soil is a critical resource needed for plant growth. The amount of water available at any given time is determined by the *soil water balance*, which includes the gain, loss, and storage of soil water. Figure 20.9 illustrates the components of the soil water balance. Water held in storage in the soil water zone is increased by recharge during precipitation, but decreased by use through evapotranspiration. Surplus water is disposed of by downward percolation to the groundwater zone or by overland flow.

The rate of water vapour return to the atmosphere from the ground and its plant cover is called *actual evapotranspiration (Ea)*.

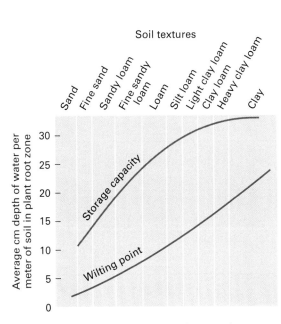

20.8 Field capacity and wilting point according to soil texture Fine-textured soils hold more water. They also hold water more tightly, so plants wilt more quickly.

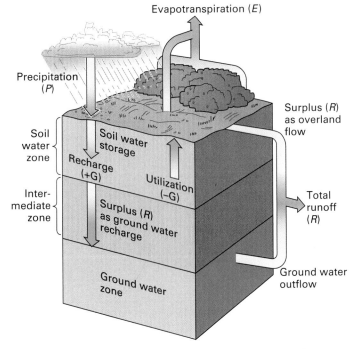

20.9 Schematic diagram of the soil water balance in a soil column (Copyright © A.N. Strahler)

Potential evapotranspiration (Ep) represents the water vapour loss under ideal conditions. These ideal conditions include a complete cover of uniform vegetation consisting of fresh green leaves and no bare ground, and an adequate water supply to maintain the soil's storage capacity at all times. This second condition can be fulfilled naturally by abundant and frequent precipitation, or artificially by irrigation. These terms can be simplified as follows:

Actual evapotranspiration *(Ea)* is **water use**.
Potential evapotranspiration *(Ep)* is **water need**.

Need refers to the quantity of soil water required to maximize plant growth in the given conditions of solar radiation, air temperature, wind speed, and the available supply of nutrients. Of these, the most important factor in determining water need is temperature. The difference between water use and water need is the *soil water shortage*, or *deficit*. This is the quantity of water that irrigation must supply to achieve maximum crop growth within an agricultural system.

A SIMPLE SOIL WATER BUDGET

The amounts of water needed to satisfy each process in the soil water balance comprise the soil *water budget.* They are calculated by adding and subtracting the mean monthly values at a given observing station. All terms of the soil water budget are stated in centimetres of water depth, as is precipitation.

Figure 20.10 shows a simplified soil water budget. The terms needed to complete the budget are the following:

Precipitation, *P*
Water need, *Ep*
Water use, *Ea*

Storage withdrawal, $-G$
Storage recharge, $+G$
Soil water shortage, *D*
Water surplus, *R*

Points on the graph represent average monthly values of precipitation and water need. In this example, precipitation *(P)* is much the same in all months. In contrast, water need *(Ep)* shows a strong seasonal cycle, with low values in winter and a high summer peak.

At the start of the year, precipitation greatly exceeds water use, and a large water surplus *(R)* exists. Runoff disposes of this surplus. By May, water use exceeds precipitation, creating a water deficit. In this month, plants begin to withdraw soil water from storage.

Storage withdrawal $(-G)$ is represented by the difference between the water-use curve and the precipitation curve. As storage withdrawal continues, however, plants obtain soil water with increasing difficulty. Thus, water use *(Ea)* is less than water need *(Ep)* during this period. Storage withdrawal continues throughout the summer, and the deficit period lasts through September. The area labelled soil water shortage *(D)* is the difference between water need and water use. It represents the total quantity of water needed by irrigation to ensure maximum growth throughout the deficit period.

In October, precipitation *(P)* begins to exceed water need *(Ep)* again, but the soil must first absorb an amount equal to the summer storage withdrawal. So a period of *storage recharge* $(+G)$ follows, which lasts through November. In December, the soil reaches its full storage capacity. Now, a water surplus *(R)* sets in again, lasting through the winter. *Working It Out 20.1 • Calculating a Simple Soil Water Budget* explains the procedures used to calculate a water budget.

SOIL DEVELOPMENT

SOIL HORIZONS

Most soils possess **soil horizons**—distinctive layers that differ in physical and chemical composition, organic content, or structure (Figure 20.11). Soil horizons usually develop by either selective removal or accumulation of certain ions, colloids, and chemical compounds caused by water moving through the soil. A **soil profile** is the full set of horizons exposed in a soil pit that is excavated down to the parent material.

Figure 20.11 illustrates the idea of the *pedon*—a soil column extending from the surface to a lower limit in regolith or bedrock—which is the smallest distinctive division of the soil of a given area. The pedon exhibits all the features needed to properly classify and describe the soil at a location. An area of soil of the same type, which is made up of many pedons, is called a *polypedon*.

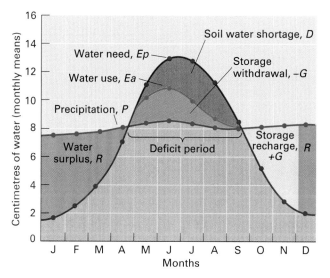

20.10 A simplified soil water budget This soil water budget is typical of a midlatitude moist climate.

WORKING IT OUT | 20.1   **Calculating a Simple Soil Water Budget**

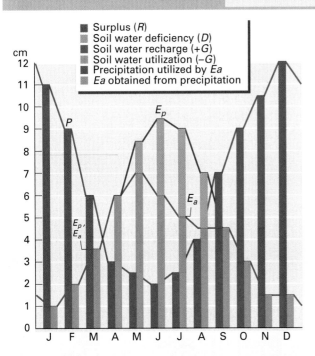

A model soil water budget Bars are scaled to the values in the table.

The soil water balance is a system in which an input of precipitation flows through soil pathways to be output as evapotranspiration or runoff. The amount of water returned to the atmosphere through evapotranspiration *(Ea)* is determined largely by temperature and the amount of vegetation cover. The amount of runoff will depend, in the long term, on how water use *(Ea)* compares with precipitation *(P)*. If precipitation is greater, then runoff will occur, both as overland flow and as groundwater outflow.

Water storage *(S)* in the soil layer is an important characteristic of the system. This storage provides a reserve of water for plants to draw upon, so that water use can exceed precipitation for part of the year. Later, when precipitation exceeds water use, the reserve can be recharged, and only after the reserve is replenished can runoff occur again in the annual cycle.

In this example, the monthly soil water budget is calculated for a station

in the marine west-coast climate, which exhibits wet, cool winters and warm, drier summers. For each month, the basic data required are precipitation *(P)*, water use *(Ea)*, and water need *(Ep)*, from which the other quantities are derived. In any month, precipitation can follow only three pathways: to the atmosphere as water use *(Ea)*, to streams and rivers as runoff *(R)*, or to storage *(+G)* within the soil layer. In January, precipitation is 11 centimetres, and water use is 1 centimetre, leaving 10 centimetres for storage or runoff. At this time of the year, storage is full, however, so the entire amount goes to runoff.

The flow of runoff persists until April, when water use (6 centimetres) exceeds precipitation (3 centimetres) by 3 centimetres. The needed 3 centimetres is withdrawn from storage, giving the value of −3 for −G. Storage

withdrawals continue through August, yielding a total withdrawal of 14.5 centimetres.

In September, precipitation (7 centimetres) again exceeds water use (4.5 centimetres), leaving a surplus (2.5 centimetres) for +G. This surplus begins to recharge the soil water storage. In October, soil water storage continues to increase into November, when it is 9 centimetres. However, only 6 centimetres is required to balance the total summer withdrawal of 14.5 centimetres, leaving 9 − 6 = 3 centimetres for runoff *(R)*. In December, the surplus goes entirely to runoff.

Soil water shortage *(D)* is the difference between water need *(Ep)* and water use *(Ea)*. The two quantities are equal until May, when water use drops below water need as significant water storage withdrawals begin to occur. This condition persists until September, when abundant precipitation allows water use to equal water need.

Note, for practical applications in agriculture, the depth of storage water in the soil that is available to plants is normally limited to 30 centimetres for field crops.

Simplified example of a soil water budget

Equation	P =	Ea	−G	+G	+R	Ep	(Ep − Ea) = D
January	11.0 =	1.0			+10.0	1.0	0.0
February	9.0 =	2.0			+7.0	2.0	0.0
March	6.0 =	3.5			+2.5	3.5	0.0
April	3.0 =	6.0	−3.0			6.0	0.0
May	2.5 =	7.0	−4.5			8.5	1.5
June	2.0 =	6.0	−4.0			9.5	3.5
July	2.5 =	5.0	−2.5			9.0	4.0
August	4.0 =	4.5	−0.5			7.0	2.5
September	7.0 =	4.5		+2.5		4.5	0.0
October	9.0 =	3.0		+6.0		3.0	0.0
November	10.5 =	1.5		+6.0	+ 3.0	1.5	0.0
December	12.0 =	1.5			+10.5	1.5	0.0
Totals	78.5 =	45.5	− 14.5	+14.5	+33.0	57.0	11.5
	78.5 =	78.5					

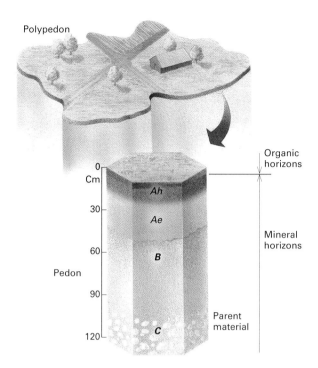

20.11 Soil horizons A column of soil will normally show a series of horizons, which are horizontal layers with different properties.

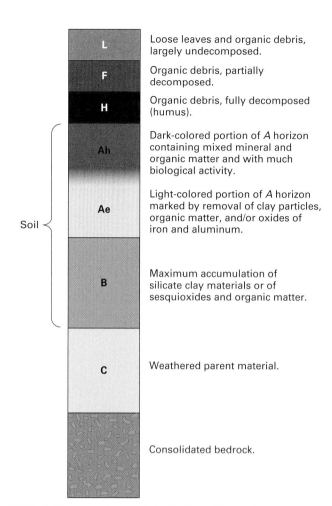

20.12 Soil horizons characteristic of a boreal forest climate A sequence of horizons that might appear in a forest soil developed in a cool, moist climate.

Figure 20.12 shows the main types of horizons found in a representative moist forest soil. Organic horizons overlie the mineral horizons and are formed from accumulations of material derived from plants and animals. Several types of organic layers are distinguished, depending on the state of decay. The Canadian System of Soil Classification recognizes three organic horizons that contain decayed matter ranging from slightly *(L)* to moderately *(F)* to very decomposed *(H)*. In addition, horizons composed of peat are designated *O*.

The mineral horizons below the organic layers of the surface are distinguished as the *A*, *B*, and *C* horizons. Plant roots readily penetrate the *A* and *B* horizons and influence soil development within them. In its scientific usage, soil refers only to the *A* and *B* horizons. The *A* horizon is the uppermost mineral horizon. Its upper portion *(Ah)* is rich in organic matter, consisting of numerous plant roots and down-washed humus from the organic horizons above. In the lower part of the *A* horizon *(Ae)*, clay particles and oxides of aluminum and iron, as well as plant nutrients, are removed by downward-seeping water that leaves behind grains of sand or coarse silt.

The *B* horizon receives the clay particles and aluminum and iron sesquioxides, as well as organic matter washed down from the *A* horizon. It is made dense and hard by the filling of natural spaces with clays and sesquioxides. Transitional horizons are present at some locations.

Beneath the *B* horizon is the *C* horizon, which consists of the parent mineral matter of the soil, or weathered regolith. Below the regolith lies unaltered bedrock or accumulated sediments.

SOIL-FORMING PROCESSES

There are four classes of soil-forming processes (Table 20.2). The first includes *soil enrichment* processes, which add material to the soil. For example, inorganic enrichment occurs when sediment is brought from higher to lower areas by overland flow. Stream flooding also deposits fine mineral particles on low-lying soil surfaces. Wind is another source of fine material that can accumulate on the soil surface. Organic enrichment occurs when humus, accumulating in organic surface layers, is carried downward to the *A* horizon below.

The second class of soil-forming processes includes processes that remove material from the soil body. Such *removal* occurs when surface erosion carries sediment away from the soil's uppermost layer. Another important process is *leaching*, in which percolating water dissolves soil materials and moves them below the soil profile or into the groundwater.

In moist climates, surplus soil water moves downward to the groundwater zone. This water movement leaches calcium carbonate from the entire soil profile in a process called *decalcification*. Soils that have lost most of their calcium are also usually

Table 20.2 Soil-forming processes

Enrichment	Addition of material to the soil, for example, by deposition of mineral matter by water or wind action.
Removal	Removal of material from the soil, for example, by erosion of uppermost layers or by leaching of dissolved matter to lower layers or to groundwater.
Decalcification	Leaching of calcium carbonate from the soil to the groundwater below by large amounts of infiltrating precipitation in moist climates.
Translocation	Movement of materials upward or downward within the soil body.
Eluviation	Downward transport of fine materials from the upper part of the soil.
Illuviation	Accumulation of fine materials in a lower part of the soil.
Calcification	Accumulation of calcium carbonate by dissolution in upper layers and precipitation in the *B* horizon.
Salinization	Upward wicking of salt-laden groundwater toward the soil surface with evaporation to produce a layer of salt accumulation.
Transformation	Transformation of material in the soil body; for example, conversion of primary to secondary minerals.
Humification	Decomposition of organic matter to produce humus.

acidic and correspondingly low in bases. The addition of lime or pulverized limestone not only corrects the acid condition, but also restores the calcium, which is a plant nutrient.

The third class of soil-forming processes involves *translocation*, in which materials are moved within the soil body, usually from one horizon to another. Two translocation processes that operate simultaneously are eluviation and illuviation. **Eluviation** consists of the downward transport of fine particles, particularly clays and colloids, from the uppermost part of the soil. Eluviation leaves behind grains of sand or coarse silt, forming the *Ae* horizon. **Illuviation** is the accumulation of materials that are brought downward, normally from the *Ae* horizon to the *B* horizon. The materials that accumulate may be clay particles, humus, or iron and aluminum sesquioxides.

The translocation of calcium carbonate is another important process. In many areas, the parent material of the soil contains a substantial proportion of calcium carbonate derived from the disintegration of limestone, a common variety of bedrock. Carbonic acid, which forms when carbon dioxide gas dissolves in rainwater or soil water, readily reacts with calcium carbonate, which then goes into the solution as Ca^{++} and CO_3^- ions.

In dry climates, such as in the prairie grasslands of Saskatchewan and Alberta, calcium carbonate dissolves in the upper layers of the soil during periods of rain or snowmelt, when soil water recharge is taking place. The dissolved carbonates are carried down to the *B* horizon, where water percolation reaches its limit. Here, the carbonate is precipitated through a process called *calcification*. Calcium carbonate deposition takes the form of white or pale-coloured grains, nodules, or plates in the *B* or *C* horizons.

Another form of translocation associated with arid climates is *salinization*. Where groundwater lies close to the surface in a poorly drained area, evaporation draws up a continual flow of moisture by capillary tension. When evaporation occurs, any dissolved salts precipitate and accumulate as a distinctive *salic horizon*. Most of the salts are compounds of sodium, which, when present in large amounts, is associated with highly alkaline conditions and is toxic to many kinds of plants. When salinization occurs in irrigated lands in a desert climate, the soil can be ruined for further agricultural use.

The last class of soil-forming processes involves the *transformation* of material within the soil body. One example is the conversion of minerals from primary to secondary types. Another is the decomposition of organic matter to produce humus, a process termed *humification*. In warm, moist climates, humification can reduce most of the organic matter to carbon dioxide and water, leaving virtually no humus in the soil.

SOIL TEMPERATURE

Soil temperature acts as a control over biologic activity and also influences the intensity of chemical processes affecting soil minerals. Below 10°C, chemical activity is slowed, and at or below the freezing point (0°C), it mostly ceases.

The temperature of the uppermost soil layer and the soil surface strongly affects the rate at which micro-organisms decompose organic matter. Thus, in cold climates where decomposition is slow, organic matter in the form of fallen leaves and stems tends to accumulate to form a thick organic horizon, which is slowly transformed into humus and carried downward to enrich the *A* horizon.

In warm, moist climates of low latitudes, the decomposition rate of plant material is rapid; bacterial activity disposes of nearly all the fallen leaves and stems. Under these conditions, organic horizons may be missing and the entire soil profile will contain very little organic matter (Figure 20.13).

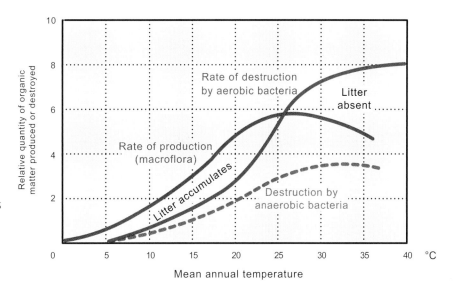

20.13 Production and decomposition rates of plant materials Plant production increases with temperature to a maximum of about 25–30°C and then declines due to high rates of plant respiration, which accelerate consumption of products formed in photosynthesis. Decomposition typically increases with temperature. The net effect is that litter tends to accumulate on the soil surface in colder climates, but not in warmer climates.

SURFACE CONFIGURATION

The *slope* and *aspect* of the ground surface are important factors in soil formation. Typically, soil horizons are thick on gentle slopes, but thin on steep slopes where material is more rapidly removed by erosion. In low-lying areas, accumulated drainage water may saturate the soil. Under extreme conditions, this can lead to oxygen deficits in the soil, which can affect chemical and biological processes (Figure 20.14). Bog soils and mucks formed under such conditions exhibit distinct horizons of gley—a blue, gray, or olive-coloured sticky clay and an accompanying smell of hydrogen sulphide when disturbed.

Slope aspect affects the soil temperature and water regime. Slopes facing away from the Sun are sheltered from direct insolation and tend to have cooler, moister soils. Slopes facing toward the Sun are exposed to more intense solar radiation, raising soil temperature and increasing evapotranspiration.

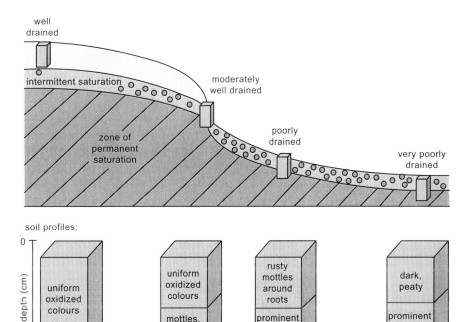

20.14 Soil catena On undulating terrain, soil properties change in relation to topographic position, mainly because of drainage conditions and the influence they have on plant growth and other soil-forming processes. Mottles are a pattern of light and dark "fingers" of soil colour that develop under anaerobic conditions in some soils.

BIOLOGICAL PROCESSES

The presence and activities of living plants and animals, as well as their non-living organic products, are essential for good soil development. Organic matter in the form of humus is important for soil fertility. The colloidal structure of humus holds bases, which are needed for plant growth. Humus reduces leaching and helps to keep nutrients cycling through plants and soils. It also helps to create good soil structure by binding mineral particles together. Good soil structure allows water and air to penetrate the soil freely, providing a healthy environment for plant roots.

The types of animals living in the soil range from bacteria to burrowing mammals. The role animals play in soil formation is extremely important especially when conditions are warm and moist enough to support large populations. For example, earthworms continually rework the soil not only by burrowing, but also by passing soil through their intestinal tracts. They ingest large amounts of decaying leaf matter, carrying it down from the surface, and incorporating it into the mineral horizons. Many forms of insect larvae perform a similar function. Larger animals, such as gophers, rabbits, and badgers, are also noted for their excavations and soil disturbance.

Many agricultural soils have been cultivated for centuries. As a result, both the structure and composition of agricultural soils have undergone great changes. These altered soils are often recognized as distinct soil classes and are given the same status in soil classifications as natural soils.

THE GLOBAL SCOPE OF SOILS

An important aspect of soil science for physical geography is the classification of soils into major types and subtypes as they are distributed over the lands of the Earth. Mapping of soil types reveals linkages among climate, natural vegetation, parent material, time, biologic processes, and landforms that differentiate the Earth's surface. The geography of soils is thus essential in determining the nature and quality of global environments.

Because they are the product of many processes acting at different rates, soils exist as a continuum. As a result, specific criteria are used to group soils into discrete classes. Typically, soil classification systems developed for large regions group soils by their basic properties rather than their usage. One such global classification system, called the World Reference Base for Soil Resources, was developed by the United Nations Food and Agriculture Organization with the support of the UN Environment Program and the International Society of Soil Science. The World Reference Base uses soil morphology, including soil structures, assemblages of soil constituents, soil horizons, and their vertical sequence. It recognizes 30 soil groups at the highest level of the classification.

The World Reference Base was developed over several decades by successive international congresses of soil scientists. Its goal was to provide a framework through which existing soil classification systems developed by individual nations could be correlated, and ongoing soil classification efforts could be harmonized. Its final objective was to identify a set of major soil groupings at the global level, as well as the criteria for defining and identifying them. The World Reference Base serves not only as a framework for soil scientists from many countries and regions to use in their work, but also as a common ground for the inventory of land and natural resources.

Many countries have developed unique classification systems—Canada and the United States are two examples. The Canadian System of Soil Classification pays special attention to the soils of cold regions and permafrost terrains. The American system, which is called the Comprehensive Soil Classification System (CSCS), is more general and can be applied globally.

Soil classification systems typically use the **soil order** as the highest-level grouping. In the Canadian system, the next subdivision is the *great group*, while the American system uses the term *suborder*. Soil orders and suborders, or great groups, are often distinguished by the presence of a *diagnostic horizon*. Each diagnostic horizon has some unique combination of physical properties (such as colour, structure, and texture) or chemical properties (for example, an abundance of calcium). The two basic kinds of diagnostic horizons are a horizon formed at the surface, called an *epipedon*, and a subsurface horizon formed by processes occurring at various depths in the soil.

SOILS OF CANADA

Although classification systems such as the World Reference Base or the American CSCS cover the global range of soils, Canada has evolved a unique soil classification system that is especially suited to its own soils. Because Canada is located entirely north of lat. 40° N, it has no tropical and equatorial soil types. Moreover, nearly all of Canada experienced glaciation during the last ice age. Ice sheets and glaciers largely removed pre-existing soils, replacing them with unsorted glacial debris, transported sands and gravels, and lake-bottom deposits. As a result, the Canadian system emphasizes young soils of cold regions in more detail than other systems do.

Like the World Resources Base and the American CSCS, the Canadian Soil Classification System uses a system of classes based on the properties of the soils themselves, rather than interpretations of various uses of the soils. Thus, the classes are based on generalized properties of real, not idealized, soils. Although the Canadian system's classes are defined according to actual soil properties that can be observed and measured, the system has a genetic bias. It favours properties or combinations of properties

Table 20.3 Horizons and subhorizons of the Canadian system of soil classification

Organic Horizons

O	Organic horizon developed mainly from mosses, rushes, and woody materials (e.g., peat).
L	Organic horizon characterized by an accumulation of organic matter derived mainly from leaves, twigs, and woody materials in which the organic structures are easily discernible.
F	Same as *L* above, except that the original structures are difficult to recognize.
H	Organic horizon characterized by decomposed organic matter in which the original structures are not discernible.

Mineral Horizons

A	Mineral horizon found at or near the surface in the zone of leaching or eluviation of materials in solution or suspension, or of maximum *in situ* accumulation of organic matter, or both.
B	Mineral horizon characterized by enrichment in organic matter, sesquioxides, or clay; by the development of soil structure; or by a change of colour denoting hydrolysis, reduction, or oxidation.
C	Mineral horizon comparatively unaffected by the pedogenic processes operating in *A* and *B* horizons. Gleying processes and the accumulation of calcium, magnesium, and more soluble salts can occur in this horizon.

Subhorizons (Lowercase suffixes)

b	Buried soil horizon.
c	Irreversibly cemented pedogenic horizon, also known as a hardpan.
ca	Horizon of secondary carbonate enrichment, in which the concentration of lime exceeds that in the unenriched parent material.
cc	Horizon containing irreversibly cemented pedogenic concretions.
e	Horizon characterized by the eluviation of clay, iron, aluminum, or organic matter, alone or in combination.
f	Horizon enriched with amorphous material, principally aluminum and iron combined with organic matter; reddish near upper boundary, becoming more yellow at depth.
g	Horizon characterized by grey colours or prominent mottling, or both, indicating permanent or intense chemical reduction.
h	Horizon enriched with organic matter.
j	Used as a modifier of suffixes *e*, *f*, *g*, *n*, and *t* to denote an expression of, but failure to meet, the specified limits of the suffix it modifies.
k	Denotes the presence of carbonate as indicated by visible effervescence when dilute hydrogen chloride (HCl) is added.
m	Horizon slightly altered by hydrolysis, oxidation, or solution, or all three to give a change in colour or structure, or both.
n	Horizon in which the ratio of exchangeable calcium (Ca) to exchangeable sodium (Na) is 10 or less, as well as the following distinctive morphological characteristics: prismatic or columnar structure, dark coating on ped surfaces, and hard to very hard consistency when dry.
p	Horizon disturbed by human activities, such as cultivation, logging, and habitation.
s	Horizon of salts, including gypsum, which may be detected as crystals, veins, or surface crusts of salt crystals.
sa	Horizon with secondary enrichment of salts more soluble than calcium and magnesium carbonates; the concentration of salts exceeds that in the unenriched parent material.
ss	Horizon containing *slickensides*—shear surfaces that form when one soil mass moves over another.
t	Illuvial horizon enriched with silicate clay.
u	Horizon that is markedly disrupted by physical or faunal processes other than cryoturbation.
v	Horizon affected by disruption and mixing caused by shrinking and swelling of the soil mass.
x	Loamy subsurface horizon of high bulk density and very low organic matter content; when dry, it is hard and seems to be cemented; also known as a fragipan.
y	Horizon affected by cryoturbation, as manifested by disrupted or broken horizons, incorporation of materials from other horizons, and mechanical sorting.
z	A frozen layer.

Source: The Canadian System of Soil Classification, 3rd ed., Ottawa: NRC Research Press, 1998. Used with permission.

that reflect processes of soil formation when distinguishing between higher divisions. Thus, the soils grouped under a single soil order are considered the product of a similar set of dominant soil-forming processes resulting from broadly similar climatic conditions.

Table 20.3 provides a listing and description of the major horizons recognized by the Canadian Soil Classification System, including organic and mineral horizons. Also described are subhorizons, identified with lowercase letters; for example, the *Ah* and *Ae* subhorizons used in the stylized boreal forest soil profile

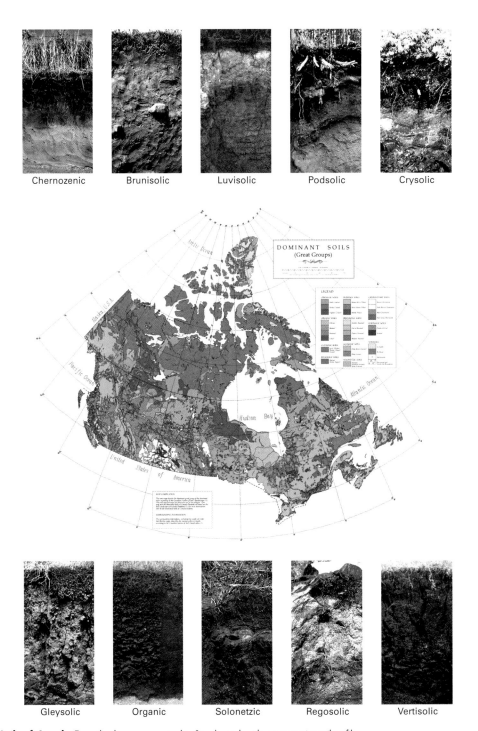

20.15 Soils of Canada Ten soil orders are recognized in Canada, each with representative soil profiles.

in Figure 20.12. The nature and order of horizons and subhorizons found in the soil profile are primary determinants of the soil order and the great group class assigned to a soil.

The Canadian System of Soil Classification includes 10 soil orders:

Chernozemic	Cryosolic	Regosolic
Brunisolic	Gleysolic	Vertisolic
Luvisolic	Organic	
Podzolic	Solonetzic	

Figure 20.15 shows the distribution of these 10 soil orders in Canada together with representative soil profiles.

Chernozemic Order

Soils of the *Chernozemic order* are the predominant soils of the agricultural regions in the Canadian Prairies. They have thick, dark, *A* horizons, rich in organic matter. This *Ah* horizon originates from decomposition of grasses that grow abundantly in these regions. Typically, the *C* horizon contains varying amounts of calcium, which is deposited through evaporation during periods of drought. The presence of a *Cca* horizon is a noted characteristic of these soils. **Chernozemic soils** form in areas that have cold winters, hot summers, and low precipitation that quickly evaporates in the summer heat. These soils range in colour from black to various shades of brown and grey, depending on how much organic matter has been incorporated into the profile. The deepest and darkest soils have developed under tall-grass prairie in more easterly regions of the Prairies, especially in Manitoba, where precipitation is a little higher. Toward the west, the soils become shallower and paler in colour. The amount of calcium in the profile increases and is found progressively closer to the surface as drought conditions become more intense and prolonged. Chernozems are rich in nutrients and have excellent structure and good water-holding capacity, so are ideal for agriculture.

Brunisolic Order

Brunisolic soils typically lack the degree of horizon development found in many of the other soil orders. They are generally associated with coniferous or deciduous forests and occur in a wide range of climates throughout Canada. A distinctive character of these soils is the presence of a *Bm* horizon that is brownish in colour and may contain small accumulations of aluminum and iron or clay. Brunisols develop under a range of drainage conditions from good to imperfect drainage. Although suitable for agriculture, brunisols tend to be somewhat acidic with pH values often as low as 5.5.

Luvisolic Order

Luvisolic soils are characteristic of forested regions that are well to imperfectly drained. The largest area of these soils occurs in the central to northern interior plains under deciduous, mixed, and coniferous forest. Luvisolic soils have a surface humus layer overlying a leached, light greyish eluvial *A* horizon that is low in clay and iron-bearing minerals. Beneath this *Ae* horizon is an illuvial *Bt* horizon that is enriched with clay and often exhibits high levels of available nutrient ions, such as calcium, magnesium, and potassium. The genesis of luvisolic soils is thought to involve the suspension of clay in the soil solution near the soil surface. The clay is then carried through the profile and deposited at a depth, where downward motion of the soil solution ceases or becomes very slow.

Podzolic Order

Podzolic soils typically form under coniferous forests in cool, moist regions where abundant precipitation causes pronounced leaching. Podzols are especially well-developed where parent materials, such as granite or sandstone, are rich in quartz. The cool, moist climates do not favour rapid breakdown of organic litter by soil organisms. Consequently, the soil surface is typically covered by coniferous needles in various stages of decomposition, which form distinctive L, H, and F layers of pronounced acidity, 5–10 centimetres thick. Organic acids, such as tannins dissolved from the litter, increase the leaching potential of percolating water.

The constant downward movement of water carries nutrients from the upper layers, and acidity characteristically is below pH 5.5. Podzols are usually distinguished by the presence of a marked eluviated horizon from which iron and aluminum sesquioxides, clays, and organic material has been removed. The resulting *Ae* horizon is characteristically pale ash-grey in colour and sandy in texture. Redeposition in the *B* horizon results in a dark-coloured *Bh* horizon that is enriched with organic matter. Beneath this is a rust-coloured *Bf* horizon, in which the iron and aluminum sesquioxides accumulate. If abundant, these minerals can form a hard, impermeable layer that restricts drainage of water through the soil. Where present, such hardpans are designated as *Bc* horizons.

Podzols are formed under conditions of severe leaching, which can leave the upper horizons virtually depleted of all soil constituents except quartz grains. Some of these constituents are re-deposited in the lower part of the soil profile, but most of the plant nutrients can be carried away in drainage waters. Consequently, podzols are not ideal for agriculture, although productivity can be improved with additions of lime and other fertilizers.

Cryosolic Order

Cryosolic soils occupy much of the northern third of Canada where permafrost remains close to the surface of both mineral and organic deposits. Cryosols predominate in arctic tundra north of the tree line, but are also common in the open subarctic forests and extend into the boreal forest, especially in some organic materials. Cryosols are occasionally found in alpine areas of mountainous regions (Figure 20.16a)

Formation of cryosols is predominantly influenced by severe climatic conditions. Short, cool summers and long winters favour the accumulation of organic matter without decomposition, because of limited microbial activity. A unique characteristic of cryosols is the presence of a perennially frozen layer (*Cz*), typically 40–80 centimetres below the surface. This invariably leads to intense seasonal freeze–thaw activity. Above the *permafrost layer*, the upper portion of the soil thaws in the summer to form the *active layer*. The thickness of the active layer is controlled by factors such as soil texture, soil moisture regime, and thickness of the insulating organic surface layer.

At the end of the summer, freezing in the active layer occurs upward from the frost table and downward from the soil surface. Soil material trapped between these freezing fronts comes under *cryostatic pressure*, resulting in contorted and displaced soil horizons (Figure 20.16b). Coarser rock fragments are also heaved and sorted, which further disturbs the soil profile and leads to the formation of patterned ground (see Chapter 14).

20.16 Cryosolic soil order Cryosols are soils of permafrost terrains. (a) Permafrost and cryosols underlie many boreal forest landscapes, such as this one in Yukon Territory. (b) A Turbic Cryosol profile from this region. Note the flow structures, formed by frost heaving and solifluction. (From the Canadian Soil Information System)

Weak leaching and translocation of materials occur in permafrost soils because permafrost impedes drainage and cryoturbation mixes the soil materials in the active layer. In winter, the soils are frozen and firm, but during summer, they become soft and pliable and can become waterlogged, leading to gleying in the *C* horizon (*Cg*). The pH of cryosols largely depends on the composition of the parent material, and this is accentuated by cryoturbation, which continually mixes materials within the entire soil profile. The nutrient content of cryosols is low, with most of the plant nutrients released slowly from organic matter.

Cryosols are sensitive to disturbance, especially when the insulating organic layer is removed, leading to increased thawing of the underlying permafrost and the potential for severe thermokarst development. Cryosols are an important store of organic carbon under present climate conditions. Increased decomposition rates due to global warming could potentially release carbon dioxide and methane to the atmosphere and accelerate climate change.

Gleysolic Order

Gleysolic soils have features indicating periodic or prolonged saturation and are usually associated with either a high groundwater table for part of the year, or temporary saturation above a relatively impermeable layer. They are particularly abundant in the low-lying river basins but commonly occur in patches among other soils in the landscape.

Gley soils typically have a thick organic horizon (*Ah*) at the surface. The underlying *B* horizon is greyish in colour. When accompanied by streaks and patches of greenish-blue or orange-brown, the mottled horizons are designated *Bg*. Mottling develops when waterlogging restricts oxygen diffusion into the soil from the atmosphere. Under these conditions, anaerobic microorganisms extract oxygen from chemical compounds. For example, iron sesquioxide when reduced to ferrous oxide imparts a greenish-blue colour to the soil (Figure 20.17). An orange-brown mottled colour is usually indicative of localized re-oxidation of ferrous salts and is often associated with root channels or cracks that develop during dry spells.

Organic Order

Soils of the *Organic order* are composed largely of organic materials. They include most of the soils commonly known as peat, muck, or bog soils and represent accumulations of partially or completely decomposed plant materials formed under anaerobic conditions. **Organic soils** are often associated with poorly drained depressions and so are saturated with water for prolonged periods. However, they can also form under cool, wet climatic conditions, which in combination with high acidity and nutrient deficiency restrict microbiological decomposition.

Typical organic soils have greater than 30 percent organic matter content and surface organic horizons of 40–60 centimetres, although layers as thin as 10 centimetres sometimes develop directly over bedrock. Different types of organic soils are recognized according to the degree of decomposition of the organic material. For example, fibrisols are composed largely of organic matter that has not decomposed; in mesisols, decomposition is partially complete, and in humisols, most of the organic material has been broken down into a soft amorphous mass.

Organic soils possess high water-holding capacity and low load-bearing strength. They are usually acidic, with a pH higher than 5, unless supplied with nutrient-rich drainage water that has percolated through adjacent mineral soils. Organic soils supplied with nutrients in this way support *fens*. Organic soils that are supplied only by precipitation develop into highly acidic *bogs*, often dominated with Sphagnum mosses.

MUNSELL*COLOR CHART FOR GREY

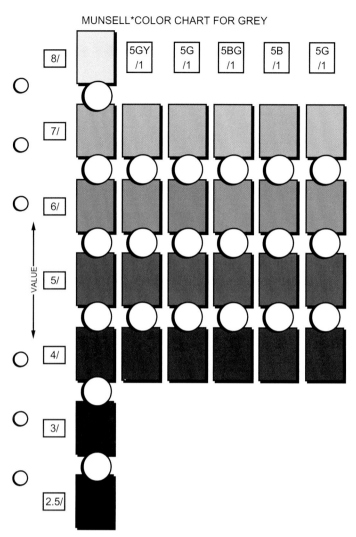

20.17 Munsell colour chart for gley Green, blue, and grey patches and streaks in a soil are characteristic of the gley conditions associated with waterlogged soil.

20.18 Solenetzic soil High sodium content gives rise to the distinctive white-capped, columnar structure seen in this dry, prairie soil.

in the parent material or transported into the soil by upward moving groundwater, produces a distinctive prismatic or columnar structure in the soil, described as a *Bnt* horizon (Figure 20.18).

The productivity of solonetzic soils is low, due to the limited depth of topsoil and hardness of the *Bnt* horizon. Moisture is usually the major limitation, but alkalinity is also a problem. Solonetzic soils tend to be low in organic matter, and reserves of nutrients are often small. Deep plowing and, more recently, subsoiling have been developed to improve these soils. In deep plowing, calcium in the form of lime is mixed into the *Bnt* horizon to replace the sodium and prevent the hardpan from re-forming. Subsoiling accomplishes little mixing of the soil horizons but may improve crop growth by shattering the hardpan.

Solonetzic Order

Soils of the *Solonetzic order* occur on saline parent materials in some areas of the semi-arid to subhumid prairies, in association with chernozemic soils and, to a lesser extent, with luvisolic and gleysolic soils. Most **solonetzic soils** are associated with a vegetation cover of grasses and forbs.

Solonetzic soils are characterized by tough, almost impermeable *hardpans* that may vary from 5 to 30 centimetres or more below the surface. This hardpan severely restricts root and water penetration into the subsoil. Although these *subsurface* horizons are typically hard when dry, they swell to a sticky, plastic mass of low permeability when wet. Sodium, present either

Regosolic Order

Regosolic soils develop on unconsolidated parent materials, such as dune sands and alluvium (Figure 20.19). They have weakly developed profiles in which *B* horizons are absent. Typically, only a thin humic *Ah* horizon and a surface litter layer overlies the unaltered parent substrate, resulting in AC profiles. This lack of development may be due to the youthfulness of the parent material, as is the case for recently deposited alluvium, or instability, as occurs in colluvium on slopes subject to mass wasting. Drainage in regosolic soils is quite variable and may be quite rapid in materials, such as dune sand, but slow in finer alluviums. Regosols occur in a wide range of vegetation and climates, but generally are of limited extent in Canada.

20.19 Regosolic soil profile Regosols are weakly developed soils of recently deposited material. This profile shows silty river alluvium from the Atlantic provinces. Horizon development is generally lacking, although the upper 20 centimetres of the profile appears slightly lighter in colour. (From the Canadian Soil Information System)

Vertisolic Order

Soils of the *Vertisolic order* occur in parent materials rich in mont-morillonite clays that expand greatly when wet and then shrink excessively when dry. Shrinking and swelling is strong enough that horizons characteristic of other soil orders have either been prevented from forming or have been severely disrupted. This is because **vertisolic soils** develop deep cracks as they dry, and material from the upper part of the soil profile falls into them. Thus, vertisolic soils are self-mixing and lack distinctive horizons (Figure 20.20). Typically they have deep *Ah* horizons that merge with an indistinct *Bv* horizon, which is disrupted by shrinking and swelling of the soil mass. This process is called *argillipedoturbation*.

The vertisolic order was only recently recognized in Canada. The major areas of occurrence are in the cool, semi-arid to sub-humid grasslands of the interior plains of western Canada. Minor areas of vertisolic soils occur in valleys in the western mountains, in parts of the southern boreal forest, and in the cool temperate regions of central Canada.

Vertisols typically form from basic parent materials in climates that are seasonally humid. In terms of colour and organic matter content, vertisols are usually dark and may have a chernozemic-like *A* horizon at the surface, or they may have mottling in the *B* horizon that is a characteristic feature of soils of the gleysolic order. However, they are distinguished from both of these by the presence in the lower *B* horizon of slickensides (*Bss*)—shear surfaces that form when one soil mass moves over another—and a vertic (*v*) horizon that has been strongly affected by shrinking and swelling.

20.20 Vertisol This deeply cracked clay-rich soil lacks distinctive horizons.

SOILS OF THE WORLD

The soils of the world present a much broader range of soil types and conditions than those found in Canada. This survey of global soils uses the U.S. Comprehensive Soil Classification System (CSCS).

Table 20.4 shows the CSCS soil orders. At the highest level, the system recognizes three groups of soil orders. The largest group includes seven orders with well-developed horizons or fully weathered minerals. A second group includes a single soil order that is rich in organic matter. The last group includes three soil orders with poorly developed or no horizons. Although each order has several suborders, this summary includes only the suborders associated with the alfisols.

The world soils map (Figure 20.21) shows the general distribution of the soil orders and representative profiles of each. Note that the alfisols have been subdivided into four important suborders that correspond well to four basic climate zones. Entisols, inceptisols, histosols, and andisols are not shown as these orders largely occur locally. In highland regions, the soil patterns are too complex to show on a global scale.

Three soil orders dominate the low latitudes: oxisols, ultisols, and vertisols. Soils of these orders have developed over long time spans in an environment of warm soil temperatures. Soil water is abundant in a wet season or lasts throughout the year.

Table 20.4 Soil orders

Group I
Soils with well-developed horizons or fully weathered minerals, resulting from long-continued adjustment to prevailing soil temperature and soil water conditions.

Oxisols	Very old, highly weathered soils of low latitudes, with a subsurface horizon of mineral oxide accumulation and very low base status.
Ultisols	Soils of equatorial, tropical, and subtropical latitude zones, with a subsurface horizon of clay accumulation and low base status.
Vertisols	Soils of subtropical and tropical zones with high clay content and high base status. Vertisols develop deep, wide cracks when dry, and the soil blocks formed by cracking move with respect to each other.
Alfisols	Soils of humid and subhumid climates with a subsurface horizon of clay accumulation and high base status. Alfisols range from equatorial to subarctic latitude zones.
Spodosols	Soils of cold, moist climates, with a well-developed *B* horizon of illuviation and low base status.
Mollisols	Soils of semi-arid and subhumid midlatitude grasslands, with a dark, humus-rich epipedon and very high base status.
Aridisols	Soils of dry climates, low in organic matter, and often having subsurface horizons of carbonate mineral or soluble salt accumulation.

Group II
Soils with a large proportion of organic matter.

Histosols	Soils with a thick upper layer very rich in organic matter.

Group III
Soils with poorly developed or no horizons that are capable of further mineral alteration.

Entisols	Soils lacking horizons, usually because their parent material has accumulated only recently.
Inceptisols	Soils with weakly developed horizons, having minerals capable of further alteration by weathering processes.
Andisols	Soils with weakly developed horizons, having a high proportion of glassy volcanic parent material produced by erupting volcanoes.

Oxisols

Oxisols have developed on stable land areas in equatorial, tropical, and subtropical regions with large water surpluses. They are principally associated with rainforests and occur throughout the wet equatorial climate zone in Africa, South America, and Asia.

Oxisols usually lack distinct horizons, except for darkened surface layers. Soil minerals are weathered to an extreme degree and are dominated by stable aluminum and iron sesquioxides. Red, yellow, and yellowish-brown colours are normal. The base status of oxisols is very low; nearly all the bases plants require have been removed from the soil profile, either through storage in the vegetation or leaching from the profile. The soil is quite easily broken apart and allows easy penetration by rainwater and plant roots.

Ultisols

Ultisols are similar to the oxisols in appearance and environment of origin. Ultisols are reddish to yellowish in colour. They have a subsurface horizon of illuviated clay, which is not found in the oxisols. Although forest is the characteristic native vege-

tation, the base status of the ultisols is low like oxisols. Hence, shallow roots of trees and shrubs must take up and recycle nutrients quickly.

Ultisols are widespread throughout Southeast Asia and the East Indies. Other important areas include South America, Africa, and northeastern Australia. Ultisols also occur in the southeastern United States, where they correspond with the area of moist subtropical climate. In lower latitudes, ultisols are identified with the wet-dry tropical climate and the monsoon and trade-wind coastal climate. Note that all these climates have a dry season, even though it may be short.

Vertisols

Vertisols are typically black in colour and have a high clay content. The predominant clay mineral is montmorillonite, which shrinks and swells with seasonal changes in soil water content. Wide, deep vertical cracks develop in the soil during the dry season (Figure 20.22). When moistened and softened by rain, some fragments of surface soil drop into the cracks before they close, so that the soil is constantly being mixed. Vertisols are equivalent to vertisolic soils in the Canadian system.

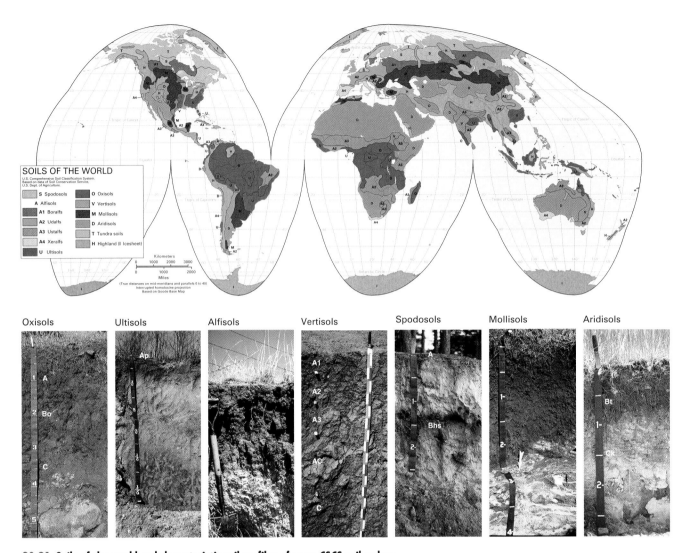

20.21 Soils of the world and characteristic soil profiles of some CSCS soil orders

Vertisols typically form under grass and savanna vegetation in subtropical and tropical climates with a pronounced dry season. Because vertisols require a particular clay mineral as a parent material, the major areas of occurrence are scattered and show no distinctive pattern on the world map. An important region of vertisols is the Deccan Plateau of western India, where basalt supplies silicate minerals that are altered into montmorillonite clay.

Alfisols

Alfisols are characterized by a pale-coloured eluviated *A* horizon that is low in bases, clay minerals, and sesquioxides. These materials are concentrated by illuviation in the *B* horizon where the clay holds bases, such as calcium and magnesium. The base status of the alfisols, therefore, is generally quite high.

Alfisols are distributed throughout the world, ranging from latitudes as high as 60° N in North America and Eurasia to the equatorial zone in South America and Africa. Because the alfisols occur in many climate types, four important suborders are noted.

Boralfs are alfisols of the cold (boreal) forest lands of North America and Eurasia. They have a grey surface horizon and a brownish subsoil. *Udalfs* are brownish alfisols of the midlatitude zone. They are closely associated with the moist continental climates in North America, Europe, and eastern Asia. These suborders are equivalent to luvisolic soils in Canada.

Ustalfs are reddish-brown alfisols of the warmer climates. They range from the subtropical zone to the equator and are associated with the wet-dry tropical climate in Southeast Asia, Africa, Australia, and South America. *Xeralfs* are alfisols of the Mediterranean climate, with its cool moist winter and dry summer. The xeralfs typically are brownish or reddish in colour.

Spodosols

Spodosols are formed in the cold boreal climate beneath coniferous forest. They are equivalent to the podzolic soils of Canada. Spodosols are distinguished by their acidic, reddish-orange *B* horizons. This horizon is called the *spodic* horizon and is com-

20.22 A Vertisol in Texas The clay minerals that are abundant in vertisols shrink when they dry out, producing deep cracks in the soil surface.

posed of organo-aluminum and iron compounds brought downward by eluviation. Intensive leaching produces a conspicuous, ash-grey sandy horizon in the upper part of the soil profile, with a dark organic layer at the soil surface.

Spodosols are closely associated with regions recently covered by the great ice sheets of the late-Cenozoic era. These soils are therefore very young. Typically, the parent material is coarse sand, consisting largely of quartz, which does not weather to form clay minerals. Because of heavy leaching, spodosols are strongly acidic.

Histosols

Throughout the northern regions where spodosols are located are countless patches of **histosols**, also known as organic soils in the Canadian system. This unique soil order has a high organic matter content in a thick, dark upper layer. Most histosols go by such common names as peat or muck. They have formed in shallow lakes and ponds by accumulation of partially decayed plant matter. In time, a layer of organic matter, or *peat*, replaces the water and becomes a *bog*. Some peat bogs are used for cultivation of cranberries (cranberry bogs). Sphagnum peat from bogs is dried and baled for sale as mulch for use on suburban lawns and shrub bery beds. Dried peat has also been used for centuries in Europe as a low-grade fuel.

Entisols

Entisols are mineral soils that lack distinct horizons. They are soils in the sense that they support plants, but they may be found in any climate and under any vegetation. Entisols lack distinct horizons for two reasons. It may be the result of a parent material, such as quartz sand, in which horizons do not readily form. Alternatively, there may have been no time for horizons to form in recent deposits of alluvium or on actively eroding slopes. In the Canadian system, these soils are known as regosolic soils.

Inceptisols

Inceptisols are soils with horizons that are weakly developed, usually because the soil is quite young. Inceptisols occur quite extensively in the same regions as ultisols and oxisols. Especially important are the inceptisols of the river flood plains and delta plains in Southeast Asia that support dense populations of rice farmers. Here, annual river floods cover low-lying plains and deposit layers of fine silt. This sediment is rich in primary minerals that yield bases as they weather chemically over time. Inceptisols are also found in glaciated areas, where young sediment is present. In Canada, these are brunsolic soils.

Andisols

Andisols are soils in which more than half of the parent mineral matter is volcanic ash. They are, for the most part, fertile soils and, in moist climates, they support a dense natural vegetation cover. A high proportion of carbon, formed by the decay of plant matter, is also typical, so that the soil usually appears very dark in colour. Andisols form over a wide range of latitudes and climates, and are found in localized patches associated with individual volcanoes.

Mollisols

Mollisols are soils associated with the semi-arid and subhumid midlatitude grasslands. Mollisols have a thick, dark brown to black surface horizon with a loose, granular structure. Calcium dominates the bases of the *A* and *B* horizons, and the soil has a high base status. Mollisols are some of the most naturally fertile soils and now produce most of the world's commercial grain crop.

Mollisols are the dominant soils in the Great Plains region of North America and occur extensively in the Pampa region of Argentina and Uruguay. In Eurasia, a great belt of mollisols stretches across the steppes of Russia. The chernozemic soils of the Prairie Provinces are the Canadian equivalent to the mollisols.

Aridisols

Aridisols, the soils of desert regions, are dry for long time periods. Because the climate supports only very sparse vegetation, humus is lacking and the soil colour ranges from pale grey to pale red. Soil horizons are weakly developed, but there may be important subsurface horizons of accumulated calcium carbonate and other soluble salts, especially sodium. The salts give the aridisols a high degree of alkalinity. The closest equivalent in the Canadian soil classification would be soils of the solonetzic order.

Locally, where water supplies permit irrigation, aridisols can be highly productive for a wide variety of crops (Figure 20.23). Great irrigation systems, such as those of the Imperial Valley of the United States, the Nile Valley of Egypt, and the Indus Valley of Pakistan, have made aridisols highly productive, but not without problems of salt buildup and waterlogging.

Tundra Soils

Tundra soils are formed largely of primary minerals, ranging in size from silt to clay, that have been broken down by frost and glacial action. Layers of peat are often present between mineral layers. Beneath the tundra soil is perennially frozen ground (permafrost). Because the annual summer thaw affects only a shallow surface layer, soil water cannot easily drain away. Thus, the soil is saturated with water over large areas. Repeated freezing and thawing of this shallow surface layer disrupts plant roots, so that only small, shallow-rooted plants can maintain a hold. In the Comprehensive Soil Classification System, tundra soils are classed as wet subtypes of inceptisols, but in the Canadian system, they are recognized at the highest level as the cryosolic soil order.

A Look Ahead

Soil is the complex outermost layer that covers most of the Earth's continents. It is influenced by many factors, including the parent materials from which it is derived, the vegetation that it harbours,

and the water regime of precipitation and evapotranspiration that it experiences. In many environments, soil-forming processes operate very slowly. Thus, some soils can be the products of complex histories involving climatic changes. By inducing soil erosion, human activities can rapidly strip the uppermost soil horizons, leaving less productive layers at the surface. Soil horizons and properties can also vary strongly over short distances. This not only affects crop production, but is an important factor controlling natural vegetation patterns and ecosystem processes, which the final three chapters that follow discuss.

20.23 Cotton farming along the Colorado River below Parker, Arizona Note the pattern of roads alternating with irrigation canals. The alternate method of irrigation using centre-pivot systems gives distinctive circles in the landscape as shown: *34° 03' N; 114° 20' W. (Fly By: 48° 17' N; 78° 12' W)*

CHAPTER SUMMARY

- The soil layer is a complex mixture of solid, liquid, and gaseous components. It is derived from parent material, or regolith, that is produced from rock by *weathering*. The major factors influencing soil and soil development are parent material, topography, climate, vegetation, and time.

- Soil texture refers to the proportions of *sand*, *silt*, and *clay* that are present. Colloids are the finest particles in soils and are important because they help retain nutrients, or bases, that plants use. Soils show a wide range of *pH values*, from acid to alkaline. Soils with granular or blocky structures are most easily cultivated.

- In soils, *primary minerals* are chemically altered to form secondary minerals, which include oxides and *clay minerals*. The nature of the clay minerals determines the soil's base status. If

base status is *high*, the soil retains nutrients. If *low*, the soil can lack fertility. When a soil is fully moistened by heavy rainfall or snowmelt and allowed to drain, it reaches its *storage capacity*. Evaporation from the surface and transpiration from plants draws down the soil water store until precipitation occurs again to recharge it.

- The soil water balance describes the gain, loss, and storage of soil water. It depends on water need *(potential evapotranspiration)*, water use *(actual evapotranspiration)*, and precipitation. Monitoring these values on a monthly basis provides a *soil water budget* for the year.

- Most soils possess distinctive horizontal layers called horizons. These layers are developed by processes of *enrichment*, *removal*, *translocation*, and *transformation*. In downward translocation, materials such as humus, clay particles, and mineral oxides are removed by eluviation from an upper horizon and accumulate by illuviation in a lower one. In *salinization*, salts are moved upward by evaporating water to form a *salic horizon*. In the *humification* transformation process, organic matter is broken down by bacterial decay. Where soil temperatures are warm, this process can be highly effective, leaving a soil low in organic content. Animals, such as earthworms, can be very important in soil formation.

- Soil classification systems typically group soils together according to soil form, properties, and formation process rather than by usage. The classification systems of Canada and the United States are examples. Canada's system emphasizes soils of cold regions and permafrost terrains, while the American system is more global in scope. These systems use the presence and nature of *diagnostic horizons* to distinguish *soil orders* and *suborders* or *great groups*. The Canadian system recognizes 10 soil orders, the American 11 orders.

- Soils of Canada's *chernozemic order* have a dark, rich, loose *A* horizon with carbonate accumulation in the *B* horizon. These prairie soils develop on the semi-arid grasslands of the Prairie provinces.

- Soils of the *brunisolic order* lack strong horizon development. They underlie northern vegetation covers ranging from boreal forest to heath and tundra.

- Soils of the *luvisolic order* have a light-coloured *A* horizon from which clay has eluviated and a *B* horizon where clay has accumulated. They are found under forest, south of the permafrost zone.

- Soils of the *podzolic order* typically have an organic layer overlying a pale grey *A* horizon, with humus and aluminum and iron oxides accumulating in the *B* horizon. They are found under coniferous forest in cool to cold, moist environments.

- *Cryosolic order* soils occupy permafrost terrains. Horizon development is typically disrupted by freeze-thaw mixing and soil flowage.

- Soils of the *gleysolic order* are typically saturated with water during some or most of the annual cycle. They often show mottling in the *B* horizon, brought about by chemical-reducing conditions during saturation.

- *Organic order* soils are composed largely of organic materials that decay very slowly in cold, wet environments.

- Soils of the *solonetzic order* have *B* horizons that are very hard when dry and sticky and plastic when wet. They develop in semi-arid saline environments in interior plains.

- *Regosolic order* soils have weakly developed horizons. Often they are fresh deposits of sediment; for example, recent alluvium.

- Soils of the *vertisolic order* develop on clays that expand and shrink on wetting and drying, leading to mechanical mixing of the soil. They lack distinct horizons. In Canada, they are found at some locations in the interior plains under grassland.

- The soils of the world can be classified using the U.S. Comprehensive Soil Classification System.

- Oxisols are old, highly weathered soils of low latitudes. They have a horizon of mineral oxide accumulation and a low base status. Ultisols are also found in low latitudes. They have a horizon of clay accumulation and are also of low base status. Vertisols are rich in a type of clay mineral that expands and contracts with moistening and drying, and have a high base status.

- Alfisols have a horizon of clay accumulation like ultisols, but they are of high base status. They are found in moist climates from equatorial to subarctic zones.

- Spodosols, found in cold, moist climates, exhibit a horizon of illuviation and low base status. Mollisols have a thick upper layer rich in humus. They are soils of midlatitude grasslands. Aridisols are soils of arid regions, marked by horizons of accumulation of carbonate minerals or salts.

- Histosols have a thick upper layer formed almost entirely of organic matter.

- Three soil orders have poorly developed or no horizons: entisols, inceptisols, and andisols. Entisols have no horizons and often consist of fresh parent material. The horizons of inceptisols are only weakly developed. Andisols are weakly developed soils occurring on young volcanic deposits.

KEY TERMS

alfisols	entisols	parent material	solonetzic soils
andisols	gleysolic soils	podzolic soils	spodosols
aridisols	histosols	regosolic soils	ultisols
bases	illuviation	secondary minerals	vertisolic soils
brunisolic soils	inceptisols	soil	vertisols
chernozemic soils	luvisolic soils	soil horizons	water need
colloids	mollisols	soil orders	water use
cryosolic soils	organic soils	soil profile	
eluviation	oxisols	soil texture	

REVIEW QUESTIONS

1. What important factors condition the nature and development of the soil?

2. Soil colour, texture, and structure are used to describe soils and soil horizons. Identify each of these three terms, showing how they are applied.

3. Explain the concepts of acidity and alkalinity as they apply to soils.

4. Identify two important classes of secondary minerals in soils and provide examples of each class.

5. How does the ability of soils to hold water vary, and how does this ability relate to soil texture?

6. Define water need (potential evapotranspiration) and water use (actual evapotranspiration). How are they used in the soil water balance?

7. Identify the following terms as used in the soil water budget: storage, withdrawal, storage recharge, soil water shortage, water surplus.

8. What is a soil horizon? How are soil horizons named? Provide two examples.

9. Identify four classes of soil-forming processes and describe each.

10. What are translocation processes? Identify and describe four translocation processes.

11. What is a diagnostic horizon? Explain how soils are identified by diagnostic horizons and provide several examples.

12. How many soil orders are there in the Canadian system? Name them all.

13. Chernozemic and brunisolic soils are both found in interior Canada. Compare these two soil types.

14. Describe the process by which luvisolic soils attain their distinctive properties.

15. What are the unique features of podzolic soils? What environments are associated with podzolic soils?

16. Where are cryosolic soils located? What are their key features?

17. Gleysolic and organic soils are both soils of wet environments. How do they differ?

18. What is unique about soils of the solonetzic order? Where do they occur?

19. What factors contribute to the weak development of horizons in regosolic soils?

20. What is unique about soils of the vertisolic order? Where do they occur?

21. How many soil orders are there in the Comprehensive Soil Classification System (CSCS)? Name them all.

22. Name three CSCS soil orders that are especially associated with low latitudes. For each order, provide at least one distinguishing characteristic and explain it.

23. Compare alfisols and spodosols. What features do they share? What features differentiate them? Where are they found?

24. Where are mollisols found? How do the properties of mollisols relate to climate and vegetation cover?

25. Desert and tundra are extreme environments. Which soil order is characteristic of each environment? Briefly describe desert and tundra soils.

26. Identify five Canadian soil types that have good equivalents in the CSCS global system and name these equivalents.

VISUALIZATION EXERCISES

1. Sketch the profile of a cool, moist forest soil, labelling *L, F, H, Ah, Ae,* and *C* horizons. Indicate the movement of materials from the zone of eluviation to the zone of illuviation.

2. Examine the Canadian soils map (Figure 20.15) and identify three soil types that are found near your location. Develop a short list of characteristics that would help you tell them apart.

ESSAY QUESTIONS

1. Document the important role of clay particles and clay mineral colloids in soils. What is meant by the term *clay*? What are colloids? What are their properties? How does the type of clay mineral influence soil fertility? How does the amount of clay influence the water-holding capacity of the soil? What is the role of clay minerals in horizon development?

2. Using the world maps of global soils and global climate, compare the pattern of soils on a north-south transect along the 20° E longitude meridian with the corresponding climate patterns. What conclusions can you draw about the relationship between soils and climate? Be specific.

PROBLEM   WORKING IT OUT 20.1 Calculating a Simple Soil Water Budget

1. To the right are monthly values and annual totals for precipitation, water need *(Ep)*, and water use *(Ea)* at Urbana, Illinois, in the moist continental climate. Prepare and plot a soil water budget similar to the figure in *Working It Out 20.1* for this station. This will involve determining soil water shortage *(D)*, soil water use *(−G)*, soil water recharge *(+G)*, and water surplus *(R)* for each month following the method described in feature box.

Find answers at the back of the book

Soil water budget for Urbana, Illinois

Month	P	= Ea	− G	+ G	+ R	Ep	D
Jan	5.7	0.0				0.0	
Feb	4.5	0.0				0.0	
Mar	8.2	1.4				1.4	
Apr	10.0	4.4				4.4	
May	9.9	8.8				8.8	
Jun	8.4	12.4				12.6	
Jul	8.0	13.4				14.9	
Aug	9.0	11.6				13.0	
Sep	8.3	7.8				7.8	
Oct	6.6	4.8				4.8	
Nov	5.7	1.4				1.4	
Dec	5.5	0.0				0.0	
Total	89.8	66.0	−12.0	+12.1	23.7	69.1	3.1

EYE ON THE LANDSCAPE |

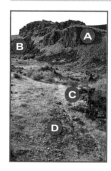

Chapter Opener Horsethief Butte State Park, Washington. The basaltic lava flows (**A**) in this region of Washington weather to a fine-textured fertile soil. Soil mantles the ground on all but the steepest slopes where it is continually lost through mass wasting (**B**). Grasses and herbs (**C**) are the dominant plants on the deeper soils, with mosses commonly growing over exposed bedrock (**D**).

Autumn in the Alberta foothills. Various plant communities are associated with the different soil conditions that occur in response to the topographic diversity in this region. Grasses and shrubs occur on the well-drained, coarse textured colluvium (**A**) on the lower valley slopes. Moister conditions favor the growth of aspen and white spruce (**B**) and in the valley floor where drainage is slower (**C**) standing water and boggy soils prevail which support sedges and groups of black spruce. Podzols are often present on the forested mountain slopes (**D**) while above the tree-line (**E**) thin mountain soils and bare rock will sustain alpine tundra and lichens. The golden color of the aspen foliage (**F**) suggests that the first frosts of the fall have occurred and that the trees will soon be entering their leafless winter state.

Chapter 21

EYE ON THE LANDSCAPE

A papaya tree in a tropical microcosm. What else would the geographer see? . . . Answers at the end of the chapter.

SYSTEMS AND CYCLES OF THE BIOSPHERE

Energy Flow in Ecosystems
The Food Web
Photosynthesis and Respiration
Net Photosynthesis
Net Primary Production
Net Production and Climate
Biomass as an Energy Source
Biogeochemical Cycles in the Biosphere
Geographers at Work • Joel Mortyn
Nutrient Elements in the Biosphere
The Carbon Cycle
The Oxygen Cycle
The Nitrogen Cycle
Eye on Global Change 21.1 • Human Impact on
    the Carbon Cycle
The Sulphur Cycle
Sedimentary Cycles
A Closer Look:
Focus on Remote Sensing 21.2 • Monitoring
    Global Productivity from Space

This chapter begins a section on **biogeography**, a branch of geography that focuses on the distribution of the Earth's plants and animals. It identifies and describes the processes that influence plant and animal distribution patterns on varying scales of space and time. Biogeography encompasses two major themes. *Ecological biogeography* is concerned with how the environment affects the distribution patterns of organisms. *Historical biogeography* focuses on the origins of present spatial patterns of organisms. It studies the processes of evolution, migration, and extinction of species. Chapter 21 examines how ecosystems cycle energy and matter. Chapter 22 discusses how organisms live and interact with their environment from the perspectives of ecological and historical biogeography. The final chapter of the book describes the major biomes of the world.

ENERGY FLOW IN ECOSYSTEMS
THE FOOD WEB

Energy transformations in an ecosystem occur through a series of *trophic levels*, or feeding levels, that are collectively referred to as a **food chain** or **food web** (Figure 21.1).

The plants and algae in a food web are the **primary producers** and comprise the first trophic level. These organisms use light energy to convert carbon dioxide and water into carbohydrates, and eventually into other biochemical molecules needed to support life. The primary producers support the **consumers**—organisms that ingest other organisms as their food source. The *primary consumers* or herbivores are the lowest level of consumers. Next are the *secondary consumers* or carnivores that feed on the primary consumers. More than one level of carnivore is usually present in an ecosystem, as larger predators normally feed on smaller carnivores. Some animals are **omnivores** and can feed on plant materials as well as animals. Omnivores tend to be more selective in their plant food requirements than herbivores, and will normally select fruits and grains to supplement their intake of animal tissue. The **decomposers** feed on *detritus*, or decaying organic matter, derived from all feeding levels. These might include insect larvae and worms, but are mainly micro-organisms and bacteria.

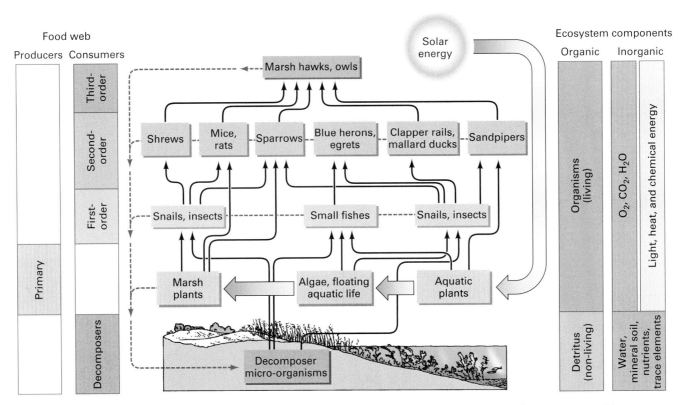

21.1 Energy flow diagram of a salt-marsh ecosystem in winter The arrows show how energy flows from the Sun to producers, consumers, and decomposers. (Food chain from R. L. Smith, *Ecology and Field Biology*, New York: Harper and Row.)

The food web is an energy flow system, tracing the path of solar energy through the ecosystem. Solar energy is absorbed by the primary producers and stored in the chemical products of photosynthesis. As consumers eat and digest these organisms, chemical energy is released. This chemical energy is used to power new biochemical reactions, which again produce stored chemical energy in the consumers' bodies.

At each level of energy flow in the food web, energy is expended in respiration, and ultimately lost as waste heat. This means that both the numbers of organisms and their total biomass generally decrease at each successive level in a food web. Typically, between 10 and 50 percent of the energy stored in organic matter at one level can be passed to the next level, which consequently limits the system to four levels of consumers.

For each species, the number of individuals present in an ecosystem ultimately depends on the level of resources available to support that species' population. If these resources provide a steady supply of energy, the population size will stabilize. Normally, the population will expand exponentially until the resources become increasingly scarce. As this happens, an increasing proportion of the population is unable to sustain itself, and population growth slows and eventually stops. This type of population growth, in which the rate slows and eventually reaches zero, is called *logistic growth*. At this point, the habitat's *carrying capac-*

ity has been reached. Carrying capacity is the number of individuals a particular habitat can sustain without significant negative impact on the population or its environment.

PHOTOSYNTHESIS AND RESPIRATION

Photosynthesis is the production of carbohydrate—a general term for a class of organic compounds consisting of the elements carbon (C), hydrogen (H), and oxygen (O). Common sugar or sucrose ($C_{12}H_{22}O_{11}$) is one example. Photosynthesis requires a series of complex biochemical reactions using water (H_2O) and carbon dioxide (CO_2) as well as light energy. A simplified chemical reaction for photosynthesis can be written as

$$H_2O + CO_2 + \text{light energy} \longrightarrow -CHOH- + O_2$$

Oxygen in the form of gas molecules (O_2) is a byproduct of photosynthesis.

There are three different photosynthesis processes. About 95 percent of plants use the C_3 **pathway**, so designated because the first stable product manufactured from CO_2 is a 3-carbon compound called *phosphoglycerate* (PGA). About 1 percent of plants use the C_4 **photosynthetic pathway**. In C_4 plants, CO_2 is converted into oxaloacetic acid, which is a 4-carbon compound. Like C_4 plants, plants that use the **CAM** (Crassulacean acid metabolism) **pathway** initially manufacture oxaloacetic acid. However,

CAM plants take in carbon dioxide at night and store it for later processing during daylight hours.

Each photosynthesis pathway has adaptive value. C_3 plants are more efficient than C_4 and CAM plants under cool and moist conditions and under normal light intensities. C_4 plants have an adaptive advantage under high light intensity and high temperatures. They also tend to use water more efficiently. Many tropical grasses, as well as sugar cane and maize, are C_4 species. CAM plants include succulents, such as cacti and agaves. They open their stomata at night when it is cooler. Their transpiration rates are low, and they use water sparingly. This gives cacti and other CAM plants an advantage in desert environments.

Respiration is the opposite of photosynthesis in that carbohydrate is broken down and combined with oxygen to yield carbon dioxide and water. The reaction can be simplified as follows:

$$-CHOH- + O_2 \longrightarrow CO_2 + H_2O + \text{chemical energy}$$

Photosynthesis and respiration act in a continuous cycle, which in its simplest case can be represented by a primary producer and a decomposer (Figure 21.2). The cycle involves one loop for hydrogen, another for carbon, and two loops for oxygen.

A living plant draws water up from the soil. Photosynthesis takes place in the green leaves of the plant, while the leaf cells absorb light energy. Carbon dioxide is brought in from the atmosphere, and oxygen is liberated and begins its atmospheric cycle. The plant tissue then dies and falls to the ground, where the decomposer acts on it. Through respiration, oxygen is taken out of the atmosphere or soil air and combined with the decomposing carbohydrate. Energy is liberated as heat. Both carbon dioxide and water enter the atmosphere as gases during decomposition.

An important concept emerges from this flow diagram. Energy passes through the system. It comes from the Sun and eventually returns to outer space. However, the material components—hydrogen, oxygen, and carbon—are recycled within the total system. Many other material components, including plant nutrients, are recycled in a similar way. Because the Earth as a planet is a closed system, the material components never leave the total system. However, they can be stored in other ways and forms where they are not available for use by organisms for prolonged periods.

NET PHOTOSYNTHESIS

Because both photosynthesis and respiration occur simultaneously in a plant, the amount of new carbohydrate placed in storage is less than the total carbohydrate being synthesized. A distinction, therefore, is made between gross photosynthesis and net photosynthesis. **Gross photosynthesis** is the total amount of carbohydrate produced by photosynthesis. **Net photosynthesis** is the amount of carbohydrate remaining after respiration has broken down enough carbohydrate to meet the plant's own needs. Thus,

Net photosynthesis = Gross photosynthesis − Respiration

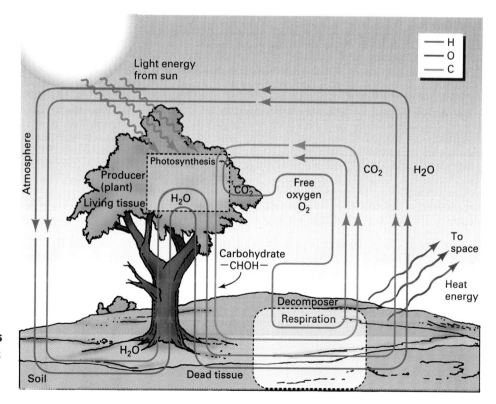

21.2 Photosynthesis and respiration cycles
A simplified flow diagram of the essential components of photosynthesis and respiration through the biosphere.

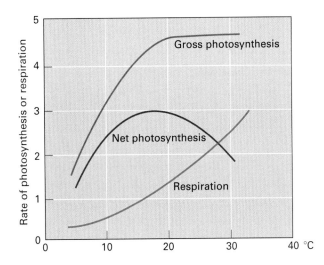

21.3 Photosynthesis and energy flow Photosynthesis varies with light intensity and temperature, but respiration is temperature dependent. As a result, gross photosynthesis tends to reach a maximum plateau, but net photosynthesis declines at higher temperatures.

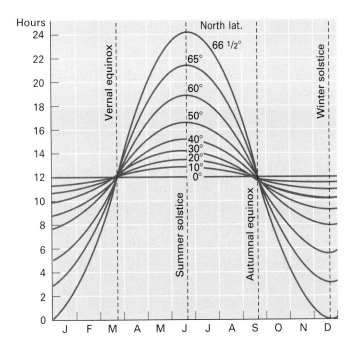

21.4 Day length variation Duration of the daylight period at various latitudes throughout the year. The vertical scale gives the number of hours the Sun is above the horizon.

The rate of photosynthesis is strongly dependent on air temperature and the available light energy (Figure 21.3). At first, the photosynthesis rate rises rapidly as both temperature and light intensity increases. The rate then slows and reaches a maximum value. It then decreases because high light intensity also heats the leaves. This heating increases the rate of respiration, which offsets gross production and decreases net photosynthesis.

The necessary light intensity to allow maximum net photosynthesis is only 10 to 30 percent of full summer sunlight for most green plants. Additional light energy is simply ineffective. Duration of daylight then becomes the important factor in the rate at which products of photosynthesis accumulate as plant tissues. The seasonal contrast in day length increases with latitude (Figure 21.4). At higher latitudes, duration of daylight changes markedly with the seasons. Conversely, at low latitudes, the daylight period remains close to 12 hours throughout the year. In sub-arctic latitudes, photosynthesis can take place during most of the 24-hour day in summer, a factor that can compensate significantly for the shortness of the growing season.

NET PRIMARY PRODUCTION

Accumulated net production by photosynthesis is measured in terms of **biomass**, which is usually expressed as the dry weight of organic matter per unit of surface area within the ecosystem; that is, kilograms per square metre or tonnes per hectare (1 hectare=10^4 square metres). Of all ecosystems, forests have the greatest biomass because of the large amount of wood that the trees accumulate over time. For fresh water bodies and the oceans, biomass is small—about one percent of the biomass in the grass-lands and croplands.

Although the amount of biomass present per unit area is an important indicator of the amount of photosynthetic activity, it can be misleading. In some ecosystems, consumers and decomposers quickly break down biomass, so the amount maintained is low. From the viewpoint of ecosystem productivity, what is important is the annual yield of useful energy produced by the ecosystem, or the **net primary production** (Table 21.1). The highest values are in two quite different environments: forests and

Table 21.1 Net primary production for various ecosystems

	Grams per Square Metre per Year	
	Average	**Typical Range**
Lands		
Rain forest of the equatorial zone	2,000	1,000–5,000
Freshwater swamps and marshes	2,500	800–4,000
Midlatitude forest	1,300	600–2,500
Midlatitude grassland	500	150–1,500
Agricultural land	650	100–4,000
Lakes and streams	500	100–1,500
Alpine tundra	100	80–500
Arctic tundra	50	5–300
Extreme desert	3	0–10
Oceans		
Algal beds and reefs	2,000	1,000–3,000
Estuaries (tidal)	1,800	500–4,000
Continental shelf	360	300–600
Open ocean	125	1–400

wetlands (estuaries). Net primary production in agricultural land is extremely variable due to factors such as availability of soil water, soil fertility, and use of fertilizers and machinery.

Productivity of the oceans is generally low. The deep water oceanic zone, which comprises about 90 percent of the world ocean area, is the least productive of the marine ecosystems. Continental shelf areas and zones of upwelling are much more productive. Upwelling of cold water from ocean depths brings nutrients to the surface and greatly increases the growth of microscopic *phytoplankton*. These, in turn, serve as food sources for marine animals in the food chain. An example is the Peru Current off the west coast of South America. Zones of upwelling and continental shelves together support about 99 percent of the world's fish production.

NET PRODUCTION AND CLIMATE

In addition to temperature and intensity and duration of light, water availability also influences net photosynthesis. Annual net primary production increases rapidly with precipitation in drier regions, from desert through semi-arid to subhumid climates (Figure 21.5). Production tends to level off in the humid range, possibly because a large soil water surplus has some counteractive influence, such as removal of plant nutrients by leaching. Combining the effects of light intensity, temperature, and precipitation, approximate values of productivity can be assigned to various climates as follows (units are grams of carbon per square metre per year—g C m^{-2} a^{-1}):

Highest productivity (more than 800)	Wet equatorial
Very high productivity (600–800)	Monsoon and trade-wind coastal Wet-dry tropical
High productivity (400–600)	Wet-dry tropical Moist subtropical Marine west coast
Moderate productivity (200–400)	Mediterranean Moist continental
Low productivity (100–200)	Dry tropical, semi-arid Dry midlatitude, semi-arid Boreal forest
Very low productivity (0–100)	Dry tropical, desert Dry midlatitude, desert Tundra

For natural ecosystems, productivity is largely dependent on climate and soils. For agricultural ecosystems, however, productivity is strongly influenced by the flow of energy provided to the crop in the form of agricultural chemicals and irrigation water. Much if not all of this energy is derived from the burning of fossil fuels and so represents a conversion of fossil fuel energy to human foodstuffs that is not always efficient.

Within the past decade or so, remote sensing has come into use as a tool for mapping primary productivity on a global scale. Based

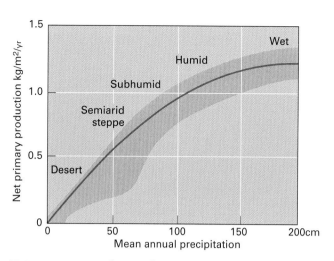

21.5 Precipitation and net production Net primary production increases rapidly with increasing precipitation, but levels off in the higher values. Observed values fall mostly within the shaded zone. (Data from Whittaker, 1970)

on remote sensing data, terrestrial productivity has increased since about 1980, while ocean productivity has decreased. These trends are most likely linked to global climate changes, including global warming at higher latitudes, reductions in cloud cover in equatorial regions, and decreasing winds over oceans. *A Closer Look: Focus on Remote Sensing 21.2 • Monitoring Global Productivity from Space* at the end of the chapter provides more information on this.

BIOMASS AS AN ENERGY SOURCE

Net primary production by photosynthesis represents a source of renewable energy derived from the Sun that can be exploited to fill human energy needs. The use of biomass—the measure of net primary production—as an energy source involves releasing solar energy that has been fixed in plant tissues through photosynthesis. This process can take place in a number of ways—the simplest is direct burning of plant matter as fuel. Other approaches involve the generation of intermediate fuels from plant matter; for example, methane gas, charcoal, and alcohol. Biomass energy conversion is not highly energy efficient. Typical values of net annual primary production of plant communities range from 1 to 3 percent of available solar energy. However, the abundance of terrestrial biomass is so great that, with proper development, biomass use could provide the energy equivalent of several million barrels of oil per day.

One important use of biomass energy is the burning of firewood for cooking in developing nations. The annual growth of wood in the forest of developing countries totals about half the world's energy production. However, fuelwood use exceeds production in many areas, creating local shortages and severe strains on some forest ecosystems. The forest-desert transition areas of thorntree, savanna, and desert scrub in central Africa south of the Sahara Desert are examples.

Even in closed stoves, wood burning is not very efficient, ranging from 10 to 15 percent for cooking. However, the conversion of wood to charcoal or gas can boost efficiencies to 70 or 80 percent with appropriate technology. In this process, called *pyrolysis*, controlled partial burning in an oxygen-deficient environment reduces carbohydrate to free carbon (charcoal) and yields flammable gases, such as carbon monoxide and hydrogen. Charcoal is more energy efficient than wood, burns more cleanly, and is easier to transport. In addition, charcoal can be made from waste fibres and agricultural residues that would normally be discarded. Thus, charcoal is an efficient fuel that can help extend the firewood supply in areas where wood is in high demand.

A second method of extracting energy from biomass uses anaerobic digestion to produce *biogas*. In this process, animal and human wastes are fed into a closed digesting chamber, where anaerobic bacteria break down the waste to produce a gas that is a mixture of methane and carbon dioxide. The biogas can be easily burned for cooking or heating, or it may be used to generate electric power. The digested residue is a sweet-smelling fertilizer. China now maintains a vigorous program of construction of biogas digesters for small family units to use. The benefits include better sanitation and reduced air and water pollution, as well as more efficient fuel usage.

Another use of biomass that is increasing in importance is the conversion of agricultural wastes to alcohol. In this process, yeast micro-organisms convert the carbohydrate to alcohol through fermentation. An advantage of alcohol is that it can serve as a substitute and extender for gasoline. Gasohol, a mixture of up to 10 percent alcohol in gasoline, can be burned in conventional engines without adjustment; its use is promoted in Canada.

Relying on biomass energy can also yield important benefits in reducing carbon dioxide emissions. Burning of biomass does not reduce the carbon dioxide flow to the atmosphere; in fact, it releases carbon dioxide more quickly than decomposition does. However, the energy obtained from biomass burning will, in all likelihood, substitute a quantity of fossil fuel. This reduces overall carbon dioxide emissions, since the equivalent amount of fossil fuel no longer needs to be burned.

BIOGEOCHEMICAL CYCLES IN THE BIOSPHERE

Solar energy flows through ecosystems, passing from one part of the food chain to the next, until ultimately the biosphere loses it as energy radiated to space. Matter also moves through ecosystems, but because of gravity, matter cannot be lost to space. As molecules are formed and reformed by chemical and biochemical reactions within ecosystems, the atoms that compose them are neither changed nor lost. Thus matter is conserved, and the constituent atoms and molecules can be used and reused, or cycled, within ecosystems.

Atoms and molecules move through ecosystems under the influence of both physical and biological processes. The pathways of a particular type of matter through the Earth's ecosystem comprise a **biogeochemical cycle** (sometimes referred to as a *material cycle* or *nutrient cycle*). Figure 21.6 shows the major features of a biogeochemical cycle. Any area or location of material concentration is a pool. There are two types of pools: **active pools**, where materials are in forms and places easily accessible to life processes; and **storage pools**, where materials are more or less inaccessible to life. A system of material flow pathways connects the various active and storage pools within the cycle. Life processes usually control pathways between active pools, while physical processes usually control pathways between storage pools.

There are two types of biogeochemical cycles: gaseous and sedimentary. In a *gaseous cycle*, the element or compound can be converted directly into a gas. The gas diffuses throughout the atmosphere and thus arrives over land or sea, to be reused by the biosphere, in a much shorter time. The primary constituents of living matter—carbon, hydrogen, oxygen, and nitrogen—all move through gaseous cycles. In a *sedimentary cycle*, weathering releases the compound or element from rock. It then follows the movement of running water, either in solution or as sediment, to the sea. Eventually, through precipitation and sedimentation, these materials convert into rock. The cycle completes when the rock is uplifted and exposed to weathering.

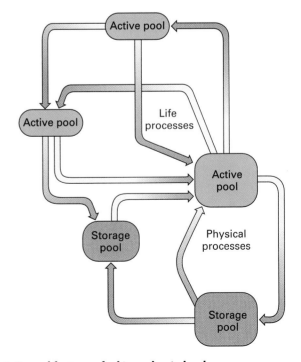

21.6 General features of a biogeochemical cycle

GEOGRAPHERS AT WORK

Balancing Forest Values
by Joel Mortyn, University of British Columbia

Forests are vital to the health of our planet. They mitigate the effects of global warming by replacing atmospheric carbon with oxygen. They provide a home to plants, animals and micro-organisms, making them a rich source of biodiversity. They regulate local climates, enrich soils, recycle nutrients and play an important role in the global water cycle.

Forests are also vital to our societies. They supply us with building materials, firewood, paper, food, water and some medicinal products. We use forests for recreational activities such as hiking, camping and fishing. Forests underpin many of the cultural and spiritual beliefs of Aboriginal peoples.

We often have conflicting views over the importance of each of these values. It is the job of forest scientists, such as Joel Mortyn to devise strategies that balance these needs.

Our society's demand for forest products is not decreasing. At the same time, society is expressing a wish to conserve more forests. If we are to achieve each of these goals, we must make better use of the trees we remove. Mortyn's research focuses on this important issue. He is studying how the economics of lumber recovery for Western hemlock (Tsuga heterophylla) varies throughout British Columbia. Hemlock logs often contain an internal core of rotten wood which is unsuitable for structural timber. As a result, most of these logs are used for paper production. However, the outer portions of these logs often contain high quality wood. Mortyn aims to find out whether it is more profitable to cut lumber from the outer portions while still using the inner

portion for paper production. If so, we will make better use of the different components of the logs. Over time, fewer forests will need to be logged to satisfy our demand for structural timber. (See *Eye on Global Change 21.1* for more of the human impact on our forests.)

Joel Mortyn conducting a forest assessment in southern British Columbia.

The magnitudes of the total storage and total active pools can be very different. In some cases, the active pools are much smaller than storage pools, and materials move more rapidly between active pools than between storage pools or in and out of storage. For example, in the carbon cycle, photosynthesis and respiration will cycle all the carbon dioxide in the atmosphere (active pool) through plants in about 10 years. But it may be many millions of years before the carbonate sediments (storage pool) now forming as rock will be uplifted and decomposed to release carbon dioxide.

NUTRIENT ELEMENTS IN THE BIOSPHERE

Table 21.2 lists the 15 elements that are most abundant in living matter. The three principal components of a carbohydrate—hydrogen, carbon, and oxygen—account for 99.5 percent of all living matter and are called **macronutrients**. Macronutrients are elements that substantial quantities of organic life require to thrive. The remaining 0.5 percent is divided among 12 elements. Six of these are also macronutrients: nitrogen, calcium, potassium, magnesium, sulphur, and phosphorus. Nitrogen is the fourth most abundant element in the composition of living matter and is taken up as nitrates from the soil. Calcium, potassium, and magnesium are elements derived from silicate rocks through mineral weathering. The sup-

ply of sulphur and phosphorus also depends on rock weathering. The other elements listed in Table 21.2, such as iron and manganese, are considered important for life, but are needed in

Table 21.2 Elements composing global living matter

Basic Carbohydrate	Percent*
Hydrogen (M)	49.74
Carbon (M)	24.90
Oxygen (M)	24.83
	Subtotal 99.47
Other Nutrients	
Nitrogen (M)	0.272
Calcium (M)	0.072
Potassium (M)	0.044
Silicon	0.033
Magnesium (M)	0.031
Sulphur (M)	0.017
Aluminum	0.016
Phosphorus (M)	0.013
Chlorine	0.011
Sodium	0.006
Iron	0.005
Manganese	0.003
	M—macronutrient

* Based on the 15 most abundant elements.

Source: E. S. Deevey, Jr., *Scientific American*, Vol. 223.

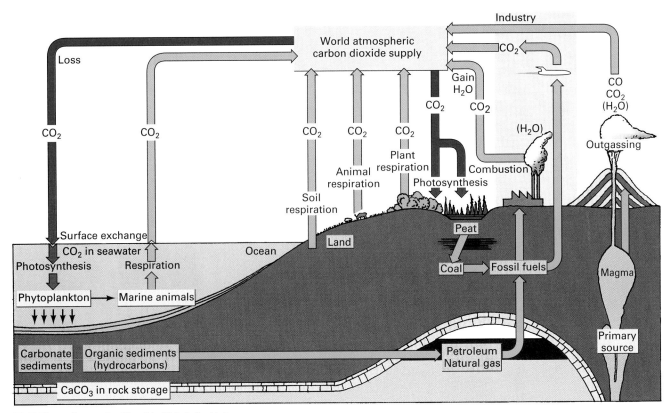

21.7 The carbon cycle (Copyright © A. N. Strahler)

lesser amounts. They are usually referred to as micronutrients. Additional elements that life requires only in minute quantities, including boron, copper, molybdenum, and zinc, are usually referred to as trace elements.

THE CARBON CYCLE

The movements of carbon through the life layer are of great importance because all life is composed of carbon compounds of one form or another. Of the total carbon available, most lies in storage pools as carbonate sediments below the Earth's surface. Only about 0.2 percent is readily available to organisms as carbon dioxide or as decaying biomass in active pools.

In the gaseous portion of the cycle, carbon moves largely as carbon dioxide, in the form of a free gas in the atmosphere or as a dissolved gas in fresh and salt water (Figure 21.7). In the sedimentary portion of the cycle, carbon is in the form of carbohydrate molecules in organic matter, hydrocarbon compounds in rock (petroleum, coal), and mineral carbonate compounds, such as calcium carbonate ($CaCO_3$). The world supply of atmospheric carbon dioxide is about 2 percent of the carbon in active pools. This atmospheric pool is supplied by plant and animal respiration in the oceans and on the lands. Under natural conditions, some new carbon enters the atmosphere each year from volcanoes by out-gassing in the form of carbon dioxide and carbon monoxide (CO). Industry injects substantial amounts of carbon into the atmosphere through combustion of fossil fuels.

Carbon dioxide leaves the atmospheric pool to enter the oceans, where phytoplankton uses it in photosynthesis. These organisms are primary producers in the ocean ecosystem and are consumed by marine animals in the food chain. Phytoplankton also build skeletal structures of calcium carbonate. This mineral matter settles to the ocean floor to accumulate as sedimentary strata, such as chalk and limestone, and forms an enormous storage pool not available to organisms until released later by rock weathering. Organic compounds synthesized by phytoplankton also settle to the ocean floor and eventually transform into the hydrocarbon compounds making up petroleum and natural gas. On land, plant matter accumulating over geologic time forms layers of peat that ultimately transform into coal. Petroleum, natural gas, and coal comprise the fossil fuels and represent huge storage pools of carbon.

Human activity is currently having a significant effect on the carbon cycle. The burning of fossil fuels is releasing carbon dioxide to the atmosphere at a rate far beyond that of any natural process. *Eye on Global Change 21.1 • Human Impact on the Carbon Cycle* documents how human activity has influenced the major flows within the carbon cycle.

THE OXYGEN CYCLE

Figure 21.8 shows details of the *oxygen cycle*. The active pool is mostly found in the atmosphere, but a small active pool is also present in the oceans. The complete picture of the cycling of oxygen includes its movements and storage when combined with car-

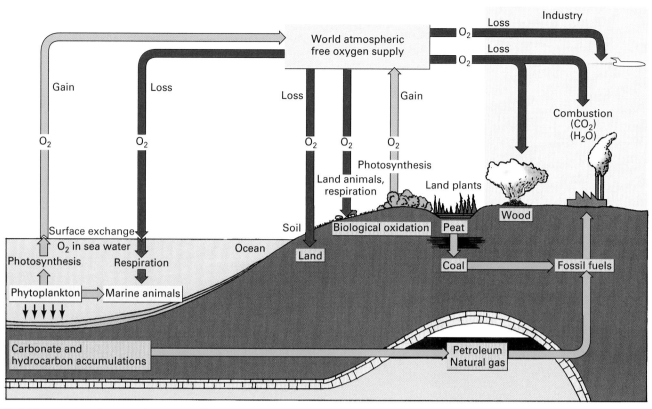

21.8 The oxygen cycle (Copyright © A. N. Strahler)

bon as carbon dioxide and as organic and inorganic compounds. These were covered in the discussion of the carbon cycle.

Oxygen enters the active pool through release in photosynthesis, both in the oceans and on land. Each year, a small amount of new oxygen comes from volcanoes through out-gassing, principally as carbon dioxide and water. Loss through organic respiration and mineral oxidation balances the input to the atmospheric pool. Adding to the withdrawal from the atmospheric oxygen pool is industrial activity through the combustion (oxidation) of wood and fossil fuels. Forest fires and grass fires are another means of oxygen consumption. Some oxygen from the small, active pool dissolved in the oceans is continuously placed in storage in mineral carbonate form in ocean-floor sediments.

Human activity reduces the amount of oxygen in the air by (1) burning fossil fuels; (2) clearing and draining land, which speeds the oxidation of soil minerals and soil organic matter; and (3) reducing photosynthesis by clearing forests for agriculture and by paving and covering previously productive surfaces. Fortunately, the oxygen pool is so large that the human impact or potential impact is small at this time.

THE NITROGEN CYCLE

Nitrogen moves through the biosphere in a gaseous *nitrogen cycle* in which the atmosphere, containing 78 percent nitrogen as N_2 by volume, is a vast storage pool (Figure 21.9). Nitrogen in the atmosphere, in the form of N_2, is an inert gas, and most plants or animals cannot assimilate it directly. The process by which nitrogen is converted into nitrogen compounds, such as ammonia and nitrates, is called *nitrogen fixation*. Through this process, nitrogen can be used in various biochemical processes. Only certain micro-organisms possess the ability to use nitrogen directly. Some of these are species of free-living soil bacteria; other nitrogen-fixing micro-organisms live in symbiotic association on the roots of higher plants. Symbiotic bacteria of the genus *Rhizobium* are associated with about 200 species of trees and shrubs, as well as almost all members of the legume family, which includes important agricultural species such as clover, alfalfa, soybeans, peas, beans, and peanuts. The blue-green algae are an important group as they can fix nitrogen, both in soil and in the oceans. Nitrogen fixation by all biological processes totals approximately 175 million metric tonnes.

Nitrogen fixation also occurs as a result of lightning, which can dissociate nitrogen molecules, enabling the atoms to combine with oxygen in the air to form nitrogen oxides. These dissolve in rain, and are carried to the Earth as nitrates. Atmospheric nitrogen fixation contributes about 5 percent of the total nitrogen fixed. The Haber-Bosch Process carries out industrial conversion of nitrogen to ammonia using an iron-based catalyst, very high pressure, and high temperature. This process converts about 50 million metric tonnes of nitrogen annually, and accounts for about 25 percent of nitrogen fixation by all processes.

EYE ON GLOBAL CHANGE | 21.1   Human Impact on the Carbon Cycle

Carbon cycles continuously among the land surface, atmosphere, and ocean in many complex pathways. However, human activity now strongly influences these flows. The most important human impact on the carbon cycle is the burning of fossil fuels. Another important impact arises from changing the Earth's land covers; for example, in clearing forests or abandoning agricultural areas.

About half of the output of carbon by fossil fuel burning is taken up by the atmosphere. Of the remaining amount, about 65 percent is absorbed by the oceans and 35 percent flows into the biosphere. Thus, ecosystems are a *sink* for carbon dioxide, accepting about 0.7 gigatonnes per year of carbon. Ecosystems cycle carbon in photosynthesis, respiration, decomposition, and uncontrolled combustion through burning.

If the value of 0.7 gigatonnes per year in carbon uptake is correct, the amount of terrestrial biomass must be increasing at that rate. However, forests are diminishing in area as they are logged or converted to farmland or grazing land. This conversion is primarily occurring in tropical and equatorial regions; it is estimated that this will release about 1.6 gigatonnes per year of carbon to the atmosphere. Since this release is included in the net land ecosystem uptake of 0.7 gigatonnes per year, the Earth's remaining forest cover must be taking up the 1.6 gigatonnes per year released by deforestation as well as an additional 0.7 gigatonnes per year. Thus, mid- and high-latitude forests are estimated to be increasing in area or biomass at a rate of 2.3 gigatonnes per year.

Independent evidence seems to confirm this conclusion. In Europe, for example, forest statistics show an approximate 25 percent increase in the volume of living trees in the past two decades. This increase has been sustained despite damage to forests by air pollution, especially in eastern Europe. In North America, forest areas are increasing in many regions, and there is an increasing trend in reforestation following timber removal.

Some of the increase in global biomass may also be the result of enhancement of photosynthesis by warmer temperatures and increased atmospheric carbon dioxide concentrations. Another possible reason for increased ecosystem productivity is nitrogen fertilization of soils by washout of nitrogen-pollutant gases in the atmosphere.

Some foresters have observed that harvesting mature forests and replacing them with young, fast-growing timber should increase the rate of withdrawal of carbon dioxide from the atmosphere. Since the lumber of the mature forests goes into semi-permanent storage in dwellings and structures where it is protected from decay and oxidation to carbon dioxide, it represents a withdrawal of carbon dioxide from the atmosphere. The young forests that replace the mature ones grow quickly, fixing carbon at a much faster rate than the older, mature forest, in which annual growth has slowed.

However, some reports indicate that the conversion of old-growth forests to young, fast-growing forests will not significantly decrease atmospheric carbon dioxide. While 42 percent of the harvested timber goes into comparatively long-term storage (greater than five years) in building structures, much of the remainder is directly discarded on the logging site, where it is burned or rapidly decomposes. In addition, some biomass becomes waste in factory processing of the lumber, where sawdust and scrap are burned as fuel. Similarly, the manufacture of paper also results in short-term conversion of a large proportion of the harvested trees to carbon dioxide. Thus, harvesting of old-growth forests as now practised appears to contribute substantially to atmospheric carbon dioxide.

Some environmentalists have advocated increased tree planting as a way of enhancing carbon dioxide fixation. To take up the quantity of carbon now being released by fossil fuel burning would require some 7 million square kilometres of new closed-crown broad-leaved deciduous forest—an area about the size of Australia.

Higher carbon dioxide concentration in the atmosphere might enhance photosynthesis and thus increase the rate of carbon fixation. The enhancement of photosynthesis by increased carbon dioxide concentrations has been observed for many plants and demonstrated as a way of increasing yields of some crops. However, carbon dioxide is only one factor in photosynthesis. Light, heat, nutrients, and water are also needed, and restrictions in any of these would affect productivity.

While the dynamics of forests are important in the global carbon cycle, soils may be even more important. Recent inventories estimate that about four times as much carbon resides in soils than in above-ground biomass. The largest reservoir of soil carbon is in the boreal forest. In fact, there is about as much carbon in boreal forest soils as in all above-ground vegetation. This soil carbon has accumulated over thousands of years under cold conditions that have slowed its decay. However, there is now great concern that global warming, which is acting more strongly at high latitudes, will increase the rate of decay of this vast carbon pool, and that boreal forests, which are presently a sink for carbon dioxide, will become a source.

Reducing the rate of carbon dioxide buildup in the atmosphere is a matter of great international concern and the subject of several international treaties designed to limit emissions of greenhouse gases. The latest was signed by about 180 countries at Kyoto, Japan, in December 1997 (see map below). Since that time, nations have been struggling with the Kyoto Protocol's implementation, including Canada. While a lot of good progress has been made, more work is necessary, specifically an effective global commitment to reduce carbon dioxide releases and control of global warming.

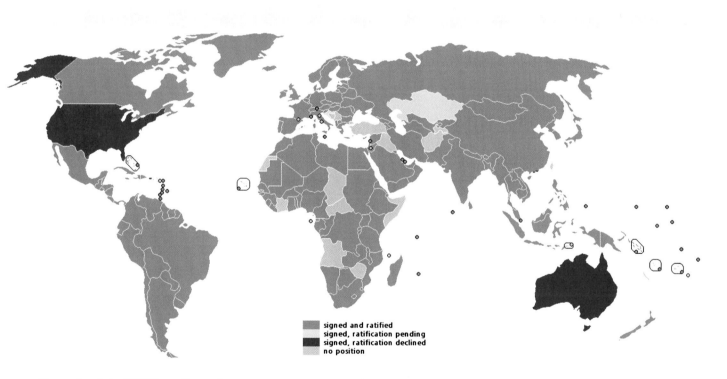

signed and ratified
signed, ratification pending
signed, ratification declined
no position

Signatories of the 1997 Kyoto Protocol

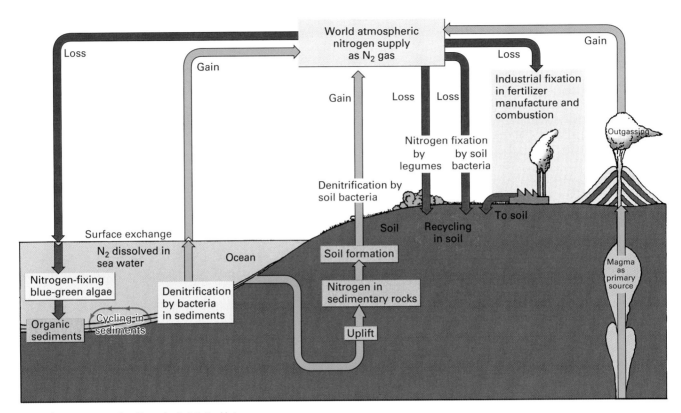

21.9 The nitrogen cycle (Copyright © A.N. Strahler)

Nitrogen is lost to the biosphere by *denitrification*, a process in which certain soil bacteria, such as species of *Pseudomonas* and *Thiobacillus*, convert nitrogen from usable forms back to N_2. Denitrification depletes the soil of available nitrogen and reduces plant productivity. By returning nitrogen to the atmosphere, denitrification completes the organic portion of the nitrogen cycle.

Currently, nitrogen fixation is far exceeding denitrification, and usable nitrogen is accumulating in the life layer. Human activities have almost entirely produced this excess of fixation. Human activity fixes nitrogen in the manufacture of nitrogen fertilizers and by oxidizing nitrogen in the combustion of fossil fuels. Widespread cultivation of legumes has also greatly increased worldwide nitrogen fixation. At present rates, nitrogen fixation attributable to human activity nearly equals all natural biological fixation.

Much of the nitrogen fixed by human activities is carried from the soil into rivers and lakes where it can cause major water pollution problems by stimulating the growth of algae and phytoplankton to create algal blooms. Ultimately, the nitrogen reaches the oceans. The long-term impact of large amounts of nitrogen on the Earth's marine ecosystems remains uncertain, although nitrogen is implicated as the cause of the "*dead zones*" that have been reported in some oceans (Figure 21.10).

Dead zones are characterized by *hypoxia*, a condition in which oxygen is almost entirely depleted. Hypoxia develops because of high biological demand from decomposer bacteria that flourish on the abundant but short-lived algal blooms caused by excess nutrients in the water. Dead zones have been reported in the North, Adriatic, Baltic, and Black seas, and Japan's Seto Inland Sea. However, the dead zones in these seas are small compared with the area of nearly lifeless ocean in the Gulf of Mexico. The waters of the Gulf are so depleted of oxygen that fish and other forms of sea life are killed or forced to move away (Figure 21.11).

The rivers that flow into the Gulf of Mexico drain from about 40 percent of all U.S. land area and account for nearly 90 percent of the freshwater runoff into the Gulf. They carry excess nutrients generated from agricultural runoff containing chemical fertilizers and animal manure, both of which are rich in nitrogen. The dead zone in the Gulf of Mexico was first reported in the 1980s. Its size changes every year. In drought years, the dead zone is relatively small, but when river discharge is high, it can exceed 20,000 square kilometres. This type of ocean pollution is predicted to become more acute because industrial fixation of nitrogen in fertilizer manufacture is currently doubling about every six years.

THE SULPHUR CYCLE

Most of the Earth's sulphur is tied up in rocks and ocean sediments. A relatively small amount is also present in the atmosphere (Figure 21.12). Sulphur originates from igneous rocks, such as

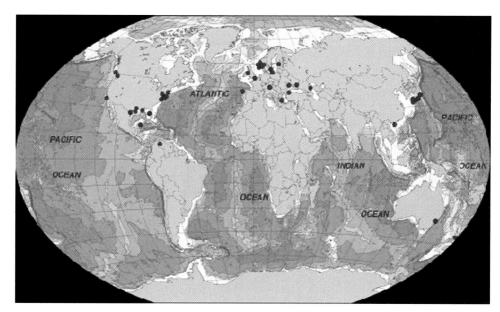

21.10 The Earth's marine dead zones Dead zones are areas where concentrations of dissolved oxygen in the bottom waters are so reduced that the organisms that normally inhabit the region are unable to survive. Dead zones are associated with areas of high phytoplankton production caused by agricultural fertilizers carried to the sea by rivers.

pyrite (FeS_2), and is also found in gypsum ($CaSO_4.2H_2O$) and other sedimentary deposits. Long-term storage of sulphur occurs in both organic and inorganic forms, from which it is released by weathering and decomposition. Sulphur in mineral form can be mobilized through oxidation of sulphides to sulphate, which may then go into solution and be transported to the ocean in runoff.

Sulphur can also enter the atmosphere as sulphur dioxide (SO_2) and hydrogen sulphide (H_2S).

Volcanic activity releases sulphur gases to the atmosphere. In addition, hydrogen sulphide (H_2S), dimethyl sulphide (DMS), and carbonyl sulphide (COS) enter the atmosphere through biological activity. Fine particles of gypsum are also carried into the

21.11 Gulf of Mexico dead zone (a) Dissolved oxygen in the Gulf Coast waters. Reds and oranges represent low oxygen concentrations. (b) Summer phytoplankton conditions along the Gulf Coast. Reds and oranges represent high concentrations of phytoplankton and river sediment. (Image acquired July 2002 by the MODIS/Aqua platform NASA/Goddard Space Flight Center)

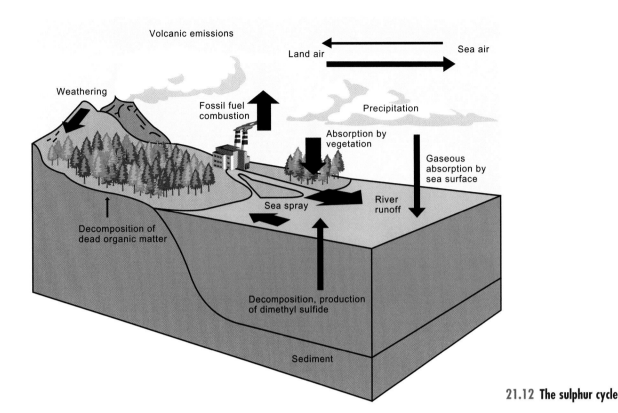

21.12 The sulphur cycle

atmosphere from desert soils. However, the largest input is from human activity; this enters the atmosphere mainly as sulphur dioxide (SO_2) or hydrogen sulphide through combustion of coal and oil, petroleum refining, and smelting of sulphide ores of copper, lead, zinc, and other metals. Most of these terrestrial emissions return to the land in the form of dry deposition or as acid rain, but some transfer from the land to the sea.

Sulphur occurs in the ocean mainly in the form of sulphates that are derived from rivers and atmospheric redeposition of sulphate-containing aerosols evaporated from the ocean surface. The principal gas originating from the oceans is dimethyl sulphide, produced by decomposing phytoplankton cells. Dimethyl sulphide represents the largest natural source of gaseous sulphur in the atmosphere. It combines with water droplets to form acidic aerosols that act as cloud condensation nuclei. This has been suggested as a mechanism that might offset global warming (see Chapter 6).

A small component of the sulphur in river flows comes from the natural weathering of pyrite and gypsum. However, most of the sulphur transported by rivers comes from human activity associated with mining, erosion, and air pollution. Acid mine drainage is a particular problem and arises where mining operations expose rocks containing ferrous sulphide (FeS_2). Reaction with air and water produces sulphuric acid (H_2SO_4) and other acidifying substances. This, in turn, can increase solubility of heavy metal contaminants, such as cobalt and arsenic. The effect of acid mine drainage on aquatic ecosystems can be devastating.

An important distinction between cycling of sulphur and cycling of nitrogen and carbon is that sulphur is present in the environment in a fixed or available form as sulphate anions (SO_4^-), which living organisms can directly use. Although sulphur is not required on the same scale as nitrogen and other nutrients, it is an essential component of amino acids, such as cystine, and is important for the proper functioning of proteins and enzymes in plants and animals. In addition, many bacteria in anaerobic environments use sulphate as a source of oxygen for respiration. Some green and purple sulphur-reducing bacteria are able to combine hydrogen sulphide with carbon dioxide to form carbohydrates. This anaerobic process may be the method of photosynthesis that sustained the Earth's earliest life forms.

SEDIMENTARY CYCLES

The carbon, oxygen, nitrogen, and sulphur cycles are all referred to as gaseous cycles because they possess a gaseous phase in which the element involved is present in significant quantities in the atmosphere. Many other elements move in sedimentary cycles; that is, from the land to ocean in running water, returning after millions of years in uplifted terrestrial rocks. These elements are not present in the atmosphere, except in small quantities as blowing dust or condensation nuclei in precipitation.

Figure 21.13 shows how some important macronutrients move in sedimentary cycles. Mineral nutrients derive mainly from weathering of soil minerals and decomposition of organic residues. Some are stored in nutrient pools within the soil, from which they

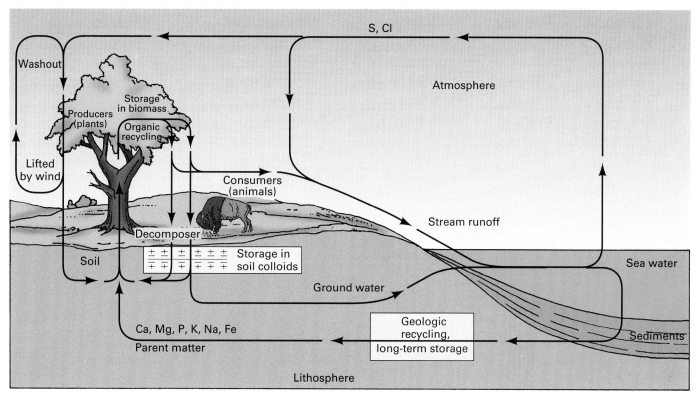

21.13 Sedimentary cycles A flow diagram of the sedimentary cycle of materials in and out of the biosphere and within the inorganic realm of the lithosphere, hydrosphere, and atmosphere

may, to varying degrees, be extracted by plants or lost by leaching. Nutrients that are held as ions on the surfaces of soil colloids are readily available to plants; however, in other forms, they may be relatively insoluble and become available only slowly over long periods. Although nutrients in solution can be taken up immediately by plants, in this form, they also are susceptible to leaching and removal by percolating water. In addition, some nutrients are also lost in particulate form through overland flow and erosion. Other particles are lifted into the atmosphere by winds sometimes falling back to the Earth great distances from their source.

Exchangeable cations, such as phosphorus (P) and potassium (K), are generally held in a short-term storage pool and rapidly replenish nutrient ions as they are removed from the soil solution. The rate at which this transfer of soil minerals proceeds depends on their solubility. Minerals in the form of nitrates, carbonates, sulphates, and chlorides are readily taken up in the soil solution. Other forms may require a complex series of reactions before they are released from the mineral particles. For example, phosphates that are strongly bound to clays are released quite slowly. Soil organic matter also releases nutrients slowly as it decomposes. Most readily available nutrients are held on the negatively charged surfaces of clay particles and organic matter (see Chapter 20).

Cations are continually exchanged with other soluble ions until their supply is depleted. Cation exchange is a major source of nutrients like calcium and magnesium, as well as micronutrient trace metals like zinc, manganese, and copper.

Cycling of many plant nutrients closely follows the carbon cycle. When nutrient-rich plant residues return to the soil, this pool of carbon compounds is acted upon by decomposer organisms. As organic matter breaks down to simpler compounds, plant nutrients are released in available forms to growing plants. Rapid cycling of nutrients in this way is important in ecosystems, such as the tropical rain forest, where heavy rain could quickly leach essential cations from the soil.

Considerable element recycling occurs among the producers, consumers, and decomposers in an ecosystem. However, the elements used in the biosphere are continually escaping to the sea as ions dissolved in stream runoff and ground water flow. Here, enormous storage pools hold nutrient elements, making them unavailable to organisms. As well as sea water, these storage pools include sediments on the sea floor and enormous accumulations of sedimentary rock beneath both land and oceans. Eventually, weathering releases elements held in the geologic storage pools into the soil.

As human activity drives carbon dioxide concentrations in the atmosphere to ever higher levels, it becomes ever more important to understand and model the biosphere's photosynthetic activity. Analysis of the global carbon budget shows that the biosphere must be a *carbon sink*; that is, for carbon flows to balance, global photosynthesis must exceed global respiration by a significant amount. This means that fixed carbon is accumulating in the biosphere, reducing the amount of carbon dioxide buildup in the atmosphere. However, this conclusion is not certain without some sort of direct measurement. Until recently, there was no way to measure the Earth's photosynthetic activity, but new techniques using remote sensing, meteorological observations, and models of biological productivity now make it possible.

The factors that control primary production at any point on the Earth include the following:

- *The amount of photosynthetic material present.* For terrestrial plants, this is measured by the surface area of leaves above a square metre of ground. For oceanic phytoplankton, it is the chlorophyll concentration within a cubic metre at the ocean surface. Chlorophyll is a plant pigment that converts light energy into the chemical energy needed to fix carbon.

- *Light.* Photosynthesis requires light. What is important is the amount of light absorbed by the photosynthetic material, which depends on two things: (1) the amount of illumination from the Sun and sky, determined by such factors as season, latitude, and cloud cover; and (2) the amount of photosynthetic material present to absorb the available light energy.

- *Temperature.* The biochemical process of fixing carbon is sensitive to temperature. Warmer temperatures favour photosynthesis, but if temperatures are too high, photosynthesis shuts down.

- *Water.* Terrestrial plants transport water from their roots to their leaves in the process of transpiration. As the water evaporates from leaf pores, it cools the leaves and allows them to maintain high levels of photosynthesis, even under intense sunlight. Of course, water is freely available to aquatic phytoplankton at all times.

- *Nutrients.* Nutrients can play an indirect role by reducing the health of organisms or limiting their development. In many cases, the productivity of phytoplankton is limited by a scarcity of nutrients, especially iron.

Land Productivity

The "greenness" of the surface detected by remote sensing can be related to the amount of photosynthetic material. Land plants strongly absorb red light and strongly reflect near infrared light. Thus, locations that appear darker in a red band image and brighter in a near-infrared band image will have more leaf area.

Figure 1a shows an image of leaf-area index for an area of North America that includes the United States and a portion of southeastern Canada. Leaf-area index (LAI) is the ratio of the area of the upper surface of a leaf to the area of ground below. Thus, a leaf-area index of two indicates that each square metre of ground surface is covered by two square metres of leaf area. Typical leaf-area indices can range from zero in barren deserts to six or seven in dense forest. The map of leaf-area index is derived from MODIS images, acquired May 1–10, 2003. At this time of year, the southeast, mid-west, and west-coast regions are abundantly green.

Figure 1b converts the leaf-area measurement into another measurement—the fraction of photosynthetically active radiation (FPAR) absorbed by the leaf area. As shown on the scale, this fraction ranges from 0 to 100 percent. It expresses the proportion of useful solar radiation (visible light) that the leaf canopy actually absorbs. By comparing this value with the solar energy falling on each point during each day, it is possible to estimate the energy available for photosynthesis.

The amount of photosynthesis that actually occurs depends on both temperature and water availability, as noted above. Another factor is the efficiency of the type of vegetation cover; given the same environmental conditions and the same leaf area, some vegetation types are more productive than others. Gross primary productivity can be calculated by combining temperature and rainfall from meteorological data sources with efficiency based on a map of vegetation cover type (Figure 1c). The map shows the same general patterns as the others, with maximum gross primary productivity in the southeast and west-coast regions, but the fine detail shows how rainfall, cloudiness, temperature, and type of vegetation cover affect areas.

The last step is to estimate respiration—the rate at which carbon is released to the atmosphere in plant metabolism. This will depend on many of the same factors as photosynthesis,

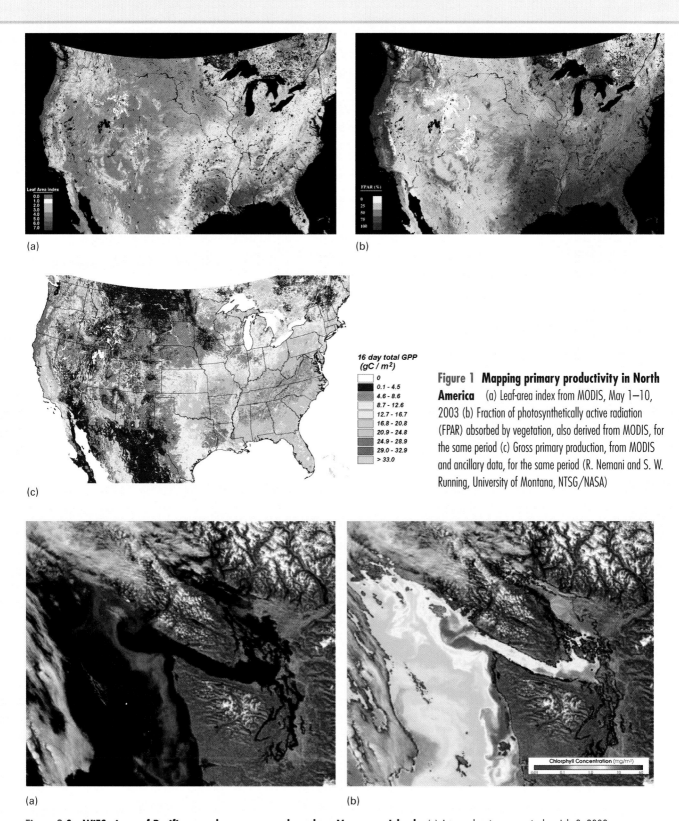

(a)

(b)

(c)

16 day total GPP
(gC / m²)

	0
	0.1 - 4.5
	4.6 - 8.6
	8.7 - 12.6
	12.7 - 16.7
	16.8 - 20.8
	20.9 - 24.8
	24.9 - 28.9
	29.0 - 32.9
	> 33.0

Figure 1 Mapping primary productivity in North America (a) Leaf-area index from MODIS, May 1–10, 2003 (b) Fraction of photosynthetically active radiation (FPAR) absorbed by vegetation, also derived from MODIS, for the same period (c) Gross primary production, from MODIS and ancillary data, for the same period (R. Nemani and S. W. Running, University of Montana, NTSG/NASA)

(a)

(b)

Figure 2 SeaWiFS views of Pacific coastal waters around southern Vancouver Island (a) A true-colour image acquired on July 9, 2003 (b) Chlorophyll concentration on the same day (SeaWiFS Project, NASA/GSFC/ORBIMAGE)

(continued)

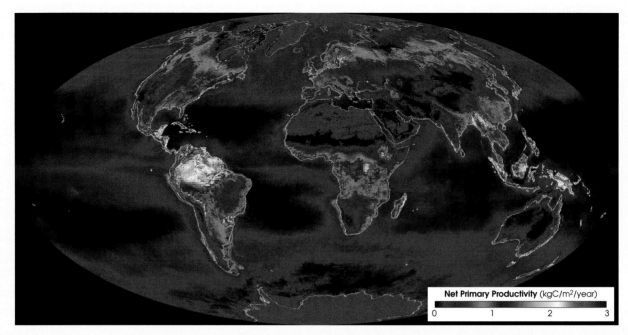

Figure 3 Global net primary productivity from MODIS for 2002 This image maps global productivity, including both land and oceans, as viewed by the MODIS instruments on NASA's Terra and Aqua satellite platforms. (MODIS Science Team/NASA)

including leaf area, temperature, and moisture. The result is net primary productivity, which is shown globally in figures 3 and 4.

Ocean Productivity

For oceans, the amount of photosynthetic material present at a location is measured in terms of chlorophyll concentration in ocean water. This quantity is expressed in units of milligrams of chlorophyll per cubic metre of sea water. As with terrestrial plants, this quantity is determined through colour, detected by satellite instruments using special narrow spectral bands.

Figure 2a shows a colour image of the waters around southern Vancouver Island and the Puget Sound area acquired by the SeaWiFS (Sea-viewing Wide Field-of-view Sensor) instrument, which is designed specifically to image the colour of ocean water. The band of greenish water along the coast of Washington State indicates the presence of phytoplankton.

By carefully analyzing the spectral information acquired by SeaWiFS, it is possible to estimate chlorophyll concentration across the image (Figure 2b). Chlorophyll concentration, like leaf-area index, is not sufficient to model photosynthesis accurately. The presence of chlorophyll doesn't mean that active photosynthesis is proceeding at a maximum rate. For example, lack of nutrients such as iron may slow the photosynthetic process.

The efficiency of photosynthesis by phytoplankton can be assessed from MODIS satellite data. This satellite instrument has special spectral bands that measure chlorophyll fluorescence, which causes chlorophyll to emit light at particular wavelengths. Chlorophyll fluorescence occurs primarily when absorbed sunlight is not being used to fix carbon. In other words, when chlorophyll fluorescence is strong, photosynthesis is weak. Therefore, the strength of the fluorescence leaving ocean waters provides a measure of the efficiency of the chlorophyll that is present. With this missing link in the chain of information, mapping primary productivity of the ocean becomes possible.

Global Productivity

Figure 3 shows a global map of net primary production of both land and oceans for 2002. Net primary production is generally higher on land than in oceans. However, there is a lot more ocean surface than land surface. Taking this into account, terrestrial and oceanic net primary production are about equal.

Large central areas of the oceans are unproductive (dark purple). This is largely due to lack of essential nutrients in surface waters. Near land and along coasts where nutrient-rich deep water rises from below, nutrients are in greater supply and oceans are more productive. For example, upwelling is particularly important along the west coast of South America from Peru southward and shows dark and light blue tones.

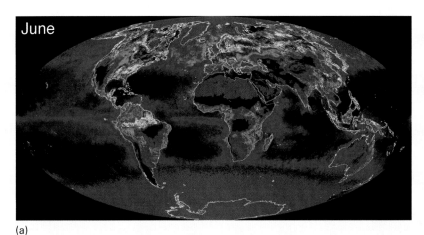

(a)

(b)

Figure 4 Global net primary productivity with the seasons (a) Net primary productivity for June 2002, as derived from MODIS (b) Net primary productivity for December 2002 (MODIS Science Team/NASA)

On land, the highest values of net primary production are shown by the yellow and red tones of the Amazon basin and equatorial Asia. Lesser, but still high, values are evident in eastern North America, central Africa, eastern Asia, and Scandinavia. Low values characterize arid and semi-arid regions, such as the interior deserts of Australia and semi-arid western North America (grey areas show regions for which data are not available.)

Figure 4 shows net primary productivity for June 2002 (a) and December 2002 (b). Comparing the two images, ocean productivity patterns are much the same at low and midlatitudes, but quite different at high latitudes. Here, oceans are noticeably more productive in summer in each hemisphere. Over land, the June pattern shows the highest values of net primary productivity in the boreal and eastern continental zones of the northern hemisphere. Long days and warm temperatures produce these values. Productivity remains high in the Amazon and equatorial regions. In December, net primary production is high in South America and sub-Saharan Africa,

as well as along northern and eastern coasts of Australia. Meanwhile, most of northern North America and Eurasia are dormant.

Recent Changes in Global Productivity

Analysis of satellite images acquired by NOAA's Advanced Very High Resolution Radiometer (AVHRR) between 1982 and 1999 shows that terrestrial net primary productivity increased by about 6 percent during that period (Figure 5). The greatest increases occurred in northwestern North America, equatorial and subtropical South America, and India. Large increases were also noted in Africa and northern Russia. By comparing these results with meteorological data, it was concluded that the increase at low latitudes was due to reduced cloud cover, which allowed more light for photosynthesis. At high latitudes, the causes were increased temperature and, to some extent, increased water availability.

Over a similar time period, large changes in ocean phytoplankton concentration took place, as well (Figure 6). The colour scale is logarithmic, with deeper tones of blue and red showing much larger changes than their lighter tones. The most striking feature observed was the decline in phytoplankton in northern oceans (blue), which amounted to about 30 percent in the North Pacific and 14 percent in the North Atlantic. In the equatorial zones, increases of up to 50 percent (red tones) were observed at some locations, but the increases were not large enough to account for the high-latitude decreases. Thus, it was concluded that global concentrations of phytoplankton had decreased overall.

One explanation for this decline is that warmer sea-surface temperatures are increasing the duration and strength of the thermocline at high latitudes. This inhibits the mixing of nutrient-rich deep water with nutrient-poor surface water and thus keeps the phytoplankton population in a nutrient-limited condition. Another possibility is that wind speeds are decreasing, which will also reduce mixing. Both of these changes have already been observed. Summer sea-surface temperatures in northern regions increased by 0.4°C between the early 1980s and 2000, and average spring wind stresses on the sea surface have decreased by about 8 percent. However, it is not certain that these changes are the result of global warming or originate from a multi-year ocean cycle yet to be discovered.

(continued)

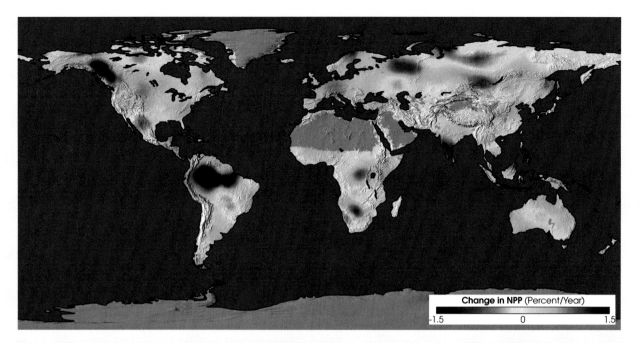

Figure 5 Change in net primary productivity for land, 1982–99   This image, constructed from images of NOAA's AVHRR instruments acquired from 1982 to 1999, shows the change in net primary productivity in units of percent per year. (R. Nemani and S. W. Running, University of Montana, NTSG/NASA)

Implications for the Global Carbon Budget

What are the implications of these changes in global productivity for the global carbon budget? First, the balance in global productivity between land and ocean seems to be shifting toward land. Although this may increase the amount of terrestrial biomass and decrease the rate of carbon dioxide buildup, it also means that ocean productivity is probably declining and altering oceanic ecosystems.

Second, soil respiration must also be considered. Soil respiration is analogous to the decay of organic matter in soils, which releases carbon dioxide to the atmosphere. Soil respiration is quite temperature dependent, and small increases in temperature can stimulate large increases in respiration. Considering that boreal forest soils are rich in organic matter and that global temperatures are increasing most rapidly at high latitudes, increased release of carbon dioxide by soil respiration could exceed increased biomass production and result in higher levels of atmospheric carbon dioxide.

Third, it is not known if the increase in land productivity will continue. For example, with more sunlight and

A Look Ahead

The organizing principles of energy flow and matter cycling help to clarify some of the processes acting in the biosphere. Just as solar energy is the driving force for the circulation of the global atmosphere and ocean, it also provides the power source for photosynthesis, on which all the world's organisms, including humans, ultimately depend for sustenance.

Organisms also rely on the smooth functioning of material cycles. Without the tiny fraction of the Earth's atmosphere that is carbon dioxide, no terrestrial photosynthesis would occur. If human activity enhances this component through carbon dioxide release, it is sure to impact the productivity of the biosphere directly, as well as indirectly through climate change. Similarly, humans influence the nitrogen and oxygen cycles without fully knowing the consequences.

The final two chapters examine the biosphere from the perspective of the spatial patterns of plants and animals on scales ranging from local to global. Chapter 22 discusses the processes that determine the distributions of individuals and species, including organism–environment relationships and dynamic processes such as species dispersal, migration, and extinction. The book closes with a survey of the world's major biome types and how human activities affect them.

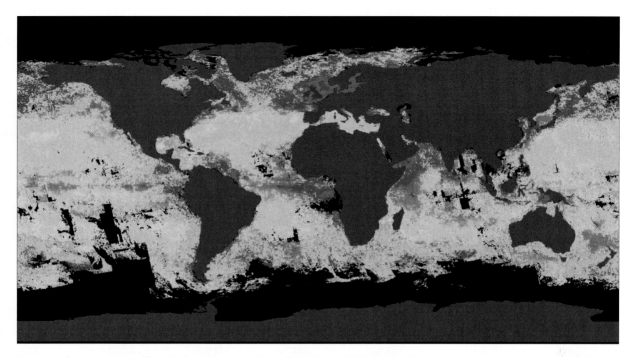

Figure 6 Change in ocean chlorophyll concentration from 1979–86 to 1997–2000   This image shows how summer ocean chlorophyll concentrations have changed over the past two decades. Data are for July–September, as acquired by NOAA's Coastal Zone Color Scanner (1979–86) and SeaWiFS instrument (1997–2000). (NASA/NOAA)

higher temperatures, moisture may become limited in equatorial and tropical forests, causing net primary production to plateau or even decrease. Moreover, the large changes in temperature forecast for high-latitude regions will ultimately lead to reduced productivity as boreal forests come under increasing stress, and trees at their southern limits die.

All these uncertainties emphasize the importance of the ability to map and monitor global production using remote sensing. While it may not yet be possible to predict exactly how the carbon cycle will behave in the future, at least current trends can be monitored. This will allow carbon cycle models to be refined over time, so that the full impact of human activity on the global carbon cycle can be assessed.

CHAPTER SUMMARY

- The food web of an ecosystem details how food energy flows from primary producers through consumers and on to decomposers. Because energy is lost at each level, only a relatively few top-level consumers are normally present.
- Photosynthesis is the production of carbohydrate from water, carbon dioxide, and light energy by primary producers. Respiration is the opposite process, in which carbohydrate is broken down into carbon dioxide and water to yield chemical energy and thus power organisms. Net photosynthesis is the amount of carbohydrate remaining after respiration has reduced gross

photosynthesis. Net photosynthesis increases with increasing light and temperature, up to a point.
- Forests and estuaries are ecosystems with high rates of net primary production, while grasslands and agricultural lands are generally lower. Oceans are most productive in coastal and upwelling zones near continents. Among climate types, those with abundant rainfall and warm temperatures are most productive.
- Biomass is a form of solar-powered energy. Charcoal, biogas, and alcohol are biomass products that can be used as fuels.

- There are two types of biogeochemical cycles; *gaseous*, in which the element has an important gaseous phase and moves within the atmosphere, and *sedimentary*, when no important gaseous phase is involved. Biogeochemical cycles consist of *active pools* and *storage pools* linked by flow paths. Of most concern are biogeochemical cycles of the macronutrients, which include carbon, hydrogen, oxygen, nitrogen, as well as calcium, potassium, phosphorus, and magnesium.
- The *carbon cycle* includes an active pool of biospheric carbon and atmospheric carbon dioxide, with a large storage pool of carbonate in sediments. Human activities have provided a pathway from storage to active pools by the burning of fossil fuel. The *oxygen cycle* features an active pool of atmospheric O_2, which is increased by photosynthesis and reduced by respiration, combustion, and mineral oxidation.
- The *nitrogen cycle* also has an important gas phase, but nitrogen is largely held in the form of N_2, which most organisms cannot use directly. *Nitrogen fixation* occurs when bacteria or blue-green algae convert N_2 to more useful forms, often in symbiosis with higher plants. Human activity has doubled the rate of nitrogen fixation, largely through fertilizer manufacture. Human activity has also greatly influenced the *sulphur cycle*, mainly through the deleterious effects of acid precipitation and acid mine drainage.
- Sedimentary cycles involve macronutrients that do not have an important gas phase. These elements are held in active pools in living and decaying organisms and in soils. Storage pools include sea water, sediments, and sedimentary rocks.

KEY TERMS

active pool	C_4 pathway	food web	omnivores
biogeochemical cycle	CAM pathway	gross photosynthesis	photosynthesis
biogeography	consumers	macronutrients	primary producers
biomass	decomposers	net photosynthesis	respiration
C_3 pathway	food chain	net primary production	storage pool

REVIEW QUESTIONS

1. What is a food web or food chain? What are its essential components? How does energy flow through the food web of an ecosystem?
2. How is net primary production related to biomass? Identify some types of terrestrial ecosystems that have a high rate of net primary production and some with a low rate.
3. What is a biogeochemical cycle? What are its essential features? Identify and compare two types of biogeochemical cycles.
4. List nine macronutrients and identify those associated with gaseous and sedimentary cycles.
5. What are the essential features and flow pathways of the carbon cycle? How have human activities affected the carbon cycle?
6. What are the essential features and flow pathways of the nitrogen cycle? What role do bacteria play? How has human activity modified the nitrogen cycle?
7. What is a "dead zone"? How does it form, and how might it change over time?
8. What are the essential features and flow pathways of macronutrients in sedimentary cycles?

EYE ON THE ENVIRONMENT 21.1   Human Impact on the Carbon Cycle

1. How is change in land use affecting the global carbon balance? Will replacing old-growth forests with younger faster-growing forests help remove carbon dioxide from the atmosphere?
2. Will increasing levels of atmospheric carbon dioxide have an effect on the rate of carbon fixation by ecosystems? How will increases in temperature affect the release of boreal soil carbon?

A CLOSER LOOK

Focus on Remote Sensing 21.2 Monitoring Global Productivity from Space

1. What factors control primary production at any point on the globe? Identify each factor and relate it to the process of photosynthesis.
2. What is a leaf-area index? How and why can it be sensed remotely? How does it relate to the fraction of photosynthetically active radiation absorbed by the plants on the surface?
3. What factor is most important in determining ocean primary productivity? How is it mapped using remote sensing?
4. What are the main patterns of primary productivity over the oceans? Over land?
5. What changes in net primary productivity over land have occurred over the past two decades? How are they related to the controlling factors identified in Question 1?
6. What changes in net primary productivity over oceans have occurred over the past two decades? What are the possible causes of these changes?
7. What are the implications of productivity changes for the carbon cycle? How will remote sensing be useful in assessing the impact of human activity on the global carbon cycle?

VISUALIZATION EXERCISE

1. Draw a diagram of the general features of a biogeochemical cycle, in which life processes and physical processes link storage pools and active pools.

ESSAY QUESTIONS

1. Select one of the cycles described in the text (carbon, oxygen, nitrogen, sulphur, or sedimentary). Identify and describe the power sources for each of the major pathways in the cycle.
2. Suppose atmospheric carbon dioxide concentration doubles. What will be the effect on the carbon cycle? How will flows change? Which pools will increase? decrease?

EYE ON THE LANDSCAPE |

Chapter Opener   Tropical microcosm. This papaya (*Carica sp.*) tree is growing out of an abandoned termite mound (**A**) and benefits from the higher nutrient status of the soil that has been reworked by the insects. The smooth bark (**B**) is characteristic of many tropical woody species, as is the general lack of branching, which is often confined to the higher parts of the stem (**C**). Large leaves (**D**) are not uncommon on the smaller plants that are generally confined to the understory of tropical forests, and their waxy surfaces help to shed water in this wet climate. The ripening fruit (**E**) is borne in clusters near the top of the stem. Part of the land has recently been cleared and debris left on the surface (**F**) can add nutrients to the soil and reduce surface runoff and erosion.

Web of life. The grizzly bear (**A**) is an omnivore that feeds on berries and meat as opportunities arise. Salmon (**B**) face a hazardous journey when they return to their spawning grounds. Fast flowing currents, waterfalls and rapids are a daunting prospect, to which is added the perils of confronting a large and voracious top carnivore. The river (**C**) itself plays a role in the movement of matter and energy through the ecosystem, as it carries sediment and nutrients to the sea under the influence of gravity.

Chapter 22

> > > >

EYE ON THE LANDSCAPE

Wetlands at Chicko River, British Columbia.
What else would the geographer see? . . . Answers at the end of the chapter.

BIOGEOGRAPHIC PROCESSES

Ecological Biogeography
Water Need
Temperature
Other Climatic Factors
Bioclimatic Frontiers
Geomorphic and Edaphic (Soil) Factors
Disturbance
Interactions among Species
Ecological Succession
Geographers at Work • Dr. Xulin Guo
Succession, Change, and Equilibrium
Historical Biogeography
Evolution
Speciation
Extinction
Dispersal
Distribution Patterns
*Working It Out 22.1 • Island Biogeography:
   The Species–Area Curve*
Biogeographic Realms
Biodiversity

The preceding chapter focused on ecosystems from the perspective of energy capture and global cycling of carbon, nitrogen, and other nutrients. But ecosystems are composed of individual organisms that use and interact with their environment in different ways. Each organism has a range of environmental conditions that limits its survival, as well as a set of characteristic adaptations that it exploits to obtain the energy it needs to live.

This chapter begins with a discussion of ecological biogeography and examines how relationships between organisms and the environment help determine where they are found. Ecological biogeography explains the spatial pattern of organisms on local and regional scales and on a time scale of a few generations.

Historical biogeography examines the distribution of organisms on continental or global scales over longer time periods. This branch of biogeography describes processes such as evolution, dispersal, and extinction of species through time. Together, ecological and historical biogeography provide a comprehensive framework for understanding and appreciating the diversity of the biosphere, and how human activity threatens that diversity.

ECOLOGICAL BIOGEOGRAPHY

The relationship between organisms and their physical environments is strongly influenced by landforms and soils. For example, upland soils are often stony and well-drained, while the soils in valleys may be finer, richer in organic matter, and wetter. Such variations provide a range of **habitats** that may be more or less suited to the needs and preferences of organisms or groups of organisms.

A concept related to the habitat, the **ecological niche**, includes the functional role an organism plays as well as the physical space it inhabits. An organism's ecological niche describes how and where it obtains its energy and how it influences other species and the environment around it. Included in the ecological niche are the organism's tolerances and responses to changes in moisture, temperature, soil chemistry, and other factors. Although many different species may occupy the same habitat, no two species can occupy identical ecological niches. Thus, the theoretical limit for a species, as determined by its physiological requirements for all environmental factors, establishes its **fundamental niche**. The reduced environmental space that it actually occupies because of competition with other species is called its **realized niche**.

Each habitat is home to a distinct group of species, each occupying specific, but interrelated, ecological niches. A group of species that live in a particular habitat and interact with each other form a *community*. Because similar habitats in a region often contain similar assemblages of species, communities can be grouped together as *associations*. Such associations are defined by dominant species, as in the hemlock-white pine-northern hardwoods association found in the mixed-wood forests of Ontario and Quebec.

In general, moisture and temperature are the most important environmental factors that determine where species are found. Although under conditions of extreme temperature or dryness, organisms are sometimes present as spores or cysts, nearly all organisms have limits that are exceeded at least somewhere on the Earth at some time. On the global scale, temperature and moisture patterns translate into climate. For this reason, there is a strong relationship between climate and vegetation.

WATER NEED

The availability of water to terrestrial organisms at a particular point in time or space is determined by the balance between precipitation, evaporation, runoff, and infiltration. This balance is, in turn, affected by organisms, especially the plant cover. Through transpiration, plants return much of the soil water to the atmosphere. This process is important in tropical rainforests, where transpiration directly contributes as much as 50 percent of the rain. By obstructing overland flow and increasing soil porosity, plants reduce runoff and increase infiltration. Burrowing animals enhance infiltration, as well. Although these local water flows are important within individual habitats, their effects are generally small compared with those of the physical processes that control the major features of the water cycle. Thus, the overall dynamics of the atmosphere and oceans in the form of global climate still determine the major pattern of variation in water from one place to another.

Both plants and animals show a variety of adaptations that enable them to cope with a scarcity or abundance of water. Plants that are adapted to drought conditions are called **xerophytes** (Figure 22.1). While most xerophytes are associated with desert habitats with scarce rainfall, some are adapted to habitats that dry quickly following rapid drainage of precipitation; for example, sand dunes, beaches, and bare rock surfaces. Many species adapt to drought by closing the stomatal cells in the leaf epidermis to limit transpiration. In some xerophytes, a thick layer of wax-like material on leaves and stems reduces water loss. The wax helps to seal water vapour inside the plant tissues. Others adapt by greatly reducing their leaf area or by bearing no leaves at all. Cacti, for example, produce spines rather than leaves.

(a)

(b)

(c)

22.1 Plant adaptations to drought (a) Leaf succulents such as *Aloe lateritia* can store water in their swollen leaves. (b) The creosote bush (*Larrea tridentata*) has small leaves that are naturally varnished to conserve water. (c) Leaves on the ocotillo (*Fouqueria splendens*) develop quickly after a rainstorm, but soon wither and fall from the stem as the soil dries out. In this way, the ocotillo can produce several crops of leaves each year.

(a)

(b)

(c)

22.2 Plants of wet habitats (a) Tall cattails (*Typha latifolia*) are able to conduct oxygen to their submersed roots. (b) The small leaves of Labrador tea (*Ledum groenlandicum*) have characteristics similar to those of desert shrubs; in both environments, the leaves have adapted to limit transpiration. (c) The floating leaves of water lilies (*Nymphaea odorata*) allow them to be fully exposed to air and sunlight.

Adaptations of plants to water-scarce environments also include improved abilities to obtain and store water. Roots may extend several metres to reach soil moisture far below the surface. If the roots reach the groundwater zone, they will have a steady supply of water. Other desert plants develop a widespread, but shallow, root system. This enables them to absorb water from brief storms that saturate only the uppermost soil layer. Leaves of desert plants, such as aloes and agaves and stems of cacti, are greatly thickened by spongy tissue in which they can store water. Plants with this adaptation are called *succulent*. This spongy tissue also plays a role in CAM photosynthesis, where plants use it to store intermediate products formed by gas exchange during hours of darkness.

Another adaptation to extreme dryness is a very short life cycle. Many small desert plants are *ephemeral annuals*. Following germination, they leaf out, bear flowers, and produce seeds in a few weeks immediately following a heavy rain shower. They complete their life cycle when soil moisture is available, and the species survive the dry period as seeds that require no moisture.

In the wet-dry tropical climate, which has a yearly cycle with a pronounced dry season, trees and shrubs will often respond by dropping their leaves at the end of the moist season, becoming dormant during the dry season. When water is available again, they leaf out and grow quickly. In the Mediterranean climate, which experiences hot, dry summers, the plants typically retain their tough, leathery leaves all year. Plants that retain their leaves through a dry or cold season have the advantage of being able to resume photosynthesis immediately when growing conditions become favourable, whereas deciduous plants have to grow a new set of leaves each growing season. This is particularly valuable for conifers faced with a short growing season at high latitudes.

Plants that grow in lakes, marshes, and bogs, called **hydrophytes**, are adapted to cope with excessive moisture (Figure 22.2). This generally means that the plants can tolerate low concentrations of soil oxygen. Oxygen diffuses very slowly into saturated soils, which results in anaerobic conditions and unusual soil chemistry. In particular, iron becomes very soluble and, potentially, could become toxic to plants. Plants that grow in shallow water environments, such as cattails (*Typha latifolia*), typically have abundant air space tissue that allows them to take in air through the leaves and quickly conduct it down the stems to the roots where it passes into the mud immediately surrounding the roots. Through this process of *radial oxygen loss*, the iron is oxidized to insoluble forms and precipitated externally to the roots. In this way, the plants can take up other essential nutrients preferentially. Another way to reduce iron toxicity is to limit uptake of dissolved cations by reducing the flow of water needed for transpiration. This is achieved in species such as Labrador tea (*Ledum groenlandicum*) by reducing leaf size. Other adaptations in aquatic plants include floating leaves, such as water lilies (*Nymphaea odorata*) and water hyacinths (*Eichhornia crassipes*). The leaves' waxy, water-repellent surfaces keep them dry, and their seeds float so they can disperse in the currents.

To cope with water shortages, some invertebrate animals have evolved methods that are similar to those of ephemeral annual plants; they avoid the dry period by becoming dormant. When rain falls, they emerge to take advantage of the moisture and the newly developed, short-lived vegetation. Similarly, many bird species regulate their behaviour to nest only when the rains occur, as this is the time of most abundant food for their offspring. Tiny brine shrimp may wait many years in dormancy until normally dry lake beds fill with water, an event that occurs perhaps three or four times a century. The shrimp then emerge and complete their life cycles before the lake evaporates.

Mammals are by nature poorly adapted to desert environments, but many survive through a variety of mechanisms that enable them to avoid water loss. Just as plants reduce transpiration to conserve water, many desert mammals do not sweat through skin glands. Instead they rely on other methods of cooling, such as avoiding the sun and becoming active only at night. In this respect, they join most of the rest of the desert fauna, spending their days in cool burrows in the soil and their nights foraging for food. However, the camel is well-adapted to the desert environment, mainly because its blood doesn't thicken through dehydration. It can still circulate freely and prevent the animal's core temperature from becoming dangerously high. Kangaroo rats (*Dipodomys spp.*) can metabolize water from a diet of dry seeds, so they are essentially immune to prolonged drought.

Most animal groups include numerous species that have adapted to wet environments. Some, like amphibians, are able to exploit both terrestrial and aquatic habitats. However, problems can arise for some species when land areas are periodically flooded. Worms, for example, are often forced out of the soil during periods of saturation, and rodents in savanna grasslands can be drowned in their burrows during the wet season. The beaver (*Castor canadensis*), meanwhile, works industriously building dams to flood the land.

TEMPERATURE

The temperature of the air and soil acts directly on organisms by influencing the rates at which physiological processes take place in plant and animal tissues. Thus, each plant species has an optimum temperature associated with processes such as photosynthesis, flowering, and seed germination. There are also limiting lower and upper temperatures for these functions and for the total survival of the plant itself. In general, the colder the climate, the fewer the species that are capable of surviving. In the severely cold arctic and alpine environments of high latitudes and high altitudes, relatively few plant and animal species are found. A plant's tolerance to cold is closely tied to the physical disruption that accompanies the growth of ice crystals inside cells. Cold-tolerant plant species can expel excess water from cells so that it freezes in the intercellular spaces where it does no damage.

The effects of temperature variations on animals are moderated by their physiology and their ability to seek sheltered environments. Most animals lack a physiological mechanism for internal temperature regulation. These animals, including reptiles, invertebrates, and fish, are *cold-blooded animals*—their body temperatures passively follow the environment. With a few exceptions (notably fish), these animals are active only during the warmer parts of the year. They survive the cold, winter weather of the midlatitudes by becoming dormant. Some vertebrates enter a dormant state called *hibernation*, in which metabolic processes virtually stop and body temperatures closely parallel those of the surroundings. Most hibernators seek out burrows, nests, or other environments where winter temperatures do not reach extremes or fluctuate rapidly. Because the annual range of soil temperatures is much smaller below the uppermost layers, burrows are particularly suited to hibernation.

Other animals maintain a relatively constant body temperature by internal metabolism. This group includes the birds and mammals. These *warm-blooded animals* possess a variety of adaptations to regulate their body temperature. Fur, hair, and feathers act as insulation by trapping air next to the skin surface, thereby reducing heat loss. A thick layer of fat also provides excellent insulation. Other adaptations are for cooling; for example, sweating or panting uses the high latent heat of vaporization of water to dissipate body heat. Heat loss is also facilitated by exposing blood-circulating tissues to the cooler surroundings. A seal's flippers, a bird's feet, and the long ears of a jackrabbit serve this function.

OTHER CLIMATIC FACTORS

On the global scale, light available for plant growth varies by latitude. Duration of daylight in summer increases rapidly with higher latitude and reaches its maximum within the Arctic and Antarctic circles, where the sun may be above the horizon for 24 hours (see Figure 21.4). Although frost greatly shortens the growing season for plants at high latitudes, the prolonged daylight greatly accelerates the rate of plant growth during the brief summer. In midlatitudes, the deciduous species endure an annual rhythm of increasing and decreasing periods of daylight, which determines the timing of budding, flowering, fruiting, and leaf shedding.

The diurnal and seasonal cycles of illumination also influence animal behaviour. The day–night cycle controls the activity patterns of many animals. Birds, for example, are generally active during the day, whereas small mammals often forage at night. The daylight length, or *photoperiod*, influences seasonal activity in animals. In midlatitudes, as autumn days grow shorter, squirrels and other rodents hoard food for the coming winter season, and birds begin their migrations. In the spring, a larger photoperiod will trigger such activities as mating and reproduction.

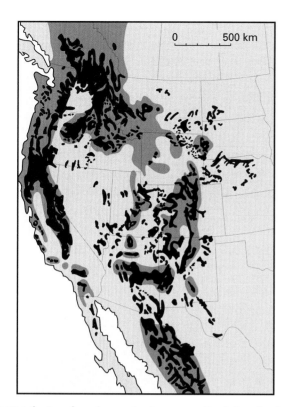

22.3 Distribution of ponderosa pine in western North America Areas of ponderosa pine (*Pinus ponderosa*) are shown in black. The edge of the shaded area marks the 500-millimetre rainfall isohyet.

Wind is an important environmental factor that can affect vegetation structure in highly exposed areas. Close to the timberline in high mountains and along the northern limits of tree growth in the arctic zone, wind has deformed trees in such a way that the branches project from the lee side of the trunk only. The wind causes excessive drying, damaging the exposed side of the plant. The tree limit on mountainsides thus varies in elevation with the degree of exposure to strong prevailing winds and will extend higher on lee slopes and in sheltered pockets.

BIOCLIMATIC FRONTIERS

Climatic factors, such as moisture and temperature, often limit the distribution of plant and animal species. Biogeographers recognize that there is a critical level of climatic stress beyond which a species cannot survive. A geographic boundary can thus mark the limits of the potential distribution of a species. This boundary is sometimes referred to as a **bioclimatic frontier**. Although the frontier is usually marked by a variety of climatic elements, it is sometimes possible to single out one climatic element that approximately coincides with it.

The distribution of ponderosa pine (*Pinus ponderosa*) in western North America is an example (Figure 22.3). In this mountainous region, annual rainfall varies sharply with elevation. The 500 millimetre total annual precipitation isohyet encloses

most of the upland areas with ponderosa pine. The parallel of the isohyet with the forest boundary, rather than actual degree of co-incidence, is significant. The sugar maple *(Acer saccharum)* is a more complex example (Figure 22.4). Here, the boundaries on the north, west, and south coincide roughly with selected values of annual precipitation, mean annual minimum temperature, and mean annual snowfall.

Although bioclimatic limits must exist for all species, no plant or animal is necessarily found within its frontier. Many other factors, such as diseases or predation, can limit the distribution of a species. Alternatively, a species (especially a plant species) may migrate slowly and may still be radiating outward from the location where it originated. Some species may be dependent on another species and are therefore limited by the latter's distribution.

GEOMORPHIC AND EDAPHIC (SOIL) FACTORS

Geomorphic factors influencing ecosystems include slope steepness, slope aspect, and relief. Slope steepness influences the rate at which precipitation drains from the surface. On steep slopes, surface runoff is rapid, and soil water recharge by infiltration decreases. On gentle slopes, much of the precipitation can penetrate the soil

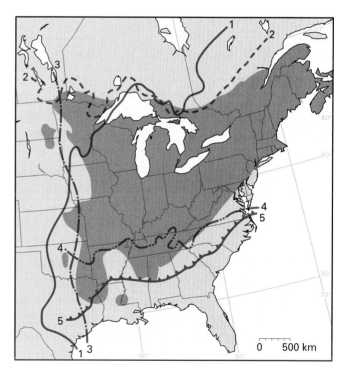

22.4 Bioclimatic limits of sugar maple The shaded area shows the distribution of sugar maple (*Acer saccharum*) in eastern North America. Line 1 represents annual precipitation of 750 millimetres. Line 2 represents a mean annual minimum temperature of −40°C. Line 3 represents the eastern limit of the boundary between arid and humid climates. Line 4 represents a mean annual snowfall of 25 centimetres. Line 5 represents a mean annual minimum temperature of −10°C.

22.5 Slope orientation and habitat In this photo of the Chisos Mountains of Big Bend National Park in Texas, dry south-facing slopes on the left support a community of low, xerophytic shrubs, while moister north-facing slopes on the right support an open forest cover of piñon pine and juniper.

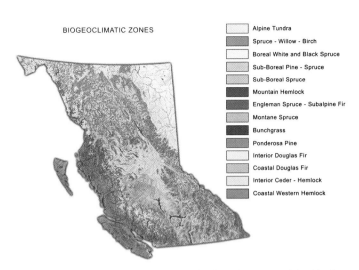

BIOGEOCLIMATIC ZONES

	Alpine Tundra
	Spruce - Willow - Birch
	Boreal White and Black Spruce
	Sub-Boreal Pine - Spruce
	Sub-Boreal Spruce
	Mountain Hemlock
	Engleman Spruce - Subalpine Fir
	Montane Spruce
	Bunchgrass
	Ponderosa Pine
	Interior Douglas Fir
	Coastal Douglas Fir
	Interior Ceder - Hemlock
	Coastal Western Hemlock

22.6 Biogeoclimatic zones of British Columbia The zones are defined by the dominant forest species associated with general macroclimatic conditions.

and be retained. More rapid erosion on steep slopes can result in thinner soils compared with those found on gentler slopes.

Slope aspect has a direct influence on plants by increasing or decreasing the exposure to sunlight and prevailing winds. Slopes facing the sun have a warmer, drier environment than slopes that are shaded for much of the day. In midlatitudes, these slope-aspect contrasts can be so strong as to produce quite different biotic communities on north-facing and south-facing slopes (Figure 22.5).

Geomorphic factors are partly responsible for the dryness or wetness of the habitat within a region that has the same general climate. On divides, peaks, and ridge crests, the soil tends to dry out because of rapid drainage and because the surfaces are more exposed to sunlight and wind. By contrast, valley floors are wetter because surface runoff over the ground and into streams causes water to collect there. In humid climates, the groundwater table in the valley floors may lie close to or at the ground surface to produce marshes, swamps, ponds, and bogs.

Edaphic (soil) factors can be considered on two scales. On the broadest scale, terrestrial ecosystem types can be associated with soil distribution, which is controlled by general climatic regimes. Alternatively, on a local scale, edaphic factors are important in differentiating habitat conditions. Properties, such as soil texture and structure, humus content, and nutrient status, all contribute to a habitat.

The influence of topography and soils on habitat has been used to develop provincial forest ecosystem classifications in Canada. This approach was first developed in British Columbia, where *biogeoclimatic zones* were established according to dominant tree species growing under broadly defined macroclimatic

conditions (Figure 22.6). In the original scheme, the potential growth of forest species within each zone was described according to soil nutrient status and soil moisture conditions. In modified schemes, sites are described according to the vegetation communities they would support at maturity under specified soil moisture and soil nutrient regimes.

The average amount of soil water annually available for evapotranspiration defines soil moisture regimes, which range from very dry (xeric) to very wet (subhydric). The amount of essential soil nutrients that are available to plants define soil nutrient regimes, which range from very poor to very rich. This information is represented in a two-dimensional diagram known as an **edatopic grid**, on which representative plant communities are plotted (Figure 22.7).

DISTURBANCE

Another environmental factor affecting ecosystems is *disturbance*, which includes fire, flood, volcanic eruption, storm waves, high winds, and other infrequent catastrophic events that damage or destroy ecosystems and modify habitats. Although disturbance can greatly alter the nature of an ecosystem, it is often part of a natural cycle of regeneration that provides opportunities for short-lived or specialized species to grow and reproduce. In this way, disturbance is often a natural process to which many ecosystems have adapted.

In converting biomass to ash, *fire* mobilizes nutrients, such as potassium and phosphorus, making them available for a new generation of plants. However, other elements, such as nitrogen and sulphur, may be lost through volatilization and removed as gases. Fire affects ecosystems by changing species composition

Nutrient Regime

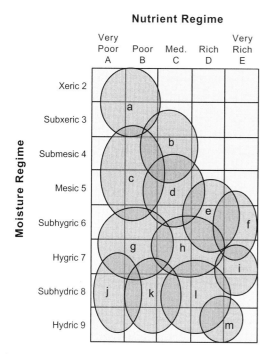

22.7 Edatopic grid The edatopic grid is a convenient method of placing sites (in this example, simply labelled a–m) into a two-dimensional ecological space, which has direct implications for forest growth, composition, and management.

and creating a patchy structure of diversity on the landscape. It can also increase runoff and soil erosion where it has destroyed a significant layer of vegetation.

Fire strikes most forests at some point and, in many instances, is beneficial. It cleans out the understory and consumes dead and decaying organic matter, while leaving some of the overstory trees untouched. On the forest floor, mineral soil is exposed and fertilized with new ash, providing a productive environment for dormant seeds. Sunlight is also abundant, with shrubs and forbs no longer shading the soil. Among tree species, pines are typically well-adapted to germinate under these conditions. For example, the jack pine of eastern Canada and the lodgepole pine of the intermountain west have cones that remain tightly closed until the heat of a fire opens them, releasing the seeds. These species are directly dependent on fire to maintain their geographic range and importance in the ecosystem. In the Rockies and boreal forests, there are many patches, large and small, of jack and lodgepole pine of different ages that document a long history of fire. These stands also serve as specialized habitats for particular insects, birds, and mammals.

Fires are also important to the preservation of grasslands. Grasses have extensive root systems with buds located at or just below the surface, making them quite fire-resistant. However, woody plants are not so resistant and are usually killed by grass-fires. Most grasslands are dependent to some degree on fire for their maintenance. In Mediterranean climates, chaparral vegetation is also adapted to regular burning.

Fires can be remotely sensed in several ways (Figure 22.8). Thermal imagers detect active fires as bright spots because they emit more heat energy than normal surfaces. Smoke plumes can also identify the location of fires, but are hard to distinguish from clouds in some images. Fire scars can be detected, especially using mid-infrared bands.

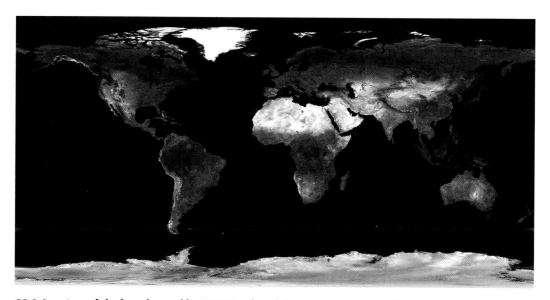

22.8 Locations of the fires detected by MODIS on board the Terra and Aqua satellites from January 11–20, 2007 Each coloured dot indicates a location where MODIS detected at least one fire during the compositing period. Colour ranges from red, where the fire count is low, to yellow, where number of fires is large.

22.9 Wind damage Trees felled by wind create openings that allow understory species and seedlings to flourish and are an important regeneration mechanism in some forests.

Fire is one of several forms of disturbance that can disrupt plants and animals. Strong winds and ice storms can bring down large areas of forest (Figure 22.9). High winds also bring destructive waves to coral reefs, as well as sand and scouring to bays and coastal marshes. On a fine spatial scale, disturbance includes the fall of individual trees or large limbs within forests, creating light openings for understory species to fill. Even animal burrows and wallows can provide unique habitats of disturbance.

In addition to natural agents, the modern world is increasingly subject to human disturbance. The earliest societies mostly affected the environment through hunting. In some cases, this led to the extinction of species. Particularly susceptible were large flightless birds and other megafauna, such as the wooly mammoth. Humans have been able to dominate every environment on the Earth and have wrought inexorable change. Often this change has been direct and deliberate, for example, through introduction of new species—often ones that have been bred specifically for foodstuffs. But equally often, humans have introduced species with devastating results, for example rabbits in Australia. In many cases, humans have inadvertently introduced species, such as zebra mussels, to regions beyond their natural range.

Direct and indirect human-induced environmental change is immense. Earlier chapters have covered some of these topics; for example, air pollution and global warming. The list can be extended to include soil erosion, water quality, desertification, loss of wetlands, logging, plantation agriculture, agrochemicals, and so on, all of which can and have had an impact on the biosphere. The pervasive impact of human disturbance is difficult and, in many cases, impossible to rectify. For many species, it is too late regardless of conservation efforts that are being mounted worldwide.

INTERACTIONS AMONG SPECIES

Species interactions can be an important determinant of distribution patterns of plants and animals. Two species that are part of the same ecosystem can interact with one another in three ways: interaction may be negative to one or both species; the two species may be neutral, not affecting each other; or interaction may be positive, benefiting at least one of the species.

Competition, a negative interaction, occurs whenever two species require a common resource that is in short supply. Because neither species has full use of the resource, both populations suffer, with lower growth rates than would normally occur if only one of the species were present. Competition is an unstable situation that may lead to the elimination of one of the species.

Predation and parasitism are also negative interactions between species. *Predation* occurs when one species feeds on another. The benefits are obviously positive to the predator species, which obtains energy for survival, and negative to the prey species. *Parasitism* occurs when one species gains nutrition from another, typically when the parasite organism invades or attaches to the body of the host in some way (Figure 22.10).

Although predation and parasitism are usually regarded as negative processes that benefit one species at the expense of the other, these interactions may really be beneficial in the long term to the prey populations. Predation helps to maintain prey populations at levels that are in harmony with the environment's ability to support them. In addition to maintaining equilibrium population levels, predation and parasitism differentially remove the weaker individuals and can improve the genetic composition of the species.

22.10 Parasitism Parasitic dwarf mistletoe (*Arceuthobium americanum*) causes the clumped growth deformations known as witches' brooms on many conifer species.

A third type of interaction is *herbivory*. Some plant species have adapted well to grazing and can maintain themselves in the face of increased grazing pressure; others may be quite sensitive to this process. When overgrazing occurs, these differing sensitivities can produce significant changes in the structure and composition of plant communities.

Another type of negative interaction is *allelopathy*, a phenomenon in which chemical toxins produced by one plant species serve to inhibit the growth of others. Several shrub species in the California chaparral are allelopathic and exude toxins from their foliage. The allelopathic toxins accumulate in the soil until periodic fires break them down.

The term **symbiosis** includes three types of positive interactions between species: commensalism, protoco-operation, and mutualism. In commensalism, one of the species benefits and the other is unaffected. Examples of commensals include epiphytic plants, such as orchids or Spanish moss, which live on the branches of larger plants. These epiphytes depend on their hosts for physical support only. In the animal kingdom, small commensal crabs or fish seek shelter in the burrows of sea worms; or the commensal remora fish attaches itself to a shark.

When the relationship benefits both parties but is not essential to their existence, it is called protoco-operation. The attachment of a stinging coelenterate, such as a jelly fish or sea anemone, to a crab is an example of protoco-operation. The crab gains camouflage and an additional measure of defence, while the coelenterate eats bits of stray food that the crab misses. A similar arrangement is seen between species of acacia and ants (Figure 22.11). The ants defend the plants against herbivorous animals and also chew away any encroaching plants, clearing areas as much as four metres in radius. In return, the plant provides proteins and fats from specialized Beltian bodies and nectar. The ants hollow out thorns for their nests.

22.11 Ant guards The bullhorn acacia (*Acacia cornigera*) is one of several acacia species that is occupied by colonies of stinging ants (*Pscudomyrmcx fcrruginca*).

Where protoco-operation has progressed to the point that one or both species cannot survive alone, the result is mutualism. The association of the nitrogen-fixing bacterium *Rhizobium* with the root tissue of legumes is an example. The bacteria convert nitrogen gas to a form directly usable by the plant. The association is mutualistic because *Rhizobium* cannot survive alone.

ECOLOGICAL SUCCESSION

The Earth is a dynamic planet in which landscapes change continually. The phenomenon of change in plant and animal communities through time is referred to as **ecological succession**. If succession begins on newly deposited sediment or bare rock, it is called *primary succession*. If succession occurs on a previously vegetated area that has been recently disturbed by fire, flood, windstorm, or humans, it is called *secondary succession*.

Sites on which primary succession occurs can originate in several ways; for example, a sand dune or beach, the surface of a new lava flow or freshly deposited volcanic ash, the deposits of silt on the inside of a river bend, or a recently exposed glacial moraine. These sites will not likely have a true soil with horizons, but will consist of little more than a deposit of coarse mineral fragments.

The first stage of a succession—the *pioneer stage*—includes a few plant and animal species unusually well-adapted to adverse conditions, such as rapid water drainage and desiccation. As pioneer plants grow, their roots penetrate the substrate, and their subsequent death and decay adds humus to the rudimentary soil. Fallen leaves and stems add an organic layer to the ground surface.

Soon conditions are favourable for other species to invade the area and displace the pioneers. The new arrivals may be larger plant forms providing more extensive cover. In this case, the microclimate near the ground alters, becoming one of less extreme air and soil temperatures, high humidity, and less intense insolation. Still other species now invade and thrive in the modified environment. When the succession has finally run its course, a **climax community** of plant and animal species of more or less stable composition will exist.

The colonization of a coastal sand dune provides an example of primary succession. Growing foredunes bordering the ocean present a sterile habitat. The dune sand usually lacks nitrogen, phosphorus, and other nutrients, and its water-holding ability is very low. Under intense solar radiation, the dune surface becomes hot and dry. At night, radiation cooling in the absence of moisture produces low surface temperatures.

One of the first pioneers of this extreme environment is beach grass, which reproduces by sending out rhizomes (creeping underground stems). The plant thus slowly spreads over the dune. Unlike less specialized plants, beach grass does not die when moving sand buries it; instead, it sends up new shoots.

Climate change plays a major role in landscape change and ecosystem degradation along with anthropogenic factors. Global warming has resulted in increased temperature and uneven distribution of precipitation. What effects are we expecting from predicted rising temperatures and more frequent extreme weather events in Canada? Analyzing historical vegetation response to climate variation will help understand the interaction between climate and vegetation.

National parks in different ecosystems are designed for protecting and presenting significant examples of Canada's natural heritage and to ensure the ecological integrity of the landscapes. National parks provide ideal conditions with limited human influence in which to investigate how climate change affects ecosystem dynamics.

Using multi-temporal and multi-spatial resolution satellite imagery to study ecosystem responses to climate change in the Northern hemisphere is a major focus of the scientific community. During the past two decades, many studies of vegetation distribution and condition at global and regional scales have been based on satellite imagery data. Applications include spring green up time variation (phenology), maximum productivity, and changes associated with temperature and precipitation fluctuations.

Dr. Xulin Guo has conducted research in several Canadian national parks. In Grasslands National Park, she and her research group have used different spatial resolution satellite sensors including IKONOS, SPOT, Landsat, AVHRR, and MODIS, to characterize vegetation biophysical and spectral features. Understanding the landscape heterogeneity helps in mapping habitat change and understanding species at risk challenges. In addition, her studies have focused on evaluating relationships between vegetation growth, satellite signal function, and climate variation in 11 northern national parks (1983-2006).

Looking forward to the field campaign in front of the visiting centre of Grasslands National Park of Canada.

After colonization, the beach grass shoots begin to suppress sand movement, and the dune becomes more stable. With increasing stabilization and better water-holding capacity, other plants can establish. On beach and dune ridges in Nova Scotia, for example, the species that follow beach grass include beach pea, poison ivy, and sedge. Where active deposition of sand has ceased, the beach grass is gradually replaced by other species, such as scouring rush, bayberry, and lichens. With the establishment of other grass species, beach grass disappears. Over time, heath and scrub take over; bearberry, blueberry, and juniper are common. Trees, such as aspen and red oak, may also be present. Eventually a forest of white pine, white spruce, oak, and maple develops (Figure 22.12).

Although this example has stressed the changes in plant cover, animal species are also changing as succession proceeds (Table 22.1). The developmental stages shown in the table for these inland dunes are somewhat different from those described for the coastal environment.

Where disturbance alters an existing community, secondary succession can occur. Examples include areas that have been disturbed by logging, fire, and insects, or simply through the collapse of a tree (Figure 22.13). Secondary succession also occurs on abandoned farmland as it reverts to its former natural state. Secondary succession usually occurs more rapidly than primary succession. One reason is that there is already a supply of seeds and roots in the soil from which new growth can quickly germinate or sprout. Likewise, not all of the vegetation is necessarily destroyed. Patches of plants that survived the disturbance and those in nearby sites can act as local seed sources. In addition, previous organisms have substantially modified the fertility and structure of the soil, compared with freshly deposited substrates like dune sand. This makes it more amenable for growth and colonization, especially if some of the other micro-environmental conditions, such as shade from surviving trees or dead snags, persist in the post-disturbance site.

SUCCESSION, CHANGE, AND EQUILIBRIUM

Successional change, which arises from the actions of the plants and animals themselves, is called *autogenic succession*. One group of inhabitants modifies the environment in a way that makes it more suitable for other species to establish. In many cases, however, autogenic succession does not run its full course. Environmental disturbances, such as wind, fire, flood, or renewed clearing for agriculture, may divert the course of succession temporarily or even permanently.

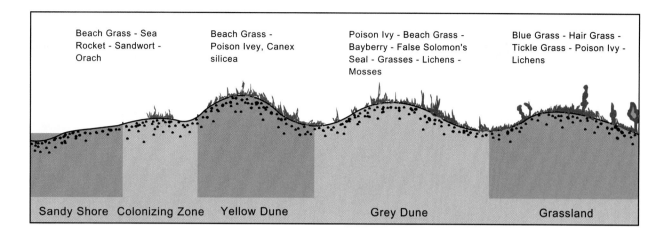

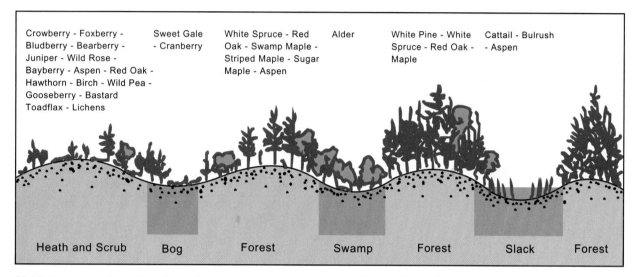

22.12 Dune succession in Nova Scotia Beach grass is a pioneer on beach dunes and helps stabilize the dune against wind erosion.

Table 22.1 Invertebrate succession on the Lake Michigan dunes

| | Successional Stages | | | | |
Invertebrate	Beach grass–Cottonwood	Jack Pine Forest	Black Oak Dry Forest	Oak and Oak–Hickory Moist Forest	Beech–Maple Forest Climax
White tiger beetle	x				
Sand spider	x				
Long-horn grasshopper	x	x			
Burrowing spider	x	x			
Bronze tiger beetle		x			
Migratory locust		x			
Ant lion			x		
Flat bug			x		
Wireworms			x	x	x
Snail			x	x	x
Green tiger beetle				x	x
Camel cricket				x	x
Sow bugs				x	x
Earthworms				x	x
Wood roaches				x	x
Grouse locust					x

Source: V. E. Shelford, as presented in E. P. Odum, Fundamentals of Ecology, Philadelphia: W. B Saunders Co., p. 259.

(a)

(b)

22.13 Secondary succession in mixed wood forest in Saskatchewan (a) Small-scale disturbance, such as the collapse of a tree, creates a gap in the overstory that light-demanding species, which are normally excluded from a mature forest stand, may initially occupy. (b) Fire killed most of the above-ground vegetation at this site, but new growth has developed from seeds and roots that survived in the soil.

Introduction of a new species can also greatly alter existing ecosystems and successional pathways. The parasitic chestnut blight, introduced from Europe to New York in about 1910, decimated populations of the American chestnut tree within about 40 years. This tree species, which may have accounted for as much as 25 percent of the mature trees in eastern forests, is now found only as small blighted stems sprouting from old root systems.

This example shows that, while succession is a reasonable model to explain many of the changes seen in ecosystems, it may be more realistic to view the pattern of ecosystems on the landscape as a reflection of a spatial dynamic equilibrium between autogenic forces of self-induced change and external forces of disturbance that reverse or redirect vegetation change temporarily or permanently. The biotic landscape is a mosaic of distinctive biotic communities with different biological potentials and different histories.

This view of the landscape assumes that all successional species are available to colonize new space or establish in existing communities. In this case, the nature of the biotic communities is determined by varying environmental and ecological factors that act within a new space created by physical and human processes. But not all species are available to colonize new spaces, particularly on continental and global scales, and especially over long time spans. On these broader spatial and temporal scales, the processes of migration, dispersal, evolution, and extinction are more important in determining the spatial patterns. These processes are within the realm of historical biogeography.

HISTORICAL BIOGEOGRAPHY

Historical biogeography examines four key processes that influence the distribution of species: evolution, speciation, extinction, and dispersal.

EVOLUTION

An astonishing number of organisms exist on the Earth—about 40,000 species of micro-organisms, 350,000 species of plants, and 2.2 million species of animals, including some 800,000 insect species, have been identified and described. However, many organisms remain unclassified. Estimates suggest that species of plants will ultimately number about 540,000, and at present perhaps only a third of all insects have been classified. This great diversity has been achieved through *evolution*, as a result of the interaction between organisms and the environment. The process of evolution, as enunciated by Charles Darwin, is based on the premise that all life possesses *variation;* that is, differences arise between parent and offspring. The environment acts on variation in organisms in a way that ultimately favours individuals who are best suited to their environment. Darwin called this survival and reproduction of the fittest **natural selection**. Over time, this would bring about the formation of new species, whose individuals differed greatly from their ancestors.

A *species* is defined as all individuals capable of interbreeding to produce fertile offspring. A *genus* is a collection of closely related species that share a similar genetic evolutionary history. Each species has a scientific name, composed of a generic name and a specific name in combination. Thus, white spruce, a

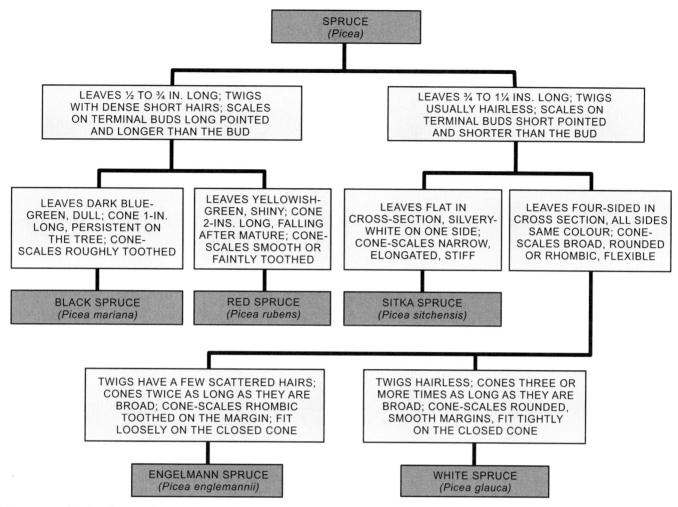

22.14 Key to the identification of Canadian spruces Six species of native spruces are recognized according to the size and shape of their leaves and cones and related characteristics.

common conifer throughout the boreal forest of North America, is *Picea glauca*. The related black spruce is *Picea mariana*.

Although the true test of a species is the ability of all of its individuals to reproduce successfully, this criterion is not always easily applied. Instead, a species is usually defined by *morphology*—the outward form and appearance of its individuals. The *phenotype* of an individual is the morphological expression of its genetically inheritable information, or *genotype*, and includes all the physical aspects of its structure that are readily seen. Species, then, are usually defined by a characteristic phenotype and are identified by criteria, such as leaf shape (Figure 22.14).

SPECIATION

Speciation refers to the process by which species are differentiated and maintained. It often occurs when populations become isolated from one another. This **geographic isolation** can happen in several ways. For example, plate tectonics may uplift a mountain range that subsequently separates a population into two subpopulations isolated by a climatic or physical barrier. Chance

long-distance dispersal can also establish a new population far from the original population. These are examples of *allopatric speciation*, in which populations are geographically isolated. With time, the populations gradually diverge and eventually lose the ability to interbreed.

In contrast, *sympatric speciation* occurs within a larger population. This may result from mutations that allow some members of a species to exploit the environment in slightly different ways. When subject to natural selection, these mutants will begin to evolve into different subpopulations, each adapted to its own environmental niche. Eventually, the subpopulations may become separate species. This type of evolution, in which there is the opportunity for the formation of many new species adjusted to different habitats, is called *adaptive radiation*. A good example is the 300 or more species of cichlid fishes that have evolved in the lakes of the African Rift Valley.

Another mechanism of sympatric speciation that is quite important in plants is **polyploidy**. Normal organisms have two sets of genes and chromosomes; that is, they are *diploid*. Through accidents in the reproduction process, two closely related species can cross in such a way that the offspring have both sets of genes from both parents. These *tetraploids* are fertile but cannot reproduce with the populations from which they arose, and so are instantly isolated as a new species. By some estimates, 70 to 80 percent of higher plant species have developed in this way.

EXTINCTION

Over geologic time, the fate of all species is *extinction*. When conditions change more quickly than populations can evolve new adaptations, population size decreases. When that occurs, the population becomes increasingly vulnerable to chance occurrences, such as a fire, a rare climatic event, or an outbreak of disease. Ultimately, the population succumbs, and the species becomes extinct.

Some extinctions are very rapid, particularly those induced by human activity. A classic example is that of the passenger pigeon, a dominant bird of eastern North America in the late nineteenth century. Flying in huge flocks and feeding on seeds and fruits, these birds were easily captured in nets and shipped to markets for food. By 1890, they were virtually gone. The last known passenger pigeon died in the Cincinnati Zoo in 1914.

Rare but extreme events can also cause extinctions. Many lines of evidence suggest that the Earth was struck by an asteroid about 65 million years ago. The impact, which occurred on the continental shelf near the Yucatan Peninsula, raised a global dust cloud that blocked sunlight from the surface, cooling the Earth's climate intensely for a period of perhaps several years. Dinosaurs and many other groups of terrestrial and marine organisms vanished. Less affected were organisms that were not as sensitive to a brief, but intense, period of cold. These included birds and mammals that have internal metabolic temperature regulation, as well as seed plants and insects that spend part of their life cycle in a dormant state.

DISPERSAL

Nearly all types of organisms have some **dispersal** mechanism; that is, a capacity to move from a location of origin to new sites. Often dispersal is confined to one stage in the life cycle, as in the dispersal of higher plants as seeds. Even in animals that are inherently mobile, there is often a developmental stage when movement from one site to the next is more likely to occur.

Normally, dispersal does not change the geographic range of a species. Seeds fall near their source, and animals seek out nearby habitats to which they are adjusted. However, when new areas become available, dispersal moves colonists into the available habitat. For example, dispersal is a fundamental aspect of succession. But species also disperse by *diffusion*, the slow extension of range from year to year. Rare, long-distance dispersal events can also occur, and these are significant in establishing global distribution patterns.

Some species have propagation modes that are especially well-adapted to long-distance dispersal. Mangrove species, which line coastal estuaries in equatorial and tropical regions, have seeds that ocean currents carry thousands of kilometres to populate far distant shores. Another example of a plant well-adapted to oceanic dispersal is the coconut palm. Its large seed, housed in a floating husk, has made it a universal occupant of tropical beaches. Among the animals, birds, bats, and insects are frequent long-distance travellers. Generally, non-flying mammals, freshwater fishes, and amphibians are less likely to make long migrations, with rats and tortoises the exceptions.

The case of the cattle egret demonstrates both long-distance dispersal and diffusion. This small heron crossed the Atlantic, arriving in northeastern South America from Africa in the late 1880s. One hundred years later, it had become one of the most abundant herons of the Americas (Figure 22.15). Another example of diffusion is the northward colonization of trees following the retreat of the ice sheets at the end of the Late-Cenozoic Ice Age (Figure 22.16). The oak reached their present northern limit in Canada about 8,000 years ago, beech achieved a similar range about 4,000 years ago, and chestnut spread to the southern shores of Lake Ontario only about 2,000 years ago.

Dispersal often means surmounting *barriers;* that is, regions a species is unable to occupy even for a short period. Oceans have proven to be effective barriers to long-distance dispersal for most species. But other barriers are not so obvious. For example, the

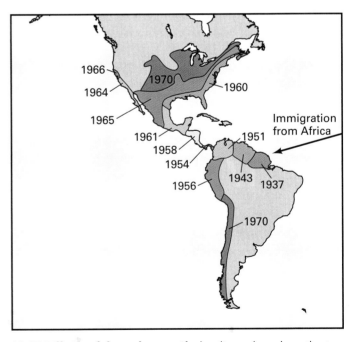

22.15 Diffusion of the cattle egret After long-distance dispersal to northeastern South America, the cattle egret spread to Central and North America as well as to coastal regions of western South America. (From J. H. Brown and M. V. Lomolino, *Biogeography*, second ed., Sunderland, Massachusetts: Sinauer, 1998, used with permission)

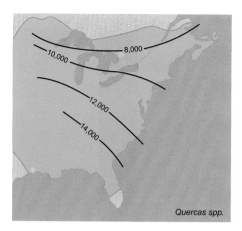
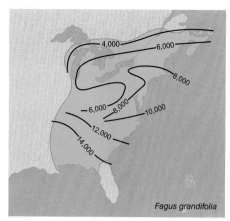

22.16 Post-glacial migration of deciduous trees in eastern North America The oak (*Quercus spp.*) migrated quickly northward and reached its present limit about 8,000 years ago. Beech (*Fagus grandifolia*) was well-established in southern Canada about 4,000 years ago, and chestnut (*Castanea dentata*) migrated more slowly, reaching its northern limit only about 2,000 years ago. Apart from a few small individuals, chestnut has since disappeared from eastern North America because of chestnut blight.

basin and range country of Utah, Nevada, and California presents an expanse of desert with small patches of forest. While birds and bats may have no difficulty moving from one patch to the next, a small mammal may never be able to cross the desert. In this case, the barrier is one in which the physiological limits of the species are exceeded. But there may be ecological barriers, as well; for example, a zone of intense predation or a region occupied by a vigorous competitor.

Just as there are barriers to dispersal, there also are corridors that facilitate dispersal. Central America forms a present-day *land bridge* connecting North and South America. It has been in place for about 3.5 million years. Other corridors of great importance to present-day species distribution patterns have existed in the recent past. For example, the Bering Strait region between Alaska and eastern Siberia was dry land during the early Cenozoic Era and during the Ice Age, when sea level dropped by more than 100 metres. Many plant and animal species of Asia crossed this bridge and then spread southward into the Americas. One notable migrant species of the last continental glaciation was the aboriginal human, and there is

substantial evidence to support the hypothesis that the skilled hunters who crossed the Bering land bridge were responsible for the extinction of many of the large animals that disappeared from the Americas about 10,000 years ago.

Ecologists and biogeographers have intensively studied the dispersal process, coupled with extinction, and the result has been the development of mathematical models explaining the number of species that might be expected within a region of a given size. *Working It Out 22.1 • Island Biogeography: The Species–Area Curve* discusses this.

DISTRIBUTION PATTERNS

The processes of evolution, speciation, extinction, and dispersal have, over time, produced the many spatial patterns of species distribution seen on the Earth. One of the simplest patterns is that of the **endemic** species, which is restricted to one area and nowhere else. An endemic distribution can arise in one of two ways; as the result of a contraction of a broader range, or as the origin location of a species that has not dispersed widely. Some

WORKING IT OUT | 22.1   Island Biogeography: The Species–Area Curve

The basic principle of island biogeography is simply that the larger the island, the more species it contains. That is, the number of species systematically increases with island area. This finding is true whether studying lists of all plant and animal species or just those species within a particular genus, family, or other group of related organisms.

The table shows the areas (in square kilometres) of some islands of the Lesser and Greater Antilles in the Caribbean, as well as the number of species of amphibians and reptiles found on each island.

Island	Area km²	Species*
Redonda	1.3	3
Saba	13	5
Montserrat	102	9
Puerto Rico	9,104	40
Jamaica	10,991	39
Hispaniola	76,480	84
Cuba	110,860	76

*Data of P.J. Darlington, 1967

These data are plotted in the figure below. In graph a, as the area of the island increases, the species count increases steeply at first. But for larger islands, the number of species increases more gradually. This *species–area curve* follows a power function; the number of species is related to the area raised to a power. Expressed as an equation,

$$S = cA^z$$

where S is the number of species, A is the area of the island, and c is a constant. The variable z, the power to which A is raised, lies between zero and one to provide a curve of this shape.

When a power function is plotted on logarithmic axes, it follows a straight line, as shown in part b of the figure. For these data the resulting best-fit equation for the power function is

$$S = 2.47A^{0.30}$$

The second example, shown in the graph on the opposite page, plots area against species counts of land birds and freshwater birds on 23 islands of the Sunda group, the Republic of the Philippines, and New Guinea. These islands are located in a broad arc off the southeast coast of Asia. As seen for the Antilles, there is a strong straight-line relationship between species and area. For these data, the best-fit values of c and z are

$$S = 4.61A^{0.34}$$

Note that the values of z are somewhat similar in these two examples. A number of studies of species–area curves for different islands and different species groups show that values of z are generally in the range of 0.20 to 0.35. Mathematical ecologists have determined that z, theoretically, is expected to be about 0.26 if species follow a log-normal rule of abundance. This rule states that, in any given region, only a few species are relatively common, while most species are moderately rare or very rare. The rule fits a broad range of data and holds for many situations.

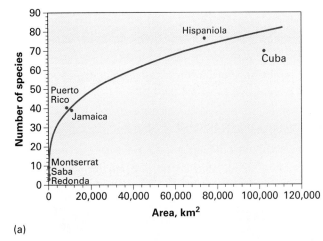

(a)

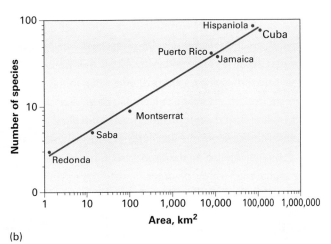

(b)

Species counts of reptiles and amphibians on selected islands of the Lesser and Greater Antilles (a) Data plotted on linear axes (b) Data plotted on logarithmic axes

The values of c, however, are less similar in these examples—4.61 for the Asian data, and 2.47 for the Antilles. These values depend on the number of species of a particular type that the island environment can support. For example, for a fixed area, say 1,000 square kilometres, the expected number of amphibian and reptile species is 19.6 compared with 48.3 species of land and freshwater birds. This shows that, given islands of comparable size, the Sunda Islands-Philippines-New Guinea group supports more bird species than the Antilles islands support amphibian and reptile species.

Species migrate to islands from the mainland or other islands. Over hundreds or thousands of years, individuals find their way to islands by chance—a typhoon, for example, may carry a bird to an island hundreds of kilometres from its normal habitat. Or a flood may sweep an animal out to sea on floating debris that comes to rest on a far island shore. In this way, islands receive new immigrants.

The immigration rate depends on the area of the island—a large island is more likely to receive more immigrants, simply because it has more area to intercept accidental travellers. Also, a large island is more likely to have a greater diversity of habitats, so more immigrants can successfully establish themselves once they arrive. So if the rate of immigration (I) is defined as the number of new species per year that an unpopulated island could support, that rate will be low for small islands and high for large islands.

At the same time, species on islands suffer extinction, which reduces the total number of species present. The extinction rate also depends on the island's area. On a small island, the population of a species will also tend to be small, and a catastrophic event, like a typhoon or volcanic eruption, is more likely to wipe out that population. On a larger island, there is a better chance for survival of enough individuals to keep the species in existence there. The extinction rate (E) is the number of species per year that are lost (given that all possible species are present); it will be high for small islands and low for large islands. These situations are shown in the table below.

	Small Island	Large Island
Immigration rate	low	high
Extinction rate	high	low
$I/(I + E)$	small	large

Combining the immigration and extinction rates gives the *turnover rate* ($I + E$). This rate expresses the number of species that either immigrate or become extinct in a given year. The expression $I/(I + E)$ is then the proportion of the turnover rate that arises from immigration; it will be small for small islands and large for large islands. This proportion determines the average size of the species pool present on the island. Thus,

$$S = P[I/(I + E)]$$

where S is the expected number of species and P is the number of species in the pool available for colonization of the island (*i.e.*, on the mainland).

This expression shows that the number of species depends on the balance between immigration and extinction—a balance that in turn depends on the size of the island. Small islands have low immigration rates and high extinction rates, so have few species. Larger islands have higher immigration rates and lower extinction rates, so have more species.

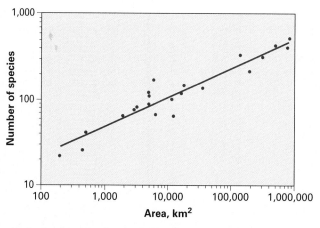

Species–area plot for land and freshwater bird species Data are shown for 23 islands of the Sunda group, the Republic of the Philippines, and New Guinea. (Data are from sources as cited in R. H. MacArthur and E. O. Wilson, *Evolution*, vol. 17, 1963, p. 374.)

22.17 Gingko tree The gingko tree is an eastern Chinese endemic that has survived millions of years with little evolutionary change. It has been introduced in many parts of the world. The fan-shaped leaves show an uncommon arrangement of parallel veins radiating from the leaf stem.

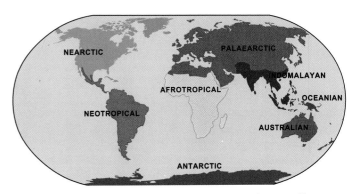

22.19 Biogeographic realms Each biogeographic realm is characterized by a distinctive set of plant and animal species that has arisen because of past and present barriers to dispersal.

endemic species are ancient relics of species that have otherwise gone extinct. An example is the gingko tree (Figure 22.17), which was widespread throughout the Mesozoic Era but until recently was restricted to a small region in eastern China. It is now widely planted as an urban street tree, known for its hardiness.

In contrast to endemics are *cosmopolitan* species, such as dandelions, which are distributed widely. Most cosmopolitan species either are very small, or have efficient propagating forms that are readily dispersed by wind or oceanic currents.

Another interesting pattern is **disjunction**, in which one or more closely related species are found in widely separated regions. An example is the distribution of the tinamous and flightless ratite birds, which include the ostrich, emu, cassowary, and kiwi (Figure 22.18). This disjunct pattern is

thought to result from an ancestral species that was widespread across the ancient continent of Gondwana. As plate tectonics split Gondwana into North and South America, Africa, Australia, and New Zealand, isolation and evolution differentiated the ancestral lineage into the diverse array of related species that now inhabit these continents.

BIOGEOGRAPHIC REALMS

As spatial distributions of species are examined on a global scale, certain common patterns emerge. From this, it is possible to delineate **biogeographic realms**, characterized by a distinct assemblage of plants and animals (Figure 22.19). The largest realm is the Palearctic. It covers 54.1 million square kilometres and includes most of Eurasia and North Africa. The Nearctic realm, which includes most of North America, covers 22.9 million square kilometres and is similar in size to the Afrotropical realm (22.1 million square kilometres), which extends from the southern margin of the Sahara Desert to the southern tip of Africa. South America and the Caribbean are included in the Neotropical realm (19.0 million square kilometres). The remaining realms range from 7.7 million

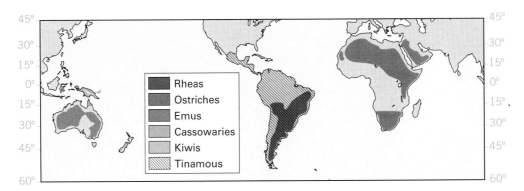

22.18 Disjunct distribution The distribution pattern of ratite birds and tinamous shows disjunctions resulting from isolation by continental movements. (From J. H. Brown and M. V. Lomolino, *Biogeography*, second ed., Sunderland, Massachusetts: Sinauer, 1998, used with permission.)

square kilometres for Australasia to 0.3 million square kilometres for the Antarctic.

The similarity in plant and animal distribution patterns suggests they have experienced related histories of evolution and environmental affinity. However, these patterns are not identical; therefore, there are distinct floristic and faunal regions (Figure 22.20). The boundaries of each region are related to sig-

nificant barriers across which dispersion to adjacent realms has been limited. The Australasian realm is effectively isolated by oceans. The Palearctic realm is separated from the Afrotropical realm by the hostile environment of the Sahara. Similarly, the Himalayas and associated mountain ranges separate the Palearctic and Indomalayan realms. These barriers have been particularly restrictive to animal dispersal.

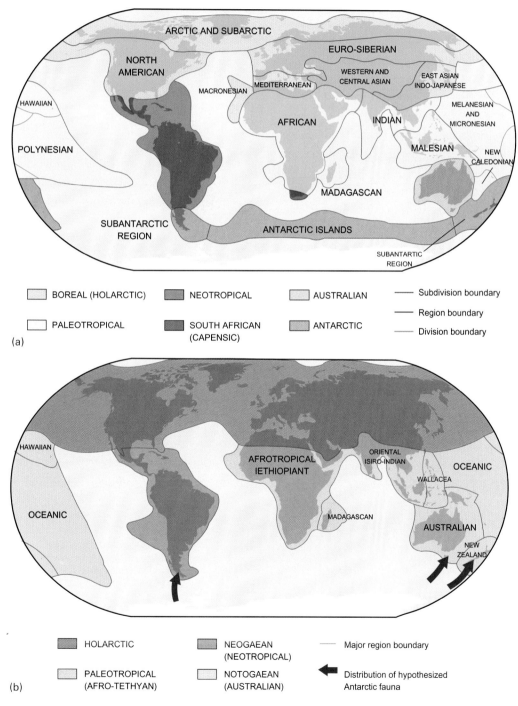

22.20 Floristic and faunal regions of the world (a) Floristic regions are distinguished mainly on the basis of distinctiveness and how endemic families and genera of flowering plants are. (b) Faunal regions are distinguished by their distinctive animal assemblages, particularly their mammals.

The number of plant and animal groups common to each realm increases where the boundaries between regions are less effective. The degree of similarity between regions also reflects how recently the barriers were formed or broken. For example, India drifted into Asia about 45 million years ago, and North and South America became linked through Central America about 3.5 million years ago. More recently, North America and Asia were joined by the Bering land bridge, a 1,500-kilometre wide tract of steppe grassland, as recently as 12–15,000 years ago.

The boundary that has been the most contentious is the one separating Southeast Asia from Australia. Plant and animal groups on the various islands of the East Indies are transitional and have greater or lesser affinity with the adjacent continents. Original work by Alfred Russel Wallace in 1876 noted a clear distinction in the bird families that occurred on two of these islands—Bali and Lombok—that were separated by a distance of only 35 kilometres. On Bali, the birds were related to those of Java, Sumatra, and mainland Malaysia. On Lombok, they more closely resembled those of New Guinea and Australia. Based on this, the Oriental and Australasian faunal regions were separated by Wallace's Line, drawn between Bali and Lombok and extended northward between Borneo and Sulawesi (Figure 22.21). Like most biogeographic regions, the faunal and floral assemblages of the Oriental and Australian regions are clearly distinct at their centre, but species merge at the periphery. This blurring can be traced to late-Cenozoic sea level changes, which alternately exposed and submerged extensive areas of land in this transition zone.

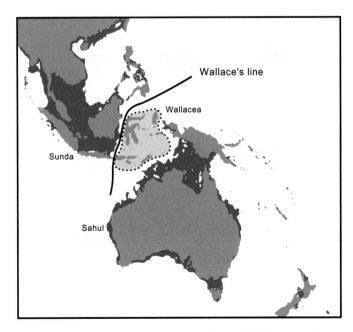

22.21 Wallace's Line Wallace's Line separates the Oriental and Australasian faunal regions. The light blue areas are the Sunda and Sahul shelves, which were exposed when sea level dropped in the Late-Cenozoic Ice Age. The grey area, known as Wallacea, consists of a group of islands that are separated from these continental shelves by deep water.

BIODIVERSITY

Biodiversity—the variety of biological life on the Earth—depends on the variations of the Earth's environments and the processes of evolution, dispersal, and extinction through geologic time. Currently, the Earth's biodiversity is rapidly decreasing as a result of human activity. Humans now use some 20 to 40 percent of global primary productivity, as well as exploit 70 percent of the marine fisheries to provide food. This environmental exploitation has doubled the natural rate of nitrogen fixation, used more than half of the Earth's supply of surface water, and transformed more than 40 percent of the land surface.

In addition, recent human activity has ushered in a wave of extinctions unlike any that has been seen for millions of years. In the last 40 years, biologists have documented the disappearance of several hundred land animal species, including 58 mammals, 115 birds, 100 reptiles, and 64 amphibians. Aquatic species have also been severely affected, with 40 species or subspecies of freshwater fish lost in North America alone in the last few decades. Botanists estimate that more than 600 plant species have become extinct in the past four centuries.

In 2001, the Committee on the Status of Endangered Wildlife in Canada (COSEWIC) identified 358 species or populations as endangered, threatened, or of special concern (Table 22.2). The majority of these species designated to be at risk are in the southern parts of Canada, particularly southwestern and interior British Columbia, the southern Prairies, and southern Ontario.

The numerous extinctions that have been documented globally may represent only a fraction of the species that have been lost. Many species have not yet been discovered and so may become extinct without ever being identified. Some of these species could be important to a better understanding of evolutionary history, because they may include representatives of new biological divisions of the highest level. In the past decade, three new families of flowering plants and two new phyla of animals were

Table 22.2 COSEWIC status designations as of November 2001

	Extirpated or Extinct	Endangered	Threatened	Special Concern	Not at Risk	Data Deficient
Mammals	6	16	14	25	47	8
Birds	5	20	8	21	35	3
Amphibians and reptiles	3	10	13	13	13	—
Fish	8	10	21	37	31	6
Invertebrates	5	10	3	3	1	2
Plants	2	49	34	44	16	4
Mosses	—	3	1	—	—	—
Lichens	—	1	—	3	—	—
Total	29	118	94	146	143	23

discovered. Among animal groups, insects, spiders, and other invertebrates are the most poorly known. Fungi and microbes are other groups in which a large proportion of species have yet to be identified.

Human activity has caused extinctions in several ways. One way has been to disperse new organisms that out-compete or predate existing organisms. Island populations have especially suffered from this process. Developing in isolation, island species have often not evolved defence mechanisms to protect themselves from predators, including humans. Many islands were subjected to waves of invading species, ranging from rats to weeds, brought first by prehistoric humans and later by explorers and conquerors. Hunting by prehistoric humans alone was enough to exterminate many species, not only from islands, but from large continental regions, as well. Another mechanism is the use of fire. As humans learned to use fire in hunting and to clear and maintain open land, large areas became subject to periodic burning. Habitat alteration and fragmentation is yet another cause of extinctions. By isolating plant and animal populations and altering their environment, human activities cause populations to shrink, making extinction more likely.

Biodiversity is not uniform over the Earth's surface. In general, tropical and equatorial regions have more species and more variation in species composition between different habitats. In isolated areas, such as islands or mountaintops, species diversity tends to increase with the size of the isolated area and decrease with the degree of isolation from surrounding sources of colonists. Much of this diversity is contributed by endemic species. Geographic areas in which biodiversity is especially high are referred to as **hotspots**. Hence, an important strategy for preservation of global biodiversity is to identify hotspots and take conservation measures to protect them.

There are currently 25 hotspots (Figure 22.22). These contain 44 percent of all vascular plant species and 35 percent of terrestrial vertebrates, but account for only 1.4 percent of the Earth's surface. However, these regions have collectively lost almost 90 percent of their original primary vegetation, with direct consequences on plant species diversity and the animals that rely on them.

An area qualifies as a hotspot if it contains endemic vascular plant species composing at least 0.5 percent of all the world's plant species. This is equivalent to at least 1,500 endemic plant species. In addition, the area must have lost at least 70 percent of its primary plant cover. In many hotspots, the numbers of endemic plant and vertebrate species are strongly correlated. More than half of all threatened plants and terrestrial vertebrates are endemic to biodiversity hotspots. Conservation of these important areas is therefore essential to minimize the threat of mass extinctions.

Nature, operating over millennia, has provided an incredibly rich array of organisms that interact with one another in a seamless web of organic life. Humans are an integral part of that web.

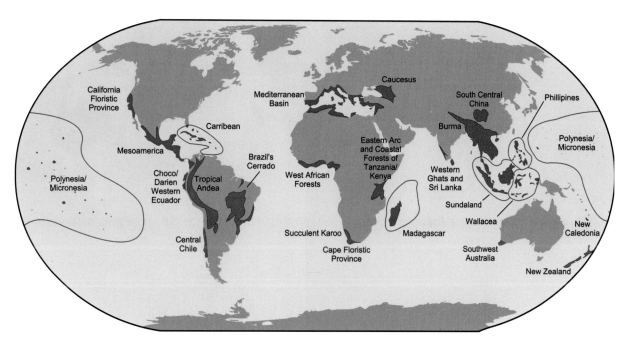

22.22 Biodiversity hotspots These areas exhibit unusually rich biodiversity, but are also threatened with destruction.

When human activity causes the extinction of a species, a link in the web is broken. Ultimately, the web will become so impoverished that it could threaten the future of the human species and many other life forms. There is no way of knowing which organisms future humans will rely on. Thus, it seems prudent to keep as many of them around for as long as possible.

One approach to promoting a balanced relationship between humans and the biosphere has been the development of **biosphere reserves**, designed to conserve examples of characteristic ecosystems. Biosphere reserves combine both conservation and sustainable use of natural resources. To date there are more than 500 biosphere reserves in more than 100 countries around the world (Figure 22.23). Each consists of a core area that is protected by law; for example, a nature reserve or national park.

The core area is designed to protect sensitive and valuable species. It may include original ecosystems, as well as those that are valuable to humans, such as grazing land. Core areas may also include characteristic cultural features. Land and water use may occur in the core area, provided it is consistent with the aims of nature conservation. Surrounding a core area is a buffer zone in which activities and resource use are consistent with protection

of the core. Restrictions in buffer zones are based on local voluntary agreements. Beyond the buffer zone is a transition area where sustainable development is a priority. This is the margin of the biosphere reserve; it is important for the economic development of the region. To be successful, a biosphere reserve therefore depends on an appropriate mix of research, monitoring, education, and training. Perhaps most importantly, each biosphere reserve depends on voluntary co-operation to conserve and use resources for the well-being of all.

A Look Ahead

In this chapter, the focus has been on the processes that determine the spatial patterns of biota on the Earth. It has discussed how organisms adjust to their environments and how natural selection works in response to environmental pressures. It also noted how evolution, dispersal, and extinction generate patterns of species distribution and determine biodiversity. The final chapter takes a more functional view of the life layer by reviewing the global biomes—major divisions of ecosystems that are based largely on the dominant vegetation life forms.

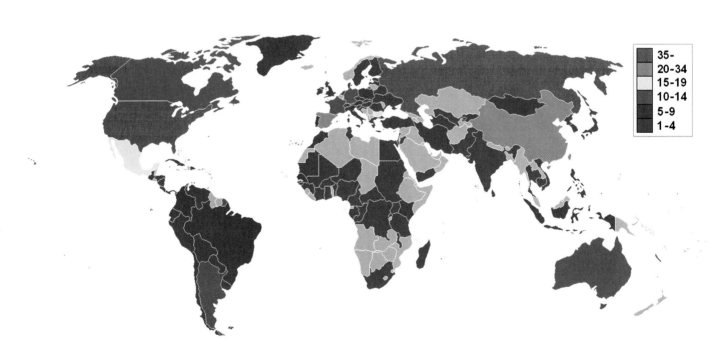

22.23 Global network of biosphere reserves Biosphere reserves are designed to create and promote innovative approaches to conservation and development.

CHAPTER SUMMARY

- *Ecological biogeography* examines how relationships between organisms and environment help determine when and where organisms are found. *Historical biogeography* examines how, where, and when species have evolved and how they are distributed over longer times and broader scales.

- A *community* of organisms occupies a particular environment, or habitat. *Associations* are often defined by characteristic species or vegetation life forms. Environmental factors influencing the distribution patterns of organisms include moisture, temperature, light, and wind.

- Organisms require water to live, so are limited by the availability of water. Xerophytes are adapted to dry habitats. They reduce water loss by having waxy leaves, spines instead of leaves, or no leaves at all. Some develop extensive root systems. Hydrophytes are adapted to wet environments such as bogs and swamps. Xeric animals include vertebrates that are nocturnal and have various adaptations to conserve water. Invertebrates such as brine shrimp can adjust their life cycle to prolonged drought.

- Temperature triggers and controls stages of plant growth, as well as limits growth at temperature extremes. Survival below freezing requires special adaptations; only a small proportion of plants are frost tolerant. *Cold-blooded animals* have body temperatures that follow the environment; however, they can moderate these temperatures by seeking warm or cool places. Mammals and birds are warm-blooded animals that maintain constant internal temperatures through a variety of adaptation mechanisms.

- The light available to a plant depends on its position in the structure of the community. Duration and intensity of light vary with latitude and season and serve as a cue to initiate growth stages in many plants. The day–night cycle regulates a lot of animal behaviour, as does the *photoperiod* (seasonal change in the duration of daylight). Wind deforms plant growth by desiccating buds and young growth on the windward side of the plant.

- A bioclimatic frontier marks a species' potential distribution boundary. Other factors may also limit the distribution of a species.

- *Geomorphic factors* of slope, steepness, and orientation affect both the moisture and temperature environment of the habitat and serve to differentiate the microclimate of each community.

- Soil, or *edaphic*, *factors*, such as soil texture, structure, acidity, alkalinity, and salinity, can also limit the distribution patterns of organisms or affect community composition.

- *Disturbance* includes catastrophic events that damage or destroy ecosystems. Fire is a common type of disturbance that influences forests, grasslands, and shrublands. Floods, high winds, and storm waves are others. Many ecosystems include specialized species that are well-adapted to disturbance.

- Species interact in a number of ways, including competition, *predation* and *parasitism*, and *herbivory*. In *allelopathy*, plant species literally poison the soil environment against competing species. Positive (beneficial) interactions between species are called *symbiosis*.

- Ecological succession comes about as ecosystems change through time. *Primary succession* occurs on new soil substrate, while *secondary succession* occurs in disturbed habitats. Although succession is a natural tendency for ecosystems to change with time, it is opposed by natural disturbances and limited by local environmental conditions.

- *Historical biogeography* focuses on the influence of evolution, speciation, extinction, and dispersal on the distribution patterns of species.

- A species is best defined as a population of organisms that is capable of interbreeding successfully, but is usually defined by a typical *morphology* or *phenotype*.

- Speciation is the process by which species are differentiated and maintained. It includes mutation and natural selection. Geographic isolation involves the isolation of subpopulations of a species, allowing genetic divergence and speciation to occur. In sympatric speciation, adaptive pressures force a breeding population to separate into different subpopulations that may become species. Sympatric speciation of plants has included polyploidy.

- Extinction occurs when populations become very small and thus are vulnerable to chance occurrences of fire, disease, or climate anomaly. Rare but extreme events can cause mass extinctions. An example is the asteroid impact that the Earth suffered about 65 million years ago.

- Species expand their ranges by dispersal. Seeds generally disperse plants, while animals often disperse through their own movements and migrations. Long-distance dispersal, though rare, is important in establishing biogeographic patterns. *Barriers*, often climatic or topographic, inhibit dispersal and induce geographic isolation.

- Endemic species are found in one region or location and nowhere else. They arise by either a contraction of the range of a species or a recent speciation event. *Cosmopolitan* species are widely dispersed and nearly universal. Disjunction occurs when one or more closely related species appear in widely separated regions.

• Biogeographic realms capture patterns of species occurrence that arise from common histories and similar environmental preferences. The boundaries between realms are often quite marked where significant change in habitat occurs. For this reason, oceans are effective barriers to terrestrial organisms. Less distinct boundaries occur where adjacent land areas were joined comparatively recently.

• Biodiversity is rapidly decreasing as human activity progressively affects the Earth. Extinction rates for many groups of plants and animals are as high or higher today than they have been at any time in the past. Humans act to disperse predators, parasites, and competitors widely, disrupting long-established evolutionary adjustments of species with their environments. Hunting and burning have exterminated many species. Habitat alteration and fragmentation also lead to extinctions. Preservation of global biodiversity includes a strategy of protecting hotspots where diversity is greatest. Biosphere reserves provide a functional compromise between conservation and sustainable land use.

KEY TERMS

bioclimatic frontier
biodiversity
biogeographic realm
biosphere reserve
climax community
competition

disjunction
dispersal
ecological niche
ecological succession
edatopic grid
endemic

fundamental niche
geographic isolation
habitat
hotspot
hydrophyte
natural selection

polyploidy
realized niche
speciation
symbiosis
xerophyte

REVIEW QUESTIONS

1. What is a habitat? What are some of the characteristics that differentiate habitats? Compare *habitat* with *niche*.

2. Contrast the terms *community* and *association*.

3. Although water is a necessity for terrestrial life, many organisms have adapted to dry environments. Describe some of the adaptations that plants and animals have evolved to cope with desert environments.

4. Terrestrial temperatures vary widely. How does the annual variation in temperature influence plant growth, development, and distribution? How do animals cope with variation in temperature?

5. How does the ecological factor of light affect plants and animals?

6. What does the term *bioclimatic frontier* mean? Provide an example.

7. How do primary succession and secondary succession differ?

8. Distinguish between allopatric and sympatric speciation. Provide examples of each.

9. Describe the effects of barriers and corridors in the dispersal process.

10. How are biogeographic realms differentiated?

11. What is biodiversity? How has human activity affected biodiversity?

VISUALIZATION EXERCISE

1. Compare parts (a) and (b) of Figure 22.20. Which boundaries are similar, and which are different? Speculate on possible reasons for the similarities and differences.

ESSAY QUESTION

1. Imagine yourself as a biogeographer discovering a new group of islands. Select a global location for your island group, including climate, geologic substrate, and indicate proximity to nearby continents or land masses. What types of organisms would you expect to find within your island group and why?

1. A zoogeographer assembles the following data for area and species counts of termites and butterflies on a set of (fictitious) islands several hundred kilometres off the coast of Brazil.

Island	Area (km²)	Termites	Butterflies
Alhambra	156	11	23
Bonarote	845	16	41
Carlo	1,746	26	47
Delore	3,550	31	64
Edmundo	14,323	42	101
Fonseca	71,420	68	151

Analysis shows that the species–area curve for termites best fits the expression $S = 2.07A^{0.31}$, while the curve for butterflies best fits the expression $S = 4.56A^{0.32}$. Plot the data in two ways: on arithmetic (linear) axes and on logarithmic axes. Show data for both termites and butterflies on each graph. Sketch smooth curves to fit the data plotted on the arithmetic graph and straight lines for the data plotted using the logarithmic axes.

2. Use the expressions for the two species–area curves to find the predicted number of species of termites and butterflies for Carlo Island. Compare these with the observed values in the table. Are the observed values higher or lower than predicted? Identify some possible reasons why this difference might occur.

Find answers at the back of the book

EYE ON THE LANDSCAPE |

Chapter Opener Wetlands at Chicko River, British Columbia. Permanent water in the low-lying area favours the growth of sedges (**A**), the remains of which slowly accumulate as peaty deposits allowing the plants to advance into the depression as it gradually fills in. Shrubs (**B**), such as willows, typically grow in the moister soils adjacent to wetlands, which in turn are replaced by aspens (**C**) and conifers (**D**) in the better-drained soils beyond. Young conifers are establishing at the edge of the pine stand (**E**), and over time, will replace the aspens and willows as the vegetation cover changes through succession.

Jackpine, regeneration after fire, Ontario, Jackpine (*Pinus banksiana*) (**A**) is a fire-dependent species that requires the heat of a fire to open its serotinous cones before the seeds can be released. It is also a shade-intolerant species and will germinate and establish successfully only in open sites where competition is reduced and mineral soil is exposed through removal of the forest litter (**B**). Fireweed (*Epilobium angustifolium*) (**C**) is also an early successional species; it establishes immediately in burned sites where its seeds have been transported by wind. Forest fires burn with varying intensity. Fallen trees (**D**) often indicate a more intense fire. The standing dead trees (**E**) will eventually decay and

fall, within a few years, but nonetheless the dead trees provide essential habitat for insectivorous species such as woodpeckers. The woody debris typically adds little nutrients to the soil, but is used by ants and other decomposer organisms that are essential for the health of the forest ecosystem.

Chapter 23

EYE ON THE LANDSCAPE

What else would the geographer see? . . . Answers at the end of the chapter.

THE EARTH'S TERRESTRIAL BIOMES

Natural Vegetation

Structure and Life Form of Plants

Terrestrial Ecosystems—The Biomes

Geographers at Work • Dr. Paul Treitz

Forest Biome

Focus on Remote Sensing 23.1 • Mapping Global Land Cover by Satellite

Eye on the Environment 23.2 • Exploitation of the Low-Latitude Rain Forest Ecosystem

Savanna Biome

Grassland Biome

Geographers at Work • Dr. Will Wilson

Desert Biome

Tundra Biome

Climatic Gradients and Vegetation Types

Altitude Zones of Vegetation

The previous two chapters focused on ecology and biogeography, examining the principles and processes that determine the distribution of plants and animals on the Earth's surface. This concluding chapter describes the broad global distribution patterns and characteristics of the Earth's major biomes. It emphasizes vegetation because it is a visible and obvious part of the landscape. Also, the largest division of ecosystems—the biome—is defined primarily by the characteristics of the vegetation cover. Vegetation tends to be related to climate, so the occurrence of major biomes coincides, in many cases, with broad climate types.

NATURAL VEGETATION

Over the last few thousand years, human societies have come to dominate much of the land area and, in many regions, have changed the natural vegetation—sometimes drastically, other times more subtly. **Natural vegetation** is a plant cover that develops with little or no human interference. It is subject to natural forces of modification and destruction, such as storms or fires. Natural vegetation still occurs in some remote areas, such as the arctic and alpine tundra or parts of the wet equatorial rain forests.

In contrast, much of the land surface in midlatitudes has been totally affected by human activities, through intensive farming or urbanization. Some areas of natural vegetation appear to be untouched but are actually dominated by human activity in a subtle way. For example, most national parks and forests were protected from fire for many decades. In the past, when lightning started a forest fire, it was extinguished as quickly as possible. However, periodic burning is part of the natural cycle in many regions, and, in recent years, suppressing wildfires has been stopped in some parks, allowing the return to more natural processes.

Humans have influenced vegetation by moving plant species from their original habitats to foreign lands and different environments. The eucalyptus tree is a striking example. From Australia, species of eucalyptus have been transplanted in various places, including California, North Africa, and India. Sometimes exported plants thrive like weeds, forcing out natural species and becoming a major nuisance. For example, brome grass (*Bromus inermis*), originally imported to North America to improve forage quality, has spread aggressively in many prairie habitats, reducing the diversity of native species.

Nevertheless, all plants have limited tolerance to the environmental conditions imposed by factors such as temperature, soil water availability, and soil nutrients. Consequently, the structure and appearance of the plant cover conforms to basic environmental controls, and each vegetation type, whether forest, grassland, or desert, is associated with a characteristic geographical region.

STRUCTURE AND LIFE FORM OF PLANTS

Botanists recognize and classify plants by species. However, plant geographers are often less concerned with individual species and more concerned with the vegetation as a whole. In describing the vegetation, plant geographers refer to the **life form** of the plants, which emphasizes their physical structure, size, and shape. This method of classifying plants can provide information on the relationship between the species and their environments. Thus, vegetation in widely separated regions is often remarkably similar in appearance, even though there are few if any common species. For example, the trees that grow in the boreal forests of Canada look very similar to those growing in Scandinavia and Russia, but in fact no tree species occurs naturally in both regions. They are, however, similarly adapted to cope with the rigorous climate and thin soils that characterize northern Canada and northern Eurasia. Likewise, the shrub life form that is characteristically associated with Mediterranean climate region of California is almost indistinguishable from shrubs growing in the southern tip of Africa, Greece, or central Chile.

Both trees and shrubs are erect, *perennial* woody plant forms that endure from year to year. Most have lifespans of many years. *Trees* typically have a single upright main trunk, often with few branches in the lower part but many in the upper part to form a crown. *Shrubs* usually have several stems branching from a base near the soil surface so their foliage is close to the ground.

Lianas are also woody plants, but they take the form of vines supported on trees and shrubs. Lianas include not only the tall, heavy vines of the wet equatorial and tropical rain forests, but also some woody vines of midlatitude forests.

Herbs compose a major class of plant life forms. They lack woody stems so are usually smaller plants. Some are *annuals*, living only for a single season; others are *perennials* and live for multiple seasons. Some herbs are broad-leaved, and others are narrow-leaved, such as grasses. Herbs usually form a low ground layer.

In forests, the trees grow close together. Crowns are in contact, so the foliage shades the ground. Many forests in moist climates have at least three life-form layers (Figure 23.1). Tree crowns form the uppermost layer, shrubs an intermediate layer, and herbs a lower layer. There is sometimes a fourth, lowest layer that consists of mosses and other small plants. In **woodland**, the

23.1 Vertical structure of a forest Most forests consist of one or more layers, including an understory of shrubs and a ground layer of herbs and grasses. Dead and decaying plants are also an important component of this forest ecosystem in northern Saskatchewan.

trees are widely spaced so that their crowns are separated by open areas that usually support grasses, low herbs, or a shrub layer.

Lichens are another life form that can be abundant in the ground layer. They are plant forms in which algae and fungi live together to form a single plant structure. In some alpine and arctic environments, lichens grow in profusion and dominate the vegetation (Figure 23.2).

TERRESTRIAL ECOSYSTEMS— THE BIOMES

From the viewpoint of human use, ecosystems are natural resource systems. Food, fibre, fuel, and structural material are products of ecosystems and are manufactured by organisms using

23.2 Lichen Despite its name, the reindeer moss photographed here is actually a common variety of lichen.

GEOGRAPHERS AT WORK

Mapping Forest Ecosystems and Their Biophysical Characteristics with Remote Sensing
by Dr. Paul Treitz, Queen's University

Did you ever hug a tree? If you've done any field work in forestry, you've probably hugged a lot of them. To measure the size of a tree trunk, you have to put your arms around the tree in order to pass a special measuring tape from one hand to the other. Pulling the tape tight, you read off the diameter indicated where the tape's end crosses the remainder of the tape.

How about the height of a tree? Well, you could climb the tree, carrying the end of a measuring tape to the top. Or you could cut the tree down and lay the tape along its length. Of course, there is an easier way—go a fixed distance away from the tree and then sight the angle to its top. From trigonometry, it's easy to find the height.

The height and trunk diameter of a tree are basic characteristics that are related both to the tree's biomass and the amount of photosynthetic material it bears in its leaves. Thus, these measure-ments are very important in assessing the health and productivity of forest ecosystems and are linked to many ecosystem processes. But measuring height, diameter, and related quantities, such as tree spacing or crown diameter, is very tedious. Wouldn't it be useful if we could measure these characteristics indirectly using remote sensing?

That's where Paul Treitz's research comes in. He's developing techniques for using airborne laser scanners (i.e., light detection and ranging (lidar) to survey large areas of Canadian forests. His goal is to provide the information on forest structure (i.e., height, volume, biomass) that forest managers need for sustainable development of forest resources while conserving the biodiversity and long-term health of forest ecosystems. For example, biomass is an important variable that plays a significant role in assessing carbon stocks; is an important element in global change and productivity models; and is a measure of vegetation community structure which influences biodiversity. You can find out more about his research and see photos of his students making tree measurements at our website.

Paul Treitz's research focuses on the spectral, spatial, and temporal analysis of remote sensing data of forest ecosystems.

energy derived from the sun. **Terrestrial ecosystems** are directly influenced by climate and interact with the soil. In this way, they are closely woven into the fabric of physical geography. On the global scale, the largest recognizable subdivision within terrestrial ecosystems is the **biome**. Although the biome includes the total assemblage of plant and animal life interacting within the life layer, green plants dominate the biome physically because of their enormous biomass compared with that of other organisms.

There are five principal biomes. The **forest biome** is dominated by trees, which form a closed or nearly closed canopy. Forests require an abundance of soil water, so they are found in moist climates. Temperatures must also be suitable, requiring at least a warm season, if not warm temperatures year round. The **savanna biome** is transitional between forest and grassland. It supports an open cover of trees with grasses and herbs beneath. The **grassland biome** develops in regions with moderate shortages of soil water. Temperatures must provide adequate warmth during the growing season. The semi-arid regions of the dry tropical, dry subtropical, and dry midlatitude climates are associated with the grassland biome. The **desert biome** includes organisms that have adapted to the moderate to severe water shortages that occur for most, if not all, of the year. The characteristic plants are xerophytes, species that can survive with limited water. Temperatures in deserts range from very hot to cool. The **tundra biome** is limited by cold temperatures. Only small plants that grow quickly when temperatures rise above freezing during the short growing season can survive.

Biogeographers subdivide biomes into smaller vegetation units, called *formation classes*, based on plant life forms. For example, different types of forests, such as evergreen needleleaf, broadleaf deciduous, and broadleaf evergreen, are easily distinguished within the forest biome. There are at least three kinds of temperate grasslands, including tall grass prairie, mixed prairie, and steppes. Deserts, too, span a wide range in terms of plant

FOCUS ON REMOTE SENSING | 23.1   Mapping Global Land Cover by Satellite

One of the striking things about the Earth's surface when viewed from space is its colour. Deserts appear in shades of brown, dotted with white salty playas. Equatorial forests are green, dissected by branching lines of dark rivers. Shrublands are marked with a greenish tinge. Some regions show substantial change throughout the year. In the midlatitude zone, deciduous forests go from intense green in the summer to brown as leaves drop from the trees. In the tropical zones, grasslands and savannas change from brown to green as the rainy season comes and goes. Thus, the colour of the land surface can provide information about the type of plant cover present at a location.

Ever since they received the first satellite images of the Earth, scientists have used colour—that is, the spectral reflectance of the surface—as an indicator of land cover type. They have produced land cover maps using individual Landsat images for more than 30 years. Detailed maps can be generated from

the spectral signatures of known plant types, and these are subsequently used to classify the vegetation in an area.

Most satellite-derived vegetation maps now use the normalized difference vegetation index (NDVI). This index is based on the absorption and reflectance characteristics of plant tissues for different wavelengths of light. The pigment chlorophyll strongly absorbs visible light (from 0.4 to 0.7 micrometres), whereas leaf surfaces and other plant tissues strongly reflect near-infrared light (from 0.7 to 1.1 micrometres). The difference in the amount of energy absorbed and reflected in these wavelengths is used to calculate NDVI.

$$NDVI = \frac{NIR - Red}{NIR + Red}$$

NDVI provides an assessment of plant density and vigour and is extremely useful in comparing seasonal changes in vegetation cover. In Figure 1, it is evident that vegetation in Canada is mostly dormant in the

winter months, but grows actively in the summer.

Global mapping of land cover requires instruments that can observe the surface on a daily or near-daily basis, thus maximizing the opportunity to image the surface in areas that are frequently covered in cloud. These instruments have a coarser ground resolution than Landsat, with pixels in the range of 500 to 1,100 metres on a side; however, that is not a deficiency when working on a global scale. Shown in Figure 2 is a map of global land cover produced from MODIS images acquired mainly in 2001. The legend recognizes 17 types of land cover, including forests, shrublands, savannas, grasslands, and wetlands.

Evergreen broadleaf forest dominates the equatorial belt, stretching from South America, through Central Africa, to south Asia. Adjoining the equatorial forest belt are regions of savannas and grasslands, which have seasonal wet-dry climates. The vast desert region running from the Sahara to the Gobi is barren or sparsely vegetated. It is flanked by grasslands on the west, north, and east. Broadleaf deciduous forests are prominent in eastern North America, western Europe, and eastern Asia. Evergreen needleleaf forests span the boreal zone from Alaska and northwest Canada to Siberia. Croplands are found throughout most regions of human habitation, except for dry desert regions and cold boreal zones.

The map was constructed using both spectral and temporal information. The graph in Figure 3 shows how these

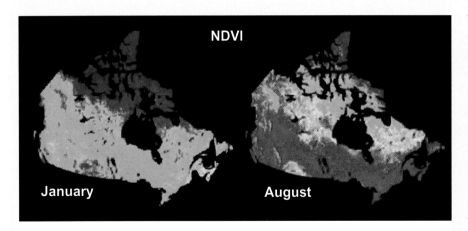

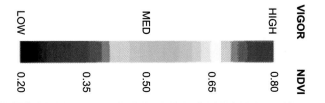

Figure 1 January and August normalized difference vegetation indices for Canada The NDVI provides a measure of the density and vigour (photosynthetic activity) of the vegetation cover. It uses data derived from NOAA's AVHRR sensor.

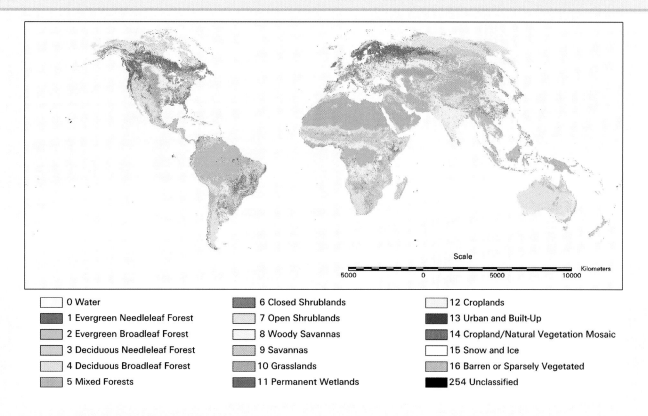

☐ 0 Water	■ 6 Closed Shrublands	☐ 12 Croplands
■ 1 Evergreen Needleleaf Forest	☐ 7 Open Shrublands	■ 13 Urban and Built-Up
■ 2 Evergreen Broadleaf Forest	☐ 8 Woody Savannas	■ 14 Cropland/Natural Vegetation Mosaic
☐ 3 Deciduous Needleleaf Forest	■ 9 Savannas	☐ 15 Snow and Ice
☐ 4 Deciduous Broadleaf Forest	■ 10 Grasslands	☐ 16 Barren or Sparsely Vegetated
■ 5 Mixed Forests	■ 11 Permanent Wetlands	■ 254 Unclassified

Figure 2 Global land cover from MODIS This map of global land cover types was constructed from MODIS data acquired mainly during 2001. The map has a spatial resolution of one square kilometre; that is, each square kilometre of the Earth's land surface is independently assigned a land cover type label. (A. H. Strahler, Boston University/NASA)

information sources are used. It depicts reflectance values in red and near-infrared (NIR) spectral bands for three land cover types, as observed in the southeastern United States: evergreen needleleaf forest, cropland/natural vegetation mosaic, and deciduous broadleaf forest. The reflectances are shown as they change over the course of about a year and a half, from 31 October 2000 (2000305) to 23 April 2002 (2002113).

Figure 3 Spectral and temporal reflectance patterns
The graph shows how the spectral reflectance in red and near-infrared bands of three land cover types in the southeastern United States varies during the period from 31 October 2000 to 23 April 2002. Dates are shown in Julian date format: the first four digits indicate the year, and the remaining three indicate the day of the year, with 001 representing 1 January, and so on. (A. H. Strahler, Boston University/NASA)

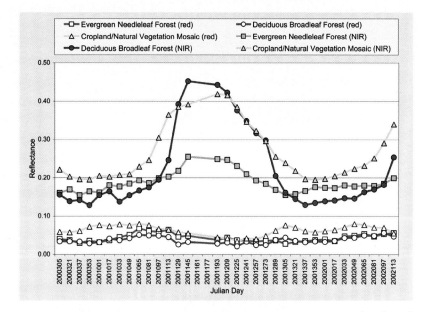

(continued)

The top three curves, which are near-infrared values, show the patterns of the three types most clearly. During the winter, values are generally low, with deciduous broadleaf forest, evergreen needleleaf forest, and cropland in increasing order of reflectance. As spring begins, cropland reflectance rises before deciduous broadleaf forest, but deciduous broadleaf forest reaches a higher peak. Evergreen needleleaf forest also shows a spring green-up, but it is later than the others and peaks at a lower value. Reflectance gradually drops during the fall, and the three types reach about the same reflectance levels as in the prior year. The three lower curves, which display the red band, also have distinctive features, but they are less obvious on this graph.

The process by which each global pixel is given a label is referred to as *classification*. In short, a computer program is presented with many examples of each type of land cover. It then "learns" the examples and uses them to classify pixels, depending on their spectral and temporal pattern. The MODIS global land cover map shown was prepared with more than 1,500 examples of the 17 land cover types. It is estimated to be about 75–80 percent accurate.

Land cover mapping is a common application of remote sensing. Given the ability of space-borne instruments to image the Earth consistently and repeatedly, classification of remotely sensed data is a natural way of extending our knowledge from the specific to the general to provide valuable new geographic information.

23.3 Natural vegetation of the world

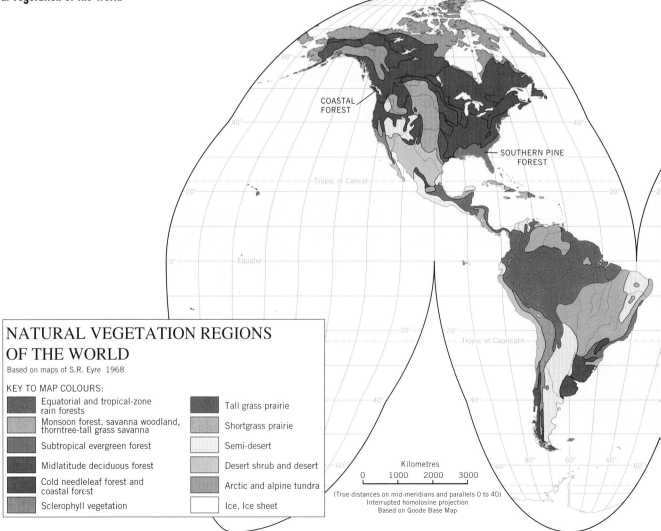

NATURAL VEGETATION REGIONS OF THE WORLD

Based on maps of S.R. Eyre 1968

KEY TO MAP COLOURS:

- Equatorial and tropical-zone rain forests
- Monsoon forest, savanna woodland, thorntree-tall grass savanna
- Subtropical evergreen forest
- Midlatitude deciduous forest
- Cold needleleaf forest and coastal forest
- Sclerophyll vegetation
- Tall grass prairie
- Shortgrass prairie
- Semi-desert
- Desert shrub and desert
- Arctic and alpine tundra
- Ice, Ice sheet

Kilometres

0   1000   2000   3000

(True distances on mid-meridians and parallels 0 to 40)
Interrupted homolosine projection
Based on Goode Base Map

abundance and representative life forms, which, in addition to cacti, include drought-tolerant trees, shrubs, and herbs. The formation classes this chapter describes are major, widespread types that are clearly associated with specific climate types. Figure 23.3 shows their general distribution. With remote sensing, it is possible to map global land cover more accurately. *Focus on Remote Sensing 23.1 • Mapping Global Land Cover by Satellite* describes how data from NASA's MODIS instrument is used for this purpose.

FOREST BIOME

Within the forest biome, there are six major formations: low-latitude rain forest, monsoon forest, subtropical evergreen forest, midlatitude deciduous forest, needleleaf forest, and sclerophyll forest.

Low-Latitude Rain Forest

Low-latitude rain forest, found in the equatorial and tropical latitude zones, consists of tall, closely set trees. The crowns of the trees are generally arranged in two or three layers (Figure 23.4). The highest layer consists of scattered "emergent" crowns that protrude from the closed canopy below, often rising to 40 metres. Below the layer of emergents is a second, continuous layer, which is 15 to 30 metres high. A third, lower layer consists of small, slender trees 5 to 15 metres high, with narrow crowns. Together the trees form a continuous canopy of foliage and provide shade to the understory (Figure 23.5a). Some of the tall emergent species develop wide buttress roots to help support themselves (Figure 23.5b). The trees are characteristically smooth-barked and have no branches on the lower two thirds of their trunks. Tree leaves

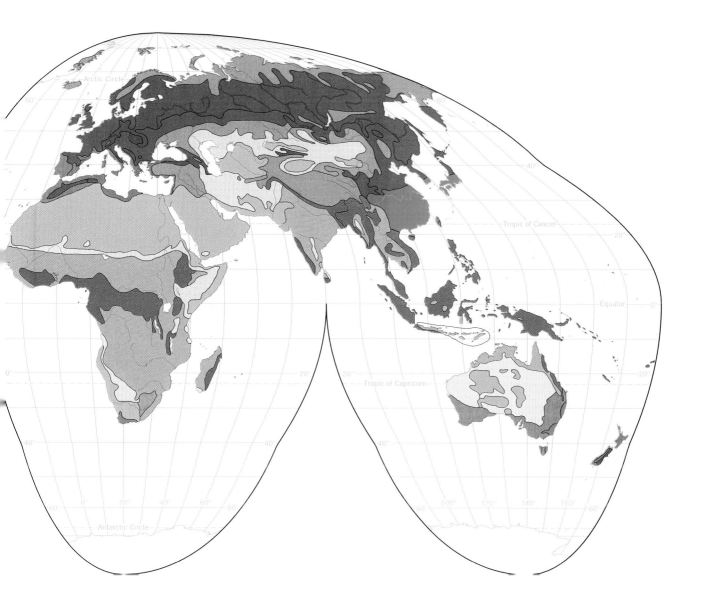

23.4 Rain forest layers This diagram shows the typical structure of equatorial rain forest. (From J. S. Beard, *The Natural Vegetation of Trinidad*, Oxford: Clarendon Press)

are large and evergreen—thus, equatorial rain forest is often described as *broadleaf evergreen forest*.

Typical of the low-latitude rain forest are thick, woody lianas supported by the trunks and branches of trees. Some are slender, like ropes, while others reach thicknesses of 20 centimetres. They climb high into the trees to the upper canopy, where light is available, and develop numerous branches of their own. *Epiphytes* are also common in low-latitude rain forest. These plants attach themselves to the trunks, branches, or foliage of trees and lianas, which they use solely as a means of physical support. Epiphytes include many different types of plants: ferns, orchids, mosses, and lichens (Figure 23.5c).

A particularly important characteristic of the low-latitude rain forest is the large number of species of trees that coexist. In some regions, as many as 3,000 species may be found in a few square kilometres. Individuals of a given species are often widely separated. It is speculated that many species of plants and animals in this diverse ecosystem are still undocumented.

Equatorial and tropical rain forests are not jungles of impenetrable plant thickets. Rather, the floor of the low-latitude rain forest is usually so shaded that plant foliage is sparse close to the ground. This opens up the forest, making it easy to travel within its interior. The ground surface is covered only by a thin litter layer of leaves. Dead plant matter rapidly decomposes because the warm temperatures and abundant moisture promote its breakdown by bacteria and termites. Roots quickly absorb the nutrients released by decay. As a result, the organic matter content of the soil is low.

Large herbivores are uncommon in the low-latitude rain forest. However, those that do live in this environment include the African okapi and the tapir of South America and Asia. Most herbivores living in low-latitude rain forests are climbers and include many primates—monkeys and apes. Tree sloths spend their life

hanging upside down as they browse the forest canopy. Toucans and parrots join fruit-eating bats as flying grazers of the forest. There are few large predators. Notable are the leopards of African and Asian forests and jaguars and ocelots in South America.

Low-latitude rain forest develops in a climate that is continuously warm, frost-free, and has abundant precipitation in all months of the year (or, at most, has only one or two drier months). These conditions occur in the wet equatorial climate and the monsoon and trade-wind coastal climate. In the absence of a cold or dry season, plant growth continues throughout the year. In this uniform environment, some plant species grow new leaves and shed old ones continuously. Still other plant species shed their leaves according to their own physiological schedule, responding to the slight changes in daylight length that occur with the seasons.

Figure 23.6 shows the world distribution of the low-latitude rain forest. *Equatorial rain forest*, which lies astride the equator, occurs extensively in South America where it occupies much of the Amazon lowland. In Africa, equatorial rain forest is found in the Congo lowland and in a coastal zone extending westward from Nigeria to Guinea. In Indonesia, the island of Sumatra marks the boundary of a region of equatorial rain forest that stretches eastward to the islands of the western Pacific.

The low-latitude rain forest extends poleward through the tropical zone (lat. 10° to 25° N and S) along monsoon and trade-wind coasts. This *tropical-zone rain forest* thrives in the monsoon and trade-wind coastal climates that experience a short dry season. However, the dry season is not intense enough to deplete soil water.

In the northern hemisphere, trade-wind coasts that have rain forests are found in the Philippine Islands and along the eastern coasts of Central America and the West Indies. These highlands re-

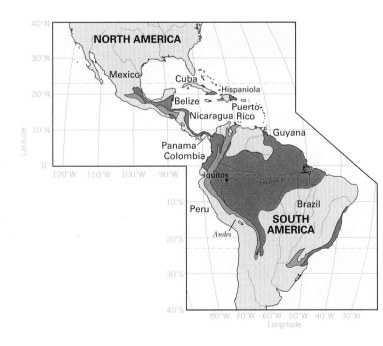

(a)

(c)

(b)

23.5 Tropical rain forest (a) This general view of tropical forest shows the dense canopy and straight, branchless trunks of the tallest trees. (b) Buttress roots provide support at the base of a large tree in Daintree National Park, Queensland, Australia. (c) Red-flowering epiphytes adorn the trunks of sierra palms in the El Yunque rain forest, Caribbean National Forest, Puerto Rico.

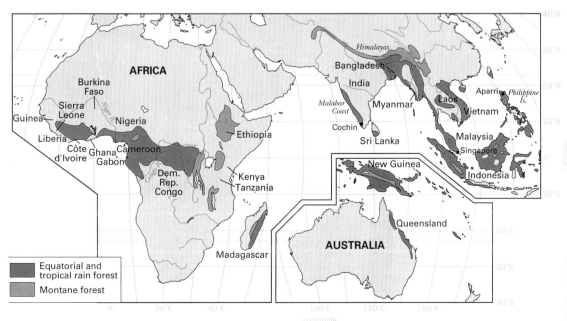

23.6 Low-latitude rain forest
World map of low-latitude rain forest, showing equatorial and tropical rain forest types (Based on maps of S.R. Eyre, 1968)

EYE ON THE ENVIRONMENT | 23.2 — Exploitation of the Low-Latitude Rain Forest Ecosystem

Many of the world's equatorial and tropical regions are home to the rain forest ecosystem. This ecosystem is perhaps the most diverse on the Earth; it possesses more species of plants and animals than any other. Large tracts of rain forest still exist in South America, south Asia, and some parts of Africa. Ecolo-gists regard this ecosystem as a genetic reservoir of many species of plants and animals. But as human populations expand and the quest for agricultural land continues, low-latitude rain forests are being threatened with clearing, logging, cultivation of cash crops, and animal grazing.

In the past, native peoples farmed low-latitude rain forests using the *slash-and-burn* method—cutting down all the vegetation in a small area, then burning it (see photo at left). In a rain forest ecosystem, most of the nutrients are held within living plants rather than in the soil. Burning the vegetation on the site releases the trapped nutrients, returning a portion of them to the soil. Here, the nutrients are available to growing crops. The supply of nutrients derived from the original vegetation cover is small, however, and the harvesting of crops rapidly depletes the nutrients. After a few seasons of cultivation, a new field is cleared, and the old field is abandoned. Rain forest plants re-establish their hold on the abandoned area, and eventually, the rain forest returns to its original state. This cycle shows that primitive, slash-and-burn agriculture can maintain the rain forest ecosystem.

Modern intensive agriculture, on the other hand, uses large areas of land and is not compatible with the rain forest ecosystem. When large areas are abandoned, seed sources are so far away that the original forest species cannot take hold. Instead, secondary species dominate, often accompanied by species from other vegetation types. These species are good invaders, and once they enter an area, they tend to stay. The dominance of these secondary species is permanent, at least on the human time scale. Thus, we can regard the rain forest ecosystem as a resource that, once cleared, may never return. The loss of low-latitude rain forest will result in the disappearance of thousands of species of organisms from the rain forest environment—a loss of millions of years of evolution, together with the destruction of the most complex ecosystem on the Earth.

In Amazonia, the transformation of large areas of rain forest into agricultural land has included the use of heavy machinery to carve out major highways,

Slash-and-burn clearing This rain forest in Maranhão, Brazil has been felled and burned in preparation for cultivation.

---

ceive abundant orographic rainfall in the belt of the trade winds. A good example of tropical-zone rain forest is the eastern mountains of Puerto Rico. In southeast Asia, tropical-zone rain forest occurs in Vietnam and Laos, southeastern China, and on the western coasts of India and Myanmar. In the southern hemisphere, belts of tropical-zone rain forest extend down the coast of Madagascar, and the northeastern coastal region of Australia. The corresponding belt of coastal rain forest in eastern Brazil has largely been cleared for agriculture.

Within the regions of low-latitude rain forest are many highland regions where climate is cooler and the orographic effect increases rainfall. Here, rain forest extends upward on the rising mountain slopes. Between 1,000 and 2,000 metres, the rain forest gradually changes in structure and becomes *montane forest* (Figure 23.6). The canopy of montane forest is more open and tree heights are lower than in the rain forest. Tree ferns and bamboos are numerous, and epiphytes are particularly abundant. As elevation increases, mist and fog become persistent, giving high-elevation montane forest the name *cloud forest*.

Low-latitude rain forest is under increasing human pressure. Slowly but surely, the rain forest is being conquered by logging, clearcutting, and conversion to grazing and farmland. *Eye on the Environment 23.2 • Exploitation of the Low-Latitude Rain Forest Ecosystem* describes this process in more detail.

Monsoon Forest

Monsoon forest of the tropical latitude zone differs from tropical rain forest because it is deciduous. Most of the monsoon forest

such as the Trans-Amazon Highway in Brazil. Once the initial route was established, innumerable secondary roads and trails were developed, as shown in the photo sequence below. Large fields for cattle pasture or commercial crops have been created by cutting, bulldozing, clearing, and burning the vegetation. In some regions, the great broadleaved rain forest trees are being removed for commercial lumber.

A recent computer simulation indicated that if the Amazon rain forest is entirely removed and replaced with pasture, surface and soil temperatures will increase by 1 to 3°C. Precipitation in the region will decline by 26 percent and evaporation by 30 percent. The deforestation will change weather and wind patterns so that less water vapour enters the Amazon Basin from outside sources, making the basin even drier. In areas with a marked dry season, that season will be longer. Although these models contain simplifications and are subject to error, the results confirm the pessimistic conclusion that once large-scale deforestation has occurred, artificial restoration of a rain forest

Landsat photos of Rondonia, Brazil   These photos show the amount of deforestation that has occurred as highways, secondary roads, and trails have been developed through the rain forest over the course of just 17 years.

comparable to the original one may be impossible to achieve.

According to a report by the United Nations Food and Agriculture Organization, about 0.6 percent of the world's rain forest is lost annually by conversion to other uses. More rain forest land, 2.2 million hectares, is lost annually in Asia than in Latin America and the Caribbean, where 1.9 million hectares are converted every year. Africa's loss of rain forest is estimated at about 470,000 hectares per year. Among individual countries, for-

est conversion in Brazil and Indonesia accounts for nearly half of this loss. Deforestation in low-latitude moist deciduous forests, dry deciduous forests, and hill and montane forests in these regions is also very serious.

Although deforestation rates are rapid in some regions, many nations are working to reduce the loss of rain forest environment. However, because the rain forest can provide agricultural land, minerals, and timber, the pressure to allow deforestation continues.

trees shed their leaves during the dry season as a result of moisture stress. In the dry season, the forest resembles the deciduous forests of the midlatitudes in winter.

Monsoon forest is typically open, with patches of shrubs and grasses interspersed among the trees (Figure 23.7). Because of its open nature, light easily penetrates to the understory, which, consequently, is better developed than in the rain forest. Tree heights are also lower. Typically, as many as 30 to 40 tree species are present

23.7 Monsoon woodland This woodland is in the Bandipur National Park wildlife sanctuary in the Nilgiri Hills of southern India. The photo was taken in the rainy season, with trees in full leaf.

23.8 Monsoon forest World map of monsoon forest and related types — savanna woodland and thorntree-tallgrass savanna (Based on maps of S.R. Eyre, 1968)

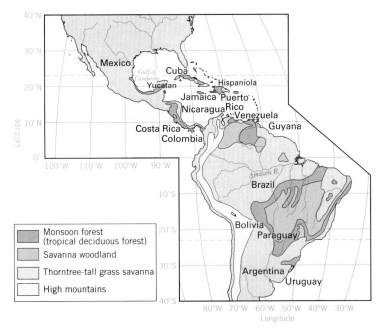

in a small tract of monsoon forest, which is less than half of those present in most rain forests. Tree trunks are massive, often with thick, rough bark. Branching starts at a comparatively low level and produces large, round crowns.

Monsoon forest develops in the wet-dry tropical climate in which a long rainy season alternates with a dry, rather cool season. These conditions are most strongly developed in the Asiatic monsoon climate, but are not limited to that area. The typical regions of monsoon forest are in Myanmar, Thailand, and Cambodia (Figure 23.8). In the monsoon forest of southern Asia, the teakwood tree was once abundant and was widely exported to the Western world to make furniture, panelling, and decks. Now this great tree is logged out, and the Indian elephant, once trained to carry out this logging work, is unemployed. Large areas of monsoon forest also occur in south central Africa and in Central and South America, bordering the equatorial and tropical rain forests.

Subtropical Evergreen Forest

Subtropical evergreen forest is generally found in regions of moist subtropical climate, where winters are mild and there is ample rainfall throughout the year. This forest occurs in two forms: broadleaf and needleleaf. The *subtropical broadleaf evergreen forest* has fewer and shorter species of trees than the low-latitude broadleaf evergreen rain forest. Subtropical broadleaf evergreen leaves tend to be smaller and more leathery, and the leaf canopy less dense. This forest often has a well-developed lower layer of vegetation. Depending on the location, this layer may include tree ferns, small palms, bamboos, shrubs, and herbaceous plants. Lianas and epiphytes are abundant.

In the northern hemisphere, subtropical evergreen forest consists of broad-leaved trees such as evergreen oaks and trees of the laurel and magnolia families. These forests are called "laurel forest." They are associated with the moist subtropical climate in

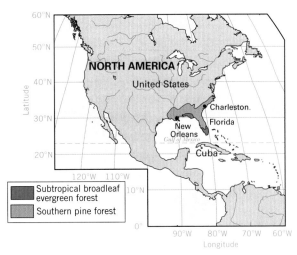

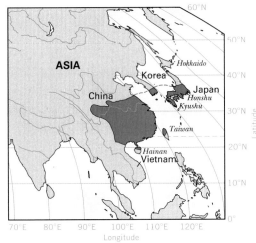

23.9 Subtropical evergreen forest Northern hemisphere map of subtropical evergreen forests, including the southern pine forest (Based on maps of S.R. Eyre, 1968)

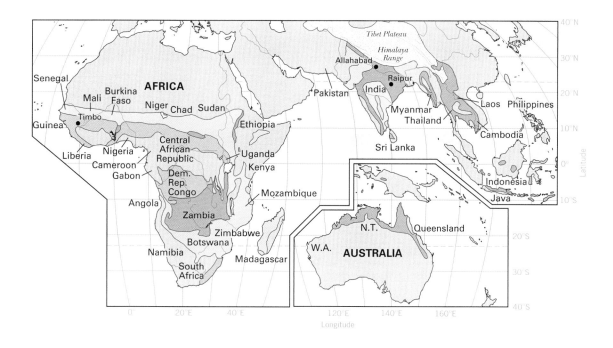

the southeastern United States, southern China, and southern Japan (Figure 23.9). However, these regions are under intense crop cultivation because of their favourable climate. The land has been cleared of natural vegetation for centuries, and little natural laurel forest remains.

The *subtropical needleleaf evergreen forest* occurs only in the southeastern United States (Figure 23.9). Here it is called the southern pine forest, since it is dominated by species of pine. It is found on the wide belt of sandy soils that borders the Atlantic and Gulf coasts. Because the soils are coarse-textured, water drains away quickly, leaving them quite dry. During infrequent drought years, these forests may burn. Since pines are well adapted to droughts and fires, they form a stable vegetation cover over large parts of the region. Timber companies have taken advantage of this natural preference for pines, creating many plantations that yield valuable lumber and pulp (Figure 23.10).

Midlatitude Deciduous Forest

Midlatitude deciduous forest is the native forest type of Western Europe, eastern Asia, and eastern North America. It is dominated by tall, broadleaf trees that provide a continuous and dense canopy in summer but shed their leaves completely in the winter (Figure 23.11). Small trees and shrubs provide a distinctive understory in some areas. The trees resume growth in the spring in response to warmer temperatures and longer days. Many herbaceous species flower early while the forest floor remains bright and sunny. Ferns and other shade-loving plants that thrive beneath the leafy canopy develop in midsummer. The end of the growing season is marked by vivid autumn colours, and soon after the leaves fall to the ground.

23.10 Pine plantation This plantation of longleaf pine grows on the sandy soil of the southeastern coastal plain near Waycross, Georgia.

23.11 Temperate deciduous forest The leafy summer foliage falls to the ground in autumn and the trees go through winter in a leafless state.

Midlatitude deciduous forests are found almost entirely in the northern hemisphere (Figure 23.12). Throughout much of its range, this forest type is associated with the moist continental climate. It receives adequate precipitation in all months, normally with a summer maximum. There is a strong annual temperature cycle with cool winters and warm summers. The length of the growing season is about 120 days in northerly regions, increasing southward to more than 250 days.

In Western Europe, the midlatitude deciduous forest is associated with the marine west-coast climate. Oaks have been planted extensively and are the dominant species in many areas, although birch, beech, and ash are also common. Elm was particularly widespread in southern Britain but has been drastically reduced by Dutch elm disease. Because of intense land-use pressures, the deciduous forests of eastern Asia are mostly found in hilly areas that are not easily cultivated. The forests are composed of oak, elm, maple, lime, and ash. Numerous flowering shrubs are found in undisturbed areas. A small area of deciduous forest is found in the drier parts of Patagonia near the tip of South America. This forest is composed almost entirely of southern beech.

In North America, deciduous forests extend from the Great Lakes region south to the Gulf of Mexico. The greatest diversity is in the southern Appalachians and includes beech, tulip-tree, basswood, sugar maple, buckeye, and oaks that can grow to heights of 35 to 40 metres (Figure 23.13a). The number of tree species decreases westward until, in the driest areas bordering the prairie grasslands, only oaks and hickories are common. Beech and sugar maple are the dominant species in the deciduous forests of south-

ern Ontario and Quebec (Figure 23.13b). Other species include elm, basswood, ash, and walnut. The northern limit of the deciduous forests occurs in a mixed-wood forest region that runs from northern Minnesota to the Maritime provinces (Figure 23.13c). Climate and soils are less favourable for deciduous species in this transitional region and conifers increase in abundance. Here, pure deciduous stands, composed mainly of trembling aspen and birch, are generally limited to sites that have been disturbed by fire.

The deciduous forest includes a variety of animal life, much of it stratified according to canopy layers. There can be as many as five layers: the upper canopy, lower canopy, understory, shrub layer, and ground layer. Because the ground layer presents a more uniform environment in terms of humidity and temperature, it contains the largest concentration of organisms and the greatest diversity of species. Many small mammals burrow in the soil for shelter or food in the form of soil invertebrates. Among this burrowing group are ground squirrels, mice, and shrews, as well as some larger animals—foxes, woodchucks, and rabbits. Most of

23.12 Midlatitude deciduous forest Northern hemisphere map of midlatitude deciduous forests (Based on maps of S.R. Eyre, 1968)

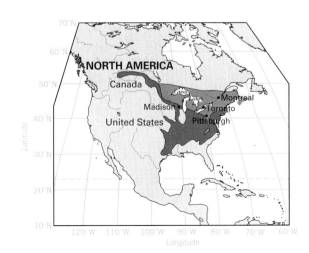

(a)

(b)

(c)

23.13 Deciduous forests (a) The most diverse forest, and often the biggest trees, occur in the southern Appalachian region of the United States. (b) Beech-maple forests occur extensively in Ontario and Quebec. (c) A zone of mixed-wood forest marks the transition between the deciduous and the boreal forest to the north.

the larger mammals feed on ground and shrub layer vegetation. However some, such as the brown bear, are omnivorous and prey upon the small animals, as well.

Large herbivores that graze in the deciduous forest include the red deer of Eurasia and the white-tailed deer of North Amer-

ica. Voles, mice, and squirrels are common species of smaller herbivores. Predators include bears, lynx, and wolves, as well as owls.

Even though birds possess the ability to move through the layers at will, many restrict themselves to one or two layers. For example, the wood peewee is found in the lower canopy, and the

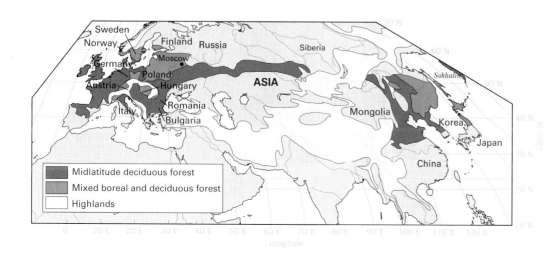

23.14 Needleleaf forest Northern hemisphere map of cold-climate needleleaf forests, including coastal forest (Based on maps of S.R. Eyre, 1968)

(a)

(b)

(c)

red-eyed vireo is found in the understory. Above them, species such as the scarlet tanagers dwell in the upper canopy. Below are the ground dwellers, such as grouse and warblers. Flying insects often show similar patterns of stratification.

Needleleaf Forest

Needleleaf forest refers to a forest composed largely of conifer trees with short trunks, relatively short branches and small, narrow, needlelike leaves. Most are evergreen, retaining their needles for several years before shedding them. When needleleaf forest is dense, it provides the ground with continuous and deep shade. Lower layers of vegetation are sparse or absent, except for a possible thick carpet of mosses. Variations in species are few—in fact, large tracts of needleleaf forest consist almost entirely of only one or two species.

Boreal forest is the cold-climate needleleaf forest of high latitudes. It occurs in two great continental belts one in North America and one in Eurasia (Figure 23.14). These belts correspond closely to the boreal forest climate region. The structure of the boreal forest changes as growing conditions become increasingly severe at higher latitudes. The *forest–tundra transition* extends beyond the Arctic tree line, where it is represented by small patches of conifers that survive in sites protected from the rigorous climate (Figure 23.15a). Further south lies the open *lichen woodland* in which trees are several metres apart and a thick carpet of feathery lichen, commonly known as *reindeer moss*, covers the intervening ground (Figure 23.15b, also see Figure 23.2). The dense stands of closely spaced trees that form a continuous cover in the southernmost section of the boreal forest are called *close forest* (Figure 23.15c).

The Eurasian boreal forests, or *taiga*, extend from Scandinavia across Siberia to the Pacific Ocean. Scots pine is the dominant

23.15 Boreal forest (a) At the northernmost edge of the boreal forest, trees are present only in sheltered patches. (b) Open lichen woodland consists of scattered trees, mostly black spruce, with reindeer moss forming the predominant cover in the adjacent openings. (c) In close forest, little light penetrates the canopy and the understory is predominantly hardy shrubs and mosses.

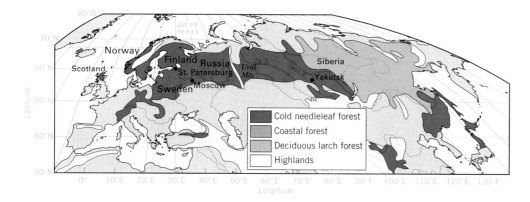

species in the west, but eastward it is replaced successively by Norway spruce, Siberian spruce, and Siberian larch. East of the Ural Mountains, Dahurian larch increases in abundance. Larches shed their needles in winter and thus are deciduous needleleaf trees. Broadleaf deciduous trees, such as aspen, willow, and birch, are usually restricted to burned areas and in time are replaced by conifers.

The boreal forests of North America form a continuous belt across the continent from Alaska to Newfoundland but, like the taiga, they are composed of relatively few species. White spruce and black spruce are widely distributed, but other tree species have more limited ranges. In Alaska and the Yukon, the forests are composed mainly of white spruce, growing in pure stands or mixed with birch and aspen. Black spruce and tamarack are common in wetter settings. In central Canada, drier, sandy soils support stands of jack pine, and balsam fir appears in response to moister climatic conditions. The diversity of the boreal forest increases to the east of the Great Lakes. Although white spruce and balsam fir remain the dominant species over much of eastern

Canada, other conifers, such as red pine, eastern white pine, eastern hemlock, and eastern white cedar, are common. Black spruce regains its importance in the rugged, windswept areas of Newfoundland and Labrador, where the character of the forest resembles the open lichen woodlands of the subarctic.

Needleleaf evergreen forest extends into lower latitudes along mountain ranges and high plateaus. For example, in western North America montane forests are found on the Rocky Mountains and the Sierra Nevada. There are four relatively distinct regions of montane needleleaf forest based on species distributions. White spruce is the dominant species in the north, but the presence of lodgepole pine and subalpine fir distinguishes the montane forest from adjacent boreal forest. Engelmann spruce largely replaces white spruce at about lat. 53°N. Further to the south, Engelmann spruce and alpine fir persist at higher elevations (Figure 23.16a), but below about 2,500 metres, they are replaced by Douglas fir and ponderosa pine. Giant sequoias and ancient bristlecone pines are distinctive montane species in the south (Figure 23.16b, c).

(a)

(b)

(c)

23.16 Montane needleleaf forest (a) Engelmann spruce and alpine fir are common in high-elevation sites in the Canadian Rockies. (b) Giant sequoias occur on the mid-elevation slopes of the Sierra Nevada. (c) Bristlecone pine grow to a considerable height in the harsh climate near the tree line.

Mammals of the boreal needleleaf forest in North America include deer, moose, elk, black bear, marten, mink, wolf, and wolverine. Common birds include jays, ravens, chickadees, nuthatches, and a number of warblers. The caribou, lemming, and snowshoe rabbit inhabit both needleleaf forest and the adjacent tundra biome. The boreal forest often experiences large fluctuations in animal species populations, a result of the low diversity and highly variable environment.

Coastal forest extends in a narrow zone from southern Alaska to northern California (Figure 23.14). Here, in a band of heavy orographic precipitation, mild temperatures, and high humidity, are perhaps the densest of all conifer forests. They contain the world's largest trees, which commonly grow 50 to 80 metres in height, with massive trunks 2 to 3 metres in diameter. Hemlock and Sitka spruce are the dominant species in Alaska and northern British Columbia. The coastal forests of southern British Columbia support hemlock, Douglas fir, and western red cedar (Figure 23.17), with Douglas fir increasing in importance in the forests of Washington and Oregon. The southern limit of the coastal forest is marked by stands of redwoods in California, which depend on coastal fogs to supplement precipitation during the dry summer months.

Sclerophyll Forest

The native vegetation of the Mediterranean climate is adapted to survival through the long summer drought. Shrubs and trees that can survive this drought are characteristically equipped with small,

23.17 Giants of the coastal forest These huge Douglas firs thrive in the cool, moist environment of the coastal forest in southern British Columbia.

hard, or thick leaves that resist water loss through transpiration. These plants are called sclerophylls.

Sclerophyll forest consists of trees with small, hard, leathery leaves. The trees are often low-branched and gnarled, with thick bark. The formation class includes *sclerophyll woodland*, an open forest in which only 25 to 60 percent of the ground is covered by trees. Also included are extensive areas of *scrub*, a plant formation type consisting of shrubs covering somewhat less than half of the

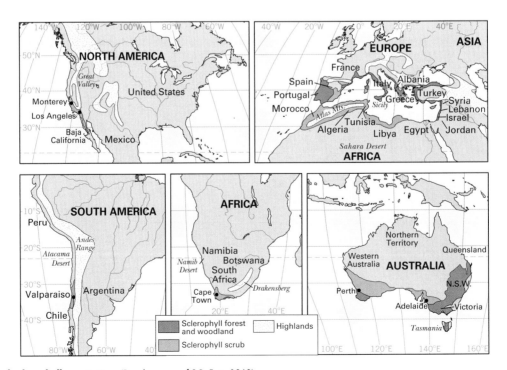

23.18 World map of sclerophyll vegetation (Based on maps of S.R. Eyre, 1968)

ground. The trees and shrubs are evergreen, retaining their thickened leaves despite a severe annual drought.

The map of sclerophyll vegetation (Figure 23.18) includes forest, woodland, and scrub types. Sclerophyll forest is closely associated with the Mediterranean climate and is narrowly limited to west coasts between lat. 30° and 45° N and S. In the Mediterranean lands, the sclerophyll forest forms a narrow, coastal belt around the Mediterranean Sea. Here, the forest consists of cork oak, live oak, Aleppo pine, stone pine, and olive. Over the centuries, human activity has reduced the sclerophyll forest to woodland or destroyed it entirely. Today, large areas of this former forest consist of dense scrub.

The other northern hemisphere region of sclerophyll vegetation is the California coast. Here, the sclerophyll forest or woodland is typically dominated by live oak and white oak (Figure 23.19a). Grassland occupies the open ground between the scattered oaks. Much of the remaining vegetation is sclerophyll scrub known as *chaparral* (Figure 23.19b), which varies in composition with elevation and exposure. Chaparral may contain wild lilac, manzanita, mountain mahogany, poison oak, and evergreen live oak.

In central Chile and the Cape region of South Africa, sclerophyll vegetation has a similar appearance, but the dominant species are quite different (Figure 23.19c). Important areas of sclerophyll forest, woodland, and scrub are also found in southeast, south central, and southwest Australia, including many species of eucalyptus and acacia (Figure 23.19d).

SAVANNA BIOME

The savanna biome is usually associated with the tropical wet-dry climate of Africa and South America. It includes vegetation formation classes ranging from woodland to grassland. In *savanna woodland*, the trees are spaced rather widely apart because there is not enough soil moisture during the dry season to support a full

(a)

(c)

(b)

(d)

23.19 Sclerophyll vegetation (a) Trees in the sclerophyll woodland in California are mainly species of oak and often grow into rather twisted forms. (b) Shrub-dominated communities, such as the chaparral of California and (c) the fynbos of South Africa, occur extensively in all Mediterranean climate regions. (d) Species of Eucalyptus compose forest, woodland, and scrub formations in the Mediterranean regions of Australia.

(a)

(b)

(c)

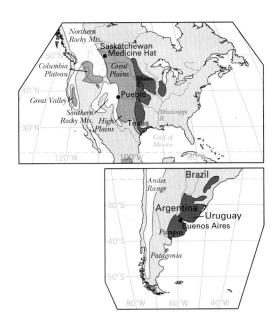

23.20 Tropical savannas (a) The tropical grasslands which form the extensive undulating plains of East Africa are home to large herds of grazing animals. (b) Most of the grasses grow more than one metre in height and provide good cover for the predators that follow the herds. (c) Trees, mainly acacias, with their distinctive spreading canopy are often present and give the ecosystem an open park-like structure.

tree cover. The open spacing allows for the development of a dense lower layer, which usually consists of grasses. The woodland has an open, park-like appearance (Figure 23.20a) and typically lies in a broad belt adjacent to equatorial rain forest.

In the tropical savanna woodland of Africa, the trees are of medium height. Tree crowns are flattened or umbrella-shaped, and

the trunks have thick, rough bark (Figure 23.20b). Some species of trees are xerophytic; their small leaves and thorns are a distinctive characteristic of species that have adapted to drought. Others are broad-leaved deciduous species that shed their leaves in the dry season. In this respect, savanna woodland resembles monsoon forest.

Fire is a frequent occurrence in the savanna woodland during the dry season; however, savanna tree species are particularly resistant to fire. Periodic burning of the savanna grasses helps to protect the grassland from the invasion of forest. Fire does not kill the underground parts of grass plants, but it limits tree growth to fire-resistant species. Animal grazing, which kills many young trees, also maintains grassland.

Figure 23.8 shows the regions of savanna woodland, along with monsoon forest (discussed earlier). In Africa, the savanna woodland grades into a belt of *thorntree-tall grass savanna*, a formation class that provides a transition to the desert biome. The thorntrees are predominantly species of acacia. These leguminous, pod-bearing plants enhance soil fertility through the activity of nitrogen-fixing microbes associated with their roots. In the thorntree-tallgrass savanna, trees are more widely scattered, and the open grassland is more extensive than in the savanna woodland. Elephant grass (*Pennisetum purpureum*) is a common species. It can grow as high as seven metres to form an impenetrable thicket.

The thorntree-tall grass savanna is closely identified with the semi-arid subtype of the dry tropical and subtropical climates. In the semi-arid climate, soil water storage is adequate for the needs of plants only during the brief rainy season. The onset of the rains is quickly followed by the greening of the trees and grasses. For this reason, vegetation of the savanna biome is described as rain-green, a term that also applies to the monsoon forest.

The African savanna is widely known for the diversity of its large grazing mammals, which include numerous species of an-

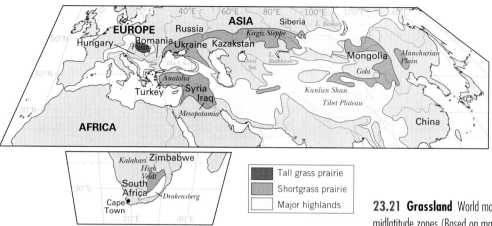

23.21 Grassland World map of the grassland biome in subtropical and midlatitude zones (Based on maps of S.R. Eyre, 1968)

Legend:
- Tall grass prairie
- Shortgrass prairie
- Major highlands

telopes, wildebeest, zebra, and giraffe. Careful studies have shown that each species has a particular preference for different parts of the grasses—blade, sheath, and stem. They also feed on these grasses at different times of the year during their regular seasonal migrations. Grazing stimulates the grasses to continue to grow, and so the ecosystem is more productive when grazed than when left alone. With these grazers comes a large variety of predators—lions, leopards, cheetahs, hyenas, and jackals (Figure 23.20c). Elephants are the largest animals of the savanna and adjacent woodland regions. However, as in the tropical forests, termites and other soil-dwelling organisms are the most abundant animal groups in the savanna ecosystem.

GRASSLAND BIOME

The grassland biome includes two major formation classes—tallgrass prairie and steppe—as well as a broad transition between the two, known as mixed prairie (Figure 23.21). *Tall grass prairie* consists largely of grasses that can exceed two metres in height at the end of the growing season. *Forbs*, which are broad-leaved herbs,

---

GEOGRAPHERS AT WORK Actively Working to Conserve our Grasslands
by Dr. Will Wilson, Lakehead University

Grasslands are one of the most ecologically important vegetation communities in areas with warm climates. In fact, at least one type of grassland, the tallgrass prairie of central and eastern North America, was created, in part, during the last great period of climate warming in the Earth's history, the Hypsithermal. With climate change already occurring and projected to increase in severity, understanding the dynamics of grassland plants, animals, and human cultures will be vital to facilitate their spread into areas now covered by other types of vegetation, such as forests.

Unfortunately, grasslands are also the most ecologically devastated terrestrial biome. In fact, the only place left in the world with intact grassland landscapes is the country of Mongolia. In North America, the shortgrass prairies were destroyed in less than 50 years, ending in the first decade of the 20th century when the last wild grassland grizzly bears and bison were killed, and the last nomads were forcibly settled. The tallgrass prairies of the East were mostly plowed by the 1880s, although many of the tallgrass plants and animals remained until the use of chemicals in agriculture became common after the 1940s.

Dr. Will Wilson works with local communities and conservation organizations to develop and implement conservation plans to protect and reintroduce grasslands. Some of his projects include supervising thesis students as they find and document remnant prairies and prairie plants in Ontario, living with nomads in Central and Inner Asia while learning how to use their traditional ecological knowledge in grassland management, working with farmers in northwestern Ontario to develop new pasture management techniques based on native grassland plants and endangered heritage breeds of livestock, and reading historical accounts of early explorers and plant hunters to reconstruct forgotten grassland ecologies. He has even started his own sheep farm so he can better understand the implications and impacts of his work on grasslands!

Dr. Will Wilson in Grassland National Park, Saskatchewan

(a)

(b)

(c)

23.22 Temperate grasslands (a) The moist tall grass prairie contains a rich diversity of herbs, such as the purple-flowered blazing star (*Liatris pycnostachya*). In drier mixed prairie, herbs such as the yellow prairie cone flower (*Ratibida columnifera*) add occasional colour. (b) Sloughs, small water bodies associated with the rolling topography created by the continental ice sheets, provide excellent waterfowl habitat in the northern mixed prairie. (c) Shorter grasses of the steppe grasslands often grow in association with small shrubs, such as the silver sage (*Artemisia Cana*).

are also present. There are no trees or shrubs in much of the prairie, but they may occur as narrow patches of woodland in stream valleys. The grasses are deeply rooted and form a thick and continuous turf (Figure 23.22).

Prairie grasslands develop best in regions of the midlatitude and subtropical zones with significant winter and summer seasons. The grasses flower in spring and early summer, and the forbs flower in late summer. Tall grass prairies are closely associated with the drier parts of the moist continental climate region, where soil water is in short supply during the summer months.

When European settlers first arrived in North America, the tallgrass prairies extended in a belt from the Texas Gulf coast northward to southern Manitoba (Figure 23.22a). A broad peninsula of tallgrass prairie extended eastward into Illinois, where conditions are somewhat moister. Since this time, these prairies have been converted almost entirely to agricultural land. Another major area of tall grass prairie is the Pampa region of South America, which occupies parts of Uruguay and eastern Argentina. The Pampa region falls into the moist subtropical climate, with mild winters and abundant precipitation.

Between tall grass prairie and steppe is a band of *mixed prairie* (Figure 23.22b). In North America, mixed prairie occupies much of the central Great Plains and stretches from Saskatchewan to Texas. It is the most extensive grassland type in Canada and extends across southeastern Alberta, southern Saskatchewan, and

eastern Manitoba. Mixed prairie is composed of medium-height and shortgrass species, and the relative dominance of each life form reflects variations in growing conditions. On well-drained soils throughout the Canadian Prairies, the dominant species include wheat grasses (*Agropyron spp.*) and spear grasses (*Stipa spp.*). In drier areas, drought-resistant species, such as blue grama (*Bouteloua gracilis*), increase in abundance, while June grass (*Koeleria cristata*) is characteristically associated with the heavy clays of glacial lake beds. Herbaceous plants, such as prairie crocuses and lilies, add diversity to the grasslands, but shrubs and trees are generally restricted to moister valleys and depressions. Groves of aspen increase toward the north and ultimately form a zone of woodland adjacent to the boreal mixed-wood ecosystem. In addition, the northern glacial plains are marked with water-filled depressions (sloughs) that form valuable waterfowl habitat (Figure 23.22c). Various aquatic plants grow in the sloughs, which are often bordered by small willows and aspen.

Steppe, also called *shortgrass prairie*, is a vegetation type dominated by sparse clumps or bunches of grasses that typically grow to a maximum height of about 30 centimetres (Figure 23.22d). Scattered shrubs and low trees may also be found in a steppe. The plant cover is poor, and a lot of bare soil is exposed. Many species of grasses and forbs grow. A typical grass of the American steppe is buffalo grass (*Buchloe dactyloides*). Other typical plants are the sunflower and loco weed. Steppe transforms

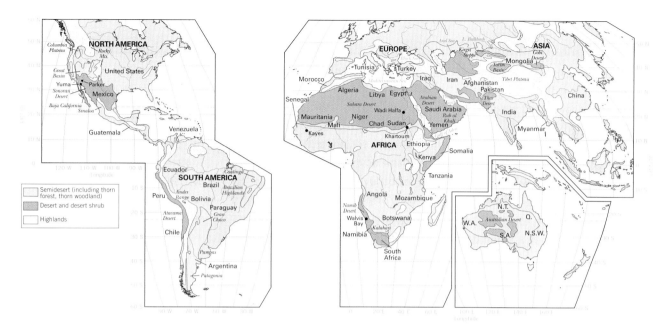

23.23 Desert World map of the desert biome, including desert and semi-desert formation classes (Based on maps of S.R. Eyre, 1968)

into semi-desert in dry environments and into prairie where there is more rainfall.

The map of the grassland biome (Figure 23.21) shows that steppe grassland is concentrated largely in the midlatitude areas of North America and Eurasia. The only southern hemisphere occurrence is the "veldt" region of South Africa.

Steppe grasslands correspond well with the semi-arid subtype of the dry continental climate. Spring rains nourish the grasses, which grow rapidly until early summer. By midsummer, the grasses are usually dormant. Occasional summer rainstorms cause periods of revived growth.

Grassland animals are distinctive. Before the exploitation of the North American grassland for cattle ranching, large grazing mammals were abundant. These included pronghorn antelope, elk, and bison, which ranged widely from tall grass prairie to steppe. Now nearly extinct, bison once numbered 60 million. By 1889, however, this herd had been reduced to 800. Several small herds of bison remain in the Prairies, and their reintroduction into the Grasslands National Park of Saskatchewan in 2006 adds an important element to the regional ecosystem. For much of the prairie, rodents and rabbits join cattle as the major grazers in the grasslands.

The grassland ecosystem supports some rather unique adaptations to life. A common adaptive mechanism is jumping or leaping, assuring an unimpeded view of the surroundings. Jackrabbits and jumping mice are examples. The pronghorn antelope combines the leap with great speed, which allows it to avoid predators and fire. Burrowing is also another common life habit; the soil provides the only shelter in the exposed grasslands. Examples are burrowing rodents, including prairie dogs, gophers, and field mice. Rabbits use old burrows for nesting or shelter. Invertebrates also seek shelter in the soil, and many are adapted to living with the burrows of rodents, where moisture and temperatures are more moderate. The burrowing owl, an endangered species in Canada, has also adapted to the prairie environment in this way. These little birds feed voraciously on mice and insects, and this has been promoted as an asset to landowners, who are being encouraged to maintain the species on their lands.

DESERT BIOME

The desert biome includes several formation classes that are transitional between grassland and savanna biomes and the arid desert. Only two basic formation classes—semi-desert and dry desert—are described here (Figure 23.23).

Semi-desert is a transitional formation class that is distributed over a wide latitude range, from the tropical zone to midlatitudes. Semi-desert consists of sparse xerophytic shrubs. One example is the sagebrush vegetation of the middle and southern Rocky Mountain region and Colorado Plateau (Figure 23.24a). Recently, as a result of overgrazing and trampling by livestock, semi-desert shrub vegetation has expanded widely into areas of the western United States that were formerly steppe grasslands.

Thorntree semi-desert of the tropical zone consists of xerophytic trees and shrubs that have adapted to a climate with a long, hot dry season and a brief, but intense, rainy season. These conditions are found in the semi-arid and arid subtypes of the dry tropical and dry subtropical climates. The thorny trees and shrubs are mainly deciduous plants that shed their leaves in the dry season. The shrubs may grow closely together to form dense thickets. Cacti are present in some locations.

Dry desert is a formation class of xerophytic plants that disperse sparsely over the ground. The permanent vegetation of dry desert consists of small, hard-leaved, or spiny shrubs, succulent plants (such

(a)　　　　　　　　　　　　　　　　　(b)　　　　　　　　　　　　　　(c)

23.24 Desert vegetation (a) Sagebrush (*Artemisia tridentata*) dominates the landscape in many semi-desert regions in North America. (b) The giant saguaro cactus (*Cereus giganteus*) is a distinctive species in parts of the American southwest. (c) Welwitschia (*W. mirabilis*) is an ancient and rare plant of Namibia.

as cacti), or hardy grasses. In addition, many species of small annual plants may appear sporadically, after a rare, but heavy, downpour. In many parts of arid deserts, there is no plant cover at all because the surfaces consist of shifting dune sands or sterile salt flats.

Desert plants differ greatly in appearance from one part of the world to another. In the Mojave and Sonoran deserts of the southwestern United States, some species are large and prominent in the landscape, such as the tree-like saguaro cactus (Figure 23.24b). In Namibia, the ancient and unique Welwitschia (Figure 23.24c) and the small flowering stones (*Lithops spp.*) are distinctive plants.

Desert animals, like the plants, are typically adapted to the dry conditions of the desert. Important herbivores in American deserts include kangaroo rats, jackrabbits, and grasshopper mice. Insects are abundant, as are insect-eating bats and birds, such as the cactus wren. Reptiles, especially lizards and snakes, are also common. The best adapted mammal is the camel, whose legendary survival skill is largely due to the way it can dissipate body heat by maintaining good blood circulation even when dehydrated.

TUNDRA BIOME

In the *Arctic tundra* plants grow during the long days of the brief summer and then endure the long (or continuous) darkness of winter. In the summer months, air temperatures rise above freezing, and a shallow layer of ground ice thaws at the surface. The permafrost beneath, however, remains frozen, restricting meltwater drainage. This creates a saturated environment over wide areas for at least part of the growing season. Because plant remains decay slowly in the cold meltwater, layers of organic matter can build up in the moist ground. Frost in the soil fractures and breaks roots, thus restricting the size of tundra plants. In winter, wind-driven snow and extreme cold also injure plant parts that project above the snow.

Arctic tundra grows over a wide range of latitudes. Its extreme northern limit is at 85.5° N. The decrease in summer warmth northward is marked by progressive changes in the vegetation. In North America, the most southerly division is known as the *low arctic*, which is characterized by a continuous plant cover in all but the driest or most exposed sites. Northward, this division merges with the polar semi-desert of the *middle arctic*. Beyond this is the *high arctic*—a polar "desert" where vascular plants are restricted to a few favourable sites.

The rolling terrain of the low arctic tundra supports a thin cover of short grasses and sedges (Figure 23.25a). Small shrubs, such as crowberry and Labrador tea, may grow to heights of 15 to 30 centimetres, and mosses and lichens form a ground cover. Taller dwarf birches and willows form scrub communities in sheltered sites. In drier upland sites, mosses and lichens are more conspicuous, and patches of hardy, drought-resistant evergreen shrubs, such as arctic heather, are present (Figure 23.25b). The distribution of the shrubs is dependent on the depth and duration of the winter snowpack. In the most exposed sites, the shrubs are replaced by low-growing cushion plants, such as moss campion and purple saxifrage. Cotton grass is common in moist locations.

Similar plants grow in the middle arctic, but their distribution and productivity are more limited and the vegetation cover is much more open. In the high arctic, the vegetation cover is sparse (Figure 23.25c). Small mats of Arctic willow take advantage of the warmer temperatures near the ground surface, and patches of heath occasionally grow on warmer slopes. Lichens are common in exposed areas (Figure 23.25d).

Tundra also occurs in the southern hemisphere. In the south polar region, tundra is limited to ice-free areas in the Antarctic peninsula and rocky coastlines. Only two species of flowering plants are native to Antarctica, but more species occur in the subantarctic islands, where the characteristic cover is tussock grasses.

Alpine tundra is found at high elevations, above the limit of tree growth and below the vegetation-free zone of bare rock and perpetual snow (Figure 23.26a). In North America, alpine tundra occurs discontinuously from Alaska to Mexico and less extensively in the mountains of eastern Canada. The elevation of the tree line is related to temperature during the growing season. For example, trees grow at elevations 500 to 600 metres higher in the Rockies than in the Pacific Coast Ranges because summer temperatures in

(a)

(b)

(c)

(d)

23.25 Arctic tundra (a) Short grasses and sedges dominate much of the tundra landscape. The tall grasses seen here have taken advantage of higher soil nitrogen levels around an animal burrow. (b) The dwarf shrub arctic heather (*Cassiope tetragona*) survives only where it is protected by snow during the severe winter. (c) The high arctic landscape affords little opportunity for plant growth, although purple saxifrage (*Saxifraga oppositifolia*) is well adapted to these conditions. (d) In extreme locations, only lichens survive.

(a)

(b)

(c)

23.26 Alpine tundra (a) Above the tree line, mostly sedges, grasses, and perennial herbs replace woody plants. (b) At higher elevations in the Rockies, the closed coniferous forests are replaced by tree islands mainly comprising Engelmann spruce and subalpine fir. (c) Wind-blown ice crystals can abrade the branches on the upwind side of exposed tree trunks, resulting in the flagged krummholz form seen in this subalpine fir near Banff, Alberta.

the inland continental climate are warmer than on the coast. As it approaches the tree line, the forest cover begins to open and the trees become arranged in clumps known as tree islands (Figure 23.26b). Still higher, near the limit of tree growth, harsh winter conditions cause the trees to develop misshapen flagged krummholz forms, mostly because blowing ice crystals abrade the branches on the upwind side of the trunk (Figure 23.26c).

Many alpine tundra species also grow in the Arctic, but some species are endemic to specific mountain areas. This is especially the case in Europe and Asia, where mountain ranges, such as the Alps, Pyrenees, and Carpathians, are separated geographically. Vegetation patterns in alpine tundra are closely related to the complex mountain topography and develop in response to snow depth, timing of snow melt, exposure, and drainage. At high elevations, solar radiation can be intense, and many of the plants must cope with periods of summer drought in combination with drying winds. The plants are mostly perennial herbs and dwarf shrubs, with grasses, sedges, and mosses found in wetter locations. Vegetation becomes increasingly important as growing conditions deteriorate; flowers are conspicuous in the alpine meadows. Some species are pollinated by the wind, but most are pollinated by insects that spend their days in the alpine meadows and then descend into the valleys at night. These excursions are facilitated by diurnal air movements in the mountains (see Chapter 7).

The alpine flora of tropical mountains is very distinctive. The tree line occurs at 3,500 to 4,000 metres, with the upper limit of the alpine zone at about 4,800 metres. The tropical alpine zone experiences marked diurnal temperature fluctuations, with strong daytime heating and rapid heat loss at night. Tropical alpine vegetation consists mainly of tussock grasses and small-leaved shrubs, but the most distinctive species are the giant rosette plants that grow on the high volcanoes in East Africa and the Andes (Figure 23.27).

23.27 Giant rosette plants of tropical alpine tundra The giant groundsel (*Senecio keniodendron*), shown her near Mount Kenya, can reach a height of 5 metres with a one-metre long spike of yellow flowers.

As is often the case in particularly dynamic environments, species diversity in the tundra is low, but the number of individuals is high. Among the animals, vast herds of caribou (in North America) and reindeer (in Eurasia) roam the tundra, lightly grazing on the lichens and plants and constantly moving. A smaller number of musk-oxen are also primary consumers of the tundra vegetation. Wolves and wolverines, as well as arctic foxes and polar bears, are predators, although some of these animals may feed directly on plants as well. Among the smaller mammals, snowshoe rabbits and lemmings are important herbivores. Invertebrates are scarce in the tundra, except for a small number of insect species. Black flies, deerflies, mosquitoes, and "no-see-ums" (tiny biting midges) are all abundant and can make July on the tundra most uncomfortable. Reptiles and amphibians are rare. The boggy tundra, however, presents an ideal summer environment for many migratory birds such as waterfowl, sandpipers, and plovers.

The food web of the tundra ecosystem is simple and direct. The important producer is reindeer moss, the lichen *Cladonia rangifera*. In addition to caribou and reindeer, lemmings, ptarmigan (arctic grouse), and snowshoe rabbits are lichen grazers. During the summer, the abundant insects provide food for the migratory waterfowl populations.

CLIMATIC GRADIENTS AND VEGETATION TYPES

Climate is an important control of vegetation, especially on the global scale represented by the major formation classes. As climate changes with latitude or longitude, vegetation will also change (corresponding changes also occur in the soils). Figure 23.28 shows three transects across portions of continents that illustrate this principle. Note, for these transects, the effects of highland regions on climate and vegetation is not considered.

The upper transect stretches from the equator to the Tropic of Cancer in Africa. Across this region, climate ranges through all four low-latitude climates: wet equatorial, monsoon and tradewind coastal, wet-dry tropical, and dry tropical. Vegetation ranges from equatorial rain forest, savanna woodland, and savanna grassland to tropical scrub and tropical desert.

The middle transect is a composite from the Tropic of Cancer to the Arctic Circle in Africa and Eurasia. Climates include many of the mid- and high-latitude types: dry subtropical, Mediterranean, moist continental, boreal forest, and tundra. The vegetation cover ranges from tropical desert through subtropical steppe to sclerophyll forest in the Mediterranean. Further north is the midlatitude deciduous forest in the moist continental climate region, which transforms into boreal needleleaf forest, subarctic woodland, and finally tundra.

The lower transect ranges across the United States, from Nevada to Ohio. On this transect, the climate begins as dry midlatitude but changes to moist continental as precipitation gradu-

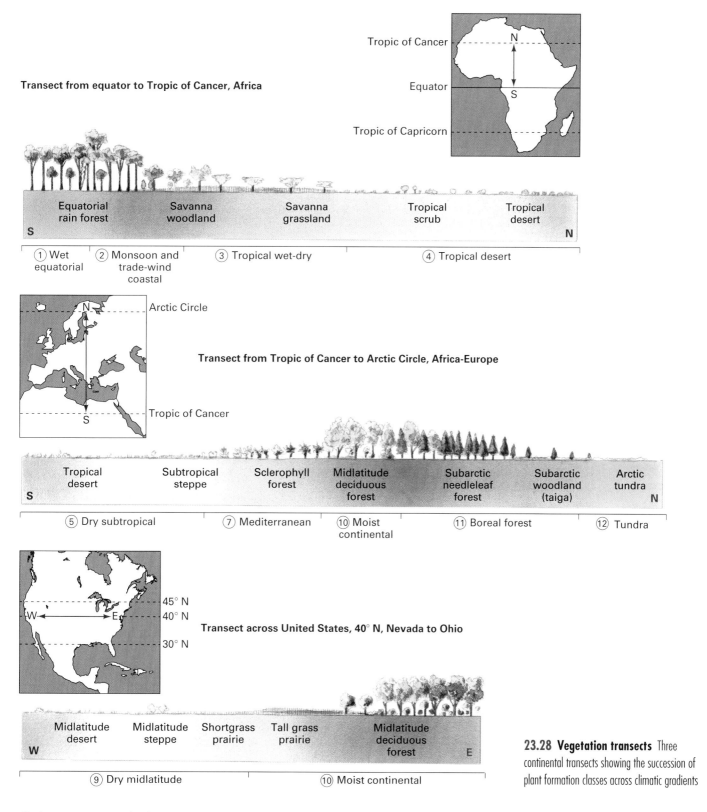

Transect from equator to Tropic of Cancer, Africa

Tropic of Cancer

Equator

Tropic of Capricorn

| Equatorial rain forest | Savanna woodland | Savanna grassland | Tropical scrub | Tropical desert |

S   N

① Wet equatorial ② Monsoon and trade-wind coastal ③ Tropical wet-dry ④ Tropical desert

Arctic Circle

Transect from Tropic of Cancer to Arctic Circle, Africa-Europe

Tropic of Cancer

| Tropical desert | Subtropical steppe | Sclerophyll forest | Midlatitude deciduous forest | Subarctic needleleaf forest | Subarctic woodland (taiga) | Arctic tundra |

S   N

⑤ Dry subtropical ⑦ Mediterranean ⑩ Moist continental ⑪ Boreal forest ⑫ Tundra

45° N
40° N
30° N

W   E

Transect across United States, 40° N, Nevada to Ohio

| Midlatitude desert | Midlatitude steppe | Shortgrass prairie | Tall grass prairie | Midlatitude deciduous forest |

W   E

⑨ Dry midlatitude ⑩ Moist continental

23.28 Vegetation transects Three continental transects showing the succession of plant formation classes across climatic gradients

ally increases eastward. The vegetation changes from midlatitude desert and steppe to shortgrass prairie, tall grass prairie, and midlatitude deciduous forest.

The changes on these transects are largely gradational rather than abrupt. Yet, the global maps of both vegetation and climate show distinct boundaries from one region to the next. Which is correct? The changes on the transects are more accu-

rate. Maps must necessarily have boundaries to communicate information. But climate and vegetation know no specific boundaries. Instead, they are classified into specific types for convenience in studying their spatial patterns. When studying any map of natural features, keep in mind that boundaries are always approximate and gradational. For vegetation, these transitional zones are called ecotones.

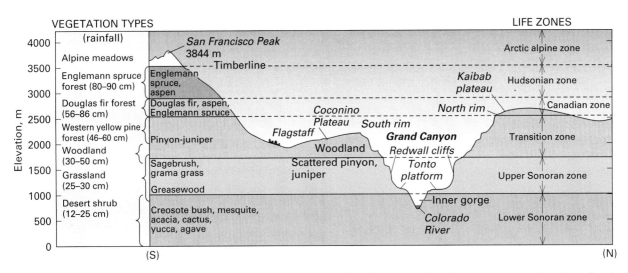

23.29 Altitude zone of vegetation in the arid southwestern United States The profile shows the Grand Canyon–San Francisco Peak district of northern Arizona. Life zones are distinguished from the vegetation types because they include representative animals, in much the same way as biomes include all life forms. (Based on data of G. A. Pearson, C. H. Merriam, and A. N. Strahler)

ALTITUDE ZONES OF VEGETATION

Generally, temperatures decrease and precipitation increases with elevation, which is reflected in a systematic change in the vegetation cover. The changes that occur with altitude are similar to those related to latitude, and many of the vegetation forms seen in the major biomes recur in the bands of vegetation associated with mountain areas. The concept of altitudinal zonation originated from studies carried out in Arizona, where, over a short distance, elevation ranges from about 700 metres at the bottom of the Grand Canyon to over 3,800 metres at the summit of the San Francisco Peaks, about 100 kilometres away (Figure 23.29). Annual precipitation ranges from 12 to 25 centimetres in the desert scrub vegetation type, to 80 to 90 centimetres in the Engelmann spruce forest that grows up to the tree line. Because corresponding trends were noted between plants and animals, the concept of **life zones** was proposed. The life zones that occur along geographically restricted elevational gradients are therefore analogous to the biomes that occur much more extensively in similar environments.

There are six different zones in this altitudinal transect, ranging from the Lower Sonoran Life Zone to the Arctic-Alpine Life Zone. In the lowest elevation sites of the Lower Sonoran Life Zone, the dominant plants are drought-resistant shrubs, such as the creosote bush, and succulent plants typical of the Sonoran desert. At the highest elevations on the mountain peaks, two main habitat types are found: rocky outcrops and boulder fields where lichens predominate; and areas of alpine meadow with herbs, grasses, sedges, rushes, and mosses. Of the approximately 50 species of plants found in the Arctic-Alpine Life Zone, about half occur as disjunct populations; that is, populations that are disconnected from their main distribution area in high-latitude tundra.

Between these extreme habitats are the Upper Sonoran, Transition, Canadian, and Hudsonian life zones. With the exception of the Transition Life Zone, they take their names from regions with characteristic vegetation cover in the major biomes. For example, the Hudsonian Life Zone, at 2,900 to 3,500 metres, is dominated by needleleaf forest that resembles (in appearance, but not in species composition) the boreal forest bordering Hudson Bay.

The Upper Sonoran Life Zone supports a diversity of plant communities, including desert scrub, mainly composed of sagebrush; woodlands of evergreen oaks, juniper, and pinyon pine; areas of chaparral, composed of scrub oaks, manzanita, and mountain mahogany; and grasslands. Open stands of ponderosa pine forest are representative of the Transition Life Zone. In the Canadian Life Zone, the composition of the forests change to Douglas fir and white fir, interspersed with deciduous trees, such as Gambel oak and trembling aspen. Rising above this to the tree line is the Hudsonian Life Zone, where common species include Engelmann spruce, Alpine fir, and ancient bristlecone pines.

A concept similar to altitude zones has been proposed for tropical regions, based on mean annual biotemperature. Biotemperature refers to all temperatures above freezing, with all temperatures below freezing adjusted to 0° C; it is based on the length of the growing season, as well as temperature. Subdivisions of the major vegetation zones are calculated from total annual precipitation and the ratio of mean annual potential evapotranspiration to mean total annual precipitation. The derived value is used to define humidity provinces (Figure 23.30). The secondary axes include altitudinal and latitudinal descriptors. The scheme was intended to be used globally, but has seen only limited application outside of Central America. It helps to

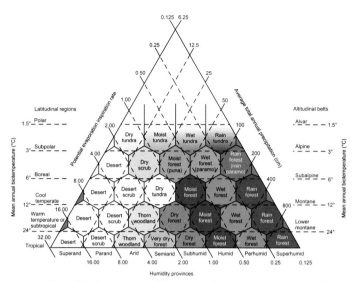

23.30 The Holdridge system of life zones Thirty community types are defined according to growing season temperatures and available moisture. In this system, a potential evapotranspiration (PE) ratio of 1.00 indicates that potential moisture demand equals precipitation; the climate becomes more humid as PE ratio drops below 1.00 and more arid as PE ratio increases above 1.00.

emphasize the relationship between vegetation and climate; however, such an approach is a gross oversimplification of the complex environment of the plant world.

The diverse world of plants and animals is an expression of the factors and forces that operate on all scales in the natural world. Thus, it provides a fitting end to our study of physical geography.

A Look Back

This chapter concludes our overview of physical geography as the science of global environments. The overview began with an examination of weather and climate systems, powered largely by solar energy. This was followed by a description of the systems of the solid Earth, and the dramatic forces associated with it. The primary land forms created by tectonic forces are sculpted more finely into a diversity of landscapes by geomorphic agents that derive their energy from gravity and solar power. Lastly, we covered systems of soils and ecosystems, grouped together by their dependence on climate and substrate, and powered largely by solar energy.

As a discipline, physical geography stresses the interrelationships among its many component sciences and focuses on the environment in an integrated way. Originally, physical geography emphasized natural processes, but today few of these are completely immune from the impacts of human activity. As the human population continues to grow, its influence on the Earth will become more and more profound. This book has provided the tools to understand and assess the essential features of human impact on natural systems and to evaluate the implications of environmental change. Hopefully what you have learned will serve you well as you face the challenges of world citizenship in the twenty-first century.

CHAPTER SUMMARY

- Natural vegetation is a plant cover that develops with little or no human interference. Although much vegetation appears to be in a natural state, humans influence the vegetation cover through fire suppression and the introduction of new species.
- The life form of a plant refers to its physical structure, size, and shape. Life forms include *trees, shrubs, lianas, herbs,* and *lichens.*
- The largest unit of terrestrial ecosystems is the biome: forest, grassland, savanna, desert, and tundra. The forest biome includes a number of important forest formation classes. The *low-latitude rain forest* exhibits a dense canopy and open floor with a large number of species. *Subtropical evergreen forest* occurs in *broadleaf* and *needleleaf* forms in the moist subtropical climate. Monsoon forest is largely deciduous, with most species shedding their leaves after the wet season.
- *Midlatitude deciduous forest* is associated with the moist continental climate. Its species shed their leaves before the cold season. *Needleleaf forest* consists largely of evergreen conifers. It includes the *coastal forest* of the Pacific Northwest, the *boreal forest* of high latitudes, and needleleaf mountain forests. *Sclero*

phyll forest is composed of trees with small, hard, leathery leaves and is found in the Mediterranean climate region.
- The savanna biome consists of widely spaced trees with an understory, often of grasses. Dry-season fire is frequent in the savanna biome, limiting the number of trees and encouraging the growth of grasses.
- The grassland biome of midlatitude regions includes *tall grass prairie,* in moister environments, and *shortgrass prairie,* or steppe, in semi-arid areas.
- Vegetation of the desert biome ranges from thorny shrubs and small trees to dry desert vegetation composed of drought-adapted species.
- Tundra biome vegetation is limited largely to low herbs that are adapted to the severe drying cold experienced on the fringes of the Arctic Ocean.
- Since climate changes with altitude, vegetation typically occurs in altitudinal life zones. Climate also changes gradually with latitude, and so biome changes are typically gradual, without abrupt boundaries.

KEY TERMS

biome	grassland biome	natural vegetation	terrestrial ecosystem
desert biome	life form	savanna biome	tundra biome
forest biome	life zones	steppe	woodland

REVIEW QUESTIONS

1. What is natural vegetation? How do humans influence vegetation?

2. Plant geographers describe vegetation by its overall structure and by the life forms of individual plants. Define and differentiate the following terms: forest, woodland, tree, shrub, herb, liana, perennial, deciduous, evergreen, broadleaf, and needleleaf.

3. What are the five main biome types that ecologists and biogeographers recognize? Describe each briefly.

4. Low-latitude rain forests occupy a large region of the Earth's land surface. What are the characteristics of these forests? Include forest structure, types of plants, diversity, and climate.

5. Monsoon forest and midlatitude deciduous forest are both deciduous but for different reasons. Compare the characteristics of these two formation classes and their climates.

6. Subtropical broadleaf evergreen forest and tall grass prairie are two vegetation formation classes that have been greatly altered by human activities. How was this done and why?

7. Distinguish among the types of needleleaf forest. What characteristics do they share? How are they different? How do their climates compare?

8. Which type of forest, with related woodland and scrub types, is associated with the Mediterranean climate? What are the features of these vegetation types? How are they adapted to the Mediterranean climate?

9. How do traditional agricultural practices in the low-latitude rain forest compare to present-day practices? What are the implications for the rain forest environment?

10. What are the effects of large-scale clearcutting on the rain forest environment?

11. Describe the formation classes of the savanna biome. Where is this biome found and in what climate types? What role does fire play in the savanna biome?

12. Compare the two formation classes of the grassland biome. How do their climates differ?

13. What is meant by the term "mixed prairie"? Where is it found? What are its characteristics?

14. Describe the vegetation types of the desert biome.

15. What are the features of arctic and alpine tundra? How does the cold tundra climate influence the vegetation cover? How does alpine tundra differ from arctic tundra?

16. How does elevation influence vegetation? Provide an example of how vegetation zones are related to elevation.

VISUALIZATION EXERCISE

1. Forests often contain plants of many different life forms. Sketch a cross-section of a forest, including typical life forms, and label them.

ESSAY QUESTION

1. Figure 23.28 shows some vegetation transects along with associated climates. Consulting this figure and Figure 8.8 (world climate map), construct and describe a similar transect starting at Calgary, then passing through Winnipeg, Sudbury, Montreal, and Halifax. Discuss how the vegetation pattern is related to climate.

EYE ON THE LANDSCAPE |

Chapter Opener   Near the headwaters of the Thelon River, Northwest Territories. Black spruce (*Picea mariana*) is a the principal tree species at the Arctic tree-line where it typically grows in small groups (**A**) where conditions are more favorable. Dwarf birch (*Betula nana*) (**B**) is also common and usually acquires a shrubby form in this harsh environment; here it displays its fall foliage. The soils which blanket the ancient bedrock of the Canadian Shield are thin and poorly developed. Drainage is often limited because of low relief (**C**), permafrost or bedrock and small lakes are common in low-lying areas. Arctic tundra is dominated by grasses, sedges and moss (**D**) depending on soil thickness and drainage.

The fragile forest. As well as natural hazards such as fire; freezing rain and windstorms, the forests are subject to a variety of human impacts ranging from extensive logging to recreation. In this stand of conifers even a lightly used trail have a long-lasting effect through compaction and removal of topsoil and plant litter (**A**) and damage to root systems through breakage and exposure (**B**). The closed canopy of a conifer stand filters out much of the incoming

radiation (**C**), and the typical ground cover is of shade-tolerant species such as ferns (**D**). Many conifers adapt to the shade by shedding their lower branches (**E**).

Appendix

1 | CLIMATE DEFINITIONS AND BOUNDARIES

The following table summarizes the definitions and boundaries of climates and climate subtypes based on the soil-water balance, as described in Chapter 15 and shown on the world climate map, Figure 8.8. All definitions and boundaries are provisional.

Ep Water need (potential evapotranspiration)
D Soil water shortage (deficit)
R Water surplus (runoff)
S Storage (limited to 30 cm)

Group I: Low-Latitude Climates

1. **Wet equatorial climate**
   Ep ≥ 10 cm in every month, and
   S ≥ 20 cm in 10 or more months.

2. **Monsoon and trade-wind coastal climate**
   Ep ≥ 4 cm in every month, or
   Ep > 130 cm annual total, or both, and
   S ≥ 20 cm in 6, 7, 8, or 9 consecutive months, or, if
   S > 20 cm in 10 or more months, then Ep ≤ 10 cm in 5 or more consecutive months.

3. **Wet-dry tropical climate**
   D ≥ 20cm, and
   R ≥ 10 cm, and
   Ep ≥ 130 cm annual total, or Ep ≥ 4 cm in every month, or both, and
   S ≥ 20 cm in 5 months or fewer, or minimum monthly S < 3 cm.

4. **Dry tropical climate**
   D ≥15 cm, and
   R = 0, and
   Ep ≥ 130 cm annual total, or Ep ≥ 4 cm in every month, or both.

Subtypes of dry climates
   s Semiarid subtype (Steppe subtype) At least 1 month with S > 2 cm.
   a Desert subtype
      No month with S > 2 cm.

Group II: Midlatitude Climates

5. **Dry subtropical climate**
   D ≥ 15 cm, and
   R = 0, and Ep < 130 cm annual total, and
   Ep ≥ 0.8 cm in every month, and
   Ep < 4 cm in 1 month.

6. **Moist subtropical climate**
   D < 15 cm when R = 0, and
   Ep < 4 cm in at least 1 month, and
   Ep ≥ 0.8 cm in every month.

7. **Mediterranean climate**
   D ≥ 15 cm, and
   R ≥ 0, and
   Ep ≥ 0.8 cm in every month, and storage index > 75%, or P/Ea × 100 < 40%.

8. **Marine west-coast climate**
   D < 15 cm, and
   Ep < 80 cm annual total, and
   Ep ≥ 0.8 cm in every month.

9. **Dry midlatitude climate**
   D ≥ 15 cm, and
   R = 0, and
   Ep ≤ 0.7 cm in at least 1 month, and
   Ep > 52.5 cm annual total.

10. **Moist continental climate**

D < 15 cm when R = 0, and

Ep ≤ 0.7 cm in at least 1 month, and

Ep > 52.5 cm annual total.

Group III: High-Latitude Climates

11. **Boreal forest climate**

52.5 cm > Ep > 35 cm annual total, and

Ep = 0 in fewer than 8 consecutive months.

12. **Tundra climate**

Ep < 35 cm annual total, and

Ep = 0 in 8 or more consecutive months.

13. **Ice-sheet climate**

Ep = 0 in all months.

Appendix

2 | CONVERSION FACTORS

Metric to English

Metric Measure	Multiply by*	English Measure
LENGTH		
Millimetres (mm)	0.0394	Inches (in.)
Centimetres (cm)	0.394	Inches (in.)
Metres (m)	3.28	Feet (ft)
Kilometres (km)	0.621	Miles (mi)
AREA		
Square centimetres (cm^2)	0.155	Square inches (in^2)
Square metres (m^2)	10.8	Square feet (ft^2)
Square metres (m^2)	1.12	Square yards (yd^2)
Square kilometres (km^2)	0.386	Square miles (mi^2)
Hectares (ha)	2.47	Acres
VOLUME		
Cubic centimetres (cm^3)	0.0610	Cubic inches (in^3)
Cubic metres (m^3)	35.3	Cubic feet (ft^3)
Cubic metres (m^3)	1.31	Cubic yards (yd^3)
Millilitres (ml)	0.0338	Fluid ounces (fl oz)
Litres (l)	1.06	Quarts (qt)
Litres (l)	0.264	Gallons (gal)
MASS		
Grams (g)	0.0353	Ounces (oz)
Kilograms (kg)	2.20	Pounds (lb)
Kilograms (kg)	0.00110	Tons (2000 lb)
Tonnes (t)	1.10	Tons (2000 lb)

English to Metric

English Measure	Multiply by*	Metric Measure
LENGTH		
Inches (in.)	2.54	Centimetres (cm)
Feet (ft)	0.305	Metres (m)
Yards (yd)	0.914	Metres (m)
Miles (mi)	1.61	Kilometres (km)
AREA		
Square inches (in^2)	6.45	Square centimetres (cm^2)
Square feet (ft^2)	0.0929	Square metres (m^2)
Square yards (yd^2)	0.836	Square metres (m^2)
Square miles (mi^2)	2.59	Square kilometres (km^2)
Acres	0.405	Hectares (ha)
VOLUME		
Cubic inches (in^3)	16.4	Cubic centimetres (cm^3)
Cubic feet (ft^3)	0.0283	Cubic metres (m^3)
Cubic yards (yd^3)	0.765	Cubic metres (m^3)
Fluid ounces (fl oz)	29.6	Millilitres (ml)
Pints (pt)	0.473	Litres (l)
Quarts (qt)	0.946	Litres (l)
Gallons (gal)	3.79	Litres (l)
MASS		
Ounces (oz)	28.4	Grams (g)
Pounds (lb)	0.454	Kilograms (kg)
Tons (2000 lb)	907	Kilograms (kg)
Tons (2000 lb)	0.907	Tonnes (t)

* Conversion factors shown to 3 decimal-digit precision.

ANSWERS TO PROBLEMS

Working It Out 2.1 ● Distances from Latitude and Longitude

1. If the two cities are separated by 44° latitude and each latitude is equal to 111 km, then:

$$\text{Distance} = 44° \text{ lat} \times \frac{111 \text{ km}}{1° \text{ lat}} = 4884 \text{ km}$$

2. The two cities are on the 52° parallel. Therefore the number of kilometres equal to one degree longitude is:

$$\cos(52)° \times 111 \text{ km} = 0.62 \times 111 \text{ km} = 68.3 \text{ km}$$

Therefore the distance between the two cities is:

$$\text{Distance} = (106° - 1°) \times \frac{68.3 \text{ km}}{1° \text{ long}} = 7176 \text{ km}$$

3. In order to find the area of the map, we must convert 1° latitude and 1° longitude to kilometres. At the equator, both 1° latitude and 1° longitude are equal to 111 kilometres. The size of the area at the equator is therefore 111 km × 111 km = 12321 km^2. For the area near Churchill, 1° latitude is still 111 kilometres, however, the number of kilometres equal to 1° longitude must be calculated:

$$\cos(60°) \times 111 \text{ km} = 0.5 \times 111 \text{ km} = 55.5 \text{ km}$$

4. The area of this area near Churchill is then:

$$111 \text{ km} \times 55.5 \text{ km} = 6161 \text{ km}^2$$

which is exactly half of the area at the equator.

5. Toronto is one and a half hours behind St. Johns. Therefore, the flight left at 11:30 am Newfoundland Standard Time. With a flight time of 2 hours, the flight will arrive at 1:30 pm, local time.

6. Beijing is 15 hours ahead of Vancouver. Therefore, when the flight leaves Vancouver the time in Beijing is 5:20 am Sunday, June 15th. The flight duration is 12 hours and 20 minutes. Therefore, the flight arrives in Beijing at 5:40 pm Sunday, June 15th.

Working It Out 3.2 ● Radiation Laws

1. To calculate the flow rate of energy coming from the Sun, we must use the Stefan-Blotzmann Law, which states that the flow rate is a function of the surface temperature of a blackbody. The Sun's surface temperature is 5950 K.

$$M = \sigma T^4 = (5.67 \times 10^{-8} \text{ W/m}^2\text{K}^4) \times (5950 \text{ K})^4$$

$$M = \sigma T^4 = (5.67 \times 10^{-8} \text{ W/m}^2\text{K}^4)1.25 \times 10^{15} \text{ K}^4$$

$$M = 7.11 \times 10^7 \text{ W/m}^2$$

The wavelength of greatest radiance is given by Wien's Law:

$$\lambda_{max} = b/T = \frac{(2892 \, \mu\text{mK})}{(5950 \text{ K})} = 0.49 \, \mu\text{m}$$

2. Again the Stefan-Boltzmann law must be used here, using the surface temperature of the Earth in kelvin, not Celsius: 15.4 °C + 273.15 = 288.55 K.

$$M = \sigma T^4 = (5.67 \times 10^{-8} \text{ W/m}^2\text{K}^4) \times (288.55)^4$$

$$M = \sigma T^4 = (5.67 \times 10^{-8} \text{ W/m}^2\text{K}^4)6.9 \times 10^9 \text{ K}^4$$

$$M = 392 \text{ W/m}^2$$

The Wavelength of maximum irradiance for the Earth's Surface:

$$\lambda_{max} = b/T = \frac{(2892 \, \mu\text{mK})}{(288.15 \text{ K})} = 10 \, \mu\text{m}$$

3. The Ratio of the flow rate of energy emitted by the Sun's surface to that of Earth's:

$$\frac{(M_{Sun})}{(M_{Earth})} = \frac{(7.11 \times 10^7 \text{ W/m}^2)}{(392 \text{ W/m}^2)} = 1.8 \times 10^5$$

4. Since the radiation flow to Venus is 1.92 times greater than the radiation flow to the Earth, the solar constant for Venus will be 1.92 × 1.37 kW/m^2 = 2.63 kW/m^2. If the radius of Venus is 6050 km, the area the planet presents to the Sun will be that of a disk with area $\pi r^2 = 3.14 \times (6050 \text{ km})^2 = 3.14 \times 3.65 \times 10^7$ km$^2 = 1.15 \times 10^8$ km$^2 = 1.15 \times 10^{14}$ m^2. The total energy flow intercepted is equal to Venus's solar constant times the area it presents to the Sun, or = 2.63 kW/m$^2 \times 1.15 \times 10^{14}$ m$^2 = 3.03 \times 10^{14}$ kW. For outflows, 65 percent of the incoming solar radiation is directly reflected back, providing a shortwave radiation flow rate to space of 0.65 × 3.03 × 10^{14} kW = 1.97 × 10^{14} kW. The remaining portion, 35 percent, or 0.35 × 3.03 × 10^{14} kW = 1.06 × 10^{14} kW, flows outward to space as longwave radiation emitted by the planet and its atmosphere.

Working It Out 4.2 ● Exponential Growth

1. For the 2 percent rate, the multiplier will be

$$M = e^{(R \times T)} = e^{(0.02 \times 50)} = 2.718^{1.00} = 2.72,$$

so $2.72 \times 360 = 979$ ppm. For the 3 percent rate,

$$M = e^{(0.03 \times 50)} = 2.718^{1.50} = 4.48,$$

and $4.48 \times 360 = 1613$ ppm.

2. Doubling time for Singapore is $70 \div 1.3 = 53.8$ yrs, and for Republic of Congo is $70 \div 3.0 = 23.3$ yrs. For Singapore,

$$M = e^{(0.013 \times 25)} = 2.718^{0.325} = 1.38,$$

and so the population will be $2.8 \times 1.38 = 3.88$ million. For Congo,

$$M = e^{(0.03 \times 25)} = 2.718^{0.75} = 2.12,$$

and the population will be $2.12 \times 2.4 = 5.08$ million.

Working It Out 5.1 ● Pressure and Density in the Oceans and Atmosphere

1. Since P (in t/m^2) $= D$ (in m) for the ocean, the pressure at the bottom of the diving pool will be $5\ t/m^2$. Adding atmospheric pressure ($1\ t/m^2$) gives $6\ t/m^2$. The fraction due to water is then 5/6, while the fraction due to the atmosphere is 1/6. For the deep-sea diver, the pressure will be $100 + 1 = 101$, with fractions 100/101 for ocean water and 1/101 for the atmosphere.

2. For Yellowhead Pass, $Z = 1{,}110$ m $= 1.11$ km. The pressure at the pass is the calculated as:

$$P_Z = 101.4\ \text{kPa} \times [1 - (0.0226 \times 1.11\ \text{km})]^{5.26}$$

$$P_Z = 101.4\ \text{kPa} \times [1 - 0.0251]^{5.26}$$

$$P_Z = 101.4\ \text{kPa} \times 0.975^{5.26}$$

$$P_Z = 101.4\ \text{kPa} \times 0.875$$

$$P_Z = 88.73\ \text{kPa}$$

For the summit of Mt. Robson, $Z = 3{,}954$ m $= 3.954$ km. The pressure for the summit is then:

$$P_Z = 101.4\ \text{kPa} \times [1 - (0.0226 \times 3.954\ \text{km})]^{5.26}$$

$$P_Z = 62.0\ \text{kPa}$$

3. The equation for atmospheric pressure must be re-arranged in order to find the altitude at which a pressure of 97.9 kPa would be found under normal conditions. Before the passage of the hurricane, the normal conditions exhibited a sea level pressure of 101.6 kPa.

$$P_Z = 101.6\ \text{kPa} \times [1 - (0.0226 \times Z)]^{5.26}$$

$$\frac{P_Z}{101.6\ \text{kPa}} = [1 - (0.0226 \times Z)]^{5.26}$$

$$\left[\frac{P_Z}{101.6\ \text{kPa}}\right]^{1/5.26} = [1 - (0.0226 \times Z)]$$

$$0.0226 \times Z = 1 - \left[\frac{P_Z}{101.6\ \text{kPa}}\right]^{1/5.26}$$

$$Z = \frac{1 - \left[\dfrac{P_Z}{101.6\ \text{kPa}}\right]^{1/5.26}}{0.0226}$$

Now plug in pressure value of 97.9 kPa for P_Z:

$$Z = \frac{1 - \left[\dfrac{97.9\ \text{kPa}}{101.6\ \text{kPa}}\right]^{1/5.26}}{0.0226} = \frac{1 - 0.993}{0.0226} = 0.311\ \text{km} = 311\ \text{m}$$

Alternative Calculation:

$$P_Z = 101.6\ \text{kPa} \times [1 - (0.0226 \times Z)]^{5.26}$$

$$97.9\ \text{kPa} = 101.6\ \text{kPa} \times [1 - (0.0226 \times Z)]^{5.26}$$

$$0.964 = [1 - (0.0226 \times Z)]^{5.26}$$

$$(0.964)^{1/5.26} = [1 - (0.0226 \times Z)]$$

$$0.993 = 1 - (0.0226 \times Z)$$

$$-0.007 = -0.0226 \times Z$$

$$0.311\ \text{km} = Z$$

Therefore, under normal conditions before the passage of the hurricane, a pressure 97.9 kPa would be found at an altitude of 311 metres.

Working It Out 6.1 ● Energy and Latent Heat

1. Following the example, we can easily find that $4.19 \times (100 - 15) + 2260 = 4.19 \times 85 + 2260 = 2616$ kJ/kg are required. Thus, we have

$$490\ \text{km}^3 \times \left[\frac{10^3\ \text{m}}{1\ \text{km}}\right]^3 \times \frac{10^3\ \text{kg}}{1\ \text{m}^3} \times \frac{2616\ \text{kJ}}{1\ \text{kg}} = 1.28 \times 10^{18}\ \text{kJ}.$$

Working It Out 6.3 ● The Lifting Condensation Level

1.

$$H = 1000 \times \frac{25 - 18}{8.2} = 1000 \times \frac{7}{8.2} = 854\ \text{m}$$

$$T = 25 - 854 \times \frac{10}{1000} = 25 - 8.5 = 16.5°\text{C}$$

2. Equation (1) states

$$T_0 - H \times R_{DRY} = T_{DEW} - H \times R_{DEW}$$

Placing terms with T on the left and terms with H on the right,

$$T_0 - T_{DEW} = H \times R_{DRY} - H \times R_{DEW}$$

Factoring,

$$T_0 - T_{DEW} = H (R_{DRY} - R_{DEW})$$

Solving for H, we have

$$H = \frac{T_0 - T_{DEW}}{R_{DRY} - R_{DEW}} = \frac{T_0 - T_{DEW}}{\dfrac{10}{1000} - \dfrac{1.8}{1000}} = 1000 \frac{T_0 - T_{DEW}}{8.2}$$

which is Equation (2).

Working It Out 8.2 ● Averaging in Time Cycles

1. Average precipitation for 2001–2005 in Dryden, Ontario in mm:

Jan.	29.82	May	101.96	Sept.	108.58
Feb.	13.56	June	119.82	Oct.	59.16
March	26.34	July	77.06	Nov.	29.36
April	46.74	Aug.	102.38	Dec.	31.46

2. During the Autumn and Winter months, the values for the 2001–2005 average and the long-term average are similar. However, within the Spring, the 2001–2005 data shows higher precipitation rates. The 2001–2005 data also shows large variability within the summer months, with a significant local minimum in July, while the long-term mean shows less variability within the summer months and a local minimum in August. However, both data sets show a maximum value in June.

Working It Out 9.1 ● Cycles of Rainfall in the Low Latitudes

1. The mean is 164 mm, and the mean deviation is 43 mm.

2. The relative variability is 0.26, which makes San Juan about the same as Padang, but less variable than Abbassia and Bombay.

Working It Out 10.1 • Standard Deviation and Coefficient of Variation

1. For Gander Newfoundland:

Statistic	January	July
Sample Std. Dev. (mm)	42.47	28.12
Mean (mm)	108.9	88.8
Coeff. of variation	0.39	0.32

Precipitation in January has both the largest standard deviation and the largest coefficient of variation.

Working It Out 12.1 • Radioactive Decay

1. The half-life is 1.28 b.y., $k = 0.693/1.28 = 0.542$; so

$$P(t = 1) = e^{-0.542 \times 1} = e^{-0.542} = 0.582 = 58.2\%$$
$$P(t = 3) = e^{-0.542 \times 3} = e^{-1.626} = 0.197 = 19.7\%$$

2. Here, the half-life is 14.1 b.y., $k = 0.693/14.1 = 0.0491$, and

$$P(t = 5) = e^{-0.0491 \times 5} = e^{-0.245} = 0.782 = 78.2\%$$
$$P(t = 10) = e^{-0.0491 \times 10} = e^{-0.491} = 0.612 = 61.2\%$$
$$P(t = 15) = e^{-0.0491 \times 15} = e^{-0.737} = 0.479 = 47.9\%$$

3. For ^{14}C, $k = 0.693/5730 = 1.21 \times 10^{-4}$. Then

$$P(t) = e^{-1.21 \times 10^{-4} t}.$$

In this case, $P(t)$ is known and equal to 0.1, while t is unknown. Solving for t, we start with

$$P(t) = e^{-kt}$$

and taking the log to the base e of both sides, we obtain

$$\ln(P(t)) = -kt$$

Then we can simply solve for t:

$$t = -\frac{\ln(P(t))}{k}$$

Substituting values for $P(t)$ and k in this formula yields

$$t = \frac{\ln(P(t))}{k} = -\frac{\ln(0.1)}{1.21 \times 10^{-4}} = -\frac{-2.30}{1.21 \times 10^{-4}} = 1.90 \times 10^{-4}$$
$$= 19,000 \text{ yrs}$$

4. $t = \frac{1}{k} \ln\left[\frac{D}{M} + 1\right] = \frac{1}{0.155} \ln[0.448 + 1] = \frac{0.370}{0.155} = 2.39$ b.y.

5. For this decay sequence, $H = 7.04 \times 10^8$ yr $= 0.704$ b.y., and $k = 0.693/H = 0.693/0.704 = 0.985$ for time in b.y. Thus,

$$t = \frac{1}{0.985} \ln[9.56 + 1] = \frac{2.36}{0.985} = 2.40 \text{ b.y.}$$

The results of the two analyses are therefore quite consistent.

Working It Out 14.1 • The Power of Gravity

1. To apply the formula, we first need to convert the velocity of the rock mass from km/hr to m/sec:

$$\frac{150 \text{ km}}{\text{hr}} \times \frac{1 \text{ hr}}{60 \text{ min}} \times \frac{1 \text{ min}}{60 \text{ s}} \times \frac{10^3 \text{ m}}{\text{km}} = 41.7 \text{ m/s}$$

Then we have

$$E = \frac{1}{2}mv^2 = \frac{1}{2} \times 7.56 \times 10^{10} \text{ kg} \times \left(\frac{41.7 \text{ m}}{\text{s}}\right)^2$$
$$= 6.57 \times 10^{13} \text{ J}$$

The total energy released is 2.82×10^{14} J, so the ratio of kinetic to total energy is

$$\frac{6.57 \times 10^{13} \text{ J}}{2.82 \times 10^{14} \text{ J}} = 0.223 = 22.3\%$$

2. The free-fall velocity will be

$$v = \sqrt{2gd} = \sqrt{2 \times \frac{9.8 \text{ m}}{\text{s}^2} \times 500 \text{ m}} = 99.0 \text{ m/s}$$

The ratio of the velocity of the Madison Slide to the free-fall velocity is then

$$\frac{41.7 \text{ m/s}}{99.0 \text{ m/s}} = .421 = 42.1\%$$

Thus, the slide moves less than half as fast as a free-falling body. The slide moves more slowly because it encounters friction in moving.

Working It Out 15.2 • Magnitude and Frequency of Flooding

1.

Rank	Discharge (m³/s)	Recurrence Interval (years)
1	856	36
2	733	18
3	698	12
4	637	9
5	570	7.2
6	565	6
7	561	5.14
8	532	4.5
9	530	4
10	501	3.6
11	479	3.27
12	477	3
13	470	2.77
14	447	2.57
15	446	2.4
16	432	2.25
17	431	2.12
18	430	2
19	429	1.89
20	425	1.8
21	380	1.71
22	368	1.64
23	358	1.57
24	355	1.5
25	318	1.44
26	289	1.38
27	265	1.33
28	256	1.29
29	255	1.24
30	251	1.2
31	247	1.16
32	235	1.13
33	229	1.09
34	224	1.06
35	96	1.03

2. In this data set, a flow of 251 m³/s has a recurrence interval of 1.2, while a flow of 247 m³/s has a recurrence interval of 1.16. The probability that these two flows will be equaled or exceeded in a given year is $100/1.2 = 83.3\%$ and $100/1.16 = 86.2\%$, respectively. Therefore, the probability will of the flow exceeding 250 m³/s will be somewhere between 83.3% and 86.2%. A good guess would be 85.4%.

We can find the recurrence interval that is associated with a particular probability by rearranging the formula: probability = 100/I, so that: I = 100/probability. Therefore, the recurrence interval associated with a probability of exeedence of 25% is $100/25 = 4$. Looking at the data, we see that the ninth highest flow rate recorded, 530 m³/s has a recurrence interval of exactly 4. Therefore, a flow of 530 m³/s has a 25% chance of being exceeded in any given year.

Working It Out 16.1 • River Discharge and Suspended Sediment

1. Applying the suspended load-discharge relationship for Oldman River and setting Q to 50 m³/s:

$S = 8.72\,Q^{3.87} = 8.72 \times (50\,\text{m}^3/\text{s})^{3.87} = 8.72 \times \text{times}; 3{,}758{,}000 = 32{,}800{,}000$ t/d

(Should be around 70 t/d)
If Q = 500 m/s then:

$$S = 8.72 \times (500\,\text{m}^3/\text{s})^{3.87} = 2.43 \times 10^{11} \text{ t/d}$$

(Should be around 70,000 t/d)
If we have an original discharge, X, the suspended load, S_1 would be:

$$S_1 = 8.72\,Q^{3.87} = 8.72 \times (X)^{3.87}$$

If the discharge is then doubled to 2X, then we can calculate the effect on the suspended load quantity without needing to know the actual value of X. We need to substitute 2X for Q:

$$S_2 = 8.72 \times (2X)^{3.87}$$

Separating out the factor of 2, we get:

$$S_2 = 8.72 \times (X)^{3.87}\,(2)^{3.87}$$

Simplifying, we get:

$$S_2 = S_1 \times (2)^{3.87} = S_1 \times 14.6$$

Therefore, if the discharge is doubled, then the suspended sediment load increases by a factor 14.6.

2. Let's assume that the discharge is 1 m³/s and doubles to 2 m³/s. The ratio of suspended sediment transported at 2 m³/s to that at 1 m³/s is then

$$\frac{S(2)}{S(1)} = \frac{178(2)^{1.75}}{178(1)^{1.75}} = \frac{2^{1.75}}{1^{1.75}} = \frac{2^{1.75}}{1} = 3.36$$

Thus, we would expect that the sediment load would more than triple with a doubling of discharge.

Working It Out 17.2 • Properties of Stream Networks

1. From the problem statement, we have $R_b = 3.5$ and $k = 6$. Thus, the formula is
$$N_u = R_b^{(k-u)} = 3.5^{(6-u)},$$
so we have

$$N_1 = 3.5^{(6-1)} = 3.5^5 = 525$$
$$N_2 = 3.5^{(6-2)} = 3.5^4 = 150$$
$$N_3 = 3.5^{(6-3)} = 3.5^3 = 42.9$$
$$N_4 = 3.5^{(6-4)} = 3.5^2 = 12.3$$
$$N_5 = 3.5^{(6-5)} = 3.5^1 = 3.5$$
$$N_6 = 3.5^{(6-6)} = 3.5^0 = 1$$

2. Substituting $L_1 = 0.12$ and $R_L = 3.2$ into the formula for expected cumulative mean length from the text gives

$$L_u^* = L_1\,R_L^{(u-1)} = 0.12 \times 3.2^{(u-1)},$$

so we have

$$L_1^* = 0.12 \times 3.2^{(1-1)} = 0.12 \times 3.2^0 = 0.12 \text{ km}$$
$$L_2^* = 0.12 \times 3.2^{(2-1)} = 0.12 \times 3.2^1 = 0.384 \text{ km}$$
$$L_3^* = 0.12 \times 3.2^{(3-1)} = 0.12 \times 3.2^2 = 1.23 \text{ km}$$
$$L_4^* = 0.12 \times 3.2^{(4-1)} = 0.12 \times 3.2^3 = 3.93 \text{ km}$$

Using the formula $L_u = L_u^* - L_{u-1}^*$ to obtain the (noncumulative) mean lengths, we have

$$L_1 = L_1^* = 0.12$$
$$L_2 = L_2^* - L_1^* = 0.384 - 0.12 = 0.264 \text{ km}$$
$$L_3 = L_3^* - L_2^* = 1.23 - 0.384 = 0.846 \text{ km}$$
$$L_4 = L_4^* - L_3^* = 3.93 - 1.23 = 2.70 \text{ km}$$

Working It Out 20.1 • Calculating a Simple Soil Water Budget

1.

Soil-water Budget for Urbana, Illinois

Month	p	Ea	−G	+G	R	Ep	D
Jan	5.7	0.0			+5.7	0.0	0.0
Feb	4.5	0.0			+4.5	0.0	0.0
Mar	8.2	1.4			+6.8	1.4	0.0
Apr	10.0	4.4			+5.6	4.4	0.0
May	9.9	8.8			+1.1	8.8	0.0
Jun	8.4	12.4	−4.0			12.6	0.2
Jul	8.0	13.4	−5.4			14.9	1.5
Aug	9.0	11.6	−2.6			13.0	1.4
Sep	8.3	7.8		+0.5		7.8	0.0
Oct	6.6	4.8		+1.8		4.8	0.0
Nov	5.7	1.4		+4.3		1.4	0.0
Dec	5.5	0.0		+5.5		0.0	0.0
Total	89.8	66.0	−12.0	+12.1	23.7	69.1	3.1

Working It Out 22.1 • Island Biogeography: The Species–Area Curve

1.

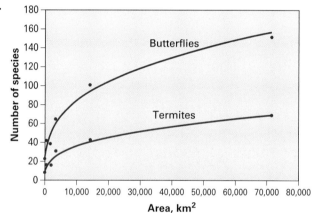

2. I

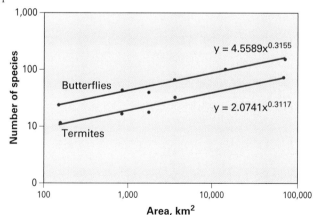

Termites: $S = 2.07\,A^{0.31}$
$= 2.07(1746)^{0.31}$
$= 2.07(10.12)$
$= 2.09$

Butterflies: $S = 4.56\,A^{0.32}$
$= 4.56(1746)^{0.32}$
$= 4.56(10.90)$
$= 49.7$

The two predicted values are somewhat higher than those actually observed. Perhaps Charleston Island has a less diverse environment than the other islands, given their sizes, and so has fewer termite and butterfly species. Or it may be that the island has not been fully explored and all species present might not have been found. Or perhaps a recent catastrophic event has wiped out some species and they have not yet returned by migration.

GLOSSARY

This glossary contains definitions of terms shown in the text in italics or boldface. Terms that are *italicized* within the definitions will be found as individual entries elsewhere in the glossary.

A horizon   *mineral* horizon of the *soil*, overlying the *E* and *B horizons*.

ablation   a wastage of glacial ice by both *melting* and *evaporation*.

abrasion   erosion of *bedrock* of a *stream channel* by impact of particles carried in a stream and by rolling of larger *rock* fragments over the stream bed; abrasion is also an activity of glacial ice, waves, and wind.

abrasion platform   sloping, nearly flat *bedrock* surface extending out from the foot of a *marine cliff* under the shallow water of breaker zone.

absolute instability   a condition that develops when the rising air remains warmer and less dense than the surrounding air.

absolute stability   a condition that develops when the *temperature* of the surrounding air is warmer than the air parcel.

absorption   transfer of *electromagnetic energy* into heat *energy* within a *gas* or *liquid* through which the radiation is passing or at the surface of a *solid* struck by the radiation.

abyssal plain   large expanse of very smooth, flat ocean floor found at depths of 4600 to 5500 m (15,000 to 18,000 ft).

abyssal hills   small hills rising to heights from a few tens of metres to a few hundred metres above the deep ocean floor.

accelerated erosion   *soil erosion* occurring at a rate much faster than *soil horizons* can be formed from the parent *regolith*.

accretion of lithosphere   production of new *oceanic lithosphere* at an active *spreading* plate boundary by the rise and solidification of *magma* of basaltic composition.

accretionary prism   mass of deformed trench *sediments* and ocean floor *sediments* accumulated in wedgelike slices on the underside of the overlying plate above a plate undergoing *subduction*.

acid deposition   the *deposition* of acid raindrops and/or dry acidic dust particles on vegetation and ground surfaces.

acid mine drainage   sulfuric acid effluent from *coal* mines, mine tailings, or spoil ridges made by *strip mining*.

acid action   solution of minerals by acids occurring in soil and ground water.

acid rain   rainwater having an abnormally low *pH*, between 2 and 5, as a result of air pollution by sulfur oxides and nitrogen oxides.

active continental margins   continental margins that coincide with tectonically active plate boundaries. (See also *continental margins, passive continental margins.*)

active layer   shallow surface layer subject to seasonal thawing in *permafrost* regions.

active pool   type of pool in the *biogeochemical cycle* in which the materials are in forms and places easily accessible to life processes. (See also *storage pool.*)

active systems   *remote sensing* systems that emit a beam of wave *energy* at a source and measure the intensity of that *energy* reflected back to the source.

actual evapotranspiration (water use)   Actual rate of *evapotranspiration* at a given time and place.

adiabatic lapse rate   (See *dry adiabatic lapse rate, wet adiabatic lapse rate.*)

adiabatic process   change of temperature within a *gas* because of compression or expansion, without gain or loss of heat from the outside.

advection fog   *fog* produced by *condensation* within a moist basal air layer moving over a cold land or water surface.

aeolian   the processes collectively referring to the *geomorphic* work of wind in terrestrial environments. This includes erosion, transportation, and deposition of Earth materials.

aerosols   tiny particles present in the *atmosphere*, so small and light that the slightest movements of air keep them aloft.

aggradation   raising of *stream channel* altitude by continued *deposition* of *bed load*.

air   a mixture of gases that surrounds the Earth.

air mass   extensive body of air within which upward gradients of temperature and moisture are fairly uniform over a large area.

air pollutant   an unwanted substance injected into the *atmosphere* from the Earth's surface by either natural or human activities; includes *aerosols, gases*, and *particulates*.

air temperature   temperature of air, normally observed by a *thermometer* under standard conditions of shelter and height above the ground.

albedo   percentage of downwelling solar *radiation* reflected upward from a surface.

albic horizon   pale, often sandy *soil horizon* from which *clay* and free iron oxides have been removed. Found in the profile of the *Spodosols*.

Alfisols   *soil order* consisting of *soils* of humid and subhumid climates, with high *base status* and an *argillic horizon.*

allele   specific version of a particular gene.

allelopathy   interaction among *species* in which a plant secretes substances into the soil that are toxic to other organisms.

allopatric speciation   type of *speciation* in which populations are geographically isolated and gene flow between the populations does not take place.

alluvial fan   gently sloping, conical accumulation of coarse *alluvium* deposited by a *braided stream* undergoing *aggradation* below the point of emergence of the channel from a narrow *gorge* or *canyon.*

alluvial meanders   sinuous bends of a *graded stream* flowing in the alluvial deposit of a *floodplain.*

alluvial river   *stream* of low *gradient* flowing upon thick deposits of *alluvium* and experiencing approximately annual overbank flooding of the adjacent *floodplain.*

alluvial terrace   benchlike landform carved in *alluvium* by a *stream* during *degradation.*

alluvium   any stream-laid *sediment* deposit found in a *stream channel* and in low parts of a stream valley subject to flooding.

alpine chains   high mountain ranges that are narrow belts of *tectonic activity* severely deformed by *folding* and thrusting in comparatively recent geologic time.

alpine debris avalanche   *debris flood* of steep mountain *slopes,* often laden with tree trunks, limbs, and large boulders.

alpine glacier   long, narrow, mountain *glacier* on a steep downgrade, occupying the floor of a troughlike valley.

alpine permafrost   *permafrost* occurring at high altitudes equatorward of the normal limit of *permafrost.*

alpine tundra   a plant *formation class* within the *tundra biome,* found at high altitudes above the limit of *tree* growth.

amphibole group   *silicate minerals* rich in calcium, magnesium, and iron, dark in colour, high in *density,* and classed as *mafic minerals.*

amplitude   for a smooth wavelike curve, the difference in height between a crest and the adjacent trough.

andesite   *extrusive igneous rock* of diorite composition, dominated by *plagioclase feldspar;* the extrusive equivalent of *diorite.*

Andisols   a *soil order* that includes *soils* formed on volcanic ash; often enriched by organic matter, yielding a dark soil colour.

anemometer   weather instrument used to indicate *wind* speed.

aneroid barometer   *barometer* using a mechanism consisting of a partially evacuated air chamber and a flexible diaphragm.

angle of repose   natural surface inclination *(dip)* of a *slope* consisting of loose, coarse, well-sorted *rock* or *mineral fragments;* for example, the *slip face* of a *sand dune,* a *talus slope,* or the sides of a *cinder cone.*

annuals   plants that live only a single growing season, passing the unfavorable season as a seed or spore.

annular drainage pattern   a stream network dominated by concentric (ringlike) major *subsequent streams.*

antarctic circle   *parallel of latitude* at 66 1/2°S.

antarctic front zone   frontal zone of interaction between antarctic *air masses* and polar air masses.

antarctic zone   *latitude* zone in the latitude range 60° to 75°S (more or less), centred on the *antarctic circle,* and lying between the *subantarctic zone* and the *polar zone.*

anticlinal valley   valley eroded in weak *strata* along the central line or axis of an eroded *anticline.*

anticline   upfold of *strata* or other layered *rock* in an archlike structure; a class of *folds.* (See also *syncline.*)

anticyclone   centre of high *atmospheric pressure.*

aphelion   point on the Earth's elliptical orbit at which the Earth is farthest from the Sun.

aquatic ecosystem   *ecosystem* of a *lake, bog,* pond, river, *estuary,* or other body of water.

Aquepts   *suborder* of the *soil order Inceptisols;* includes Inceptisols of wet places, seasonally saturated with water.

aquiclude   *rock* mass or layer that impedes or prevents the movement of *groundwater.*

aquifer   *rock* mass or layer that readily transmits and holds *groundwater.*

arc   curved line that forms a portion of a circle.

arc-continent collision   collision of a volcanic arc with *continental lithosphere* along a *subduction* boundary.

arctic circle   *parallel of latitude* at 66 1/2°N.

arctic front zone   frontal zone of interaction between arctic *air masses* and polar air masses.

arctic tundra   a plant *formation class* within the *tundra biome,* consisting of low, mostly herbaceous plants, but with some

very small stunted *trees,* associated with the *tundra climate.*

arctic zone   *latitude* zone in the latitude range 60° to 75°N (more or less), centred about on the *arctic circle,* and lying between the *subarctic zone* and the *polar zone.*

arête   sharp, knifelike divide or crest formed between two *cirques* by alpine glaciation.

argillic horizon   *soil horizon,* usually the B *horizon,* in which *clay minerals* have accumulated by *illuviation.*

arid (dry climate subtype)   subtype of the dry climates that is extremely dry and supports little or no vegetation cover.

Aridisols   *soil order* consisting of soils of dry climates, with or without *argillic horizons,* and with accumulations of *carbonates* or soluble salts.

artesian well   drilled well in which water rises under hydraulic pressure above the level of the surrounding *water table* and may reach the surface.

aspect   compass orientation of a *slope* as an inclined element of the ground surface.

association   plant-animal community type identified by the typical organisms that are likely to be found together.

asthenosphere   soft layer of the upper *mantle,* beneath the rigid *lithosphere.*

astronomical hypothesis   explanation for glaciations and interglaciations making use of cyclic variations in the form of solar *energy* received at the Earth's surface.

atmosphere   envelope of gases surrounding the Earth, held by *gravity.*

atmospheric pressure   pressure exerted by the atmosphere because of the force of *gravity* acting upon the overlying column of air.

atoll   circular or closed-loop *coral reef* enclosing an open *lagoon* with no island inside.

atomic mass number   total number of protons and *neutrons* within the nucleus of an atom.

atomic number   number of protons within the nucleus of an atom; determines element name and chemical properties of the atom.

autogenic succession   form of *ecological succession* that is self-producing—that is, results from the actions of plants and animals themselves.

autumnal equinox   *equinox* occurring on September 22 or 23.

average deviation   difference between a single value and the mean of all values, taken without respect to sign.

axial rift   narrow, trenchlike depression situated along the centre line of the *mid-oceanic ridge* and identified with active seafloor spreading.

axis of rotation centre line around which a body revolves, as the Earth's axis of rotation.

B horizon mineral *soil horizon* located beneath the A *horizon,* and usually characterized by a gain of *mineral matter* (such as *clay minerals* and oxides of aluminum and iron) and organic matter *(humus).*

backswamp area of low, swampy ground on the *floodplain* of an *alluvial river* between the *natural levee* and the *bluffs.*

backwash return flow of *swash* water under influence of gravity.

badlands rugged land surface of steep *slopes,* resembling miniature mountains, developed on weak *clay* formations or clay-rich *regolith* by fluvial erosion too rapid to permit plant growth and soil formation.

bar low ridge of *sand* built above water level across the mouth of a *bay* or in shallow water paralleling the shoreline. May also refer to embankment of sand or gravel on floor of a *stream channel.*

bar (pressure) unit of pressure equal to 10^5 Pa *(pascals);* approximately equal to the pressure of the Earth's *atmosphere* at sea level.

barchan dune *sand dune* of crescentic base outline with a sharp crest and a steep lee *slip face,* with crescent points (horns) pointing downwind.

barometer instrument for measurement of *atmospheric pressure.*

barrier (to dispersal) a zone or region that a *species* is unable to colonize or perhaps even occupy for a short time, thus halting *diffusion.*

barrier island long narrow island, built largely of beach *sand* and dune sand, parallel with the mainland and separated from it by a *lagoon.*

barrier reef *coral reef* separated from mainland *shoreline* by a *lagoon.*

barrier-island coast *coastline* with broad zone of shallow water offshore (a *lagoon*) shut off from the ocean by a *barrier island.*

basalt *extrusive igneous rock* of *gabbro* composition; occurs as *lava.*

base flow that portion of the *discharge* of a *stream* contributed by *ground water* seepage.

base level lower limiting surface or level that can ultimately be attained by a *stream* under conditions of stability of the Earth's crust and sea level; an imaginary surface equivalent to sea level projected inland.

base status of soils quality of a *soil* as measured by the presence or absence of *clay minerals* capable of holding large numbers of *bases.* Soils of high *base status* are rich in base-holding *clay minerals;* soils of low *base status* are deficient in such minerals.

bases certain positively-charged *ions* in the *soil* that are also plant nutrients; the most important are calcium, magnesium, potassium, and sodium.

batholith large, deep-seated body of *intrusive igneous rock,* usually with an area of surface exposure greater than 100 km^2 (40 mi^2).

bauxite mixture of several *clay minerals,* consisting largely of aluminum oxide and water with impurities; a principal ore of aluminum.

bay a body of water sheltered from strong wave action by the configuration of the *coast.*

beach thick, wedge-shaped accumulation of *sand, gravel,* or cobbles in the zone of breaking waves.

beach drift transport of *sand* on a beach parallel with a *shoreline* by a succession of landward and seaward water movements at times when *swash* approaches obliquely.

bearing direction angle between a line of interest and a reference line, which is usually a line pointing north.

bed load that portion of the *stream load* moving close to the stream bed by rolling and sliding.

bedrock solid *rock* in place with respect to the surrounding and underlying *rock* and relatively unchanged by *weathering* processes.

bedrock slump *landslide* of *bedrock* in which most of the *bedrock* remains more or less intact as it moves.

Bergeron process the formation of precipitation in the cold *clouds* of the mid and upper latitudes by ice crystal growth.

bioclimatic frontier geographic boundary corresponding with a critical limiting level of climate stress beyond which a *species* cannot survive.

biodiversity the variety of biological life on Earth or within a region.

biogas mixture of methane and *carbon dioxide* generated by action of anaerobic bacteria in animal and human wastes enclosed in a digesting chamber.

biogeochemical cycle total system of *pathways* by which a particular type of *matter* (a given element, compound, or ion, for example) moves through the Earth's *ecosystem* or *biosphere;* also called a *material cycle* or *nutrient cycle.*

biogeographic realm region in which the same or closely related plants and animals tend to be found together.

biogeography the study of the distributions of organisms at varying spatial and temporal *scales,* as well as the processes that produce these distribution patterns.

biomass dry weight of living organic matter in an *ecosystem* within a designated surface area; units are kilograms of organic matter per square metre.

biome largest recognizable subdivision of *terrestrial ecosystems,* including the total assemblage of plant and animal life interacting within the *life layer.*

biosphere all living organisms of the Earth and the environments with which they interact.

biosphere reserve a *biosphere* designed to conserve examples of characteristic *ecosystems*

bitumen combustible mixture of hydrocarbons that is highly viscous and will flow only when heated; considered a form of petroleum.

bituminous sand (See *bitumen.*)

blackbody ideal object or surface that is a perfect radiator and absorber of *energy;* absorbs all radiation it intercepts and emits radiation perfectly according to physical theory.

block mountains class of mountains produced by block faulting and usually bounded by *normal faults.*

block separation separation of individual joint blocks during the process of *physical weathering.*

blowout shallow depression produced by continued *deflation.*

bluffs steeply rising ground slopes marking the outer limits of a *floodplain.*

bog a shallow depression filled with organic matter, for example a glacial lake or pond basin filled with *peat.*

Boralfs *suborder* of the *soil order Alfisols;* includes Alfisols of *boreal forests* or high mountains.

boreal forest variety of *needleleaf forest* found in the *boreal forest climate* regions of North America and Eurasia.

boreal forest climate cold climate of the *subarctic zone* in the northern *hemisphere* with long, extremely severe winters and several consecutive months of zero *potential evapotranspiration (water need).*

Borolls *suborder* of the *soil order Mollisols;* includes Mollisols of cold-winter semiarid plants *(steppes)* or high mountains.

braided stream *stream* with shallow channel in coarse *alluvium* carrying multiple threads of fast flow that subdivide and rejoin repeatedly and continually shift in position.

breaker sudden collapse of a steepened water wave as it approaches the shoreline.

broadleaf deciduous forest *forest* type consisting of broadleaf *deciduous trees* and found in the *moist subtropical climate* in parts of the *marine west-coast climate.* (See also *mid-latitude deciduous forest.*)

broadleaf evergreen forest *forest* type consisting of broadleaf *evergreen trees* and found in the wet equatorial and tropical climates. (See also *low latitude rainforest.*)

Brunisolic soils   a class of *forest soils* in the Canadian soil classification system with brownish *B horizon*.

budget   in flow systems, an accounting of *energy* and *matter* flows that enter, move within, and leave a system.

bush-fallow farming   agricultural system practised in the African *savanna woodland* in which *trees* are cut and burned to provide cultivation plots.

butte   prominent, steep-sided hill or peak, often representing the final remnant of a resistant layer in a region of flat-lying *strata*.

C horizon   *soil horizon* lying beneath the *B horizon*, consisting of *sediment* or *regolith* that is the *parent material* of the soil.

C₃ pathway   one of the three processes of *photosynthesis*. This process converts *carbon dioxide* and ribulose bisphosphate.

C₄ pathway   one of the three processes of *photosynthesis*. In this process, *carbon dioxide* is converted into oxaloacetic acid, which is a 4-carbon compound.

calcification   accumulation of *calcium carbonate* in a soil, usually occurring in the *B* or *C horizons*.

calcite   mineral having the composition *calcium carbonate*.

calcium carbonate   compound consisting of calcium (Ca) and carbonate (CO₃) *ions*, formula CaCo₃, occurring naturally as the mineral *calcite*.

caldera   large, steep-sided circular depression resulting from the explosion and subsidence of a *stratovolcano*.

CAM pathway   one of the three processes of *photosynthesis*. In this process, like the *C4 pathway*, the plant initially manufactures oxaloacetic acid. What makes this process different is that at night, *carbon dioxide* is taken in, and stored for later processing during daylight hours.

canyon   (See *gorge*.)

capillary action   process by which *capillary tension* draws water into a small opening, such as a *soil* pore or a *rock joint*.

capillary tension   a cohesive force among surface molecules of a *liquid* that gives a droplet its rounded shape.

carbohydrate   class of organic compounds consisting of the elements carbon, hydrogen and oxygen.

carbon cycle   *biogeochemical cycle* in which carbon moves through the *biosphere*; includes both *gaseous cycles* and *sedimentary cycles*.

carbon dioxide   the chemical compound CO₂, formed by the union of two atoms of oxygen and one atom of carbon; normally a gas present in low concentration in the *atmosphere*.

carbon fixation   (See *photosynthesis*.)

carbonates (carbonate minerals, carbonate rocks)   *minerals* that are carbonate compounds of calcium or magnesium or both, i.e., *calcium carbonate* or magnesium carbonate. (See also *calcite*.)

carbonic acid   a weak acid created when CO₂ gas dissolves in water.

carbonic acid action   chemical reaction of *carbonic acid* in rainwater, *soil water*, and *ground water* with *minerals*; most strongly affects carbonate minerals and *rocks*, such as limestone and marble; an activity of *chemical weathering*.

cartography   the science and art of making maps.

Celsius scale   temperature scale in which the *freezing* point of water is 0°, and the boiling point is 100°.

Cenozoic Era   last (youngest) of the *eras* of geologic time.

channel   (See *stream channel*.)

chaparral   sclerophyll scrub and dwarf *forest* plant *formation class* found throughout the coastal mountain ranges and hills of central and southern California.

chemical energy   *energy* stored within an organic molecule and capable of being transformed into *heat* during metabolism.

chemical weathering   chemical change in *rock-forming* minerals through exposure to atmospheric conditions in the presence of water; mainly involving *oxidation*, *hydrolysis*, *carbonic acid action*, or direct solution.

chemically precipitated sediment   *sediment* consisting of *mineral matter* precipitated from a water solution in which the matter has been transported in the dissolved state as *ions*.

chernozem   type of *soil order* closely equivalent to *Mollisol*; an order of the Canadian Soil Classification System.

Chernozemic soils   a class of grassland *soils* in the Canadian soil classification system with a thick *A horizon* rich in organic matter.

chert   *sedimentary rock* composed largely of silicon dioxide and various impurities, in the form of nodules and layers, often occurring with *limestone* layers.

chinook wind   a *local wind* occurring at certain times to the lee of the Rocky Mountains; a very dry wind with a high capacity to evaporate *snow*.

chlorofluorocarbons (CFCs)   synthetic chemical compounds containing chlorine, fluorine, and carbon atoms that are widely used as coolant fluids in refrigeration systems.

cinder cone   conical hill built of coarse *tephra* ejected from a narrow volcanic vent; a type of *volcano*.

circle of illumination   great circle that divides the globe at all times into a sunlit *hemisphere* and a shadowed hemisphere.

circum-Pacific belt   chains of andesite *volcanoes* making up mountain belts and *island arcs* surrounding the Pacific Ocean basin.

cirque   bowl-shaped depression carved in *rock* by glacial processes and holding the *firn* of the upper end of an *alpine glacier*.

clast   rock or mineral fragment broken from a parent *rock* source.

clastic sediment   *sediment* consisting of particles broken away physically from a parent *rock* source.

clay   *sediment* particles smaller than 0.004 mm in diameter.

clay minerals   class of *minerals* produced by alteration of *silicate minerals*, having plastic properties when moist.

claystone   *sedimentary rock* formed by lithification of *clay* and lacking *fissile* structure.

cliff   sheer, near-vertical *rock* wall formed from flat-lying resistant layered *rocks*, usually *sandstone*, *limestone*, or *lava* flows; may refer to any near-vertical rock wall. (See also *marine cliff*.)

climate   generalized statement of the prevailing *weather* conditions at a given place, based on statistics of a long period of record and including mean values, departures from those means, and the probabilities associated with those departures.

climate change   the variation in the Earth's global climate or in regional *climates* over time.

climatic frontier   a geographical boundary that marks the limit of survival of a plant *species* subjected to climatic stress.

climatology   the science that describes and explains the variability in space and time of the heat and moisture states of the Earth's surface, especially its land surfaces.

climax community   stable community of plants and animals reached at the end point of *ecological succession*.

climograph   a graph on which two or more climatic variables, such as monthly mean temperature and monthly mean precipitation, are plotted for each month of the year.

closed flow system   flow system that is completely self-contained within a boundary through which no *matter* or *energy* is exchanged with the external environment. (See also *open flow system*.)

cloud forest   a type of low evergreen rainforest that occurs high on mountain slopes, where *clouds* and *fog* are frequent.

clouds   dense concentrations of suspended water or ice particles in the diameter range 20 to 50 μm. (See *cumuliform clouds, stratiform clouds*.)

coal *rock* consisting of hydrocarbon compounds, formed of compacted, lithified, and altered accumulations of plant remains *(peat)*.

coarse textured (rock) having *mineral* crystals sufficiently large that they are at least visible to the naked eye or with low magnification.

coast (See *coastline.*)

coastal and marine geography the study of the geomorphic processes that shape shores and coastlines and their application to coastal development and marine resource utilization.

coastal blowout dune high *sand dune* of the *parabolic dunes* class formed adjacent to a beach, usually with a deep *deflation* hollow *(blowout)* enclosed within the dune ridge.

coastal foredunes a narrow belt of dunes in the form of irregularly shaped hills and depressions, typically found on the land side of sand beaches. They normally bear a cover of beach grass and a few other species of plants capable of survival in the severe environment.

coastal forest subtype of *needleleaf evergreen forest* found in the humid coastal zone of the northwestern United States and western Canada.

coastal plain coastal belt, emerged from beneath the sea as a former *continental shelf*, underlain by *strata* with gentle *dip* seaward.

coastline (coast) zone in which coastal processes operate or have a strong influence.

coefficient of variation in statistics, the ratio of the standard deviation to the mean.

cold front moving weather *front* along which a cold *air mass* moves underneath a warm air mass, causing the latter to be lifted.

cold-blooded animal animal whose body temperature passively follows the temperature of the environment.

cold-core ring circular eddy of cold water, surrounded by warm water and lying adjacent to a warm, poleward-moving *ocean current*, such as the Gulf Stream. (See also *warm-core ring.*)

collision-wake capture during precipitation, the process by which larger water droplets drag smaller droplets down behind them, which arises because of the range of droplet sizes naturally present in clouds.

colloids particles of extremely small size, capable of remaining indefinitely in suspension in water. May be mineral or organic in nature.

colluvium deposit of *sediment* or *rock* particles accumulating from overland flow at the base of a *slope* and originating from higher slopes where *sheet erosion* is in progress. (See also *alluvium.*)

community an assemblage of organisms that live in a particular *habitat* and interact with one another.

competition form of interaction among plant or animal *species* in which both draw resources from the same *pool*.

component in flow systems, a part of the system, such as a *pathway*, connection, or flow of *matter* or *energy*.

composite volcano volcano composed of layers of ash and lava. See *stratovolcano*.

compression (tectonic) squeezing together, as horizontal compression of crustal layers by *tectonic* processes.

condensation process of change of *matter* in the gaseous state *(water vapour)* to the liquid state (liquid water) or solid state (ice).

condensation nucleus a tiny bit of solid *matter (aerosol)* in the *atmosphere* on which *water vapour* condenses to form a tiny water droplet.

conditional instability arises when the environmental temperature lapse rate (ETLR) is between the dry adiabatic lapse rate (DALR) and the saturated adiabatic lapse rate (SALR) and is associated with moist air.

conduction transmission of *sensible heat* through *matter* by transfer of *energy* from one atom or molecule to the next in the direction of decreasing temperature.

cone of depression conical configuration of the lowered *water table* around a well from which water is being rapidly withdrawn.

conformal projection *map* projection that preserves without shearing the true shape or outline of any small surface feature of the Earth.

conglomerate a *sedimentary rock* composed of pebbles in a matrix of finer *rock* particles.

conic projections a group of *map projections* in which the *geographic grid* is transformed to lie on the surface of a developed cone.

consequent stream *stream* that takes its course down the slope of an *initial landform*, such as a newly emerged *coastal plain* or a *volcano*.

consumers animals in the *food chain* that live on organic matter formed by *primary producers* or by other *consumers*. (See also *primary consumers, secondary consumers.*)

consumption (of a lithospheric plate) destruction or disappearance of a subducting *lithospheric plate* in the *asthenosphere*, in part by *melting* of the upper surface, but largely by softening because of heating to the temperature of the surrounding *mantle rock*.

continental collision event in *plate tectonics* in which subduction brings two segments of the *continental lithosphere* into contact, leading to formation of a *continental suture*.

continental crust crust of the continents, of felsic composition in the upper part; thicker and less dense than *oceanic crust*.

continental drift hypothesis, introduced by Alfred Wegener and others early in the 1900s, of the breakup of a parent continent, *Pangea*, starting near the close of the *Mesozoic Era*, and resulting in the present arrangement of *continental shields* and intervening *ocean-basin floors*.

continental lithosphere *lithosphere* bearing *continental crust* of *felsic igneous rock*.

continental margins (1) Topographic: one of three major divisions of the ocean basins, being the zones directly adjacent to the continent and including the *continental shelf, continental slope*, and *continental rise*. (2) Tectonic: marginal belt of continental crust and lithosphere that is in contact with *oceanic crust* and *lithosphere*, with or without an active plate boundary being present at the contact. (See also *active continental margins, passive continental margins.*)

continental rise gently sloping seafloor lying at the foot of the *continental slope* and leading gradually into the *abyssal plain*.

continental rupture crustal spreading apart affecting the *continental lithosphere*, so as to cause a *rift valley* to appear and to widen, eventually creating a new belt of *oceanic lithosphere*.

continental scale scale of observation at which we recognize continents and other large Earth surface features, such as ocean currents.

continental shelf shallow, gently sloping belt of seafloor adjacent to the continental shoreline and terminating at its outer edge in the *continental slope*.

continental shields ancient crustal *rock* masses of the continents, largely *igneous rock* and *metamorphic rock*, and mostly of *Precambrian age*.

continental slope steeply descending belt of seafloor between the *continental shelf* and the *continental rise*.

continental suture long, narrow zone of crustal deformation, including underthrusting and intense *folding*, produced by a *continental collision*. Examples: Himalayan Range, European Alps.

continuous permafrost *permafrost* that underlies more than 90 percent of the surface area of a region.

convection a process by which a fluid is heated by a warm surface, expands, and rises, creating an upward flow. This flow moves heat away from the surface.

convection (atmospheric) air motion consisting of strong updrafts taking place within a *convection cell*.

convection cell individual column of strong updrafts produced by atmospheric *convection*.

convection loop circuit of moving *fluid,* such as *air* or water, created by unequal heating of the *fluid.*

convectional precipitation a form of *precipitation* induced when warm, moist air is heated at the ground surface, rises, cools, and condenses to form water droplets, raindrops, and eventually, rainfall.

convergent precipitation occurs when moist air flows come together and then rise, and is associated with hurricanes.

converging boundary boundary between two crustal plates along which *subduction* is occurring and *lithosphere* is being consumed.

Coordinated Universal Time the legal time standard recognized by all nations and administered by the Bureau International de l'Heure, located near Paris. It is a high-precision atomic time standard that is periodically adjusted by *leap seconds* to compensate for discrepancies in the Earth's rotation.

coral reef rocklike accumulation of *carbonates* secreted by corals and algae in shallow water along a marine shoreline.

coral-reef coast *coast* built out by accumulations of *limestone* in *coral reefs.*

core spherical central mass of the Earth composed largely of iron and consisting of an outer liquid zone and an interior solid zone.

Coriolis effect effect of the Earth's rotation tending to turn the direction of motion of any object or *fluid* toward the right in the northern *hemisphere* and to the left in the southern hemisphere.

corrosion erosion of *bedrock* of a *stream channel* (or other *rock* surface) by chemical reactions between solutions in stream water and *mineral* surfaces.

cosmopolitan species *species* that are found very widely.

counter-radiation *longwave radiation* of atmosphere directed downward to the Earth's surface.

covered shields areas of *continental shields* in which the ancient *rocks* are covered beneath a thin layer of sedimentary *strata.*

crater central summit depression associated with the principal vent of a *volcano.*

creep very slow movement of soil or rock material over a period of several years.

crescentic dune (See *barchan dunes.*)

crevasse gaping crack in the brittle surface ice of a *glacier.*

crude oil liquid fraction of *petroleum.*

crust of Earth outermost solid shell or layer of the Earth, composed largely of *silicate minerals.*

Cryaquepts great group within the soil *suborder* of *Aquepts;* includes Aquepts of

cold climate regions and particularly the *tundra climate.*

Cryosolic soils a class of *soils* in the Canadian soil classification system associated with strong frost action and underlying *permafrost.*

cryoturbation movement of mineral particles of any size by freezing and thawing of ice.

cuesta *erosional landform* developed on resistant *strata* having low to moderate *dip* and taking the form of an asymmetrical low ridge or hill belt with one side a steep slope and the other a gentle slope; usually associated with a *coastal plain.*

cultural energy *energy* in forms exclusive of solar *energy* of *photosynthesis* that is expended on the production of raw food or feed crops in agricultural *ecosystems.*

cumuliform clouds *clouds* of globular shape, often with extended vertical development.

cumulonimbus cloud large, dense *cumuliform cloud* yielding *precipitation.*

cumulus cloud type consisting of low-lying, white cloud masses of globular shape well separated from one another.

curvature effect limits the growth of small, pure water droplets to conditions of supersaturation. It is related to the size of the droplet and arises because the surface of a small droplet has greater curvature than the surface of a large droplet.

cutoff cutting-through of a narrow neck of land, so as to bypass the stream flow in an *alluvial meander* and cause it to be abandoned.

cycle in flow systems, a closed flow system of *matter.* Example: *biogeochemical cycle.* (See *closed flow system.*)

cycle of rock transformation or rock cycle total cycle of changes in which *rock* of any one of the three major *rock* classes—*igneous rock, sedimentary rock, metamorphic rock*—is transformed into *rock* of one of the other classes.

cyclone centre of low *atmospheric pressure.* (See *tropical cyclone, wave cyclone.*)

cyclonic precipitation a form of *precipitation* that occurs as warm moist air is lifted by air motion occurring in a *cyclone.*

cyclonic storm intense weather disturbance within a moving *cyclone* generating strong winds, cloudiness, and *precipitation.*

cylindric projections group of *map projections* in which the *geographic grid* is transformed to lie on the surface of a developed cylinder.

data acquisition component component of a *geographic information system* in which data are gathered together for input to the system.

data management component component of a *geographic information system* that creates, stores, retrieves, and modifies data layers and *spatial objects*

daughter product new *isotope* created by decay of an *unstable isotope.*

daylight saving time time system under which time is advanced by one hour with respect to the *standard time* of the prevailing *standard meridian.*

debris flood (debris flow) streamlike flow of muddy water heavily charged with *sediment* of a wide range of size grades, including boulders, generated by sporadic torrential rains upon steep mountain watersheds.

decalcification removal of *calcium carbonate* from a *soil horizon* as *carbonic acid* reacts with *carbonate mineral matter.*

December solstice (See *winter solstice.*)

deciduous plant *tree* or *shrub* that sheds its leaves seasonally.

declination of Sun latitude at which the Sun is directly overhead; varies from $-23\frac{1}{2}°$ ($23\frac{1}{2}°$ S lat.) to $+23\frac{1}{2}°$ N lat.)

décollement detachment and extensive sliding of a *rock* layer, usually *sedimentary,* over a near-horizontal basal *rock* surface; a special form of low-angle thrust *faulting.*

decomposers organisms that feed on dead organisms from all levels of the *food chain;* most are microorganisms and bacteria that feed on decaying organic matter.

deep sea cone a fan-shaped accumulation of undersea *sediment* on the *continental rise* produced by sediment-rich currents flowing down the *continental slope.*

deficit (soil-water shortage) in the soil-water budget, the difference between *water use* and *water need;* the quantity of irrigation water required to achieve maximum growth of agricultural crops.

deflation lifting and transport in *turbulent suspension* by wind of loose particles of *soil* or *regolith* from dry ground surfaces.

deglaciation widespread recession of *ice sheets* during a period of warming global climate, leading to an interglaciation. (See also *glaciation, interglaciation.*)

degradation lowering or downcutting of a *stream channel* by stream *erosion* in *alluvium* or *bedrock.*

degree of arc measurement of the angle associated with an *arc,* in degrees.

delta *sediment* deposit built by a stream entering a body of standing water and formed of the *stream load.*

delta coast *coast* bordered by a *delta.*

delta kame flat-topped hill of *stratified drift* representing a glacial *delta* constructed adjacent to an *ice sheet* in a marginal glacial lake.

dendritic drainage pattern   *drainage pattern* of treelike branched form, in which the smaller streams take a wide variety of directions and show no parallelism or dominant trend.

dendrochronology   a method of scientific dating based on the analysis of tree-ring growth patterns.

denitrification   biochemical process in which nitrogen in forms usable to plants is converted into molecular nitrogen in the gaseous form and returned to the atmosphere—a process that is part of the *nitrogen cycle.*

density of matter   quantity of mass per unit of volume, stated in kg m^{-2}.

denudation   total action of all processes whereby the exposed *rocks* of the continents are worn down and the resulting *sediments* are transported to the sea by the *fluid agents;* includes also *weathering* and *mass wasting.*

deposition (atmosphere)   the change of state of a substance from a *gas (water vapour)* to a *solid* (ice); in the science of *meteorology,* the term sublimation is used to describe both this process and the change of state from solid to vapour. (See *sublimation.*)

deposition (of sediment)   (See *stream deposition.*)

depositional landform   *landform* made by *deposition* of *sediment.*

depression   a term that describes a low-pressure centre, particularly, those that bring cloudy conditions and precipitation in mid-latitudes.

desert biome   *biome* of the dry climates consisting of thinly dispersed plants that may be *shrubs,* grasses, or perennial *herbs,* but lacking in *trees.*

desert pavement   surface layer of closely fitted pebbles or coarse *sand* from which finer particles have been removed.

desertification   (See *land degradation.*)

detritus   decaying organic matter on which *decomposers* feed.

dew-point lapse rate   rate at which the dew point of an air mass decreases with elevation; typical value is 1.8°C/1000 m (1.0°F/1000 ft)

dew-point temperature   temperature of an *air mass* at which the air holds its full capacity of water vapour.

diagnostic horizons   *soil horizons,* rigorously defined, that are used as diagnostic criteria in classifying *soils.*

diffuse radiation   solar radiation that has been *scattered* (deflected or reflected) by minute dust particles or cloud particles in the *atmosphere.*

diffuse reflection   solar *radiation* scattered back to space by the Earth's atmosphere.

diffusion   the slow extension of the range of a *species* by normal processes of dispersal.

digital image   numeric representation of a picture consisting of a collection of numeric brightness values (pixels) arrayed in a fine grid pattern.

dike   thin layer of *intrusive igneous rock,* often near-vertical or with steep *dip,* occupying a widened fracture in the surrounding *rock* and typically cutting across older *rock* planes.

dimictic   in *limnology,* mixing twice each year, as a dimictic *lake.*

diorite   *intrusive igneous rock* consisting dominantly of plagioclase feldspar and pyroxene; a *felsic igneous rock.*

dip   acute angle between an inclined natural *rock* plane or surface and an imaginary horizontal plane of reference; always measured perpendicular to the *strike.* Also a verb, meaning to incline toward.

diploid   having two sets of chromosomes, one from each parent organism.

discharge   volume of flow moving through a given cross section of a stream in a given unit of time; commonly given in cubic metres (feet) per second.

discontinuous permafrost   *permafrost* that underlies from 10 to 90 percent of the surface of a region.

disjunction   geographic distribution pattern of *species* in which one or more closely related species are found in widely separated regions.

dispersal   the capacity of a *species* to move from a location of birth or origin to new sites.

distributary   branching *stream channel* that crosses a *delta* to discharge into open water.

diurnal   adjective meaning "daily."

Dobson Unit   the standard way to express ozone amounts in the atmosphere.

doldrums   belt of calms and variable winds occurring at times along the *equatorial trough.*

dolomite   carbonate mineral or *sedimentary rock* having the composition calcium magnesium carbonate.

dome   (See *sedimentary dome.*)

drainage basin   total land surface occupied by a *drainage system,* bounded by a *drainage divide* or watershed.

drainage divide   imaginary line following a crest of high land such that overland flow on opposite sides of the line enters different *streams.*

drainage pattern   the plan of a network of interconnected *stream channels.*

drainage system   a branched network of *stream channels* and adjacent land *slopes,* bounded by a *drainage divide* and converging to a single channel at the outlet.

drainage winds   *winds,* usually cold, that flow from higher to lower regions under the direct influence of *gravity.*

drawdown (of a well)   difference in height between base of cone of depression and original water table surface.

drought   occurrence of substantially lower-than-average *precipitation* in a season that normally has ample precipitation for the support of food-producing plants.

drumlin   hill of glacial *till,* oval or elliptical in basal outline and with smoothly rounded summit, formed by plastering of till beneath moving, debris-laden glacial ice.

dry adiabatic lapse rate   rate at which rising air is cooled by expansion when no *condensation* is occurring; 10°C per 1000 m (5.5°F per 1000 ft).

dry deposition   the process by which atmospheric gases and particles are transferred to the surface as a result of random turbulent air motions.

dry desert   plant *formation class* in the *desert biome* consisting of widely dispersed xerophytic plants that may be small, hard-leaved or spiny *shrubs,* succulent plants (cacti), or hard grasses.

dry lake   shallow basin covered with salt deposits formed when stream input to the basin is subjected to severe *evaporation;* may also form by evaporation of a saline lake when climate changes; see also *salt flat.*

dry midlatitude climate   dry climate of the *midlatitude zone* with a strong annual cycle of *potential evapotranspiration (water need)* and cold winters.

dry subtropical climate   dry climate of the *subtropical zone,* transitional between the *dry tropical climate* and the *dry midlatitude climate.*

dry tropical climate   climate of the *tropical zone* with large total annual *potential evapotranspiration (water need).*

dune   (See *sand dune.*)

dust bowl   western Great Plains of the United States, which suffered severe wind deflation and soil drifting during the drought years of the middle 1930s.

dust storm   heavy concentration of dust in a turbulent *air mass,* often associated with a *cold front.*

E horizon   soil mineral horizon lying below the *A horizon* and characterized by the loss of *clay minerals* and oxides of iron and aluminum; it may show a concentration of *quartz* grains and is often pale in colour.

Earth's crust   (See *crust of Earth.*)

earth hummock   low mound of vegetation-covered earth found in *permafrost* terrain, formed by cycles of *ground ice* growth and melting (see also *mud hummock*).

earth flow moderately rapid downhill flowage of masses of water-saturated *soil, regolith*, or weak *shale*, typically forming a step-like terrace at the top and a bulging toe at the base.

earth rotation refers to the counter-clockwise turning of the Earth on its axis, an imaginary straight line passing through the centre of the planet and joining the North and South poles.

earthquake a trembling or shaking of the ground produced by the passage of *seismic waves.*

earthquake focus point within the Earth at which the *energy* of an *earthquake* is first released by rupture and from which *seismic waves* emanate.

easterly wave weak, slowly moving trough of low pressure within the belt of *tropical easterlies;* causes a weather disturbance with rain showers.

ebb current oceanward flow of *tidal current* in a *bay* or tidal stream.

ecological niche a concept that includes the role an organism occupies and the function it performs in an ecosystem.

ecological succession time-succession (sequence) of distinctive plant and animal communities occurring within a given area of newly formed land or land cleared of plant cover by burning, clear cutting, or other agents.

ecology science of interactions between life forms and their environment; the science of *ecosystems.*

economic geography the study of the location, distribution and spatial organization of economic activities across the Earth.

ecosystem group of organisms and the environment with which the organisms interact.

edaphic factors factors relating to soil that influence a terrestrial ecosystem.

edatopic grid a method of placing sites into a two-dimensional ecological space, which has direct implications for forest growth, composition, and management.

El Niño episodic cessation of the typical *upwelling* of cold deep water off the coast of Peru; literally, "The Christ Child," for its occurrence in the Christmas season once every few years.

electromagnetic radiation (electromagnetic energy) wavelike form of *energy* radiated by any substance possessing heat; it travels through space at the speed of light.

electromagnetic spectrum the total *wavelength* range of *electromagnetic energy.*

eluviation soil-forming process consisting of the downward transport of fine particles, par-

ticularly the *soil colloids* (both mineral and organic), carrying them out of an upper *soil horizon.*

emergence exposure of submarine landforms by a lowering of sea level or a rise of the crust, or both.

endemic species a *species* found only in one region or location.

energy the capacity to do work, that is, to bring about a change in the state or motion of *matter.*

energy balance (global) balance between *shortwave* solar *radiation* received by the Earth-atmosphere system and *radiation* lost to space by *shortwave* reflection and *longwave radiation* from the Earth-atmosphere system.

energy balance (of a surface) balance between the flows of *energy* reaching a surface and the flows of *energy* leaving it.

energy flow system *open system* that receives an input of *energy*, undergoes internal *energy* flow, *energy* transformation, and *energy* storage, and has an *energy* output.

Entisols *soil order* consisting of mineral soils lacking *soil horizons* that would persist after normal plowing.

entrenched meanders winding, sinuous valley produced by *degradation* of a *stream* with trenching into the *bedrock* by downcutting.

environmental pollution unchecked human activity, such as fuel consumption, fertilizer runoff, toxic industrial production waste, and acid mine drainage, which degrades environmental quality.

environmental temperature lapse rate rate of temperature decrease upward through the *troposphere;* standard value is 6.4 C°/km (3 F°/1000 ft).

epilimnion warm, less dense, upper layer of a *lake* that forms by solar heating and wind-induced circulation.

epicentre the point on the Earth's surface directly above the focus of an earthquake.

epipedon *soil horizon* that forms at the surface.

epiphytes plants that live above ground level out of contact with the soil, usually growing on the limbs of *trees* or *shrubs;* also called "air plants."

epoch a subdivision of geologic time.

equal-area projections class of *map projections* on which any given area of the Earth's surface is shown to correct relative areal extent, regardless of position on the globe.

equator *parallel of latitude* occupying a position midway between the Earth's poles of *rotation;* the largest of the parallels, designated as *latitude* 0°.

equatorial current westward-flowing *ocean current* in the belt of the *trade winds.*

equatorial easterlies upper-level easterly air flow over the *equatorial zone.*

equatorial rainforest plant *formation class* within the *forest biome*, consisting of tall, closely set broadleaf *trees* of evergreen or semideciduous habit.

equatorial trough atmospheric low-pressure trough centred more or less over the *equator* and situated between the two belts of *trade winds.*

equatorial zone *latitude* zone lying between lat. 10° S and 10° N (more or less) and centred upon the *equator.*

equilibrium in flow systems, a state of balance in which flow rates remain unchanged.

equilibrium line marks the boundary between the accumulation zone and the ablation zone.

equilibrium tide the hypothetical tide that would be produced by the lunar and solar tidal forces in the absence of ocean constraints and dynamics.

equinox instant in time when the *subsolar point* falls on the Earth's equator and the *circle of illumination* passes through both poles. *Vernal equinox* occurs on March 20 or 21; *autumnal equinox* on September 22 or 23.

era major subdivision of geologic time consisting of a number of geologic periods. The three *eras* following *Precambrian time* are *Paleozoic, Mesozoic*, and *Cenozoic.*

erg large expanse of active *sand dunes* in the Sahara Desert of North Africa.

erosional landforms class of the *sequential landforms* shaped by the removal of *regolith* or *bedrock* by agents of erosion. Examples: *gorge,* glacial *cirque, marine cliff.*

esker narrow, often sinuous embankment of coarse gravel and boulders deposited in the bed of a meltwater *stream* enclosed in a tunnel within stagnant ice of an *ice sheet.*

estuary *bay* that receives fresh water from a river mouth and salt water from the ocean.

Eurasian-Indonesian belt mountain arc system extending from southern Europe across southern Asia and Indonesia.

eustatic referring to a true change in sea level, as opposed to a local change created by upward or downward *tectonic* motion of land.

eutrophication excessive growth of algae and other related organisms in a *stream* or *lake* as a result of the input of large amounts of nutrient *ions*, especially phosphate and nitrate.

evaporation process in which water in liquid state or solid state passes into the vapour state.

evaporites class of *chemically precipitated sediment* and *sedimentary rock* composed of soluble salts deposited from saltwater bodies.

evapotranspiration combined water loss to the atmosphere by *evaporation* from the *soil* and *transpiration* from plants.

evergreen plant *tree* or *shrub* that holds most of its green leaves throughout the year.

evolution the creation of the diversity of life forms through the process of natural selection.

exfoliation see *unloading*.

exfoliation dome smoothly rounded *rock* knob or hilltop bearing *rock* sheets or shells produced by spontaneous expansion accompanying *unloading*.

exotic river *stream* that flows across a region of dry climate and derives its *discharge* from adjacent uplands where a *water surplus* exists.

exponential growth increase in number or value over time in which the increase is a constant proportion or percentage within each time unit.

exposed shields areas of *continental shields* in which the ancient basement *rock,* usually of *Precambrian* age, is exposed to the surface.

extension (tectonic) drawing apart of crustal layers by *tectonic activity* resulting in *faulting*.

extinction the event that the number of organisms of a *species* shrinks to zero so that the species no longer exists.

extreme events catastrophic events, such as floods, fires, hurricanes, and earthquakes, that have great and long-lasting impacts on human and natural systems.

extrusion release of molten *rock magma* at the surface, as in a flow of *lava* or shower of volcanic ash.

extrusive igneous rock *rock* produced by the solidification of *lava* or ejected fragments of *igneous rock (tephra)*.

eye the circular region of relatively light winds and fair weather found at the centre of a severe tropical cyclone.

Fahrenheit scale temperature scale in which the *freezing* point of water is 32°, and the boiling point 212°.

fair weather system a traveling *anticyclone,* in which the descent of air suppresses clouds and precipitation and weather is typically fair.

fallout *gravity* fall of atmospheric particles of *particulates* reaching the ground.

fault sharp break in *rock* with a displacement (slippage) of the block on one side with respect to an adjacent block. (See *normal fault, overthrust fault, strike-slip fault, transform fault.*)

fault coast *coast* formed when a *shoreline* comes to rest against a *fault scarp*.

fault creep more or less continuous slippage on a *fault plane,* relieving some of the accumulated strain.

fault plane surface of slippage between two Earth blocks moving relative to each other during faulting.

fault scarp clifflike surface feature produced by faulting and exposing the *fault plane;* commonly associated with a *normal fault*.

fault-line scarp erosion scarp developed upon an inactive *fault* line.

feedback in flow systems, a linkage between flow paths such that the flow in one *pathway* acts either to reduce or increase the flow in another *pathway*.

feldspar group of *silicate minerals* consisting of silicate of aluminum and one or more of the metals potassium, sodium, or calcium. (See *plagioclase feldspar, potash feldspar.*)

felsenmeer expanse of large blocks of *rock* produced by *joint* block separation and shattering by *frost action* at high altitudes or in high latitudes; from the German for "rock sea."

felsic *quartz* and *feldspars* treated as a mineral group of light colour and relatively low *density*. (See also *mafic minerals.*)

felsic igneous rock *igneous rock* dominantly composed of *felsic minerals*.

fetch distance over water that wind blows, creating wind waves.

fine textured (rock) having *mineral* crystals too small to be seen by eye or with low magnification.

fiord narrow, deep ocean embayment partially filling a *glacial trough*.

fiord coast deeply embayed, rugged coast formed by partial *submergence* of *glacial troughs*.

firn granular old *snow* forming a surface layer in the zone of accumulation of a *glacier*.

fissile adjective describing a *rock,* usually *shale,* that readily splits up into small flakes or scales.

fjord a deep, steep-walled, U-shaped valley formed by erosion by a glacier and submerged with seawater.

flash flood flood in which heavy rainfall causes a stream or river to rise very rapidly.

flood stream flow at a stream *stage* so high that it cannot be accommodated within the *stream channel* and must spread over the banks to inundate the adjacent *floodplain*.

flood basalts large-scale outpourings of basalt *lava* to produce thick accumulations of *basalt* over large areas.

flood current landward flow of a *tidal current*.

flood stage designated stream-surface level for a particular point on a *stream,* higher than which overbank flooding may be expected.

flood plain belt of low, flat ground, present on one or both sides of a *stream channel,* subject to inundation by a *flood* about once annually and underlain by alluvium.

flow system a physical *system* in which *matter, energy,* or both move through time from one location to another.

fluid substance that flows readily when subjected to unbalanced stresses; may exist as a *gas* or a *liquid*.

fluid agents *fluids* that erode, transport, and deposit *mineral matter* and organic matter; they are running water, waves and currents, glacial ice, and *wind*.

fluvial landforms *landforms* shaped by running water.

fluvial processes geomorphic processes in which running water is the dominant *fluid* agent, acting as *overland flow* and *stream flow*.

focus (See *earthquake focus*.)

fog cloud layer in contact with land or sea surface, or very close to that surface. (See *advection fog, radiation fog*.)

folding process by which *folds* are produced; a form of *tectonic activity*.

folds wavelike corrugations of *strata* (or other layered *rock* masses) as a result of crustal *compression*.

food chain (food web) organization of an *ecosystem* into steps or levels through which *energy* flows as the organisms at each level consume *energy* stored in the bodies of organisms of the next lower level.

food web (See *food chain*.)

forb broad-leaved *herb,* as distinguished from the grasses.

forearc trough in plate tectonics, a shallow trough between a *tectonic arc* and a continent; accumulates *sediment* in a basinlike structure.

foredunes ridge of irregular *sand dunes* typically found adjacent to *beaches* on low-lying *coasts* and bearing a partial cover of plants.

foreland folds *folds* produced by *continental collision* in *strata* of a *passive continental margin*.

forest assemblage of *trees* growing close together, their crowns forming a layer of foliage that largely shades the ground.

forest biome *biome* that includes all regions of *forest* over the lands of the Earth.

formation classes subdivisions within a *biome* based on the size, shape, and structure of the plants that dominate the vegetation.

fossil fuels naturally occurring hydrocarbon compounds that represent the altered remains of organic materials enclosed in *rock;* examples are, *coal, petroleum (crude oil),* and *natural* gas.

fractional scale (See *scale fraction*.)

freezing change from liquid state to solid state accompanied by release of *latent heat,* becoming *sensible heat.*

freezing front location at which freezing is occurring in the *active layer* of *permafrost* during the annual freeze-over; fronts may move downward from the top or upward from the bottom.

fringing reef *coral reef* directly attached to land with no intervening *lagoon* of open water.

front surface of contact between two unlike *air masses.* (See *cold front, occluded front, polar front, warm front.*)

frost action *rock* breakup by forces accompanying the *freezing* of water.

frost hollows areas that are susceptible to imperceptible cold air drainage.

frost point when the dew-point temperature is below freezing.

fundamental niche the theoretical limit for a species, as determined by its physiological requirements for all environmental factors.

gabbro *intrusive igneous rock* consisting largely of pyroxene and *plagioclase feldspar,* with variable amounts of *olivine;* a *mafic igneous rock.*

gas (gaseous state) *fluid* of very low density (as compared with a liquid of the same chemical composition) that expands to fill uniformly any small container and is readily compressed.

gaseous cycle type of *biogeochemical cycle* in which an element or compound is converted into *gaseous* form, diffuses through the *atmosphere,* and passes rapidly over land or sea where it is reused in the *biosphere.*

gene flow speciation process in which evolving populations exchange alleles as individuals move among populations.

genetic drift *speciation* process in which chance mutations change the genetic composition of a breeding population until it diverges from other populations.

genotype the gene set of an individual organism or *species.*

genus a collection of closely related *species* that share a similar genetic evolutionary history.

geographic cycle a theory that views landscapes as stages in a cycle, beginning with rapid uplift and followed with erosion by streams.

geographic grid complete network of parallels and meridians on the surface of the globe, used to fix the locations of surface points.

geographic information system (GIS) a system for acquiring, processing, storing, querying, creating, and displaying *spatial data;* normally computer-based.

geographic isolation *speciation* process in which a breeding population is split into parts by an emerging geographic barrier, such as an uplifting mountain range or a changing climate.

geography the study of the evolving character and organization of the Earth's surface.

geography of soils the study of the distribution of soil types and properties and the processes of soil formation.

geoid refers to the shape that the Earth would assume if it were entirely covered with water and responding to the forces acting upon it.

geologic norm stable natural condition in a moist climate in which slow *soil erosion* is paced by maintenance of *soil horizons* bearing a plant community in an equilibrium state.

geology science of the solid Earth, including the Earth's origin and history, materials comprising the Earth, and the processes acting within the Earth and upon its surface.

geomorphology science of Earth surface processes and *landforms,* including their history and processes of origin.

geostrophic wind *wind* at high levels above the Earth's surface blowing parallel with a system of straight, parallel *isobars.*

geyser periodic jetlike emission of hot water and steam from a narrow vent at a geothermal locality.

glacial abrasion *abrasion* by a moving *glacier* of the *bedrock* floor beneath it.

glacial delta *delta* built by meltwater streams of a *glacier* into standing water of a marginal glacial lake.

glacial drift general term for all varieties and forms of *rock* debris deposited in close association with *ice sheets* of the *Pleistocene Epoch.*

glacial plucking removal of masses of *bedrock* from beneath an *alpine glacier* or *ice sheets* as ice moves forward suddenly.

glacial trough deep, steep-sided *rock* trench of U-shaped cross section formed by *alpine glacier* erosion.

glaciation (1) general term for the total process of glacier growth and *landform* modification by *glaciers.* (2) single episode or time period in which *ice sheets* formed, spread, and disappeared.

glacier large natural accumulation of land ice affected by present or past flowage. (See *alpine glacier.*)

Gleysolic soils a class of *soils* in the Canadian soil classification system characterized by indicators of periodic or prolonged water saturation.

global energy balance a full accounting of all the energy flows among the Sun, the atmosphere, the Earth's surface, and space.

Global Positioning System (GPS) a satellite-based system constantly sending radio signals to Earth with information that allows a GPS receiver to calculate its position on the Earth's surface.

global radiation balance the energy flow process by which the Earth absorbs shortwave solar radiation and emits longwave radiation. In the long run, the two flows must balance.

global scale scale at which we are concerned with the Earth as a whole, for example in considering Earth-Sun relationships.

global warming potential a measure of how much a given mass of greenhouse gas is estimated to contribute to global warming.

gneiss variety of *metamorphic rock* showing banding and commonly rich in *quartz* and *feldspar.*

Gondwana a *supercontinent* of the Permian *Period* including much of the regions that are now South America, Africa, Antarctica, Australia, New Zealand, Madagascar, and peninsular India.

Goode projection an equal-area *map projection,* often used to display areal thematic information, such as *climate* or *soil* type.

gorge (canyon) steep-sided *bedrock* valley with a narrow floor limited to the width of a *stream channel.*

graben trenchlike depression representing the surface of a crustal block dropped down between two opposed, infacing *normal faults.* (See *rift valley.*)

graded profile smoothly descending profile displayed by a *graded stream.*

graded stream *stream* (or *stream channel*) with *stream gradient* so adjusted as to achieve a balanced state in which average *bed load* transport is matched to average bed load input; an average condition over periods of many years' duration.

gradient degree of *slope,* as the gradient of a river or a flowing glacier.

granite *intrusive igneous rock* consisting largely of *quartz, potash feldspar,* and *plagioclase feldspar,* with minor amounts of biotite and hornblende; a *felsic igneous rock.*

granitic rock general term for *rock* of the upper layer of the *continental crust,* composed largely of *felsic igneous* and *metamorphic rock; rock* of composition similar to that of *granite.*

granular disintegration grain-by-grain breakup of the outer surface of coarse-grained *rock,* yielding *sand* and gravel and leaving behind rounded boulders.

graphic scale *map scale* as shown by a line divided into equal parts.

grassland biome *biome* consisting largely or entirely of *herbs,* which may include grasses, grasslike plants, and *forbs.*

gravitation mutual attraction between any two masses.

gravity gravitational attraction of the Earth upon any small mass near the Earth's surface. (See *gravitation.*)

gravity gliding the sliding of a *thrust sheet* away from the centre of an *orogen* under the force of *gravity.*

great circle circle formed by passing a plane through the exact centre of a perfect sphere; the largest circle that can be drawn on the surface of a sphere.

great group subdivision in soil classification systems; a first subdivision in the Canadian Soil Classification System and a second subdivision in the U.S. Comprehensive Soil Classification System.

greenhouse effect accumulation of heat in the lower *atmosphere* through the absorption of *longwave radiation* from the Earth's surface.

greenhouse gases atmospheric gases such as CO_2 and *chlorofluorocarbons (CFCs)* that absorb outgoing *longwave radiation,* contributing to the *greenhouse effect.*

groin wall or embankment built out into the water at right angles to the *shoreline.*

gross photosynthesis total amount of *carbohydrate* produced by *photosynthesis* by a given organism or group of organisms in a given unit of time.

ground ice frozen water within the pores of *soils* and *regolith* or as free bodies or lenses of solid ice.

ground inversion an air layer with its base at the ground surface and in which temperature increases with height.

ground moraine moraine formed of till distributed beneath a large expanse of land surface covered at one time by an ice sheet.

groundwater *subsurface water* occupying the *saturated zone* and moving under the force of *gravity.*

growth rate (of a population) rate at which a population growths or shrinks with time; usually expressed as a percent or proportion of increase or decrease in a given unit of time.

gullies deep, V-shaped trenches carved by newly formed *streams* in rapid headward growth during advanced stages of *accelerated soil erosion.*

guyot sunken remnant of a volcanic island.

gyres large circular *ocean current* systems centred upon the oceanic subtropical *high-pressure cells.*

habitat subdivision of the environment according to the needs and preferences of organisms or groups of organisms.

Hadley cell atmospheric circulation cell in low latitudes involving rising air over the *equatorial trough* and sinking air over the *subtropical high-pressure belts.*

hail form of *precipitation* consisting of pellets or spheres of ice with a concentric layered structure.

hairpin dune a long, narrow dune with parallel sides.

half-life time required for an initial quantity at time-zero to be reduced by one-half in an exponential decay system.

hammada a desert surface from which wind has removed most of the regolith, leaving only bedrock surfaces scattered with large rocks.

hanging valley stream valley that has been truncated by marine erosion so as to appear in cross section in a *marine cliff,* or truncated by glacial erosion so as to appear in cross section in the upper wall of a *glacial trough.*

haze minor concentration of *pollutants* or natural forms of *aerosols* in the atmosphere causing a reduction in visibility.

hazards assessment a field of study blending *physical* and *human geography* to focus on the perception of risk of natural hazards and on developing public policy to mitigate that risk.

heat (See *sensible heat, latent heat.*)

heat energy a form of energy in which the atoms and molecules within a solid (or liquid, or gas) are in rapid motion.

heat island persistent region of higher air temperatures centred over a city.

hemisphere half of a sphere; that portion of the Earth's surface found between the *equator* and a pole.

herb tender plant, lacking woody stems, usually small or low; may be annual or perennial.

herbivory form of interaction among *species* in which an animal (herbivore) grazes on herbaceous plants.

heterosphere region of the *atmosphere* above about 100 km in which *gas* molecules tend to become increasingly sorted into layers by molecular weight and electric charge.

hibernation dormant state of some vertebrate animals during the winter season.

high base status (See *base status of soils.*)

high-latitude climates group of climates in the *subarctic zone, arctic zone,* and *polar zone,* dominated by arctic *air masses* and polar air masses.

high-level temperature inversion condition in which a high-level layer of warm air overlies a layer of cooler air, reversing the normal trend of cooling with altitude.

high-pressure cell centre of high barometric pressure; an *anticyclone.*

highland climates cool to cold, usually moist, climates that occupy mountains and high plateaus.

Histosols *soil order* consisting of *soils* with a thick upper layer of organic matter.

hogbacks sharp-crested, often sawtooth ridges formed of the upturned edge of a resistant *rock* layer of *sandstone, limestone,* or *lava.*

Holocene Epoch last *epoch* of geologic time, commencing about 10,000 years ago; it followed the *Pleistocene Epoch* and includes the present.

homosphere the lower portion of the *atmosphere,* below about 100 km altitude, in which atmospheric *gases* are uniformly mixed.

horse latitudes *subtropical high-pressure belt* of the North Atlantic Ocean, coincident with the central region of the Azores high; a belt of weak, variable winds and frequent calms.

horst crustal block uplifted between two *normal faults.*

hot springs springs discharging heated *groundwater* at a temperature close to the boiling point; found in geothermal areas and thought to be related to a *magma* body at depth.

hotspot (biogeography) geographic region of high biodiversity.

hotspot (plate tectonics) centre of intrusive *igneous* and *volcanic* activity thought to be located over a rising *mantle plume.*

human habitat the lands of the Earth that support human life.

human geography the part of *systematic geography* that deals with social, economic and behavioral processes that differentiate *places.*

human-influenced vegetation vegetation that has been influenced in some way by human activity, for example through cultivation, grazing, timber cutting, or urbanization.

humidity general term for the amount of *water vapour* present in the air. (See *relative humidity, specific humidity.*)

humification *pedogenic process* of transformation of plant tissues into *humus.*

humus dark brown to black organic matter on or in the *soil,* consisting of fragmented plant tissues partly digested by organisms.

hurricane *tropical cyclone* of the western North Atlantic and Caribbean Sea.

hydraulic action *stream erosion* by impact force of the flowing water upon the bed and banks of the *stream channel.*

hydrograph graphic presentation of the variation in *stream discharge* with elapsed time, based on data of stream gauging at a given station on a stream.

hydrologic cycle   total plan of movement, exchange, and storage of the Earth's free water in gaseous state, liquid state, and solid state.

hydrology   science of the Earth's water and its motions through the *hydrologic cycle.*

hydrolysis   chemical union of water molecules with *minerals* to form different, more stable mineral compounds.

hydrophyte   a plant that grows in lakes, marshes and bogs, and is adapted to cope with excessive moisture.

hydrosphere   total water realm of the Earth's surface zone, including the oceans, surface waters of the lands, *groundwater,* and water held in the *atmosphere.*

hygrometer   instrument that measures the *water vapour* content of the *atmosphere;* some types measure *relative humidity* directly.

hypolimnion   cold, dense lower layer of a *lake.*

ice age   span of geologic time, usually on the order of one to three million years, or longer, in which glaciations alternate with interglaciations repeatedly in rhythm with cyclic global climate changes. (See also *interglaciation, glaciation.*)

Ice Age (Late-Cenozoic Ice Age)   the present ice age, which began in late Pliocene time, perhaps 2.5 to 3 million years ago.

ice lens   more-or-less horizontal layer of *segregated ice* formed by capillary movement of soil water toward a freezing front.

ice lobes (glacial lobes)   broad tonguelike extensions of an *ice sheet* resulting from more rapid ice motion where terrain was more favorable.

ice nuclei   particles which act as the nuclei for the formation of ice crystals in the atmosphere.

ice sheet   large thick plate of glacial ice moving outward in all directions from a central region of accumulation.

ice shelf   thick plate of floating glacial ice attached to an *ice sheet* and fed by the ice sheet and by *snow* accumulation.

ice storm   occurrence of heavy glaze of ice on solid surfaces.

ice wedge   vertical, wall-like body of ground ice, often tapering downward, occupying a shrinkage crack in *silt* of *permafrost* areas.

ice sheet climate   severely cold climate, found on the Greenland and Antarctic *ice sheets,* with *potential evapotranspiration (water need)* effectively zero throughout the year.

ice-wedge polygons   polygonal networks of *ice wedges.*

iceberg   mass of glacial ice floating in the ocean, derived from a *glacier* that extends into tidal water.

iceberg   mass of glacial ice floating in the ocean, derived from a glacier that extends into tidal water.

igneous rock   *rock* solidified from a high-temperature molten state; *rock* formed by cooling of *magma.* (See *extrusive igneous rock, felsic igneous rock, intrusive igneous rock, mafic igneous rock, ultramafic igenous rock.*)

illuviation   accumulation in a lower *soil horizon* (typically, the *B horizon*) of materials brought down from a higher horizon; a soil-forming process.

image processing   mathematical manipulation of digital images, for example, to enhance contrast or edges.

Inceptisols   *soil order* consisting of soils having weakly developed *soil horizons* and containing weatherable *minerals.*

individual-scale   the finest level of scale, including such landscape features as a grassy sand dune on a beach.

induced deflation   loss of *soil* by wind erosion that is triggered by human activity such as cultivation or overgrazing.

induced mass wasting   *mass wasting* that is induced by human activity, such as creation of waste *soil* and *rock* piles or undercutting of *slopes* in construction.

infiltration   absorption and downward movement of *precipitation* into the *soil* and *regolith.*

infrared imagery   images formed by *infrared radiation* emanating from the ground surface as recorded by a remote sensor.

infrared radiation   *electromagnetic energy* in the *wavelength* range of 0.7 to about 200 μm.

initial landforms   *landforms* produced directly by internal Earth processes of *volcanism* and *tectonic activity.* Examples: *volcano, fault scarp.*

inner lowland   on a *coastal plain,* a shallow valley lying between the first *cuesta* and the area of older *rock* (oldland).

input   flow of *matter* or *energy* into a system.

inselbergs   an isolated hill, knob, or small mountain that rises abruptly from a gently sloping or virtually level surrounding plain.

insolation   interception of solar *energy (shortwave radiation)* by an exposed surface.

inspiral   horizontal inward spiral or motion, such as that found in a *cyclone.*

interglaciation   within an *ice age,* a time interval of mild global climate in which continental *ice sheets* were largely absent or were limited to the Greenland and Antarctic ice sheets; the interval between two glaciations. (See also *deglaciation, glaciation.*)

interlobate moraine   *moraine* formed between two adjacent lobes of an *ice sheet.*

International Date Line   the 180° *meridian of longitude,* together with deviations east and west of that meridian, forming the time boundary between adjacent *standard time zones* that are 12 hours fast and 12 hours slow with respect to Greenwich standard time.

interrupted projection   projection subdivided into a number of sectors (gores), each of which is centred on a different central meridian.

intertropical convergence zone (ITCZ)   zone of convergence of *air masses* of *tropical easterlies (trade winds)* along the axis of the *equatorial trough.*

intrusion   body of *igneous rock* injected as *magma* into preexisting crustal *rock;* example: *dike* or *sill.*

intrusive igneous rock   *igneous rock* body produced by solidification of *magma* beneath the surface, surrounded by preexisting *rock.*

inversion   (See *temperature inversion.*)

ion   atom or group of atoms bearing an electrical charge as the result of a gain or loss of one or more electrons.

island arcs   curved lines of volcanic islands associated with active *subduction* zones along the boundaries of *lithospheric plates.*

isobars   lines on *map* passing through all points having the same *atmospheric pressure.*

isohyet   line on a *map* drawn through all points having the same numerical value of *precipitation.*

isopleth   line on a *map* or globe drawn through all points having the same value of a selected property or entity.

isostasy   principle describing the flotation of the *lithosphere,* which is less dense, on the plastic *asthenosphere,* which is more dense.

isostatic compensation   crustal rise or sinking in response to unloading by *denudation* or loading by sediment deposition, following the principle of *isostasy.*

isostatic rebound   local crustal rise after the melting of ice sheets, following the principle of *isostasy.*

isotherm   line on a *map* drawn through all points having the same air temperature.

isotope   form of an element with a unique *atomic mass number.*

jet stream   high-speed air flow in narrow bands within the *upper-air westerlies* and along certain other global *latitude* zones at high levels.

joints   fractures within *bedrock,* usually occurring in parallel and intersecting sets of planes.

Joule   unit of work or energy in the metric system; symbol, J.

June solstice   (See *summer solstice.*)

karst landscape or topography dominated by surface features of *limestone* solution and underlain by a *limestone cavern* system.

Kelvin scale (K) temperature scale on which the starting point is absolute zero, equivalent to −273°C.

kilopascal the standard unit, for meteorological purposes, that pressure is measured in.

kinetic energy form of *energy* represented by *matter* (mass) in motion.

knob and kettle terrain of numerous small knobs of *glacial drift* and deep depressions usually situated along the *moraine* belt of a former *ice sheet.*

kopje rocky outcrops of old granite which, because of erosion and weathering, have broken up into a rough and jumbled surface.

lag time interval of time between occurrence of precipitation and peak discharge of a *stream.*

lagoon shallow body of open water lying between a *barrier island* or a *barrier reef* and the mainland.

lahar rapid downslope or downvalley movement of a tonguelike mass of water-saturated *tephra* (volcanic ash) originating high up on a steep-sided volcanic cone; a variety of mudflow.

lake body of standing water that is enclosed on all sides by land.

laminar flow smooth, even flow of a *fluid* shearing in thin layers without *turbulence.*

land breeze local wind blowing from land to water during the night.

land degradation *degradation* of the quality of plant cover and *soil* as a result of overuse by humans and their domesticated animals, especially during periods of *drought.*

landforms configurations of the land surface taking distinctive forms and produced by natural processes. Examples: hill, valley, plateau. (See *depositional landforms, erosional landforms, initial landforms, sequential landforms.*)

landmass large area of *continental crust* lying above sea level (base level) and thus available for removal by *denudation.*

landmass rejuvenation episode of rapid fluvial *denudation* set off by a rapid crustal rise, increasing the available *landmass.*

landslide rapid sliding of large masses of *bedrock* on steep mountain slopes or from high *cliffs.*

lapse rate rate at which temperature decreases with increasing altitude (See *environmental temperature lapse rate, dry adiabatic lapse rate, wet adiabatic lapse rate.*)

large-scale map *map* with *fractional scale* greater than 1:100,000; usually shows a small area.

Late-Cenozoic Ice Age the series of *glaciations, deglaciations* and *interglaciations* experienced during the late *Cenozoic Era.*

latent heat heat absorbed and held in storage in a *gas* or *liquid* during the processes of *evaporation,* or *melting,* or *sublimation;* distinguished from *sensible heat.*

latent heat transfer flow of *latent heat* that results when water absorbs heat to change from a *liquid* or *solid* to a *gas* and then later releases that heat to new surroundings by *condensation* or *deposition.*

lateral moraine *moraine* forming an embankment between the ice of an *alpine glacier* and the adjacent valley wall.

laterite rocklike layer rich in *sequioxides* and iron, including the minerals *bauxite* and *limonite,* found in low latitudes in association with *Ultisols* and *Oxisols.*

latitude arc of a *meridian* between the *equator* and a given point on the globe.

Laurasia a *supercontinent* of the Permian *Period* including much of the regions that are now North America and western Eurasia.

lava *magma* emerging on the Earth's solid surface, exposed to air or water.

leaching *pedogenic process* in which material is lost from the *soil* by downward washing out and removal by percolating surplus soil water.

leads narrow strips of open ocean water between ice floes.

level of condensation elevation at which an upward-moving parcel of moist air cools to the *dew point* and *condensation* begins to occur.

liana woody vine supported on the trunk or branches of a *tree.*

lichens plant forms in which algae and fungi live together (in a symbiotic relationship) to create a single structure; they typically form tough, leathery coatings or crusts attached to *rocks* and tree trunks.

life cycle continuous progression of stages in a growth or development process, such as that of a living organism.

life form characteristic physical structure, size, and shape of a plant or of an assemblage of plants.

life layer shallow surface zone containing the *biosphere;* a zone of interaction between *atmosphere* and land surface, and between atmosphere and ocean surface.

life zones series of vegetation zones describing vegetation types that are encountered with increasing elevation, especially in the southwestern U.S.

lifting condensation level the altitude at which the air will become saturated because of adiabatic cooling (caused by expansion).

limestone nonclastic *sedimentary rock* in which *calcite* is the predominant *mineral,* and with varying minor amounts of other minerals and *clay.*

limestone caverns interconnected subterranean cavities formed in *limestone* by *carbonic acid action* occurring in slowly moving *groundwater.*

limnology study of the physical, chemical, and biological processes of *lakes.*

limonite mineral or group of *minerals* consisting largely of iron oxide and water, produced by *chemical weathering* of other iron-bearing minerals.

line type of *spatial object* in a *geographic information system* that has starting and ending *nodes;* may be directional.

liquid *fluid* that maintains a free upper surface and is only very slightly compressible, as compared with a *gas.*

lithosphere strong, brittle outermost *rock* layer of the Earth, lying above the *asthenosphere.*

lithospheric plate segment of *lithosphere* moving as a unit, in contact with adjacent lithospheric plates along plate boundaries.

littoral drift transport of *sediment* parallel with the *shoreline* by the combined action of *beach drift* and *longshore current* transport.

loam soil-texture class in which no one of the three size grades *(sand, silt, clay)* dominates over the other two.

local scale scale of observation of the Earth in which local processes and phenomena are observed.

local winds general term for *winds* generated as direct or immediate effects of the local terrain.

loess accumulation of yellowish to buff-coloured, fine-grained *sediment,* largely of *silt* grade, upon upland surfaces after transport in the air in *turbulent suspension* (i.e., carried in a *dust storm*).

logistic growth growth according to a mathematical model in which the *growth rate* eventually decreases to near zero.

longitude arc of a *parallel* between the *prime meridian* and a given point on the globe.

longitudinal dunes class of *sand dunes* in which the dune ridges are oriented parallel with the prevailing wind.

longshore current current in the breaker zone, running parallel with the *shoreline* and set up by the oblique approach of waves.

longshore drift *littoral drift* caused by action of a *longshore current.*

long-wave radiation *electromagnetic energy* emitted by the Earth, largely in the range from 3 to 50 μm.

low base status (See *base status of soils.*)

low-angle overthrust fault   *overthrust fault* in which the *fault plane* or fault surface has a low angle of *dip* or may be horizontal.

low-latitude climates   group of climates of the *equatorial zone* and *tropical zone* dominated by the subtropical high-pressure belt and the *equatorial trough.*

low-latitude rainforest   evergreen broadleaf forest of the wet equatorial and tropical climate zones.

low-latitude rainforest environment   low-latitude environment of warm temperatures and abundant *precipitation* that characterizes rainforest in the *wet equatorial* and *monsoon and trade-wind coastal* climates.

low-level temperature inversion   atmospheric condition in which temperature near the ground increases, rather than decreases, with elevation.

low-pressure trough   zone of low pressure between two *anticyclones.*

lowlands   broad, open valleys between two *cuestas* of a *coastal plain.* (The term may refer to any low areas of land surface.)

Luvisolic soils   a class of *forest soils* in the Canadian soil classification system in which the *B horizon* accumulates *clay.*

macronutrients   essential chemical elements needed by all life in large quantities for it to function normally. The three principal elements are hydrogen, carbon, and oxygen.

mafic   *minerals,* largely *silicate minerals,* rich in magnesium and iron, dark in colour, and of relatively great density.

mafic igneous rock   *igneous rock* dominantly composed of *mafic minerals.*

magma   mobile, high-temperature molten state of *rock,* usually of *silicate mineral* composition and with dissolved *gases.*

manipulation and analysis component   component of a *geographic information system* that responds to spatial queries and creates new data layers.

mantle   *rock* layer or shell of the Earth beneath the *crust* and surrounding the *core,* composed of *ultramafic igneous rock* of *silicate mineral* composition.

mantle plume   a columnlike rising of heated *mantle rock,* thought to be the cause of a *hot spot* in the overlying *lithospheric plate.*

map   a paper representation of space showing point, line, or area data.

map projection   any orderly system of parallels and meridians drawn on a flat surface to represent the Earth's curved surface.

marble   variety of *metamorphic rock* derived from *limestone* or dolomite by recrystallization under pressure.

marine cliff   *rock* cliff shaped and maintained by the undermining action of breaking waves.

marine geography   (See *coastal* and *marine geography.*)

marine scarp   steep seaward *slope* in poorly consolidated *alluvium, glacial drift,* or other forms of *regolith,* produced along a coastline by the undermining action of waves.

marine terrace   former *abrasion platform* elevated to become a steplike coastal *landform.*

marine west-coast climate   cool moist climate of west coasts in the *midlatitude zone,* usually with a substantial annual *water surplus* and a distinct winter *precipitation* maximum.

marl   soft, white, carbonate-rich mud produced when calcium carbonate precipitates from a *lake* and accumulates in beds and banks on and near the lake shore.

mass number   (See *atomic mass number.*)

mass wasting   spontaneous downhill movement of *soil, regolith,* and *bedrock* under the influence of *gravity,* rather than by the action of *fluid* agents.

massive icy beds   layers of ice-rich sediment, found in *permafrost* regions, formed by upwelling *ground water* that flows to a freezing front.

material cycle   a closed matter flow system, in which matter flows endlessly, powered by energy inputs (See *biogeochemical cycle.*)

mathematical modeling   using variables and equations to represent real processes and systems.

matter   physical substance that has mass and density.

matter flow system   total system of *pathways* by which a particular type of *matter* (a given element, compound, or ion, for example) moves through the Earth's *ecosystem* or *biosphere.*

mean annual temperature   mean of daily air temperature means for a given year or succession of years.

mean daily temperature   sum of daily maximum and minimum air temperature readings divided by two.

mean monthly temperature   mean of daily air temperature means for a given calendar month.

mean velocity   mean, or average, speed of flow of water through an entire stream cross section.

meanders   (See *alluvial meanders.*)

mechanical energy   *energy* of motion or position; includes *kinetic energy* and *potential energy.*

mechanical weathering   (See *physical weathering.*)

medial moraine   long, narrow deposit of fragments on the surface of a *glacier;* created by the merging of *lateral moraines* when two glaciers join into a single stream of ice flow.

Mediterranean climate   climate type of the *subtropical zone,* characterized by the alternation of a very dry summer and a mild, rainy winter.

melting   change from solid state to liquid state, accompanied by absorption of *sensible heat* to become *latent heat.*

Mercator projection   conformal *map projection* with horizontal parallels and vertical meridians and with *map scale* rapidly increasing with increase in *latitude.*

mercury barometer   *barometer* using the Torricelli principle, in which *atmospheric pressure* counterbalances a column of mercury in a tube.

meridian   north–south line on the surface of the global *oblate ellipsoid,* connecting the *north pole* and *south pole.*

meridional transport   flow of *energy* (heat) or *matter* (water) across the *parallels of latitude,* either poleward or equatorward.

mesa   table-topped *plateau* of comparatively small extent bounded by *cliffs* and occurring in a region of flat-lying *strata.*

mesopause   Upper limit of the *mesosphere.*

mesosphere   atmospheric layer of upwardly diminishing temperature, situated above the stratopause and below the mesopause.

Mesozoic Era   second of three geologic *eras* following *Precambrian time.*

metalimnion   layer of a *lake,* between the *epilimnion* and *hypolimnion,* in which temperature decreases with depth; contains the thermocline.

metamorphic rock   *rock* altered in physical structure and/or chemical *(mineral)* composition by action of heat, pressure, *shearing stress,* or infusion of elements, all taking place at substantial depth beneath the surface.

meteorology   science of the *atmosphere;* particularly the physics of the lower or inner atmosphere.

mica group   aluminum-silicate *mineral* group of complex chemical formula having perfect cleavage into thin sheets.

microburst   brief onset of intense *winds* close to the ground beneath the downdraft zone of a *thunderstorm* cell.

microcontinent   fragment of *continental crust* and its *lithosphere* of subcontinental dimensions that is embedded in an expanse of *oceanic lithosphere.*

micrometre   metric unit of length equal to one-millionth of a metre (0. 000001 m); abbreviated μm.

microwaves waves of the *electromagnetic radiation* spectrum in the *wavelength* band from about 0.03 cm to about 1 cm.

mid-oceanic ridge one of three major divisions of the ocean basins, being the central belt of submarine mountain topography with a characteristic *axial rift*.

midlatitude climates group of climates of the *midlatitude zone* and *subtropical zone*, located in the *polar front zone* and dominated by both tropical *air masses* and polar air masses.

midlatitude deciduous forest plant *formation class* within the *forest biome* dominated by tall, broadleaf deciduous *trees*, found mostly in the *moist continental climate* and *marine west-coast climate*.

midlatitude zones latitude zones occupying the *latitude* range 35° to 55° N and S (more or less) and lying between the *subtropical zones* and the *subarctic (subantarctic) zones*.

millibar unit of *atmospheric pressure*; one-thousandth of a bar. *Bar* is a force of one million dynes per square centimetre.

mineral naturally occurring inorganic substance, usually having a definite chemical composition and a characteristic atomic structure. (See *felsic minerals, mafic minerals, silicate minerals*.)

mineral alteration chemical change of *minerals* to more stable compounds upon exposure to atmospheric conditions; same as *chemical weathering*.

mineral matter (soils) component of *soil* consisting of weathered or unweathered mineral grains.

mineral oxides (soils) secondary *minerals* found in *soils* in which original minerals have been altered by chemical combination with oxygen.

minute (of arc) 1/60 of a degree.

mistral local drainage wind of cold air affecting the Rhone Valley of southern France.

Moho contact surface between the Earth's *crust* and *mantle*; a contraction of Mohorovic, the name of the seismologist who discovered this feature.

moist continental climate moist climate of the *midlatitude zone* with strongly defined winter and summer seasons, adequate *precipitation* throughout the year, and a substantial annual *water surplus*.

moist subtropical climate moist climate of the *subtropical zone*, characterized by a moderate to large annual *water surplus* and a strongly seasonal cycle of *potential evapotranspiration (water need)*.

mollic epipedon relatively thick, dark-coloured surface *soil horizon*, containing substantial amounts of organic matter *(humus)* and usually rich in *bases*.

Mollisols *soil order* consisting of *soils* with a *mollic horizon* and high *base status*.

moment magnitude scale a scale that measures the total energy released by an earthquake. It is based on the movement that occurs on a fault, not on how much the ground shakes during an earthquake.

monadnock prominent, isolated mountain or large hill rising conspicuously above a surrounding *peneplain* and composed of a *rock* more resistant than that underlying the peneplain; a *landform* of *denudation* in moist climates.

monomictic in *limnology*, mixing once each year, as a monomictic *lake*.

monsoon and trade-wind coastal climate moist climate of low latitudes showing a strong rainfall peak in the season of high sun and a short period of reduced rainfall.

monsoon forest *formation class* within the *forest biome* consisting in part of deciduous *trees* adapted to a long dry season in the *wet dry tropical climate*.

monsoon system system of low-level *winds* blowing into a continent in summer and out of it in winter, controlled by *atmospheric pressure* systems developed seasonally over the continent.

montane forest plant *formation class* of the *forest biome* found in cool upland environments of the *tropical zone* and *equatorial zone*.

moraine accumulation of *rock* debris carried by an *alpine glacier* or an *ice sheet* and deposited by the ice to become a *depositional landform*. (See *lateral moraine, terminal moraine*.)

morphology the outward form and appearance of individual organisms or *species*.

mottles spots or blotches of varying colour found in a *soil profile*; usually indicating acid or chemically reducing conditions persisting in a wet soil environment.

mountain arc curving section of an *alpine chain* occurring on a *converging boundary* between two crustal plates.

mountain roots erosional remnants of deep portions of ancient *continental sutures* that were once *alpine chains*.

mountain winds daytime movements of air up the *gradient* of valleys and mountain slopes; alternating with nocturnal *valley winds*.

mucks organic *soils* largely composed of fine, black, sticky organic matter.

mud *sediment* consisting of a mixture of *clay* and *silt* with water, often with minor amounts of *sand* and sometimes with organic matter.

mud hummock low mound of earth found in *permafrost* terrain, formed by cycles of *ground ice* growth and melting, with centre of bare ground; vegetation may occur at edges (see also *earth hummock*).

mudflow a form of *mass wasting* consisting of the downslope flowage of a mixture of water and *mineral* fragments (*soil, regolith*, disintegrated *bedrock*), usually following a natural drainage line or *stream channel*.

mudstone *sedimentary rock* formed by the lithification of *mud*.

multipurpose map *map* containing several different types of information.

multispectral image image consisting of two or more images, each of which is taken from a different portion of the spectrum (e.g., blue, green, red, infrared).

multispectral scanner *remote sensing* instrument, flown on an aircraft or spacecraft, that simultaneously collects multiple *digital images (multispectral images)* of the ground. Typically, images are collected in four or more spectral bands.

mutation change in genetic material of a reproductive cell.

nappe overturned recumbent *fold* of *strata*, usually associated with *thrust sheets* in a collision *orogen*.

natural bridge natural *rock* arch spanning a *stream channel*, formed by cutoff of an *entrenched meander* bend.

natural flow systems flow systems of *energy* or naturally-occurring substances that are powered largely or completely by natural power sources.

natural gas naturally occurring mixture of hydrocarbon compounds (principally methane) in the gaseous state held within certain porous *rocks*.

natural levee belt of higher ground paralleling a meandering *alluvial river* on both sides of the *stream channel* and built up by *deposition* of fine *sediment* during periods of overbank flooding.

natural selection selection of organisms by environment in a process similar to selection of plants or animals for breeding by agriculturalists.

natural vegetation stable, mature plant cover characteristic of a given area of land surface largely free from the influences and impacts of human activities.

neap tides when the tidal range between high and low water is at its lowest.

needleleaf evergreen forest *needleleaf forest* composed of evergreen tree species, such as spruce, fir, and pine.

needleleaf forest plant *formation class* within the *forest biome*, consisting largely of needleleaf *trees*. (See also *boreal forest*.)

needleleaf tree tree with long, thin or flat leaves, such as pine, fir, larch, or spruce.

negative exponential mathematical form of a curve that smoothly decreases to approach a steady value, usually zero.

negative feedback in flow systems, a linkage between flow paths such that the flow in one *pathway* acts to reduce the flow in another *pathway*. (See also *feedback, positive feedback*.)

net photosynthesis *carbohydrate* production remaining in an organism after *respiration* has broken down sufficient *carbohydrate* to power the metabolism of the organism.

net primary production rate at which *carbohydrate* is accumulated in the tissues of plants within a given *ecosystem*; units are kilograms of dry organic matter per year per square metre of surface area.

net radiation difference in intensity between all incoming *energy* (positive quantity) and all outgoing *energy* (negative quantity) carried by both *shortwave radiation* and *longwave radiation*.

neutral stability occurs when the environmental temperature lapse rate equals the dry adiabatic lapse rate or the saturated adiabatic lapse rate.

neutron atomic particle contained within the nucleus of an atom; similar in mass to a proton, but without a magnetic charge.

nitrogen cycle *biogeochemical cycle* in which nitrogen moves through the *biosphere* by the processes of *nitrogen fixation* and *denitrification*.

nitrogen fixation chemical process of conversion of *gaseous* molecular nitrogen of the *atmosphere* into compounds or ions that can be directly utilized by plants; a process carried out within the *nitrogen cycle* by certain microorganisms.

node point marking the end of a *line* or the intersection of *lines* as spatial objects in a *geographic information system*

noon (See *solar noon.*)

noon angle (of the Sun) angle of the Sun above the horizon at its highest point during the day.

normal fault variety of *fault* in which the *fault plane* inclines (*dips*) toward the downthrown block and a major component of the motion is vertical.

normal temperature lapse rate measures the drop in temperature in the stationary air, averaged for the entire Earth over a long period.

north pole point at which the northern end of the Earth's *axis of rotation* intersects the Earth's surface.

northeast trade winds surface *winds* of low latitudes that blow steadily from the northeast. (See also *trade winds.*)

nuclei (atmospheric) minute particles of solid *matter* suspended in the *atmosphere* and serving as cores for *condensation* of water or ice.

nutrient cycle (See *biogeochemical cycle.*)

O₁ horizon surface *soil horizon* containing decaying organic matter that is recognizable as leaves, twigs, or other organic structures.

Oₐ horizon *soil horizon* below the O₁ horizon containing decaying organic matter that is too decomposed to recognize as specific plant parts, such as leaves or twigs.

oasis desert area where *groundwater* is tapped for crop irrigation and human needs.

oblate ellipsoid geometric solid resembling a flattened sphere, with polar axis shorter than the equatorial diameter.

occluded front weather *front* along which a moving *cold front* has overtaken a *warm front*, forcing the warm *air mass* aloft.

ocean basin floors one of the major divisions of the ocean basins, comprising the deep portions consisting of *abyssal plains* and low hills.

ocean current persistent, dominantly horizontal flow of ocean water.

ocean tide periodic rise and fall of the ocean level induced by gravitational attraction between the Earth and Moon in combination with Earth *rotation*.

oceanic crust crust of basaltic composition beneath the ocean floors, capping *oceanic lithosphere*. (See also *continental crust.*)

oceanic lithosphere *lithosphere* bearing *oceanic crust*.

oceanic trench narrow, deep depression in the seafloor representing the line of *subduction* of an oceanic *lithospheric plate* beneath the margin of a continental lithospheric plate; often associated with an *island arc*.

oceanography the study of the Earth's oceans and seas.

oil sand (See *bituminous sand.*)

old-field succession form of *secondary succession* typical of an abandoned field, such as might be found in eastern or central North America.

olivine *silicate mineral* with magnesium and iron but no aluminum, usually olive-green or grayish-green; a *mafic mineral*.

omnivores animals that feed on plant materials as well as animals.

open flow system system of interconnected flow paths of *energy* and/or *matter* with a boundary through which that *energy* and/or *matter* can enter and leave the system.

organic matter (soils) material in *soil* that was originally produced by plants or animals and has been subjected to decay.

Organic soils a class of *soils* in the Canadian soil classification system that is composed largely of organic materials.

organic sediment *sediment* consisting of the organic remains of plants or animals.

orogen the mass of tectonically deformed *rocks* and related *igneous rocks* produced during an *orogeny*.

orogeny major episode of *tectonic activity* resulting in *strata* being deformed by folding and faulting.

orographic pertaining to mountains.

orographic precipitation *precipitation* induced by the forced rise of moist air over a mountain barrier.

outcrop surface exposure of *bedrock*.

output the flow of *matter* or *energy* out of a system.

outspiral horizontal outward spiral or motion, such as that found in an *anticyclone*.

outwash glacial deposit of stratified drift left by *braided streams* issuing from the front of a *glacier*.

outwash plain flat, gently sloping plain built up of *sand* and gravel by the *aggradation* of meltwater *streams* in front of the margin of an *ice sheet*.

overburden *strata* overlying a layer or *stratum* of interest, as overburden above a *coal* seam.

overland flow motion of a surface layer of water over a sloping ground surface at times when the *infiltration* rate is exceeded by the *precipitation* rate; a form of *runoff*.

overthrust fault *fault* characterized by the overriding of one crustal block (or *thrust sheet*) over another along a gently inclined *fault plane*; associated with crustal *compression*.

overturn in *limnology*, mixing of water from the surface to the bottom in a *lake*; occurs when a lake has a uniform temperature profile.

oxbow lake crescent-shaped lake representing the abandoned channel left by the *cutoff* of an *alluvial meander*.

oxidation chemical union of free oxygen with metallic elements in *minerals*.

oxide chemical compound containing oxygen; in *soils,* iron oxides and aluminum oxides are examples.

Oxisols *soil order* consisting of very old, highly weathered *soils* of low latitudes, with an oxic horizon and low *base status*.

oxygen cycle *biogeochemical cycle* in which oxygen moves through the *biosphere* in both *gaseous* and sedimentary forms.

ozone a form of oxygen with a molecule consisting of three atoms of oxygen, O₃.

ozone layer layer in the *stratosphere*, mostly in the altitude range 20 to 35 km (12 to 31 mi), in which a concentration of *ozone* is produced by the action of solar *ultraviolet radiation*.

P waves produced by earthquakes, they are the highest velocity of all seismic waves. They can travel through both solid and liquid rock material, as well as the water of the oceans..

pack ice floating *sea ice* that completely covers the sea surface.

Paleozoic Era first of three geologic *eras* comprising all geologic time younger than *Precambrian time*.

Pangaea hypothetical parent continent, enduring until near the close of the *Mesozoic Era*, consisting of the *continental shields* of *Laurasia* and *Gondwana* joined into a single unit.

parabolic dunes isolated low *sand dunes* of parabolic outline, with points directed into the prevailing *wind*.

parallel east-west circle on the Earth's surface, lying in a plane parallel with the *equator* and at right angles to the *axis of rotation*.

parasitism form of negative interaction between *species* in which a small species (parasite) feeds on a larger one (host) without necessarily killing it.

parent material inorganic, *mineral* base from which the *soil* is formed; usually consists of *regolith*.

particulates *solid* and *liquid* particles capable of being suspended for long periods in the *atmosphere*.

pascal metric unit of pressure, defined as a force of one newton per square metre (1 N/m²); symbol, Pa; 100 Pa = 1 mb, 10⁵ Pa = 1 bar.

passive continental margins continental margins lacking active plate boundaries at the contact of *continental crust* with *oceanic crust*. A passive margin thus lies within a single *lithospheric plate*. Example: Atlantic continental margin of North America. (See also *continental margins, active continental margins*.)

passive systems electromagnetic remote sensing systems that measure radiant *energy* reflected or emitted by an object or surface.

pathway in an *energy flow system*, a mechanism by which *matter* or *energy* flows from one part of the system to another.

patterned ground general term for a ground surface that bears polygonal or ring-like features, including stone circles, nets, polygons, steps, and stripes; includes *ice wedge polygons*; typically produced by *frost action* in cold climates.

peat partially decomposed, compacted accumulation of plant remains occurring in a *bog* environment.

ped individual natural *soil* aggregate.

pediment gently sloping, rock-floored land surface found at the base of a mountain mass or *cliff* in an arid region.

pedogenic processes group of recognized basic soil-forming processes, mostly involving the gain, loss, *translocation*, or transformation of materials within the *soil* body.

pedology science of the *soil* as a natural surface layer capable of supporting living plants; synonymous with *soil science*.

pedon soil column extending down from the surface to reach a lower limit in some form of *regolith* or *bedrock*.

peneplain land surface of low elevation and slight relief produced in the late stages of *denudation* of a *landmass*.

perched water table surface of a lens of *ground water* held above the main body of ground water by a discontinuous impervious layer.

percolation slow, downward flow of water by *gravity* through *soil* and subsurface layers toward the *water table*.

perennials plants that live for more than one growing season.

peridotite *igneous rock* consisting largely of olivine and pyroxene; an *ultramafic igneous rock* occurring as a pluton, also thought to compose much of the upper *mantle*.

periglacial in an environment of intense *frost action*, located in cold climate regions or near the margins of *alpine glaciers* or large *ice sheets*.

periglacial system a distinctive set of landforms and land-forming processes that are created by intense frost action.

perihelion point on the Earth's elliptical orbit at which the Earth is nearest to the Sun.

period in *limnology*, the time for a full wave of a water surface to occur at a point, as in the period of a *seiche*.

period of geologic time time subdivision of the *era*, each ranging in duration between about 35 and 70 million years.

permafrost *soil, regolith*, and *bedrock* at a temperature below 0°C (32°F), found in cold climates of arctic, subarctic, and alpine regions.

permafrost table in *permafrost*, the upper surface of perennially frozen ground; lower surface of the *active layer*

petroleum (crude oil) natural liquid mixture of many complex hydrocarbon compounds of organic origin, found in accumulations (oil pools) within certain *sedimentary rocks*.

pH measure of the concentration of hydrogen ions in a solution. (The number represents the logarithm to the base 10 of the reciprocal of the weight in grams of hydrogen ions per litre of water.) Acid solutions have pH values less than 6, and basic solutions have pH values greater than 6.

phenotype the morphological expression of the *genotype* of an individual. It includes all the physical aspects of its structure that are readily perceivable.

photochemical reaction a chemical reaction which is induced by light.

photoperiod duration of daylight on a given day of the year at a given latitude.

photosynthesis production of carbohydrate by the union of water with *carbon dioxide* while absorbing light *energy*.

phreatophytes plants that draw water from the *ground water table* beneath *alluvium* of dry stream channels and valley floors in desert regions.

phylum highest division of higher plant and animal life.

physical geography the part of *systematic geography* that deals with the natural processes occurring at the Earth's surface that provide the physical setting for human activities; includes the broad fields of *climatology, geomorphology, coastal and marine geography, geography of soils*, and biogeography.

physical weathering breakup of massive *rock (bedrock)* into small particles through the action of physical forces acting at or near the Earth's surface. (See *weathering*.)

phytoplankton microscopic plants found largely in the uppermost layer of ocean or lake water.

pingo conspicuous conical mound or circular hill, having a core of ice, found on plains of the arctic tundra where *permafrost* is present.

pioneer stage first stage of an *ecological succession*.

pioneer plants plants that first invade an environment of new land or a *soil* that has been cleared of vegetation cover; often these are annual *herbs*.

place in geography, a location on the Earth's surface, typically a settlement or small region with unique characteristics.

plagioclase feldspar aluminum-silicate *mineral* with sodium or calcium or both.

plane of the ecliptic imaginary plane in which the Earth's orbit lies.

plant ecology the study of the relationships between plants and their environment.

plant nutrients *ions* or chemical compounds that are needed for plant growth.

plate tectonics theory of *tectonic activity* dealing with *lithospheric plates* and their activity.

plateau upland surface, more or less flat and horizontal, upheld by resistant beds of *sedimentary rock* or *lava* flows and bounded by a steep *cliff*.

playa flat land surface underlain by fine *sediment* or evaporite minerals deposited from shallow lake waters in a dry climate in the floor of a closed topographic depression.

Pleistocene Epoch *epoch* of the *Cenozoic Era*, often identified as the Ice Age; it preceded the *Holocene Epoch*.

plinthite iron-rich concentrations present in some kinds of *soils* in deeper *soil horizons* and capable of hardening into rocklike material with repeated wetting and drying.

plucking (See *glacial plucking.*)

pluton any body of *intrusive igneous rock* that has solidified below the surface, enclosed in preexisting *rock*.

pocket beach *beach* of crescentic outline located at a *bay* head.

podzol type of *soil order* closely equivalent to *Spodosol*; an order of the Canadian Soil Classification System.

Podzolic soils a class of *forest* and heath *soils* in the Canadian soil classification system in which an amorphous material of humified organic matter with Al and Fe accumulates.

point *spatial object* in a *geographic information system* with no area.

point bar deposit of coarse bed-load *alluvium* accumulated on the inside of a growing *alluvial meander*.

polar easterlies system of easterly surface winds at high latitude, best developed in the southern *hemisphere*, over Antarctica.

polar front *front* lying between cold polar *air masses* and warm tropical air masses, often situated along a *jet stream* within the *upper-air westerlies*.

polar front jet stream *jet stream* found along the *polar front*, where cold polar air and warm tropical air are in contact.

polar front zone broad zone in midlatitudes and higher latitudes, occupied by the shifting *polar front*.

polar high persistent low-level centre of high *atmospheric pressure* located over the *polar zone* of Antarctica.

polar outbreak tongue of cold polar air, preceded by a *cold front*, penetrating far into the *tropical zone* and often reaching the *equatorial zone*; it brings rain squalls and unusual cold.

polar projection *map projection* centred on Earth's *north pole* or *south pole*.

polar zones *latitude* zones lying between 75° and 90° N and S.

poleward heat transport movement of heat from equatorial and tropical regions toward the poles, occurring as *latent* and *sensible heat transfer*.

pollutants in air pollution studies, foreign matter injected into the lower *atmosphere* as *particulates* or as chemical pollutant *gases*.

pollution dome broad, low dome-shaped layer of polluted air, formed over an urban area at times when winds are weak or calm prevails.

pollution plume (1) The trace or path of pollutant substances, moving along the flow paths of *groundwater*. (2) Trail of polluted air carried downwind from a pollution source by strong winds.

polygon type of *spatial object* in a *geographic information system* with a closed chain of connected *lines* surrounding an area.

polymictic in *limnology*, mixing throughout the year, as a monomictic *lake*.

polyploidy mechanism of *speciation* in which entire chromosome sets of organisms are doubled, tripled, quadrupled, etc.

polypedon smallest distinctive geographic unit of the *soil* of a given area.

pool in flow systems, an area or location of concentration of *matter*. (See also *active pool, storage pool.*)

positive feedback in flow systems, a linkage between flow paths such that the flow in one *pathway* acts to increase the flow in another *pathway*. (See also *feedback, negative feedback.*)

potash feldspar aluminum-silicate *mineral* with potassium the dominant metal.

potential energy *energy* of position; produced by *gravitational* attraction of the Earth's mass for a smaller mass on or near the Earth's surface.

potential evapotranspiration (water need) ideal or hypothetical rate of *evapotranspiration* estimated to occur from a complete canopy of green foliage of growing plants continuously supplied with all the *soil water* they can use; a real condition reached in those situations where *precipitation* is sufficiently great or irrigation water is supplied in sufficient amounts.

pothole cylindrical cavity in hard *bedrock* of a *stream channel* produced by *abrasion* of a rounded *rock* fragment rotating within the cavity.

power source flow of *energy* into a *flow system* that causes *matter* to move.

prairie plant f*ormation class* of the *grassland biome*, consisting of dominant tall grasses and

subdominant *forbs*, widespread in subhumid continental climate regions of the *subtropical zone* and *midlatitude zone*. (See *short-grass prairie, tall-grass prairie.*)

Precambrian time all of geologic time older than the beginning of the Cambrian Period, i.e., older than 600 million years.

precession A change in the direction of the Earth's axis of rotation in space.

precipitation particles of *liquid* water or ice that fall from the atmosphere and may reach the ground. (See *orographic precipitation, convectional precipitation, cyclonic precipitation.*)

predation form of negative interaction among animal *species* in which one species (predator) kills and consumes the other (prey).

preprocessing component component of a *geographic information system* that prepares data for entry to the system.

pressure gradient change of *atmospheric pressure* measured along a line at right angles to the *isobars*.

pressure gradient force force acting horizontally, tending to move air in the direction of lower *atmospheric pressure*.

prevailing westerly winds (westerlies) surface winds blowing from a generally westerly direction in the *midlatitude zone*, but varying greatly in direction and intensity.

primary consumers organisms at the lowest level of the *food chain* that ingest *primary producers* or *decomposers* as their *energy* source.

primary minerals in *pedology (soil science)*, the original, unaltered *silicate minerals* of *igneous rocks* and *metamorphic rocks*.

primary producers organisms that use light *energy* to convert *carbon dioxide* and water to *carbohydrates* through the process of *photosynthesis*.

primary succession *ecological succession* that begins on a newly constructed substrate.

prime meridian reference meridian of zero *longitude*; universally accepted as the Greenwich meridian.

proton positively charged particle within the nucleus of a atom.

product generation component component of a *geographic information system* that provides output products such as maps, images, or tabular reports.

progradation shoreward building of a *beach, bar*, or *sandspit* by addition of coarse *sediment* carried by *littoral drift* or brought from deeper water offshore.

pyroxene group complex aluminum-silicate *minerals* rich in calcium, magnesium, and iron, dark in colour, high in density, classed as *mafic minerals*.

quartz mineral of silicon dioxide composition.

quartzite *metamorphic rock* consisting largely of the mineral *quartz.*

quick clays *clay* layers that spontaneously change from a solid condition to a near-liquid condition when disturbed.

radar an active *remote sensing* system in which a pulse of radiation is emitted by an instrument, and the strength of the echo of the pulse is recorded.

radial drainage pattern stream pattern consisting of *streams* radiating outward from a central peak or highland, such as a *sedimentary dome* or a *volcano.*

radiant energy the form of energy that is emitted from the Sun and is the Earth's principal form of energy.

radiant energy transfer net flow of radiant *energy* between an object and its surroundings.

radiation (See *electromagnetic radiation.*)

radiation balance condition of balance between incoming *energy* of solar *shortwave radiation* and outgoing *longwave radiation* emitted by the Earth into space.

radiation fog *fog* produced by radiation cooling of the basal air layer.

radioactive decay spontaneous change in the nucleus of an atom that leads to the emission of *matter* and *energy.*

radiogenic heat heat from the Earth's interior that is slowly released by the *radioactive decay* of *unstable isotopes.*

radiometric dating a method of determining the geologic age of a *rock* or *mineral* by measuring the proportions of certain of its elements in their different isotopic forms.

rain form of *precipitation* consisting of falling water drops, usually 0.5 mm or larger in diameter.

rain gauge instrument used to measure the amount of *rain* that has fallen.

rain-green vegetation vegetation that puts out green foliage in the wet season, but becomes largely dormant in the dry season; found in the *tropical zone,* it includes the *savanna biome* and *monsoon forest.*

rainshadow belt of arid climate to lee of a mountain barrier, produced as a result of adiabatic warming of descending air.

raised shoreline former *shoreline* lifted above the limit of wave action; also called an elevated *shoreline.*

rapids steep-*gradient* reaches of a *stream channel* in which *stream* velocity is high.

realized niche The reduced environmental space that a species actually occupies because of competition with other species.

Realms the major components of the planet, each with its own unique properties. The four great realms are the atmosphere, lithosphere, hydrosphere, and biosphere.

reclamation in *strip mining,* the process of restoring *spoil* banks and ridges to a natural condition.

recombination source of variation in organisms arising from the free interchange of *alleles* of genes during the reproduction process.

recumbent overturned, as a folded sequence of *rock* layers in which the folds are doubled back upon themselves.

reflection outward scattering of *radiation* toward space by the *atmosphere* and/or Earth's surface.

reg desert surface armored with a pebble layer, resulting from long-continued *deflation;* found in the Sahara Desert of North Africa.

regional geography that branch of *geography* concerned with how the Earth's surface is differentiated into unique *places.*

regional scale the scale of observation at which subcontinental regions are discernable.

regolith layer of *mineral* particles overlying the *bedrock;* may be derived by *weathering* of underlying bedrock or be transported from other locations by *fluid* agents. (See *residual regolith, transported regolith.*)

Regosolic soils a class of *soils* in the Canadian soil classification system that exhibits weakly developed *horizons.*

relative humidity ratio of *water vapour* present in the air to the maximum quantity possible for *saturated air* at the same temperature.

relative variability ratio of a variability measure, such as the average deviation, to the mean of all observations.

remote sensing measurement of some property of an object or surface by means other than direct contact; usually refers to the gathering of scientific information about the Earth's surface from great heights and over broad areas, using instruments mounted on aircraft or orbiting space vehicles.

remote sensor instrument or device measuring *electromagnetic radiation* reflected or emitted from a target body.

removal in soil science, the set of processes that result in the removal of material from a *soil horizon,* such as surface erosion or *leaching.*

representative fraction (R.F.) (See *scale fraction.*)

residual regolith *regolith* formed in place by alteration of the *bedrock* directly beneath it.

resolution on a *map,* power to resolve small objects present on the ground.

respiration the oxidation of organic compounds by organisms that powers bodily functions.

retrogradation cutting back (retreat) of a *shoreline, beach, marine cliff,* or *marine scarp* by wave action.

retrogressive thaw slump slump and flowage of overlying sediment occurring where erosion exposes ice-rich *permafrost* or massive *ground ice* to thawing.

reverse fault type of *fault* in which one fault block rides up over the other on a steep *fault plane.*

revolution motion of a planet in its orbit around the Sun, or of a planetary satellite around a planet.

rhyolite *extrusive igneous rock* of *granite* composition; it occurs as *lava* or *tephra.*

ria coastal embayment or *estuary.*

ria coast deeply embayed *coast* formed by partial *submergence* of a *landmass* previously shaped by fluvial *denudation.*

Richter magnitude scale scale of magnitude numbers describing the quantity of *energy* released by an *earthquake.*

ridge-and-valley landscape assemblage of *landforms* developed by *denudation* of a system of open *folds* of *strata* and consisting of long, narrow ridges and valleys arranged in parallel or zigzag patterns.

rift valley trenchlike valley with steep, parallel sides; essentially a *graben* between two *normal faults;* associated with crustal spreading.

rill erosion form of *accelerated erosion* in which numerous, closely spaced miniature channels (rills) are scored into the surface of exposed *soil* or *regolith.*

rock natural aggregate of *minerals* in the solid state; usually hard and consisting of one, two, or more mineral varieties.

rockslide *landslide* of jumbled *bedrock* fragments.

rock terrace terrace carved in *bedrock* during the *degradation* of a *stream channel* induced by the crustal rise or a fall of the sea level. (See also *alluvial terrace, marine terrace.*)

Rodinia early *supercontinent,* predating *Pangea,* that was fully formed about 700 million years ago.

Rossby waves horizontal undulations in the flow path of the *upper-air westerlies;* also known as upper-air waves.

rotation spinning of an object around an axis.

runoff flow of water from continents to oceans by way of *stream flow* and *groundwater* flow; a term in the water balance of the *hydrologic cycle.* In a more restricted sense, runoff refers to surface flow by *overland flow* and channel flow.

S waves produced by earthquakes, they move slower than P Waves, and shear through rock sideways, at right angles to the direction they are travelling. They cannot propagate in liquid materials.

Sahel (Sahelian zones) belt of *wet–dry tropical* and *semiarid dry tropical* climate in Africa in which *precipitation* is highly variable from year to year.

salic horizon *soil horizon* enriched by soluble salts.

salinity degree of "saltiness" of water; refers to the abundance of such ions as sodium, calcium, potassium, chloride, fluoride, sulfate, and carbonate.

salinization precipitation of soluble salts within the *soil*.

salt flat shallow lake basin covered with salt deposits formed when stream input to the basin is evaporated to dryness; may also form by evaporation of a saline lake when climate changes; see also *dry lake*.

salt marsh *peat*-covered expanse of *sediment* built up to the level of high tide over a previously formed tidal mud flat.

salt water intrusion occurs in a coastal well when an upper layer of fresh water is pumped out, leaving a salt water layer below to feed the well.

salt-crystal growth a form of *weathering* in which *rock* is disintegrated by the expansive pressure of growing salt crystals during dry weather periods when *evaporation* is rapid.

saltation leaping, impacting, and rebounding of sand grains transported over a *sand* or pebble surface by *wind*.

sample standard deviation in statistics, the square root of the average squared deviation from the mean for a sample.

sample statistics numerical values that give basic information about a sample and its variability.

sand *sediment* particles between 0.06 and 2 mm in diameter.

sand dune hill or ridge of loose, well-sorted *sand* shaped by *wind* and usually capable of downwind motion.

sand sea field of *transverse dunes*.

sandspit narrow, fingerlike embankment of *sand* constructed by *littoral drift* into the open water of a *bay*.

sandstone variety of *sedimentary rock* consisting largely of mineral particles of sand grade size.

Santa Ana easterly *wind*, often hot and dry, that blows from the interior desert region of southern California and passes over the coastal mountain ranges to reach the Pacific Ocean.

saturated adiabatic lapse rate (See *wet adiabatic lapse rate*.)

saturated air air holding the maximum possible quantity of *water vapour at* a given temperature and pressure.

saturated zone zone beneath the land surface in which all pores of the *bedrock* or *regolith* are filled with *groundwater*.

saturation vapour pressure (SVP) the maximum amount of water vapour that can be present at a given temperature.

savanna a vegetation cover of widely-spaced *trees* with a grassland beneath.

savanna biome *biome* that consists of a combination of *trees* and grassland in various proportions.

savanna woodland plant *formation class* of the *savanna biome* consisting of a *woodland* of widely spaced *trees* and a grass layer, found throughout the *wet–dry tropical climate* regions in a belt adjacent to the *monsoon forest* and *low-latitude rainforest*.

scale the magnitude of a phenomenon or system, as for example, global scale or local scale.

scale fraction ratio that relates distance on the Earth's surface to distance on a *map* or surface of a globe.

scale of globe ratio of size of a globe to size of the Earth, where size is expressed by a measure of length or distance.

scale of map ratio of distance between two points on a *map* and the same two points on the ground.

scanning systems *remote sensing* systems that make use of a scanning beam to generate images over the frame of surveillance.

scarification general term for artificial excavations and other land disturbances produced for purposes of extracting or processing mineral resources.

scattering turning aside of radiation by an atmospheric molecule or particle so that the direction of the scattered ray is changed.

schist foliated *metamorphic rock* in which mica flakes are typically found oriented parallel with foliation surfaces.

sclerophyll forest plant *formation class* of the *forest biome*, consisting of low sclerophyll *trees*, and often including sclerophyll woodland or *scrub*, associated with regions of *Mediterranean climate* .

sclerophyll woodland plant *formation class* of the *forest biome* composed of widely-spaced sclerophyll *trees* and *shrubs*.

sclerophylls hard-leaved evergreen *trees* and *shrubs* capable of enduring a long, dry summer.

scoria *lava* or *tephra* containing numerous cavities produced by expanding gases during cooling.

scrub plant *formation class* or subclass consisting of *shrubs* and having a canopy coverage of about 50 percent.

sea arch arch-like *landform* of a rocky, cliffed coast created when waves erode through a narrow headland from both sides.

sea breeze local wind blowing from sea to land during the day.

sea cave cave near the base of a *marine cliff*, eroded by breaking waves.

sea fog *fog* layer formed at sea when warm moist air passes over a cool ocean current and is chilled to the *condensation* point.

sea ice floating ice of the oceans formed by direct *freezing* of ocean water.

second of arc 1/60 of a minute, or 1/3600 of a degree.

secondary consumers animals that feed on *primary consumers*.

secondary minerals in *soil science*, minerals that are stable in the surface environment, derived by *mineral alteration* of the *primary minerals*.

secondary succession *ecological succession* beginning on a previously vegetated area that has been recently disturbed by such agents as fire, flood, windstorm, or humans.

sediment finely divided *mineral matter* and organic matter derived directly or indirectly from preexisting *rock* and from life processes. (See *chemically precipitated sediment, organic sediment*.)

sediment yield quantity of sediment removed by *overland flow* from a land surface of given unit area in a given unit of time.

sedimentary cycle type of *biogeochemical cycle* in which the compound or element is released from *rock* by *weathering*, follows the movement of running water either in solution or as *sediment* to reach the sea, and is eventually converted into *rock*.

sedimentary dome up-arched *strata* forming a circular structure with domed summit and flanks with moderate to steep outward *dip*.

sedimentary rock *rock* formed from accumulation of *sediment*.

segregated ice lenses of ice occurring as free masses in soil or regolith of *permafrost* terrain.

seiche standing wave on a *lake* causing oscillation in the water level during a period of minutes to hours; initiated by wind forcing water across the surface.

seismic moment a quantity used to measure the size of an earthquake. It is calculated by measuring the total length of fault rupture and then factoring in the depth of rupture, total slip along the rupture, and the strength of the faulted rocks.

seismic sea wave (tsunami)   train of sea waves set off by an *earthquake* (or other seafloor disturbance) traveling over the ocean surface.

seismic waves   waves sent out during an *earthquake* by faulting or other crustal disturbance from an *earthquake focus* and propagated through the solid Earth.

semiarid (steppe) dry climate subtype   subtype of the dry climates exhibiting a short wet season supporting the growth of grasses and *annual* plants.

semidesert   plant *formation class* of the *desert biome*, consisting of xerophytic *shrub* vegetation with a poorly developed herbaceous lower layer; subtypes are semidesert scrub and *woodland*.

semidiurnal tide   a tide that has two high and two low tides each tidal day.

sensible heat   heat measurable by a *thermometer*; an indication of the intensity of *kinetic energy* of molecular motion within a substance.

sensible heat transfer   flow of heat from one substance to another by direct contact.

sequential landforms   *landforms* produced by external Earth processes in the total activity of *denudation*. Examples: *gorge, alluvial fan, floodplain.*

seral stage   stage in a *sere.*

sere   in an *ecological succession*, the series of biotic communities that follow one another on the way to the stable stage, or *climax.*

sesquioxides   oxides of aluminum or iron with a ratio of two atoms of aluminum or iron to three atoms of oxygen.

set-down   temporary lowering of *lake* level when wind blows water toward the opposite side of the lake.

set-up   temporary raising of *lake* level when wind blows water toward the lake shore.

shadow zone   an area of the Earth from which waves do not emerge or cannot be recorded.

shale   fissile, *sedimentary rock* of *mud* or *clay* composition, showing lamination.

shearing (of rock)   slipping motion between very thin *rock* layers, like a deck of cards fanned with the sweep of a palm.

sheet erosion   type of *accelerated soil erosion* in which thin layers of *soil* are removed without formation of rills or *gullies.*

sheet flow   overland flow taking the form of a continuous thin film of water over a smooth surface of *soil, regolith,* or *rock.*

sheeting structure   thick, subparallel layers of massive *bedrock* formed by spontaneous expansion accompanying *unloading*

shield volcano   low, often large, domelike accumulation of basalt lava flows emerging from long radial fissures on flanks.

shoreline   shifting line of contact between water and land.

short-grass praire   plant *formation class* in the *grassland biome* consisting of short grasses sparsely distributed in clumps and bunches and some *shrubs,* widespread in areas of semiarid climate in continental interiors of North America and Eurasia; also called *steppe.*

shortwave infrared   *infrared radiation* with wavelengths shorter than 3 μm.

short-wave radiation   *electromagnetic energy* in the range from 0.2 to 3 μm, including most of the *energy* spectrum of solar radiation.

shrubs   woody perennial plants, usually small or low, with several low-branching stems and a foliage mass close to the ground.

silica   silicon dioxide in any of several mineral forms.

silicate minerals (silicates)   *minerals* containing silicon and oxygen atoms, linked in the crystal space lattice in units of four oxygen atoms to each silicon atom.

sill   *intrusive igneous rock* in the form of a plate where *magma* was forced into a natural parting in the *bedrock,* such as a bedding surface in a sequence of *sedimentary rocks.*

silt   *sediment* particles between 0.004 and 0.06 mm in diameter.

sinkhole   surface depression in *limestone,* leading down into *limestone caverns.*

slash-and-burn   agricultural system, practised in the *low-latitude rainforest,* in which small areas are cleared and the *trees* burned, forming plots that can be cultivated for brief periods.

slate   compact, fine-grained variety of *metamorphic rock,* derived from *shale,* showing well-developed cleavage.

sleet   form of *precipitation* consisting of ice pellets, which may be frozen raindrops.

slickensides   soil surfaces with grooves or striations made by movement of one mass of soil against the other while the soil is in a moist, plastic state; typical of *Vertisolic soils* and *Vertisols*

sling psychrometer   form of *hygrometer* consisting of a wet-bulb thermometer and a dry-bulb thermometer.

slip face   steep face of an active *sand dune,* receiving sand by *saltation* over the dune crest and repeatedly sliding because of oversteepening.

slope   (1) Degree of inclination from the horizontal of an element of ground surface, analogous to *dip* in the geologic sense. (2) Any portion or element of the Earth's solid surface. (3) Verb meaning "to incline."

small circle   circle formed by passing a plane through a sphere without passing through the exact centre.

small-scale map   *map* with *fractional scale* of less than 1 : 100,000; usually shows a large area.

smog   mixture of *aerosols* and chemical *pollutants* in the lower atmosphere, usually found over urban areas.

snow   form of *precipitation* consisting of ice particles.

soil   natural terrestrial surface layer containing living matter and supporting or capable of supporting plants.

soil colloids   mineral particles of extremely small size, capable of remaining suspended indefinitely in water; typically they have the form of thin plates or scales.

soil creep   extremely slow downhill movement of *soil* and *regolith* as a result of continued agitation and disturbance of the particles by such activities as *frost action,* temperature changes, or wetting and drying of the soil.

soil enrichment   additions of materials to the *soil* body; one of the *pedogenic processes.*

soil erosion   erosional removal of material from the *soil* surface.

soil horizon   distinctive layer of the *soil,* more or less horizontal, set apart from other soil zones or layers by differences in physical and chemical composition, organic content, structure, or a combination of those properties, produced by soil-forming processes.

soil order   a soil class at the highest category in the classification of *soils.*

soil orders   those eleven *soil* classes forming the highest category in the classification of soils.

soil profile   display of *soil horizons* on the face of a freshly cut vertical exposure through the *soil.*

soil science   (See *pedology.*)

soil solum   that part of the *soil* made up of the A, E, and B *soil horizons;* the soil zone in which living plant roots can influence the development of soil horizons.

soil structure   presence, size, and form of aggregations (lumps or clusters) of *soil* particles.

soil texture   descriptive property of the *mineral* portion of the *soil* based on varying proportions of *sand, silt,* and *clay.*

soil water   water held in the *soil* and available to plants through their root systems; a form of *subsurface water.*

soil water balance   balance among the component terms of the *soil-water budget;* namely, *precipitation, evapotranspiration,* change in soil water storage, and water surplus.

soil water belt   *soil* layer from which plants draw *soil water.*

soil water budget   accounting system evaluating the daily, monthly, or yearly amounts of *precipitation, evapotranspiration,* soil-water storage, water deficit, and water surplus.

soil water recharge   restoring of depleted *soil water* by *infiltration* of *precipitation.*

soil water shortage   (See *deficit.*)

soil water storage   actual quantity of water held in the *soil water belt* at any given instant; usually applied to a soil layer of given depth, such as 300 cm (about 12 in.).

solar constant   intensity of solar radiation falling upon a unit area of surface held at right angles to the Sun's rays at a point outside the Earth's *atmosphere;* equal to an *energy* flow of about 1400 W/m^2.

solar day   average time required for the Earth to complete one *rotation* with respect to the Sun; time elapsed between one solar noon and the next, averaged over the period of one year.

solar noon   instant at which the *subsolar point crosses* the *meridian of longitude* of a given point on the Earth; instant at which the Sun's shadow points exactly due north or due south at a given location.

solar power tower   a type of solar furnace using a tower to receive focused sunlight.

solids   substances in the solid state; they resist changes in shape and volume, are usually capable of withstanding large unbalanced forces without yielding, but will ultimately yield by sudden breakage.

solifluction   tundra (arctic) variety of *earthflow* in which sediments of the *active layer* move in a mass slowly downhill over a water-rich plastic layer occurring at the top of *permafrost;* produces *solifluction terraces* and *solifluction lobes.*

solifluction lobe   bulging mass of saturated *regolith* with steep curved front moved downhill by *solifluction.*

solifluction terrace   mass of saturated *regolith* formed by *solifluction* into a flat-topped terrace.

Solonetzic soils   a class of *soils* in the Canadian soil classification system with a *B horizon* of sticky *clay* that dries to a vary hard condition.

solute effect   allows water droplets to develop when air is below saturation. It occurs because cloud droplets form around condensation nuclei, and are rarely composed of pure water.

sorting   separation of one grade size of *sediment* particles from another by the action of currents of air or water.

source region   extensive land or ocean surface over which an *air mass* derives its temperature and moisture characteristics.

south pole   point at which the southern end of the Earth's *axis of rotation* intersects the Earth's surface.

southeast trade winds   surface *winds* of low latitudes that blow steadily from the southeast. (See also *trade winds.*)

southern pine forest   pine forest that is typically found on sandy soils of the Atlantic and Gulf Coast coastal plains.

Southern Oscillation   episodic reversal of prevailing barometric pressure differences between two regions, one centred on Darwin, Australia, in the eastern Indian Ocean, and the other on Tahiti in the western Pacific Ocean; a precursor to the occurrence of an El Niño event. (See also *El Nino.*)

southern pine forest   subtype of *needleleaf forest* dominated by pines and occurring in the *moist subtropical climate.*

spatial data   information associated with a specific location or area of the Earth's surface.

spatial object   a geographic area, *line* or *point* to which information is attached.

speciation   the process by which *species* are differentiated and maintained.

species   a collection of individual organisms that are capable of interbreeding to produce fertile offspring.

specific heat   physical constant of a material that describes the amount of heat energy in joules required to raise the temperature of one gram of the material by one Celsius degree.

specific humidity   mass of *water vapour* contained in a unit mass of air.

spit   (See *sandspit.*)

splash erosion   soil erosion caused by direct impact of falling raindrops on a wet surface of *soil* or *regolith.*

spodic horizon   *soil horizon* containing precipitated amorphous materials composed of organic matter and *sesquioxides* of aluminum, with or without iron.

Spodosols   *soil order* consisting of *soils* with a *spodic horizon,* an *albic horizon,* with low *base status,* and lacking in *carbonate* materials.

spoil   *rock* waste removed in a mining operation.

spreading plate boundary   *lithospheric plate* boundary along which two plates of *oceanic lithosphere* are undergoing separation, while at the same time new lithosphere is being formed by *accretion.* (See also *transform plate boundary.*)

spring tide   a tide that has the greatest range between high and low water.

stable air mass   *air mass* in which the *environmental temperature lapse rate* is less than the *dry adiabatic lapse rate,* inhibiting *convectional* uplift and mixing.

stack (marine)   isolated columnar mass of *bedrock* left standing in front of a retreating *marine cliff.*

stage   height of the surface of a river above its bed or a fixed level near the bed.

standard meridians   *standard time* meridians separated by 15° of *longitude* and having values that are multiples of 15°. (In some cases meridians are used that are multiples of $7\frac{1}{2}°$.)

standard time system   time system based on the local time of a *standard meridian* and applied to belts of *longitude* extending $7\frac{1}{2}°$ (more or less) on either side of that meridian.

standard time zone   zone of the Earth in which all inhabitants keep the same time, which is that of a *standard meridian* within the zone.

star dune   large, isolated *sand dune* with radial ridges culminating in a peaked summit; found in the deserts of North Africa and the Arabian Peninsula.

statistics   a branch of mathematical sciences that deals with the analysis of numerical data.

Stefan-Boltzmann's Law   describes the relationship between the flow of energy from an object and that object's surface temperature.

steppe   semiarid grassland occurring largely in dry continental interiors. (See *short-grass prairie.*)

steppe climate   (See *semiarid (steppe) dry climate subtype.*)

stone polygons   linked ringlike ridges of cobbles or boulders lying at the surface of the ground in arctic and alpine tundra regions.

storage capacity   maximum capacity of *soil* to hold water against the pull of *gravity.*

storage pool   type of pool in a *biogeochemical cycle* in which materials are largely inaccessible to life. (See also *active pool.*)

storage recharge   restoration of stored soil water during periods when *precipitation* exceeds *potential evapotranspiration (water need).*

storage withdrawal   depletion of stored *soil water* during periods when *evapotranspiration* exceeds *precipitation,* calculated as the difference between *actual evapotranspiration (water use)* and *precipitation.*

stored energy   (See *potential energy.*)

storm surge   rapid rise of coastal water level accompanying the onshore arrival of a *tropical cyclone.*

strata   layers of *sediment* or *sedimentary rock* in which individual beds are separated from one another along bedding planes.

stratified drift   *glacial drift* made up of sorted and layered *clay, silt, sand,* or gravel deposited from meltwater in *stream channels,* or in marginal lakes close to the ice front.

stratiform clouds   clouds of layered, blanketlike form.

stratopause   upper limit of the stratosphere.

stratosphere   layer of *atmosphere* lying directly above the *troposphere*.

stratovolcano   volcano constructed of multiple layers of *lava* and *tephra* (volcanic ash).

stratus   cloud type of the low-height family formed into a dense, dark gray layer.

stream   long, narrow body of flowing water occupying a *stream channel* and moving to lower levels under the force of *gravity*. (See *consequent stream, graded stream, subsequent stream.*)

stream capacity   maximum *stream load* of solid matter that can be carried by a *stream* for a given *discharge*.

stream channel   long, narrow, troughlike depression occupied and shaped by a *stream* moving to progressively lower levels.

stream deposition   accumulation of transported particles on a *stream* bed, upon the adjacent *floodplain*, or in a body of standing water.

stream erosion   progressive removal of mineral particles from the floor or sides of a *stream channel* by drag force of the moving water, or by *abrasion*, or by *corrosion*.

stream flow   water flow in a *stream channel*; same as channel flow.

stream gradient   rate of descent to lower elevations along the length of a *stream channel*, stated in m/km, ft/mi, degrees, or percent.

stream load   solid matter carried by a *stream* in dissolved form (as *ions*), in *turbulent suspension,* and as *bed load*.

stream ordering   a system for studying the properties of a stream, in which streams are organized by their general status in the drainage basin. This process progressively unites the channel segments that comprise a branching river network from it's headwaters to its mouth.

stream profile   a graph of the elevation of a *stream* plotted against its distance downstream.

stream transportation   downvalley movement of eroded particles in a *stream channel* in solution, in *turbulent suspension,* or as *bed load*.

strike   compass direction of the line of intersection of an inclined *rock* plane and a horizontal plane of reference. (See *dip*.)

strike-slip fault   variety of *fault* on which the motion is dominantly horizontal along a near-vertical *fault plane*.

strip mining   mining method in which overburden is first removed from a seam of *coal*, or a sedimentary ore, allowing the coal or ore to be extracted.

structure (of a system)   the pattern of the pathways and their interconnections within a flow system.

subantarctic low-pressure belt   persistent belt of low *atmospheric pressure* centred about at lat. 65°S over the Southern Ocean.

subantarctic zone   *latitude* zone lying between lat. 55° and 60°S (more or less) and occupying a region between the *midlatitude zone* and the *antarctic zone*.

subarctic zone   *latitude* zone between lat. 55° and 60° N (more or less), occupying a region between the *midlatitude zone* and the *arctic zone*.

subduction   descent of the downbent edge of a *lithospheric plate* into the *asthenosphere* so as to pass beneath the edge of the adjoining plate.

sublimation   process of change of ice (solid state) to *water vapour* (gaseous state); in *meteorology*, sublimation also refers to the change of state from water vapour (liquid) to ice (solid), which is referred to as *deposition* in this text.

submergence   inundation or partial drowning of a former land surface by a rise of sea level or a sinking of the *crust* or both.

suborder   a unit of *soil* classification representing a subdivision of the *soil order*.

subsea permafrost   *permafrost* lying below sea level, found in a shallow offshore zone fringing the arctic seacoast.

subsequent stream   *stream* that develops its course by *stream erosion* along a band or belt of weaker *rock*.

subsolar point   point on the Earth's surface at which solar rays are perpendicular to the surface.

subsurface water   water of the lands held in *soil*, *regolith*, or *bedrock* below the surface.

subtropical broadleaf evergreen forest   a formation class of the forest biome composed of broadleaf evergreen *trees*; occurs primarily in the regions of the *moist subtropical climate* .

subtropical evergreen forest   a subdivision of the *forest biome* composed of both broadleaf and needleleaf evergreen *trees*.

subtropical high-pressure belts   belts of persistent high *atmospheric pressure* trending east–west and centred about on lat. 30° N and S.

subtropical jet stream   *jet stream* of westerly winds forming at the *tropopause*, just above the *Hadley cell*.

subtropical needleleaf evergreen forest   a *formation class* of the *forest biome* composed of needleleaf evergreen *trees* occurring in the *moist subtropical climate* ⑥ of the southeastern U.S. ; also referred to as the southern pine forest.

subtropical zones   *latitude* zones occupying the region of lat. 25° to 35° N and S (more or less) and lying between the *tropical zones* and the *midlatitude zones*.

succulents   plants adapted to resist water losses by means of thickened spongy tissue in which water is stored.

summer monsoon   inflow of maritime air at low levels from the Indian Ocean toward the Asiatic low pressure centre in the season of high Sun; associated with the rainy season of the *wet–dry tropical climate* and the Asiatic monsoon climate.

summer solstice   solstice occurring on June 21 or 22, when the *subsolar point* is located at 23 1/2°N.

Sun-synchronous orbit   satellite orbit in which the orbital plane remains fixed in position with respect to the Sun.

supercontinent   single world continent, formed when *plate tectonic* motions move continents together into a single, large land mass. (See also *Pangea*.)

supercooled water   water existing in the liquid state at a temperature lower than the normal *freezing* point.

surface   the very thin layer of a substance that received and radiates energy and conducts heat to and away from the substance.

surface energy balance equation   equation expressing the balance among *heat* flows to and from a surface.

surface water   water of the lands flowing freely (as *streams*) or impounded (as ponds, *lakes*, marshes).

surface wave   a wave that transmits energy from an earthquake's epicentre along the Earth's surface. Along with S Wave's, these produce the strongest vibrations and are the source of the most damage.

surges   episodes of very rapid downvalley movement within an *alpine glacier*.

suspended load   that part of the *stream load* carried in *turbulent suspension*.

suspension   (See *turbulent suspension*.)

suture   (See *continental suture*.)

swash   surge of water up the *beach* slope (landward) following collapse of a *breaker*.

symbiosis   form of positive interaction between *species* that is beneficial to one of the species and does not harm the other.

sympatric speciation   type of *speciation* in which speciation occurs within a larger population.

synclinal mountain   steep-sided ridge or elongate mountain developed by erosion of a syncline.

synclinal valley   valley eroded on weak *strata* along the central trough or axis of a *syncline*.

syncline   downfold of *strata* (or other layered *rock*) in a troughlike structure; a class of *folds*. (See also *anticline*.)

system   (1) a collection of things that are somehow related or organized; (2) a scheme for naming, as in a classification system; (3) a flow system of *matter* and *energy*.

systematic geography   the study of the physical, economic, and social processes that differentiate the Earth's surface into *places*.

systems approach   the study of the inter-connections among natural processes by focusing on how, where, and when *matter* and *energy* flow in natural systems.

systems theory   body of knowledge explaining how systems work.

taiga   plant *formation class* consisting of *woodland* with low, widely spaced *trees* and a ground cover of lichens and mosses, found along the northern fringes of the region of *boreal forest climate*; also called cold woodland.

tailings   (See *spoil.*)

talik   pocket or region within permafrost that is unfrozen; ranges from small inclusions to large "holes" in permafrost under lakes.

tall-grass prairie   a *formation class* of the *grassland biome* that consists of tall grasses with broad-leaved *herbs*.

talus   accumulation of loose *rock* fragments derived by fall of *rock* from a *cliff*.

talus slope   slope formed of *talus*.

tar sand   (See *bitumin*.)

tarn   small *lake* occupying a *rock* basin in a *cirque* of *glacial trough*.

tectonic activity   process of bending (folding) and breaking (faulting) of crustal mountains, concentrated on or near active *lithospheric plate* boundaries.

tectonic arc   long, narrow chain of islands or mountains or a narrow submarine ridge adjacent to a *subduction* boundary and its trench, formed by *tectonic processes,* such as the construction and rise of an *accretionary prism*.

tectonic crest   ridgelike summit line of a *tectonic arc* associated with an *accretionary prism*.

tectonics   branch of *geology* relating to tectonic activity and the features it produces. (See also *plate tectonics, tectonic activity*.)

temperature gradient   rate of temperature change along a selected line or direction.

temperature inversion   upward reversal of the normal *environmental temperature lapse rate,* so that the air temperature increases upward. (See *low-level temperature inversion, high-level temperature inversion*.)

temperature regime   distinctive type of annual temperature cycle.

tephra   collective term for all size grades of solid *igneous rock* particles blown out under gas pressure from a volcanic vent.

terminal moraine   *moraine* deposited as an embankment at the terminus of an *alpine glacier* or at the leading edge of an *ice sheet*.

terrane   continental crustal *rock* unit having a distinctive set of lithologic properties, reflecting its geologic history, that distinguish it from adjacent or surrounding *continental crust*.

terrestrial ecosystems   *ecosystems* of land plants and animals found on upland surfaces of the continents.

tetraploid   having four sets of chromosomes instead of a normal two sets.

thematic map   *map* showing a single type of information.

theme   category or class of information displayed on a *map*.

thermal erosion   in regions of permafrost, the physical disruption of the land surface by melting of *ground ice,* brought about by removal of a protective organic layer.

thermal infrared   a portion of the *infrared radiation wavelength* band, from approximately from 3 to 20 μm, in which objects at temperatures encountered on the Earth's surface (including fires) emit *electromagnetic radiation*.

thermal pollution   form of water pollution in which heated water is discharged into a *stream* or *lake* from the cooling system of a power plant or other industrial heat source.

thermally direct   an increase in density that results from a loss of energy.

thermistor   electronic device that measures (air) temperature.

thermocline   water layer of a lake or the ocean in which temperature changes rapidly in the vertical direction.

thermokarst   in arctic environments, an uneven terrain produced by thawing of the upper layer of *permafrost,* with settling of sediment and related water erosion; often occurs when the natural surface cover is disturbed by fire or human activity.

thermokarst lake   shallow lake formed by the thawing and settling of permafrost, usually in response to disturbance of the natural surface cover by fire or human activity.

thermometer   instrument measuring temperature.

thermometer shelter   louvered wooden cabinet of standard construction used to hold *thermometers* and other weather-monitoring equipment.

thermosphere   atmospheric layer of upwardly increasing temperature, lying above the *mesopause*.

thorntree semidesert   *formation class* within the *desert biome*, transitional from *grassland biome* and *savanna biome* and consisting of xerophytic *trees* and *shrubs*.

thorntree-tall-grass savanna   plant *formation class*, transitional between the *savanna biome* and the *grassland biome*, consisting of widely scattered *trees* in an open grassland.

threshold velocity   the velocity required to entrain a particle of a given diameter and increases with the square root of the particle size.

thrust sheet   sheetlike mass of *rock* moving forward over a *low-angle overthrust fault*.

thunderstorm   intense, local convectional storm associated with a *cumulonimbus cloud* and yielding heavy *precipitation*, also with lightning and thunder, and sometimes the fall of *hail*.

tidal current   current set in motion by the *ocean tide*.

tidal-generating force   the cyclically varying resultant effect of lunar and solar gravitational forces and their directions.

tidal inlet   narrow opening in a *barrier island* or baymouth *bar* through which *tidal currents* flow.

tide   (See *ocean tide*.)

tidewater glacier   glaciers that end in the ocean or in a lake.

tide curve   graphical presentation of the rhythmic rise and fall of ocean water because of *ocean tides*.

till   heterogeneous mixture of *rock* fragments ranging in size from *clay* to boulders, deposited beneath moving glacial ice or directly from the *melting* in place of stagnant glacial ice.

till plain   undulating, plainlike land surface underlain by glacial *till*.

time cycle   in flow systems, a regular alternation of flow rates with time.

time zones   zones or belts of given east–west (*longitudinal*) extent within which *standard time* is applied according to a uniform system.

topographic contour   *isopleth* of uniform elevation appearing on a *map*.

tornado   small, very intense wind vortex with extremely low air pressure in centre, formed beneath a dense *cumulonimbus cloud* in proximity to a *cold front*.

traction   transport of sediment by wind or water in which the sediment remains in contact with the ground or bed of the stream, moving by rolling or sliding.

trade winds (trades)   surface winds in low latitudes, representing the low-level airflow within the *tropical easterlies*.

transcurrent fault   *fault* on which the relative motion is dominantly horizontal, in the direction of the *strike* of the fault; also called a *strike-slip fault.*

transform fault   special case of a *strike-slip fault* making up the boundary of two moving *lithospheric plates;* usually found along an offset of the *mid-oceanic ridge* where seafloor spreading is in progress.

transform plate boundary   *lithospheric plate* boundary along which two plates are in contact on a *transform fault;* the relative motion is that of a *strike-slip fault.*

transform scar   linear topographic feature of the ocean floor taking the form of an irregular scarp or ridge and originating at the offset *axial rift* of the *mid-oceanic ridge;* it represents a former *transform fault* but is no longer a plate boundary.

transformation (soils)   a class of soil-forming processes that transform materials within the soil body; examples include *mineral alteration* and *humification.*

translocation   a soil-forming process in which materials are moved within the soil body, usually from one horizon to another.

transpiration   evaporative loss of water to the *atmosphere* from leaf pores of plants.

transportation   (See *stream transportation.*)

transported regolith   *regolith* formed of *mineral matter* carried by *fluid* agents from a distant source and deposited upon the *bedrock* or upon older regolith. Examples: floodplain silt, lake clay, beach sand.

transverse dunes   field of wavelike *sand dunes* with crests running at right angles to the direction of the prevailing *wind.*

traveling anticyclone   centre of high pressure and *outspiraling* winds that travels over the Earth's surface; often associated with clear, dry weather.

traveling cyclone   centre of low pressure and *inspiraling* winds that travels over the Earth's surface; includes *wave cyclones, tropical cyclones,* and *tornadoes.*

travertine   *carbonate mineral matter,* usually *calcite,* accumulating upon *limestone cavern* surfaces situated in the *unsaturated zone.*

tree   large erect woody perennial plant typically having a single main trunk, few branches in the lower part, and a branching crown.

trellis drainage pattern   *drainage pattern* characterized by a dominant parallel set of major *subsequent streams,* joined at right angles by numerous short tributaries; typical of *coastal plains* and belts of eroded *folds.*

tropic of cancer   *parallel of latitude* at 23 1/2°N.

tropic of capricorn   *parallel of latitude* at 23 1/2°S.

tropical cyclone   intense *traveling cyclone* of tropical and subtropical latitudes, accompanied by high *winds* and heavy rainfall.

tropical easterlies   low-latitude wind system of persistent air flow from east to west between the two *subtropical high-pressure belts.*

tropical easterly jet stream   upper-air *jet stream* of seasonal occurrence, running east to west at very high altitudes over Southeast Asia.

tropical high-pressure belt   a high-pressure belt occurring in tropical latitudes at a high level in the *troposphere;* extends downward and poleward to form the *subtropical high-pressure* belt, located at the surface.

tropical zones   *latitude* zones centred on the *tropic of cancer* and the *tropic of capricorn,* within the latitude ranges 10° to 25° N and 10° to 25°S, respectively.

tropical-zone rainforest   plant *formation class* within the *forest biome* similar to *equatorial rainforest,* but occurring farther poleward in tropical regions.

tropopause   boundary between *troposphere* and *stratosphere.*

tropophyte   plant that sheds its leaves and enters a dormant state during a dry or cold season when little soil water is available.

troposphere   lowermost layer of the *atmosphere* in which air temperature falls steadily with increasing altitude.

tsunami   (See *seismic sea wave.*)

tundra biome   *biome* of the cold regions of *arctic tundra* and *alpine tundra,* consisting of grasses, grasslike plants, flowering *herbs,* dwarf *shrubs,* mosses, and *lichens.*

tundra climate   cold climate of the *arctic zone* with eight or more consecutive months of zero *potential evapotranspiration (water need).*

tundra soils   soils of the arctic *tundra climate* regions.

turbulence   in *fluid* flow, the motion of individual water particles in complex eddies, superimposed on the average downstream flow path.

turbulent flow   mode of *fluid* flow in which individual *fluid* particles (molecules) move in complex eddies, superimposed on the average downstream flow path.

turbulent suspension   *stream transportation* in which particles of *sediment* are held in the body of the *stream* by turbulent eddies. (Also applies to wind transportation.)

typhoon   *tropical cyclone* of the western North Pacific and coastal waters of Southeast Asia.

Udalfs   suborder of the *soil order Alfisols;* includes Alfisols of moist regions, usually in the *midlatitude zone,* with deciduous forest as the natural vegetation.

Udolls   suborder of the *soil order Mollisols;* includes Mollisols of the moist soil water regime in the *midlatitude zone* and with no horizon of *calcium carbonate* accumulation.

Ultisols   *soil order* consisting of *soils* of warm soil temperatures with an *argillic horizon* and low *base status.*

ultramafic igneous rock   *igneous rock* composed almost entirely of *mafic minerals,* usually olivine or *pyroxene group.*

ultraviolet radiation   *electromagnetic energy* in the *wavelength* range of 0.2 to 0.4 μm.

unloading   process of removal of overlying *rock* load from *bedrock* by processes of *denudation,* accompanied by expansion and often leading to the development of *sheeting structure.*

unsaturated zone   *subsurface water* zone in which pores are not fully saturated, except at times when *infiltration* is very rapid; lies above the *saturated zone.*

unstable air   air with substantial content of *water vapour,* capable of breaking into spontaneous convectional activity leading to the development of heavy showers and *thunderstorms.*

unstable isotope   elemental isotope that spontaneously decays to produce one or more new isotopes. (See also *daughter product.*)

upper-air westerlies   system of westerly winds in the upper *atmosphere* over middle and high latitudes.

upwelling   upward motion of cold, nutrient-rich ocean waters, often associated with cool equatorward currents occurring along *continental margins.*

Ustalfs   suborder of the *soil order Alfisols;* includes Alfisols of semiarid and seasonally dry climates in which the *soil* is dry for a long period in most years.

Ustolls   suborder of the *soil order Mollisols;* includes Mollisols of the semiarid climate in the *midlatitude zone,* with a horizon of *calcium carbonate* accumulation.

valley winds   air movement at night down the *gradient* of valleys and the enclosing mountainsides; alternating with daytime *mountain winds.*

vapour pressure   refers to the contribution that water vapour makes to the pressure exerted by the atmosphere.

variability   measure of the variation in an series of observations that centre around a mean.

variation   in the study of evolution, natural differences arising between parents and offspring as a result of *mutation* and *recombination.*

varve   annual layer of *sediment* on the bottom of a *lake* or the ocean marked by a change in colour or texture of the *sediment.*

veins   small, irregular, branching network of *intrusive rock* within a preexisting *rock* mass.

vernal equinox   *equinox* occurring on March 20 or 21, when the *subsolar point* is at the *equator.*

Vertisolic soils   a class of soils in the Canadian Soil Classification System with a high clay content, developing deep, wide cracks when dry and showing evidence of movement between aggregates; similar to Vertisols.

Vertisols   *soil order* consisting of *soils* of the *subtropical zone* and the *tropical zone* with high *clay* content, developing deep, wide cracks when dry, and showing evidence of movement between aggregates.

visible light   *electromagnetic energy* in the *wavelength* range of 0.4 to 0.7 μm.

void   empty region of pore space in sediment; often occupied by water or water films.

volcanic bombs   boulder-sized, semisolid masses of *lava* that are ejected from an erupting *volcano.*

volcanic neck   isolated, narrow steep-sided peak formed by erosion of *igneous rock* previously solidified in the feeder pipe of an extinct *volcano.*

volcanism   general term for *volcano* building and related forms of extrusive igneous activity.

volcano   conical, circular structure built by accumulation of *lava* flows and *tephra.* (See *stratovolcano, shield volcano.*)

volcano coast   *coast* formed by *volcanoes* and *lava* flows built partly below and partly above sea level.

warm front   moving weather *front* along which a warm *air mass* is sliding up over a cold air mass, leading to production of *stratiform clouds* and *precipitation.*

warm-blooded animal   animal that possesses one or more adaptations to maintain a constant internal temperature despite fluctuations in the environmental temperature.

warm-core ring   circular eddy of warm water, surrounded by cold water and lying adjacent to a warm, poleward moving ocean current, such as the Gulf Stream. (See also *cold-core ring.*)

washout   downsweeping of atmospheric *particulates* by *precipitation.*

water gap   narrow transverse *gorge* cut across a narrow ridge by a *stream,* usually in a region of eroded *folds.*

water need   (See *potential evapotranspiration.*)

water resources   a field of study that couples basic study of the location, distribution, and movement of water with the utilization and quality of water for human use.

water surplus   water disposed of by *runoff* or percolation to the groundwater zone after the *storage capacity* of the *soil* is full.

water table   upper boundary surface of the *saturated zone;* the upper limit of the *groundwater* body.

water use   (See *actual evapotranspiration.*)

water vapour   the gaseous state of water.

waterfall   abrupt descent of a *stream* over a *bedrock* step in the *stream channel.*

waterlogging   rise of a *water table* in *alluvium* to bring the zone of saturation into the root zone of plants.

watershed   a land area that drains water into a river system or other body of water.

watt   unit of power equal to the quantity of work done at the rate of one joule per second; symbol, W.

wave cyclone   *traveling cyclone* of the midlatitudes involving interaction of cold and warm *air masses* along sharply defined *fronts.*

wave-cut notch   *rock* recess at the base of a *marine cliff* where wave impact is concentrated.

wavelength   distance separating one wave crest from the next in any uniform succession of traveling waves.

weak equatorial low   weak, slowly moving low-pressure centre *(cyclone)* accompanied by numerous convectional showers and *thunderstorms;* it forms close to the *intertropical convergence zone* in the rainy season, or *summer monsoon.*

weather   physical state of the *atmosphere* at a given time and place.

weather system   recurring pattern of atmospheric circulation associated with characteristic weather, such as a *cyclone* or *anticyclone.*

weathering   total of all processes acting at or near the Earth's surface to cause physical disruption and chemical decomposition of *rock.* (See *chemical weathering, physical weathering.*)

west-wind drift   ocean drift current moving eastward in zone of *prevailing westerlies.*

westerlies   (See *prevailing westerly winds, upper-air westerlies.*)

wet adiabatic lapse rate (saturated adiabatic lapse rate)   reduced *adiabatic lapse rate* when *condensation* is taking place in rising air; value ranges between 4 and 9°C per 1000 m (2.2 and 4.9°F per 1000 ft).

wet equatorial climate   moist climate of the *equatorial zone* with a large annual *water surplus,* and with uniformly warm temperatures throughout the year.

wetlands   land areas of poor surface drainage, such as marshes and swamps.

wet-dry tropical climate   climate of the *tropical zone* characterized by a very wet season alternating with a very dry season.

whiting   precipitation of calcium carbonate in lake water, causing the water to have a milky appearance.

Wien's Law   states that there is an inverse relationship between the wavelength of the peak of the emission of a blackbody and its temperature.

Wilson Cycle   *plate tectonic* cycle in which continents rupture and pull apart, forming oceans and *oceanic crust,* then converge and collide with accompanying subduction of *oceanic crust.*

wilting point   quantity of stored *soil water,* less than which the foliage of plants not adapted to *drought* will wilt.

wind   air motion, dominantly horizontal relative to the Earth's surface.

wind abrasion   mechanical wearing action of wind-driven *mineral* particles striking exposed *rock* surfaces.

wind erosivity   the ability of wind to erode.

wind vane   weather instrument used to indicate *wind* direction.

winter monsoon   outflow of continental air at low levels from the Siberian high, passing over Southeast Asia as a dry, cool northerly *wind.*

winter solstice   solstice occurring on December 21 or 22, when the *subsolar point* is at 23 1/2°S.

Wisconsinan Glaciation   last glaciation of the *Pleistocene Epoch.*

woodland   plant *formation class,* transitional between *forest biome* and *savanna biome,* consisting of widely spaced *trees* with canopy coverage between 25 and 60 percent.

Xeralfs   suborder of the *soil order Alfisols;* includes Alfisols of the *Mediterranean climate.*

xeric animals   animals adapted to dry conditions typical of a *desert* climate.

Xerolls   suborder of the *soil order Mollisols;* includes Mollisols of the *Mediterranean climate.*

xerophytes   plants adapted to a dry environment.

zone of ablation   the area in which annual loss of snow through melting, evaporation, iceberg calving and sublimation exceeds annual gain of snow and ice on the surface.

zone of accumulation   the area above the firn line, where snowfall accumulates and exceeds the losses from ablation.

zooplankton   microscopic animals found largely in the uppermost layer of ocean or lake water.

PHOTO CREDITS

Chapter 1 Part Opener: J. D. Ives. Chapter Opener: Liaison Agency, Inc. Fig. 1.2: Robin Karpan/Parkland. Fig. 1.3: Victor Last. Fig. 1.4a: National Resources Canada, Centre for Topographic Information. © 2006. Produced under licence from Her Majesty the Queen in Right of Canada, with permission of Natural Resources Canada. Fig. 1.4b: ESRI Canada. Fig. 1.4c: Provided by the SeaWiFS Project, NASA/Goddard Space Flight Center, and ORBIMAGE. Fig. 1.5: Frank Slide Interpretive, Crowsnest Pass, Alberta. Fig. 1.6: CP/Larry MacDougal. Fig. 1.7: Simon Kriby. Fig.1.8: iStockphoto. Fig. 1.9: NASA. Fig. 1.12: iStockphoto. Fig. 1.13: iStockphoto. Fig. 1.14: iStockphoto. Eye on the Landscape: Victor Last.

Chapter 2 Part Opener: Stephen Sage. Chapter Opener: D. Archibold. 2.4: © National Maritime Museum, Greenwich, London. Fig. 2.5: Garmin.com. Fig. 2.6: Aquarius.NET®. Fig. 2.8: Committee on Geographical Studies, University of Chicago. Fig. 2.9: U.S. Navy Oceanographic Office. Fig. 2.10a&b: National Research Council of Canada, http://inms-ienm.nrc-cnrc.gc.ca/time_services/daylight_saving_e.html. Printed with permission from the National Research Council Canada (NRC), 2007. Fig. 2.12: First Light/Olivier Mackay. Fig. 2.14: Courtesy of Charles Graves. A Closer Look, Fig. 1: INCA/Global Chemical Weather Forecast. A Closer Look, Fig. 2: A.N. Strahler. A Closer Look, Fig. 3a&b: National Resources Canada, Centre for Topographic Information. © 2006. Produced under licence from Her Majesty the Queen in Right of Canada, with permission of Natural Resources Canada. A Closer Look, Fig. 5: D. Archibold. A Closer Look, Fig. 7: Adapted from: Alberta Agriculture and Food, 2005. A Closer Look, Fig. 8: Adapted from: Alberta Agriculture and Food, 2005. Eye on the Landscape: Photo Richard McGuire - richardmcguire.ca.

Chapter 3 Fig. 3.1: Fab Lab, Center for Bits and Atoms, Massachusetts Institute of Technology. Fig. 3.3: Courtesy Daedalus Enterprises, Inc. & National Geographic Magazine. Fig. 3.4: After W.D. Sellers, Physical Climatology, University of Chicago Press. Fig. 3.6: A.N. Strahler. Fig. 3.9: University of Cambridge Centre for Atmospheric Science. Fig. 3.10: Data received by the Canada Centre for Remote Sensing. Image provided by RADARSAT International. Fig. 3.11: A.N. Strahler. Eye on Global Change, Fig. 3.3a&b: NASA. Fig. 3.12: Victor Last (Image #43720). Focus on Re-

mote Sensing, Fig. 3.4: NASA. Fig. 3.16: Joe Flores, Southern California Edison/NREL. Focus on Remote Sensing, Fig. 2: Courtesy Bow River Irrigation District, Land Division. Focus on Remote Sensing, Fig. 3a,b,&c: NASA/GSFC/MITI/ERSDA C/JAROS and U.S./Japan ASTER Science Team. Focus on Remote Sensing, Fig. 4: SAR image courtesy of Intera Technologies Corporation, Calgary, Alberta, Canada. Focus on Remote Sensing, Fig. 5: Courtesy of Dr. Diana Engel, Data Discovery Hurricane Science Center. Focus on Remote Sensing, Fig. 6: Philips Maps Co. UK. Eye on the Landscape: Photo Archive (Gino Donato/Sudbury Star).

Chapter 4 Chapter Opener: CORBIS/Lowell Georgia. Fig. 4.2: Environment Canada Weather Office. Fig. 4.5: Earth Observatory, NASA. Fig. 4.7: European Centre for Medium-Range Weather Forecasts. Fig 4.8a&b: V.B. Mendes, University of Lisbon. Fig. 4.11: D. Archibold. Fig. 4.14: European Centre for Medium-Range Weather Forecasts. Fig. 4.21a&b: Center for Coastal Physical Oceanography, Old Dominion University, Norfolk, Virginia. Eye on Global Change 4.3, Fig. 2: A.N. Strahler. Fig. 4.23: After Hansen et al., 2000, Proc. Nat. Acad. Sci. Fig. 4.24: James Hansen/NASA Goddard Institute for Space Studies. Fig. 4.25: Shawn Henry/SABA. Fig. 4.26: Dr. Henri D. Grissino-Mayer. Fig. 4.27: Courtesy of Gordon C. Jacoby of the Tree-Ring Laboratory of the Lamont-Doherty Geological Observatory of Columbia University. Eye on Global Change 4.4: IPCC, Climate Change Report 2001: Synthesis Report, copyright IPCC 2001, Cambridge University Press. Fig. 4.28: Department of Fisheries & Oceans, Government of Canada; from J. Shaw et al., "Climate Change and the Canadian Coast," 1996. Reproduced with the permission of Her Majesty the Queen in Right of Canada, 2007. Eye on the Landscape: iStockphoto.

Chapter 5 Chapter Opener: CP Photo Archive/Adrian Wyld. Fig. 5.8: San Francisco State University, Department of Geosciences, California Regional Weather Server (http://virga.sfsu.edu). Fig. 5.10: Winter Science Curriculum Project, State University of New York at Oswego. Adapted from Val Eichenlaub (1979), *Weather and Climate of the Great Lakes Region*, page 60. Notre Dame, Indiana: University of Notre Dame Press. Fig. 5.14: iStockphoto. Fig. 5.16: North Pacific Seaplanes, Prince Rupert, BC. Fig. 5.17: NASA. Fig. 5.19: Wikipedia; usage free under

GNU Free Documentation License. Eye on the Environment, Fig. 1: iStockphoto image ID 107900. Eye on the Environment, Fig 2a&b: Canadian Wind Energy Atlas, Environment Canada. Reproduced with the permission of the Minister of Public Works and Government Services Canada, 2007. Fig. 5;23: A.N. Strahler. Fig. 5.24: A.N. Strahler. Figs. 5.25: National Oceanic and Atmospheric Administration. Fig. 5.26b: Environment Canada Weather Office. Fig. 5.27a: Kevin Ambrose, Washingtonprints.com. Fig. 5.28: A.N. Strahler. Fig. 5.30: A.N. Strahler. Fig. 5.31: iStockphoto. Eye on Global Change, Fig. 2a&b: (El Niño) Ropelewski, C. F. and M. S. Halpert, 1987: Global and regional scale precipitation patterns associated with the El Niño/Southern Oscillation., Mon. Wea. Rev., 115, 1606 1626. Eye on Global Change, Fig. 3a&b: (La Niña) Ropelewski, C. F. and M. S. Halpert, 1989: Precipitation patterns associated with the high index phase of the Southern Oscillation. Jour. of Climate., 2, 268 284. Eye on Global Change, Fig. 4: A.N. Strahler. Eye on the Landscape: Dr. Alexey Sergeev.

Chapter 6 Chapter Opener: Wolfgang Kaehler/CORBIS. Fig. 6.5: iStockphoto. Fig. 6.7: Robert Body (www.RobertBody.com). Fig. 6.8: A.N. Strahler. Fig. 6.9: A.N. Strahler. Focus on Remote Sensing Fig a: NASA GSFC GOES Project. Focus on Remote Sensing Fig b: Unidata Program Center, http://www.unidata.ucar.edu/. Fig. 6.11a,b,c,&d: (a) Darryl Torckler/Stone/Getty Images; (b) Keese van den Berg/Photo Researchers; (c) Trevor Mein/Stone/Getty Images; (d) Larry Ulrich/Stone/Getty Images. Fig 6.12: Ron Sanford/Stone/Getty Images. Fig. 6.13: "Mean Number of Days/Freezing Rain, Canada" from *The Climates of Canada*, 1990, pp. 60–61. Reproduced with the permission of the Minister of Public Works and Government Services Canada, 2007. Fig. 6.14: Public Safety and Emergency Preparedness Canada, *Keeping Canadians Safe. Average number of days with hail.* www.psepc.gc.ca/images/hail_map_e.gif. Reproduced with the permission of the Minister of Public Works and Government Services Canada, 2007. Fig. 6.15: Courtesy of John Cassano, University of Colorado. Fig 6.16a&b: Original map data provided by The Atlas of Canada http://atlas.gc.ca/ © 2007. Produced under licence from Her Majesty the Queen in Right of Canada, with permission of Natural Resources Canada. Fig. 6.17: "Mean Number of Days/Blowing Snow, Canada" from *The Climates of Canada*, 1990, pp. 60–61. Reproduced with the permission of the Minister of Public

Works and Government Services Canada, 2007. Fig. 6.23: Stu Hood (nkife) / flickr.com. Eye on the Landscape: CP Photo Archive/Kaz Novak.

Chapter 7 Chapter Opener: Dusty Davis. Fig 7.1: Meteorological Directorate, Ministry of Transportation, Kingdom of Bahrain; Bahrain TradeNet WLL. Fig. 7.2: U.S. Dept. of Commerce. Fig. 7.3: The Image Bank/Getty Images. Fig. 7.6: A.N. Strahler. Fig. 7.7: A.N. Strahler. Fig 7.8: Sierra College Geography Department. Fig. 7.9: A.N. Strahler. Fig. 7.11d: CBC/Radio-Canada. Fig. 7.11e: CBC/Radio-Canada. Fig. 7.11f: CBC/Radio-Canada. Fig 7.17: © Christopher Smith/Getty Images News & Sport Services. Fig. 7.18: Data from H. Riehl, Tropical Meterology, NY McGraw-Hill. Fig 7.19: GOES-12 satellite, NASA, NOAA. Fig. 7.20: The Green Lane™, Environment Canada's World Wide Web site. Fig 7.21: Redrawn from NOAA, National Weather Service. Fig. 7.22: NASA. Fig 7.23: USGS Coastal & Marine Geology Program. Eye on the Environment, Fig1a,b&c: CP Photo Archives (David J. Phillips/AP). Eye on the Environment, Fig. 2: Justin R. Glenn, Southeast Regional Climate Center, Columbia, SC. Fig. 7.25: A.N. Strahler. Eye on the Landscape: © Jim Reed Photography.

Chapter 8 Chapter Opener: © Todd Tiffan, Tiffan Images. Fig 8.3: Simplified and modified from Plate 3, World Climatology, Volume 1, The Time Atlas, Editor John Bartholomew, The Times Publishing Company, Ltd. London, 1958. Fig. 8.8: A.N. Strahler. Eye on the Landscape: Richard Wear/Design Pics.

Chapter 9 Chapter Opener: Courtesy of Town Council of Alice Springs. Fig. 9.2: iStockphoto. Fig. 9.6a: Spelbrink.org. Fig. 9.9: © Will & Deni McIntyre/Photo Researchers. Fig. 9.12: Calvin Jones. Fig. 9.14a&b: Bill Archibold. Fig. 9.15: Bill Archibold. Fig. 9.16a&b: Earth Observatory, NASA. Eye on Global Change, Fig. 2: US Geographic Service. Eye on Global Change, Fig. 3: Sharon E. Nicholson, Department of Meteorology, Florida State University, Tallahassee. Fig. 9.17a&b: Visible Earth, NASA. Fig 9.21a,b,&c: (a) Michael Fogden/DRK Photo (b) Walter Schmidt/Peter Arnold Inc. (c) Art Wolfe/Photo Researchers. Fig. 9.23a: Government of Australia. Fig. 9.25: Michele Burgess/Corbis. Eye on the Landscape: Christian Puff, Faculty Center of Botany, University of Vienna, Austria.

Chapter 10 Chapter Opener: Craig Tuttle/Corbis. Fig. 10.2: © Fred Hirschmann. Fig. 10.3: © Royal Botanic Gardens & Domain Trust, Sydney, Australia. Fig. 10.6a&b: NOAA Southeast Regional Climate Center. Fig. 10.8: Biodiversity Conservation and Sustainable Development, University of Wisconsin. Fig. 10.11a: Department of Labour, New Zealand, http://www.osh.dol.govt.nz. Fig. 10.11b: Mark L. Kaufman. Fig. 10.12a: Porterfield-Chickering/Photo Researchers. Fig. 10.12b: The Society for Growing Australian Plants. Fig. 10.12c: Copyright © 2004, Point Reyes National Seashore, US National Park Service. Fig. 10.14: David Muench Photography. Fig. 10.15: School of Natural Resources & Agricultural Science, University of Fairbanks, Alaska. Fig. 10.16: Nechako Fisheries Conservation Program. Fig. 10.17: Government of British Columbia. Eye on Environment: Agriculture and Agri-Food Canada, http://agr.gc.ca/pfra/drought/maps/dr02fig1e.pdf. Fig. 10.19b: Unimaps.com. Fig. 10.23: ImageShack.com. Fig. 10.26: NASA. Fig. 10.27a: Canadian Wildland Fire Information System, Natural Resources Canada. Fig. 10.27b: Canadian Wildland Fire Information System, Natural Resources Canada. Fig. 10.27c: University of Toronto Fire Management Systems Laboratory. Fig. 10.28: R.N. Drummond. Eye on the Landscape: Jacqueline Pratt.

Chapter 11 Part Opener: www.gdaywa.com. Chapter Opener: Yann Arthus-Bertrand/Altitude. Fig. 11.1: U.S. Geological Survey, 345 Middlefield Road, MS 870, Menlo Park, CA 94025. Fig. 11.3: Roberto de Gugliemo/Photo Researchers. Fig. 11.4a&b: National Atlas of the United States. Fig. 11.5: Bill Archibold. Fig. 11.6: National Atlas of the United States. Fig. 11.7: National Atlas of the United States. Fig. 11.9a&b: Ward's Natural Science Establishment. Fig. 11.11a&b: Ward's Natural Science Establishment. Fig. 11.13: L.J. Maher. Fig. 11.14: Greg McLearn. Fig. 11.15: Andrew Alden, geology.about.com. Fig. 11.16: Field British Columbia, www.field.ca/. Fig. 11.17: Jason Shim, flickr.com. Fig. 11.18: iStockphoto. Fig. 11.19: Victor Englebert/Photo Researchers. Fig. 11.20: Elk Valley Coal Corporation, www.elkvalleycoal.ca/. Fig. 11.21: A.N. Strahler. Fig. 11.22: A.N. Strahler. Fig. 11.23: © Freeman Patterson/Masterfile. Focus on Systems 11.1: A.N. Strahler. Eye on the Landscape: The Mars Society, www.marssociety.org/portal.

Chapter 12 Chapter Opener: Banque de Photos SVT-Lyon. Fig. 12.3: A.N. Strahler. Working it Out 12.1: A.N. Strahler. Fig. 12.4: A.N. Strahler. Fig. 12.5: A.N. Strahler. Fig. 12.6: Wikipedia. Fig. 12.8: Dave Sandwell. Fig. 12.10: A.N. Strahler. Fig. 12.13: A.N. Strahler. Fig. 12.16: A.N. Strahler. Fig. 12.17: A.N. Strahler. Fig. 12.18: A.N. Strahler. Fig. 12.19: NASA. Fig. 12.20: From *Plate Tectonics*, copyright © 1998 by Arthur N. Strahler. Fig. 12.22: Redrawn and simplified from maps by R. S. Dietz and J. C. Holden, Jour. Geophysical Research, vol. 75, pp. 4943–4951, Figures 2 to 6. Copyrighted by the American Geophysical Union. Used with permission. Eye on the Landscape: Chris Durbin/www.sln.org.uk/geography.

Chapter 13 Chapter Opener: Credit: P. Rona, OAR/National Undersea Research Program (NURP); NOAA. Fig. 13.1: A.N. Strahler. Fig. 13.2: James Mason/Black Star. Fig. 13.3: A.N. Strahler. Fig. 13.4: K. Hamdorft/Auscape International Ltd. Fig. 13.5: Kevin West/Liaison Agency, Inc./Getty Images. Fig. 13.6: Greg Vaughn/Tom Stack & Associates. Fig. 13.7: epa/Corbis. Fig. 13.8: Werner Stoy/Camera Hawaii, Inc. Focus on Systems 13.1, Fig. 1: A.N. Strahler. Focus on Systems 13.1, Fig. 2: *Adapted with permission from Macdonald and Abbott, 1970,* Volcanoes in the Sea: The Geology of Hawaii, Honolulu, *University of Hawaii Press, p. 138, Figure 92.* Focus on Systems 13.1, Fig 3: A.N. Strahler. Fig. 13.9: Arthur N. Strahler. Fig. 13.10: Carol Evenchick, Natural Resources Canada. Reproduced with the permission of the Minister of Public Works and Government Services Canada, 2007 and Courtesy of Natural Resources Canada, Geological Survey of Canada. Fig. 13.11: ©Larry Ulrich. Fig. 13.12: Steve Vidler/Leo de Wys, Inc. Eye on the Environment 13.2: iStockphoto. Fig. 13.13: A.N. Strahler. Focusn on Remote Sensing 13.3, Fig. 1: Image courtesy NASA/GSFC/MITI/ERSDAC/JAROS and U. S./Japan ASTER Science Team. Focus on Remote Sensing 13.3, Fig. 2: Image courtesy NASA/JPL/NIMA. Focus on Remote Sensing 13.3, Fig. 3: Image courtesy Ron Beck, EROS Data Center. Fig. 13.15a: © Russ Heinl. Fig. 13.15b: © Russ Heinl. Fig. 13.16: Bill Archibold. Fig. 13.18: A. N. Strahler. Fig. 13.20: Georg Gerster/Comstock Images. Fig. 13.21: James Balog/Black Star. Fig. 13.22: Exelsior/Sipa Press. Fig. 13.23: A.N. Strahler. Fig. 13.25: National Oceanic and Atmospheric Administration, West Coast & Alaska Tsunami Warning Center. Fig. 13.26: NOAA Center for Tsunami Research. Eye on the Landscape: Tom Bean/Corbis.

Chapter 14 Part Opener: Yann Arthus-Bertrand/Altitude. Chapter Opener: A.N. Strahler. Fig. 14.1: A.N. Strahler. Fig. 14.2a: © Susan Rayfield/Photo Researchers. Fig. 14.2b: © Steve McCutcheon. Fig. 14.4: © Tom Bean. Fig. 14.6: A.N. Strahler. Fig. 14.8: A.N. Strahler. Fig. 14.10: A.N. Strahler. Fig. 14.11: A.N. Strahler. Fig 14.12: © AP/Wide World Photos. Fig. 14.13: © Bill Davis/Black Star. Fig. 14.14: Adapted from Troy L. Pewe, Geotimes, vol 29, no 2, copyright 1984, by the American Geologic Institute. Fig. 14.15: From Robert F. Black, "Permafrost," Chapter 14 of P. D. Trask's Applied Sedimentation., copyright © by John Wiley & Sons. Focus on Systems 14.2: Adapted with permission from A. H. Lachenbruch in Rhodes W. Fairbridge, Ed., The Encyclopedia of Geomorphology, New York: Reinhold Publishing Corp. Fig. 14.18: Chris Burn. Fig. 14.19: Adapted with permission from A. H. Lachenbruch in Rhodes W. Fairbridge, Ed., The Encyclopedia of Geomorphology, New York: Reinhold Publishing Corp. Fig. 14.20: Stephen J. Krasemann/DRK Photo. Fig. 14.21: Steve McCutcheon. Fig. 14.23: Chris Burn. Fig. 14.24: Chris Burn. Fig. 14.25a&b: Chris Burn. Fig. 14.26: Chris Burn. Fig. 14.27: Chris Burn. Fig. 14.28: Bernard Hallet, Periglacial Laboratory, Quaternary Research Center. Fig. 14:29: © Steve McCutcheon/Alaska Pictorial Service. Eye on the Landscape: © Douglas Peebles Photography.

Chapter 15 Chapter Opener: Yann Arthus-Bertrand/Altitude. Fig. 15.3: A.N. Strahler. Fig 15.4: A.N. Strahler. Fig. 15.5: A.N. Strahler. Fig. 15.6: © Laurence Parent. Fig 15.7: Dr. Brian Jones, Dept. of Earth & Atmospheric Sciences, University of Alberta. Fig. 15.8: A.N. Strahler. Fig. 15.9: © Bruno Barbey/Magnum Photos, Inc. Fig. 15.10: A.H. Strahler. Fig. 15.11: A.H. Strahler. Fig. 15.12: Atlas of Canada, Natural Resources Canada. Focus on Systems 15.1, Fig. 1: A.N. Strahler. Focus on System 15.1, Fig. 2: David R. Frazier/Photo Researchers. Fig. 15.14: After Hoyt and Langbein, Floods, copyright © Princeton University Press. Used by permission. Fig. 15.16: Natural Resources Canada. Reproduced with the permission of the Minister of Public Works and Government Services Canada, 2007 and Courtesy of Natural Resources Canada, Geological Survey of Canada. Fig. 15.18a: K. Francis, S. Fick/Canadian Geographic: Sources: Manitoba Centre for Remote Sensing; Royal Commission on Flood Cost-Benefit, 1958. Fig. 15.18c: Robert Gilbert. Fig. 15.18d: CP Photo Archive (Winnipeg Free Press/Ken Gigliotti). Fig. 15.18e: CP Photo Archive (Winnipeg Free Press/Ken Gigliotti). Fig. 15.19: NASA/Corbis Images. Fig. 15.20: Robert Gilbert. Eye on the Environment 15.3, Fig. 1: A.N. Strahler. Eye on the Environment 15.3, Fig. 2: A.N. Strahler. Fig. 15.24: © Tom Till Photography. Fig. 15.28: Robert Gilbert. Eye on the Landscape: Portland General Electric, Oregon.

Chapter 16 Chapter Opener: Wolfgang Weber, Canadian Arctic Gallery. Fig. 16.1: A.N. Strahler. Fig. 16.3: Dr. P.M. Haygarth, Institute of Grassland and Environmental Research, Devon, UK. Fig. 16.4: © Bernie Bauer. Fig. 16.7a: John Quinton, Lancaster Environment Centre, Lancaster University, Lancaster, United Kingdom. Fig. 16.7b: Gene Dayton, Central Queensland University. Fig. 16.8: Rob Broek. Fig. 16.9: Bill Archibold. Fig. 16.14a: Nick Gray. Fig. 16.14b: A.M. Stacey. Fig 16.14c: H.J.A. Berendsen. Fig. 16.15: Keith Taylor, Gloucester, England. Fig. 16.16: A.N. Strahler. Fig. 16.17: Mark Zanzig/www.zanzig.com. Fig. 16.18a: A.N. Strahler. Fig. 16.19: Bill Archibold. Fig. 16.20: A.N. Strahler. Fig 16.21: Dept. of Sedimentary Geology, Penn State University. Fig. 16.22: A.N. Strahler. Fig. 16.23: A.N. Strahler. Fig. 16.24: © Rocky Mountaineer Vacations® and The Armstrong Group. Fig. 16.25: Ralph R.B. von Frese. Fig. 16.26: A.N. Strahler. Focus on Systems 16.2, Fig. 1: A.N. Strahler. Fig. 16.27: L.D. Gordon/The Image Bank. Fig. 16.28: Jan Kopec/Stone/Getty Images. Fig. 16.30: Courtesy of B.C. Hydro. Fig. 16.31: © Marc Reid. Fig. 16.32: Breck Kent/Animals Animals/Earth Scenes. Fig. 16.33: Kazza. Fig. 16.34: T.A. Wiewnadt/DRK Photo. Fig. 16.35: A.N. Strahler. Fig. 16.36: Mark A. Melton. Eye on the Landscape: 4x4 Adventures, NZ.

Chapter 17 Chapter Opener: Ron Chapple/Corbis. Fig 17.1: A.N. Strahler. Fig. 17.2: A.N. Strahler. Focus on Remote Sensing 17.1, Fig. 1: Earth Satellite Corporation. Focus on Remote Sensing 17.1, Fig. 2: Alan Strahler. Focus on Remote Sensing 17.1, Fig. 3: USGS/NASA. Focus on Remote Sensing 17.1, Fig. 4: USGS/NASA. Fig. 17.3: Tom Bean. Fig. 17.4: A.N. Strahler. Fig. 17.5: A.N. Strahler. Fig. 17.6: Mishuna.image.pbase.com. Fig. 17.8: University of Washington Astronomy Department. Fig. 17.9: A.N. Strahler. Fig. 17.10: LANDSAT. Fig. 17.11: A.N. Strahler. Fig. 17.12: EPIC. Fig. 17.13: EPIC. Fig. 17.14: EPIC. Fig. 17.15: Dow Williams, dowclimbing.com. Fig. 17.16: Steve Smith. Fig. 17.17: Ludo Kuipers, OzOutback. Fig. 17.18: LANDSAT/NASA. Fig. 17.19: Alex McLean/Landslides. Fig. 17.20: Galen Rowell/Corbis. Fig. 17.21: David & Shannon Steffenson. Fig. 17.22: NASA/D. Roddy, LPI. Fig. 17.23: Dr. Jill Bechtold.

Fig. 17.24: A.N. Strahler. Eye on the Landscape: Steven Vidler/Eurasia Press/Corbis.

Chapter 18 – Chapter Opener: Copyright © Adrian Heisey. Fig. 18.3: Declan McCullagh Photograph. Fig. 18.4: M.J. Coe/Animals Animals/Earth Scenes. Fig. 18.5: Satellite image courtesy of GeoEye. Fig. 18.6: Patrick Hesp. Fig.18.7: ©Johnathan Bascom. Fig. 18.8: Andrew G. Fountain. Fig.18.9: © Atlantide Phototravel/Corbis. Fig. 18.10: Darwin Wiggett/firstlight.ca. Fig. 18.11: John S. Shelton. Fig. 18.14: © Michael Fenton, http://www.earthscience-world.org/. Fig. 18.15: Photography courtesy of Patrick Hesp. Fig. 18.16: ©Brian J. McMorrow. Fig. 18.17: Copyright © George Steinmetz. Fig. 18.18: J.A. Kraulis/Masterfile. Fig. 18.19: G.R. Roberts/The Natural Sciences Image Library. Fig. 18.21: Alex McLean/Landslides. Fig. 18.23: © Wolfgang Kaehler/CORBIS. Fig. 18.24: Andrew Stacey. Fig. 18.26: Steve Dunwell. Fig 18.27: © Yann Arthus-Bertrand/CORBIS. Fig. 18.28: With permission of The Scanbrit School of English, Bournemouth, UK. Fig. 18.33a: © John Carnemolla/Australian Picture Library/CORBIS. Fig. 18.33b: © Wolfgang Meier/zefa/Corbis. Fig. 18.34a: Roy Toft/Getty Iimages. Fig. 18.34b: Robert Harding World Imagery/Getty Images. Fig. 18.35a: M-SAT-LTD/SPL/PUBLIPHOT. Fig. 18.35c © CORBIS SYGMA. A Closer Look 18.1, Fig. 1: © Stephen Rose/Liaison Agency, Inc./Getty Images. A Closer Look 18.1, Fig. 2: Stephen Crowley/New York Times Pictures. A Closer Look 18.1, Fig. 3: Stock Images/firstlight.ca. A Closer Look 18.1, Fig. 4: photograph courtesy of Steve Solomon, Geological Survey of Canada Atlantic. Eye on the Landscape: Harald Sund/Getty Images.

Chapter 19 Chapter Opener: Dean Conger/Corbis. Fig. 19.1a: © Ron Niebrugge/www.wildnatureimages.com. Fig. 19.1b: Dr. William Shilts. Fig. 19.1c: Ann Badjura. Fig. 19.1d: Icefield Helicopter Tours, Alberta, Canada. Fig. 19.2: Dr. Lindley S. Hanson, Salem State College, Dept. Geological Sciences. Fig. 19.3: A.N. Strahler. Focus on Systems 20.1, Fig. 1: A.N. Strahler. Fig. 19.4: A.N. Strahler. Fig. 19.5a: About.com. Fig. 19.5b: Icefield Helicopter Tours, Alberta, Canada. Fig. 19.6a: H.J.A. Berendsen. Fig. 19.6b: Gifford H. Miller. Fig. 19.7a: The Lakes Gliding Club, Walney Airfield, Barrow-in-Furness, Cumbria, UK. Fig. 19.7b: Prof. T. Smith, University of California at Santa Barbara. Fig. 19.7c: Jim Carpenter. Fig. 19.7d: CORBIS. Focus on Remote Sensing 19.2, Fig. 1: NASA. Focus on Remote Sensing 19.2, Fig. 2: NASA/GSFC/MITI/ERSDAC/JAROS and U.S./Japan ASTER science team. Focus on Remote Sensing 19.2, Fig. 3: A.N. Strahler. Eye on Global Change 19.3, Fig. 2a: Carl Key/US Geological Service. Eye on Global Change 19.3, Fig. 2b: Blase Reardon/US Geological Service. Fig. 19.9: Canadian Foundation for Innovation. Fig. 19.10a,b,c&d: National Snow and Ice Data Center, Boulder, CO. Fig. 19.12: MSN Encarta. Fig. 19.13: Roberto Julio Bessin. Fig. 19.14: Prof. Steven Dutch, University of Wisconsin, Green Bay.

Fig. 19.16: J.B. Krygier. Fig. 19.17: Juerg Alean, Eglisau, Switzerland, Glaciers online

Chapter 20 Part Opener: Alan Bauer Photography. Chapter Opener: Alan Bauer Photography. Fig. 20.2: X-Rite Inc., Grand Rapids, MI. Fig. 20.4: Ontario Ministry of Food, Agriculture and Rural Affairs. Copyright © 2007 Queen's Printer for Ontario. Reproduced with permission. Fig. 20.9: A.N. Strahler. Fig. 20.15a: University of Calgary Dept. of Geography. Fig. 20.15b: Agriculture & Agri-food Canada, Canadian Soil Information System (CanSIS). Fig. 20.16c: University of Calgary Dept. of Geography. Fig. 20.15d: Agriculture & Agri-food Canada, Canadian Soil Information System (CanSIS). Fig. 20.15e: Agriculture & Agri-food Canada, Canadian Soil Information System (CanSIS). Fig. 20.15f: Agriculture & Agri-food Canada, Canadian Soil Information System (CanSIS). Fig. 20.15g: US Geological Service. Fig. 20.15h: Agriculture & Agri-food Canada, Canadian Soil Information System (CanSIS). Fig. 20.15i: Agriculture & Agri-food Canada, Canadian Soil Information System (CanSIS). Fig. 20.15j: US Dept. of Agriculture Natural Resources Conservation Service. Fig. 20.15k: US Dept. of Agriculture Natural Resources Conservation Service. Fig. 20.16: Canadian Soil Information System (CanSIS), Agriculture and Agri-Food Canada. Fig. 20.17: X-Rite Inc., Grand Rapids, MI. Fig. 20.18: University of Calgary Dept. of Geography. Fig. 20.19: Agriculture and Agri-Food Canada. Fig. 20.20: US Dept. of Agriculture Natural Resources Conservation Service. Fig. 20.21a: Henry D. Foth. Fig. 20.21b: Henry D. Foth. Fig. 20.21c: Henry D. Foth. Fig. 20.21d: Henry D. Foth. Fig. 20.21e: Henry D. Foth. Fig. 20.21f: Henry D. Foth. Fig. 20.21g: Henry D. Foth. Fig. 20.22: William E. Ferguson. Fig 20.23: M. Collier/DRK Photo. Eye on the Landscape: Fridmar Damm/zefa/Corbis.

Chapter 21 Chapter Opener: Dr. Bruce Marcot. Fig 21.7: A.N. Strahler. Eye on Global Change 21.1: Dr. Bruce Marcot. Fig. 21.8: A.N. Strahler. Fig. 21.9: A.N. Strahler. Fig. 21.10: NASA. Fig. 21.11a: NASA. Fig. 21.11b: NASA. Focus on Remote Sensing 21.2, Fig. 1a&b: Natural Resources Canada. Focus on Remote Sensing 21.2, Fig. 2a&b: SeaWiFS Project/ORBIMAGES. Focus on Remote Sensing 21.2, Fig. 3: MODIS Science Team/NASA. Focus on Remote Sensing 21.2, Fig. 4a&b: MODIS Science Team/NASA. Focus on Remote Sensing 21.2, Fig. 5: R. Nemani, University of Montana NTSG/NASA. Focus on Remote Sensing 21.2, Fig. 5: NASA/NOAA. Eye on the Landscape: First People, www.firstpeople.us.

Chapter 22 Chapter Opener: Joel W. Rogers/Corbis. Fig. 22.1a: Parks.it. Fig. 22.1b: ©2007, Doug Von Gausig, Critical Eye Photography. Fig. 22.1c: iStockphoto. Fig. 22.2a: iStockphoto. Fig. 22.2b: © Copyright 1996-2006 Earl J.S. Rook. Fig. 22.2c: iStockphoto. Fig. 22.5: Gregory

D. Dimijian/Photo Researchers. Fig. 22.6: Adapted from BC Ministry of Forests and Range. Fig. 22.8: NASA. Fig. 22.9: iStockphoto. Fig. 22.10: Daniel Mosquin. Fig. 22.11: Dan L. Perlman. Fig. 22.12: Adapted from Nova Scotia Museum of Natural History, *Natural History of Nova Scotia, Vol. 1: Habitats*, pp. 424, 424. Fig. 22.13a: Bill Archibold. Fig. 22.13b: Bill Archibold. Fig. 22.15: J.H. Brown and M.V. Lomolino, Biogeography, 2nd ed., 1998, Sinauer, Sunderland, Mass., used by permission. Fig. 22.17: S.W. Carter/Photo Researchers. Fig. 22.18: J. H. Brown and M. V. Lomolino, Biogeography, second ed., 1998, Sunderland, Massachusetts, used by permission. Fig. 22.19: Adapted from Millennium Ecosystem Assessment. Fig. 22.22: Adapted from Norman Myers, Russell A. Mittermeir, Cristina G. Mittermier, Gustavo A. B. da Fonseca, and Jennifer Kent. Biodiversity hotspots for conservation priorities. Nature, 403:853-858, 2000. Fig. 22.23: Wikipedia. Eye on the Landscape: National Forestry Database Program.

Chapter 23 Chapter Opener: © Galen Rowell/Corbis. Fig. 23.1a: Natural Resources Canada. Fig. 23.1b: Natural Resources Canada. Fig. 23.1c: D. Archibold. Fig. 23.2: © Steve McCutcheon. Fig. 23.2b: NASA/Goddard Space Flight Center, Scientific Visualization Studio. Fig. 23.2c: NASA/Goddard Space Flight Center Scientific Visualization Studio. Focus on Remote Sensing 23.1, Fig. 2: A.H. Strahler/NASA. Focus on Remote Sensing 23.1, Fig. 3: A.H. Strahler/NASA. Fig. 23.5a: Paul A. Souders/Corbis. Fig. 23.5b: © Ferrero/Labat/Auscape International Pty. Ltd. Fig. 23.5c: © Tom Bean. Eye on the Environment 23.2: © Jacques Jangoux/Peter Arnold Inc. Fig. 23.7: © John S. Shelton. Fig. 23.10: © Kenneth Murray/Photo Researchers. Fig. 23.11a: © Kevin Schafer/CORBIS. Fig. 23.11b: © Chris Taylor; Cordaiy Photo. Fig. 23.13a: © David Muench/CORBIS. Fig. 23.13b: © Raymond Gehman/CORBIS. Fig. 23.13c: © Gunter Marx Photography/CORBIS. Fig. 23.15a: © Stuart Westmorland/CORBIS. Fig. 23.15b: © Raymond Gehman/CORBIS. Fig. 23.15c: © Galen Rowell/CORBIS. Fig. 23.16a: D. Archibold. Fig. 23.16b: © David Muench/Corbis. Fig. 23.16c: © David Samuel Robbins/CORBIS. Fig. 23.17a: Fletcher & Baylis/Photo Researchers. Fig. 23.17b: © DLILLC/Corbis. Fig. 23.19a: © David Muench/Corbis. Fig. 23.19b: © Andrew Brown; Ecoscene/CORBIS. Fig. 23.19c: © Roger De La Harpe; Gallo Images/CORBIS. Fig. 23.20a: © Arthur Morris/Corbis. Fig, 23.20b: © DLILLC/Corbis. Fig. 23.20c: © Paul A. Souders/Corbis. Fig. 23.22a: © Annie Griffiths Belt/Corbis. Fig. 23.22b: © Layne Kennedy/Corbis. Fig. 23.22c: © Nik Wheeler/Corbis. Fig. 23.24a: © Tom Bean/Corbis. Fig. 23.24b: © George D. Lepp/Corbis. Fig. 23.24c: © Peter Johnson/Corbis. Fig. 23.25a: © Geray Sweeney/Corbis. Fig. 23.25b: © Scott T. Smith/Corbis. Fig. 23.25d: © David Muench/Corbis. Fig. 23.26a: © Michael T. Sedam/Corbis. Fig. 23.26b: Bill Archibold. Fig. 23.26c: © Paul A. Souders/Corbis. Fig. 23.27: M. P. Kahl/Photo Researchers. Eye on the Landscape: © Andrew Brown; Ecoscene/Corbis.

Index

A

A horizon, soil, 501, 505
Abatement, 259
Ablation, 463
Ablation till, 479
Abrasion, 465
Abrasion platform, 448
Absolute instability, 145
Absolute stability, 145
Absorption, 45
Abyssal plains, 295
Accretionary prism, 301
Acid action, 340–341
Active continental margins, 297
Active layer, 351
Active pools, 524
Active radar system, 61
Actual evapotranspiration, 498, 499
Adaptive radiation, 555
Adiabatic cooling, 135
Adiabatic process, 134–136
Advection fog, 140
Aeolian, 437
Aerial photography, 61
Aerosols, 50
African Sahel, 212–213
African savanna, 588
Aggradation, 402
Air masses, 153–155

Air pollution climatology, 172
Air temperature, 68–97
   annual range, 85
   daily insolation and net radiation, 72–73
   daily temperature, 73–74
   future scenarios, 91, 94
   global warming, 86–91
   high mountain environments, 78–79
   influencing factors, 70
   IPCC Report of 2001, 92–93
   land-water contrast, 80–82
   measuring, 70–71
   net radiation, 80
   radiation, 42, 80
   rural/urban temperature, 74–75
   stratosphere and upper layers, 77–78
   surface temperature, 70
   temperature inversion, 79
   temperature record (1886-2002), 90–91
   temperatures close to the ground, 74
   troposphere, 76–77
   world patterns, 82–85
Albedo, 54
Aleutian Low, 107
Alfisols, 512
Allelopathy, 551
Allopatric speciation, 555
Alluvial rivers, 405

Alluvial terrace, 405
Alluvial terrace formation, 405
Alluvium, 342
Alpine chains, 292
Alpine debris avalanches, 344
Alpine folds, 300
Alpine glaciers, 465–468, 489
Alpine permafrost, 348, 359
Alpine tundra, 249, 359, 592, 593
Altitude, 100
Altitude zones of vegetation, 595–597
Altocumulus clouds, 137, 138
Altostratus cloud, 137
Aluminum sesquioxide, 496
American plate, 297
Amphibole group, 271
Anabranching river, 406
Ancient mountain roots, 295
Andean alpine glacial features, 468
Andesite, 273
Andisols, 513
Anenometer, 100
Aneroid barometer, 100
Angle of repose, 442
Annual insolation, 46, 47
Annular drainage pattern, 425
Ant guards, 551
Antarctic Circle, 33

Antarctic Ice Sheet, 472, 484
Antarctic sea ice, 475
Antarctic zones, 48
Antarctic-front zone, 188
Anticlinal valley, 425
Anticlines, 321
Anticyclones, 108, 156–163
Aphelion, 31
Aquicludes, 367
Aquifer, 367
Arabian plate, 297
Aragonite, 278
Aral Sea, 241–242
Arc, 410
Arctic air mass, 155
Arctic and polar deserts, 184
Arctic Circle, 33
Arctic permafrost, 250–251
Arctic sea ice, 85, 474
Arctic shoreline, 457
Arctic tundra, 592, 593
Arctic tundra environment, 249
Arctic zone, 48
Arctic-Alpine Life Zone, 596
Arctic-front zone, 188
Arête, 468, 470
Argillipedoturbation, 510
Argon, 49
Arid climate, 192, 215
Arid climate landforms, 422
Arid landforms, 461
Aridisols, 513–514
Arizona monsoon, 105
Artesian well, 367, 368
Asiatic Low, 108
Associations, 544
ASTER, 62, 280, 281, 322
Asthenosphere, 290
Astronomical hypothesis, 486
Aswan, Egypt, 80
Athabasca tar sands, 9
Atmosphere, 10
　air pollutants, 50–52
　components, 48–49
　energy flows, 57
　ozone, 49–51
　pressure and density, 102
　primordial, 48
　solar energy losses, 53

stratosphere and upper layers, 77–78
　troposphere, 76–77
Atmospheric absorption, 45
Atmospheric dust transport, 440
Atmospheric moisture and precipitation, 128–151
　adiabatic process, 134–136
　atmospheric stability, 145–149
　clouds, 136–138, 139
　fog, 138–140
　global precipitation, 181–185
　humidity, 132–134
　hydrologic cycle, 131–132
　hydrosphere, 130–131
　microburst, 148, 149
　precipitation, 140–145
　relative humidity, 133–134
　seasonability of precipitation, 185–187
　thunderstorm, 146–147, 148
Atmospheric pressure, 99–100, 102, 103
Atmospheric pressure maps, 106
Atmospheric stability, 145–149
Atoll, 458
Atomic mass number, 292
Austral-Indian plate, 297
Australasian realm, 561
Autogenic succession, 552
Autumnal (fall) equinox, 32
Axial rift, 295
Azores High, 105, 106, 108

B

B horizon, soil, 501, 505
Backswamp, 406
Backwash, 448
Badland topography, 399
Baja California desert, 218
Banded gneiss, 280
Bar, 449
Barchan dunes, 442
Barometer, 99–100
Barrier reef, 458
Barrier-island coast, 453
Barringer Meteorite Crater, 430, 431
Basalt, 273
Basaltic shield volcanoes, 314
Base flow, 376
Base status, 496
Bauxite, 496

Beach, 448
Beach drift, 448
Beaufort scale, 445
Bed load, 400
Bedrock, 342, 494
Bedrock disintegration, 338
Bergeron process, 140
Bioclimatic frontier, 547
Biodiversity, 9, 562–564
Biogas, 524
Biogeochemical cycles, 524–533
Biogeoclimatic zones, 548
Biogeographic processes, 542–567
　bioclimatic frontier, 547
　biodiversity, 562–564
　biogeographic realms, 560–562
　dispersal, 556–557
　distribution patterns, 557, 560
　disturbance, 548–550
　ecological biogeography, 544–551
　ecological succession, 551–554
　edaphic (soil) factors, 548
　evolution, 554–555
　extinction, 556
　geomorphic factors, 547–548
　historical biogeography, 554–562
　interaction among species, 550–551
　island biogeography, 558–559
　speciation, 555–556
　species-area curve, 558–559
　temperature, 546
　water need, 544–546
Biogeographic realms, 560–562
Biogeography, 6, 519
Biomass, 522, 524
Biomass energy, 523–524
Biome, 571. See also Terrestrial biomes
Biosphere, 10
Biosphere - systems and cycles, 518–541
　biogeochemical cycles, 524–533
　biomass energy, 523–524
　carbon cycle, 526, 528–529
　energy flow in ecosystems, 519–523
　food web, 520–521
　net photosynthesis, 521–522
　net primary production, 522–523
　net production and climate, 523
　nitrogen cycle, 527–530
　nutrient elements, 525–526

oxygen cycle, 526–527
photosynthesis, 520–521
respiration, 521
sedimentary cycles, 532–533
sulphur cycle, 530–532
Biosphere reserves, 564
Biotite, 271
Bismark plate, 297
Bitumen, 279
Blackbody, 44
Block mountains, 324
Block separation, 338
Blowing snow, 142
Blowout (sand dune), 440
Bluffs, 406
Bora, 110
Boreal forest, 30, 584
Boreal forest climate, 247–249
Boreal stage, 487
Braided channel, 402, 404
Brandberg Massif, 421
Brisbane, Australia, 75
Broadleaf evergreen forest, 576
Brunisolic soils, 507
Burgess Shale fossils, 276, 277
Burgess Shale trilobites, 277
Butte, 422

C

C horizon, soil, 501, 505
C$_3$ pathway, 520
C$_4$ photosynthetic pathway, 520
cA, 154, 155
cAA, 155
Calcification, 502
Calcite, 278
Caldera, 313
Caledonides, 295
California rainfall cycle, 238
CAM pathway, 520
Cambrian period, 291
Canadian glacier types, 464
Canadian Life Zone, 596
Canadian Shield, 293
Canadian system of soil classification, 504–510
Canyon, 408
Capacitive sensors, 134
Capillary tension, 498

Carbon credits, 88
Carbon credits trading, 88
Carbon cycle, 9, 88, 89, 526, 528–529
Carbon dioxide, 49
Carbon dioxide (CO$_2$), 86–89
Carbonic acid, 274
Carboniferous period, 291
Caribbean plate, 297
Caroline plate, 297
Cartography, 7, 24
Cascade Mountain volcanoes, 300
Catastrophic events, 10
Cementation, 276
Cenozoic era, 291
CERES, 56–57
CFCs, 50, 88–89
Chalk, 278
Channel, 372
Channel forms, 402, 404
Chaparral, 587
Chaplin Lake, Saskatchewan, 277
Chemical energy, 43
Chemical weathering, 274, 340–341
Chemically precipitated sedimentary rocks, 276–278
Chernozemic soils, 506–507
Cherrapunji, 204
Chert, 278
Chinook, 111
Chittagong, Bangladesh, 186
Chlorofluorocarbons (CFCs), 50, 88–89
Cinder cones, 315
Circle of illumination, 32
Circles, 358
Circum-Pacific belt, 292
Cirque, 465
Cirque glacier, 464
Cirrocumulus cloud, 137
Cirrostratus cloud, 137
Cirrus clouds, 137, 138
Clast, 275
Clastic sedimentary rocks, 275–276
Clay, 496
Clay loam, 496
Clay minerals, 274, 496
Claystone, 276
Clearcutting, 237
Cliff, 422
Climate, 176–263

classification, 187–196
defined, 177
factors to consider, 177
global precipitation, 181–185
high-latitudes (*See* High-latitude climates)
low-latitudes (*See* Low-latitude climates)
midlatitudes (*See* Midlatitude climates)
seasonability of precipitation, 185
temperature regimes, 178–181
time cycles, 179–181
Climate change
Arctic, in, 360
coastal environments, 454
ice sheets, 484–485
monitoring, from space, 552
multiple stresses, 252
regional impacts, 252
Climate Change 2001, 92–93
Climate classification, 187–196
Climatology, 6
Climax community, 551
Climograph, 179
Close forest, 584
Closed flow system, 12
Cloud cover and global warming, 169, 172
Cloud forest, 578
Clouds, 14, 53, 57, 90, 136–138, 139
CO$_2$, 86–89
Coal, 279
Coastal blowout dune, 443
Coastal erosion, 454
Coastal foredunes, 444
Coastal forest, 586
Coastal plains, 423–424
Coastal sediment cell, 449
Coastline of submergence, 452–453
Coccolithophores, 278
Coccoliths, 278
Cocos plate, 297
Coefficient of variation, 226
Cold front, 158
Cold-blooded animals, 546
Collision-wake capture, 140
Colluvium, 342, 397
Colours, 60
Columbia Icefield, 464
Comfort index of heat, 231
Community, 544

Competition, 550

Composite volcanoes, 312

Comprehensive soil classification system (CSCS), 504, 510–514

Compressional tectonic activity, 300

Compressive flow, 466

Condensation nuclei, 49

Condensation nucleus, 136

Conditional instability, 145

Conduction, 52, 70

Cone of depression, 370

Conglomerate rock, 276, 277

Consequent stream, 424

Consumers, 519

Contamination of groundwater, 370–371

Continent-continent collisions, 302, 303

Continental antarctic air mass (cAA), 155

Continental arctic air mass (cA), 154, 155

Continental crust, 290

Continental drift, 306

Continental margin, 296–297

Continental polar air mass (cP), 155

Continental rupture, 303

Continental scale, 10

Continental shields, 293–295

Continental suture, 302

Continental tropical air mass (cT), 154, 155

Continents of the past, 304–305

Continuous permafrost, 250, 348

Contour strip mining, 348

Convection, 52, 70, 142

Convectional precipitation, 144–145

Convective wind system, 104

Convergence, 142

Convergent precipitation, 144

Converging boundary, 300

Coordinated Universal Time (UTC), 29

Coral reefs, 455–456

Coral-reef coasts, 457–458

Coriolis effect, 21, 101

Corrosion, 400

Cosmopolitan species, 560

Counter-radiation, 54–55

Covered shields, 295

cP, 155

Crater Lake, 313

Creep, 438

Cretaceous period, 291

Crevasses, 466

Crust, 288

Cryosolic soils, 507–508

CSCS, 504, 510–514

CSCS soil orders, 510–514

Cuestas, 424

Cumuliform clouds, 138

Cumulonimbus cloud, 137, 138

Cumulus cloud, 137, 138

Curvature effect, 136–137

Cutoff, 406

Cycle of rock transformation, 281–283

Cyclone tracks/cyclone families, 161–162

Cyclones, 108–109, 156–163

Cyclonic (or frontal) precipitation, 142

Cyclonic precipitation, 156

Cyclonic storm, 156

D

Daily insolation, 46–47

Daily insolation and net radiation, 72–73

Daily temperature, 73–74

Daily world weather map, 161

DALR, 135, 145

Dam, 411–412

DART buoys, 331

Darwin, Charles, 554

Daughter product, 292

Davis, William Morris, 431

Day-night cycle, 546

Daylight saving time, 29–30

Dead zones, 530, 531

Death Valley, 11

Debris flood, 344

Decalcification, 501

December solstice, 32

Deciduous forest, 581–584

Declination, 33

Decomposers, 519

Deep-sea hydrothermal vent, 333

Deeply eroded volcanoes, 429–430

Deflation, 440

Deflation hollow, 440

Deglaciation, 475

Delta coasts, 453

Deltas, 457

Dendritic drainage pattern, 423, 424

Dendrochronology, 90, 91

Denitrification, 530

Denudation, 395

Depositional landforms, 396

Depression, 108n

Depression tracks, 107

Desert animals, 592

Desert biome, 591–592

Desert climate, 195

Desert landscapes, 218

Desert pavement, 411, 440

Desert plants, 592

Desertification, 212

Devils Tower, 430

Devonian period, 291

Dew-point lapse rate, 135

Dew-point temperature, 133

Diagnostic horizon, 504

Differential GPS, 24

Diffuse radiation, 53

Diffuse reflection, 53

Diffusion, 556

Digital imaging, 62–63

Dike, 273

Dimictic, 388

Diorite, 272

Dip, 420–422

Diploid, 555

Discharge, 372

Discontinuous permafrost, 250, 348

Disjunct distribution, 560

Disjunction, 560

Dispersal, 556–557

Dissolved matter, 400

Distances from latitude and longitude, 23

Distributaries, 453

Distribution patterns, 557, 560

Disturbance, 548–550

Diurnal tides, 452

Dobson unit (DU), 49

Dolomite, 278

Dome erosion, 425

Doubling time, 87

Drainage basin, 373

Drainage divide, 373

Drainage patterns, 423

Drainage systems, 373

Drainage winds, 110

Drawdown, 370

Drifting snow, 142

Drizzle, 140

Drought on the Prairies, 244–245
Drumlin, 481, 482
Drunken trees, 355
Dry adiabatic lapse rate (DALR), 135, 145
Dry adiabatic rate, 135
Dry-bulb thermometer, 134
Dry climates, 191–192, 192
Dry deposition, 52
Dry desert, 592
Dry lake, 386
Dry midlatitude climate, 240–242
Dry subtropical climate, 226–227
Dry tropical climate, 215–219
Dryland salinization, 227
DU, 49
Dune succession, 553
Dunes, 442–444
Dunes and drifting snow, 127
Dust Bowl, 240
Dust storms, 439–440

E

Earth
    axis, 31
    crust, 267–268
    interior, 288–290
    radius, 288
    relief features, 291–295
    revolution, 31
    rotation, 21
    shape, 21
Earth hummock, 358
Earth impact sites, 431
Earth realms, 10
Earth rotation, 21, 22
Earth's crust, 267–268
Earth-observing satellites, 64
Earthquake, 325–331
    Canada, in, 330
    locations, 329, 330
    measurement, 325–329
    plate tectonics, 329–330
    seismic sea waves, 330–331
Earthquake magnitude, 326–329
Earthquake waves, 288, 289
East African Rift Valley, 324
Easterly wave, 164
Ebb current, 452
Ecological biogeography, 519, 544–551

Ecological niche, 543
Ecological succession, 551–554
Economic geography, 5
Edaphic (soil) factors, 548
Edatopic grid, 548, 549
El Niño, 120–123, 238
Electromagnetic radiation, 41–43
Electromagnetic spectrum, 42
Eluviation, 502
Emergent coastlines, 453
Emissivity, 61
Endemic species, 557
Energy, 43
Energy balance, 70
Energy flow system, 11, 13
Entisols, 513
Entrenched meanders, 412
Environmental pollution, 9
Environmental temperature lapse rate
    (ETLR), 76, 145
Eocene epoch, 291
Ephemeral annuals, 545
Epicentre, 325
Epicontinental seas, 476
Epilimnion, 387
Epiphytes, 576
Equator, 22
Equatorial current, 118
Equatorial easterlies, 115
Equatorial rain forest, 576
Equatorial trough, 102
Equatorial zone, 47
Equilibrium, 13
Equilibrium line, 465
Equinox, 32
Equinox conditions, 32–33
Ergs, 441
Eroded anticline, 428
Erosion, 477–478
    coastal, 454
    dome, 425
    glaciers, 465, 470
    ice sheets, 477–478
    marine, 446–448
    slope, 396–399
    stream, 399–400
    wind, 440–441
Erosional landforms, 396
Eruption of Mount Pinatubo, 90

Esker, 481, 482
ETLR, 76, 145
Eurasian plate, 297
Eurasian-Indonesian belt, 292
Evaporation, 130, 131
Evaporites, 276
Evapotranspiration, 74, 366, 498
Evolution, 554–555
Exfoliation, 340
Exfoliation dome, 340, 429
Exponential growth, 87
Exposed batholiths, 428–429
Exposed shields, 293
Extending flow, 466
Extensional tectonic activity, 300
Extinction, 556
Extreme events, 10
Extrusion, 272
Extrusive igneous rock, 271, 272–273
Eye, 166

F

Fair-weather systems, 156
Fallout, 52
Fault, 321
Fault coast, 458
Fault-line scarp, 428
Fault plane, 321
Fault scarp, 428
Fault scarps, 324
Faults/fault landforms, 321–325
Feedback, 13
Feldspars, 271
Felsenmeer, 338
Felsic rocks, 271
Field capacity, 498
Fields of systematic geography, 6
Finger lakes, 471
Fire, 548–550
Fire ratings, 249
Fire Weather Index, 248, 249
Firn, 465
First-order basin, 411
First-order segment, 411
Fish ladder, 393
Fissile, 276
Fjord, 471
Fjord coast, 453
Flash flood, 217, 377, 413

Flocculation, 276
Flood, 376
Flood basalts, 315
Flood current, 452
Flood expectancy graphs, 379
Flood plain, 376, 377
Flood plain development, 402–406
Flood prediction, 377–380
Flood stage, 377
Flooding, 237, 240, 376–380
Floodplain mapping, 240
Floristic and faunal regions, 561
Flotation, 400
Flow systems, 11–14
Fluid drag, 401
Fluvial processes and landforms, 394–417
    aggradation, 402
    alluvial fans, 414
    arid climate, 413–414
    dam, 411–412
    definitions, 395
    depositional landforms, 396
    entrenched meanders, 412
    erosional landforms, 396
    flood plain development, 402–406
    fluvial processes (arid climate), 413–414
    river discharge/suspended sediment, 403
    slope erosion, 396–399
    stream capacity, 402
    stream deposition, 402
    stream erosion, 399–400
    stream gradation, 406–409
    stream networks as trees, 410–411
    stream transportation, 400–402
    waterfall, 409–410
Foehn, 111
Fog, 138–140
Fold, 300
Fold belts, 317, 321, 425, 428
Folded strata, 321
Food web, 520–521
Forbs, 589
Force, 346
Fore-arc trough, 301
Forest biome, 575–587
    low-latitude rain forest, 575–579
    midlatitude deciduous forest, 581–584
    monsoon forest, 578–580
    needleleaf forest, 584–586

    sclerophyll forest, 586–587
    subtropical evergreen forest, 580–581
Forest fires, 356–357
Forest-tundra transition, 584
Forests, 525, 570, 599
Formation classes, 571
Fossil fuels, 279
Fragile forest, 599
Frank slide, 9
Freezing front, 352
Freezing rain, 140
Fringing reef, 458
Front, 155
Frontal precipitation, 142
Frost action, 338, 339
Frost hollows, 79
Frost point, 133
Fujita scale of tornado intensity, 163
Fundamental niche, 543

G

Gabbro, 272
Gamma rays, 42
Gaseous biogeochemical cycle, 524
Gasohol, 524
Genotype, 555
Genus, 554
Geodetic surveys, 21
Geographic cycle, 431–432
Geographic grid, 22
Geographic information system (GIS), 7, 28–29, 30
Geographic isolation, 555
Geography, 5
Geography of soils, 6
Geoid, 450
Geologic norm, 397
Geologic time scale, 290, 291
Geomorphic factors, 547–548
Geomorphology, 6, 311
GEOSS, 331
Geostationary satellites, 64
Geostrophic wind, 112–114
Geothermal energy sources, 320
Giant rosette plants, 594
Gibber plains, 440
Gingko tree, 560
GIS, 7, 28–29, 30
Glacial abrasion, 465

Glacial deltas, 482
Glacial drift, 479
Glacial erosion, 470
Glacial lakes, 471
Glacial moraine, 469, 470
Glacial retreat, 472–473
Glacial spillway, 482
Glacial till, 479
Glacial trough, 469–471
Glacial valleys, 471
Glaciation, 475–477
Glaciation cycles, 486
Glacier mass balance, 472
Glacier systems, 462–489
    alpine glaciers, 465–468, 489
    drumlin, 482
    erosion, 477–478
    esker, 482
    fjord, 471
    glacial retreat, 472–473
    glacial trough, 469–471
    glaciation cycles, 486
    glaciers as flow system, 466–467
    global warming, 484–485
    Great Lakes, 478–479
    Holcene environments, 486–487
    ice age, 475–477, 484–487
    ice sheets, 472–473
    iceberg, 474
    landforms, 468–469, 477–484
    marginal lakes, 482–484
    moraines, 480–481
    outwash, 482
    remote sensing of glaciers, 468–469
    sea ice, 473–475
    till plain, 482
Glaciolacustrine plains, 483
Gleyosolic soils, 508
Global carbon budget, 538
Global circulation system. See Winds and global circulation system
Global climate change, 9
Global energy balance, 55–58
Global energy system, 53–55
Global positioning system (GPS), 23–24
Global precipitation, 181–185
Global productivity, 536–537
Global scale, 10
Global tide patterns, 452

Global time, 26–31
   daylight saving time, 29–30
   international date line, 30–31
   standard time, 27, 29
   time zones, 27, 29
   world time zone, 29
Global upper-level winds, 115
Global warming, 86–91. *See also* Climate
   change
Global warning potential (GWP), 86
Global water balance, 131, 132
GMT, 29
Gneiss, 280
Gobi, 441
GOES series of geostationary satellites,
   138, 139
GOES weather satellites, 64
Gondwana, 295, 305
Goode projection, 26, 34
Gorge, 408
GPS, 23–24
Graben, 324
Graded profiles, 407
Graded stream, 407
Granite, 272
Granitic rock, 290
Granular disintegration, 338
Graphic scale, 35
Grassland biome, 589–591
Gravitation, 346
Gravity, 346–347
Great Australian desert, 218
Great circles, 22
Great Lakes, 384–385, 478–479
Greenhouse effect, 49, 54, 55, 169, 172
Greenhouse gases, 51, 86
Greenland Ice Sheet, 472, 484
Greenwich Mean Time (GMT), 29
Greenwich meridian, 22
Ground inversion, 79
Ground moraine, 479
Groundwater, 130, 366–367
Groundwater contamination, 370–371
Groynes, 449
Gulf of Mexico dead zone, 531
Gullies, 398, 399
Gullying, 14
Guyot, 314
GWP, 86

Gypsum, 276
Gyres, 118

H
Haber-Bosch process, 527
Habitats, 543
Hadley cells, 102
Hail, 141, 147
Hair hygrograph, 134
Hairpin dune, 443
Half-life, 292
Hamburg, Germany, 80
Hammada, 441
Hanging glaciers, 464
Hanging valley, 470, 471
Hawaiian High, 108
Hazards assessment, 7
Heat energy, 43
Heat index, 231
Heat island, 75
Herbivory, 551
Herbs, 570
Heterosphere, 78
Hibernation, 546
High arctic, 592
High clouds, 14
High mountain environments, 78–79
High-latitude climates, 247–251
   boreal forest climate, 247–249
   ice sheet climate, 251
   overview, 247
   tundra climate, 248–251
High-latitude coasts, 456–457
Highland climates, 184–185
Historical biogeography, 519, 554–562
Histosols, 513
Hoarfrost, 133
Hogbacks, 424
Holcene environments, 486–487
Holdridge system of life zones, 597
Holocene epoch, 486
Holocene Pleistocene epoch, 291
Homosphere, 78
Hopewell Rocks, New Brunswick, 14
Horn, 468, 470
Horsethief Butte State Park, 517
Horst, 324
Hortonian overland flow, 396
Hot springs, 315–316

Hotspot, 314
Hotspot volcano, 318–319
Hotspots, 563
Hudsonian Life Zone, 596
Human disturbance, 550
Human geography, 5
Humidity, 132–134
Humification, 502
Hummocks, 358
Humus, 494
Hurricane, 164–168
Hurricane Katrina, 165, 170–171, 175
Hydraulic action, 400
Hydrograph, 373
Hydrologic cycle, 12, 131–132
Hydrologic drought, 245
Hydrology, 365
Hydrolysis, 274, 340
Hydrophytes, 545
Hydrosphere, 10, 130–131
Hydrothermal vents, 333
Hypolimnion, 387

I
Ibyuk pingo, 355
Ice, 351–355
Ice age, 476. *See also* Glacier systems
Ice floes, 474
Ice lenses, 352
Ice lobes, 480
Ice nuclei, 137
Ice sheet, 465
Ice sheet climate, 251
Ice sheets, 472–473
Ice shelves, 473
Ice storms, 140, 243, 246
Ice wedge, 353, 354
Ice-contact deposit, 482
Ice-wedge polygons, 353, 354
Iceberg, 474
Icefield, 464
Icelandic Low, 107
Igneous rocks, 270–274
Illinoisan glaciation, 476
Illuviation, 502
Image processing, 63
Impact structures, 430–431
Inceptisols, 513
Incoming short-wave radiation, 55–56

Individual scale, 10
Induced deflation, 440
Induced earth flow, 346
Induced mass wasting, 345–346
Infiltration, 365
Infiltration excess overland flow, 396
Infrared radiation, 42
Initial landforms, 311, 312
Inner node, 410
Inselbergs, 429
Insolation, 46–47
Instrument shelters, 70
Interglaciation, 476
Intergovernmental Panel on Climate Change (IPCC), 91
Interlobate moraine, 480
International date line, 30–31
Intertropical convergence zone (ITCZ), 102, 105
Intrusive igneous rock, 271, 272–273
IPCC, 91
IPCC Report of 2001, 92–93
Irrigation salinization, 227
Island arcs, 292
Island biogeography, 558–559
Isobars, 100
Isostatic compensation, 432
Isostatic rebound, 476, 477
Isotherms, 82, 83
Isotopes, 292
ITCZ, 102, 105

J

January ocean currents, 119
Japan Meteorological Agency seismic intensity (shindo) scale, 327, 328
Jet streams, 116–117
Joint-block separation, 338
Joints, 338
Joule (J), 130
Juan de Fuca plate, 297
June solstice, 32
Jurassic period, 291

K

Kaduna, Nigeria, 186
Kangaroo rats, 546
Kansan glaciation, 476
Karst landscape, 368–369

Katabatic winds, 110
Katrina (Hurricane), 165, 170–171, 175
Kauai, 421
Kayes, Mali, 216
Kenny Dam, 237
Kettle holes, 481
Kilimanjaro, 220
Kilopascals (kPa), 99
Kinetic energy, 43
Knob-and-kettle topography, 480
Kopje, 429
Köppen, Vladimir, 187
Köppen climate system, 192–193
Köppen-Geiger system of climate classification, 194–196
kPa, 99
Krakatoa, 313
Kyoto Protocol, 89, 94

L

La Niña, 120–123
Lacustrine sedimentary record, 389
Lag deposits, 440
Lag time, 376
Lagoon, 453
Lahar, 317, 344, 430
Lake, 380–389
    circulation, 386–388
    Great Lakes, 384–385
    origin, 382
    saline lakes, 386
    salt flats, 386
    seasonal circulation, 387–388
    sediment, 388–389
    temperature, 386–387
    water balance, 382–383
Lake Agassiz, 483
Lake effect, 243, 246
Lake Eyre, 217, 219
Lake sediment corer, 389
Lambert Glacier, 469
Laminar sublayer, 437
Land and sea breezes, 109
Land bridge, 557
Land-cover alteration, 90
Land degradation, 212
Land-ocean contrasts, 107
Land productivity, 534–536
Land subsidence, 454–455

Land-water contrast, 80–82
Landforms, 311
Landforms and rock structure, 418–435
    arid regions, 422
    coastal plains, 423–424
    deeply eroded volcanoes, 429–430
    drainage patterns, 423
    exposed batholiths, 428–429
    fold belts, 425, 428
    geographic cycle, 431–432
    impact structures, 430–431
    sedimentary dome, 424–425
    strike and dip, 420–422
Landforms of tectonic activity
    faults/fault landforms, 321–325
    fold belts, 317, 321
Landslide, 344
Lapse rate, 76
Large-scale map, 35
Late-Cenozoic ice age, 475–477, 484–487. See also Glacier systems
Latent heat, 52, 129, 130
Latent heat transfer, 53, 70
Lateral moraine, 469, 470
Latitude, 22–24
Latitude zones, 47–48
Laurasia, 305
Laurel forest, 580
Laurentia, 295
Lava flows, 314
Leaching, 501
Leads, 474
Leap years, 31
Leibnitz, Gottfried, 43
Lenticular clouds, 138
Levee, 406
Lianas, 570
Lichen woodland, 584
Lichens, 570
Life layer, 10
Life zones, 596–597
Lifting condensation level, 135, 136
Lightning, 147
Limestone, 278
Limestone caverns, 368
Limnology, 380
Limonite, 496
Lithosphere, 10, 290
Lithospheric plate boundary, 300

Lithospheric plates, 290, 297, 298
Little Ice Age, 487
Littoral drift, 448–449
Loam, 495, 496
Loamy sand, 496
Local scale, 10
Local winds, 109–111
Lodgement till, 479
Loess, 444–445
Logistic growth, 520
Long-wave radiation, 45–46
Longitude, 22–24
Longitudinal dunes, 443, 444
Longshore current, 448
Longshore drift, 448
Low arctic, 592
Low clouds, 14
Low-angle overthrust fault, 325
Low-latitude climates, 200–223
    dry tropical climate, 215–219
    highland climates, 219–220
    low-latitude rainforest environment,
      207
    Monsoon and trade-wind coastal
      climate, 204–207
    overview, 201–202, 203
    rainfall, 210–211
    savanna environment, 208–214
    tropical desert environment, 217–219
    wet equatorial climate, 202–204
    wet-dry tropical climate, 207–214
Low-latitude rain forest, 575–579
Low-latitude rainforest environment, 207
Low-level temperature inversion, 79
Lower Sonoran Life Zone, 596
Lowlands, 424
Luvisolic soils, 507

M
Mackerel sky, 138
Madison Slide, 346
Mafic minerals, 271
Mafic rocks, 271
Magma, 268
Manaus, 80
Manicouagan Impact Structure, 421
Manning's equation, 400
Manning's *n*, 400
Mantle, 288

Map, 34–37
Map projections, 24–26, 34
Mapping forest ecosystems, 571
Mapping global land cover by satellite,
    572–574
Marble, 280
Mare's tail, 138
Marginal lakes, 482–484
Marine dead zones, 530, 531
Marine erosion, 446–448
Marine geography, 6
Marine scarp, 447
Marine terraces, 458
Marine transport and deposition, 448–449
Marine west-coast climate, 234–239
Maritime polar air mass (mP), 154, 155
Maritime tropical air mass (mT), 154, 155
Marl, 388
Mass wasting, 341
    debris flood, 344
    earth flow, 343
    induced, 345–346
    induced earth flow, 346
    landslide, 344
    mudflow, 344
    scarification, 346–348
    slopes, 342
    soil creep, 342–343341
Massive icy beds, 354
Material cycle, 12, 524
Mathematical modelling, 7
Matter flow system, 12
mb, 99
McFarlane Falls, Saskatchewan, 6
Mean deviation, 226
Meander, 405
Meander belt, 405
Meandering channel, 402, 404
Mechanical energy, 43
Medial moraine, 469
Mediterranean climate, 232–234
Medvedev-Sponheuer-Karnik (MSK)
    scale, 327
Mercalli scale, 327, 328
Mercator projection, 25–26
Mercury barometer, 99
Meridians, 22
Mesa, 422
Mesopause, 77

Mesosphere, 77
Mesothermal climates, 192
Mesozoic era, 291
Metalimnion, 387
Metamorphic rocks, 279–281
Meteor Crater, 430
Meteorology, 6
Methane, 51
Mica Dam, 411, 412
Microburst, 148, 149
Microthermal climates, 192
Middle arctic, 592
Middle-infrared radiation, 42
Midlatitude climates, 225–246
    dry midlatitude climate, 240–242
    dry subtropical climate, 226–227
    marine west-coast climate, 234–239
    Mediterranean climate, 232–234
    moist continental climate, 243–246
    moist subtropical climate, 228–232
    overview, 228, 229
Midlatitude deciduous forest, 581–584
Midlatitude deserts and steppes, 183
Midlatitude west coasts, 184
Midlatitude zones, 48
Midoceanic ridge, 295
Milankovitch, Milutin, 486
Milankovitch curve, 486
Mild, humid (mesothermal) climates, 192
Millibars (mb), 99
Mineral, 268. *See also* Rocks and minerals
Mineral alteration, 274, 496
Mineral horizons, 505
Mineral oxides, 496
Minute, 22
Miocene epoch, 291
Missingas black smokers, 333
Mississippi Delta, 457
Mississippi Delta marshland, 455
Mistral, 110
Mitigation, 259
Mixed prairie, 243, 590
Mixed tides, 452
Modified Mercalli Scale, 327
Moho, 288
Mohorovicic, Andrija, 288
Moist climates, 191–192
Moist continental climate, 243–246
Moist subtropical climate, 228–232

Moist subtropical regions, 183–184
Moisture. *See* Atmospheric moisture and precipitation
Mollisols, 513
Moment magnitude scale, 326, 328
Monadnocks, 429
Monomictic, 388
Monsoon, 105
Monsoon and trade-wind coastal climate, 204–207
Monsoon forest, 578–580
Monsoon wind patterns, 107
Monsoon woodland, 579
Montane forest, 578
Montane needleleaf forest, 585
Monterey, California, 233, 234
Monthly precipitation patterns, 185
Monument Valley, 423, 425
Moon, 31
Moraines, 469, 470, 480–481
Morphology, 555
Mount Angel glacier, 489
Mount Edziza, 315
Mount Fuji, 322
Mount Mayon, 313
Mount Meager, 315
Mount Nazko, 315
Mount Ruapehu, 429, 430
Mount Vesuvius, 322
Mount Waialeale, 204, 205
Mountain and valley winds, 109
Mountain arc, 292
Mountain pine beetle epidemic, 236
Mountain roots, 295
mP, 154, 155
MSK scale, 327
mT, 154, 155
Mud hummock, 358
Mudflow, 344
Mudstone, 276
Munsell colour system, 494–495

N

Nazca plate, 297
Nappes, 301
Natural bridge, 412, 413
Natural flow systems, 11
Natural gas, 279
Natural levee, 406

Natural selection, 554
Natural vegetation, 569–570
Natural vegetation regions, 574
NDVI, 572
Neap tides, 451
Near-infrared radiation, 42
Nearctic realm, 560
Nebraskan glaciation, 476
Nebula, 48
Needleleaf forest, 237, 584–586
Negative feedback, 13
Neogener period, 291
Net photosynthesis, 521–522
Net primary production, 522–523
Net production and climate, 523
Net radiation, 58, 70, 72–73, 80
Network, 410
Neutral coastlines, 453, 457
Neutral stability, 145
Newton, Sir Isaac, 43
Niagara Escarpment, 410
Niagara Falls, 409
Niche formation, 339
Nile Delta, 457
Nimbostratus cloud, 137
Nitrogen, 49
Nitrogen cycle, 527–530
Nitrogen fixation, 527, 530
Nitrous oxide, 51
Node, 410
Normal fault, 322–324
Normal temperature lapse rate, 76
Normalized difference vegetation index (NDVI), 572
North American air masses, 155
North Atlantic Drift, 119
Northeast trades, 103
NOx, 52
Nuée ardente, 313
Nutrient cycle, 524

O

Obsidian, 273
Occluded front, 158, 159
Ocean
    currents, 118–124
    pollution, 530
    pressure and density, 102
    productivity, 523, 536

temperature layers, 118
Ocean basin feature, 295–297
Ocean basin floor, 295
Ocean basis opening process, 303–304
Ocean circulation, 121
Ocean currents, 118–124
Ocean pollution, 530
Ocean productivity, 536
Ocean temperature structure, 118
Oceanic crust, 290
Oceanic heat transport, 169
Oceanic trenches, 297
Oceanography, 6
Oil deposits, 279
Oil shale, 276
Oligocene epoch, 291
Olivine, 271
Omega block, 115, 116
Omnivores, 519
Open flow system, 12
Open lichen woodland, 584
Orbiting earth satellites, 63–64
Ordovician period, 291
Organic coastlines, 457–458
Organic horizons, 505
Organic soils, 508
Orogen, 301
Orogens and collisions, 302
Orogeny, 301, 302
Orographic precipitation, 143–144
Ottawa, Ontario, 242
Outcrops, 342
Outer node, 410
Outlet glaciers, 472
Outwash, 482
Outwash plain, 482
Overflow channels, 482
Overland flow, 366, 371–372, 396
Overthrust fault, 322, 325
Overthrust faults, 301
Overturn, 387
Oxbow lake, 406
Oxidation, 274, 340
Oxisols, 511
Oxygen, 49
Oxygen cycle, 526–527
Ozone hole, 50, 51
Ozone layer, 50–51, 77

P

P wave, 288, 289
Pa, 99
Pacific Decadal Oscillation Index, 238, 239
Pacific plate, 297
Pacific Tsunami Warning Centre (PTWC), 331
Paired terraces, 405
Palearctic realm, 560
Paleocene epoch, 291
Paleoclimate studies, 356
Paleogene period, 291
Paleomagnetism, 484
Paleozoic era, 291
Pangaea, 295, 304
Pangaea's breakup, 307
Parabolic dune, 443
Parallels, 22
Parasitism, 550
Parent material, 494
Particulates, 51
Pascal (Pa), 99
Passive continental margins, 297
Passive radar system, 61
Patterned ground, 358
Pedestal rocks, 441
Perched water table, 367
Percolation, 366
Peridotite, 272
Periglacial system, 348
Perihelion, 31
Permafrost, 250–251, 348
Permafrost regions, 348
    active layer, 351
    alpine permafrost, 359
    climate change in Arctic, 360
    energy flow system, 352
    environmental problems, 359–360
    forest fires, 356–357
    ice, 351–355
    maps, 349, 350
    patterned ground, 358
    permafrost, 348
    permafrost temperature, 350–351
    retrogressive thaw slumps, 357–358
    solifluction, 359
    thermokarst lakes, 355–356
Permafrost table, 348
Permian period, 291

Petroleum, 275, 279
Phanerozoic eon, 290
Phases of the moon, 31
Phenotype, 555
Philippine plate, 297
Photoperiod, 546
Photosynthesis, 520–521
Physical geography, 5
Pine Lake Tornado, 162
Pine plantation, 581
Pingo, 354–355
Pioneer stage, 551
Plagioclase feldspar, 271
Plane of the ecliptic, 31
Planimetric, 411
Plants, structure/life form, 570
Plate motions/interactions, 297–300
Plate tectonics
    continental rupture, 303
    earthquakes, 329–330
    ocean basis opening process, 303–304
    orogens and collisions, 302
    plate motions/interactions, 297–300
    subductor tectonics, 301–302
    tectonic plate processes, 300–301
Plateau, 422
Pliocene epoch, 291
Plucking, 465
Plunging anticlines, 428
Plunging folds, 428
Plunging synclines, 428
Pluton, 273
Pocket beaches, 449
Podzolic soils, 507
Point bar, 406, 408
Polar climates, 192
Polar easterlies, 104
Polar front, 103, 115, 155, 188
Polar jet streams, 116, 117
Polar outbreak, 164
Polar outbreaks, 103
Polar projection, 25
Polar zones, 48
Poleward heat transport, 168–169
Pollution, 9
Polygons, 358
Polymictic, 388
Polyploidy, 555
Pool, 524

Popocatepetl, 322
Positive feedback, 13
Potash, 276
Potassium feldspar, 271
Potential energy, 43
Potential evapotranspiration, 499
Power, 346
Power source, 11
Prairie grasslands, 590
Prairie sloughs, 481
Precambrian time, 290
Precession, 486
Precipitation, 140–145. *See also* Atmospheric moisture and precipitation
Precipitation and global warming, 172
Precipitation fog, 138
Predation, 550
Pressure gradient, 100
Pressure gradient force, 100
Prevailing westerlies, 103
Primary consumers, 519
Primary minerals, 496
Primary producers, 519
Primary succession, 551
Prime meridian, 23
Primordial atmosphere, 48
Pro-glacial deposit, 482
Pro-glacial lakes, 478
Progradation, 449
Protoco-operation, 551
Protons, 292
PTWC, 331
Pyrolysis, 524
Pyroxene group, 271

Q

Quadrature, 451
Quartz, 271, 274
Quartz crystal, 268
Quartzite, 280
Quick clays, 343

R

Radar, 61–62
Radial drainage pattern, 429
Radial oxygen loss, 545
Radiant energy, 43
Radiation
    counter, 54–55

diffuse, 53
electromagnetic, 41–43
incoming short-wave, 55–56
long-wave, 45–46
net, 58
solar, 44–45
temperature, and, 42, 80
UV, 42
Radiation balance, 56
Radiation fog, 138
Radiation laws, 44
Radioactive decay, 292
Radiometric dating, 292–293
Rain, 140
Rain forest layers, 576
Rain gauge, 142
Rain shadow, 144
Raised beaches, 477
Realized niche, 543
Realms, 10
Recessional moraines, 469, 480
Reclamation, 347
Red River flood, 380, 381
Red Sea, 303, 304
Reg, 440
Regelation, 466
Region, 411
Regional geography, 5
Regional scale, 10
Regolith, 342, 494
Regosolic soils, 509
Reindeer moss, 584
Rejuvenation, 432
Relative humidity, 133–134
Remote sensing, 7, 60–64
    glaciers, 468–469
    mapping forest ecosystems, 571
    mapping global land cover by satellite,
        572–574
    monitoring global activity from space,
        534–538
    rock structures, 421
    volcanoes, 322
    wetland ecosystems, 456
Representative fraction, 35
Residual regolith, 342
Respiration, 521
Retrogradation, 449
Retrogressive thaw slumps, 357–358

Reverse fault, 322, 325
Revolution of earth and moon, 31
Rhyolite, 273
Ria coast, 452
Ribbon lakes, 471
Richter scale, 325, 326, 328
Richter, Charles F., 326
Ridge-and-valley landscape, 421, 425
Rift valley, 303
Rift Valley systems, 324
Rifting, 300
Rill erosion, 398, 399
Rilling, 14
River deltas, 457
River discharge, 372
River discharge/suspended sediment, 403
River drainage basin, 11
River floods, 376–377
River flow, 376
Roche mountain, 478
Rock cycle (cycle of rock transformation),
    281–283
Rock structure. *See* Landforms and rock
    structure
Rocks and minerals, 268–283
    igneous rocks, 270–274
    metamorphic rocks, 279–281
    overview, 268–270
    rock cycle (cycle of rock transforma-
        tion), 281–283
    sedimentary rocks, 274–279
Root, 410
Rooted tree, 410–411
Ross Ice Shelf, 485
Rossby wave, 115–116
Rossby wave cycle, 116
Runoff, 132, 232, 396
Rural temperatures, 74–75

S

S wave, 288, 289
Sagebrush, 591
Saguaro cactus, 591
Sahara Desert, 218
Sahelian drought, 213
Sahelian zone, 212–213
Salic horizon, 502
Saline lakes, 386
Salinization, 502

SALR, 145
Salt encrustations, 386
Salt flats, 386
Salt marsh, 452
Salt-crystal growth, 339–340
Salt-marsh ecosystem, 520
Saltation, 438
Saltwater intrusion, 371
Sample standard deviation, 226
Sample statistics, 226
San Andreas Fault, 325
San Quentin Glacier, 468
Sand, 496
Sand dunes, 442–444
Sand ripples, 439
Sand sea, 443
Sandspit, 449
Sandstone, 276
Sandy clay, 496
Sandy clay loam, 496
Sandy loam, 496
Santa Ana, 110
Santa Tecla landslide, 345
Saturated adiabatic lapse rate (SALR), 145
Saturated zone, 366
Saturation specific humidity, 133
Saturation vapour pressure (SVP), 132
Savanna biome, 587–589
Savanna environment, 208–214
Scale, 10
Scarification, 346–348
Scattering, 45
Schist, 279–280
Sclerophyll forest, 586–587
Sclerophyll vegetation, 587
Sclerophyll woodland, 586
Scoria, 273
Scotia plate, 297
Sea and land breezes, 109
Sea breeze, 109
Sea cliffs, 447
Sea fog, 140
Sea ice, 473–475
Sea-level rise, 455, 458
Seasonability of precipitation, 185
Second, 23
Second-order segment, 411
Secondary consumers, 519
Secondary minerals, 496

Secondary succession, 551, 554

Sediment, 275, 342

Sediment load, 401

Sediment transport, 401

Sediment yield, 398

Sedimentary biogeochemical cycle, 524

Sedimentary cycles, 532–533

Sedimentary dome, 424–425

Sedimentary rocks, 269, 274–279

Sedimentary strata, Utah, 285

Segregated ice, 352

Seismic moment, 326

Seismic sea waves, 330–331

Semi-arid climate, 192, 215, 217

Semi-desert, 591

Semidiurnal tides, 452

Sensible heat, 52

Sensible heat transfer, 52

Sequential landforms, 311, 312

Shadow zone, 288, 289

Shale, 276

Shanghai, China, 186

Shearing, 279

Sheet erosion, 396

Sheet flow, 371

Sheeting structure, 340

Shield volcanoes, 314–316

Ship Rock, New Mexico, 430

Short-grass prairies, 241

Short-wave infrared radiation, 42

Short-wave radiation, 45

Shortgrass prairie, 590

Shrubs, 570

Side slopes, 409

Sief dunes, 443

Silica, 276

Sill, 273

Silt loam, 496

Silty clay, 496

Silty clay loam, 496

Silurian period, 291

Simpson-Saffir scale of tropical cyclone
    intensity, 167

Sinkhole, 369

Slash-and-burn clearing, 578

Slate, 279, 280

Sleet, 140

Sling psychrometer, 134

Slope, 375

Slope erosion, 396–399

Sloughs, 481

Small circles, 22

Small-scale map, 35

Smog, 138

Snow, 142

Snow melt in mountains, 263

Snowy-forest (microthermal) climates, 192

Soil. *See* Soil systems

Soil acidity and alkalinity, 497–498

Soil catena, 503

Soil colloids, 496–497

Soil colour, 494–495

Soil configuration, 503

Soil creep, 342–343341

Soil enrichment, 501

Soil horizons, 499, 501, 505

Soil leaching, 497

Soil minerals, 496

Soil moisture, 498

Soil order, 504

Soil salinity, 227

Soil structure, 495–496

Soil systems, 492–517

    biological processes, 504

    Canadian system of classification,
        504–510

    classification systems, 504

    CSCS soil orders, 510–514

    mineral matter, 495

    soil acidity and alkalinity, 497–498

    soil colloids, 496–497

    soil colour, 494–495

    soil configuration, 503

    soil horizons, 499, 501, 505

    soil minerals, 496

    soil moisture, 498

    soil structure, 495–496

    soil temperature, 502

    soil texture, 495, 496

    soil water balance, 498–499

    soil water budget, 499, 500

    soil, defined, 493, 494

    soil-farming processes, 501–502

Soil temperature, 502

Soil texture, 495, 496

Soil water, 130

Soil water balance, 498–499

Soil water belt, 366

Soil water budget, 499, 500

Soil water recharge, 498

Soil water shortage, 499

Soil-farming processes, 501–502

Solar energy, 59

Solar I, 59

Solar II, 59

Solar noon, 26

Solar power tower, 59

Solifluction, 359

Solonetzic soils, 509

Solstice, 32

Solstice conditions, 33

Solute effect, 137

Solution weathering, 363

Sorting, 276

Source regions, 153

South Polar High, 108

Southeast trades, 103

Southern Oscillation, 122

Southern Oscillation Index, 238, 239

SO_x, 52

SPAR lab, 236

Spatial object, 28

Speciation, 555–556

Species, 554

Species-area curve, 558–559

Specific humidity, 133

Spectral signature, 60

Spillways, 482

Splash erosion, 397

Spodosols, 512–513

Spoil, 347

Spreading boundary, 300

Spring tides, 451

Spruce, 555

Squall line, 158

Stable air, 145

Stage, 374

Standard deviation, 226

Standard meridian, 27

Standard time, 27, 29

Standard time system, 27

Star dunes, 443, 444

Steam fog, 138

Stefan-Boltzman law, 44

Steppe/Steppes, 183, 590

Steppe climate, 195

Steps, 358

Stevenson screens, 70

Stoke's Law, 438

Storage pools, 524

Storage recharge, 499

Storage withdrawal, 499

Stored energy, 43

Storm surge, 167

Straight channel, 402, 404

Strata, 268

Stratified drift, 479

Stratiform clouds, 138

Stratocumulus cloud, 137

Stratopause, 77

Stratosphere and upper layers, 77–78

Stratovolcanoes, 312–314

Stream, 372

Stream capacity, 402

Stream deposition, 402

Stream discharge, 372–373

Stream erosion, 399–400

Stream flow, 373–376

Stream gradation, 406–409

Stream load, 400

Stream network properties, 426–427

Stream ordering, 409

Stream profiles, 407

Stream segment, 411

Stream transportation, 400–402

Stream tree, 411

Strike and dip, 420–422

Strike-slip fault, 324

Strip mining, 347, 348

Stripes, 358

Sub-boreal stage, 487

Sub-glacial stream, 482

Subantarctic zones, 48

Subarctic zone, 48

Subduction, 297

Subduction ARCS, 314

Subductor tectonics, 301–302

Subhorizons, 505

Sublimation, 129

Subsea permafrost, 348

Subsequent streams, 424

Subsidence, 370, 454–455

Subsolar point, 32

Subsurface water, 130

Subtropical broadleaf evergreen forest, 580

Subtropical evergreen forest, 580–581

Subtropical high-pressure belts, 102–103, 104–105

Subtropical jet streams, 117

Subtropical needleleaf evergreen forest, 581

Subtropical zones, 47

Successional change, 551–554

Succulent, 545

Sugar Creek hydrograph, 373

Sulphur cycle, 530–532

Sulphur oxides, 52

Summer monsoon, 105

Summer solstice, 32

Sun's ephemeris, 33

Sun-synchronous satellites, 63

Surface energy balance equation, 72

Surface energy flows, 56–57

Surface latent heat flux, 81

Surface ocean currents, 118–120

Surface temperature, 70

Surface water, 130, 371–373

Surface waves, 288, 289

Surface winds, 102–104

Suspended load, 400

Suspension, 400, 438

SVP, 132

Swash, 448

Swells, 446

Symbiosis, 551

Sympatric speciation, 555

Synclinal mountain, 425

Synclines, 321

System, 11

Systematic geography, 5

Systems approach, 11

T

Tailings, 347

Tall grass prairie, 589, 590

Tall-grass prairie, 242

Talus, 338

Talus cones, 363

Talus slopes, 338

Tamanrasset, Algeria, 186

Tarn, 470, 471

Taupo volcanic zone (TVZ), 430

Tectonic activity, 290

Tectonic arc, 301

Tectonic crest, 301

Tectonic landforms. *See* Landforms of tectonic activity

Tectonic plate processes, 300–301

Temperate deciduous forest, 582

Temperate grasslands, 590

Temperature. *See* Air temperature

Temperature gradients, 83

Temperature inversion, 79

Temperature regimes, 178–181

Temperatures close to the ground, 74

Tephra, 312

Terminal fall velocity, 438

Terminal moraine, 469, 470, 480

Terrestrial biomes, 568–599

    altitude zones of vegetation, 595–597

    climatic gradients/vegetation types, 594–595

    desert biome, 591–592

    forest biome, 575–587 (*See also* Forest biome)

    grassland biome, 589–591

    life zones, 596–597

    natural vegetation, 569–570

    plants, structure/life form, 570

    savanna biome, 587–589

    tundra biome, 592–594

    vegetation transects, 594–595

Terrestrial ecosystems, 571

Tetraploids, 556

Thermal conductivity, 70

Thermal infrared radiation, 42

Thermal infrared sensing, 61

Thermally direct, 168, 169

Thermistor, 71

Thermocline, 118, 387

Thermohaline circulation, 120, 123, 124

Thermokarst lakes, 355–356

Thermometer, 70

Thermosphere, 77

Thornthwaite, C.W., 187

Thorntree semi-desert, 592

Thorntree-tall grass savanna, 588

Threshold velocity, 438

Thrust sheets, 301

Thunderstorm, 146–147, 148

Tidal currents, 452

Tidal inlets, 453

Tide-generating force, 450

Tides, 449–452

Tidewater glaciers, 465
Till, 479
Till plain, 481, 482
Tilt of earth's axis, 31, 32
Tilted rock strata, 422
Time. *See* Global time
Time cycle, 14–15
Time cycles of climate, 179–181
Time zones, 27, 29
Tornado, 162–163
Tower karst, 370
Traction, 438
Tractive force, 450
Trade winds, 103
Trade-wind coastal climate, 204–207
Trade-wind coasts, 183
Transcurrent fault, 323, 324–325
Transform boundary, 300
Transform scars, 304
Transition Life Zone, 596
Translocation, 502
Transpiration, 74
Transported regolith, 342
Transverse dunes, 443
Travelling cyclones and anticyclones,
    156–163
Travertine, 368
Tree, 410
Tree-ring analysis, 90, 91, 187
Trees, 570
Trellis drainage pattern, 424
Triassic period, 291
Tropic of Cancer, 33
Tropic of Capricorn, 33
Tropical cyclone, 164–168
Tropical desert environment, 217–219
Tropical deserts, 183
Tropical easterly jet stream, 117
Tropical high-pressure belt, 115
Tropical microcosm, 541
Tropical rain forest, 576, 577
Tropical rainforest climate, 194
Tropical rainy climates, 192
Tropical savanna climate, 194
Tropical zones, 47
Tropopause, 77
Troposphere, 76–77
Tropospheric aerosols, 89–90
TROWAL, 159

Tsunami, 330–331
Tundra biome, 592–594
Tundra climate, 248–251
Tundra soils, 514
Turbidites, 301
Turbidity currents, 301
Turtle Mountain, 9
Turtle Mountain slide, 345
TVZ, 430

U

Ultisols, 511
Ultramafic rocks, 271
Ultraviolet (UV) radiation, 42
Undersea topography, 296
Unloading, 340
Unpaired terraces, 405
Unsaturated zone, 366
Upper Sonoran Life Zone, 596
Upper-air westerlies, 114
Upper-air wind map, 115
Upwelling, 119
Urban climatology, 76
Urban heat island, 75
Urban temperatures, 74–75
UTC, 29
UV radiation, 42

V

Valley fog, 138
Valley mist, 110
Valley wind, 109
Vapour pressure, 132
Varves, 389, 483
Vegetation transects, 594–595
Ventifacts, 441
Vernal (spring) equinox, 32
Vertisolic soils, 510
Vertisols, 511–512
Vestmannaeyjar, Iceland, 309
Victoria Falls, 409
Virga, 141
Visible light, 42
Volcanic activity
    environmental hazards, 316–317
    global perspective, 316
    life cycle of hotspot volcano, 318–319
    remote sensing, 322
    shield volcanoes, 314–316

    stratovolcanoes, 312–314
Volcanic eruptions, 90, 238
Volcanic glass, 273
Volcanic neck, 430
Volcanic rock formations, 274
Volcanism, 290, 312
Volcano, 312
Volcano coast, 457

W

W, 45
Walker circulation, 122
Walkerton groundwater contamination,
    371
Wallace, Alfred Russel, 562
Wallace's line, 562
Warm front, 158
Warm-blooded animals, 546
Washout, 52
Water, 129. *See also* Water/water bodies
Water gap, 425
Water need, 499, 544–546
Water resources, 7
Water table depletion, 370
Water table surface, 366–367
Water use, 499
Water vapour, 49
Water/water bodies, 364–393
    aquifer, 367
    contamination of groundwater, 370–371
    drainage systems, 373
    flooding, 376–380
    groundwater, 366–367
    karst landscape, 368–369
    lake (*See* Lake)
    limestone caverns, 368
    overland flow, 371–372
    river flow, 376
    stream discharge, 372–373
    stream flow, 373–376
    surface water, 371–373
    surface water as natural resource,
        389–390
    water table depletion, 370
    water table surface, 366–367
Waterfall, 409–410
Watershed, 373
Watt (W), 45
Wave cyclone, 157–160

Waves. *See also* Work of winds and waves

Weak equatorial low, 164

Weather recording instruments, 71

Weather systems, 152–175

   air masses, 153–155

   cloud cover and global warming, 169, 172

   cyclone tracks/cyclone families, 161–162

   defined, 153

   easterly wave, 164

   hurricane, 164–168

   Hurricane Katrina, 165, 170–171

   oceanic heat transport, 169

   polar outbreak, 164

   poleward heat transport, 168–169

   precipitation and global warming, 172

   tornado, 162–163

   travelling cyclones and anticyclones, 156–163

   tropical cyclone, 164–168

   tropical/equatorial systems, 164–168

   wave cyclone, 157–160

   weak equatorial low, 164

Weathering, 274, 338–341

Wegener, Alfred, 305–306

Welwitschia, 591

West Antarctic Ice Sheet, 484–485

West-wind drift, 119

Wet (saturated) adiabatic lapse rate, 135

Wet-and-dry-bulb thermometer, 134

Wet-dry climate, 192

Wet-dry tropical climate, 207–214

Wet equatorial belt, 183

Wet equatorial climate, 202–204

Wetland ecosystems, 456

Whiting, 388

Wien's law, 44

Wilson Cycle, 304, 305

wWilting point, 498

Wind, 100. *See also* Winds and global circulation system; Work of winds and waves

Wind deflation, 440

Wind erosivity, 438

Wind farms, 112

Wind waves, 446

Wind-transported silt, 444

Winds and global circulation system, 98–127. *See also* Work of winds and waves

   anticyclones, 108

   atmospheric pressure, 99–100

   Coriolis effect, 101

   cyclones, 108–109

   El Niño, 120–123

   geostrophic wind, 112–114

   higher latitudes, 107

   ITCZ, 105

   jet streams, 116–117

   La Niña, 120–123

   local winds, 109–111

   measurement of wind, 100

   monsoon, 105

   ocean currents, 118–124

   pressure gradient, 100

   Rossby waves, 115–116

   subtropical high-pressure belts, 104–105

   surface winds, 102–104

   thermohaline circulation, 120, 123

   upper atmosphere, 111–118

Winter monsoon, 105

Winter solstice, 32

Wisconsinan glaciation, 476

Woodland, 570

Work, 346

Work of winds and waves, 436–461. *See also* Winds and global circulation system

   Beaufort scale, 445

   coastal environments, 445–449

   coastal foredunes, 444

   coastline of submergence, 452–453

   depositional landforms, 441–442

   dust storms, 439–440

   emergent coastlines, 453

   erosion by wind, 440–441

   littoral drift and shore protection, 449

   loess, 444–445

   marine erosion, 446–448

   marine transport and deposition, 448–449

   neutral coastlines, 453, 457

   organic coastlines, 457–458

   sand dunes, 442–444

   sea-level rise, 455, 458

   terrestrial environments, 437–438

   tides, 449–452

   transport by wind, 438–440

World latitude zones, 47–48

World precipitation, 181–185

World Reference Base for Soil Resources, 504

World time zone, 29

X

X-rays, 42

Xerophytes, 544

Y

Yakutsk, Russia, 80

Z

Zenith angle, 32

Zero tillage, 245

Zonda, 111

Zone of ablation, 465

Zone of accumulation, 465

NOTES

NOTES

National Topographic System of Canada Reference

Dual highway, hard surface
Route à 2 chaussées séparées, revêtement dur

Road, hard surface, more than 2 lanes
Route, revêtement dur, plus de 2 voies

Road, hard surface, 2 lanes
Route, revêtement dur, 2 voies

Road, hard surface, less than 2 lanes
Route, revêtement dur, moins de 2 voies

Street
Rue

Road, loose or stabilized surface, all season, 2 lanes or more
Route de gravier, aggloméré, toute saison, 2 voies ou plus

Road, loose or stabilized surface, all season, less than 2 lanes
Route de gravier, aggloméré, toute saison, moins de 2 voies

Road, loose surface, dry weather
Route de gravier, temps sec

Unclassified road, street
Route non classée, rue

Vehicle track or winter road; gate
Chemin de terre ou d'hiver; barrière

Trail, cut line or portage; portage, short or position uncertain
Sentier, percée, portage; portage court ou position incertaine

Road, under construction
Route, en construction

Highway interchange with number; traffic circle
Échangeur avec numéro; rond-point

Highway route number
Numéro de route

Built-up area; street; park/sports field
Agglomération; rue; parc ou terrain de sports

Indian reserve; small
Réserve indienne, petite

Railway, single track; railway station; turntable
Chemin de fer, voie unique; gare; plaque tournante

Railway, multiple tracks
Chemin de fer, voies multiples

Railway, under construction
Chemin de fer, en construction

Railway, abandoned
Chemin de fer, abandonné

Railway on road; special track railway
Chemin de fer sur route; chemin de fer à voie spéciale

Rapid transit route: rail; road
Transport rapide : voie ferrée; route

Bridge; footbridge; snowshed
Pont; passerelle; paraneige

Bridge: swing, draw, lift; tunnel
Pont : tournant, basculant, levant; tunnel

Cut; embankment, causeway
Déblai; remblai, chaussée

Dyke or levee; with road
Digue ou levée; avec route

Ferry
Traversier

Ford
Gué

Submarine cable
Câble sous-marin

Navigation light; navigation beacon
Balise lumineuse; balise de navigation

Coast Guard station; exposed shipwreck
Station de la garde-côtière; épave émergée

Seaplane base; seaplane anchorage
Hydrobase; ancrage d'hydravions

Crib or abandoned bridge pier
Caisson ou pilier de pont abandonné

Airfield, position approximate; heliport
Terrain d'aviation, position approximative; héliport

Building(s)
Bâtiment(s)

Church; non-Christian place of worship; shrine
Église; lieu de culte non chrétien; lieu de pèlerinage

School; elevator; fire station
École; élévateur; caserne de pompiers

Sports track; stadium
Piste de course; stade

Silo; kiln; dome
Silo; séchoir; dôme

Cemetery; historic site or point of interest
Cimetière; lieu historique ou lieu d'intérêt

Landmark object (with height): tower, chimney, etc.
Objet-repère (avec hauteur) : tour, cheminée, etc.

Campground; picnic site; service centre
Terrain de camping; terrain de pique-nique; centre de service

Golf course; golf driving range; drive-in theatre
Terrain de golf; champ d'exercice-golf; ciné-parc

Wind-operated device; ruins; greenhouse
Éolienne; ruines; serre

Aerial cableway, ski lift, conveyor
Téléphérique, remontée mécanique, convoyeur

Ski area, ski jump
Station de ski, tremplin de ski

Wall; fence
Mur; clôture

Tank(s): vertical; horizontal
Réservoir(s) : vertical; horizontal

Warden, ranger station; Customs
Poste de gardien, poste de garde forestier; poste de douane

Well: oil, gas
Puits de pétrole ou de gaz

Crane: vertical; horizontal
Grue : verticale; horizontale

Rifle range with butts
Champ de tir avec buttes

Power transmission line; multiple lines
Ligne de transport d'énergie; lignes multiples

Telephone line; firebreak
Ligne téléphonique; coupe-feu

Pipeline with control valve
Pipeline; valve de contrôle

Pipeline underground; multiple pipelines underground
Pipeline souterrain; pipelines multiples souterrains

Electric facility; oil or natural gas facility
Installation électrique; installations pétrolières ou gazières

Pit: sand, gravel, clay; quarry
Sablière, gravière, glaisière; carrière

Mine; cave
Mine; caverne

International boundary with monument
Frontière internationale avec borne-repère

Boundary, first class with mile post
Limite de première classe avec borne milliaire

Boundary, first class unsurveyed
Limite de première classe non arpentée

Boundary, second class
Limite de deuxième classe

Boundary, second class in Dominion Land Survey
Limite de deuxième classe dans les régions de l'Arpentage des terres du Canada

Boundary, third class
Limite de troisième classe

Boundary, fourth class
Limite de quatrième classe

Boundary, fifth class
Limite de cinquième classe

Boundary, sixth class
Limite de sixième classe

Boundary, sixth class unsurveyed
Limite de sixième classe non arpentée

Boundary, seventh class
Limite de septième classe

Boundary, eighth class
Limite de huitième classe

Boundary, ninth class
Limite de neuvième classe

Precise elevation
Point coté, précis

Dam: small; large; carrying road
Barrage : petit; grand; portant une route

Wharf; pier or dock; seawall; breakwater
Quai; jetée; mur de protection; brise-lames

Slip; drydock; boat ramp
Cale; cale sèche; rampe de chargement

Lock; sluice gate
Écluse; vanne

Sewage disposal pond, settling pond
Bassin d'épuration des eaux d'égout, étang de décantation

Watercourse or shoreline: definite; indefinite
Cours d'eau ou rive : précis(e); imprécis(e)

Watercourse; direction of flow arrow
Cours d'eau; flèche de direction du courant

Rapids
Rapides

Falls (with height in black)
Chutes (avec hauteur en noir)

Well: water, brine; spring
Puits d'eau, puits salant; source

Navigable canal; canal, abandoned
Canal navigable; canal, abandonné

Ditch, conduit; conduit, underground
Fossé, canalisation; canalisation souterraine

Conduit bridge
Pont de canalisation

Braided stream; disappearing stream
Cours d'eau anastomosé; cours d'eau disparaissant sous terre

Fish ladder
Échelle à poissons

Lake or pond; slough, intermittent lake or pond
Lac ou étang; bourbier, lac ou étang intermittent

Flooded area
Région inondée

Reservoir, dugout, swimming pool; underground reservoir
Réservoir, abreuvoir, piscine; réservoir souterrain

Tundra: ponds; polygons
Étangs de toundra; polygones de toundra

Foreshore flats or sand in water
Estrans ou sable dans l'eau

Rocks in water or small islands
Rochers dans l'eau ou îlots

Rocky ledge; rocky reef
Barre rocheuse; récif

Artificial island, small
Îlot artificiel

Kelp area; fish pound
Varech; vivier dans l'eau

Dry river bed
Lit de cours d'eau tari

Marsh, swamp, muskeg; string bog
Marais, marécage, fondrière; fondrière à filaments

Debris-covered ice; palsa bog
Glace couverte de débris; fondrière de palse

Spot elevation, non-precise; water elevation
Point coté, imprécis; altitude de la surface de l'eau

Contours: index; intermediate
Courbes de niveau : maîtresses; intermédiaires

Approximate contours
Courbes de niveau approximatives

Auxiliary contours
Courbes de niveau intercalaires

Depression contours
Courbes de cuvette

Cliff or escarpment
Falaise ou escarpement

Sand; esker; pingo
Sable; esker; pingo

Moraine, scree
Moraine, éboulis

Glacier, ice cap, snowfield
Glacier, calotte glaciaire, champ de neige

Orchard; vineyard, hopfield; wooded area
Verger; vignoble, houblonnière; région boisée